Modern Classical Mechanics

In this modern and distinctive textbook, Helliwell and Sahakian present classical mechanics as a thriving and contemporary field with strong connections to cutting-edge research topics in physics. Each part of the book concludes with a capstone chapter describing various key topics in quantum mechanics, general relativity, and other areas of modern physics, clearly demonstrating how they relate to advanced classical mechanics, and enabling students to appreciate the central importance of classical mechanics within contemporary fields of research. Numerous and detailed examples are interleaved with theoretical content, illustrating abstract concepts more concretely. Extensive problem sets at the end of each chapter further reinforce students' understanding of key concepts, and provide opportunities for assessment or self-testing. A detailed online solutions manual and lecture slides accompany the text for instructors. Often a flexible approach is required when teaching advanced classical mechanics, and, to facilitate this, the authors have outlined several paths instructors and students can follow through the book, depending on background knowledge and the length of their course.

T. M. Helliwell taught at Harvey Mudd College for more than 40 years, as well as serving in several administrative roles, including Physics Department Chair. He received two National Science Foundation fellowships, was awarded the Henry T. Mudd Prize for service to the college in 1997, and served as a consultant at the Jet Propulsion Laboratory. He is the author or co-author of more than 40 research papers on general relativity, cosmology, and quantum theory, and the author of two textbooks on special relativity.

V. V. Sahakian is Professor of Theoretical Physics at Harvey Mudd College. He has received two National Science Foundation fellowships, and authored or co-authored more than 25 research papers on string theory, cosmology, and quantum gravity. His research focuses on understanding the small-scale structure of space, often studying black holes and exploring new frameworks that extend the Standard Model of particle physics and standard inflationary cosmology.

Modern Classical Mechanics

T. M. HELLIWELL

Harvey Mudd College, California

V. V. SAHAKIAN

Harvey Mudd College, California

CAMBRIDGE
UNIVERSITY PRESS

CAMBRIDGE
UNIVERSITY PRESS

University Printing House, Cambridge CB2 8BS, United Kingdom

One Liberty Plaza, 20th Floor, New York, NY 10006, USA

477 Williamstown Road, Port Melbourne, VIC 3207, Australia

314–321, 3rd Floor, Plot 3, Splendor Forum, Jasola District Centre, New Delhi – 110025, India

79 Anson Road, #06–04/06, Singapore 079906

Cambridge University Press is part of the University of Cambridge.

It furthers the University's mission by disseminating knowledge in the pursuit of education, learning, and research at the highest international levels of excellence.

www.cambridge.org
Information on this title: www.cambridge.org/9781108834971
DOI: 10.1017/9781108874687

First published 2021

Printed in the United Kingdom by TJ Books Limited, Padstow Cornwall 2021

A catalogue record for this publication is available from the British Library.

Library of Congress Cataloging-in-Publication Data
Names: Helliwell, T. M. (Thomas M.), 1936– author. | Sahakian, V. V. (Vatche V.), author.
Title: Modern classical mechanics / T.M. Helliwell, Harvey Mudd College,
California, V.V. Sahakian, Harvey Mudd College, California.
Description: Cambridge, United Kingdom ; New York, NY : Cambridge
University Press, [2021] | Includes bibliographical references and index.
Identifiers: LCCN 2020039628 (print) | LCCN 2020039629 (ebook)
| ISBN 9781108834971 (hardback) | ISBN 9781108874687 (epub)
Subjects: LCSH: Mechanics.
Classification: LCC QA809 .H45 2021 (print) | LCC QA809 (ebook)
| DDC 531–dc23
LC record available at https://lccn.loc.gov/2020039628
LC ebook record available at https://lccn.loc.gov/2020039629

ISBN 978-1-108-83497-1 Hardback

Additional resources for this publication at www.cambridge.org/helliwell

Dedicated to Bonnie, Gaia, and Yuliya

Contents

Part III

Preface

The branch of physics known as "classical mechanics" originated in the seventeenth century, but wasn't called that until the discovery of quantum mechanics in the 1920s. It was quantum mechanics that most profoundly changed our understanding of how and why particles move as they do, and even what a particle *is*. Quantum mechanics was so completely different that the word "classical" had to be added to the older theory to make it clear which mechanics was meant. At the same time, quantum mechanics was heavily inspired and influenced by the formulations of classical mechanics by Lagrange and Hamilton dating back to the eighteenth and nineteenth centuries.

Einstein's theories of special relativity (1905) and general relativity (1915) also had important impacts on classical mechanics, changing the laws of motion primarily by revolutionizing our understanding of the spacetime arena in which physics takes place. These theories have been viewed as either introducing a new "relativistic mechanics" or more modestly as *completing* classical mechanics, making it useful even for particles moving close to the speed of light and for particles moving in strong gravitational fields.

Quantum mechanics, special relativity, and general relativity stand together as the three pillars of modern physics. Classical mechanics integrates with all three as a robust approximation framework that is both useful in practice for problem solving – but also as an inspirational venue for developing basic intuition about physics.

In the title of the book we have endowed our exposition of classical mechanics with the word "modern," because it is a modern approach in several ways. First, we focus on the Lagrangian and Hamiltonian formulations of mechanics almost from the outset, modern of course only relative to Newton's formulation. Throughout we emphasize the connections of these newer approaches to the development of quantum mechanics – through contact with Feynman's path-integral formulation of quantum mechanics and the relations of Hamilton–Jacobi theory to Schrödinger's approach to wave mechanics. We also develop the subject of mechanics with relativity in mind early on, integrating modern differential geometry notation in the narrative and motivating the variational principle through arguments that come naturally from special relativity. In particular, immediately after a compact review of Newtonian particle mechanics in Chapter 1, special relativity is introduced already in Chapter 2. Finally, the exposition is modern in that we use a tone and physics mindset that is contemporary, often with an emphasis on the role of symmetry as a guiding principle, and we draw on many examples from modern subjects and applications such as black holes, cosmology, atomic physics, particle physics, magnetic trapping, orbital mechanics, and spaceflight.

Modern classical mechanics also stands strong on its own as a useful approximation framework that addresses physics problems in regimes where quantum mechanics and/or relativity come in as sub-leading effects. In many situations, using quantum mechanics and/or relativity to study a physical system would be tantamount to shooting a fly with a catapult. Roughly speaking, classical mechanics works very well (i.e., agrees with experiments) for macroscopic objects that are moving at speeds much less than the speed of light, and where gravity is not too strong – and also where our experimental measurements are not too precise.

Take the motions of planets around the sun and moons around their planets, for example. Motions within the solar system were the most important testing ground for classical mechanics in the first place, and for nearly all purposes classical mechanics in this domain works as well now as it ever did. We still use it to plot the motion of spacecraft on their way to distant planets, for example – it would be completely unnecessary to tackle a problem like that using the full apparatus of quantum mechanics. The same can be said for the use of special and general relativity, except for tiny but nevertheless important effects like the precession of the planet Mercury's perihelion or the rate of atomic clocks in Global Positioning System (GPS) satellites around the earth.

Our book is first and foremost a textbook on classical mechanics and its many uses, while also showing where its limitations lie – limitations as defined by quantum mechanics as well as the relativity theories, and emphasizing the inspirational role the subject played in the development of modern physics. To accomplish these goals, the book is divided into three main parts. There are five chapters in each part, where the fifth chapter is a "capstone chapter," a special unit that elaborates further on the boundaries of classical mechanics as presented in the preceding four chapters and its connections to the three pillars of modern physics.

In broad strokes, the first part of the book is about the Lagrangian formulation of mechanics; the second part is about the various forces and symmetries that present themselves on the mechanics stage; and the third part is about the Hamiltonian formulation. The capstone chapter of the first part is a pedagogical exposition of Feynman's path-integral formulation of quantum mechanics and its connections to modern classical mechanics; the capstone chapter of the second part discusses general relativity and its relations to relativistic mechanics; and the capstone chapter of the third part is about Hamilton–Jacobi theory, phase space, and the connections to the wavefunction formulation of quantum mechanics.

This layout allows for different pathways through the book, depending on the time available for a given class and the background preparation of the students. The following diagram illustrates the conceptual connections and dependencies between the various sections.

Based on this, we can identify several possible pathways that can be adopted in a typical 15-week-long class.

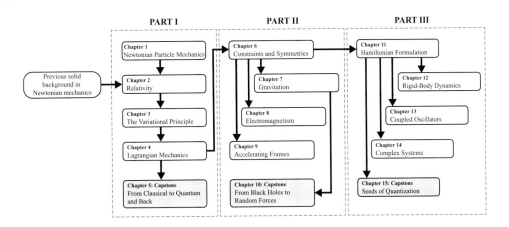

- **Basic mechanics:** For students who have had a basic calculus-based mechanics course and are looking for a basic second course. Chapters 1, 2, 3, 4, 6.1 and 6.2, 7.1 to 7.4, 9.1 to 9.4.
- **Lagrangian approach plus a bit more:** For students who have had a robust calculus-based mechanics course and are looking for a more sophisticated second course. Chapters 2, 3, 4, 6, 7, 8, and 9.
- **Traditional Lagrangian and Hamiltonian mechanics:** For students who have had a robust calculus-based mechanics course and are looking for a rather traditional course on Lagrangian and Hamiltonian mechanics. Chapters 2, 3, 4, 6, 7, 9, 11, 12 or 13.
- **Advanced mechanics:** For students who have had a robust calculus-based mechanics course and are looking for an advanced modern exposure to mechanics. Chapters 2, 3, 4, 5, 6, 7, 8, 9, 10, 11, one of 12, 13, or 14, optional reading 15.

In a 20-week timeframe, one can cover most if not all the chapters. This can also be done naturally in a 14-week graduate-level course. Chapters 5, 10, and 12 to 15 can also serve as excellent directed reading material for students who complete their second mechanics course but want to learn more advanced topics.

A Note about Notation

Throughout the book, we have attempted to accord, as much as possible, with notational conventions that are commonly used in similar textbooks. However, there is one place we have decided to adopt a notation that is instead more consistent with more advanced graduate-level textbooks: components of vectors are labeled by superscripts instead of subscripts. For example, the components of a velocity vector $\mathbf{v}$ in spherical coordinates are written as $\mathbf{v} = (v^r, v^\phi, v^\theta)$; similarly, the components of a four-velocity vector $\boldsymbol{u}$ in Cartesian coordinates take the form $\boldsymbol{u} = (u^t, u^x, u^y, u^z)$. This notation is conventional in differential geometry and graduate-level textbooks so as to distinguish vectors from co-vectors – such as the momentum co-vector and the gauge potential co-vector in electromagnetism.

Given that our modern approach to the subject of mechanics incorporates the language of special relativity from the outset, it is indeed natural to adopt the "correct" differential geometry notation from the start. This also helps the reader later on in transitioning to graduate-level coursework and research-level literature. One pitfall of this notation is that it does require a bit of an initial learning curve as the superscript might be confused with raising a variable to a power. We have addressed this issue by choosing a different font for superscripts that represent components – and generally making sure that we point out potential confusion whenever the context does not make the interpretation obvious. Because of this, we recommend that all users of this book are at least encouraged to *read* Chapter 2, which covers special relativity, even if they are already familiar with the subject. This chapter establishes the notation clearly and gets readers used to it quickly. We have tested this in the classroom over many years and found that the adoption of the notation can be rather smooth and seamless. One bonus advantage of the new notation is that subscripts can be reserved to label particles or degrees of freedom, needs that are very common in the subject of classical mechanics; and indeed, we do so throughout the book. When all is said and done, we believe it is worthwhile to introduce readers to the newer notation, and that it pays off quickly.

Each chapter ends with a list of problems arranged in the order that the topics they cover appear in the chapter. And each problem is labeled by one, two, or three stars, indicating the level of difficulty – one star being easiest and three being hardest.

Acknowledgments and Credits

Over the many years we have taught classical mechanics at Harvey Mudd College, we have benefited greatly from the questions and enthusiasm of the many students in our classes. Teaching is a learning process for the teacher and we have tried to relay this experience in the pages of this textbook. We have also benefited from suggestions and help from many of our colleagues, especially Peter Saeta, Brian Shuve, and John Townsend; also from Sami Gara and several anonymous reviewers. Their support of our work has meant a great deal.

Image of earth used throughout the text: "The Blue Marble" photograph of the earth, taken by the Apollo 17 mission on December 7, 1972 at a distance of about 29,000 km. Taken by either Harrison Schmitt or Ron Evans.

Some artwork in figures is from Wikipedia (wikipedia.org) as material in the public domain.

Notation and Conventions

$\mathbf{v}$	three-vector
$\boldsymbol{v}$	four-vector
$\hat{\mathbf{r}}$	unit vector
v^a	three-vector component
v^μ	four-vector component
$\hat{\mathcal{R}}$	matrix
r, θ	polar coordinates
ρ, φ, z	cylindrical coordinates
r, ϕ, θ	spherical coordinates: radial, azimuthal, latitude
T	kinetic energy
U	potential energy
$-+++$	spacetime signature

Useful Relations

$$T = \tfrac{1}{2}m\left(\dot{x}^2 + \dot{y}^2 + \dot{z}^2\right)$$ Cartesian

$$T = \tfrac{1}{2}m\left(\dot{\rho}^2 + \rho^2\dot{\varphi}^2 + \dot{z}^2\right)$$ cylindrical

$$T = \tfrac{1}{2}m\left(\dot{r}^2 + r^2\sin^2\theta\,\dot{\phi}^2 + r^2\dot{\theta}^2\right)$$ spherical

$$T = \tfrac{1}{2}m\mathbf{v}_{\rm rot}^2 + m\mathbf{v}_{\rm rot}\cdot(\boldsymbol{\omega}\times\mathbf{r}) + \tfrac{1}{2}m\omega^2 r^2 - \tfrac{1}{2}m\left(\boldsymbol{\omega}\cdot\mathbf{r}\right)^2$$ non-inertial

$$\mathbf{F}_{\rm rot} = \mathbf{F}_{\rm in} \underbrace{-m\,\boldsymbol{\omega}\times(\boldsymbol{\omega}\times\mathbf{r})_{\rm rot}}_{\text{centrifugal}} \underbrace{-2\,m\,(\boldsymbol{\omega}\times\mathbf{v}_{\rm rot})_{\rm rot}}_{\text{Coriolis}} \underbrace{-m\,(\dot{\boldsymbol{\omega}}\times\mathbf{r})_{\rm rot}}_{\text{Euler}}$$ fictitious forces

$$S = -m\,c^2\int dt\,\sqrt{1 - \tfrac{v^2}{c^2}} + Q\int dt\left(-\phi + \mathbf{A}\cdot\tfrac{\mathbf{v}}{c}\right)$$ charged particle

$$\frac{d}{dt}\left(\frac{\partial L}{\partial \dot{q}_k}\right) - \frac{\partial L}{\partial q_k} = \lambda_l a_{lk}, \quad a_{lk}\dot{q}_k + a_{lt} = 0$$ equations of motion

$$\delta S = \int dt\left(\frac{\partial L}{\partial q_k}\Delta q_k + \frac{\partial L}{\partial \dot{q}_k}\frac{d}{dt}\left(\Delta q_k\right) + \frac{d}{dt}(L\,\delta t)\right)$$ transformation

$$Q \equiv \frac{\partial L}{\partial \dot{q}_k}\Delta q_k + L\,\delta t$$ Noether charge

$$H = \frac{\partial L}{\partial \dot{q}_k}\dot{q}_k - L$$ Hamiltonian

$$\dot{q}_k = \frac{\partial H}{\partial p_k}, \quad \dot{p}_k = -\frac{\partial H}{\partial q_k}, \quad \frac{\partial L}{\partial t} = -\frac{dH}{dt}$$ Hamiltonian equations

$$r = \frac{\ell^2/G\,M\,m^2}{1 + \epsilon\cos\phi}$$ gravitational orbits

$$a = -\frac{G\,M\,m}{2\,E}, \quad \epsilon = \sqrt{1 + \frac{2\,E\,\ell^2}{G^2 M^2 m^3}}$$ orbit relations

Vector identities

$$\mathbf{A}\cdot(\mathbf{B}\times\mathbf{C}) = \mathbf{B}\cdot(\mathbf{C}\times\mathbf{A}) = \mathbf{C}\cdot(\mathbf{A}\times\mathbf{B})$$
$$\mathbf{A}\times(\mathbf{B}\times\mathbf{C}) = (\mathbf{A}\cdot\mathbf{C})\,\mathbf{B} - (\mathbf{A}\cdot\mathbf{B})\,\mathbf{C}$$

$$\nabla\cdot(f\mathbf{A}) = f(\nabla\cdot\mathbf{A}) + \mathbf{A}\cdot(\nabla f)$$
$$\nabla\times(f\mathbf{A}) = f(\nabla\times\mathbf{A}) + (\nabla f)\times\mathbf{A}$$
$$\nabla(\mathbf{A}\cdot\mathbf{B}) = (\mathbf{A}\cdot\nabla)\,\mathbf{B} + (\mathbf{B}\cdot\nabla)\,\mathbf{A} + \mathbf{A}\times(\nabla\times\mathbf{B}) + \mathbf{B}\times(\nabla\times\mathbf{A})$$
$$\nabla\cdot(\mathbf{A}\cdot\mathbf{B}) = (\nabla\times\mathbf{A})\cdot\mathbf{B} - \mathbf{A}\cdot(\nabla\times\mathbf{B})$$
$$\nabla\times(\mathbf{A}\times\mathbf{B}) = \mathbf{A}\,(\nabla\cdot\mathbf{B}) - \mathbf{B}\,(\nabla\cdot\mathbf{A}) + (\mathbf{B}\cdot\nabla)\,\mathbf{A} - (\mathbf{A}\cdot\nabla)\,\mathbf{B}$$
$$\nabla\times(\nabla\times\mathbf{A}) = \nabla(\nabla\cdot\mathbf{A}) - \nabla^2\mathbf{A}$$

PART I

1 Newtonian Particle Mechanics

We begin our journey of discovery by reviewing the well-known laws of Newtonian mechanics. We set the stage by introducing inertial frames of reference and the Galilean transformation that translates between them, and then present Newton's celebrated three laws of motion for both single particles and systems of particles. We review the three conservation laws of momentum, angular momentum, and energy, and illustrate how they can be used to provide insight and greatly simplify problem solving. We end by discussing the fundamental forces of nature and which of them are encountered in classical mechanics. All this is a preview to a relativistic treatment of mechanics in the following chapter.

1.1 Inertial Frames and the Galilean Transformation

Classical mechanics begins by analyzing the motion of *particles*. Classical particles are idealizations: they are point-like, with no internal degrees of freedom like vibrations or rotations. But by understanding the motion of these ideal "particles" we can also understand a lot about the motion of *real* objects, because we can often ignore what is going on inside of them. The concept of "classical particle" can in the right circumstances be used for objects all the way from electrons to baseballs to stars to entire galaxies.

In describing the motion of a particle, we first have to choose a frame of reference in which an observer can make measurements. Many reference frames could be used, but there is a special set of frames, the *non-accelerating*, **inertial frames**, in which the physics is particularly simple. Picture a set of three orthogonal meter sticks defining a set of Cartesian coordinates drifting through space with no forces applied. An inertial observer drifts with the coordinate system and uses it to make measurements of physical phenomena. This inertial frame and inertial observer are not unique, however: having established one inertial frame, any other frame moving at constant velocity relative to it is also inertial, as illustrated in Figure 1.1.

Two of these inertial observers, along with their personal coordinate systems, are depicted in Figure 1.2: observer $\mathcal{O}$ describes positions of objects through a Cartesian system labeled (x, y, z), while observer $\mathcal{O}'$ uses a system labeled (x', y', z').

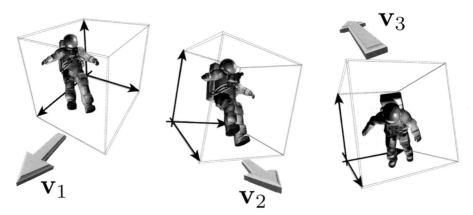

Fig. 1.1 Various inertial frames in space. If one of these frames is inertial, any other frame moving at constant velocity relative to it is also inertial.

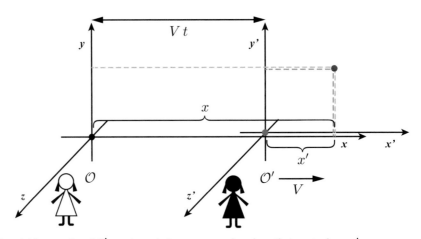

Fig. 1.2 Two inertial frames, $\mathcal{O}$ and $\mathcal{O}'$, moving relative to one another along their mutual x or x' axes.

An **event** of interest to an observer is characterized by the position in space at which the measurement is made – but also by the instant in time at which the observation occurs, according to clocks at rest in the observer's inertial frame. For example, an event could be a snapshot in time of the position of a particle along its trajectory. Hence, the event is assigned four numbers by observer $\mathcal{O}$: x, y, z, and t for time, while observer $\mathcal{O}'$ labels the same event x', y', z', and t'.

Without loss of generality, observer $\mathcal{O}$ can choose her x axis along the direction of motion of $\mathcal{O}'$, and then the x' axis of $\mathcal{O}'$ can be aligned with that axis as well, as shown in Figure 1.2. It seems intuitively obvious that the coordinates of the event are related by

$$x = x' + Vt', \quad y = y', \quad z = z', \quad t = t', \tag{1.1}$$

where we assume that the origins of the two frames coincide at time $t' = t = 0$. This is known as a **Galilean transformation**. Note that the only difference in the coordinates is in the x direction, corresponding to the distance between the two origins as each system moves relative to the other. This transformation – in spite of being highly intuitive – will turn out to be incorrect, as we shall see in the next chapter. But for now, we take it as good enough for our Newtonian purposes.

If the coordinates represent the instantaneous position of a particle, we can write

$$x(t) = x'(t') + Vt', \quad y(t) = y'(t'), \quad z(t) = z'(t'), \quad t = t'. \tag{1.2}$$

We then differentiate this transformation with respect to $t = t'$ to obtain the transformation laws of velocity and acceleration. Differentiating once gives

$$v_x = v'_x + V, \quad v_y = v'_y, \quad v_z = v'_z, \tag{1.3}$$

where, for example, $v_x \equiv dx/dt$ and $v'_x \equiv dx'/dt'$, and differentiating a second time gives

$$a_x = a'_x, \quad a_y = a'_y, \quad a_z = a'_z. \tag{1.4}$$

That is, the velocity components of a particle differ by the relative frame velocity in each direction, while the acceleration components are the same in every inertial frame. Therefore one says that the acceleration of a particle is **Galilean invariant**.

Henceforth, we assert that all statements of physics that we write are expressed from the perspective of inertial observers – unless explicitly stated otherwise. For this purpose, *any* inertial observer has a valid perspective and is no more privileged than any other. This implies that all fundamental laws of physics we will write should be unchanged between the perspectives of different inertial observers. This equivalence of physics amongst inertial frames is called the **principle of relativity**.

1.2 Newton's Laws of Motion

In his *Principia* of 1687, Newton presented his famous three laws. The first of these is the **law of inertia**:

> **I.** If there are no forces on an object, then if the object starts at rest it will stay at rest, or if it is initially set in motion, it will continue moving in the same direction in a straight line at constant speed. ∎

Since this is a statement of physics – made by definition from the perspective of any inertial observer – it should be compatible with the principle of relativity: all inertial observers can write this same statement. On the contrary, using the Galilean velocity transformation, we see that if a particle has constant velocity in one inertial frame then it has constant velocity in all inertial frames. Hence, to assure that this statement can be written by any inertial observer and is hence compatible with the principle of relativity, we use the Galilean transformations to connect the perspectives of inertial reference frames.[1] In practice, we can henceforth use this first law of Newton to *test* whether or not our frame is inertial: if we remove all interactions from a particle under observation, and if we then notice that when set at rest the particle stays put and if tossed in any direction it keeps moving in that direction with constant speed, we can conclude that the law of inertia is obeyed and our frame is inertial.

An astronaut set adrift from her spacecraft in outer space, far from earth, or the sun, or any other gravitating object, will move off in a straight line at constant speed when viewed from an inertial frame. So if her spaceship is drifting without power and is not rotating, the spaceship frame is inertial and onboard observers will see her move away in a straight line. But if her spaceship is rotating, for example, observers on the ship will see her move off in a curved path – the frame inside a rotating spaceship is not *inertial.*

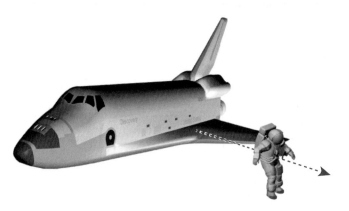

Now consider an inertial observer who observes a particle to which a force **F** is applied. Then **Newton's second law** states that

$$\mathbf{F} = \frac{d\mathbf{p}}{dt}, \tag{1.5}$$

---

[1] Alternatively, we can think of inertial frames as some yet-undefined set of reference frames for the principle of relativity, then use this first law of Newton to *define* what inertial reference frames must be, along with the associated Galilean transformations that connect them. A curious fact is that having identified an inertial frame as one in which Newton's first law is valid, which can be accomplished by purely local observations of the motion of test particles, one finds that inertial frames are also those which are neither accelerating nor rotating relative to the distant stars! It is hard to believe this is mere coincidence, but the reasons for it are not universally agreed upon.

where the momentum of the particle is $\mathbf{p} = m\mathbf{v}$, the product of its mass and velocity. That is:

> **II.** The time rate of change of a particle's momentum is equal to the net force on that particle.   ∎

Newton's second law tells us that if the momentum of a particle changes, there must be a net force causing that change. Note that the second law gives us the means to identify and quantify the effect of forces and interactions. By conducting a series of measurements of the rate of change of momenta of a selection of particles, we explore the forces acting on them in their environment. Once we understand the nature of these forces, we can use this knowledge to predict the motion of other particles in a wider range of circumstances – this time by deducing the effect of such forces on rate of change of momentum.

Note also that $d\mathbf{p}/dt = m d\mathbf{v}/dt = m\mathbf{a}$, so Newton's second law can also be written in the form $\mathbf{F} = m\mathbf{a}$, where $\mathbf{a}$ is the acceleration of the particle. The particle is taken to have a fixed mass, independent of its position or velocity. The law therefore implies that if we remove all forces from an object, neither its momentum nor its velocity will change: it will remain at rest if started at rest, and move in a straight line at constant speed if given an initial velocity. But that is just Newton's *first* law, so it might seem that the first law is just a special case of the second law! However, the second law is not true in all frames of reference. An accelerating observer will see the momentum of an object changing, even if there is no net force on it. In fact, it is only inertial observers who can use Newton's second law, so the first law is not so much a special case of the second as a means of specifying those observers for whom the second law is valid. Put differently, Newton uses the first law to implicitly *define* the concept of inertial reference frames.

Newton's second law is the most famous fundamental law of classical mechanics, and it must also be Galilean invariant according to our principle of relativity. We have already shown that the acceleration of a particle is invariant and we also take the mass of a particle to be the same in all inertial frames. So if $\mathbf{F} = m\mathbf{a}$ is to be a fundamental law, which can be used by observers at rest in any inertial frame, we must insist that the force on a particle is likewise Galilean invariant. Newton's second law itself does not specify which forces exist, but in classical mechanics any force on a particle (due to a spring, gravity, friction, or whatever) must be the same in all inertial frames.

If the drifting astronaut is carrying a wrench, by throwing it away (say) in the forward direction she exerts a force on it. During the throw the momentum of the wrench changes, and after it is released it travels in some straight line at constant speed.

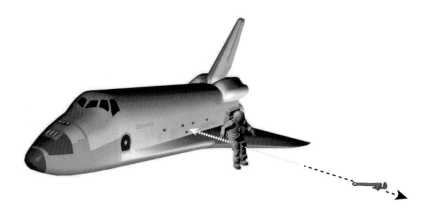

Of course, it is one thing to *know* Newton's second law; it is quite another thing to solve it to find a particle's motion in a particular case, which may range from easy to quite challenging. At the easy end of the spectrum is the case of an object of mass m moving under the influence of a constant force, such as the gravitational force $\mathbf{F} = m\mathbf{g}$ on a particle in a uniform gravitational field $\mathbf{g}$. If that is the only force, the particle's acceleration $\mathbf{a}$ will be constant, so its velocity $\mathbf{v}(t)$ can be found by integrating $\mathbf{a}$ over time, and then its position $\mathbf{r}(t)$ can be found by integrating $\mathbf{v}(t)$ over time. All this leads to the familiar equations of projectile motion.

Finally, Newton's *third* law states that

> **III.** "Action equals reaction." If one particle exerts a force on a second particle, the second particle exerts an equal but opposite force back on the first particle. ∎

We have already stated that any force acting on a particle in classical mechanics must be the same in all inertial frames, so it follows that Newton's third law is also Galilean invariant: a pair of equal and opposite forces in a given inertial frame transform to the same equal and opposite pair in another inertial frame.

While the astronaut, drifting away from her spaceship, is exerting a force on the wrench, at each instant the wrench is exerting an equal but opposite force back on the astronaut. This causes the astronaut's momentum to change as well, and if the change is large enough her momentum will be reversed, allowing her to drift back to her spacecraft in a straight line at constant speed when viewed in an inertial frame.

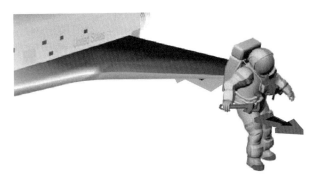

1.3 One-Dimensional Motion: Drag Forces

Before discussing the full rich possibilities of the three-dimensional motion of a particle, we will begin with the simpler case of one-dimensional motion. In fact, if the total force acting on a particle pulls or pushes it in one linear direction, say in the x direction, and if the particle begins at rest or with some initial velocity that happens also to be in this same x direction, then the particle will continue to move in the x direction.

In general, the net force on a particle moving in one dimension might depend upon the particle's position, or its velocity, or time, or any combination of these variables. In this section we will suppose that the net force on a particle depends only upon its velocity, and not its position in space or the time. Then Newton's second law takes the form

$$\mathbf{F}(\mathbf{v}) = m\mathbf{a} \equiv m\frac{d\mathbf{v}}{dt} \tag{1.6}$$

which is a *first-order* differential equation. This often makes the problem much simpler than for position-dependent forces, which lead to second-order differential equations.

Drag forces are prime examples of one-dimensional velocity-dependent forces. They include air resistance on dropped baseballs, raindrops, and skydivers; they also include the horizontal motion of automobiles or airplanes and water drag on fish or submarines. By definition, drag forces act in opposition to an object's velocity through the fluid. For small objects moving sufficiently slowly, fluid flows around an object smoothly in what is called **laminar flow**, giving rise to "viscous drag," where the drag force is proportional to the **viscosity** of the fluid, a measure of how much of the fluid is pulled along with the object as it moves. An example would be dropping a small ball into a vat of honey or molasses, both highly viscous fluids. The viscous drag force is *linear* in the velocity, so has the form $F_{\text{drag}} = -bv$, where b is a constant.

Example 1.1 **A Bacterium with a Viscous Drag Force**

The most important force on a non-swimming bacterium in a fluid is the viscous drag force $F = -bv$, where v is the velocity of the bacterium relative to the fluid and b is a constant that depends on the size and shape of the bacterium and the viscosity of the fluid – the minus sign means that the drag force is opposite to the direction of motion. If the bacterium, as illustrated in Figure 1.3, gains a velocity v_0 and then stops swimming, what is its subsequent velocity as a function of time?

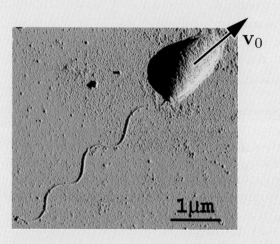

Fig. 1.3 A bacterium in a fluid. What is its motion if it begins with velocity v_0 and then stops swimming? Reprinted figure with permission from Guanglai Li, Lick-Kong Tam, and Jay X. Tang, Amplified effect of Brownian motion in bacterial near-surface swimming, PNAS, November 17, 2008 (Figure 1b). Copyright (2008) by the American Physical Society. Figure 1b. DOI: https://doi.org/10.1103/PhysRevE .84.041932

Let us assume that the fluid defines an inertial reference frame. Newton's second law then leads to the ordinary differential equation

$$m\frac{dv}{dt} = -b\,v \Rightarrow m\ddot{x} = -b\,\dot{x}, \tag{1.7}$$

where $\dot{x} \equiv dx/dt$ and $\ddot{x} \equiv d^2x/dt^2$. So Newton's second law is a second-order differential equation in position and time, but a particularly simple one that can be integrated at once to give a first-order differential equation in v and t. Separating variables and integrating:

$$\int_{v_0}^{v} \frac{dv}{v} = -\frac{b}{m} \int_{0}^{t} dt, \tag{1.8}$$

which gives $\ln(v) - \ln(v_0) = \ln(v/v_0) = -(b/m)t$. Exponentiating both sides:

$$v = v_0 e^{-(b/m)t} \equiv v_0 e^{-t/\tau} \;\Rightarrow\; a = \frac{dv}{dt} = -\frac{v_0}{\tau} e^{-t/\tau}, \tag{1.9}$$

where $\tau \equiv m/b$ is called the "time constant" of the exponential decay. In a single time constant, *i.e.*, when $t = \tau$, the velocity decreases to $1/e$ of its initial value; therefore τ is a measure of how quickly the bacterium slows down. The bigger the drag force (or the smaller the mass), the greater the deceleration.

An alternate way to solve the differential equation is to note that it is linear with constant coefficients, so the exponential form $v(t) = Ae^{\alpha t}$ is bound to work, for an *arbitrary* constant A and a *particular* constant α. In fact, the constant $\alpha = -1/\tau$, found by substituting $v(t) = Ae^{\alpha t}$ into the differential equation and requiring that it be a solution. In this first-order equation the constant A is the single required arbitrary constant. It can be determined by imposing the initial condition $v = v_0$ at $t = 0$, which tells us that $A = v_0$.

Now we can integrate once more to find the bacterium's position $x(t)$. If we choose the x direction as the $\mathbf{v_0}$ direction, then $v = dx/dt$, so

$$x(t) = v_0 \int_0^t e^{-t/\tau}\, dt = v_0\tau \left(1 - e^{-t/\tau}\right). \tag{1.10}$$

The second integration constant is fixed by the bacterium's starting position, $x(0) = 0$. As $t \to \infty$, we see that its position x asymptotically approaches the value $v_0\tau$. Note that given a starting position and an initial velocity, the path of a bacterium is determined by the forces exerted on it. Figure 1.4 shows $x(t)$ and $v(t)$ for the bacterium.

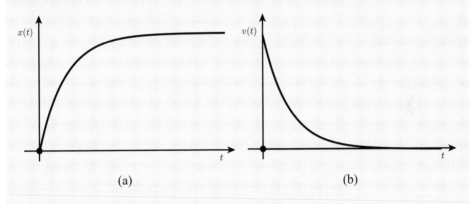

(a)   (b)

Fig. 1.4 Position (a) and velocity (b) versus time for the bacterium. ∎

For larger and more quickly moving objects there comes a point where the fluid no longer flows smoothly around the object, but becomes turbulent, churning around and shedding swirling eddies and vortices. The drag force is then approximately proportional to the *square* of the object's velocity through the fluid. This is sometimes called **inertial drag** or **Newtonian drag**. Air in front of a fast-moving baseball has no time to flow smoothly out of the way, but becomes turbulent and retains this turbulence after the ball has already passed by. This is the type of drag that normally happens all around us, including the drag force on cars and airplanes moving at typical speeds. Doubling their velocity increases the drag force by a factor of four, so in the case of automobiles, for example, designers are motivated to reduce the drag force by streamlining the shape of cars to minimize the turbulence. This helps increase fuel efficiency and also the top speed attainable.

Example 1.2   A ball of mass m and radius r is dropped from the top of a skyscraper. Find the height of the skyscraper if the ball reaches the ground at a time t later.

In this case the drag force is quadratic over essentially the entire trip, so the equation of motion is

$$m\frac{dv}{dt} = mg - cv^2, \tag{1.11}$$

where c is the drag constant and we have taken the positive direction to be downward. Note that the net force on the ball goes to zero as $v \rightarrow \sqrt{mg/c}$, so there is a terminal velocity $v_T = \sqrt{mg/c}$ which the ball never quite reaches as it falls. Initially, when v is small, the ball has downward acceleration $a \simeq g$, and then $a \rightarrow 0$ as $v \rightarrow v_T = \sqrt{mg/c}$. It is the existence of a terminal velocity that helps some cats survive when they leap out of open windows in tall apartment buildings hoping to catch a bird, or even a very few people among those whose parachutes have failed to open, or in one case a soldier who jumped without a parachute from a plane in flames, preferring to take his chances in free fall rather than getting burned alive. Thanks to the terminal velocity, the impact velocity of an object at the ground stays nearly the same no matter how high up the object begins, assuming of course that the initial altitude is sufficiently great. Using the result $v_T^2 = mg/c$, the v and t variables in $F = ma$ can be separated to give

$$gt = g\int_0^t dt = \int_0^v \frac{dv}{1 - v^2/v_T^2}. \tag{1.12}$$

A particularly simple way to carry out the integration is to use the technique of partial fractions, beginning with the identity

$$\frac{1}{1 - z^2} = \frac{1}{2}\left(\frac{1}{1+z} + \frac{1}{1-z}\right). \tag{1.13}$$

So if we let $z = v/v_T$, it follows that

$$gt = \frac{v_T}{2}\left[\ln(1+z) - \ln(1-z)\right] = \frac{v_T}{2}\ln\left(\frac{1+z}{1-z}\right), \tag{1.14}$$

which gives t in terms of v, since $v = v_T z$. We can invert this equation to find v as a function of t. The result is

$$v = v_T\left[\frac{e^{gt/v_T} - e^{-gt/v_T}}{e^{gt/v_T} + e^{-gt/v_T}}\right] = v_T \tanh(gt/v_T) \tag{1.15}$$

in terms of a hyperbolic tangent function. From this result we can verify that $v \simeq gt$ for small t, using the series expansion for exponentials $e^x = 1 + x + (1/2)x^2 + \dots$ for small x. We can also verify that $v \rightarrow v_T$ for large t, since then $e^{-gt/v_T} \rightarrow 0$.

So far so good. Now we can find how far the ball falls in a given time by integrating the last result over time, and letting y be the distance fallen. That is:

$$y = \int dy = v_T \int_0^t dt \, \tanh(gt/v_T) = \frac{v_T^2}{g} \int dq \, \tanh q = \frac{v_T^2}{g} \int dq \left(\frac{\sinh q}{\cosh q} \right), \qquad (1.16)$$

where we have defined $q = gt/v_T$ and used the identity $\tanh q = \sinh q / \cosh q$, where sinh and cosh are the hyperbolic sine and cosine functions. The differential of $\cosh q$ is $\sinh q \, dq$, so the integral is just the natural logarithm of $\cosh q$. So finally:

$$y = \left(\frac{v_T^2}{g} \right) \ln(\cosh q) = \left(\frac{v_T^2}{g} \right) \ln(\cosh gt/v_T). \qquad (1.17)$$

This is how far the ball has fallen as a function of time. One can also invert this equation to find how long it takes the ball to reach the ground in terms of its initial height.   ∎

1.4 Oscillation in One-Dimensional Motion

The drag forces we have used so far are purely velocity-dependent forces in which Newton's second law becomes a first-order differential equation. In contrast, a simple harmonic oscillator consists of a mass m attached to one end of a Hooke's-law spring exerting force $F = -kx$, where k (a positive constant) is the force constant of the spring and x is the spring stretch. For such position-dependent forces, Newton's second law becomes a *second-order* differential equation. The minus sign indicates that if x is positive, when the spring has been stretched, it will *pull* the particle back toward equilibrium, and if x is negative, the spring has been compressed, and it will *push* the particle back toward equilibrium. The importance of this linear force extends far beyond the force exerted by an actual spring, because very often it is a spring-like linear restorative force that is exerted when a particle is displaced slightly from equilibrium under the influence of a wide variety of forces. We will return to this point when we discuss energy a bit later.

If the only force on a particle moving in one dimension is due to a Hooke's-law spring, the equation of motion is

$$m\ddot{x} = -kx \quad \text{or} \quad m\ddot{x} + kx = 0, \qquad (1.18)$$

where each overdot means a time derivative, a notation due to Newton himself. This is the famous simple harmonic oscillator (SHO) equation, a linear second-order differential equation in x and t.

There are several ways to solve the equation. One way is to note that we require a couple of linearly independent functions whose second derivatives are the negatives of themselves, apart from constants; this suggests *sines* and *cosines*. The general solution can then be written

$$x(t) = A\cos(\omega t) + B\sin(\omega t) \quad \text{or} \quad x(t) = C\cos(\omega t + \varphi), \qquad (1.19)$$

where $\omega = \sqrt{k/m} = 2\pi\nu$ with ν the frequency of oscillation, and where A and B (or C and φ) are the two arbitrary constants needed in solutions of the second-order differential equation. The constants C and φ can be found in terms of A and B, or vice versa, using the trig identity $\cos(\alpha + \beta) = \cos\alpha\cos\beta - \sin\alpha\sin\beta$. The second form of the solution is depicted in Figure 1.5, illustrating the meaning of the constants C, ω, and φ.

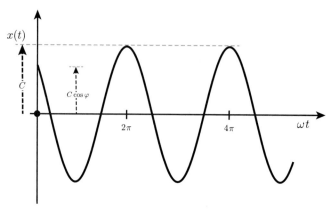

Fig. 1.5 A simple harmonic oscillation $x(t) = C\,\cos(\omega t + \varphi)$ for phase angle $\varphi = \pi/4$. Shown is the amplitude C. The period of oscillations is $P = 2\pi/\omega$, and $\omega = 2\pi\nu$, where $\nu = 1/P$ is the frequency and ω is the angular frequency of oscillation.

Another method of solving the SHO equation is more formal but also provides more insight. We can solve the equation in stages, integrating once to get a first-order differential equation, called a "first integral of motion," and then integrating a second time to get the final solution $x(t)$. This first integration can be carried out by first multiplying the equation by a so-called "integrating factor" $\dot{x}$, giving

$$m\ddot{x}\dot{x} + kx\dot{x} = 0 \quad \text{or} \quad \frac{1}{2}m\frac{d\dot{x}^2}{dt} + kx\frac{dx}{dt} = 0. \qquad (1.20)$$

Multiplying by dt, we have $(1/2)m\,d(\dot{x}^2) + kxdx = 0$, which is directly integrable because each term contains only a single variable. Integrating this last equation:

$$\frac{1}{2}m\dot{x}^2 + \frac{1}{2}kx^2 = E, \qquad (1.21)$$

where E is the constant of integration. We recognize this as a conservation of energy equation for the particle, the sum of its kinetic and potential energies. The kinetic energy $T = (1/2)m\dot{x}^2$ depends on the particle's velocity but not its position, and the potential energy $U = (1/2)kx^2$ depends on the particle's position but not its velocity. The sum is the total energy, a constant of the motion.

Now we can separate the remaining variables x and t and integrate once more:

$$\int dt = \sqrt{\frac{m}{2}} \int \frac{dx}{\sqrt{E - (1/2)kx^2}}, \tag{1.22}$$

again with only a single variable in each term. Substituting $x = \sqrt{2E/k}\cos\theta$ and integrating gives $t = -\sqrt{m/k}\,\theta + $ constant and then rearranging and using the fact that $\cos(-\theta) = \cos(\theta)$ it follows that

$$x(t) = \sqrt{\frac{2E}{k}}\cos(\omega t + \varphi), \tag{1.23}$$

where $\omega = \sqrt{k/m}$ and E and φ are the two necessary arbitrary constants. In addition to showing that energy conservation is the first integral of motion, we have found the amplitude of oscillation in terms of the energy E and force constant k.

Damped Oscillations

The simple harmonic oscillator is not damped. According to the solutions, once excited it will oscillate forever. However, real oscillations eventually die out, which means they must have additional forces exerted on them that cause them to decrease their amplitude with time. A realistic force that does this in most situations is the quadratic damping force $F_{\mathrm{drag}} = -cv^2$, where c is a constant. It will continually reduce the oscillator's amplitude.

Adding this force to the oscillating object leads to the equation $m\ddot{x} = -kx - c\dot{x}^2$, which is still a second-order differential equation, but with a new x^2 term that is *nonlinear*. Unfortunately, this nonlinearity makes the equation impossible to solve in terms of elementary functions, so the tradition is to replace quadratic damping with linear damping, which makes the full equation linear and easy to solve. Even though linear damping is usually unrealistic, it at least leads to decaying oscillations, which is more realistic than no damping at all.

A particular linearly damped oscillator consists of a mass m confined to move in the x direction attached at one end to a Hooke's-law spring of force constant k, and which is also subject to the damping force $-b\,v$ where b is a constant. That is, we assume that the damping force is linearly proportional to the velocity of the mass and in the direction opposite to its motion.

Newton's second law then gives $F = -kx - b\dot{x} = m\ddot{x}$, a second-order linear differential equation equivalent to

$$\ddot{x} + 2\beta\dot{x} + \omega_0^2 x = 0, \tag{1.24}$$

where we let $\beta \equiv b/2m$ and $\omega_0 \equiv \sqrt{k/m}$ to simplify the notation. Mathematically, we are guaranteed a solution once we fix *two* initial conditions. These can be, for example, the initial position $x(0) = x_0$ and velocity $v(0) = \dot{x}(0) = v_0$. Hence, our solution will depend on two constants to be specified in the particular problem. In general, each dynamical variable we track through Newton's second law will generate a single second-order differential equation, and so will require two initial conditions. This is the sense in which Newton's laws provide us with predictive power: fix a few constants using initial conditions, and physics will tell us the future evolution of the system. For the example at hand, Eq. (1.24) is a *linear* differential equation with constant coefficients, which can be solved by setting $x \propto e^{\alpha t}$ for some α. Substituting this form into Eq. (1.24) gives the quadratic equation

$$\alpha^2 + 2\beta\alpha + \omega_0^2 = 0 \quad \text{with solutions} \quad \alpha = -\beta \pm \sqrt{\beta^2 - \omega_0^2}. \tag{1.25}$$

There are now three possibilities: (1) $\beta > \omega_0$, the "overdamped" solution; (2) $\beta = \omega_0$, the "critically damped" solution; and (3), $\beta < \omega_0$, the "underdamped" solution, all as illustrated in Figure 1.5.

(1) In the overdamped case the exponent α is *real* and *negative*, and so the position of the mass as a function of time is

$$x(t) = A_1 e^{\gamma_1 t} + A_2 e^{\gamma_2 t}, \tag{1.26}$$

where $\gamma_1 = -\beta + \sqrt{\beta^2 - \omega_0^2}$ and $\gamma_2 = -\beta - \sqrt{\beta^2 - \omega_0^2}$. Here A_1 and A_2 are arbitrary constants. The two terms are the expected linearly independent solutions of the second-order differential equation, and the coefficients A_1 and A_2 can be determined from the initial position x_0 and initial velocity v_0 of the mass. Figure 1.6(a) shows a plot of $x(t)$.

(2) In the critically damped $\beta = \omega_0$ case the two solutions of Eq. (1.25) merge into the single solution $x(t) = A e^{-\beta t}$. However, a second-order differential equation has two linearly independent solutions, so we need one more. This additional solution is $A' t e^{-\beta t}$ for an arbitrary coefficient A', as can be seen by substituting this form into Eq. (1.24). The general solution for the critically damped case is therefore

$$x = (A + A't)e^{-\beta t}, \tag{1.27}$$

which has the two independent constants A and A' determined from the initial position x_0 and velocity v_0. Figure 1.6(b) shows a plot of $x(t)$ in this case.

(3) In the underdamped case, the quantity $\sqrt{\beta^2 - \omega_0^2} = i\sqrt{\omega_0^2 - \beta^2}$ is purely *imaginary*, so

$$x(t) = e^{-\beta t}\text{Re}\left(A_1 e^{i\omega_1 t} + A_2 e^{-i\omega_1 t}\right), \tag{1.28}$$

where $\omega_1 = \sqrt{\omega_0^2 - \beta^2}$ and we take only the *real* part of the solution, as indicated by "Re." It is mathematically legal to take only the real part of the solution since the differential equation is real and linear in x: if the complex function $x(t)$ solves the differential equation, so will the real and imaginary parts of $x(t)$ separately.[1] We can use Euler's identity

$$e^{i\theta} = \cos\theta + i\sin\theta \tag{1.29}$$

---

[1] You can convince yourself of this by plugging $x(t) = x_R(t) + i x_I(t)$ into the differential equation and extracting two identical equations for $x_R(t)$ and $x_I(t)$ from the real and imaginary parts, respectively.

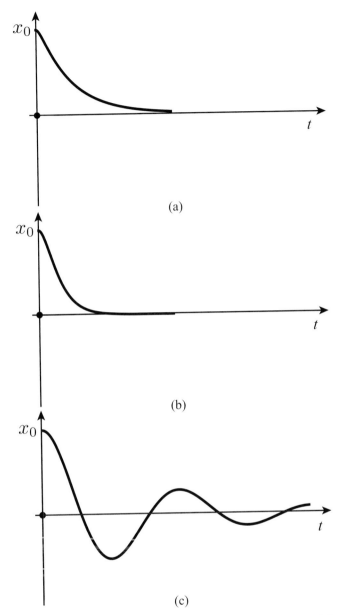

Fig. 1.6 Motion of an oscillator if it is (a) overdamped, (b) critically damped, or (c) underdamped, for the special case where the oscillator is released from rest ($v_0 = 0$) at some position x_0.

to write x in terms of purely real functions:

$$x(t) = e^{-\beta t}(\bar{A}_1 \cos \omega_1 t + \bar{A}_2 \sin \omega_1 t), \tag{1.30}$$

where $\bar{A}_1 = A_1 + A_2$ and $\bar{A}_2 = i(A_1 - A_2)$ are real coefficients. We can also use the identity $\cos(\theta + \varphi) = \cos \theta \cos \varphi - \sin \theta \sin \varphi$ to write Eq. (1.30) in the form

$$x(t) = A e^{-\beta t} \cos(\omega_1 t + \varphi), \tag{1.31}$$

where $A = \sqrt{\bar{A}_1^2 + \bar{A}_2^2}$ and $\varphi = \tan^{-1}(-\bar{A}_2/\bar{A}_1)$. That is, the underdamped solution corresponds to a decaying oscillation with amplitude $A e^{-\beta t}$. The arbitrary constants A and φ can be determined from the initial position x_0 and velocity v_0 of the mass. Figure 1.6(c) shows a plot of $x(t)$ in this case. If there is no damping at all, we have $b = \beta = 0$ (and the oscillator is obviously "underdamped"). The original Eq. (1.24) becomes the SHO equation $\ddot{x} + \omega_0^2 x = 0$ whose most general solution is

$$x(t) = A \cos(\omega_0 t + \varphi). \tag{1.32}$$

This gives away the meaning of ω_0: it is the angular frequency of oscillation of a simple harmonic oscillator, related to the oscillation frequency ν in cycles/second by $\omega_0 = 2\pi\nu$. Note that $\omega_1 < \omega_0$; i.e., the damping reduces the oscillation frequency in addition to damping the amplitude.

Whichever solution applies, it is clear that the motion of the particle is determined by (a) the initial position $x(0)$ and velocity $\dot{x}(0)$, and (b) the forces acting on it throughout its motion.

1.5 Resonance

If we "drive" a lightly damped spring–mass system with an oscillating force at the right frequency we observe the phenomenon of *resonance*. Repeated small stimulations of an oscillating system at its natural frequency of oscillation can cause the oscillation amplitude to become large, especially if the damping is small. In particular, consider adding a sinusoidal driving force $F = F_0 \sin \omega t$ to the spring force and the damping force acting upon a spring–mass system. Then Newton's law becomes

$$m\ddot{x} = F_{\text{spring}} + F_{\text{damping}} + F_{\text{driving}} = -kx - b\dot{x} + F_0 \sin \omega t. \tag{1.33}$$

We can change the driving frequency ω arbitrarily. So now we have three important frequencies, the "natural" frequency $\omega_0 = \sqrt{k/m}$ of an undamped spring–mass system; the linear damped frequency $\omega_1 = \sqrt{\omega_0^2 - \beta^2}$, where $\beta = b/2m$; and the new driving frequency ω. There are various ways to apply this sinusoidal driving force. One way is to hold the end of the spring which is *not* connected to the mass m, and move it back and forth sinusoidally in the x direction, so its position on a frictionless table as a function of time is $X = A \sin \omega t$. Then the length of the spring at any time is not x, but $(x - X)$, so the force it exerts on m is $F_{\text{spring}} = -k(x - X) = -k(x - A \sin \omega t)$. Newton's second law then gives

$$m\ddot{x} = -k(x - A \sin \omega t) - bv = -kx - b\dot{x} + kA \sin \omega t, \tag{1.34}$$

so

$$m\ddot{x} + b\dot{x} + kx = F_0 \sin \omega t \quad \text{or} \quad \ddot{x} + 2\beta\dot{x} + \omega_0^2 x = f_0 \sin \omega t, \tag{1.35}$$

where $F_0 \equiv kA$, $\beta \equiv b/2m$ is the *damping constant*, and $f_0 \equiv F_0/m$. This is the equation of a driven, linearly damped harmonic oscillator. Mathematically speaking, the differential equation is still linear and of second order, but it has been changed from a homogeneous to an inhomogeneous equation, due to the driving force term on the right. The solution of this inhomogeneous equation is the sum of the general (or "characteristic") solution $x_c(t)$ of the homogeneous equation (*i.e.*, the equation without the driving term on the right) and a particular solution $x_p(t)$ of the full inhomogeneous equation

$$x(t) = x_c(t) + x_p(t). \tag{1.36}$$

We have already found the general solution of the homogeneous equation. It is

$$x_c(t) = Ae^{-\beta t}\cos(\omega_1 t + \varphi_0), \tag{1.37}$$

where $\omega_1 = \sqrt{\omega_0^2 - \beta^2}$ and the amplitude A and phase angle φ_0 are the requisite number of arbitrary constants for the second-order differential equation. Note that this homogeneous term $x_c(t)$ gradually dies out, so it is often called the "transient" solution, as illustrated in Figure 1.7.

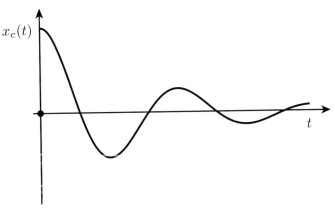

Fig. 1.7   Transient solution of a forced, damped harmonic oscillator.

It is the other, "particular" solution $x_p(t)$ that wins out in the end, and it is called the "steady state" solution

$$x_p(t) = C \sin(\omega t + \delta), \tag{1.38}$$

where C and δ are constants to be determined. The complete solution is the sum of the steady-state solution and the transient (characteristic) solution:

$$x_p(t) = Ae^{-\beta t}\cos(\omega_1 t + \varphi_0) + C\sin(\omega t - \delta). \tag{1.39}$$

The first term, the transient solution, dies away as time goes on, leaving the steady-state solution with amplitude C.

How did we know the form of $x_p(t)$? We could first try $x_p = C\sin\omega t$ for some constant C, in which the mass oscillates in synchrony with the driving force. However, that cannot work, because the first-derivative term in the differential equation converts the sine to a cosine, while every other term in the equation is the sine, so there is no value of C for which the trial solution works. Another possibility is to try the phase-shifted sine function $x_p(t) = C\sin(\omega t - \delta)$, which oscillates at the driving frequency but is phase-shifted by the angle δ.

Substituting this trial solution $x_p(t) = C\sin(\omega t - \delta)$ into the differential equation gives

$$C[(\omega_0^2 - \omega^2)\sin(\omega t - \delta) + 2\beta\omega\cos\omega t - \delta] = f_0\sin\omega t. \tag{1.40}$$

Using the trig identities

$$\sin(a \pm b) = \sin a\cos b \pm \cos a\sin b \quad\text{and}\quad \cos(a \pm b) = \cos a\cos b \mp \sin a\sin b \tag{1.41}$$

we write

$$C[(\omega_0^2 - \omega^2)(\sin\omega t\cos\delta - \cos\omega t\sin\delta)$$
$$+ 2\beta\omega(\cos\omega t\cos\delta + \sin\omega t\sin\delta)] = f_0\sin\omega t, \tag{1.42}$$

which must hold at all times. Orthogonality of the sine and cosine functions implies that the coefficients of each should independently vanish. For example, at times t such that $\omega t = 0, \pi, 2\pi$, etc., the $\sin\omega t$ terms all vanish, so the $\cos\omega t$ terms alone must satisfy the equation. That is:

$$C[(-\omega_0^2 - \omega^2)\sin\delta + 2\beta\omega\cos\delta] = 0 \tag{1.43}$$

at any one of the times mentioned above. But all of these quantities are independent of time, so this expression must *always* be zero. Therefore the quantity inside the square brackets vanishes. That is:

$$\tan\delta = \frac{2\beta\omega}{\omega_0^2 - \omega^2} = \frac{(2\beta/\omega_0)(\omega/\omega_0)}{1 - (\omega/\omega_0)^2}. \tag{1.44}$$

Notice that if the damping $\beta \to 0$, it follows that $\tan\delta \to 0$, so that the phase angle $\delta \to 0$ as well. Then the mass moves back and forth in phase with the driving force. This is also true for very low applied frequencies ω; as $\omega \to 0$, the phase angle $\delta \to 0$. This means that if the driving force causes the spring to oscillate very slowly back and forth, the mass on the other end of the spring will move back and forth in phase with the driving force.

Figure 1.8 is a graph of the phase angle δ as a function of ω/ω_0, the ratio of the driving frequency to the natural frequency of the undamped spring for a particular value of $2\beta/\omega_0$. Note that the response of the system is $\pi/2$ out of phase with the driving frequency if $\omega = \omega_0$, and out of phase by the angle π if $\omega \gg \omega_0$. It is straightforward to work out the changes in shape of this graph depending upon the value of $2\beta/\omega_0$. Equation (1.42) must also be correct for all times such that

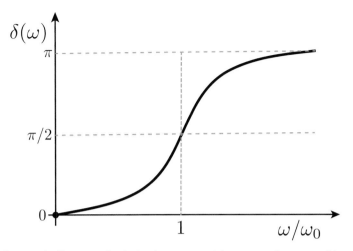

Fig. 1.8 Graph of the phase angle δ between the driving frequency and the response frequency of the oscillator, drawn for a particular value of the parameter $2\beta/\omega_0$. As the parameter is made larger, the slope of the graph becomes steeper near $\omega/\omega_0 = 1$.

$\omega t = \pi/2, 3\pi/2, 5\pi/2$, etc., when each $\cos \omega t$ term is zero. Only the $\sin \omega t$ terms survive, so it follows that

$$C[(\omega_0^2 - \omega^2)\cos\delta + 2\beta\omega\sin\delta] = f_0, \qquad (1.45)$$

so the constant C is

$$C = \frac{f_0}{(\omega_0^2 - \omega^2)\cos\delta + 2\beta\omega\sin\delta}. \qquad (1.46)$$

We have already found $\tan\delta$, so noting that

$$\tan\delta = \frac{\sin\delta}{\cos\delta} = \frac{\sin\delta}{\sqrt{1 - \sin^2\delta}}, \qquad (1.47)$$

we get

$$\sin\delta = \frac{2\beta\omega}{\sqrt{(\omega_0^2 - \omega^2)^2 + 4\beta^2\omega^2}}. \qquad (1.48)$$

Substituting these into the previous equation for C, the final result for the amplitude C as a function of the driving frequency ω is

$$C(\omega) = \frac{f_0}{(\omega_0^2 - \omega^2)\cos\delta + 2\beta\omega\sin\delta} = \frac{f_0}{\sqrt{(\omega_0^2 - \omega^2)^2 + 4\beta^2\omega^2}}. \qquad (1.49)$$

Now we have found both the amplitude and the phase angle of the "particular" (steady-state) solution $x_p(t)$ of the forced, damped oscillator:

$$x_p(t) = C(\omega)\sin(\omega t - \delta(\omega)), \qquad (1.50)$$

where $C(\omega)$ is given by Eq. (1.49) and the phase angle δ by Eq. (1.44).

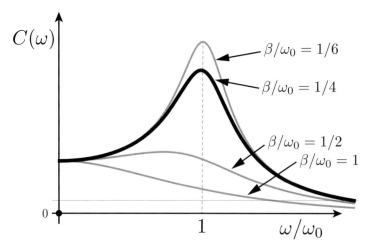

Fig. 1.9 Shape of the oscillation amplitude response of the system as a function of ω/ω_0 for various damping constants.

The shape of $C(\omega)$, the amplitude response, is especially interesting; it is displayed in Figure 1.9 as a function of the ratio ω/ω_0, for various damping constants, as characterized by the ratio β/ω_0. The curves show a resonance peak at a frequency near, but not quite at, the natural frequency ω_0 of the undamped spring–mass system. Note that the curves are sharper for small damping than for large damping. If the driving force frequency ω is close to the natural frequency ω_0 the response is large, especially if the drag is small. This is the resonance phenomenon. The resonance frequency ω_R is the frequency corresponding to the maximum in the response curve $C(\omega)$. It is then given by

$$\omega_R = \sqrt{\omega_0^2 - 2\beta^2}, \tag{1.51}$$

which is easily found by setting $dC(\omega)/d\omega = 0$. It is then easy to show that the oscillation amplitude at resonance is

$$C_R = \frac{F_0}{2m\beta\omega_1}, \tag{1.52}$$

where $\omega_1 = \sqrt{\omega_0^2 - \beta^2}$ is the frequency of the damped, undriven oscillator. Note that C_R is large if the damping β is small.

Resonance can be observed by repeated small pushes on a child on a swing at his or her natural frequency of oscillation; or by driving a car at just the right speed on a washboard road, especially when the car has no shock absorbers to damp out the motion; or when tuning a radio, where incoming radio waves striking the antenna can excite large oscillations in the radio's electrical circuits if the frequency is just right, but not otherwise, so you hear only the station you tuned for.

1.6 Motion in Two or Three Dimensions

So far all of our examples have been restricted to one-dimensional motion. When the motion is in two or three dimensions, the first step is to select an appropriate coordinate system that fits the problem. For two-dimensional motion there are Cartesian or plane polar coordinates, for example, and for three-dimensional motion there are Cartesian, spherical, or cylindrical coordinates, the most common choices among many others.

Having chosen a coordinate system, it is often convenient to express vector quantities like position, velocity, acceleration, or force using **unit vectors**. Each unit vector has unit length and points in one of the orthogonal directions corresponding to the coordinates in the system. It follows that the dot product of any unit vector with itself is unity, while the dot product of any unit vector with any other unit vector in the same system is zero.

For Cartesian coordinates in two dimensions the unit vectors are $\hat{\mathbf{x}}$ and $\hat{\mathbf{y}}$, where

$$\hat{\mathbf{x}} \cdot \hat{\mathbf{x}} = 1, \quad \hat{\mathbf{y}} \cdot \hat{\mathbf{y}} = 1, \quad \hat{\mathbf{x}} \cdot \hat{\mathbf{y}} = \hat{\mathbf{y}} \cdot \hat{\mathbf{x}} = 0. \tag{1.53}$$

The position vector of a particle is then

$$\mathbf{r} = x\hat{\mathbf{x}} + y\hat{\mathbf{y}} \tag{1.54}$$

and the particle's velocity and acceleration vectors are

$$\mathbf{v} = \frac{d\mathbf{r}}{dt} = \dot{x}\hat{\mathbf{x}} + \dot{y}\hat{\mathbf{y}} \quad \text{and} \quad \mathbf{a} = \frac{d\mathbf{v}}{dt} = \ddot{x}\hat{\mathbf{x}} + \ddot{y}\hat{\mathbf{y}}. \tag{1.55}$$

In differentiating $\mathbf{r}$ and $\mathbf{v}$ we differentiated their components, but did not have to differentiate the unit vectors, because $\hat{\mathbf{x}}$ and $\hat{\mathbf{y}}$ are constants: neither the length of these unit vectors nor their directions in space change with time. If plane polar coordinates r, θ are chosen instead, the unit vectors are $\hat{\mathbf{r}}$ and $\hat{\boldsymbol{\theta}}$, where

$$\hat{\mathbf{r}} \cdot \hat{\mathbf{r}} = 1, \quad \hat{\boldsymbol{\theta}} \cdot \hat{\boldsymbol{\theta}} = 1, \quad \hat{\mathbf{r}} \cdot \hat{\boldsymbol{\theta}} = \hat{\boldsymbol{\theta}} \cdot \hat{\mathbf{r}} = 0. \tag{1.56}$$

Now whereas Cartesian unit vectors do not change with time, the plane polar unit vectors generally *do* change as the particle moves, because their directions may change. For example, if the particle moves in a circle around the origin, both unit vectors $\hat{\mathbf{r}}$ and $\hat{\boldsymbol{\theta}}$ change direction in space. In fact, their time derivatives are

$$\frac{d\hat{\mathbf{r}}}{dt} = \dot{\theta}\hat{\boldsymbol{\theta}} \quad \text{and} \quad \frac{d\hat{\boldsymbol{\theta}}}{dt} = -\dot{\theta}\hat{\mathbf{r}}. \tag{1.57}$$

In plane polar coordinates the position vector of a particle is simply $\mathbf{r} = r\hat{\mathbf{r}}$. Therefore the velocity is

$$\mathbf{v} = \frac{d\mathbf{r}}{dt} = \dot{r}\hat{\mathbf{r}} + r\dot{\hat{\mathbf{r}}} = \dot{r}\hat{\mathbf{r}} + r\dot{\theta}\,\hat{\boldsymbol{\theta}}, \tag{1.58}$$

and the acceleration is

$$\mathbf{a} = \frac{d\mathbf{v}}{dt} = (\ddot{r} - r\dot{\theta}^2)\hat{\mathbf{r}} + (r\ddot{\theta} + 2\dot{r}\dot{\theta})\hat{\boldsymbol{\theta}}. \tag{1.59}$$

This equation contains within it the well-known results that a particle circling the origin at constant radius r and constant angular velocity $\dot{\theta} \equiv \omega$ will have an inward ("centripetal") acceleration $-r\omega^2 \hat{r} = -(v^2/r)\,\hat{r}$, and a person walking outward $\dot{r} > 0$ on a steadily rotating carousel with angular velocity $\dot{\theta} > 0$ will be accelerating *sideways*, in the $\hat{\boldsymbol{\theta}}$ direction. Much more on all of this in Chapter 9.

Of course, motion in all three dimensions requires three unit vectors, typically for Cartesian, spherical, or cylindrical coordinates. These unit vectors are given in Appendix A.

Example 1.3   A Slingshot on the Moon

Someday we may want to construct spacecraft or space colonies not on the earth or the moon but in space itself, using mined metals and other materials lifted off the moon. The moon has the advantage of a much smaller escape velocity than that of the earth, and no atmosphere to retard motion. Instead of using expensive rockets and fuel, could it be possible to achieve the escape velocity from the airless moon by *slinging* containers of material from its surface using a rapidly rotating boom? A sturdy boom of length R might swing around in a horizontal plane on the moon's surface about a central vertical axis at constant angular velocity ω. A payload container starting near the rotation axis of the boom might then slide with increasing speed out along the length of the boom and then project outward at a very high velocity when it leaves the end of the boom.

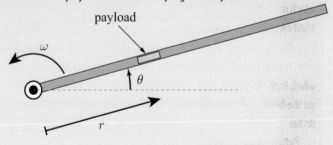

Fig. 1.10   A boom with payload on the moon's surface, rotating in a horizontal plane.

Plane polar coordinates are the obvious choice here, with r measured outward from the rotation axis and θ the angle of the boom from some initial angle $\theta = 0$ when the payload is released on the rotating boom at a small initial radius r_0, with $\dot{r}_0 = 0$. The boom keeps swinging around at constant angular velocity, so the angle of the payload is $\theta = \omega t$ until it finally flies off the end of the boom (see Figure 1.10).

We assume the payload slides frictionlessly along the boom, so the radial force $F_r = 0$. The tangential force is $F_\theta \neq 0$, which is the normal force of the boom on the payload, keeping it moving with constant angular velocity ω as it slides outward. Newton's second law is then

$$\mathbf{F} = F_\theta \hat{\boldsymbol{\theta}} = m\mathbf{a} = m[(\ddot{r} - r\omega^2)\hat{\mathbf{r}} + (r\ddot{\theta} + 2\dot{r}\dot{\theta})\hat{\boldsymbol{\theta}}], \tag{1.60}$$

so

$$\ddot{r} - \omega^2 r = 0 \quad \text{and} \quad F_\theta = m(r\ddot{\theta} + 2\dot{r}\dot{\theta}). \tag{1.61}$$

The first equation is a linear, second-order differential equation with solution $r = Ae^{\omega t} + Be^{-\omega t}$, where A and B are arbitrary constants. We can find A and B from the given initial conditions at $t = 0$, which are $r = r_0$ and $\dot{r} = 0$. This gives $A = B = r_0/2$, so

$$r = (r_0/2)(e^{\omega t} + e^{-\omega t}) \equiv r_0 \cosh \omega t \tag{1.62}$$

in terms of the hyperbolic cosine function. Then the velocity of the payload as a function of time, including both the radial and tangential components, is

$$\mathbf{v} = \dot{r}\,\hat{\mathbf{r}} + r\dot{\theta}\,\hat{\boldsymbol{\theta}} = r_0\omega \sinh \omega t\,\hat{\mathbf{r}} + r_0\omega \cosh \omega t\,\hat{\boldsymbol{\theta}}. \tag{1.63}$$

We can find the payload velocity when it reaches the end of the boom. At that point $R = r_0 \cosh \omega t_f$, where t_f is the time when this happens. Then $\cosh \omega t_f = R/r_0$ and $\sinh \omega t_f = \sqrt{\cosh^2 \omega t_f - 1} = \sqrt{(R/r_0)^2 - 1}$, where we have used the identity $1 + \sinh^2 = \cosh^2$. Substituting these results into the expression for $\mathbf{v}$:

$$\mathbf{v} = \omega \left[\sqrt{R^2 - r_0^2}\,\hat{\mathbf{r}} + R\hat{\boldsymbol{\theta}} \right] \tag{1.64}$$

and from this we can find the speed of the payload as it flies off the end:

$$v_f = \sqrt{v_r^2 + v_\theta^2} = \omega\sqrt{2R^2 - r_0^2}, \tag{1.65}$$

which must equal or exceed the moon's escape velocity. Finally, we can calculate the tangential force the boom must exert upon the payload to keep $\theta = \omega t$, as a function of time and as a function of r:

$$F_\theta = m(r\ddot{\theta} + 2\dot{r}\dot{\theta}) = m(0 + 2\omega^2 r_0 \sinh \omega t) = 2m\omega^2\sqrt{r^2 - r_0^2}, \tag{1.66}$$

which is greatest when $r = R$, at the tip of the boom. There will be an equal but opposite reaction force back on the boom due to the payload, so the boom must be strong enough to withstand this tangential force at its tip.

Putting in some numbers, the escape velocity on the moon is approximately 2.4 km/s, and we can choose $\omega = 2\pi\ \text{s}^{-1}$ and $r_0 = 1$ m. The radius of the boom must then be $R \simeq 270$ m.   ∎

1.7 Systems of Particles

Up to now we have concentrated on the dynamics of *single* particles. We will now expand our horizons to encompass systems of an arbitrary number of particles. A system of particles might be an entire solid object like a bowling ball, in which tiny parts of the ball can be viewed as individual infinitesimal particles. Or we might have a liquid in a glass, or the air in a room, or a planetary system, or a galaxy of stars, all made of constituents we treat as "particles."

The location of the ith particle of a system can be identified by a position vector $\mathbf{r}_i$ extending from the origin of coordinates to that particle, as illustrated in Figure 1.11. Using the laws of classical mechanics for each particle in the system, we can find the laws that govern the system as a whole.

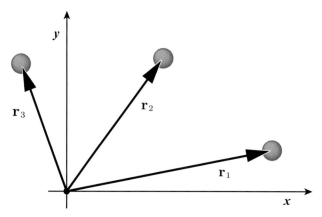

Fig. 1.11 A system of particles, with each particle identified by a position vector $\mathbf{r}_i$ with $i = 1, 2, 3$.

Define the total momentum $\mathbf{P}$ of the system as the sum of the momenta of the individual particles:

$$\mathbf{P} = \sum_i \mathbf{p}_i. \tag{1.67}$$

Similarly, define the total force $\mathbf{F}_\mathrm{T}$ on the system as the sum of all the forces on all the particles:

$$\mathbf{F}_\mathrm{T} = \sum_i \mathbf{F}_i. \tag{1.68}$$

It then follows that $\mathbf{F}_\mathrm{T} = d\mathbf{P}/dt$, just by adding up the individual $\mathbf{F}_i = d\mathbf{p}_i/dt$ equations for all the particles. If we further split up the total force $\mathbf{F}_\mathrm{T}$ into $\mathbf{F}_\mathrm{ext}$ (the sum of the forces exerted by external agents, like earth's gravity or air resistance on the system of particles that form a golfball) and $\mathbf{F}_\mathrm{int}$ (the sum of the internal forces between members of the system themselves, like the mutual forces between particles within the golfball), then

$$\mathbf{F}_\mathrm{T} = \mathbf{F}_\mathrm{int} + \mathbf{F}_\mathrm{ext} = \mathbf{F}_\mathrm{ext}, \tag{1.69}$$

because all the internal forces cancel out by Newton's third law. That is, for any two particles i and j, the force of i on j is equal but opposite to the force of j on i. Finally, we can write a grand second law for the system as a whole:

$$\mathbf{F}_\mathrm{ext} = \frac{d\mathbf{P}}{dt}, \tag{1.70}$$

showing how the system as a whole moves in response to external forces.

Now the importance of momentum is clear. For if no external forces act on the collection of particles $\mathbf{F}_{\text{ext}} = 0$, their *total* momentum cannot depend upon time, so $\mathbf{P}$ is conserved. Individual particles in the collection may move in complicated ways, but they always move in such a way as to keep the total momentum constant.

Another useful quantity characterizing a system of particles is their **center of mass** position $\mathbf{R}_{\text{CM}}$. Let the ith particle have mass m_i, and define the center of mass of the collection of particles as

$$\mathbf{R}_{\text{CM}} = \frac{\sum_i m_i \mathbf{r}_i}{M}, \tag{1.71}$$

where $M = \sum_i m_i$ is the total mass of the system. We can write the position vector of a particle as the sum $\mathbf{r}_i = \mathbf{R}_{\text{CM}} + \mathbf{r}'_i$, where $\mathbf{r}'_i$ is the position vector of the particle measured from the center of mass, as illustrated in Figure 1.12.

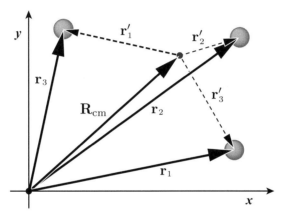

Fig. 1.12 A collection of particles, each with a position vector $\mathbf{r}_i$ from a fixed origin. The center of mass $\mathbf{R}_{\text{CM}}$ is shown, and also the position vector $\mathbf{r}'_i$ of the ith particle measured from the center of mass.

The velocity of the center of mass is

$$\mathbf{V}_{\text{CM}} = \frac{d\mathbf{R}_{\text{CM}}}{dt} = \frac{\sum_i m_i \mathbf{v}_i}{M} = \frac{\mathbf{P}}{M}, \tag{1.72}$$

differentiating term by term, and using the fact that the particle masses are constant. Again $\mathbf{P}$ is the total momentum of the particles, so we have proven that the center of mass moves at constant velocity whenever $\mathbf{P}$ is conserved – that is, whenever there is no net external force. In particular, if there is no external force on the particles, their center of mass stays at rest if it starts at rest.

This result is also very important because it shows that a real object composed of many smaller "particles" can be considered a particle itself: it obeys all of Newton's laws with a position vector given by $\mathbf{R}_{\text{CM}}$, a momentum given by $\mathbf{P}$, and the only relevant forces being the external ones. It relieves us of having to draw a distinct line between *particles* and *systems* of particles. For some purposes we think of a star as composed of many smaller particles, and for other purposes the star as a whole could be considered a single particle in the system of stars called a galaxy.

1.8 Conservation Laws

Using Newton's laws we can show that under the right circumstances there are as many as three dynamical properties of a particle that remain constant in time, *i.e.*, that are *conserved*. These properties are **momentum, angular momentum,** and **energy**. They are conserved under different circumstances, so in any particular case all of them, none of them, or only one or two of them may be conserved. As we will see, a conservation law typically leads to a *first-order* differential equation, which is generally much easier to tackle than the usual second-order equations we get from Newton's second law. This makes identifying conservation laws in a system a powerful tool for problem solving and characterizing the motion. We will see later in Chapter 6 that there are deep connections between conservation laws and symmetries in Nature.

Momentum

From Newton's second law in the form $\mathbf{F} = d\mathbf{p}/dt$ it follows that if there is no net force on a particle, its momentum $\mathbf{p} = m\mathbf{v}$ is conserved, so its velocity $\mathbf{v}$ is also constant. Conservation of momentum for a single particle simply means that a free particle (a particle with no force on it) moves in a straight line at constant speed. For a single particle, conservation of momentum is equivalent to Newton's first law.

For a *system* of particles, however, momentum conservation becomes nontrivial, because it requires the conservation of only *total* momentum $\mathbf{P}$. When there are no *external* forces acting on a system of particles, the total momentum of the individual constituents remains constant, even though the momentum of each single particle may change:

$$\mathbf{P} = \sum_i \mathbf{p}_i = \text{constant}. \tag{1.73}$$

As we saw earlier, this is the momentum of the center of mass of the system if we were to imagine the sum of all the constituent masses added up and placed at the center of mass. This relation can be very handy when dealing with several particles.

Example 1.4

A Wrench in Space

We are sitting within a spaceship watching a colleague astronaut outside holding a wrench. The astronaut-plus-wrench system is initially at rest from our point of view. The astronaut (of mass M) suddenly throws the wrench (of mass m), with some unknown force. We then see the astronaut moving with velocity $\mathbf{V}$. Without knowing anything about the force with which she threw the wrench, we can compute the velocity of the wrench. No external forces act on the system consisting of wrench plus astronaut, so its total momentum is conserved:

$$\mathbf{P} = M\mathbf{V} + m\mathbf{v} = \text{constant}, \tag{1.74}$$

where $\mathbf{v}$ is the unknown velocity of the wrench. Since the system was initially at rest, we know that $\mathbf{P} = 0$ for all time. We then deduce

$$\mathbf{v} = -\frac{M\mathbf{V}}{m} \tag{1.75}$$

without needing to use Newton's second law or any other differential equation.   ∎

Example 1.5 **Rockets**

In the preceding example the astronaut gains velocity in a direction opposite to the direction in which she throws the wrench, thereby conserving overall momentum. A rocket behaves exactly the same way, for exactly the same reason, except the single throw of a wrench is replaced by the continuous exhaust of burned fuel streaming out from the combustion chamber at the rear of the rocket. Figure 1.13 shows the rocket moving to the right in gravity-free empty space; there are no external forces, so the total momentum of the rocket plus expelled combustion gases must be conserved. At time t, shown in Figure 1.13(a), the rocket (including onboard fuel) has mass m and velocity v. Slightly later, at time $t + \Delta t$, as shown in Figure 1.13(b), the rocket has mass $m + \Delta m$ (where Δm is *negative*, since the rocket has expelled some fuel in the exhaust) and velocity $v + \Delta v$. In addition, there is now an exhaust mass $-\Delta m = |\Delta m|$, where $-\Delta m$ is positive. Note that our system of rocket plus exhaust has constant mass, which is essential here, because it only makes sense to conserve momentum for a system in which the mass stays the same.

What is the velocity of the bit of exhaust $|\Delta m|$ in the second figure? We will suppose that its velocity is u relative to the rocket, called the *exhaust velocity*, directed in the backwards direction, and so in the inertial frame in which we are viewing the rocket the rocket has velocity v (or $v + \Delta v$) to the right – it will make no difference which we choose – so the bit of exhaust has velocity $u - v$ to the *left* from our point of view. (Note that if at some instant the rocket happens to be moving to the right at speed u relative to us, then the bit of exhaust will be at rest in our frame; if the rocket is moving faster than u, the bit of exhaust will actually be moving to the right, since $u - v$ will be negative.)

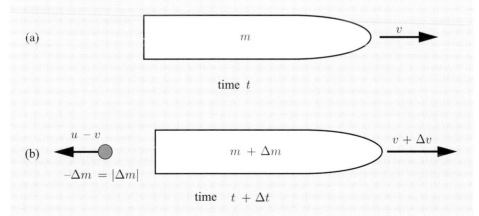

Fig. 1.13   A rocket and expelled exhaust (a) at time t and (b) at time $t + \Delta t$.

We can now conserve momentum between times t and $t + \Delta t$. That is:

$$(m + \Delta m)(v + \Delta v) - (-\Delta m)(u - v) = mv. \tag{1.76}$$

So

$$m\,\Delta v + \Delta m\,\Delta v + \Delta m\,u = 0. \tag{1.77}$$

Dividing by the brief time interval Δt and taking the limit $\Delta t \to 0$, the doubly small term $\Delta m \Delta v$ goes away in the limit, so we find that the equation of motion is

$$m(t)\frac{dv}{dt} = -u\frac{dm}{dt}. \tag{1.78}$$

This looks very similar to Newton's second law in the form $mdv/dt = F$, except that here the mass of the rocket changes with time. The "force" term on the right is called the *thrust* of the rocket:

$$\text{Thrust} \equiv -u\frac{dm}{dt}, \tag{1.79}$$

which is *positive* because the rocket mass is decreasing with time as its fuel is burned. The equation makes intuitive sense: the thrust is proportional to both the exhaust velocity and the rate at which the fuel is burned.

We can now integrate the rocket's equation of motion if we assume that the exhaust velocity u is constant. First, multiply Eq. (1.78) by dt and divide by m: this removes t as a variable, and we are left with $dv = -udm/m$. The remaining variables v and m have been separated, so we can integrate both sides:

$$\int_{v_0}^{v} dv = -u \int_{m_0}^{m} \frac{dm}{m}, \tag{1.80}$$

giving

$$v = v_0 + u\ln(m_0/m), \tag{1.81}$$

which is often called the *rocket equation*. If, for example, 90% of the initial mass of the rocket consists of fuel, while only 10% is "payload," then when all the fuel has burned the rocket has only 10% of its original mass, so its velocity has increased by

$$v - v_0 = u\ln\left(\frac{m_0}{m_{\text{payload}}}\right) = u\ln\left(\frac{m_0}{0.1m_0}\right) \simeq 2.30\,u. \tag{1.82}$$

By the end, the rocket is traveling faster than the fuel speed relative to the rocket. ∎

Finding the motion of a rocket is an example of a "variable mass" problem, called that because the mass of the object of interest (the rocket in this case) changes mass as time goes on. There are dozens of analogous problems, including for example (i) a hailstone that gains mass with time, freezing and accreting water molecules in the air as it falls; (ii) a jet aircraft whose mass increases as its wings ice up while its mass *decreases* as fuel is burned; (iii) a railroad boxcar moving along a horizontal track, open at the top and gaining mass as rain falls in, while losing mass due to a hole in the bottom of the boxcar through which water is leaking. Note that the total mass of the system does not change; it simply moves from one part of the system to another.

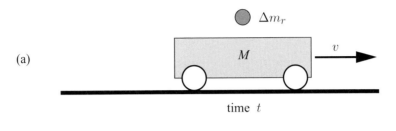

(a)

time t

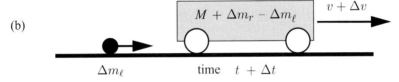

(b)

Δm_ℓ time $t + \Delta t$

Fig. 1.14   A leaky open boxcar in a rainstorm. (a) At time t the boxcar is moving at velocity v and some raindrops of mass Δm_r are about to fall in, with no horizontal component of velocity. (b) At time $t + \Delta t$ the boxcar is moving at velocity $v + \Delta v$. The raindrops Δm_r have fallen in, and a quantity of water Δm_ℓ has leaked out, still moving with horizontal velocity v.

The technique for solving such problems is to use Newton's second law $F = dp/dt$ in the form $\Delta p = F \Delta t$ over the short time interval Δt *for a system whose mass is the same at time $t + \Delta t$ as it was at time t.* That is, we can only be confident that $F = dp/dt$ is valid if the system has fixed mass. So in the case of the boxcar, for example, we draw two pictures (see Figure 1.14). The first at time t shows a boxcar of mass M moving to the right at speed v plus a small quantity of rain of mass Δm_r falling with no horizontal velocity (its vertical velocity is irrelevant here). Thus, the horizontal momentum of the system at time t is simply $p_0 = Mv$. The second picture is at time $t + \Delta t$, and shows a boxcar of mass $M + \Delta m_r - \Delta m_\ell$, indicating that the boxcar has gained mass Δm_r due to the falling rain, while losing mass Δm_ℓ due to the leak. In this picture there is also a mass Δm_ℓ, the leaked mass, moving to the right at speed v, because it "remembers" the speed it had just before it leaked out by the law of inertia, Newton's first law. The momentum of the entire system at $t + \Delta t$ is $p_1 = (M + \Delta m_r - \Delta m_\ell)(v + \Delta v) + \Delta m_\ell v$. Now if we pretend there is no horizontal force on the system due to air resistance or friction with the tracks, the total momentum of the fixed-mass system is the same at $t + \Delta t$ as it was at time t. Therefore, setting $p_1 = p_0$:

$$(M + \Delta m_r - \Delta m_\ell)(v + \Delta v) + \Delta m_\ell v = Mv. \qquad (1.83)$$

Now cancel the Mv terms, divide by Δt, and take the limit $\Delta t \to 0$. The result is the differential equation of motion of the boxcar:

$$M\frac{dv}{dt} = -\lambda_r v, \quad \text{where} \quad \lambda_r = \frac{dm_r}{dt}. \tag{1.84}$$

Here, λ_r is the rate at which rain is falling in. Note that this equation *looks* just like the equation for the bacterium subject to a linear drag force. The cause of the "drag" here is that the boxcar has to speed up the horizontal velocity of the raindrops that fall in, and the rain reacts back upon the boxcar tending to slow it down. Appearances may be deceiving, however, because in the boxcar problem M changes with time unless the rate of rainfall happens to be exactly the same as the rate of leakage. Nevertheless, we can solve the problem completely for $v(t)$ and then $x(t)$ if we assume the rates of rainfall and leaking are both constants, λ_r and λ_ℓ (see the Problems section at the end of this chapter). We can also find the differential equation of motion if there is air resistance or friction by adding nonzero forces to $\Delta p = F\Delta t$, and perhaps solve the equation exactly if F has a sufficiently simple form.

Angular Momentum

Let a position vector $\mathbf{r}$ extend from an origin of coordinates to a particle, as shown in Figure 1.15. The angular momentum of the particle is defined to be

$$\boldsymbol{\ell} = \mathbf{r} \times \mathbf{p}, \tag{1.85}$$

the vector cross product of $\mathbf{r}$ with the particle's momentum $\mathbf{p}$. Note that in a given inertial frame the angular momentum of the particle depends not only on properties of the particle itself, namely its mass and velocity, *but also upon our choice of origin*.

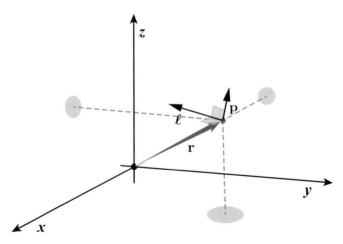

Fig. 1.15 The position vector for a particle. Angular momentum is always defined with respect to a chosen point from where the position vector originates.

Using the product rule, the time derivative of ℓ is

$$\frac{d\boldsymbol{\ell}}{dt} = \frac{d\mathbf{r}}{dt} \times \mathbf{p} + \mathbf{r} \times \frac{d\mathbf{p}}{dt}. \tag{1.86}$$

The first term on the right is $\mathbf{v} \times m\mathbf{v}$, which vanishes because the cross product of two parallel vectors is zero. In the second term, we have $d\mathbf{p}/dt = \mathbf{F}$ using Newton's second law, where $\mathbf{F}$ is the net force acting on the particle. It is therefore convenient to define the **torque** $\mathbf{N}$ on the particle due to $\mathbf{F}$ as

$$\mathbf{N} = \mathbf{r} \times \mathbf{F}, \tag{1.87}$$

so that

$$\mathbf{N} = \frac{d\boldsymbol{\ell}}{dt}. \tag{1.88}$$

That is, the net *torque* on a particle is responsible for any change in its angular momentum, just as the net *force* on the particle is responsible for any change in its momentum. The angular momentum of a particle is conserved if there is no net torque on it.

Sometimes the momentum $\mathbf{p}$ is called the "linear momentum" to distinguish it from the angular momentum $\boldsymbol{\ell}$. They have different units and are conserved under different circumstances. The momentum of a particle is conserved if there is no net external *force* and the angular momentum of the particle is conserved if there is no net external *torque*. It is easy to arrange forces on an object so that it experiences a net force but no net torque, and equally easy to arrange them so there is a net torque but no net force. For example, if the force $\mathbf{F}$ is parallel to $\mathbf{r}$, we have $\mathbf{N} = 0$; yet there is a nonzero force.

There is another striking difference between momentum and angular momentum. In a given inertial frame, the value of a particle's momentum $\mathbf{p}$ is independent of where we choose to place the origin of coordinates. But because the angular momentum $\boldsymbol{\ell}$ of the particle involves the position vector $\mathbf{r}$, the value of $\boldsymbol{\ell}$ does depend on the choice of origin. This makes angular momentum more abstract than momentum, in that in the exact same problem different people at rest in the same inertial frame may assign it different values depending on where they choose to place the origin of their coordinate system.

The angular momentum of *systems* of particles is sufficiently complex and sufficiently interesting to devote much of Chapter 12 to it. For now, we can simply say that as with linear momentum, angular momentum can be exchanged between particles in the system. The total angular momentum of a system of particles is conserved if there is no net *external* torque on the system.

Example 1.6   **A Particle in Two Dimensions Attached to a Spring**

A block of mass m is free to move on a frictionless tabletop under the influence of an attractive Hooke's-law spring force $\mathbf{F} = -k\mathbf{r}$, where the vector $\mathbf{r}$ is the position vector of the particle measured from the origin. We will find the motion $x(t), y(t)$ of the ball and show that the angular momentum of the ball about the origin is conserved.

The vector $\mathbf{r} = x\,\hat{\mathbf{x}} + y\,\hat{\mathbf{y}}$, where x and y are the Cartesian coordinates of the ball and $\hat{\mathbf{x}}$ and $\hat{\mathbf{y}}$ are unit vectors pointing in the positive x and positive y directions, respectively. Newton's second law $-k\mathbf{r} = m\ddot{\mathbf{r}}$ becomes

$$-k(x\hat{\mathbf{x}} + y\hat{\mathbf{y}}) = m(\ddot{x}\hat{\mathbf{x}} + \ddot{y}\hat{\mathbf{y}}), \tag{1.89}$$

which separates into the two simple harmonic oscillator equations

$$\ddot{x} + \omega_0^2 x = 0 \quad \text{and} \quad \ddot{y} + \omega_0^2 y = 0, \tag{1.90}$$

where $\omega_0 = \sqrt{k/m}$. It is interesting that the x and y motions are completely independent of one another in this case; the two coordinates have been decoupled, so we can solve the equations separately. The solutions are

$$x = A_1 \cos(\omega_0 t + \varphi_1) \quad \text{and} \quad y = A_2 \cos(\omega_0 t + \varphi_2), \tag{1.91}$$

showing that the ball oscillates simple harmonically in both directions.

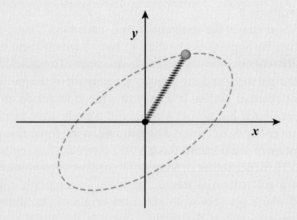

Fig. 1.16 A two-dimensional elliptical orbit of a ball subject to a Hooke's-law spring force, with one end of the spring fixed at the origin. The spring's rest length is zero.

The four constants $A_1, A_2, \varphi_1, \varphi_2$ can be evaluated in terms of the four initial conditions $x_0, y_0, v_{x_0}, v_{y_0}$. The oscillation frequencies are the same in each direction, so orbits of the ball are all closed. In fact, the orbit shapes are ellipses centered at the origin, as shown in Figure 1.16.[a] Note that in this two-dimensional problem, the motion of the ball is determined by *four* initial conditions (the two components of the position vector and the two components of the velocity vector), together with the known force throughout the motion. This is what is expected for two second-order differential equations.

The spring exerts no torque on the ball about the origin, since the cross product of any vector with itself vanishes, so $\mathbf{N} = \mathbf{r} \times \mathbf{F} = \mathbf{r} \times -k\mathbf{r} = 0$. Therefore the angular momentum of the ball is conserved about the origin. In this case, this angular momentum is given by

$$\boldsymbol{\ell} = (x\hat{\mathbf{x}} + y\hat{\mathbf{y}}) \times (m\dot{x}\hat{\mathbf{x}} + m\dot{y}\hat{\mathbf{y}}) = (mx\dot{y} - my\dot{x})\hat{\mathbf{z}}, \tag{1.92}$$

so the special combination $m\,x\,\dot{y} - m\,y\,\dot{x}$ remains constant for all time. That is certainly a highly nontrivial statement.

The angular momentum is *not* conserved about any other point in the plane, because then the position vector and the force vector would be neither parallel nor antiparallel. The angular momentum of a particle is *always* conserved if the force is purely central, *i.e.*, if it is always directly toward or away from a fixed point, as long as that same point is chosen as origin of the coordinate system.

We still have not used the conservation of angular momentum in this problem to our advantage, because we solved the full second-order differential equation. To see how we can tackle this problem without ever needing to invoke Newton's second law or *any* second-order differential equation, we need to first look at another very useful conservation law, the conservation of *energy*.   ∎

---

[a] Remember that the equation of an ellipse in the x–y plane can be written as

$$\frac{(x - x_0)^2}{a^2} + \frac{(y - y_0)^2}{b^2} = 1, \tag{1.93}$$

where (x_0, y_0) is the center of the ellipse, and a and b are the minor and major radii. One can show that Eq. (1.91) indeed satisfies this equation for appropriate relations between $\varphi_1, \varphi_2, A_1, A_2$ and x_0, y_0, a, b.

Energy

Energy is the third quantity that is sometimes conserved. Of momentum, angular momentum, and energy, energy is the most subtle and most abstract, yet it is often the most useful.

We begin by writing Newton's law for a particle in the form $\mathbf{F}_T = m d\mathbf{v}/dt$, where $\mathbf{F}_T$ is the total force on the particle. Dotting this equation with the particle's velocity $\mathbf{v}$:

$$\mathbf{F}_T \cdot \mathbf{v} = m\mathbf{v} \cdot \frac{d\mathbf{v}}{dt} = \frac{d}{dt}\left(\frac{1}{2}mv^2\right) \equiv \frac{dT}{dt}, \tag{1.94}$$

where we have defined

$$T = \frac{1}{2}mv^2 \tag{1.95}$$

as the *kinetic energy* of the particle.[4] If $\mathbf{F}$ is the force of gravity, for example, then if the particle is falling vertically its velocity is parallel to $\mathbf{F}$, so $\mathbf{F} \cdot \mathbf{v}$ is positive, causing the kinetic energy of the particle to *increase*; and if the particle is rising, its velocity is antiparallel to $\mathbf{F}$, so $\mathbf{F} \cdot \mathbf{v}$ is negative, causing the kinetic energy of the particle to *decrease*. If $\mathbf{F_T}$ is the total force acting on the particle, the time rate of change

---

[4] In deriving Eq. (1.94), we have used the identity

$$\mathbf{v} \cdot \frac{d\mathbf{v}}{dt} = v_x\frac{dv_x}{dt} + v_y\frac{dv_y}{dt} + v_z\frac{dv_z}{dt} = \frac{1}{2}\frac{d}{dt}(v_x^2 + v_y^2 + v_z^2) = \frac{1}{2}\frac{d(v^2)}{dt}. \tag{1.96}$$

$$\frac{dT}{dt} = \mathbf{F_T} \cdot \mathbf{v} \tag{1.97}$$

is called the *net power input* to the particle.

Example 1.7   **Charged Particle in a Magnetic Field**

The force exerted by a magnetic field **B** on a particle of electric charge q moving with velocity **v** is given by

$$\mathbf{F_B} = q\mathbf{v} \times \mathbf{B}. \tag{1.98}$$

What is the change in a particle's kinetic energy if this is the only force acting on it?

    Using the fact that the cross product of *any* two vectors is perpendicular to both vectors, it follows that $\mathbf{v} \cdot (\mathbf{v} \times \mathbf{B}) = 0$. Therefore, the kinetic energy of a particle moving in a magnetic field is constant in time. Seen another way, the particle generally accelerates, but its acceleration $\mathbf{a} = q(\mathbf{v} \times \mathbf{B})/m$ is always perpendicular to **v**, so the *magnitude* of **v** remains constant, and therefore the kinetic energy $T = (1/2)mv^2$ remains constant as well. The particle may move along very complicated paths, but its kinetic energy never changes. ∎

We can integrate Eq. (1.94) over time to find the change in a particle's kinetic energy as it moves from some point a to another point b. The result is

$$\Delta T \equiv T_b - T_a = \int_a^b \mathbf{F_T} \cdot \mathbf{v} \, dt = \int_a^b \mathbf{F_T} \cdot d\mathbf{s}, \tag{1.99}$$

since $\mathbf{v} \equiv d\mathbf{s}/dt$, where $d\mathbf{s}$ is the instantaneous displacement vector. At each point on the path the vector $d\mathbf{s}$ is directed along the path, and its magnitude is an infinitesimal distance along the path.

Now define the **work** W done by any one of the forces **F** acting on the particle, as it moves from a to b, as the line integral (or *path* integral)

$$W = \int_a^b \mathbf{F} \cdot d\mathbf{s}. \tag{1.100}$$

Note from the dot product that it is only the component of **F** *parallel* to the path at some point that does work on the particle. Figure 1.17 illustrates the setup.

We can then define the total work done on the particle by all of the forces $\mathbf{F}_1$, $\mathbf{F}_2, \ldots$ to be

$$W_T = W_1 + W_2 + \cdots = \int_a^b \mathbf{F}_1 \cdots d\mathbf{s} + \int_a^b \mathbf{F}_2 \cdot d\mathbf{s} + \cdots \tag{1.101}$$

so it follows from Eq. (1.99) that

$$W_T = T_b - T_a, \tag{1.102}$$

which is known as the **work–energy theorem**: *the change in kinetic energy of a particle is equal to the total work done upon it.* If we observe that the kinetic energy of a particle has changed, there must have been a net amount of work done upon it.

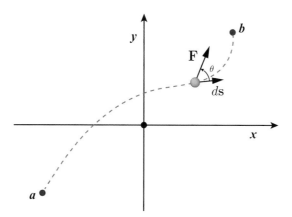

Fig. 1.17 The work done by a force on a particle is the line integral $\int_a^b \mathbf{F} \cdot d\mathbf{s}$ along the path traced by the particle.

Often the work done by a particular force $\mathbf{F}$ depends upon which path the particle takes as it moves from a to b. The frictional work done by air resistance on a ball as it flies from the bat to an outfielder depends upon how high it goes, that is, whether its total path length is short or long. There are other forces, however, like the static force of gravity, for which the work done is independent of the particle's path. For example, the work done by earth's gravity on the ball is the same no matter how it gets to the outfielder. For such forces the work depends only upon the endpoints a and b. That implies that the work can be written as the difference[5]

$$W_{\mathrm{a}\to\mathrm{b}} = -U_b + U_a \tag{1.103}$$

between a **potential energy** function U evaluated at the final point b and the initial point a.

A force $\mathbf{F}$ for which the work $W = \int_a^b \mathbf{F} \cdot d\mathbf{s}$ between any two points a and b is independent of the path is said to be **conservative**. There are several tests for conservative forces that are mathematically equivalent, in that if any one of them is true the others are true as well. The conditions are:

1 $W = \int_a^b \mathbf{F} \cdot d\mathbf{s}$ is path independent.
2 The work done around *any* closed path is $\oint \mathbf{F} \cdot d\mathbf{s} = 0$.
3 The curl of the force function vanishes: $\nabla \times \mathbf{F} = 0$.
4 The force function can always be written as the negative gradient of some scalar function U: $\mathbf{F} = -\nabla U$.

Often the third of these conditions makes the easiest test. For example, the curl of the uniform gravitational force $\mathbf{F} = -mg\,\hat{\mathbf{z}}$ is, using the determinant expression for the curl:

$$\nabla \times \mathbf{F} = \begin{vmatrix} \hat{\mathbf{x}} & \hat{\mathbf{y}} & \hat{\mathbf{z}} \\ \partial/\partial x & \partial/\partial y & \partial/\partial z \\ F_x & F_y & F_z \end{vmatrix} = 0, \tag{1.104}$$

---

[5] The reason for this choice of signs will soon become clear.

since each component of $\mathbf{F}$ is zero or a constant. Therefore this force is conservative. That means it must have a potential energy given by the indefinite integral

$$U = -\int \mathbf{F} \cdot d\mathbf{s} \tag{1.105}$$

$$= -\int (-mg\hat{\mathbf{z}}) \cdot d\mathbf{s} = mg \int dz = mgz.$$

The work done by a conservative force is equal to the *difference* between two potential energies, so it follows that the physics is exactly the same for a particle with potential energy $U(\mathbf{r})$ as it is for a potential energy $U(\mathbf{r}) + C$, where C is any constant. For example, the potential energy of a particle of mass m in a uniform gravitational field g is $U_{\text{grav}} = mgh$, where h is the altitude of the particle. The fact that any constant can be added to U in this case is equivalent to the fact that it doesn't matter from what point the altitude is measured, as long as this is done consistently throughout a problem. The motion of a particle is the same whether we measure altitude from the ground or from the top of a building.

Not all forces are conservative: for example, the curl of the hypothetical force $\mathbf{F} = \alpha xy\hat{\mathbf{z}}$, where α is a constant, is

$$\nabla \times \mathbf{F} = \begin{vmatrix} \hat{\mathbf{x}} & \hat{\mathbf{y}} & \hat{\mathbf{z}} \\ \partial/\partial x & \partial/\partial y & \partial/\partial z \\ 0 & 0 & \alpha xy \end{vmatrix}$$

$$= \hat{\mathbf{x}} \frac{\partial}{\partial y}(\alpha xy) - \hat{\mathbf{y}} \frac{\partial}{\partial x}(\alpha xy) = \alpha(x\hat{\mathbf{x}} - y\hat{\mathbf{y}}) \neq 0, \tag{1.106}$$

so this force is not conservative, and does not possess a potential energy function.

Typically both conservative ($\mathbf{F}_{\text{C}}$) and nonconservative forces ($\mathbf{F}_{\text{NC}}$) act on a particle, so the total work done on it is

$$W_{\text{T}} = W_{\text{C}} + W_{\text{NC}} = -U_{\text{b}} + U_{\text{a}} + W_{\text{NC}} = T(b) - T(a) \tag{1.107}$$

from the work–energy theorem equation (1.102), where now the potential energies U_{a} and U_{b} are the total potential energies due to all of the conservative forces. Rewriting this equation in the form

$$[T_{\text{b}} + U_{\text{b}}] - [T_{\text{a}} + U_{\text{a}}] = W_{\text{NC}}, \tag{1.108}$$

we can finally define the **energy** E of the particle as the sum of the kinetic and potential energies:

$$E \equiv T + U. \tag{1.109}$$

The change in a particle's energy as it travels from a to b is therefore

$$\Delta E = E_{\text{b}} - E_{\text{a}} = W_{\text{NC}}, \tag{1.110}$$

the total work done by nonconservative forces. The energy is *conserved*, with $E_b = E_a$, if only conservative forces act on the particle (that is, if $W_{NC} = 0$).[6]

Example 1.8 **A Child on a Swing**

A child of mass m is being pushed on a swing. Suppose there are just four forces acting on her: (i) the normal force of the seat; (ii) the hands of the pusher; (iii) air resistance; and (iv) gravity. What is the work done by each?

(i) As long as the normal force of the swing seat is perpendicular to the instantaneous displacement, the work it does must be *zero* at all times, $\mathbf{F}_N \cdot d\mathbf{s} = 0$.

(ii) While the pusher is pushing, the force is in the direction of the displacement and $\mathbf{F} \cdot d\mathbf{s} > 0$, so the work it does is *positive*. The net work done over a complete cycle is also positive, $\oint \mathbf{F} \cdot d\mathbf{s} > 0$.

(iii) The work done by air resistance is *negative*, because air resistance is opposite to the direction of motion, and hence $\mathbf{F} \cdot d\mathbf{s} < 0$. The net work done by air resistance is therefore negative, $\oint \mathbf{F} \cdot d\mathbf{s} < 0$.

(iv) The work done by gravity is positive while she is descending, and negative while she is ascending; they exactly cancel out over a complete cycle. That is, gravity is a conservative force, or $\oint \mathbf{F} \cdot d\mathbf{s} = 0$.

The only two forces that do a net amount of work on her over a complete cycle are the hands pushing (positive) and air resistance (negative). Neither force is conservative, so $\Delta E = E_b - E_a = W_{NC} = W_{hands} + W_{air}$. If the right-hand side is positive (the net work done by the pusher exceeds the magnitude of the (negative) net work done by air resistance), her energy increases; but if $W_{hands} < |W_{air}|$, her energy decreases. If the pusher stops pushing, and if we could remove air resistance, then her energy would be conserved, continually oscillating between kinetic energy (maximum at her lowest point) and gravitational potential energy (maximum at her highest points).   ∎

It is useful to expand the concept of energy beyond kinetic and potential energies by regarding the work done by nonconservative forces as external sources or sinks of the total energy. For example, in the case of the friction force, a decrease in the "mechanical energy" $T + U$ shows up in some other external form, such as heat. That is, conservation of energy is more general than one might expect from classical mechanics alone; in addition to kinetic and potential energies, there is thermal energy, the energy of deformation, energy in the electromagnetic field, and many other forms as well. Energy is a useful concept across many disparate physical systems.

Example 1.9 **A Particle Attached to a Spring Revisited**

We want to demonstrate the power of conservation laws in solving the previous problem of a particle of mass m confined to a two-dimensional plane and attached to a spring of force constant k (see Figure 1.16). The only force law is Hooke's law $\mathbf{F} = -k\mathbf{r}$. We can check that $\nabla \times \mathbf{F} = 0$, and then find that the potential energy for this conservative force is

---

[6] That, of course, is responsible for the term "conservative forces."

$$U_b - U_a = -\int_a^b \mathbf{F} \cdot d\mathbf{r} = -k \int_a^b \mathbf{r} \cdot d\mathbf{r} \Rightarrow U = \frac{1}{2}k r^2. \tag{1.111}$$

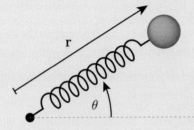

Fig. 1.18 A ball free to move in two dimensions subject to the spring force $\mathbf{F} = -k\mathbf{r}$. We assume the spring has natural length $r = 0$.

The total energy is therefore

$$E = \frac{1}{2}m v^2 + \frac{1}{2}k r^2. \tag{1.112}$$

The problem has rotational symmetry, so it is helpful to use polar coordinates. The velocity of the particle is

$$\mathbf{v} = \dot{r}\hat{\mathbf{r}} + r\dot{\theta}\,\hat{\boldsymbol{\theta}}, \tag{1.113}$$

where r and θ are the polar coordinates (see Appendix A for a review of coordinate systems). We then have

$$E = \frac{1}{2}m\left(\dot{r}^2 + r^2\dot{\theta}^2\right) + \frac{1}{2}k r^2. \tag{1.114}$$

Since E is a constant, this would be a very nice *first-order* differential equation for $r(t)$ if we could get rid of the pesky $\dot{\theta}$ term. Angular momentum conservation comes to the rescue. We know that

$$\boldsymbol{\ell} = \mathbf{r} \times (m\mathbf{v}) = m r\hat{\mathbf{r}} \times (\dot{r}\hat{\mathbf{r}} + r\dot{\theta}\,\hat{\boldsymbol{\theta}}) = m r^2 \dot{\theta}\,\hat{\mathbf{z}} = \text{constant}. \tag{1.115}$$

We can then write

$$m r^2 \dot{\theta} = \ell \Rightarrow \dot{\theta} = \frac{\ell}{m r^2} \tag{1.116}$$

with ℓ a constant. Putting this back into Eq. (1.114):

$$E = \frac{1}{2}m\dot{r}^2 + \frac{\ell^2}{2 m r^2} + \frac{1}{2}k r^2, \tag{1.117}$$

which is a first-order differential equation from which $r(t)$ can be determined; after that we can find $\theta(t)$ using Eq. (1.116). We have thus solved the problem without ever dealing with the second-order differential equation arising from Newton's second law. This is not particularly advantageous here, given that the original second-order differential equations corresponded to harmonic oscillators. In general, however, tackling only first-order differential equations is likely to be a huge advantage.

It is instructive to analyze the boundary conditions and conservation laws of this system. Newton's second law provides two second-order differential equations in two dimensions. Each differential equation requires two boundary conditions to yield a unique solution, for a total of *four* required constants. If we use conservation laws instead, we know that both energy and angular momentum are conserved. Energy conservation provides us with a single first-order differential equation requiring a single boundary condition. But the value of energy E is another constant to be specified, so there are altogether two constants to fix using energy conservation. Angular momentum conservation gives us another first-order differential equation, with a single boundary condition plus the value ℓ of the angular momentum itself, so there are another two constants. The energy and angular momentum conservation equations together thus again require a total of *four* constants to yield a unique solution. The four boundary conditions of Newton's second law are directly related to the four constants required to solve the problem using conservation equations.   ∎

Example 1.10 | **Newtonian Central Gravity and its Potential Energy**

Newton's law of gravity for the force on a "probe" particle of mass m due to a "source" particle of mass M is $\mathbf{F} = -\,(GMm/r^2)\hat{\mathbf{r}}$, where $\hat{\mathbf{r}}$ is a unit vector pointing *from* the source particle *to* the probe in spherical coordinates. The minus sign means that the force is attractive, in the negative $\hat{\mathbf{r}}$ direction. We can check to see whether this force is conservative by taking its curl.

In spherical coordinates, the curl of a vector $\mathbf{F}$ in terms of unit vectors in the r, θ, and ϕ directions is

$$\nabla \times \mathbf{F} = \frac{1}{r^2 \sin\theta} \begin{vmatrix} \hat{\mathbf{r}} & r\hat{\boldsymbol{\theta}} & r\sin\theta\,\hat{\boldsymbol{\phi}} \\ \partial/\partial r & \partial/\partial\theta & \partial/\partial\phi \\ F_r & rF_\theta & r\sin\theta F_\phi \end{vmatrix}, \tag{1.118}$$

so the curl of **F** is

$$\nabla \times \left(-\frac{GMm}{r^2}\hat{\mathbf{r}}\right) = \frac{1}{r^2 \sin\theta} \begin{vmatrix} \hat{\mathbf{r}} & r\hat{\boldsymbol{\theta}} & r\sin\theta\,\hat{\boldsymbol{\phi}} \\ \partial/\partial r & \partial/\partial\theta & \partial/\partial\phi \\ -GMm/r^2 & 0 & 0 \end{vmatrix} = 0. \tag{1.119}$$

Therefore, Newton's inverse-square gravitational force is conservative, and must have a corresponding potential energy function

$$U(r) = -\int \mathbf{F} \cdot d\mathbf{r} = GMm \int \frac{dr}{r^2} = -\frac{GMm}{r} + \text{constant}, \tag{1.120}$$

where by convention we ignore the constant of integration, which in effect makes $U \to 0$ as $r \to \infty$.   ∎

Example 1.11 | **Dropping a Particle in Spherical Gravity**

Armed with the potential energy expression due to a spherical gravitating body of mass M, we write the total energy of a probe particle of mass m as

$$E = T + U(r) = \frac{1}{2}mv^2 - \frac{GMm}{r}, \tag{1.121}$$

which is conserved. Suppose that the probe particle is dropped from rest some distance r_0 from the center of M, which we assume is so large, $M \gg m$, that it does not move appreciably as the small mass m falls toward it. The particle has no initial tangential velocity, so it will fall radially with $v^2 = \dot{r}^2$. Energy conservation gives

$$E = \frac{1}{2}m\dot{r}^2 - \frac{GMm}{r}. \tag{1.122}$$

The initial conditions are $r = r_0$ and $\dot{r} = 0$, so it follows that

$$E = -GMm/r_0. \tag{1.123}$$

Equation (1.122) is a first-order differential equation in $r(t)$. It is said to be a "first integral" of the second-order differential equation $\mathbf{F} = m\mathbf{a}$, which in this case is

$$-\frac{GMm}{r^2} = m\ddot{r}. \tag{1.124}$$

That is, if we want to find the motion $r(t)$ it is a great advantage to begin with energy conservation, because that equation already represents one of the necessary two integrations of $\mathbf{F} = m\mathbf{a}$. Solving Eq. (1.122) for $\dot{r}$, we get

$$\dot{r} = \pm\sqrt{\frac{2}{m}\left(E + \frac{GMm}{r}\right)} = \pm\sqrt{2GM\left(\frac{1}{r} - \frac{1}{r_0}\right)}. \tag{1.125}$$

We have to choose the minus sign, because when the particle is released from rest it will subsequently fall toward the origin with $\dot{r} < 0$. Separating the variables r and t and integrating both sides:

$$\int_{r_0}^{r} \frac{dr\sqrt{r}}{\sqrt{1 - r/r_0}} = -\sqrt{2GM}\int_0^t dt = -\sqrt{2GM}\,t. \tag{1.126}$$

At this point we say that the problem has been **reduced to quadrature**, an old-fashioned phrase which simply means that all that remains to find $r(t)$ (or in this case $t(r)$) is to evaluate an indefinite integral, which in the problem at hand is the integral on the left. If we are lucky, the integral can be evaluated in terms of known functions, in which case we have an *analytic* solution. If we are not so lucky, the integral can at least be evaluated numerically to any level of accuracy we need. See Chapter 14 on techniques of numerical integration.

An analytic solution of the integral in Eq. (1.126), using the substitution $r = r_0 \sin^2\theta$, gives

$$t(r) = \sqrt{\frac{r_0^3}{2GM}}\left[\frac{\pi}{2} - \sin^{-1}\sqrt{\frac{r}{r_0}} + \sqrt{\frac{r}{r_0}}\sqrt{1 - \frac{r}{r_0}}\right] \tag{1.127}$$

from which we can find the time it takes to fall to r given some initial value r_0. We cannot solve explicitly for $r(t)$ in this case, because the right-hand side is a transcendental function of r. Note that the constant r_0 in this equation is directly related to the energy E through Eq. (1.123).

The problem is much simplified if the particle falls from a great altitude to a much smaller altitude, so that $r \ll r_0$, in which case the first term in Eq. (1.127) is much bigger than the others. For example, the time it takes an astronaut to fall from rest at radius r_0 to the surface of an asteroid of radius R, where $r_0 \gg R$, is essentially

$$t = \frac{\pi}{2}\sqrt{\frac{r_0^3}{2GM}}, \tag{1.128}$$

which is independent of R! This insensitivity to the asteroid radius is due to the fact that nearly all of the travel time is spent at large radii, during which the astronaut is moving slowly. Changes in the asteroid radius R affect the overall travel time very little, because the astronaut is falling so fast near the end. On the contrary, the travel time is clearly quite sensitive to the initial position r_0. ∎

Example 1.12

Potential Energies for Positive Power-Law Forces

A particle moves in one dimension subject to the power-law force $F = -kx^n$, where the coefficient k is positive, and n is a positive integer. Let us find the potential energy of the particle and also the maximum distance x_{max} it can reach from the origin, in terms of its maximum speed v_{max}. The maximum distance is the **turning point** of the particle, because as the particle approaches this position it slows down, stops at x_{max}, and turns around and heads in the opposite direction.

The potential energy of the particle is the indefinite integral

$$U = -\int F(x)\,dx = -\int (-kx^n)\,dx = \frac{k}{n+1}x^{n+1} \tag{1.129}$$

plus an arbitrary constant of integration, which we choose to be zero. Two of these potential energy functions, one with odd n and one with even n, illustrate the range of possibilities, as shown in Figure 1.19. The case $n = 1$, corresponding to a linear restoring force, corresponds to a Hooke's-law spring, where k is the spring constant and the potential energy is $U = (1/2)kx^2$. In this case the lowest possible energy is $E = 0$, when the particle is stuck at $x = 0$. There are two turning points for energies $E > 0$, one at the right and one at the left.

Energy is conserved for any value of n, where

$$E = \frac{1}{2}mv^2 + \left(\frac{k}{n+1}\right)x^{n+1}. \tag{1.130}$$

The potential energy increases with increasing positive x, so the maximum speed of the particle is at the origin, where $x = 0$ and $E = (1/2)mv_{max}^2$. The speed goes to zero at the maximum value of x attainable, i.e., where $E = kx_{max}^{n+1}/(n+1)$. Eliminating E and solving for x_{max}, we find

$$x_{max} = \left[\frac{n+1}{2}\left(\frac{m}{k}\right)\right]^{1/(n+1)}(v_{max})^{2/(n+1)}. \tag{1.131}$$

For the spring force, which corresponds to $n = 1$, x_{max} is directly proportional to v_{max}, so if we double the particle's velocity at the origin we double the maximum x it can achieve.

Note that the conservation of energy equation (1.130) can also be solved for $v \equiv \dot{x}$ to give

$$\dot{x} = \pm\sqrt{\frac{2}{m}\left(E - \left(\frac{k}{n+1}\right)x^{n+1}\right)}, \tag{1.132}$$

which is a first-order differential equation. Dividing by the right-hand side and integrating over time yields

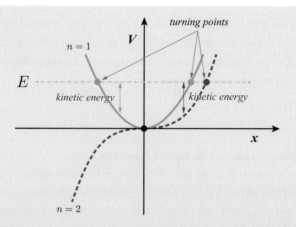

Fig. 1.19 Potential energy functions for selected positive powers n. A possible energy E is drawn as a horizontal line, since E is constant. The difference between E and $U(x)$ at any point is the value of the kinetic energy T. The kinetic energy is zero at the *turning points*, where the E line intersects $U(x)$. Note that for $n = 1$ there are two turning points for $E > 0$, but for $n = 2$ there is only a single turning point. The quadratic force with $n = 2$ has a cubic potential $U = (1/3)kx^3$ which is positive for $x > 0$ and negative for $x < 0$. Note that the slope of this potential is everywhere positive except at $x = 0$, so the force on any particle at $x \neq 0$ is toward the left, since $F = -dU/dx$ is then negative. So particles at positive x are pulled toward the origin, while particles at negative x are pushed away from the origin.

$$\int \frac{dx}{\sqrt{E - [k/(n+1)]x^{n+1}}} = \pm\sqrt{\frac{2}{m}} \int dt = \pm\sqrt{\frac{2}{m}}\, t + C, \qquad (1.133)$$

where C is a constant of integration. The problem has been reduced to quadrature.

For some values of n, the integral on the left can be evaluated in terms of standard functions; this includes the cases $n = 0$ and $+1$, for example. For other values of n the integral can be evaluated numerically; that is, there are algorithms such as "Simpson's Rule" that can be implemented on a computer to provide a numerical value for the integral, given numerical values of E, k, n, and the limits of integration. Note that conservation of energy results in a first-order differential equation, so specifying the constant of integration C is equivalent to specifying a single initial condition.

Rather than integrating Eq. (1.130), which leads to Eq. (1.133), we can *differentiate* the equation instead. The time derivative of Eq. (1.130) is

$$0 = m\dot{x}\ddot{x} + \left(\frac{k}{n+1}\right)(n+1)x^n\dot{x} = 0, \qquad (1.134)$$

since $dE/dt = 0$. The velocity $\dot{x}$ is not generally zero, so we can divide it out, leaving

$$m\ddot{x} = -kx^n, \tag{1.135}$$

which we recognize as $m\,a = F$ for the given force $F = -kx^n$. That is, the time derivative of the energy conservation first-order differential equation is simply $F = m\,a$, which is a second-order differential equation. Often, energy conservation serves as a **first integral of motion**, halfway toward a complete solution of the second-order equation $F = m\,a$.   ∎

1.9 Collisions

Collisions are commonplace: billiard balls on a billiard table, nitrogen molecules in the air, protons in a synchrotron, cars on the highway. Typically, colliding objects exert very strong equal but opposite forces on one another during a short time interval Δt, before and after which they hardly interact at all. It is true that there are usually also external forces acting on the objects during this brief time interval, such as gravity or the normal and frictional forces exerted by a pool table or road surface. However, during the brief collision times Δt such external forces are negligible compared with the internal smashing forces of one object on the other, so we can safely neglect them. Therefore, to an excellent approximation the total momentum of the colliding objects is conserved during the collision. And since their momentum is conserved, the center of mass (CM) of the colliding objects moves in a straight line at constant speed during the time just before, during, and after the collision. There is therefore an inertial frame in which the CM of the system stays at rest, called the center-of-mass (CM) frame. Analyzing the collision in the CM frame can be particularly useful.

The velocity of the CM frame in the original frame, which we will call the "lab frame," is

$$\mathbf{V}_{\mathrm{CM}} = \frac{m_0 \mathbf{v}_0 + m_1 \mathbf{v}_1}{m_0 + m_1} = \frac{\mathbf{P}}{M}, \tag{1.136}$$

where $\mathbf{P}$ is the total momentum and M is the total mass. Here m_0 and m_1 are the masses of the initial particles, and $\mathbf{v}_0$ and $\mathbf{v}_1$ are their velocities in the original lab frame. It is sometimes convenient to analyze the collision in the CM frame first, then transform results to the lab frame, or vice versa, using this relative velocity to transform between them.

In addition to momentum conservation, kinetic energy is sometimes also conserved in collisions, at least to a good approximation. Such kinetic-energy-conserving collisions are said to be *elastic*. Proton–proton collisions or ideal billiard-ball collisions may be nearly elastic, for example. We think of the billiard balls deforming slightly during such a collision, and then springing back to their original shape; that is, their initial kinetic energy is temporarily converted into a spring-like potential energy, and then returned to kinetic energy as soon as the balls separate. However, this is often just an approximation, although sometimes a pretty good one, because some of their initial energy is turned into oscillations within the balls themselves, which turns eventually into heat, robbing the balls of their macroscopic kinetic or potential energies. If a collision does not conserve macroscopic kinetic energy, it is said to be *inelastic*. And if the incident and target particles in a collision stick together during the collision, so two particles become one, that collision is said to be *totally inelastic*. A meteorite strikes the earth in a totally inelastic collision; the sum of their macroscopic kinetic energies decreases in the collision and the overall system becomes warmer to compensate.

There is an interesting special case, the elastic collision between two protons or two billiard balls of *equal mass*, where there is a "target" ball m_0 initially at rest, and an "incident" ball m_1 moving at velocity $\mathbf{v}_1$ toward its target in the "forward" direction, as shown in Figure 1.20. After they collide, and relative to the forward direction, ball m_1 bounces off at angle θ with velocity $\mathbf{v}'_1$, while ball m_0 moves off at angle φ with velocity $\mathbf{v}'_0$. Conservation of momentum tells us that

$$m\mathbf{v}_1 = m\mathbf{v}'_0 + m\mathbf{v}'_1 \quad \text{so} \quad \mathbf{v}_1 = \mathbf{v}'_0 + \mathbf{v}'_1, \tag{1.137}$$

while conservation of kinetic energy (for such an elastic collision) gives

$$\frac{1}{2}m(v_1)^2 = \frac{1}{2}m(v'_0)^2 + \frac{1}{2}m(v'_1)^2 \quad \text{so} \quad (v_1)^2 = (v'_0)^2 + (v'_1)^2. \tag{1.138}$$

before   *after*

Fig. 1.20 A collision of equal-mass balls with ball 0 initially at rest. For an elastic collision, the two balls move at right angles to one another after the collision.

Squaring the conservation of momentum equation (i.e., dotting it with itself) gives

$$\mathbf{v}_1 \cdot \mathbf{v}_1 \equiv (v_1)^2 = (v'_0)^2 + 2\mathbf{v}'_0 \cdot \mathbf{v}'_1 + (v'_1)^2. \tag{1.139}$$

Comparing this last equation with the conservation of kinetic energy equation, clearly $\mathbf{v}'_0 \cdot \mathbf{v}'_1 = 0$, so the two balls must emerge from the collision in directions perpendicular to one another, with $\theta + \varphi = 90°$. The only exception occurs for an absolutely head-on collision in which the incident ball stops dead (with $\mathbf{v}'_1 = 0$) and all of its momentum and kinetic energy are transferred to ball m_0.

1.10 Forces of Nature

The hallmark of Newtonian mechanics – the relationship $\mathbf{F} = m\,\mathbf{a}$ – is only one part of a mechanics problem. To determine the dynamics of a particle, we also need to know the left-hand side of the equation. That is, we need to specify the forces. This is a separate requirement: we need to discover and learn about what forces are present through experimentation and additional theoretical considerations. We may then be tempted to ask the bold question: *what are all of the possible forces that can arise on the left-hand side of Newton's second law?* Surprisingly, this question has a complete answer at the fundamental level, an exhaustive and finite catalogue of possibilities.

To date, depending upon what one counts as a force, there are at most four *fundamental* forces in Nature, and only two of the four can be used in classical Newtonian mechanics. For the sake of completeness, let us list these four:

1 The **electromagnetic force** can be attractive or repulsive, and acts only on particles that carry a certain mysterious attribute we call "electric charge." This force is relevant from subatomic length scales to planetary length scales, and plays a role in virtually every physical setting.

2 The **gravitational force** is an omnipresent force in classical physics, which acts on anything that has mass or energy. Gravity is by far the weakest of the four forces, but at macroscopic length scales it is very noticeable nonetheless if objects are essentially electrically neutral – so that the much stronger electromagnetic force vanishes. To make things especially mysterious, our best and current theory of gravity is Einstein's theory of general relativity, and in this theory gravity is not a force at all, but an effect of the curvature of space and time. We will discuss this theory further in Chapter 10.

3 The **weak force** is subatomic in nature, acting only over very short distances, around 10^{-15} m – a regime where it is essential to use quantum mechanics. The weak force therefore plays no role in typical classical mechanics problems. The weak force is important for understanding radioactivity, neutrinos, and the Higgs boson particle. We have also learned that the weak force is closely related to electromagnetism. The electromagnetic and weak forces collectively are sometimes referred to as the **electroweak** force.

4 The **strong force**, which is also a force of subatomic relevance at around 10^{-18} m, binds quarks together and underlies nuclear energy. This is the strongest of all the forces, but in spite of its great importance it is not directly relevant to classical mechanics, since it arises in contexts requiring the use of quantum mechanics.

In summary, if we consider electromagnetism and the weak force to be two aspects of a single electroweak force, and if we take Einstein's point of view that gravity is not in fact a force at all, then we are left with only two truly fundamental forces, the electroweak and strong forces. If, however, we look at physics from the point of view of the large-scale, classical world, the forces that matter in our day-to-day experience can be taken to be gravity and electromagnetism. That is, in a setting where the strong and weak forces play a relevant dynamical role, the framework of classical mechanics itself is typically already faltering and a full extension to quantum mechanics is needed. And if we need to take account of gravitational effects more subtle or much more exotic than Newtonian gravity, the classical laws of motion have to be modified as well.

Hence, our classical mechanics world will deal primarily with Newtonian gravity and electromagnetic forces. But what can we say about the friction and spring forces encountered already in many examples, like the normal force, the tension force in a rope, and a myriad of other force laws that make prominent appearances on the left-hand side of Newton's second law? The answer is that these are all *macroscopic effective forces*, and are not fundamental. Microscopically, they originate entirely from the electromagnetic force law. For example, when two surfaces in contact rub against one another, the atoms at the interface interact microscopically through Coulomb's law of electrostatics. When we add a large number of these tiny forces, we have an effective macroscopic force that we call **friction**. The microscopic details can often, to a good approximation, be tucked into one single parameter, the coefficient of friction. Similarly, the effect of a large number of liquid molecules on a bacterium averages out into a simple force law, $F = -bv$, where b is the only parameter left over from the detailed microscopic interactions – which are once again electromagnetic in origin. **Contact forces**, as they are called, are again not fundamental; they originate with the electromagnetic force law.

The reader may rightfully be surprised that complicated microscopic dynamics can lead to rather simple effective force laws – often described by a few macroscopic parameters. This is a rather general feature of the natural laws. When microscopic complexity is averaged over a large number of particles and length scales, it is expected that the resulting macroscopic system is described through simpler laws with fewer parameters. This is not supposed to be obvious, although it may feel intuitive. Realization of its significance and implications in physics underlies several physics Nobel prizes in the late twentieth century.[1]

1.11 Summary

So much for our brief survey of Newtonian particle mechanics. Particles obey Newton's laws of motion, and depending upon the nature of the forces on a particle, one or another of momentum, angular momentum, and energy may be conserved.

---

[1] The Nobel prize for the development of the renormalization group was awarded to Kenneth G. Wilson in 1982. Wilson described most concisely and elegantly the idea that physics at large length scales can be sensitive to physics at small length scales only through a finite number of parameters. However, the idea pervades other major benchmarks of theoretical physics, such as the Nobel prizes of 1999 to Gerardus 't Hooft and Martinus J. G. Veltman and of 1965 to Sin-Itiro Tomonaga, Julian S. Schwinger, and Richard P. Feynman.

The momentum of a particle is conserved if there is no net force on it, while the angular momentum of the particle is conserved if there is no net torque on it. Energy is conserved if all the forces acting are conservative and time independent; *i.e.*, if the work done by each force is independent of the path of the particle. Similar laws apply to systems of particles.

Given the forces on a particle together with its initial position and velocity, a classical particle moves along a single, precise path. That is the vision of Isaac Newton: particles follow deterministic trajectories. When viewed from an inertial frame, a particle moves in a straight line at constant speed unless a net force is exerted on it, in which case it accelerates according to $\mathbf{a} = \mathbf{F}/m$.

We have required that the fundamental laws of mechanics obey what is called the principle of relativity, which means that if a fundamental law is valid in one inertial frame it is valid in all inertial frames. According to the principle, there is no preferred inertial frame: the fundamental laws can be used by observers at rest in any one of them. This physical statement can be translated into a mathematical statement that given a mathematical transformation of coordinates and other quantities from one frame to another, the fundamental equations should look the same in all inertial frames. We have assumed that the Galilean transformation is the correct transformation of coordinates, and have shown that Newton's laws are invariant under that transformation (provided that any particular force considered is the same in all inertial frames). It is therefore consistent to take Newton's laws as fundamental laws of mechanics.

Then what is left to do in classical mechanics? First of all, since the time of Newton extremely useful and elegant mathematical methods have been developed that give us deep insights into mechanics and may allow us to solve whole classes of problems more easily than with the methods discussed so far. These include Chapter 3 on variational methods culminating in Lagrange's approach to mechanics in Chapter 4; also the relation between symmetries and conservation laws as summarized by Noether's theorem in Chapter 6; Hamilton's equations as presented in Chapter 11; and the Hamilton–Jacobi equation in Chapter 15. Then there are a number of chapters on special cases and applications of classical mechanics, including motion in central-force gravity in Chapter 7 and in electromagnetic fields in Chapter 8; motion as viewed in non-inertial frames of reference in Chapter 9, rigid-body rotation in Chapter 12, motion of coupled oscillators in Chapter 13, and chaotic motion in Chapter 14. Finally, to illustrate how classical mechanics fits inside the larger world of physics, the path-integral approach to quantum mechanics is discussed in capstone Chapter 5, and how Newtonian physics emerges from quantum mechanics in a certain limit; also how Einstein's general theory of relativity describes the motion of particles subject to gravity in capstone Chapter 10; and then how Schrödinger discovered his famous equation of quantum mechanics using the Hamilton–Jacobi equation of classical mechanics as a guide, in final capstone Chapter 15. But before all of this, we first introduce special relativity in Chapter 2 and show how Einstein's very simple postulates have modified classical mechanics, especially for high-energy, fast-moving particles, and the revolution the theory has brought to our understanding of the arena in which physics takes place.

Problems

★   **Problem 1.1** A meter stick is at rest in a primed frame of reference, with one end at the origin and the other at $x' = 1.0$ m. (a) Using the Galilean transformation find the location of each end of the stick in the unprimed frame at a particular time t, and then find the length of the meter stick in the unprimed frame. (b) Repeat for the case that the stick is laid out along the positive y' axis, with one end at the origin and the other at $y' = 1.0$ m. What is the length of the stick in the unprimed frame?

★   **Problem 1.2** A river of width D flows uniformly at speed V relative to the shore. A swimmer swims always at speed $2V$ relative to the water. (a) If the swimmer dives in from one shore and swims in a direction perpendicular to the shoreline in the reference frame of the flowing river, how long does it take her to reach the opposite shore, and how far downstream has she been swept relative to the shore? (b) If instead she wants to swim to a point on the opposite shore directly across from her starting point, at what angle should she swim relative to the direction of the river flow, and how long would it take her to swim across?

★   **Problem 1.3** The crews of two eight-man sculls decide to race one another on a river of width D that flows at uniform velocity V_0. The crew of scull A rows downstream a distance D and then back upstream, while the crew of scull B rows to a point on the opposite shore directly across from the starting point, and then back to the starting point. They begin simultaneously, and each crew rows at the same speed V relative to the water, with $V > V_0$. Who wins the race, and by how much time?

★   **Problem 1.4** Passengers standing in a coasting spaceship observe a distant star at the zenith, *i.e.*, directly overhead. If the spaceship then accelerates to speed $c/100$ where c is the speed of light, at what angle to the zenith (to three significant figures) do the passengers now see the star?

★★   **Problem 1.5** (a) Snow is falling vertically toward the ground at speed v. (a) A bus driver is driving through the snowstorm on a horizontal road at speed $v/3$. At what angle to the vertical are the snowflakes falling as seen by the driver? (b) Suppose that the large windshield in the flat, vertical front of the bus has been knocked out, leaving a hole of area A in the vertical plane. Given that N is the number of falling snowflakes per unit horizontal area per unit time, if the bus moves at constant speed $v/3$ to reach a destination at distance d, how many snowflakes fall into the bus before the destination is reached? (c) To minimize the total number of snowflakes that fall in, the driver considers driving faster or slower. What would be the best speed to take?

★★   **Problem 1.6** The jet stream is flowing due east at velocity v_J relative to the ground. An aircraft is traveling at velocity v_C in the northeast direction relative to the air.

(a) Relative to the ground, find the speed of the aircraft and the angle of its motion relative to the east. (b) Keeping the same speed v_C relative to the air, at what angle would the plane have to move through the air relative to the east so that it would travel northeast relative to the ground?

★ **Problem 1.7** The earth orbits the sun once per year in a nearly circular orbit of radius 150×10^6 km. The speed of light is $c = 3 \times 10^5$ km/s. Looking through a telescope, we observe that a particular star is directly overhead. If the earth were quickly stopped and made to move in the opposite direction at the same speed, at what angle to the vertical would the same star now be observed?

★ **Problem 1.8** A long chain is tied tightly between two trees and a horizontal force F_0 is applied at right angles to the chain at its midpoint. The chain comes to equilibrium so that each half of the chain is at angle θ from the straight line between the chain endpoints. Neglecting gravity, what is the tension in the chain?

★★ **Problem 1.9** An object of mass m is subject to a drag force $F = -kv^n$, where v is its velocity in the medium, and k and n are constants. If the object begins with velocity v_0 at time $t = 0$, find its subsequent velocity as a function of time.

★★ **Problem 1.10** A small spherical ball of mass m and radius R is dropped from rest into a liquid of high viscosity η, such as honey, tar, or molasses. The only appreciable forces on it are gravity mg and a linear drag force given by Stokes's law, $F_{\text{Stokes}} = -6\pi\eta Rv$, where v is the ball's velocity, and the minus sign indicates that the drag force is opposite to the direction of v. (a) Find the velocity of the ball as a function of time. Then show that your answer makes sense for (b) small times; (c) large times.

★★★ **Problem 1.11** We showed in Example 1.2 that the distance a ball falls as a function of time, starting from rest and subject to both gravity g downward and a quadratic drag force upward, is

$$y = (v_T^2/g) \ln(\cosh(gt/v_T)),$$

where v_T is its terminal velocity. (a) Invert this equation to find how long it takes the ball to reach the ground in terms of its initial height h. (b) Check your result in the limits of *small h* and *large h*. (For part (b) it is useful to know the infinite series expansions of the functions e^x, $(1 + x)^n$, and $\ln(1 + x)$ for small x.)

★ **Problem 1.12** For objects with linear size between a few millimeters and a few meters moving through air near the ground, and with speed less than a few hundred meters per second, the drag force is close to a quadratic function of velocity, $F_D = (1/2)C_D A\rho v^2$, where ρ is the mass density of air near the ground, A is the cross-sectional area of the object, and C_D is the drag coefficient, which depends upon the shape of the object. A rule of thumb is that in air near the ground (where $\rho = 1.2$ kg/m^3), then $F_D \simeq \frac{1}{4}Av^2$. (a) Estimate the terminal velocity v_T of a skydiver of mass m and cross-sectional area A. (b) Find v_T for a skydiver with $A = 0.75$ m^2 and mass 75 kg. (The result is large, but a few lucky people have survived a fall without a parachute. An example is 21-year-old Nicholas Alkemade, a British Royal Air Force tail gunner during World War II. On March 24, 1944 his plane caught fire over Germany and his parachute was destroyed. He had the choice of burning to death or jumping out. He jumped and fell about 6 km, slowed at the end by falling through pine trees and landing in soft snow, ending up with nothing but a sprained leg. He was captured by the Gestapo, who at first did not believe his story, but when they found his plane they changed their minds. He was imprisoned, and at the end of the war set free, with a certificate signed by the Germans corroborating his story.)

★   **Problem 1.13** A damped oscillator consists of a mass m attached to a spring k, with frictional damping forces. If the mass is released from rest with amplitude A, and after 100 oscillations the amplitude is $A/2$, what is the total work done by friction during the 100 oscillations?

★   **Problem 1.14** The solution of the underdamped harmonic oscillator is $x(t) = Ae^{-\beta t}\cos(\omega_1 t + \varphi)$, where $\omega_1 = \sqrt{\omega_0^2 - \beta^2}$. Find the arbitrary constants A and φ in terms of the initial position x_0 and initial velocity v_0.

★★   **Problem 1.15** An overdamped oscillator is released at location $x = x_0$ with initial velocity v_0. What is the maximum number of times the oscillator can subsequently pass through $x = 0$?

★   **Problem 1.16** There are thought to be three types of the particles called *neutrinos*: electron-type (ν_e), muon-type (ν_μ), and tau-type (ν_τ). If they were all massless they could not spontaneously convert from one type into a different type. But if there is a mass difference between two types, call them types ν_1 and ν_2, the probability that a neutrino starting out as a ν_1 becomes a ν_2 is given by the oscillating probability $P = S_{12}\sin^2(L/\lambda)$, where S_{12} is called the *mixing strength parameter*, which we take to be constant, L is the distance traveled by the neutrino, and λ is a characteristic length, given in kilometers by

$$\lambda = \frac{E}{1.27\Delta(m)^2},$$

where E is the energy of the neutrino in units of GeV (1 GeV = 10^9 eV) and $\Delta(m)^2$ is the difference in the *squares* of the two masses in units of (eV)^2. Neutrinos are formed in earth's atmosphere by the collision of cosmic-ray protons from outer space with atomic nuclei in the atmosphere. The giant detector *Super Kamiokande*, located deep underground in a mine west of Tokyo, saw equal numbers of electron-type neutrinos coming (1) from the atmosphere above the detector; (2) from the atmosphere on the other side of the earth, which pass through our planet on their way to the detector. However, Super K saw more muon-type neutrinos coming down from above than those coming up from below. This was strong evidence that muon-type neutrinos oscillated into tau-type neutrinos (which Super K could not detect) as they penetrated the earth, since it requires more time to go 13,000 km through the earth than 20 km through the atmosphere above the mine. (a) Suppose

$(\Delta m)^2 = 0.01 \text{ eV}^2$ between ν_μ and ν_τ type neutrinos, and that the neutrino energy is $E = 5 \text{ GeV}$. What is λ? How would this explain the fewer number of muon neutrinos seen from below than from above? (b) The best experimental fit is $(\Delta m)^2 = 0.0022 \text{ eV}^2$. Again assuming $E = 5 \text{ GeV}$, what is λ? Make a crude estimate of the ratio one might expect for the number of muon neutrinos from below and from above.

Problem 1.17 The "quality factor" Q of an underdamped oscillator can be defined as

$$Q = 2\pi \frac{E}{|\Delta E|}, \tag{1.140}$$

where at some time E is the total energy of the oscillator and $|\Delta E|$ is the energy loss in one cycle. (a) Show that $Q \simeq \pi/\beta P$, where β is the damping constant and P is the period of oscillation. Therefore if the damping increases, Q decreases. (b) What is Q for a simple pendulum that loses 1% of its energy during each cycle? (c) The quality factor also describes the sharpness of the resonance curve of a driven, lightly damped oscillator. Show that to a good approximation $Q \simeq \omega/(\Delta\omega)$, where $\Delta\omega$ is the angular frequency difference between the two locations on the amplitude resonance curve for which the amplitude is $1/\sqrt{2}$ that at peak resonance.

Problem 1.18 Consider the unit vectors $\hat{\mathbf{x}}$, $\hat{\mathbf{y}}$, $\hat{\mathbf{r}}$, and $\hat{\boldsymbol{\theta}}$ in a plane. (a) Find $\hat{\mathbf{r}}$ and $\hat{\boldsymbol{\theta}}$ in terms of any or all of $\hat{\mathbf{x}}$, $\hat{\mathbf{y}}$, x, and y. (b) Find $\hat{\mathbf{x}}$ and $\hat{\mathbf{y}}$ in terms of any or all of $\hat{\mathbf{r}}$, $\hat{\boldsymbol{\theta}}$, r, and θ.

Problem 1.19 The mass and mean radius of the moon are $m = 7.35 \times 10^{22} \text{ kg}$ and $R = 1.74 \times 10^6 \text{ m}$. (a) From these parameters, along with Newton's constant of gravity $G = 6.674 \times 10^{-11} \text{m}^3/\text{kg/s}^2$, find the moon's escape velocity in m/s. (b) For a slingshot boom of length 50 m, what must be the minimum rotation frequency ω to sling material off the moon, as described in Example 1.3? Take into account both the radial and tangential components of the payload velocity when it comes off the end of the boom. Assume payloads are initially set upon the boom at radius $r = 3 \text{ m}$ and with $\dot{r} = 0$.

Problem 1.20 Ninety percent of the initial mass of a rocket is in the form of fuel. If the rocket starts from rest and then moves in gravity-free empty space, find its final velocity v if the speed u of its exhaust is (a) 3.0 km/s (typical chemical burning), (b) 1000 km/s, (c) $c/10$, where c is the speed of light. (d) If the exhaust velocity is 3.0 km/s, for how long can the rocket maintain the acceleration $a = 10 \text{ m/s}^2$?

Problem 1.21 A space traveler pushes off from his coasting spaceship with relative speed v_0; he and his spacesuit together have mass M, and he is carrying a wrench of mass m. Twenty minutes later he decides to return, but his thruster doesn't work. In another 40 minutes his oxygen supply will run out, so he immediately throws the wrench away from the ship direction at speed v_w relative to himself prior to the throw. (a) What then is his speed relative to the ship? (b) In terms of given parameters, what is the minimum value of v_w required so he will return in time?

★★   **Problem 1.22** An astronaut of mass M, initially at rest in some inertial frame in
gravity-free empty space, holds n wrenches, each of mass $M/2n$. (a) Calculate her
recoil velocity v_1 if she throws all the wrenches at once in the same direction with
speed u relative to her original inertial frame. (b) Find her final velocity v_2 if she
first throws half of the wrenches with speed u relative to her original inertial frame,
and then the other half with speed u relative to the frame she reached after the first
throw. Compare v_2 with v_1 from part (a). (c) Then find her total recoil velocity v_n
if she throws all n wrenches, one at a time and in the same direction, and each
with speed u relative to her instantaneous inertial frame just before she throws it.
(d) Find her total recoil velocity in the limit $n \to \infty$, and compare with the rocket
equation.

★★   **Problem 1.23** We are planning to travel in a rocket for 6 months with acceleration 10
m/s^2, and with a final payload mass of 1000 tonnes (1 tonne $= 1000$ kg). (a) Using
a chemically fueled rocket with exhaust speed 3160 m/s, what must be the original
ship mass m_0? Compare m_0 with the mass of the observed universe. (Including so-
called "dark matter," the mass density is approximately 6×10^{-30} g/cm^3 and the
observed radius is of order 10^{10} light years.) (b) Redo part (a) if instead we use a
fuel that can be ejected at 3.16×10^7 m/s, about 10% the speed of light. (c) How
fast would this ship be moving at the end of 6 months? (d) How far will the ship
have gone by this time? Compare this distance with the distance to the star Alpha
Centauri, about 4 light years away.

★★★   **Problem 1.24** A single-stage rocket rises vertically from its launchpad by burning
liquid fuel in its combustion chamber; the gases escape with a net momentum
downward, while the rocket, in reaction, accelerates upward. The gravitational field
is g. (a) Pretending that air resistance is negligible, show that the rocket's equation
of motion is

$$m\frac{dv}{dt} = -u\frac{dm}{dt} - mg,$$

where m is the instantaneous mass of the rocket at time t, v is its upward velocity,
and u is the speed of the exhaust relative to the rocket. (b) Assume that g and u
remain constant while the fuel is burning, and that fuel is burned at a constant rate
$|dm/dt| = \alpha$. Integrate the rocket equation to find $v(m)$. (c) Suppose that $u = 4.4$
km/s and that all the fuel is burned up in 1 minute. If the rocket achieves the escape
velocity from earth of 11.2 km/s, what percentage of the original launchpad mass
was fuel?

★★   **Problem 1.25** A rocket in gravity-free empty space has fueled mass M_0 and exhaust
velocity u equal to that of a first-stage Saturn V rocket (as used in sending men to
the moon): $M_0 = 3100$ tons $= 28 \times 10^6$ kg and $u = 2500$ m/s. The ship's acceleration
is kept constant at 10 m/s^2. (a) Find the initial rate of fuel ejection $|dM/dt|_{t=0}$. (b)
After how many minutes will the ship mass be reduced to $1/e$ of its initial value?
(c) Suppose the ship accelerates as described for 20 minutes. What percentage of

its initial mass is left? How many kilograms is this? What is the ship's velocity at this time?

★★ **Problem 1.26** Beginning at time $t = 0$, astronauts in a landing module are descending toward the surface of an airless moon with a downward initial velocity $-|v_0|$ and altitude $y = h$ above the surface. The gravitational field g is essentially constant throughout this descent. An onboard retrorocket can provide a fixed downward exhaust velocity u. The astronauts need to select a fixed exhaust rate $\lambda = |dm/dt|$ in order to provide a soft landing with velocity $v = 0$ when they reach the surface at $y = 0$. (a) Explain briefly why Newton's second law for the module during its descent has the form

$$m(t)\frac{dv}{dt} = u\left|\frac{dm}{dt}\right| - m(t)g.$$

(b) Find the velocity v of the module as a function of time, in terms of $|v_0|, u, m_0, \lambda$, and g. (c) During the descent its velocity is $v = dy/dt$, negative because it is downward. Find an expression for $y(t)$ in terms of $|v_0|, g, u, \lambda, m_0$, and h.

★★★ **Problem 1.27** A space probe of mass M is propelled by light fired continuously from a bank of lasers on the moon. A mirror covers the rear of the probe; light from the lasers strikes the mirrors and bounces directly back. In the rest frame of the lasers, n_γ photons are fired per second, each with momentum $p_\gamma = h\nu_\gamma/c$, where h is Planck's constant, c is the speed of light, and ν is the photon's frequency. (a) Show that in a short time interval Δt the change in the probe's momentum is $2n'_\gamma p'_\gamma \Delta t$, where n'_γ is the number of photons striking the mirror per second, and p'_γ is the momentum of each photon, both in the *probe's* frame of reference. (b) The photons are Doppler-shifted in the probe's frame, so their frequency is only $\nu' \approx \nu(1 - v/c)$, where v is the velocity of the probe. Show also that $n'_\gamma = n_\gamma(1 - v/c)$, and then show that the ship's acceleration has the form $a = \alpha(1 - v/c)^2$, where α is a constant. Express α in terms of M, n_γ, and p_γ. (c) Find an expression for the probe's velocity as a function of time. Briefly discuss the nature of this result as the probe travels faster and faster.

★★ **Problem 1.28** A proposed interstellar ram-jet would sweep up deuterons in space, burn them in an onboard fusion reactor, and expel the reaction products out the tail of the ship. In a reference frame instantaneously at rest relative to the ship, deuterons, each of mass m, approach the ship at relative velocity v. They are burned, and the burn products, with essentially the same total mass, are ejected from the rear of the ship at velocity $v+u$. The ship mass M stays constant, the cross-sectional area of the ship is A, and the number of deuterons per unit volume is n. (a) Find dN/dt, the number of deuterons swept up per unit time, in terms of n, A, and v. (b) Find dP/dt, the change in total momentum of the ship per unit time. (c) Show that the velocity of the ship increases exponentially, with $v = v_0 e^{\alpha t}$, where v_0 is the ship's initial velocity and α is a constant, which can be expressed in terms of given parameters. Assume that u is constant.

★ **Problem 1.29** (a) An open railroad coal car of mass M is rolling along a horizontal track at velocity v_0 when a coal chute suddenly dumps a load of coal of mass m into the coal car, vertically in the frame of the ground. When the load of coal has come to rest relative to the coal car, how fast is the coal car moving? (b) A similar coal car, of mass M and velocity v_0, has a covered hole in its bottom; when the cover is suddenly opened, a mass m of coal falls onto the tracks. How fast is the coal car then moving?

★ **Problem 1.30** Half of a chain of total mass M and length L is placed on a frictionless tabletop, while the other half hangs over the edge. If the chain is released from rest, what is the speed of the last link just as it leaves the tabletop?

★ **Problem 1.31** A particle of mass m is free to move in one dimension between the coordinates $x = 0$ and $x = 2\pi/k$, where k is a positive constant. Within this range the particle is subject to the force $F = \alpha \sin(kx)$, where α is a constant. (a) If the maximum value of the corresponding potential energy is α/k, what are the turning points of the particle if its energy is $E = \alpha/2k$? (b) Find the speed of the particle as a function of x.

★★ **Problem 1.32** One end of a string of length ℓ is attached to a small ball, and the other end is tied to a hook in the ceiling. A nail juts out from the wall, a distance d $(d < \ell)$ below the hook. With the string straight and horizontal, the ball is released. When the string becomes vertical it meets the nail, and then the ball swings upward until it is directly above the nail. (a) What speed does the ball have when it reaches this highest point? (b) Find the minimum value of ℓ such that the ball can reach this point at all.

★ **Problem 1.33** A rope of mass/length λ is in the shape of a circular loop of radius R. If it is made to rotate about its center with angular velocity ω, find the tension in the rope. *Hint*: Consider a small slice of the rope to be a "particle."

★ **Problem 1.34** A particle is attached to one end of an unstretched Hooke's-law spring of force constant k. The other end of the spring is fixed in place. If now the particle is pulled so the spring is stretched by a distance x, the potential energy of the particle is $U = (1/2)kx^2$. (a) Now suppose there are *two* springs with the same force constant that are laid end-to-end in the y direction, with a particle attached between them. The other ends of the springs are fixed in place. Now the particle is pulled in the *transverse* direction a distance x. Find its potential energy $U(x)$. (b) $U(x)$ is proportional to what power of x in the limit of small x, and to what power of x in the limit of large x?

★★ **Problem 1.35** A spherical pendulum consists of a bob of mass m on the end of a light string of length R hung from a point on the ceiling, and with a uniform gravitational field g downward. The position of the bob can be specified by the polar angle θ of the string (the angle of the string and bob from the vertical) and the azimuthal angle φ (the angle of the string and bob from, say, the north as projected down onto a horizontal base plane). (a) Show that the square of the velocity of the bob at any moment is $v^2 = R^2(\dot{\theta}^2 + \sin^2\theta\,\dot{\varphi}^2)$. Then, in terms of any or all of m, R, g, and the two coordinates θ and φ and their first time derivatives: (b) find an expression for the energy E of the bob and explain why it is conserved; (c) find an expression for the angular momentum ℓ of the bob about the vertical axis passing through the point of support, and explain why it is conserved.

Problem 1.36 Consider an arbitrary power-law central force $\mathbf{F}(\mathbf{r}) = -kr^n\hat{\mathbf{r}}$, where k and n are constants and r is the radius in spherical coordinates. Prove that such a force is conservative, and find the associated potential energy of a particle subject to this force.

Problem 1.37 The potential energy of a mass m on the end of a Hooke's-law spring of force constant k is $(1/2)kx^2$. If the maximum speed of the mass with this potential energy is v_0, what are the turning points of the motion?

Problem 1.38 Planets have roughly circular orbits around the sun. Using the table below of the orbital radii and periods of the inner planets, how does the centripetal acceleration of the planets depend upon their orbital radii? That is, find the exponent n in $a = \text{con} \times r^n$. (Note that 1 A.U. = 1 astronomical unit, the mean sun–earth distance.)

Planet	Mean orbital radius (A.U.)	Period (years)
Mercury	0.387	0.241
Venus	0.723	0.615
Earth	1.000	1.000
Mars	1.523	1.881

Problem 1.39 Four mathematically equivalent conditions for a force to be conservative are given in the chapter. One condition is that a conservative force can always be written as $F = -\nabla U$. Show then that each of the other three conditions is a necessary consequence.

Problem 1.40 A rock of mass m is thrown radially outward from the surface of a spherical, airless moon of radius R. From Newton's second law its acceleration is $\ddot{r} = -GM/r^2$, where M is the moon's mass and r is the distance from the moon's center to the rock. The energy of the rock is conserved, so $(1/2)m\dot{r}^2 - GMm/r = E = \text{constant}$. (a) Show by differentiating this equation that energy conservation is a first integral of $F = m\ddot{r}$ in this case. (b) What is the minimum value of E, in terms of given parameters, for which the rock will escape from the moon? (c) For this case what is $\dot{r}(t)$, the velocity of the rock as a function of time since it was thrown?

Problem 1.41 Consider a point mass m located a distance R from the origin, and a spherical shell of mass ΔM, radius a, and thickness Δa, centered on the origin. The shell has uniform mass density ρ. (a) Find ΔM in terms of the other parameters given, assuming $\Delta a \ll a$. Show that the gravitational potential energy of the point mass m due to the shell's gravity is (b) $-G\Delta Mm/R$ for $R > a$; (c) a constant for $R < a$. (d) Then show that if a mass distribution is spherically symmetric, the gravitational field inside it is directed radially inward, and its magnitude at radius R from the center is simply $G M(R)/R^2$, where $M(R)$ is the mass *within* the sphere whose radius is R. That is, a shell whose radius is greater than R exerts no net gravitational force on m.

★★   **Problem 1.42** A tunnel is drilled straight through a uniform-density nonrotating spherically symmetric airless asteroid of radius R. The tunnel is oriented along the x axis, with $x = 0$ at the center of the asteroid and of the tunnel. Using the results of the preceding problem, (a) show that if an astronaut of mass m steps into one side of the tunnel she will experience a spring-like force $F = -kx$ as she falls through the tunnel. (b) Find k in terms of any or all of G and the mass M and radius R of the asteroid. (c) Find the time it would take for her to oscillate from one end of the tunnel to the other and back again, in terms of the same parameters.

★★   **Problem 1.43** Referring to the preceding problem, if a different straight tunnel is drilled through the same asteroid, where this time the tunnel misses the asteroid's center by a distance $R/2$, (a) how long would it take the astronaut to fall from one end of the tunnel to the other and back, assuming no friction between the sides of the tunnel and the astronaut? (b) Suppose that instead of falling through the tunnel, she is given an initial tangential velocity of just the right magnitude to insert her into a circular orbit just above the surface. How long will it take her to return to the starting point in this case?

★   **Problem 1.44** Estimate the radius of the largest spherical asteroid an astronaut could escape from by jumping.

★★★   **Problem 1.45** A particle of mass m is subject to the central attractive force $\mathbf{F} = -k\mathbf{r}$, like that of a Hooke's-law spring of zero unstretched length, whose other end is fixed to the origin. The particle is placed at a position $\mathbf{r_0}$ and given an initial velocity $\mathbf{v_0}$ that is not colinear with $\mathbf{r_0}$. (a) Explain why the subsequent motion of the particle is confined to a plane containing the two vectors $\mathbf{r_0}$ and $\mathbf{v_0}$. (b) Find the potential energy of the particle as a function of r. (c) Explain why the particle's angular momentum is conserved about the origin, and use this fact to obtain a first-order differential equation of motion involving r and dr/dt. (d) Show that the particle has both an inner and an outer turning point, and solve the equation for $t(r)$, where the particle is located at an outer turning point at time $t = 0$. (e) Invert the result to find $r(t)$ in this case.

★   **Problem 1.46** A water molecule consists of an oxygen atom with a hydrogen atom on each side. The smaller of the two angles between the two OH bonds is $108°$. Find the distance of the center of mass of a water molecule from the oxygen atom in terms of the distance d between the oxygen atom and either hydrogen atom. The O has 16 times the mass of each H.

★ **Problem 1.47** A solid semicircle of radius R and mass M is cut from sheet aluminum. Find the position of its center of mass, measured from the midpoint of the straight side of the semicircle.

★ **Problem 1.48** Star α, of mass m, is headed directly toward star β, of mass $3m$, with velocity v_0 as measured in β's rest frame. (a) What is the velocity of their mutual center of mass, measured in β's frame? (b) How fast is each star moving in the CM frame? (c) If the two stars merge upon colliding, how fast is the new star moving in the CM frame?

★ **Problem 1.49** (a) A neutron in a nuclear reactor makes a head-on elastic collision with a carbon nucleus, which is initially at rest and has 12 times the mass of a neutron. What fraction of the neutron's initial speed is lost in the collision? (b) Repeat part (a) if instead the neutron collides head-on with a deuteron (mass twice that of the neutron) within a heavy-water (D_2O) molecule? (Carbon nuclei and deuterons can both be used as moderators in a reactor, whose purpose is to *moderate* their speeds, *i.e.*, slow down neutrons, as slower neutrons are more likely to cause nuclear fission.)

★ **Problem 1.50** A neutron of mass m and velocity v_0 collides head-on with a ^{235}U nucleus of mass M at rest in a nuclear reactor, and the neutron is absorbed to form a ^{236}U nucleus. (a) Find the velocity v_A of the ^{236}U nucleus in terms of m, M, and v_0. (b) The ^{236}U nucleus subsequently fissions into two nuclei of equal mass, each emerging at angle θ to the forward direction. Find the speed v_B of each final nucleus in terms of given parameters. Use classical Newtonian physics to solve the problem.

★ **Problem 1.51** Two balls, with masses m_1 and m_2, both moving along the same straight line, strike one another head-on in a one-dimensional elastic collision. (a) Show that the magnitude of the relative velocity between the two balls is the same before and after the collision. (b) Also show that if a video were made of such a collision, and then shown to an audience, the viewers could not be sure from the motion of the balls whether the video were being run forward or backward in time. That is, such collisions are said to be time-reversal invariant. (c) Would this time-reversal invariance still be true if the collision were inelastic? Give an example.

★★ **Problem 1.52** Three perfectly elastic superballs are dropped simultaneously from rest at height h_0 above a hard floor. They are arranged vertically, in order of mass, with $M_1 \gg M_2 \gg M_3$, where M_1 is at the bottom. There are small separations between the balls. When M_1 strikes the floor it bounces back up elastically, striking M_2 on its way down. M_2 then bounces back up, striking M_3 on its way down. What is the subsequent maximum altitude achieved by each ball? *Hint*: Analyze each collision in the CM frame of that collision, which is essentially the rest frame of the heavier ball because the other ball is so much lighter. Neglect the balls' radii and the small separations between them.

★★★   **Problem 1.53** *Classical big-bang cosmological models.* Consider a very large sphere of uniform-density dust of mass density $\rho(t)$. That is, at any given time the density is the same everywhere within the sphere, but the density decreases with time if the sphere expands, or increases with time if the sphere contracts, so that the total mass of the sphere remains fixed. At time $t = 0$ the sphere is all gathered at the origin, with infinite density and infinite outward velocity, so it is undergoing a "big bang" explosion. At some instant t_0 after the big bang the density everywhere within the sphere is ρ_0 and the outward speed of a particular dust particle at radius r_0 is v_0. Use Newton's gravitational constant G and also the result found in Problem 1-41, that in spherical symmetry only those mass shells whose radius r is *less* than the radius of the particular dust particle exert a net force on the particle. Let M_r be the (time-independent) total mass within radius r. (a) Find an expression for $r(t)$, the radius of the particle as a function of time, *supposing that the particle has the escape velocity.* That is, the particle, in common with all particles in the sphere, keeps moving outward but slows down, approaching zero velocity as $r \to \infty$. (b) Then consider the same model of spherically symmetric dust, except that instead of having the escape velocity, each dust particle has at any moment a velocity *less* than the escape velocity for that particle. This means that the energy of a particle of mass m is negative, where

$$E = \frac{1}{2}m\dot{r}^2 - \frac{GM_r m}{r} \tag{1.141}$$

and where $r(t)$ is its distance from the center of the sphere. Show that in this case, the time $t(r)$ expressed in terms of r can be written

$$t(r) = \sqrt{\frac{m}{2|E|}} \int^r dr \sqrt{\frac{r}{\alpha - r}} \tag{1.142}$$

and find the constant α in terms of G, M_r, m, and $|E|$. (c) To perform the integration, substitute $r = \alpha \sin^2(\eta/2) \equiv (\alpha/2)(1-\cos\eta)$, where η is a new variable, and show that

$$t = \left(\frac{\alpha^3}{8GM_r}\right)^{1/2}(\eta - \sin\eta). \tag{1.143}$$

(d) Make a table of t and r for a few values of η between 0 and 2π, and plot $r(t)$ for these values. The resulting shape is a *cycloid*, and the equations for $t(\eta)$ and $r(\eta)$ are in fact the parametric equations for a cycloid. Note that this negative-energy cosmological model begins with a "big bang" and ends with a "big crunch." (e) Finally, consider the same model of spherically symmetric dust, except that instead of having the escape velocity or less, dust particles have at any moment a velocity *greater* than the escape velocity. This means that the energy of a dust particle of mass m is positive. Using an approach analogous to that just used for negative-energy cosmologies, show that in this case

$$t(r) = \sqrt{\frac{m}{2E}} \int^{r} dr \sqrt{\frac{r}{\alpha + r}} \qquad (1.144)$$

and find the constant α in terms of $G, M_r, m,$ and E. (f) Perform the integration by substituting $r = \alpha \sinh^2(\eta/2) \equiv (\alpha/2)(\cosh \eta - 1)$, where η is a new variable and sinh and cosh are hyperbolic sine and hyperbolic cosine functions, respectively. (g) Then write the solution in parametric form, analogous to that just carried out for negative-energy cosmologies. That is, give formulas for both $t(\eta)$ and $r(\eta)$ for positive-energy cosmologies. (h) Make a table of t and r for several values of η, and plot $r(t)$ for these values. What is the ultimate fate of such a classical model universe?

★★★　**Problem 1.54** The Friedmann equations have played an important role in relativistic big-bang cosmologies. They feature a "scale factor" $a(t)$, proportional to the distance between any two points (such as the positions of two galaxies) that are sufficiently remote from one another that local random motions can be ignored. If a increases with time, the distance between galaxies increases proportionally, corresponding to an expanding universe. If we model for simplicity the universe as filled with pressure-free dust of uniform density ρ, the Friedmann equations for $a(t)$ are

$$\ddot{a} = -\frac{4\pi G\rho}{3}a \quad \text{and} \quad \dot{a}^2 = \frac{8\pi G\rho}{3}a^2 - \frac{kc^2}{R_0^2},$$

where G is Newton's gravitational constant, c is the speed of light, R_0 is the distance between two dust particles at some particular time t_0, and $k = +1, -1,$ or 0. The density of the dust is inversely proportional to the cube of the scale factor $a(t)$, *i.e.*, $\rho = \rho_0 (a_0/a)^3$, where ρ_0 is the density when $a = a_0$. Therefore

$$\ddot{a} = -\frac{4\pi G\rho_0 a_0^3}{3a^2} \quad \text{and} \quad \dot{a}^2 = \frac{8\pi G\rho_0 a_0^3}{3a} - \frac{kc^2}{R_0^2}.$$

(a) Show that if we set the origin to be at one of the two chosen dust particles, then if M is the total mass of dust within a sphere surrounding this origin out to the radius of the other chosen particle, and if at arbitrary time t_0 we set $a_0 = 1$, then the equations can be written

$$\ddot{a} = -\frac{(GM/R_0^3)}{a^2} \quad \text{and} \quad \frac{1}{2}\dot{a}^2 - \frac{(GM/R_0^3)}{a} = -\frac{kc^2}{2R_0^2} \equiv \epsilon,$$

where ϵ and M are constants.

　(b) Show that the second equation is a first integral of the first equation. (c) Compare these equations to the $F = ma$ and energy conservation equations of a particle moving radially under the influence of the gravity of a spherical moon of mass M. (d) Einstein hoped that his general-relativistic equations would lead to a static solution for the universe, since he (like just about everyone before him) believed that the universe was basically at rest. The Friedmann equations resulting from his theory show that the universe is generally expanding or contracting, however, just as a rock far from the earth is not going to stay there, but will generally be either falling inward or on its way out. So Einstein modified his theory with the addition of a "cosmological constant" Λ, which changed the Friedmann equations for pressure-free dust to

$$\ddot{a} = -\frac{(GM/R_0^3)}{a^2} + \frac{\Lambda}{3}a \quad \text{and} \quad \frac{1}{2}\dot{a}^2 - \frac{(GM/R_0^3)}{a} - \frac{\Lambda}{6}a^2 = \epsilon.$$

Show that these equations *do* have a static solution, and find the value of Λ for which the solution is static. (e) Show however (by sketching the effective potential energy function in the second equation) that the static solution is *unstable*, so that if the universe is kicked even slightly outward it will accelerate outward, or if it is kicked even slightly inward it will collapse. A static solution is therefore physically unrealistic. (Einstein failed to realize that his static solution was unstable, and later, when Edwin Hubble showed from his observations at the Mount Wilson Observatory that the universe is in fact expanding, Einstein declared that introducing the cosmological constant was "my biggest blunder.") (f) Suppose the cosmological constant is retained in the equations, but the dust is removed so that $M = 0$. Solve the equations for $a(t)$ in this case. The solution is the **de Sitter model**, an "inflationary" model of the expanding universe. What is the constant ϵ for the de Sitter model? (g) Make a qualitative sketch of $a(t)$ if both M and Λ are positive constants. Of the terms containing M and Λ, which dominates for small times? For large times?

2 Relativity

In this chapter we extend our review of mechanics to include Einstein's special theory of relativity. We will see that our previous Newtonian framework is a useful description of the mechanical world only when speeds are much less than that of light. We also use this chapter to introduce index notation and general technical tools that will help us throughout the rest of the book. Then, in the following chapter, we will show how relativity provides insights for an entirely different formulation of mechanics – the so-called **variational principle**.

2.1 Einstein's Postulates and the Lorentz Transformation

The most beautiful concepts in physics are often the simplest ones as well. In fact, the stunningly beautiful, revolutionary insights of special relativity are based on just two simple postulates proposed by Albert Einstein in 1905.

1. **The principle of relativity**: *The fundamental laws of physics are valid in all inertial frames of reference.*

We already introduced this principle in Chapter 1: it applies equally well to both Newtonian and relativistic physics.

There is a second, rather frugal postulate inspired by electromagnetism – combined with the first, it leads to astounding conclusions that stretch one's imagination and intuition to the limit.

2. **The universal speed of light**: *The speed of light is the same in all inertial frames.*

This second postulate follows from the assertion that Maxwell's equations of electromagnetism state a fundamental law of physics. Using the first postulate, this implies that these equations are valid in all inertial frames. But we note that Maxwell's equations lead to the wave equation

$$\frac{\partial^2 \phi}{\partial x^2} + \frac{\partial^2 \phi}{\partial y^2} + \frac{\partial^2 \phi}{\partial z^2} - \frac{1}{c^2}\frac{\partial^2 \phi}{\partial t^2} = 0 \tag{2.1}$$

for the electric potential $\phi(t, x, y, z)$ in vacuum, where c is the speed of light; this is in fact the wave equation for *light*.[1] A convenient solution to the equation is given by the plane wave

$$\phi = \phi_0 \cos(k(x - ct)), \tag{2.2}$$

where ϕ_0 is the amplitude and k is the wave number. Since Eq. (2.1) follows directly from a fundamental law of physics, c is a fundamental physical scale in Nature. If we track the position of a particular wavefront in this plane wave, say one of the crests $\phi = \phi_0$ with $x - ct = 0$, the wavefront then evolves according to $x(t) = ct$ from the point of view of an observer at rest in frame $\mathcal{O}$. Now consider a different inertial frame $\mathcal{O}'$, moving in the positive x direction according to observers in $\mathcal{O}$, as shown in Figure 2.1 – $\mathcal{O}$ and $\mathcal{O}'$ are the exact same reference frames we introduced in Chapter 1. According to the second postulate, the same wavefront would be seen by $\mathcal{O}'$ as moving with the *same* speed c along x': $x'(t') = ct'$.

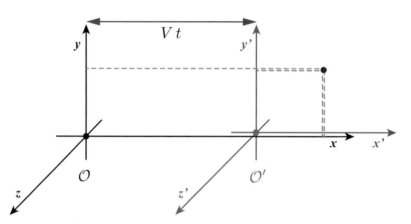

Fig. 2.1 Inertial frames $\mathcal{O}$ and $\mathcal{O}'$.

Panic ensues when we put these statements together with the Galilean transformation $x = x' + Vt$ between $\mathcal{O}$ and $\mathcal{O}'$ using Eq. (1.1); this gives

$$x = ct = x' + Vt = ct' + Vt = ct + Vt \tag{2.3}$$

since $t = t'$ from Eq. (1.1). This can be true only if the relative frame velocity V is zero!

To focus on the problem at hand, let us rephrase things in a slightly more general language. Say observer $\mathcal{O}$ is tracking a particle along a general trajectory $x(t)$. The same particle is seen by $\mathcal{O}'$ to evolve along $x'(t')$. A Galilean transformation tells us that $x(t) = x'(t') + Vt'$. Taking the time derivative of both sides of this equation, we find the usual velocity addition rule (1.3)

---

[1] There is much more on the electric potential in Chapter 8.

$$\frac{d}{dt} = \frac{d}{dt'} : \left[x(t) = x'(t') + Vt'\right] \Rightarrow \frac{dx(t)}{dt} = \frac{dx'(t')}{dt'} + V$$

$$\Rightarrow v_x = v'_x + V, \tag{2.4}$$

where $v_x = dx/dt$, $v'_x = dx'/dt'$, and we used $t = t'$ from Eq. (1.1). So if $v'_x = c$, then $v_x = c + V \neq c$ for $V \neq 0$, which contradicts the postulate, and we have a problem: *Galilean transformations are incompatible with a universal speed of light.* The second postulate can instead be seen as a restriction on the transformation rules relating the coordinate systems of inertial observers.

The critical question is then: What are the *correct* transformation equations relating the coordinates of $\mathbb{O}'$ and $\mathbb{O}$ that replace the Galilean transformation? Since the Galilean transformation arises intuitively from our basic sense of the world around us, it had better be the case that it can still be viewed as a decent approximation to the correct transformation, which we now set out to find.

As shown already in Figure 2.1, frames $\mathbb{O}$ and $\mathbb{O}'$ are assigned coordinate labels (t, x, y, z) and (t', x', y', z'), respectively, such that $\mathbb{O}'$ moves with velocity $\mathbf{V}$ in the positive x direction as seen by observers at rest in frame $\mathbb{O}$, with the x and x' axes aligned, and with the y' axis parallel to the y axis and the z' axis parallel to the z axis. According to the **Galilean** transformation, the coordinates in $\mathbb{O}$ are related to those in $\mathbb{O}'$ by (1.1)

$$x = x' + Vt', \quad y = y', \quad z = z', \quad t = t', \tag{2.5}$$

while the coordinates in $\mathbb{O}'$ are related to those in $\mathbb{O}$ by the inverse transformation

$$x' = x - Vt, \quad y' = y, \quad z' = z, \quad t' = t, \tag{2.6}$$

which can be obtained from the first set simply by interchanging primed and unprimed coordinates and letting $V \to -V$.

Now the task at hand is to derive a replacement for the Galilean transformation, one that is consistent with a universal speed of light. We begin by assuming that the new transformation for x, y, z and inverse transformation for x', y', z' have the somewhat more general, but still linear, form

$$x = \gamma x' + \zeta t', \quad x' = \gamma' x + \zeta' t, \quad y = y', \quad z = z', \tag{2.7}$$

where γ, γ', ζ, and ζ' are constants, independent of position or time. That is, we have assumed for simplicity that the y and z transformations are the same as in the Galilean transformation, that the origins of the two coordinate systems overlap at zero time, and that the equations for $x(x', t')$ and for $x'(x, t)$ are still linear. We will have to see if these assumptions are consistent with the speed of light postulate; if not, we will have to try something more general still. Note that we have *not* assumed that t is necessarily equal to t'.

Our goal now is to evaluate the constants γ, γ', ζ, and ζ'. We have four constants to determine, hence we need four physical conditions. *We can determine three of the constants in terms of the fourth without ever invoking the second postulate.*

First of all, from the meaning of the relative frame velocity V, the origin of the primed frame, $(x', y', z') = (0, 0, 0)$, must move with velocity V in the positive x direction as measured in the unprimed frame; i.e., if $x' = 0$, then $x = Vt$ (**condition 1**). This forces $\zeta' = -V\gamma'$ in the second of Eqs. (2.7). We also want the origin of the unprimed frame to move in the *opposite* direction with speed V as measured in the primed frame; i.e., if $x = 0$, then $x' = -Vt'$ (**condition 2**). This gives $\zeta = V\gamma$ in the first of Eqs. (2.7). Therefore we can write

$$x = \gamma(x' + Vt'), \quad x' = \gamma'(x - Vt), \quad y = y', \quad z = z'. \tag{2.8}$$

Now the first postulate asserts that there is no preferred inertial frame of reference, so from the symmetry this implies we must have $\gamma' = \gamma$. Why is that?

Consider a clock A' at rest at the origin of $\mathcal{O}'$; it reads time t' and it always sits at $x' = 0$. When it reads $t' = 1$ s, its distance from the origin of $\mathcal{O}$, according to $\mathcal{O}$ observers, is $x = \gamma V(1 \text{ s})$, from the first equation above. Consider another clock A at rest at the origin of $\mathcal{O}$; it reads time t and it always sits at $x = 0$. When it reads $t = 1$ s, its distance from the origin of $\mathcal{O}'$, from the point of view of $\mathcal{O}'$, is $x' = -\gamma' V(1 \text{ s})$, from the second equation above: the minus sign simply reflects the fact that $\mathcal{O}$ moves in the *negative* x' direction from the point of view of $\mathcal{O}'$. However, except for this minus sign, which is related to the direction of travel, the distances moved by A and B when each reads 1 s *should be exactly the same*, according to the egalitarian first postulate. If they were different, it would allow us to say that one frame (say the frame in which the distance moved was greater) was fundamentally "better" than the other frame. This forces us to the conclusion that $\gamma' = \gamma$ (**condition 3**).

The transformation now becomes

$$x = \gamma(x' + Vt'), \quad x' = \gamma(x - Vt), \quad y = y', \quad z = z' \tag{2.9}$$

for some still undetermined value of γ. The Galilean transformation assumes $\gamma = 1$, but as we have seen, this choice is inconsistent with a universal speed of light.

We now finally require that if $x = ct$ then also $x' = ct'$, corresponding to a beam of light emitted from the mutual coordinate origins at the instant $t = t' = 0$ when the origins coincide (**condition 4**, i.e., postulate 2). The beam moves in the x directions of both frames with the same speed c, in which case the first two of Eqs. (2.9) become

$$t = \gamma(1 + V/c)t' \quad \text{and} \quad t' = \gamma(1 - V/c)t. \tag{2.10}$$

We can eliminate t' by substituting the second equation into the first; this gives

$$t = \gamma^2(1 + V/c)(1 - V/c)t = \gamma^2(1 - V^2/c^2)t, \tag{2.11}$$

so we must choose $\gamma^2 = (1 - V^2/c^2)^{-1}$. Finally, we need the *positive* square root so that $\gamma \to 1$ as $V \to 0$, because as $V \to 0$ the two sets of axes coincide. Therefore the final results for $x(t', x')$ and $x'(t, x)$ are

$$x = \gamma(x' + Vt') \quad \text{and} \quad x' = \gamma(x - Vt), \quad \text{where} \quad \gamma = \frac{1}{\sqrt{1 - V^2/c^2}}. \tag{2.12}$$

We can now find the transformation equations for t and t'. Substitute $x' = \gamma(x - Vt)$ into the right-hand side of $x = \gamma(x' + Vt')$; the resulting equation can be solved for t' to give

$$t' = \gamma\left(t - \frac{Vx}{c^2}\right). \tag{2.13}$$

We can instead eliminate x between the two equations and then solve for t to give

$$t = \gamma\left(t' + \frac{Vx'}{c^2}\right), \tag{2.14}$$

which is the same as Eq. (2.13) if we interchange primed and unprimed coordinates and let $V \to -V$. Thus we have the amazing and profound result that *there is no longer an absolute time, the same in all frames.* Relativity shows that time and space have become closely intertwined.

The entire set of transformations from primed to unprimed coordinates can be written in the compact form

$$\begin{aligned}
ct &= \gamma(ct' + \beta x'), \\
x &= \gamma(x' + \beta ct'), \\
y &= y', \\
z &= z',
\end{aligned} \tag{2.15}$$

where

$$\beta \equiv V/c \quad \text{and} \quad \gamma = \frac{1}{\sqrt{1 - \beta^2}}. \tag{2.16}$$

These equations are collectively called the **Lorentz transformation** or colloquially **Lorentz boost**. We have used the product ct in the equations, instead of just t by itself, so that the four coordinates ct, x, y, z all have the same dimension of length. The *inverse* Lorentz transformation, for (ct', x', y', z') in terms of (ct, x, y, z), is the same, with primed and unprimed coordinates interchanged and with $\beta \to -\beta$. We began with four constants $\gamma, \gamma', \zeta, \zeta'$ and found four conditions they must obey, which determined all four in terms of the relative frame velocity V.

Having found the transformation by invoking the speed of light only in the x direction, we can verify that the transformation works also for light moving in *any* direction. Let a flashbulb flash from the mutual origins of frames $\mathcal{O}$ and $\mathcal{O}'$ just as the origins pass by one another. In the unprimed frame, the square of the distance moved by the wavefront of light in time t is

$$x^2 + y^2 + z^2 = c^2 t^2. \tag{2.17}$$

That is, the light flash spreads out spherically at speed c in all directions. Now, using the Lorentz transformation of Eqs. (2.15), we can see how the flash moves in the primed frame. Rewriting Eq. (2.17) in terms of primed coordinates, we have

$$[\gamma(x' + \beta ct')]^2 + y'^2 + z'^2 = [\gamma(ct' + \beta x')]^2, \tag{2.18}$$

which, with a little algebra, yields

$$x'^2 + y'^2 + z'^2 = c^2 t'^2. \tag{2.19}$$

That is, the light flash also moves in all directions at speed c in frame $\mathcal{O}'$. Therefore the second postulate is obeyed for light moving in *any* direction, if we use the Lorentz transformation to transform coordinates.

Let us observe the Lorentz transformation equation (2.15) for a while and note some of its features.

- For $V \ll c$, *i.e.*, when the two observers are moving with respect to one another at a speed much less than that of light, we have $\beta \ll 1$ and $\gamma \sim 1$ to leading order in β, and the Lorentz transformation equation (2.15) reduces to the Galilean transformation (1.1). That's a sanity check: our intuition led us to (1.1) because our daily experiences involve mechanics at speeds much less than that of light. Hence, we may still use Galilean transformations as long as we restrict ourselves to problems involving slow speeds and as long as we don't care about high-precision measurements. Obviously, Maxwell's equations involve light and so require the use of the full and correct form of the transformation of coordinates, the Lorentz transformations. This is why electromagnetism historically seeded the development of relativity.
- There are two obvious novelties: the mixing of time and space, the coordinates t and x, t' and x'; and an interesting scale factor $\gamma \geq 1$. Figure 2.2 shows a plot of γ as a function of β, showing that γ changes no more than 1% from unity for $0 \leq \beta \leq 0.1$, or speeds up to about 10% of light. A rough rule of thumb is then to require $v < 0.1c$ for Newtonian mechanics. Note also the divergence as $V \to c$: the corresponding flip of the sign under the square root for $V > c$ implies an upper bound on speed $\beta < 1 \Rightarrow V < c$. *Nature comes with a speed limit.*

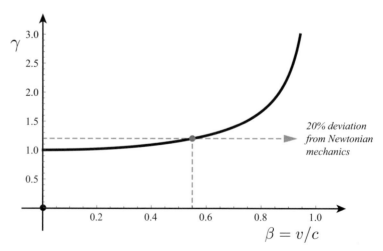

Fig. 2.2 Graph of the γ factor as a function of the relative velocity β. Note that $\gamma \cong 1$ for nonrelativistic particles, and $\gamma \to \infty$ as $\beta \to 1$.

Example 2.1   **Rotation and Rapidity**

Consider two observers $\mathcal{O}$ and $\mathcal{O}'$, stationary with respect to one another and with identical origins, but with axes $\{x, y, z\}$ and $\{x', y', z'\}$ relatively rotated. Focus on a case where observer $\mathcal{O}$'s coordinate system is rotated with respect to $\mathcal{O}'$s by a positive angle θ about the z axis:

$$x = x' \cos\theta + y' \sin\theta, \quad y = -x' \sin\theta + y' \cos\theta, \quad z = z'. \tag{2.20}$$

It is often convenient to write this transformation in matrix notation:

$$\begin{pmatrix} x \\ y \\ z \end{pmatrix} = \begin{pmatrix} \cos\theta & \sin\theta & 0 \\ -\sin\theta & \cos\theta & 0 \\ 0 & 0 & 1 \end{pmatrix} \begin{pmatrix} x' \\ y' \\ z' \end{pmatrix}. \tag{2.21}$$

In general, a rotation can be written as

$$\mathbf{r} = \hat{\mathcal{R}} \cdot \mathbf{r}', \tag{2.22}$$

for $\mathbf{r} = (x, y, z)$ and $\mathbf{r}' = (x', y', z')$, with $\hat{\mathcal{R}}$ a 3×3 rotation matrix satisfying the orthogonality condition $\hat{\mathcal{R}}^{\mathsf{T}} \cdot \hat{\mathcal{R}} = \mathbf{1}$, as well as having the determinant $|\hat{\mathcal{R}}| = 1$. Here $\hat{\mathcal{R}}^{\mathsf{T}}$ is the transpose matrix, the reflection of $\hat{\mathcal{R}}$ about its principal diagonal.

Interestingly, we can write a Lorentz boost in analogy to rotations, making its structural form more elegant and transparent. To do so, we start by introducing a *four*-component "position vector"

$$r \equiv (ct, \mathbf{r}) = (ct, x, y, z), \tag{2.23}$$

denoting an event in spacetime occurring at position (x, y, z) *and* time t. This is a natural notation, since Lorentz transformations mix space and time coordinates; again, we use ct as the time component to give it the same dimension of length as the other components.

Just as in Chapter 1, we represent *three*-vectors by *non-italicized* bold-face symbols ($\mathbf{r}$ for example). *Four*-vectors, which have a time component as well as three spatial components, we represent by *italicized* bold-faced symbols (r for example). Individual components are represented by italicized non-bold-face symbols in both cases. The time component of r is also called the "zeroth" component of the four-vector, so that the spatial components x, y, and z can still be called the first, second, and third components, just as they are for the position *three*-vector.

We can now write the Lorentz boost in Eq. (2.15) as a matrix multiplication

$$\begin{pmatrix} ct \\ x \\ y \\ z \end{pmatrix} = \begin{pmatrix} \gamma & \gamma\beta & 0 & 0 \\ \gamma\beta & \gamma & 0 & 0 \\ 0 & 0 & 1 & 0 \\ 0 & 0 & 0 & 1 \end{pmatrix} \begin{pmatrix} ct' \\ x' \\ y' \\ z' \end{pmatrix}. \tag{2.24}$$

Consider the parametrization

$$\gamma \equiv \cosh\xi \geq 1, \tag{2.25}$$

where ξ is called **rapidity**. Using the identity $\cosh^2 \xi - \sinh^2 \xi = 1$ one can easily show that

$$\gamma \beta = \sinh \xi, \tag{2.26}$$

so our Lorentz boost now becomes

$$\begin{pmatrix} ct \\ x \\ y \\ z \end{pmatrix} = \begin{pmatrix} \cosh \xi & \sinh \xi & 0 & 0 \\ \sinh \xi & \cosh \xi & 0 & 0 \\ 0 & 0 & 1 & 0 \\ 0 & 0 & 0 & 1 \end{pmatrix} \begin{pmatrix} ct' \\ x' \\ y' \\ z' \end{pmatrix}, \tag{2.27}$$

much like a rotation but with hyperbolic trigonometric functions instead, and with a sign flip. We say that Lorentz transformations rotate time and space into one another.

We can write the most general Lorentz transformation in matrix notation as well:

$$\boldsymbol{r} = \hat{\boldsymbol{\Lambda}} \cdot \boldsymbol{r}', \tag{2.28}$$

for $\boldsymbol{r} = (ct, x, y, z)$ and $\boldsymbol{r}' = (ct', x', y', z')$, with $\hat{\boldsymbol{\Lambda}}$ a *general* 4×4 matrix satisfying the condition

$$\hat{\boldsymbol{\Lambda}}^\mathsf{T} \cdot \hat{\eta} \cdot \hat{\boldsymbol{\Lambda}} = \hat{\eta} \tag{2.29}$$

as well as

$$|\hat{\boldsymbol{\Lambda}}| = 1, \tag{2.30}$$

where $\hat{\eta}$ is the 4×4 matrix

$$\hat{\eta} = \begin{pmatrix} -1 & 0 & 0 & 0 \\ 0 & 1 & 0 & 0 \\ 0 & 0 & 1 & 0 \\ 0 & 0 & 0 & 1 \end{pmatrix}. \tag{2.31}$$

For a derivation of this general statement, see the Problems section at the end of this chapter. Notice that

$$\hat{\boldsymbol{\Lambda}} = \begin{pmatrix} \cosh \xi & \sinh \xi & 0 & 0 \\ \sinh \xi & \cosh \xi & 0 & 0 \\ 0 & 0 & 1 & 0 \\ 0 & 0 & 0 & 1 \end{pmatrix} \tag{2.32}$$

satisfies Eqs. (2.29) and (2.30). Note also that $\hat{\eta}$ is *almost* the identity matrix, but not quite, because of the minus sign in the first entry. It is known as the **metric** of flat spacetime. Correspondingly, $\hat{\boldsymbol{\Lambda}}$ satisfies an "almost" orthogonality condition (2.29). We will revisit these observations in the upcoming sections as we develop our physical intuition for relativity.   ∎

2.2 Relativistic Kinematics

Kinematics deals with our *description* of motion, including the position, velocity, and acceleration of particles, while stopping short of looking for *causes* of that motion, which is the subject of *dynamics*. So we take up the essential topic of relativistic kinematics here, and then go on to relativistic dynamics in the following section.

Proper Time

Consider a particle moving in the vicinity of an observer $\mathbb{O}$ who describes its trajectory by $x(t)$, $y(t)$, $z(t)$. The observer can describe the location of the particle in time and space using a position four-vector

$$r = (c\,t, x, y, z). \tag{2.33}$$

If dt, dx, dy, and dz represent infinitesimal steps in the evolution of the particle, we can also write the infinitesimal displacement four-vector as

$$dr = (c\,dt, dx, dy, dz). \tag{2.34}$$

Observer $\mathbb{O}$ may, for some yet mysterious reason, choose to compute the quantity

$$ds^2 = dr^{\mathrm{T}} \cdot \hat{\eta} \cdot dr = -c^2 dt^2 + dx^2 + dy^2 + dz^2. \tag{2.35}$$

The first part of this expression uses matrix notation: $\hat{\eta}$ is the 4×4 matrix from Eq. (2.31), and the "T" label denotes the transpose operation on the column vector that is dr: that is, a row vector dr^{T} multiplies the matrix $\hat{\eta}$ which then multiplies the column vector dr. To see why this quantity is interesting to compute, consider the same quantity as computed by an observer $\mathbb{O}'$ at rest in the primed frame. Equation (2.28) prescribes that we must have

$$dr = \hat{\mathbf{\Lambda}} \cdot dr' \tag{2.36}$$

by taking the differential of both sides of Eq. (2.28). Substituting this into Eq. (2.35), we get

$$
\begin{aligned}
ds^2 &= dr^{\mathrm{T}} \cdot \hat{\eta} \cdot dr \\
&= dr'^{\mathrm{T}} \cdot \hat{\mathbf{\Lambda}}^{\mathrm{T}} \cdot \hat{\eta} \cdot \hat{\mathbf{\Lambda}} \cdot dr' = dr'^{\mathrm{T}} \cdot \hat{\eta} \cdot dr' \\
&= -c^2 dt'^2 + dx'^2 + dy'^2 + dz'^2,
\end{aligned}
\tag{2.37}
$$

where we used Eq. (2.29). Comparing Eqs. (2.35) and (2.37), we now see that ds^2 is an *invariant* under Lorentz transformations. Observers $\mathbb{O}$ and $\mathbb{O}'$ use the same form of the expression in their respective coordinate systems and get the same value for ds^2. In general, quantities like ds^2 that remain the same under Lorentz transformations are said to be **scalar invariants** or **Lorentz invariants**.

The quantity ds^2 has a special physical meaning. Imagine that observer $\mathbb{O}'$ happens to be "riding" with the particle at the given instant in time she measures the displacement four-vector $d\mathbf{r}'$. Observer $\mathbb{O}'$ would then see the particle momentarily at rest, with $dx' = dy' = dz' = 0$, since she is matching the particle's velocity at that instant: that is,

$$d\mathbf{r}' = (c\,dt', 0, 0, 0). \tag{2.38}$$

Now $dt' \equiv d\tau$ is an advance in time on the watch of $\mathbb{O}'$, *i.e.*, a watch in the **rest frame** of the particle. We then have from Eq. (2.37)

$$ds^2 = -c^2 d\tau^2, \tag{2.39}$$

so the value of ds^2 measures the period of an infinitesimal clock tick *as measured in the rest frame of our particle.*[2] The quantity τ is called the **proper time** of the particle. It is a four-*scalar*; that is, observers in every frame can agree on the clock's tick rate in its own rest frame. Equation (2.39) also helps us relate the proper time τ of the particle to the time t in the frame of reference of observer $\mathbb{O}$, since we know that

$$ds^2 = -c^2 d\tau^2 = -c^2 dt^2 + dx^2 + dy^2 + dz^2. \tag{2.40}$$

Dividing this equation by dt^2, we get

$$c^2 \frac{d\tau^2}{dt^2} = c^2 - \frac{dx^2}{dt^2} - \frac{dy^2}{dt^2} - \frac{dz^2}{dt^2} = c^2 - v^2, \tag{2.41}$$

where v is the speed of the particle, from which we find that

$$dt = \gamma\,d\tau, \tag{2.42}$$

with $\gamma = 1/\sqrt{1 - \beta^2}$ and $\beta \equiv v/c$. This implies that a time interval $d\tau$ in the rest frame of the particle is perceived by observer $\mathbb{O}$ as an interval $dt > d\tau$. This effect is known as **time dilation**: from the point of view of an inertial observer watching some particle move with speed v, if the observer ages by (say) 10 s, the particle may age by only 1 s from the observer's perspective. We say the particle's time slows down (i.e., time intervals stretch out or dilate) as seen by observer $\mathbb{O}$. Note that this relation holds instantaneously, even when the particle is accelerating. At every instant in time, its changing velocity results in changing time dilations with respect to observer $\mathbb{O}$.

---

[2] Note that in four-dimensional spacetime the quantity ds^2 can be positive, zero, or *negative*. It is positive for **space-like intervals**, where the spatial distance between two events is greater than their time difference, $dx^2 + dy^2 + dz^2 > c^2 dt^2$, negative for *time-like* intervals, where the time difference is greater than the spatial distance, and zero for *null* intervals, where the spatial and time differences are equal to one another, as along the path of a light ray.

We have learned that time is not at all a universal observable: it is a "malleable" quantity, with two observers in different reference frames disagreeing about its rate of advance. To talk about a notion of time that everyone agrees on in relativity, we need to refer to *proper* time – the time as measured in the rest frame of a particular reference observer or particle. For speeds that are small compared to the speed of light, we have $\gamma \sim 1$ in Eq. (2.42), and we recover the approximate Galilean statement of universal time $dt = d\tau$.

At this point we will use our current discussion to introduce a notation that will be useful throughout the rest of the book. We have already started to appreciate the elegance of lumping time and space together in a position four-vector, and we have also demonstrated the use of matrix language in compactifying the notation. Putting these technologies together, let us denote the four components of the position four-vector as

$$\boldsymbol{r} = (c\,t, x, y, z) \equiv (r^t, r^x, r^y, r^z), \tag{2.43}$$

that is, with a "zeroth" time-like component $r^t = c\,t$, as well as three spatial components $r^x = x$, $r^y = y$, and $r^z = z$. Note in particular the superscript notation we adopt, which allows us to reserve vector *subscripts* to distinguish between *different* particles. We then neatly denote the components of the position four-vector of a single particle as r^μ, where μ is an index that can take on the values, in Cartesian coordinates, t, x, y, or z. We can then write the components of the displacement four-vector $d\boldsymbol{r}$ as dr^μ. Let us now rewrite Eq. (2.35) in terms of this new "index notation":

$$ds^2 = \sum_{\substack{\mu \in \\ \{t,x,y,z\}}} \sum_{\substack{\nu \in \\ \{t,x,y,z\}}} dr^\mu \hat{\eta}_{\mu\nu} dr^\nu, \tag{2.44}$$

where we are representing the $\hat{\eta}$ matrix by its components – $\hat{\eta}_{\mu\nu}$ is the entry in the $\hat{\eta}$ matrix (2.31) in the μth row and νth column.[3] The two sums in the expression simply implement the usual matrix multiplication rule of multiplying rows against columns. Note also that this expression is now in the form of a sum over numbers: the quantities dr^μ, dr^ν, $\hat{\eta}_{\mu\nu}$ are just commutative numbers. Therefore we can write

$$ds^2 = \sum_{\substack{\mu \in \\ \{t,x,y,z\}}} \sum_{\substack{\nu \in \\ \{t,x,y,z\}}} dr^\mu \hat{\eta}_{\mu\nu} dr^\nu = \sum_{\substack{\mu \in \\ \{t,x,y,z\}}} \sum_{\substack{\nu \in \\ \{t,x,y,z\}}} dr^\mu dr^\nu \hat{\eta}_{\mu\nu}$$

$$= \sum_{\substack{\mu \in \\ \{t,x,y,z\}}} \sum_{\substack{\nu \in \\ \{t,x,y,z\}}} \hat{\eta}_{\mu\nu} dr^\mu dr^\nu. \tag{2.45}$$

---

[3] A note for readers with a more advanced background in differential geometry: in more advanced notation, a distinction is made between upper and lower Lorentz indices – corresponding to mathematical objects in the so-called tangent and co-tangent spaces of spacetime. In our notation, all vector quantities will be given in the *tangent space* and hence we use exclusively superscripts for vector indices.

As you can see, rewriting the sum symbol over and over becomes tedious. Indices such as μ and ν are often summed over the t, x, y, and z components, so we adopt the **Einstein summation convention**: any index that appears *exactly twice* in the same term is assumed to be summed over unless explicitly stated otherwise. For example, we may now write simply

$$ds^2 = \hat{\eta}_{\mu\nu} dr^\mu dr^\nu \tag{2.46}$$

with an implied sum over μ and over ν, since each index occurs exactly twice in the same term. Here, r^μ (and r^ν, where $\mu = 0, 1, 2, 3$, $\nu = 0, 1, 2, 3$) are the four components of the four-dimensional position vector.

To say that ds^2 is a Lorentz invariant means that

$$ds^2 = \hat{\eta}_{\mu\nu} dr^\mu dr^\nu = \hat{\eta}_{\mu'\nu'} dr^{\mu'} dr^{\nu'}, \tag{2.47}$$

where primed indices refer to coordinates in the coordinate system of $\mathcal{O}'$. Note that the $\hat{\eta}_{\mu'\nu'}$s are the same as the corresponding $\hat{\eta}_{\mu\nu}$s (see Eq. (2.31)).

Now we can rewrite Eq. (2.36) in our new notation

$$dr^\mu = \hat{\Lambda}^\mu_{\nu'} dr^{\nu'}. \tag{2.48}$$

The components of the Lorentz matrix $\hat{\Lambda}$ are represented by $\hat{\Lambda}^\mu_{\nu'}$ at the μth row and ν'th column. The ν' index appears twice in the expression on the right-hand side, so it is summed over: the sum implements the matrix multiplication $\hat{\Lambda} \cdot dr'$. Note that there is also a *single* index μ in each term; that index is *not* repeated – it appears only *once* in each term and so is not summed over. For every possible value of μ we have a different equation – for a total of four. These are the relations for the four components of dr. If we encounter an expression with an index that occurs *more* than twice in the same term, a mistake has been made!

Index notation takes some time to get used to, but it is worth it. It is powerful, and the physics of relativity lends itself very naturally to this notation and language. It is one of those things that requires practice to get the hang of, but once mastery is achieved, it is difficult to remember how one got by in the past without it. As we proceed with the discussion of relativity, we adopt this notation from the outset to provide the reader with as much practice and exposure as possible.

Four-Velocity

Calculus is the natural language of motion: Newton invented differential calculus to make the discussion of motion more natural and precise.[4] Similarly, four-vector notation is the natural language of relativity, because relativistic physics inherently mixes time and space. One could proceed without the use of this mathematical language of four-vectors, but that would come at the expense of unnecessarily convoluting the discussion of the physics. It is important, however, to appreciate

---

[4] Ironically, his masterpiece, the *Principia*, uses no calculus at all, because few of his readers would have understood it: the *Principia* presents instead an exposition of mechanics in a rather awkward mathematical language that often obscures the physics at hand, particularly to present-day observers familiar with calculus.

where the physics starts and where it ends. The tool of four-vectors we will use in this section is just that, a mathematical tool. It comes with layers of logic that make the symmetries underlying relativity more transparent and hence guides us to the next natural steps in the discussion. Throughout, we still need to rely on the independent statements of physics, *i.e.*, the postulates of relativity, including the universality of the speed of light.

We start by looking for an observable quantity that relates to the ordinary three-vector *velocity*, but which also fits more naturally into our new language of relativity. We want a "four-velocity," a quantity with four components whose spatial components reduce to the usual three-velocity components at small speeds. For this new quantity to be natural in relativity, it should transform under Lorentz transformations in a simple way. Let $\boldsymbol{u}$ be the velocity four-vector, with components u^μ, where $\mu = t, x, y, z$ as usual. We then require that

$$u^\mu = \hat{\Lambda}^\mu_{\ \nu'} u^{\nu'}. \tag{2.49}$$

Whatever $\boldsymbol{u}^\mu$ may be for observer $\mathbb{O}$, it relates to $\boldsymbol{u}^{\mu'}$ as seen by observer $\mathbb{O}'$ by this simple Lorentz transformation. It also needs to be related to our usual notion of velocity – the rate of change of position per unit time. However, time is not a universally invariant notion. The closest we can get to this concept is *proper time*, so the obvious candidate for four-velocity is

$$u^\mu \equiv \frac{dr^\mu}{d\tau}. \tag{2.50}$$

In this expression dr^μ is the four-displacement of a particle observed by $\mathbb{O}$, and $d\tau$ is the advance in proper time of the particle in question – which both primed and unprimed observers agree upon. Since $dr^\mu = \hat{\Lambda}^\mu_{\ \nu'} dr^{\mu'}$ and $d\tau$ is invariant, we see that this definition does lead to Eq. (2.49), as required. But how does it relate to ordinary three-vector velocity? To see this, we need to write u^μ explicitly in terms of the coordinates of a fixed observer, say $\mathbb{O}$:

$$u^\mu \rightarrow \left(\frac{dr^t}{d\tau}, \frac{dr^x}{d\tau}, \frac{dr^y}{d\tau}, \frac{dr^z}{d\tau} \right) = \left(c \frac{dt}{d\tau}, \frac{dx}{d\tau}, \frac{dy}{d\tau}, \frac{dz}{d\tau} \right)$$

$$= \left(\gamma c, \gamma \frac{dx}{dt}, \gamma \frac{dy}{dt}, \gamma \frac{dz}{dt} \right), \tag{2.51}$$

where we used the time dilation relation obtained previously in Eq. (2.42):

$$dt = \gamma \, d\tau. \tag{2.52}$$

Note that $\gamma = 1/\sqrt{1 - v^2/c^2}$, where v is the speed of the particle as seen by $\mathbb{O}$.

We now recognize the last three components of our four-vector as γ times the ordinary velocity of the particle. We may write more compactly

$$u^\mu \rightarrow (\gamma c, \gamma \mathbf{v}), \tag{2.53}$$

lumping the last three entries together. For a slow-moving particle, we have $\gamma \sim 1$ and $u^\mu \sim (c, \mathbf{v})$, so we have achieved our goal of embedding velocity into the four-vector language.

What have we gained from this exercise? Well, we now know how the ordinary velocity transforms under Lorentz transformations, as we shall see.

Example 2.2

The Transformation of Ordinary Velocity

We can now relate the ordinary velocity of the particle $\mathbf{v}$ as measured by observer $\mathcal{O}$ to the velocity $\mathbf{v}'$ as measured by $\mathcal{O}'$. To see this, we go back to Eq. (2.49) and expand it in the explicit coordinates of $\mathcal{O}$ and $\mathcal{O}'$. Let us set up the problem. A particle is moving around in spacetime, and observers $\mathcal{O}$ and $\mathcal{O}'$ are measuring its trajectory. Frame $\mathcal{O}'$ is moving with respect to $\mathcal{O}$ with speed V in the positive x direction and their spatial axes are otherwise aligned; we then have the Lorentz transformation matrix from (2.24):

$$\hat{\Lambda} = \begin{pmatrix} \gamma_V & \gamma_V \beta_V & 0 & 0 \\ \gamma_V \beta_V & \gamma_V & 0 & 0 \\ 0 & 0 & 1 & 0 \\ 0 & 0 & 0 & 1 \end{pmatrix}, \tag{2.54}$$

where $\beta_V = V/c$ and $\gamma_V = 1/\sqrt{1 - \beta_V^2}$. That is, in the context of this problem we have added the subscript V to the β and γ that describe the transformation between primed and unprimed frames with relative velocity $\mathbf{V}$, to distinguish it from the β and γ involving the velocity $\mathbf{v}$ of a particle in frame $\mathcal{O}$, and the β' and γ' involving the velocity $\mathbf{v}'$ of the particle in frame $\mathcal{O}'$. We can now write Eq. (2.49) in matrix notation:

$$\begin{pmatrix} \gamma c \\ \gamma v^x \\ \gamma v^y \\ \gamma v^z \end{pmatrix} = \begin{pmatrix} \gamma_V & \gamma_V \beta_V & 0 & 0 \\ \gamma_V \beta_V & \gamma_V & 0 & 0 \\ 0 & 0 & 1 & 0 \\ 0 & 0 & 0 & 1 \end{pmatrix} \begin{pmatrix} \gamma' c \\ \gamma' v^{x'} \\ \gamma' v^{y'} \\ \gamma' v^{z'} \end{pmatrix}. \tag{2.55}$$

All that is left is a simple matrix multiplication to obtain the relativistic velocity addition law. From the first row, we find

$$\gamma = \gamma_V \gamma' \left(1 + \beta_V \frac{v^{x'}}{c} \right), \tag{2.56}$$

which we can then use in the other three rows to get the velocity transformation

$$v^x = \frac{v^{x'} + V}{1 + V v^{x'}/c^2},$$

$$v^y = \frac{v^{y'}}{\gamma_V \left(1 + V v^{x'}/c^2 \right)},$$

$$v^z = \frac{v^{z'}}{\gamma_V \left(1 + V v^{x'}/c^2 \right)}. \tag{2.57}$$

Let us analyze the physical implications of these equations.

- As a sanity check, we should first take the small-speed limit, for which $Vv^{x'}/c^2 \ll 1$ and $\gamma_v \sim 1$; then

$$v^x = v^{x'} + V,$$
$$v^y = v^{y'},$$
$$v^z = v^{z'}. \tag{2.58}$$

These are the familiar Galilean addition of velocity rules we know very well. So far, so good.

- Now what happens when speeds become relativistic? First suppose that the particle is traveling at the speed of light in the x' direction in observer $\mathcal{O}$'s frame, $v^{x'} = c$, $v^{y'} = v^{z'} = 0$. We then have

$$v^x = \frac{c + V}{1 + V/c} = c,$$
$$v^y = \frac{0}{\gamma(1 + V/c)} = 0,$$
$$v^z = \frac{0}{\gamma(1 + V/c)} = 0. \tag{2.59}$$

At this point, we are justified in getting slightly emotional. This result is consistent with our original postulates: a particle that travels at speed c in one inertial frame travels at speed c in all other inertial frames as well.

- How about intermediate speeds, which bridge the gap between low speeds and the speed of light? The simplest way to see the implication is to plot v^x as a function of $v^{x'}$ for fixed V. Figure 2.3 shows such a plot. We now see explicitly how relativity caps speeds to be below that of light, just as we hoped.

- It is also interesting to note that the relativistic velocity addition law has nontrivial structure in the y and z directions, transverse to the relative motion of the two observers. This comes about from the ratio γ'/γ; i.e., it is due to the fact that the tick rates of the two observers' clocks are different. Even though transverse *distances* are not affected by a change of reference frame, velocity depends also on the duration of clock ticks of each observer.

Note that without the use of four-vector language the derivation of these rather involved expressions for velocities would have been more painful. The formalism helps us embed velocity into a structure that transforms in a simple way under Lorentz transformations – given by Eq. (2.49). Yet, in explicit form, this rather simple expression metamorphoses into the unfortunate result that is Eq. (2.57). The strategy now becomes obvious: try to embed any physical quantity of interest into the language of four-vectors so that we get its Lorentz transformation for free; then decompose the transformation law into its components to unravel the physical implications.

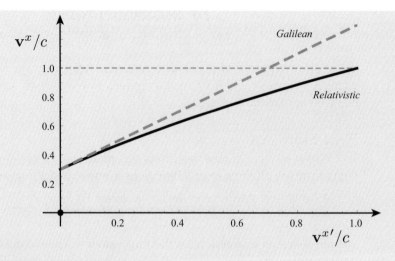

Fig. 2.3   The velocity v^x as a function of $v^{x'}$ for fixed relative frame velocity $V = 0.5c$.   ∎

Example 2.3   **Four-Velocity Invariant**

Before we proceed to a similar treatment of momentum, let us introduce a simple technical exercise. We want to compute the quantity $u^\mu u^\nu \hat{\eta}_{\mu\nu}$, in which the indices μ and ν are repeated and hence are to be summed over. It is an interesting quantity, since, writing in matrix notation:

$$\boldsymbol{u}^\mathsf{T} \cdot \hat{\eta} \cdot \boldsymbol{u} = \boldsymbol{v}'^\mathsf{T} \cdot \hat{\boldsymbol{\Lambda}}^\mathsf{T} \cdot \hat{\eta} \cdot \hat{\boldsymbol{\Lambda}} \cdot \boldsymbol{v}' = \boldsymbol{v}'^\mathsf{T} \cdot \hat{\eta} \cdot \boldsymbol{v}', \tag{2.60}$$

resulting in a Lorentz invariant, much like proper time. In index notation:

$$u^\mu u^\nu \hat{\eta}_{\mu\nu} = u^{\mu'} u^{\nu'} \hat{\eta}_{\mu'\nu'}. \tag{2.61}$$

To compute this quantity, we write it in explicit form in terms of the ordinary velocity:

$$u^\mu u^\nu \hat{\eta}_{\mu\nu} = -\gamma^2 c^2 + \gamma^2 \left((v^x)^2 + (v^y)^2 + (v^z)^2 \right) = -\gamma^2 c^2 \left(1 - \frac{v^2}{c^2} \right) = -c^2, \tag{2.62}$$

which is obviously invariant, since the speed of light is the same in all inertial frames.

Now let us compute this same expression using a different technique. Since $u^\mu u^\nu \hat{\eta}_{\mu\nu}$ is an invariant, we can evaluate it in *any* inertial frame. In particular, we can choose a frame $\mathcal{O}'$ that happens to be instantaneously at rest with respect to the particle whose velocity is represented in the expression. In that special frame we have $v^{x'} = v^{y'} = v^{z'} = 0$, and so we immediately get $u^{\mu'} u^{\nu'} \hat{\eta}_{\mu'\nu'} = -c^2$. Therefore, if we had been slightly more astute about things, we need not have done all our previous work in Eq. (2.62): by simply observing that we have an invariant, we would jump to a more convenient reference frame – the rest frame of the particle – and perform the computation there mentally. The moral: *it pays to know your invariants*.

∎

2.3 Relativistic Dynamics

We are now prepared to investigate relativistic *dynamics*, including the causes of changes in motion and how such quantities as momentum, energy, and force are changed in relativistic physics.

Four-Momentum

In classical mechanics the momentum of a particle is a three-vector, generally with components in all three spatial directions. Can we find a four-vector related to this classical momentum $\mathbf{p} = m\mathbf{v}$? Constructing it will help us learn how momentum changes when we shift perspective from one inertial observer to another in a fully relativistic context. This project turns out to be *easy*, because we have already found an expression for the four-velocity. The natural choice for the four-momentum is to multiply the four-velocity by m, the mass of the particle:

$$p^{\mu} = m\, u^{\mu}. \tag{2.63}$$

However, we need to be slightly careful. To get the required simple transformation rule

$$p^{\mu} = \hat{\mathbf{\Lambda}}^{\mu}_{\nu'} p^{\nu'} \tag{2.64}$$

from Eq. (2.49), we need the mass parameter m to be an invariant, $m = m'$. We do not want to bias ourselves toward a physical assumption that has yet to come out of the postulates of relativity. Hence, we need to justify this statement. Fortunately, we have already seen a similar situation when we encountered the transformation of time. There, we found that the notion of invariant time required a definition of proper time, the time in the rest frame of the observed particle. We can then safely adopt the same physical principle: we introduce the notion of **rest mass**, the mass of a particle measured in its rest frame. Let us denote it by m and refer to it as simply **mass** from here on; it is the only mass the particle has, the same in all reference frames. This has often been a source of confusion, since primarily for pedagogical reasons some physicists have said or written that mass increases with velocity, with a "relativistic mass" $m_{\mathrm{rel}} \equiv \gamma m$, where m is the rest mass. The advantage of using the concept of a relativistic mass that increases with velocity is that one can then still use the classical expression for the three-momentum, $\mathbf{p} = m_{\mathrm{rel}}\mathbf{v}$, and explain why a massive particle cannot be pushed to the speed of light, because its "mass" would then diverge in the limit $v \to c$, making it harder and harder to accelerate the particle. There are, however, very strong reasons not to use this idea. A danger of the simple substitution is that one might be tempted to conclude that substituting m_{rel} for m works in general for relativistic dynamics, which it most certainly does not. For example, it gives the wrong answer for kinetic energy, and also for Newton's second law in the form $\mathbf{F} = m\mathbf{a}$, except in the special case where the force exerted on the particle is perpendicular to its velocity.

The primary reason for taking mass to be a Lorentz invariant, *i.e.*, the same in all frames of reference, is that in four-dimensional spacetime, which is the natural arena for relativity, the idea of mass increasing with velocity is out of place and makes no sense. It is much better to write the three-momentum as $\mathbf{p} = \gamma m \mathbf{v}$, keeping γ in the equation *explicitly* rather than hiding it in a "relativistic mass." Then the reason it becomes harder and harder to accelerate a particle as it approaches light speeds is not that the mass is increasing, but because there is a γ factor in the true definition of momentum. The nonrelativistic definition $\mathbf{p} = m\mathbf{v}$ is only an approximation, valid for $v \ll c$. Newton's second law in the form $\mathbf{F} = d\mathbf{p}/dt$ is still valid in relativity, if we take $\mathbf{p} = \gamma m\mathbf{v}$ to be the correct expression for momentum.[5] The four-momentum is then $p^\mu = m u^\mu$, as given already in Eq. (2.63).

Let us look at the components of this new quantity p^μ and understand their physical significance. Recalling that the four-vector velocity has components $(\gamma c, \gamma \mathbf{v})$, it follows that for a particle of mass m moving with velocity $\mathbf{v}$ with $\beta = v/c$, the four-momentum is

$$p^\mu \to (\gamma mc, \gamma m\mathbf{v}), \qquad (2.65)$$

where we have collected the last three terms together into a traditional three-vector. At low speeds this has the familiar form $\boldsymbol{p} \sim (mc, m\mathbf{v})$ to linear order in v/c, with the addition of the zeroth component mc.

Note that even though the velocity $\mathbf{v}$ of a particle is restricted to be $v < c$, because of the γ factor there is no upper limit to the momentum of a particle. As the speed of the particle gets ever closer to the speed of light, the momentum grows without bound. So far, things look promising.

What is the meaning of the quantity γmc, the zeroth component of the momentum four-vector? The first clue to its meaning is the fact that in Newtonian mechanics, the momentum of a particle is conserved if there are no forces on it, and that is true in all inertial frames. By the principle of relativity we want to retain this property for relativistic particles as well, which means that the last three components (called the spatial components) of the momentum four-vector should be conserved in the absence of forces. However, when we Lorentz-transform the spatial components of a four-vector in one frame, they become a *mixture* of space and time components in another inertial frame. Therefore, to ensure conservation of the spatial components in *all* frames means that the zeroth component (also

---

[5] Einstein himself weighed in on this point. In a letter to L. Barnett (as quoted by L. B. Okun in his article "The concept of mass" in *Physics Today* 42, p. 31, June 1989), Einstein wrote: "It is not good to introduce the concept of the mass $M = m/\sqrt{1 - v^2/c^2}$ of a moving body for which no clear definition can be given. It is better to introduce no other mass concept than the 'rest mass' m. Instead of introducing M it is better to mention the expression for the momentum and energy of a body in motion."

called the time component) must be conserved as well! So the first component of the momentum four-vector must also be some conserved quantity. What quantity could it be?

A second clue to the meaning of γmc comes from evaluating it for nonrelativistic velocities. Using the binomial series (see Appendix F)):

$$(1+x)^n = 1 + nx + \frac{n(n-1)}{2!}x^2 + \frac{n(n-1)(n-2)}{3!}x^3 + \cdots, \qquad (2.66)$$

valid for $|x| < 1$, it follows that for nonrelativistic velocities $v/c \ll 1$:

$$\gamma mc = mc \left(1 - \frac{v^2}{c^2}\right)^{-1/2} \cong \frac{1}{c}\left(mc^2 + \frac{1}{2}mv^2\right), \qquad (2.67)$$

keeping the first two terms in the binomial series. This quantity is indeed conserved for a free nonrelativistic particle. We recognize the second term as the nonrelativistic **kinetic energy** of the particle, which of course is conserved in the absence of forces, while the first term is an invariant quantity proportional to the particle's mass.

Therefore we *identify* the zeroth component of the momentum four-vector as E/c, where E is the **energy** of the particle. In Newtonian mechanics we traditionally take the energy of a particle (subject to no forces or potential energies) to be *zero* if it is at rest, but we now see that a particle at rest has the **mass energy**

$$E_0 = mc^2, \qquad (2.68)$$

and if the particle is moving, it also has the kinetic energy

$$T = E - E_0 = (\gamma - 1)mc^2, \qquad (2.69)$$

which is approximately $(1/2)mv^2$ in the nonrelativistic limit $v/c \ll 1$.

In summary, the momentum four-vector is actually an "energy–momentum" four-vector, with components

$$p^\mu = \left(\frac{E}{c}, \mathbf{p}\right), \qquad (2.70)$$

where $E = \gamma mc^2 = mc^2 + (\gamma - 1)mc^2$, in which the term mc^2 is the particle's mass energy and the other term $(\gamma - 1)mc^2$ is its kinetic energy. The momentum of a particle becomes $\gamma m\mathbf{v}$ instead of the usual $m\mathbf{v}$, which for small speeds $\gamma \sim 1$ agrees with our nonrelativistic notion of momentum.

Example 2.4 | **Relativistic Dispersion Relation**

We start with a technical exercise with interesting physical implications. We want to compute the relativistic invariant

$$p^\mu p^\nu \hat\eta_{\mu\nu} = p^{\mu'} p^{\nu'} \hat\eta_{\mu'\nu'}, \tag{2.71}$$

which has a similar structure to $u^\mu u^\nu \hat\eta_{\mu\nu} = -c^2$. In fact, since the four-momentum $p^\mu = mu^\mu$, we can immediately write

$$p^\mu p^\nu \hat\eta_{\mu\nu} = -m^2 c^2. \tag{2.72}$$

Let us expand this expression in components as seen by an observer $\circleddash$. Using $p^\mu = (E/c, \mathbf{p})$, we easily get

$$-\frac{E^2}{c^2} + \mathbf{p}^2 = -m^2 c^2. \tag{2.73}$$

Alternatively, we write

$$E(p) = \sqrt{(m c^2)^2 + p^2 c^2}, \tag{2.74}$$

where we have taken $E > 0$. This is the relativistic **dispersion relation** for a particle with mass m. In general, a dispersion relation is a relation between energy and momentum, $E(p)$. The nonrelativistic limit at low speeds corresponds to $p \ll mc$, which gives, after expanding to leading order in p:

$$E(p) \simeq mc^2 \left(1 + \frac{1}{2}\frac{p^2 c^2}{(m c^2)^2} + \cdots \right) = mc^2 + \frac{p^2}{2m} + \cdots \tag{2.75}$$

Once again, we see the contribution of the mass energy $m c^2$ as well as the Newtonian kinetic energy term $T = p^2/(2m) = (1/2)mv^2$. The full relativistic form of the dispersion relation (2.74) also allows us to consider the opposite limit $m \to 0$ (that is, $p \gg mc$), the case of a light or "massless" particle

$$E(p) \simeq p c. \tag{2.76}$$

In the strict limit $m \to 0$, this expression becomes exact. *Hence, we have to entertain the possibility of a massless particle that carries energy by virtue of its momentum.* Substituting $E = \gamma mc^2$ and $p = \gamma mv$ in this expression, we also find that

$$\gamma mc = \gamma mv \to v = c. \tag{2.77}$$

Therefore we conclude that massless particles *must* travel at the speed of light. We can reverse the argument and state that a particle with $v = c \Rightarrow \gamma \to \infty$ must have zero mass if it is to have finite energy $E = \gamma mc^2$. Since light travels at speed c, there is perhaps a sense in which we can think of light as a bunch of massless particles. Historically, this simple observation helped seed the foundations of **quantum mechanics**.

We think of energy as a more fundamental physical quantity than mass. It exists irrespective of whether a particle has or does not have mass. Later on we will see, through a discussion of symmetries and conservation laws, that energy is indeed more fundamental. We will also see that even massless particles gravitate; they both cause and are affected by gravity.   ∎

Example 2.5

Decay into Two Particles

In particle and nuclear physics, many particles decay into two other particles. For example, the neutral pion π^0 decays into two photons γ: we write $\pi^0 \rightarrow 2\,\gamma$. Figure 2.4 depicts a typical scenario. The initial particle of mass m_0 is shown in its rest frame; it has energy $m_0 c^2$ and momentum zero. It subsequently decays into two particles, with masses m_1 and m_2. These two final particles must move off in opposite directions to conserve momentum. The process is then effectively one-dimensional, along the line of emission of the two particles. As we will show, given the initial and final masses, conservation of energy and momentum are sufficient to determine the energies, momenta, and speeds of each final particle. In relativity, just as in classical mechanics, we can assume that particles decay so quickly that any reasonable external forces have insufficient time to cause changes in momentum or energy during the very brief decay itself, so four-momentum is conserved. The initial four-momentum is entirely that of the particle of mass m_0:

$$p_i = \left(\frac{E_0}{c}, 0, 0, 0\right). \tag{2.78}$$

The final four-momentum is the sum of two four-momenta:

$$p_f = \left(\frac{E_1}{c}, \mathbf{p}_1\right) + \left(\frac{E_2}{c}, \mathbf{p}_2\right), \tag{2.79}$$

where we have aligned the x axis along the direction in which the two particles fly apart.

m_0

before

m_1

m_2

after

Fig. 2.4   A particle of mass m_0 decays into two particles with masses m_1 and m_2. Both energy and momentum are conserved in the decay, but mass is not conserved in relativistic physics. That is, $m_0 \neq m_1 + m_2$.

Since we need

$$p_i = p_f, \tag{2.80}$$

we immediately see that both

$$E_0 = E_1 + E_2 \tag{2.81}$$

and

$$0 = \mathbf{p}_1 + \mathbf{p}_2. \tag{2.82}$$

The mass m_0 is given and E_0 as well, since

$$E_0 = m_0 c^2 \tag{2.83}$$

for $p_0 = 0$. The conservation law implies that the two particles emerge in opposite directions. Looking back at Eqs. (2.81) and (2.82), we have four unknowns that describe the problem: E_1, E_2, and the magnitudes of the momenta p_1 and p_2. Can we then unravel the kinematics of this problem only with the given particle masses? Equation (2.81) gives us one relation. Equation (2.82) leads to $p_1 = p_2$ – the magnitudes of the momenta must be the same – which is a second relation. We then need two additional conditions. These are the relations $E^2 - p^2 c^2 = m^2 c^4$ for each of the two emerging particles from Eq. (2.74). Hence, we know the problem is solvable.[a]

It follows that

$$m_2^2 c^4 = E_2^2 - p_2^2 c^2 = (m_0 c^2 - E_1)^2 - p_1^2 c^2 \tag{2.84}$$

using energy conservation for the first term and momentum conservation for the second term. Multiplying out the right-hand side, we find that

$$m_2^2 c^4 = m_0^2 c^4 - 2 m_0 c^2 E_1 + E_1^2 - p_1^2 c^2 = m_0^2 c^4 - 2 m_0 c^2 E_1 + m_1^2 c^4. \tag{2.85}$$

We can then solve this last equation for E_1, giving

$$E_1 = \left(\frac{m_0^2 + m_1^2 - m_2^2}{2 m_0} \right) c^2. \tag{2.86}$$

Having found E_1 in terms of known quantities, we can also find E_2, both momenta, the particle velocities, and other quantities as well, using the conservation laws along with $E^2 - p^2 c^2 = m^2 c^4$.

In nuclear or particle physics, where two-particle decays are common, one usually uses *energy units* in calculations. In energy units the energy of a particle is given in MeV (10^6 electronvolts), momenta in MeV/c, and masses in MeV/c^2. As a simple example, the π^0 meson, with mass 135 MeV/c^2, decays into two photons, each massless. Therefore the energy of photon 1 as seen from the rest frame of the meson is

$$E_1 = \left(\frac{m_0^2 + 0 - 0}{2 m_0} \right) c^2 = \frac{m_0}{2} c^2 = 67.5 \text{ MeV} \tag{2.87}$$

and the magnitude of its momentum is $p_1 = E_1/c = 67.5$ MeV/c.   ∎

[a] The reader may worry about one more unknown in the full problem: the angle at which the two particles emerge back to back. But this angle is undetermined because of the spherically symmetric attributes of the system: any angle is consistent with energy and momentum conservation. Fixing the angle would require additional physical information about the natural laws underlying the decay process at hand.

Relativistic Collisions

Most of what we know about subatomic particles has come from high-energy collisions of particles in accelerators like cyclotrons, linear accelerators, and synchrotrons. In proton–proton collisions, for example, we can learn about the nature of the forces between protons and their internal structures. We can also create new particles by turning part of the kinetic energy of the initial protons into the mass and kinetic energies of newly created particles. For example, we can create single or multiple pions, proton–antiproton pairs, so-called "strange" particles like the Λ hyperon and k meson, or the famous Higgs particle that endows other particles with their masses.

Relativistic collisions are similar yet different from nonrelativistic collisions. In nonrelativistic collisions it is the total *mass* and the total *momentum* of the colliding objects that are each conserved. Mechanical energy (that is, kinetic energy plus potential energy) is not necessarily conserved; some of the kinetic energy might be converted into heat energy, for example, as in the totally inelastic collision of two objects that stick together when they collide. But in relativistic collisions it is the total *energy* and total *momentum* that are each conserved. Mass is no longer separately conserved, and is instead counted as a portion of the total energy of the system.

To be clear, let us contrast the totally inelastic collision of two particles as viewed classically and as viewed relativistically, as seen in the lab frame, the frame in which an incident particle collides with a target particle at rest. *Classically*: (i) the total mass of the particles is conserved in the collision and (ii) the sum of their momenta is also conserved. Their mechanical energy is *not* conserved: some of it is lost in the collision and converted into other forms like heat energy. *Relativistically*, each component of the energy–momentum four-vector is conserved in the collision. That is: (i) the three-vector momentum is conserved, although we must be careful to use the definition $\mathbf{p} = \gamma m \mathbf{v}$ for each particle (and not the classical approximation $\mathbf{p} = m\mathbf{v}$) and (ii) the total energy of the system is conserved as well, counting both kinetic energy and mass energy. The energy of each particle is

$$E = \gamma mc^2 \equiv (\gamma - 1)mc^2 + mc^2 \equiv \text{kinetic energy} + \text{mass energy}. \quad (2.88)$$

The total mass is *not* conserved in relativity: when the particles warm up or otherwise gain internal energy as a result of the collision their mass increases proportionally. So when two hypothetical lumps of clay collide and stick together, the net effect is that the sum of their initial momenta is equal to the final momentum of the single combined lump, and the sum of their initial energies is equal to the final energy of the single combined lump. The total kinetic energy of the individual lumps before the collision is greater than the kinetic energy of the single combined lump at the end, so kinetic energy has been lost, but the mass energy of the final lump is greater than the sum of the masses of the initial lumps, so mass energy has been gained, by just enough to compensate for the loss of kinetic energy.

Let us suppose particle 0, the "target," of mass m_0, is initially at rest, while relativistic particle 1, of mass m_1, is incident upon it. The result is a final particle of mass M. From the conservation laws we have

$$p_1 + p_0 = p_M \quad \text{so} \quad p_1 = p_M \qquad \text{three-momentum conservation} \qquad (2.89)$$

and

$$E_1 + E_0 = E_M \quad \text{so} \quad E_1 + m_0 c^2 = E_M \qquad \text{energy conservation.} \qquad (2.90)$$

In addition, each particle has three-momentum $\mathbf{p} = \gamma m \mathbf{v}$ and energy $E = \gamma m c^2$, where $\gamma = (1 - \beta^2)^{-1/2}$ for that particular particle, so the ratio of these quantities gives the velocity of the particle:

$$\beta = \frac{v}{c} = \frac{pc}{E}. \qquad (2.91)$$

Therefore, using the results of the conservation laws, the velocity of the final particle M is given by

$$\frac{v_M}{c} = \frac{p_M c}{E_M} = \frac{p_1 c}{E_1 + m_0 c^2} = \frac{V_{\mathrm{CM}}}{c}. \qquad (2.92)$$

That is, we have noted that v_M/c is also the velocity of the center of mass V_{CM}/c, since there is only one final mass, and that is its velocity. However, even though the rest frame of the final particle is the center of mass frame, this frame in relativistic physics is not literally the center of mass frame before the collision. The center of mass of the two initial particles is defined to be at position $X = (m_0 x_0 + m_1 x_1)/(m_0 + m_1)$, so the velocity of this point is $V = (m_0 v_0 + m_1 v_1)/(m_0 + m_1) = (m_1 v_1)/(m_0 + m_1)$. It is easy to show that $V \neq v_M$ except in the nonrelativistic limit. The rest frame of the final particle is, however, always the *zero momentum* frame. The momentum of the final particle is obviously zero, and so by momentum conservation the total momentum of the initial particles must also be zero. Therefore the frame of the final particle is sometimes called the "center of momentum frame" (a term invented so that one can still label it the "CM" frame), although calling it the "zero momentum frame" would be more enlightening. So we use the subscript "CM" from now on to mean the zero momentum frame, and the corresponding velocity is as given above, $V_{\mathrm{CM}}/c = p_1 c/(E_1 + m_0 c^2)$.

Example 2.6   **Threshold Energies**

An especially useful example of a totally inelastic collision is one in which two particles collide and are barely able to create one or more new particles. Given enough energy, for example, the collision of two protons can result in two protons plus a proton–antiproton pair, $p + p \rightarrow p + p + (p + \bar{p})$, where the antiproton $\bar{p}$ has the mass of a proton but the opposite electric charge. In the CM frame, the two initial protons come together with equal but opposite momenta (and therefore equal but opposite velocities, since their masses are equal). For small initial velocities they will not have enough kinetic energy to provide the mass energy of the $(p, \bar{p})$

pair, so the pair cannot be formed. But if the total energy of each initial proton is $2m_p c^2$, including both its mass energy and kinetic energy, the pair can just barely be formed, in which case the final four particles will all be at rest in the CM frame, with no energy left over to allow any of them to move. This is a totally inelastic collision, with every bit of the initial kinetic energy turned into mass energy. So in the CM frame the minimum initial kinetic energy required to create the pair is $2m_p c^2$, half of which must be provided by each initial proton. Any additional initial kinetic energy above this value will allow the final particles to have some kinetic energy in addition to their mass energy.

Now let us return to the general case in which we are barely able to create one or more new particles. We identify the final mass of the system as $\sum m_f$, the sum of all the final particle masses. We seek to find how much initial energy E_1 is required in the lab frame for an incident relativistic particle of mass m_1 to create all these final particles when it strikes the target particle of mass m_0 at rest in the lab. This is what is meant by the term **threshold energy**.

To find this energy we can use two important equalities. First, the total momentum four-vector P_T is the same before and after the collision, and that is true in both the lab and CM frames. Second, the scalar product $\eta_{\mu\nu} P_T^\mu P_T^\nu$ must be the same in all inertial frames, including the lab and CM frames. The scalar product of any two four-vectors is an invariant under Lorentz transformations.

The total momentum four-vector in the CM frame after the collision is

$$P_T^{(CM)} = \left(\frac{E_T^{(CM)}}{c}, 0, 0, 0 \right) = \left(\sum m_f c, 0, 0, 0 \right) \tag{2.93}$$

for the threshold energy, and this must be the same as it was before the collision by the conservation laws. The total four-momentum vector in the lab frame, with particle 0 at rest and particle 1 with energy E_1 before the collision, is

$$P_T^{(LAB)} = \left(\frac{E_T^{(LAB)}}{c}, p_1, 0, 0 \right) = \left(\frac{E_1 + m_0 c^2}{c}, p_1, 0, 0 \right) \tag{2.94}$$

the same after the collision according to conservation of the total four-momentum. The squares of these four-momenta are also equal to one another, because they are invariant under Lorentz transformations. So altogether we have

$$\eta_{\mu\nu} P_T^{(CM)\mu} P_T^{(CM)\nu} = - \left(\sum m_f \right)^2 c^2 \tag{2.95}$$

in the CM frame (*i.e.*, zero-momentum frame) and

$$\eta_{\mu\nu} P_T^{(LAB)\mu} P_T^{(LAB)\nu} = -(E_1 + m_0 c^2)^2/c^2 + p_1^2 = -2E_1 m_0 - m_0^2 c^2 + (E_1^2/c^2 - m_1^2 c^2) \tag{2.96}$$

in the lab frame. Here we have used the fact that for the incident particle in the lab, $p_1^2 = E_1^2/c^2 - m_1^2 c^2$. Now, setting these squared four-momenta equal to one another:

$$- \left(\sum m_f \right)^2 c^2 = -2E_1 m_1 - m_1^2 c^2 - m_0^2 c^2, \tag{2.97}$$

and then solving for E_1:

$$E_1 = E_{\text{thresh}} = \frac{[(\sum m_f)^2 - m_0^2 - m_1^2]c^2}{2m_1}.\tag{2.98}$$

This is the threshold energy in the lab frame, where the incident particle of mass m_1 strikes a target particle of mass m_0 at rest.

Now in particular, what about the creation of a $(p, \bar{p})$ pair in (p, p) collisions? In that case there are four final particles, each of mass m_p, so

$$E_1 = E_{\text{thresh}} = \frac{(4m_p)^2 - 2m_p^2)c^2}{2m_p} = 7m_p c^2,\tag{2.99}$$

which is the minimum total energy of the incident proton in the lab frame required to produce the proton–antiproton pair. The minimum *kinetic* energy of the incident proton in the lab is $6m_p c^2$, its total energy minus its mass energy. With this energy, the final four particles will all be at rest in the CM frame. Therefore, in the lab frame all four particles will be moving together at the CM velocity in the forward direction. If the incident proton has more energy than $7m_p c^2$, the final particles will have some kinetic energy in the CM frame rather than all just sitting there together at rest.

If all we are going to do is create two new particles, each of mass m_p, why isn't the kinetic energy required just $2m_p c^2$, and not $6m_p c^2$? The reason of course is that there must be enough energy to keep the total three-momentum the same after the collision as it was before. The final particles must be moving in the lab to conserve three-momentum, so kinetic energy for this motion of all four particles must be provided by the incoming proton.   ∎

Colliders

In the example just presented we found the threshold energy required in the lab to create new particles. Early accelerating machines, cyclotrons, synchrotrons, linear accelerators, and others all used beams of particles of mass m_1 and energy E_1 incident upon target particles of mass m_0 at rest. Antiprotons, antineutrons, and many "strange" particles were discovered in this way. In the case of antiprotons, for example, which had to be created with an accompanying proton,[6] a minimum energy of $E_1 = 7m_p c^2$ for the incoming proton was required in the lab frame; in practice, it required even higher energies to produce an appreciable number of $(p, \bar{p})$ pairs.

It did not escape the attention of particle physicists that if it were possible to actually perform the experiments in the CM frame instead, with two protons striking one another with equal but opposite momenta, then the minimum energy of either initial proton required would be only $E_1' = 2m_p c^2$ to produce a $(p, \bar{p})$ pair.

---

[6] The so-called "baryon number" has to be conserved. Many "heavy" particles are baryons, including the proton and neutron, each of which has baryon number +1. Their antiparticles have baryon number −1. So if an antibaryon is created, a baryon must also be created along with it so that the net baryon number before the collision is the same after the collision.

This would require a much less powerful accelerator. However, it would obviously be very challenging to accelerate two beams moving in opposite directions, and then aim the beams at one another, and finally have enough protons in the two beams so that there would be a fighting chance that some of the protons moving one way would strike an appreciable number of protons moving the other way and cause the reaction to occur.

Nonetheless, particle physicists became strongly motivated to create even heavier particles or particle pairs, so they knew they had to build "colliders," in which two beams of particles collided with one another head on. How much advantage would there be in such machines?

Suppose we want to create some new particle or particles that are much, much heavier than the colliding particles, so that the threshold energy using (p,p) collisions in the lab frame would be

$$E_1 = E_{\text{thresh}} = \frac{((\sum m_f)^2 - m_p^2 - m_p^2)c^2}{2m_p} \simeq \frac{(\sum m_f)^2 c^2}{2m_p}. \tag{2.100}$$

That is, the threshold energy would be proportional to the *square* of the total mass energy of the final particles. If the particle or particles we want to create are suspected of having a total of 50 proton masses, for example, our traditional machine would have to accelerate protons to energies of at least $1,250 m_p c^2$, a huge energy. But if we could do it in a collider instead, with two proton beams striking one another head-on, each initial proton would need a minimum energy of only about $25 m_p c^2$, 1/50th the energy required in the traditional machine. An enormous improvement in feasibility and saving in costs!

Again, for (p,p) collisions, to create highly massive particles with $\sum m_f c^2 \gg m_p c^2$ would require accelerating protons to an energy of at least $(\sum m_f)^2 c^2/(2m_p)$ in a traditional machine, but only $(\sum m_f c^2)/2$ in a collider. The ratio of these is

$$\frac{(E_1)_{\text{traditional machine}}}{(E_1)_{\text{collider}}} \simeq \frac{\sum m_f}{m_p}. \tag{2.101}$$

The larger the total mass of final particles, the more advantageous are colliders over traditional machines with stationary targets.

Photons and Compton Scattering

In his "miracle year" of 1905, Einstein not only published the theory of special relativity; among other highly original papers he also proposed that light consists of particles, later named "photons." Photons travel at the speed of light and have zero mass. In spite of being massless, they do have both energy and momentum, but we clearly cannot use the formulas $p = \gamma m v$ and $E = \gamma m c^2$, because each expression involves a product of γ, which is infinite for a particle traveling at the speed of light, and m, which is zero. This makes no sense. However, the relationship

$$\eta_{\mu\nu} p^\mu p^\nu = -E^2/c^2 + \mathbf{p} \cdot \mathbf{p} = -m^2 c^2 = 0 \tag{2.102}$$

does make sense for photons: it states in effect that the energy and momentum of a photon are related by $E = pc$, where p is the magnitude of the three-momentum. We know that a beam of light has energy; it also has momentum, as given by $p = E/c$.

Einstein proposed not only that light consists of particles we call photons, but also that photons have wave-like properties, described by quantities like wavelength and frequency as well as particle-like properties described by properties like energy and momentum. We can translate between these two worlds of particle properties and wave properties by using the relations $p = h/\lambda$, where λ is the wavelength and h is Planck's constant, and $E = h\nu$, where ν is the frequency. Then from the equation $-E^2/c^2 + p^2 = 0$ we find that $-\nu^2/c^2 + 1/\lambda^2 = 0$, or $\lambda\nu = c$, exactly what we would expect from wave theory for a wave traveling at the speed of light.[7]

An important confirmation that photons with the properties just mentioned exist came about in 1923, when A. H. Compton fired a beam of X-ray light at a graphite target, which of course contained electrons. The electrons could be treated as essentially free particles at the X-ray energies used. As X-rays scattered off these electrons, they lost some of their energy, giving it to the electrons in inelastic collisions. The energy loss meant that the X-rays would have a lower frequency and larger wavelength according to Einstein's equations $E = h\nu = hc/\lambda$, and the frequency was measured in the lab as a function of the scattering angle of the X-rays.

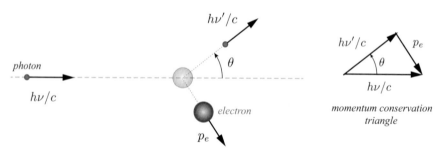

Fig. 2.5   The Compton scattering process: a photon is incident on an electron and scatters off at an angle.

We can see how λ should change as a function of the scattering angle θ, as illustrated in Figure 2.5. Let ν and ν' be the initial and final X-ray frequencies and E_e the final electron energy. Compared with the X-ray energy, the initial energies of the electrons are essentially zero. Then, in a photon–electron interaction, energy conservation gives

$$h\nu + m_e c^2 = h\nu' + E_e. \tag{2.103}$$

[7] Much later, in 1924, a French graduate student, Louis de Broglie, suggested in his doctoral dissertation that the relationships $E = h\nu$ and $p = h/\lambda$ should apply to *all* particles, not just photons. Therefore particles like electrons and protons would have wave-like properties as well as particle-like properties. Einstein was very supportive of de Broglie's idea. Much more on this in Chapter 5.

From the conservation of the total three-momentum, and using the law of cosines:

$$p_e^2 = \left(\frac{h\nu}{c}\right)^2 + \left(\frac{h\nu'}{c}\right)^2 - 2\left(\frac{h\nu}{c}\right)\left(\frac{h\nu'}{c}\right)\cos\theta. \tag{2.104}$$

The electron's momentum and energy before the interaction are related by

$$p_e^2 c^2 = E_e^2 - m_e^2 c^4 = (h\nu - h\nu' + m_e c^2)^2 - m_e^2 c^4 = h^2(\nu - \nu')^2 + 2hm_e c^2(\nu - \nu') \tag{2.105}$$

using the energy conservation equation above. Now the electron momentum p_e can be eliminated between the last two equations, resulting in

$$m_e c^2(\nu - \nu') = h\nu\nu'(1 - \cos\theta). \tag{2.106}$$

Finally, by substituting $\nu = c/\lambda$ and $\nu' = c/\lambda'$, we find the Compton scattering result

$$\lambda' - \lambda = \frac{h}{m_e c}(1 - \cos\theta) \tag{2.107}$$

for the wavelength shift of photons scattered off electrons at angle θ. The photons become redder, having transferred some of their energy to the electrons. The larger the scattering angle θ, the more energy has been transferred, and the redder the photons according to the formula. Compton's measurements agreed in detail, adding important support to the idea that light consists of photons with the properties proposed by Einstein many years before.

Four-Force

We now seek a four-vector *force* that is responsible for changes in the four-momentum of a particle. A "four-force" would allow us to reformulate Newton's second law for relativistic mechanics, since nonconservation of momentum in Newtonian physics implies the presence of a force $\mathbf{F} = d\mathbf{p}/dt$. We would like to transform this equation into four-vector language.

Let us label the four-vector force as f^μ and propose that

$$f^\mu = \frac{dp^\mu}{d\tau}, \tag{2.108}$$

where each side of the equation is a four-vector. Once again we have differentiated with respect to a four-scalar, the proper time, which is the same in all inertial frames. Therefore observer $\mathcal{O}'$ can write

$$\hat{\mathbf{\Lambda}}^{\mu'}_{\ \mu} f^\mu = \hat{\mathbf{\Lambda}}^{\mu'}_{\ \mu} \frac{dp^\mu}{d\tau} \Rightarrow f^{\mu'} = \frac{dp^{\mu'}}{d\tau} \tag{2.109}$$

to describe the same physics – with the implicit Lorentz transformation of our new four-force $f^{\mu'} = \hat{\mathbf{\Lambda}}^{\mu'}_{\ \mu} f^\mu$.

A force law describing the nature of f^μ is an independent statement of physics, so one needs to check each component of f^μ – its detailed form in terms of the parameters of the particle and its environment – to see whether the Lorentz transformation changes it beyond the expected $f^\mu = \hat{\Lambda}^\mu_{\mu'} f^{\mu'}$. All inertial observers should see the same physics, so this should not happen.

For now, we assume that whatever forces appear on the left-hand side of Eq. (2.108) are indeed consistent with the postulates of relativity. We want instead to focus on a much more urgent issue: what new physics does our reformulation of Newton's second law given by Eq. (2.108) add to the dynamics, on top of what we already know from the Newtonian realm?

To see the implications of Eq. (2.108), we can write it explicitly in component form:

$$(f^t, \mathbf{f}) = \gamma \frac{d}{dt} (\gamma m c, \gamma m \mathbf{v}), \qquad (2.110)$$

where we use the time dilation relation (2.52) to write $d\tau$ in terms of observer $\mathbb{O}$'s time differential dt, and collect the three spatial components of our four-vectors into the usual three-vector notation. We look at the easy part first: the spatial components are

$$\mathbf{f} = \gamma \frac{d}{dt} (\gamma m \mathbf{v}). \qquad (2.111)$$

Imagine that the particle is subject to no external forces, $\mathbf{f} = 0$. We then have momentum conservation

$$\frac{d}{dt} (\gamma m \mathbf{v}) = 0, \qquad (2.112)$$

where again the momentum is $\mathbf{p} = \gamma m \mathbf{v}$ (and not the nonrelativistic approximation $\mathbf{p} = m\mathbf{v}$).

Defining force $\mathbf{F}$ as the rate of change of $\mathbf{p}$, we need to write

$$\mathbf{F} \equiv \frac{d}{dt} (\gamma m \mathbf{v}). \qquad (2.113)$$

The quantity $\mathbf{F}$ corresponds to the force in Newtonian mechanics: it is the rate of change of momentum as seen by a given observer. Looking back at Eq. (2.111), we interpret the lower-case quantity $\mathbf{f}$ as

$$\mathbf{f} = \gamma \mathbf{F}. \qquad (2.114)$$

That is, the spatial components of the four-force reduce to the usual three-force $\mathbf{F}$ in the limit of small velocities.

Now, what is the meaning of the zeroth component of Eq. (2.110):

$$f^t = \gamma \frac{d}{dt} (\gamma m c)? \qquad (2.115)$$

Recall that the energy of the particle is $E = \gamma mc^2$, so

$$f^t = \frac{d}{d\tau}(\gamma mc) = \frac{1}{c}\frac{dE}{d\tau}. \tag{2.116}$$

Earlier we showed that E and p obey $E^2 = m^2c^4 + p^2c^2$, so differentiating this equation with respect to τ gives

$$2E\frac{dE}{d\tau} = 2\left(\mathrm{p}^x\frac{d\mathrm{p}^x}{d\tau} + \mathrm{p}^y\frac{d\mathrm{p}^y}{d\tau} + \mathrm{p}^z\frac{d\mathrm{p}^z}{d\tau}\right)c^2 \equiv 2\left(\mathbf{p}\cdot\frac{d\mathbf{p}}{d\tau}\right)c^2, \tag{2.117}$$

and so, using $E = \gamma mc^2$ and $\mathbf{p} = \gamma m\mathbf{v}$:

$$\frac{dE}{d\tau} = \frac{1}{\gamma mc^2}\left(\gamma m\mathbf{v}\cdot\mathbf{f}\right)c^2 = \mathbf{v}\cdot\mathbf{f} \Rightarrow \frac{dE}{dt} = \mathbf{F}\cdot\mathbf{v}, \tag{2.118}$$

which we recognize as the rate at which the force does work on the particle, *i.e.*, the *power input* to the particle. This agrees with Newtonian mechanics.

In summary, the four components of the force four-vector are

$$f^\mu \rightarrow \left(\gamma\frac{1}{c}\mathbf{v}\cdot\mathbf{F}, \gamma\mathbf{F}\right), \tag{2.119}$$

where the zeroth component of the four-force is related to the rate at which the energy flows in/out of the system. We have also learned how force must transform under Lorentz transformations, since f is a four-vector and $f^{\mu'} = \hat{\Lambda}^{\mu'}_{\mu}f^\mu$.

So far, we have been led by the postulates of relativity to a modification of the transformation rules that relate to the perspectives of inertial observers. We then developed a mathematical language that naturally lends itself to Lorentz transformations, and we discussed four-vectors and Lorentz invariants. Next, we attempted to embed various physical quantities, such as velocity, momentum, and force, into the language of four-vectors. In doing so, we wrote quantities that match the corresponding Newtonian ones at low speeds, but are packaged in a manner that easily determines how they change under Lorentz transformations. This led us to a revised velocity addition law, to a new understanding of momentum and energy, including a realization that mass is a form of energy, and finally to a revised concept of force and of Newton's second law of motion.

Dynamics in Practice

At this point it is useful to step back and think about mechanics in light of all the new revisions we have discussed. We begin by revisiting the three laws of Newton and fitting them into the postulates of relativity.

- *Unchanged first law.* There exists a class of observers – henceforth labeled **inertial observers** – for whom the laws of physics are the same. Given one inertial observer, another observer is inertial if their two frames have a constant relative velocity. In an inertial frame, in the absence of forces, a particle will move in a straight line at constant speed.

- *Revised second law.* The rate of change of four-momentum is the four-force

$$f^\mu = \frac{d}{d\tau}(mu^\mu) = \frac{dp^\mu}{d\tau}, \tag{2.120}$$

where $u^\mu = dr^\mu/d\tau$ and τ is proper time. In the absence of a four-force, energy and momentum are conserved.

- *Extended third law.* For every four-force there is an equal but opposite four-force. The spatial part of this statement ensures total momentum conservation for an isolated system: for an isolated system of particles, action–reaction pairs cancel so that the total force is zero and total momentum is conserved. We will see this in more detail in a later chapter on systems of particles. The temporal part of our new statement is about energy conservation for an isolated system: you can see this from the fact that the first component of the four-force measures rate of change of energy.

- *New fourth law.* The universal **speed of light** is a law of physics: light moves at the same speed with respect to all inertial observers. This implies that the inertial reference frames defined in the first law are connected to each other by **Lorentz transformations**.

One can use these statements to study mechanics with speeds all the way up to that of light. At low speeds we drop the new fourth law, Newton's first and third laws are unchanged, and we recover Newton's second law as an approximation. Also, Galilean transformations connect inertial reference frames. What then remains is to complete the dynamical picture by incorporating specific forces consistent with the postulates of relativity.

We can develop our physical intuitions by explicit examples, so we now proceed with a few case studies.

Example 2.7 **Uniformly Accelerated Motion**

Consider a particle of mass m moving in one spatial direction, say along the x axis of an observer $\mathbb{O}$, and suppose that this particle is subjected to an external four-force

$$(f^t, \mathbf{f}) = \left(\gamma \frac{1}{c} \mathbf{v} \cdot \mathbf{F}, \gamma \mathbf{F}\right), \tag{2.121}$$

with $\mathbf{F}$ a constant three-vector pointing in the positive x direction. Can such a constant force accelerate the particle past the speed of light?

Writing the component of $f^\mu = dp^\mu/d\tau$ in the x direction, we get

$$\gamma F^x = \gamma \frac{d}{dt}(\gamma m v^x). \tag{2.122}$$

Simplifying, we have

$$F^x = m \frac{d}{dt}(\gamma v^x), \tag{2.123}$$

which is a differential equation we can solve for v^x. The left-hand side is a constant, and the velocity $v^x(t)$ appears both explicitly and implicitly in the gamma factor $\gamma = (1 - (v^x)^2/c^2)^{-1/2}$. Integrating Eq. (2.123) with $v^x(0) = 0$, we find

$$\frac{F^x}{m}t = \frac{v^x}{\sqrt{1 - v^{x2}/c^2}},$$

(2.124)

which we can solve for $v^x(t)$:

$$\frac{v^x(t)}{c} = \frac{F^x t/mc}{\sqrt{1 + (F^x t/mc)^2}}.$$

(2.125)

We recognize $a = F^x/m$ as the Newtonian acceleration, which is a constant in this case. Therefore, in terms of a:

$$\frac{v^x(t)}{c} = \frac{at/c}{\sqrt{1 + (at/c)^2}}.$$

(2.126)

The factor at looks very familiar, but the square root in the denominator changes everything as time goes on. At early times, when the particle has not yet acquired much speed, we have $at/c \ll 1$ and we recover the Newtonian expression $v^x(t) = at$. At large times, however, the denominator ensures that we do not violate the upper speed limit

$$\frac{v^x(t)}{c} \to 1$$

(2.127)

as $t \to \infty$. Figure 2.6(a) shows a plot of $v^x(t)$ with the corresponding tapering-off feature at large speeds. We can also look at the particle's trajectory, shown in Figure 2.6(b), by integrating

$$\frac{dx}{dt} = \frac{at}{\sqrt{1 + (at/c)^2}}$$

(2.128)

with $x(0) = 0$ for a particle that starts at the origin. One finds that

$$x^2 - c^2 t^2 = \frac{c^4}{a^2},$$

(2.129)

so that in the $ct-x$ plane, the shape is that of a hyperbola. We will revisit this in the following section when we discuss Minkowski diagrams.

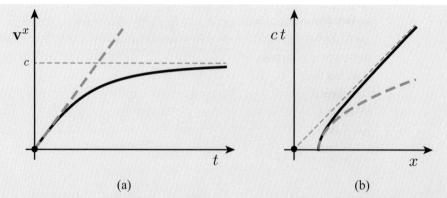

Fig. 2.6 Plots of relativistic constant-acceleration motion. (a) Shows $v^x(t)$, demonstrating that $v^x(t) \to c$ as $t \to \infty$, *i.e.*, the speed of light is a speed limit in Nature. The dashed line shows the incorrect Newtonian prediction. (b) Shows the hyperbolic trajectory of the particle on a ct–x graph. Once again the dashed trajectory is the Newtonian prediction. ∎

Example 2.8 The Doppler effect is the shifting of frequencies of sound or light waves from the perspectives of observers who are moving with respect to one another. We are most familiar with it in the context of sound (because the speed of sound is much less than the speed of light), when for example we notice a change in the pitch of the siren of an ambulance as it passes by. Sound propagates in some *medium*, whether air, liquid, or solid, so that it has a particular fixed speed given by the properties of the particular medium *in the medium's rest frame*. Its speed is therefore *not* an invariant and will be subject to the velocity addition rule. Hence, the more interesting scenario for relativity involves the Doppler effect for light, because in that case there is no medium to provide a preferred frame of reference. We want to find how the frequency of light shifts as seen by different moving observers.

Fig. 2.7 Observer $\mathcal{O}'$ shooting a laser toward observer $\mathcal{O}$ while moving toward $\mathcal{O}$.

Consider our usual setup of observer $\mathbb{O}'$ moving with a constant speed V along the positive x direction *toward* another observer $\mathbb{O}$, as shown in Figure 2.7. Observer $\mathbb{O}'$ aims a laser beam of frequency ν' (as seen from $\mathbb{O}$'s perspective) toward $\mathbb{O}$ along the x' axis, and we want to find the frequency ν perceived by $\mathbb{O}$; that is, we seek the Lorentz transformation of **frequency**.

As mentioned earlier, in 1905 Einstein showed that light consists of particles now called **photons**, and that the energy E and momentum p of a photon are each proportional to frequency, $E = h\nu$ and $p = E/c = h\nu/c$, where h is Planck's constant. This means that the four-momentum of the laser beam is

$$p^\mu = \left(\frac{E}{c}, p, 0, 0 \right) = \left(\frac{h\nu}{c}, \frac{h\nu}{c}, 0, 0 \right). \tag{2.130}$$

All that is left to do is to write the Lorentz transformation $p^\mu = \hat{\Lambda}^\mu_{\ \nu'} p^{\nu'}$ in explicit component form. That is:

$$\begin{pmatrix} h\nu/c \\ h\nu/c \\ 0 \\ 0 \end{pmatrix} = \begin{pmatrix} \gamma_V & \gamma_V \beta_V & 0 & 0 \\ \gamma_V \beta_V & \gamma_V & 0 & 0 \\ 0 & 0 & 1 & 0 \\ 0 & 0 & 0 & 1 \end{pmatrix} \begin{pmatrix} h\nu'/c \\ h\nu'/c \\ 0 \\ 0 \end{pmatrix}, \tag{2.131}$$

where $\beta_V = V/c$. This leads to

$$\nu = \gamma_V \nu' + \gamma_V \beta_V \nu', \tag{2.132}$$

where Planck's constant has dropped out of the equation. A little algebra then shows that

$$\frac{\nu}{\nu'} = \sqrt{\frac{1+\beta}{1-\beta}} > 1 \qquad \textit{approaching observers}. \tag{2.133}$$

This applies to the scenario where the laser beam is aimed from $\mathbb{O}'$ *toward* $\mathbb{O}$ as $\mathbb{O}'$ moves in the positive x direction – implicit in the fact that the x component of p^μ in Eq. (2.130) is taken to be positive and it is assumed that the beam does arrive at $\mathbb{O}$. In short, this applies when the distance between the two observers is *shrinking*. The frequency received is greater than the frequency emitted, $\nu > \nu'$, known as a **blueshift**, for obvious reasons. To see the other possibility – *i.e.*, $\mathbb{O}$ and $\mathbb{O}'$ moving *away* from each other – we can simply flip the sign of β in this expression:

$$\frac{\nu}{\nu'} = \sqrt{\frac{1-\beta}{1+\beta}} < 1 \qquad \textit{receding observers}. \tag{2.134}$$

Now the distance between the two observers is increasing, and we find that the received frequency is *less* than the emitted one: we say there has been a *redshift*.

Doppler shifts for light are an extremely useful tool in physics and technology, from determining the speeds of stars in distant galaxies leading to Hubble's discovery of the expanding universe, to the use of frequency shifts in the GPS for location and navigation. If such special-relativistic effects were not included in GPS navigation, there would be large errors in position measurements. Interestingly, it even turns out to be equally essential for GPS to include additional effects due to earth's gravity, as contained in Einstein's **general theory of relativity**.   ∎

Minkowski Diagrams

A particularly useful way to depict relativistic dynamics involves a visual tool called a **Minkowski diagram**. Simply put, it is a plot of the trajectory of a particle on a graph where the horizontal axis is one of the spatial directions and the vertical axis represents time, or actually the product ct. Figure 2.8 shows an example.

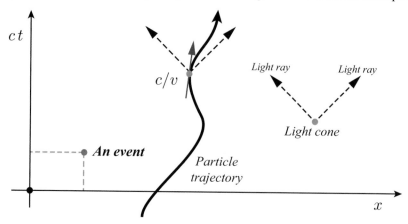

Fig. 2.8   A point on a Minkowski diagram represents an event. A particle's trajectory appears as a curve with a slope that exceeds unity everywhere.

The trajectory of the particle appears as a line moving upward, forward in time. It is sometimes referred to as the **worldline** of the particle. Light rays appear on such a diagram as straight lines at 45°, as shown in the figure. A tangent to a trajectory corresponds to c/v, the inverse relative speed of the corresponding particle. Notice that the worldline of the particle in the figure has a slope greater than unity everywhere, since $v/c < 1$.

An isolated point on a Minkowski diagram corresponds to an event, since it has a definite time and position. If two events can be connected by the physical trajectory of a particle (whose slope is everywhere greater than unity), the two events are said to be **time-like separated**. The physical implication is that the earlier event can *talk* to the later event with physical signals involving matter or light. A quick way to determine whether two events are time-like separated is to draw a forward-pointing **light cone** wedge with its apex at the earlier event, as shown in Figure 2.9. The other event should then lie within the light cone. We say that the two events are causally connected; *i.e.*, event A in the figure can cause event B. Event C lies outside the light cone for B: reaching it requires signal propagation faster than light, *i.e.*, a curve that has at least some interval where its slope is less than unity. Such events are said to be causally disconnected; we also say that events B and C are **space-like** separated. Event C in the figure lies on the light cone of A. This means that it can be reached from A with a light signal. A and C are then said to be **light-like** separated.

There is an algebraic way to determine whether two events are light-like, space-like, or time-like separated. If we look at the position four-vector $\Delta r = (c\,\Delta t, \Delta \mathbf{r})$

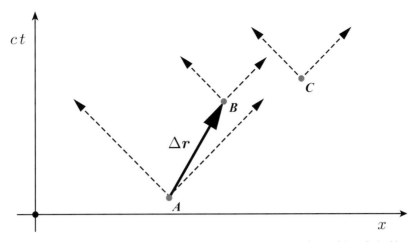

Fig. 2.9 Three events on a Minkowski diagram. Events A and B are time-like separated; A and C are light-like separated; and B and C are space-like separated.

pointing from one event to the other (see Figure 2.9), if the slope of this four-vector on the corresponding Minkowski diagram is greater than unity, then the events are time-like separated and we have

$$c\Delta t > |\Delta \mathbf{r}| \Rightarrow -c^2 \Delta t^2 + |\Delta \mathbf{r}|^2 < 0 \Rightarrow \Delta r^\mu \Delta r^\nu \hat{\eta}_{\mu\nu} < 0. \qquad (2.135)$$

Similarly, we find

$$\Delta r^\mu \Delta r^\nu \hat{\eta}_{\mu\nu} > 0 \qquad (2.136)$$

for space-like separated events, and

$$\Delta r^\mu \Delta r^\nu \hat{\eta}_{\mu\nu} = 0 \qquad (2.137)$$

for light-like separated ones. It is also useful to extend this concept to any four-vector, such as the velocity, momentum, and force four-vectors. For any such four-vector, denoted by A^μ in general, we can write

$$\begin{aligned} A^\mu A^\nu \hat{\eta}_{\mu\nu} &> 0 \quad \text{space-like} \\ A^\mu A^\nu \hat{\eta}_{\mu\nu} &< 0 \quad \text{time-like} \\ A^\mu A^\nu \hat{\eta}_{\mu\nu} &= 0 \quad \text{light-like} \end{aligned} \qquad (2.138)$$

Note in particular from Eqs. (2.62) and (2.72) that the momentum and velocity four-vectors are time-like, while the force four-vector is space-like (see the Problems section at the end of this chapter).

As an exercise in Minkowski diagram analysis, consider the trajectory of a particle under the influence of a constant four-force, as encountered in Example 2.7. From Eq. (2.129), we can now plot the profile of the worldline in Figure 2.10.

We see that the particle starts at rest with infinite slope (*i.e.*, zero speed), then speeds up and asymptotically reaches the speed of light at $45°$ slope in the figure. We note that the slope is everywhere greater than unity, as expected.

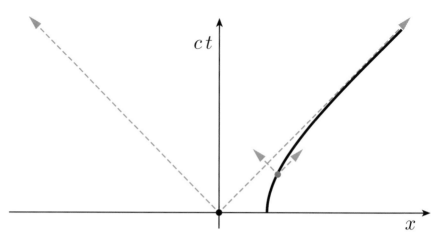

Fig. 2.10   The hyperbolic trajectory of a particle undergoing constant acceleration motion on a Minkowski diagram.

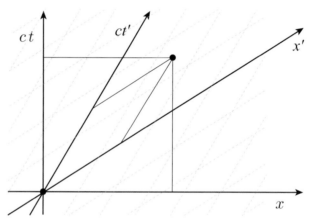

Fig. 2.11   The grid lines of two observers labeling the same event on a spacetime Minkowski diagram.

Another use of Minkowski diagrams is to picture the relation between the coordinate systems of two observers. The same set of events on a Minkowski diagram can get labeled via different coordinates by different inertial observers. Figure 2.11 shows the grid lines of observer $\mathbb{O}'$, who happens to be moving with speed V along the x axis of $\mathbb{O}$. The ct' axis is the worldline of observer $\mathbb{O}'$ as seen by $\mathbb{O}$, since it is obtained by setting $x' = 0$: after all, the ct axis of $\mathbb{O}$ is nothing but the trajectory of its origin on the Minkowski diagram at $x = 0$. In the same spirit, we see from Eq. (2.15) that the ct' axis is a straight line with slope c/V. The x' axis is given by $ct' = 0$ (as is the x axis of $\mathbb{O}$, given by the $ct = 0$ condition); from Eq. (2.15) we can see that it is a straight line with slope V/c. The ct' and x' axes are reflected images of each other across the light cone at the origin. The figure shows a geometric realization of how an event gets labeled by the two observers: each projects the event along her time and space axes, along ct and x for $\mathbb{O}$, and ct' and x' for $\mathbb{O}'$. The reader is, however, cautioned in using concepts from Euclidean geometry on such a diagram for measuring distances. The vertical axes here represent time! To measure the spacetime "distance" between two events separated by say Δt and Δx, you want to use $-c^2\Delta t^2 + \Delta x^2$, *not* $c^2\Delta t^2 + \Delta x^2$. That is, you want to use the Minkowski metric (2.31). Let us look at some examples using Minkowski diagrams to develop our visual intuition of relativity.

Example 2.9

Time Dilation

Consider once again our two observers $\odot$ and $\odot'$. The Minkowski diagrams are shown in Figure 2.12 corresponding to a relative velocity $V = (3/5)c$, i.e., observer $\odot'$ is moving in the positive x direction at $(3/5)c$ relative to $\odot$. In Figure 2.12(a), we show two events corresponding to two ticks of a clock carried by $\odot$. In Figure 2.12(b), we show two events corresponding to two ticks of a clock carried by $\odot'$ instead. Let us focus on Figure 2.12(a). $\odot$'s clock ticks are separated by Δt. Using Eq. (2.15) with $\Delta x = 0$, we have

$$c\,\Delta t' = \gamma(c\,\Delta t - \beta\Delta x) \Rightarrow \Delta t' = \gamma\Delta t. \tag{2.139}$$

The corresponding time interval $\Delta t'$ observed in the primed frame is then *greater* than Δt. To $\odot'$, this clock is moving in the negative x' direction and runs slow: this is the phenomenon of *time dilation*. Putting numbers in with $V = (3/5)c$, we have $\Delta t' = \gamma\Delta t = \Delta t/\sqrt{1 - V^2/c^2} = (5/4)\Delta t$. Figure 2.12(a) shows the same conclusion graphically.

What if we reverse our perspective? Consider a clock carried by $\odot'$ instead, which ticks with interval $\Delta t'$? Figure 2.12(b) depicts the corresponding scenario. Algebraically, the tick events of $\odot'$'s clock have $\Delta x' = 0$. Using Eq. (2.15) once again, we now get

$$c\,\Delta t = \gamma(c\,\Delta t' + \beta\Delta x') \Rightarrow \Delta t = \gamma\Delta t'. \tag{2.140}$$

Observer $\odot$ will then perceive this clock-tick separation as $\Delta t = (5/4)\Delta t' > \Delta t'$. To $\odot$, this clock is moving in the positive x direction, and once again runs slow. In summary, from the standpoint of any inertial observer, a moving clock runs slow by a factor of γ.

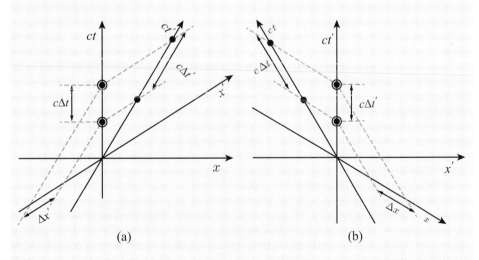

(a)   (b)

Fig. 2.12 The time dilation phenomenon. (a) Shows the scenario of a clock carried by observer $\odot$. (b) Shows the case of a clock carried by $\odot'$. ∎

Example 2.10   **Length Contraction**

Minkowski diagrams are shown in Figure 2.13 for a primed frame Ⓞ and unprimed frame Ⓞ corresponding to a relative velocity $V = (3/5)c$. Figure 2.13(a) depicts a scenario where observer $Ⓞ'$ carries a meter stick along with her. The dashed lines are the trajectories of the endpoints of the meter stick. If $Ⓞ'$ wants to measure the length of the stick, she must measure the locations of both ends *at the same time* t'. The corresponding measurement is shown in Figure 2.13(a) through two events occurring at $t' = 0$ at the endpoints. We then have $\Delta t' = 0$ and $\Delta x' = L_0$, where L_0 is the **rest length** of the stick. If observer Ⓞ is to measure the length of the same stick, he must use two events at the endpoints of the stick simultaneously in *his* reference frame, *i.e.*, two events with $\Delta t = 0$ and some value of Δx. Using Eq. (2.15) with $\Delta t' = 0$ and $\Delta x' = L_0$, one gets, after some straightforward algebra, $\Delta x = L_0/\gamma = (4/5)L_0 < L_0$! The moving stick is therefore shorter to Ⓞ. This is the phenomenon of **Lorentz contraction** or **length contraction**. If we consider a stick carried by Ⓞ instead, the scenario is shown in Figure 2.13(b). This time the rest length of the stick is given by $\Delta x = L_0$; and it is $Ⓞ'$ who perceives the stick moving, now in the negative x' direction. Once again, we can check that $Ⓞ'$ measures a length $\Delta x' = (4/5)L_0 < L_0$. Moving objects are contracted by a factor of $1/\gamma$ along the direction of motion. In fact, relativity introduces more elaborate geometric aberrations of moving objects, including a pseudorotation effect and preservation of circular shapes. We leave some of the discussion to the Problem section at the end of the chapter.

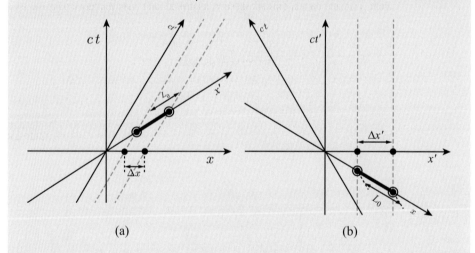

(a)   (b)

Fig. 2.13   The phenomenon of length contraction. (a) Shows the scenario of a meter stick carried by observer $Ⓞ'$. The dashed lines depict the trajectories of the two endpoints of the stick. (b) Shows the case of a stick carried by Ⓞ.   ∎

A crucial ingredient in the previous example is the realization that two events which are *simultaneous* in one reference frame are not necessarily so in another: this is known as the **relativity of simultaneity**. In the case at hand, the measurements of the locations of the two endpoints of the stick are simultaneous to one observer, and so can be used to read out the length of the stick. However, these same two measurements, as shown in the figure, are *not* simultaneous to the other observer, and so cannot be used to read out the length of the stick as measured by this other observer. *We would never judge the length of a moving stick by measuring its two endpoints at different times.*

Example 2.11 **The Twin Paradox**

Relativity abounds with so-called "paradoxes" – thought experiments that appear at first to lead to conceptual contradictions. However, they all invariably arise from one of several Newtonian traps. For example, one common pitfall is that of simultaneity: in relativity, two events that are simultaneous in one reference frame are not necessarily so in another. We saw this phenomenon at work in the previous example, leading to geometric distortions. Yet – based on our Newtonian daily experiences – we have no intuition for this, because we never encounter it in our normal experience. Often, once relativistic tinkering with the notion of time is taken into account, paradoxes are quickly resolved. And in resolving each paradox, one's intuition for relativity develops a bit more.

In this example we focus on the classic twin paradox. The scenario goes as follows. John lives on earth and tracks time with his wristwatch. His twin, Jane, is on a trek to a nearby star a distance D away. Jane will travel along a straight path at constant speed V_0 relative to John, then will turn around and come back with the same constant speed. The question is: Who has aged more when John and Jane meet? Figure 2.14(a) shows a Minkowski diagram of the setup with simultaneity lines according to John. If V_0 is large enough, time dilation effects will be important. There are three segments of the trip, two of which last for a period T_1 to John, as shown in the figure, and the middle segment lasting a period T_0. The total time of the trip will be $T_0 + 2\,T_1$ on John's wristwatch. We want to compare this to the time elapsed on Jane's wristwatch during the same period. We can immediately tell that $T_1 = D/V_0$. However, Jane's clock rate will necessarily be slow in John's rest frame, because of time dilation. For the first and third segments of the trip, Jane's speed is constant and we simply have

$$T_1 = \gamma_0 \tau_1, \tag{2.141}$$

where τ_1 is the time elapsed on Jane's wristwatch while T_1 has elapsed on John's; and $\gamma_0 = (1 - V_0^2/c^2)^{-1/2}$. Hence, $T_1 > \tau_1$ and John ages more during these segments.

The second segment is trickier, since Jane is accelerating as she turns around to return to earth. Let us assume, for simplicity, that Jane decelerates at a constant rate during the turnaround. From John's perspective, that is, from the perspective of observers at rest in his inertial frame, he can track what's happening to Jane using the relativistic form of Newton's second law. For constant acceleration, this is a problem we have already studied. We know Jane's trajectory would be hyperbolic on a Minkowski diagram, as shown in Figure 2.14. We also know that her velocity will be evolving as

$$v(t) = \frac{a\,t}{\sqrt{1 + a^2 t^2/c^2}}, \tag{2.142}$$

where a is some negative constant acceleration and $t = 0$ is taken as the moment when she has zero speed at the midpoint of the trip. Setting $v(T_0/2) = -V_0$, we can immediately deduce that

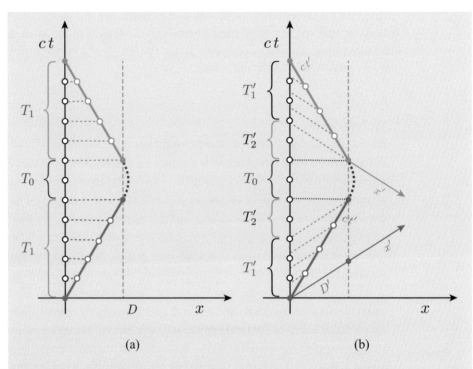

(a)   (b)

Fig. 2.14 Minkowski diagrams of the twin paradox. (a) Shows simultaneity lines according to John. During the first and third part of the trip, a time $2 \times T_1$ elapses on John's clock; during the middle part, Jane is accelerating uniformly and the time elapsed is denoted by T_0. (b) Shows simultaneity lines according to Jane, except for the two dotted lines sandwiching the accelerating segment. Jane's x' axis is also shown for two instants in time. The segment labeled T_0 is excised away and borrowed from John's perspective, since Jane is not an inertial frame during this period. T'_1 and T'_2, however, can be computed from Jane's perspective. Notice how Jane's x' axis must smoothly flip around during the time interval T_0, as she turns around. Her simultaneity lines during this period will hence be distorted and require general relativity to fully unravel.

$$\frac{a\,T_0}{2} = -\gamma_0 V_0, \tag{2.143}$$

where T_0 is the time it takes for Jane to change her speed from V_0 to $-V_0$ according to John, as shown in the figure. How much time passes on Jane's wristwatch during this period? At every instant in time, Jane is moving with some speed $v(t)$ and is subject to a time dilation effect

$$dt = \frac{d\tau}{\sqrt{1 - v(t)^2/c^2}}. \tag{2.144}$$

Substituting for $v(t)$ from Eq. (2.142) and integrating over the period $T_0/2$, we get

$$\sinh \frac{a\,\tau_0}{2\,c} = \frac{a\,T_0}{2\,c}, \tag{2.145}$$

where τ_0 is the time elapsed on Jane's wristwatch during the turnaround. Eliminating a by using Eq. (2.143), we can then write

$$T_0 = \frac{(\gamma_0 V_0/c)}{\sinh^{-1}(\gamma_0 V_0/c)}\tau_0 > \tau_0. \tag{2.146}$$

We thus have full control over the computation from John's perspective. We can tell that while Jane's wristwatch ticks for a period of

$$\tau_0 + 2\,\tau_1 \tag{2.147}$$

during the full trip, John's clock ticks

$$T_0 + 2\,T_1 = \frac{(\gamma_0 V_0/c)}{\sinh^{-1}(\gamma_0 V_0/c)}\tau_0 + 2\,\gamma_0\tau_1 \tag{2.148}$$

during the same period. Hence, John ages more, since $T_0 > \tau_0$ and $T_1 > \tau_1$. Let us summarize the result. The travel time

$$\text{on John's watch} = T_0 + 2\,T_1 = \frac{2\,\gamma_0 V_0}{|a|} + \frac{2\,D}{V_0}, \tag{2.149}$$

$$\text{on Jane's watch} = \tau_0 + 2\,\tau_1 = \frac{2\,c\,\sinh^{-1}(\gamma_0 V_0/c)}{|a|} + \frac{1}{\gamma_0}\frac{2\,D}{V_0}, \tag{2.150}$$

where we used $T_1 = D/V_0$ and $T_0 = 2\,\gamma_0 V_0/|a|$ to quote both results in terms of D (the distance of travel according to John), a (Jane's acceleration according to John), and V_0 (Jane's speed for most of the trip). Notice that for small speeds $V_0 \ll c$, the two periods become approximately the same, as expected, since $\sinh^{-1}(\gamma_0 V_0/c) \simeq (\gamma_0 V_0/c)$ and $\gamma_0 \simeq 1$.

Thus, Jane has aged less during the travel. This is fine and interesting until you try to reverse your perspective. From Jane's point of view, she was not moving. Instead, John traveled away while the star visited her. Both John and the star traveled past Jane at speed V_0 in the opposite direction, as in watching trees move past you while you are driving a car. According to Jane, is it then John's time that is dilated? Hence, by the time the trek is over and the twins meet, would Jane expect that John has aged less during her travel period? Since John and Jane can now meet and compare notes, only one of the two can be correct, and hence the paradox.

The resolution lies in the realization that Jane is not at rest in any single inertial reference frame throughout the trip, while John *is*. This is because Jane has to decelerate and turn around at the star to return to earth. During the turnaround period, Jane is not an inertial observer, and John and Jane are *not* equivalent observers as far as the laws of physics are concerned. For example, Jane can hold a pendulum and notice that it sways while she is turning around to come back home. To find out the outcome from Jane's perspective, we would need to learn how to handle the point of view of accelerating observers; we need to know how space and time are affected in Jane's reference frame when she is decelerating. However, special relativity, including the Lorentz transformation, time dilation, length contraction, and so on, stipulates that observers must be at rest in some non-accelerating, inertial frame. So what can we do? We might tackle the issue by envisioning an infinite number of inertial frames whose velocities match Jane's decelerating frame at specific times, work out what happens in those frames, and translate the results back to Jane's frame. Or we might use Einstein's general theory of relativity, which allows observers to be at rest in any frame whatsoever, accelerating or not. Neither of these approaches is necessary, however, because we can immediately deduce that John's conclusion must be the correct one, since he is indeed inertial: Jane ages less. However, we can still analyze the problem from Jane's point of view, and we will do so using graphical methods. Figure 2.14(b) shows the same setup with simultaneity lines according to Jane. The middle segment of the trek where Jane is not inertial has been excised: for this period, we still need to rely on John's perspective. We then take as given the result from Eq. (2.146):

$$T_0 = \frac{(\gamma_0 V_0/c)}{\sinh^{-1}(\gamma_0 V_0/c)} \tau_0 > \tau_0. \tag{2.151}$$

The question is now to determine T_1' and T_2' as shown in the figure. T_1' corresponds to the time elapsed on John's wristwatch during the first segment *according to Jane*. Time dilation tells us that it is given by

$$\tau_1 = \gamma_0 T_1' \Rightarrow T_1' = \frac{\tau_1}{\gamma_0}, \tag{2.152}$$

since John is doing the moving according to Jane. Note that this makes sense since Jane sees the distance D contracted to $D' = D/\gamma_0$. So, her travel time must be $\tau_1 = (D/\gamma_0)/V_0 = T_1/\gamma_0$, as we found before from John's perspective. To find T_2', we need to look at the figure and do a bit of geometry. The slope of Jane's x' axis on the figure is $\pm V_0/c$. We can then immediately read off

$$c T_2' = \frac{V_0}{c} \times D. \tag{2.153}$$

Putting things together we find that the total time of the trip on John's wristwatch from Jane's perspective is

$$2 T_1' + 2 T_2' + T_0 = \frac{2\tau_1}{\gamma_0} + \frac{2 V_0 D}{c^2} + \frac{2 \gamma_0 V_0}{|a|} = \frac{2 D}{\gamma_0^2 V_0} + \frac{2 V_0 D}{c^2} + \frac{2 \gamma_0 V_0}{|a|}$$

$$= \frac{2 D}{V_0} + \frac{2 \gamma_0 V_0}{|a|}, \tag{2.154}$$

where we used $\tau_1 = (D/\gamma_0)/V_0$ and $T_0 = (2 \gamma_0 V_0)/|a|$. We see that the conclusion is identical to John's, Eq. (2.149): John ages more. From Jane's perspective, we relied on her notion of simultaneity during the first and third segments of the trip (computations of T_1' and T_2'), during the intervals when Jane is an inertial observer. However, we did borrow John's conclusion about his and Jane's clock rates (computation of T_0), since his perspective was the inertial one – a framework where the laws of special relativity can be applied. During this acceleration phase, the laws of physics are altered from Jane's perspective, and to carry the computation of T_0 from her reference frame requires us to learn how special relativity is modified in an accelerated frame. We will see in Chapter 4 that the **principle of equivalence** plays a central role in such settings.   ∎

2.4 Summary

In this chapter we have demonstrated how a simple principle, that of the universality of the speed of light in inertial frames and in vacuum, leads to a radical reformulation of mechanics. We employed this principle to find new transformation rules, the Lorentz transformations, that connect the perspectives of inertial observers. We developed a natural mathematical language that mixes space and time, the language of four-vectors. And embedding physics in this language guided us to discover that time and space are malleable and observer-dependent notions. We also found that the concepts of momentum and energy have to be revised; and we developed a relativistic version of Newton's laws and dynamics. This framework allows us to study mechanics even when speeds are close to that of light – all along also realizing that at sufficiently slow speeds, Newtonian mechanics is a good approximation.

Problems

★   **Problem 2.1** *Time dilation and length contraction.* Clock A is placed at the origin of the primed frame; it reads time $t' = 0$ just as the origins of the primed and unprimed frames coincide. At a later time t to observers in the unprimed frame, find from the Lorentz transformation of Eqs. (2.15) (a) how far A has moved and (b) what time A reads. This is an example of the fact that *moving clocks run slow.* A stick of length L_0 is placed at rest along the x' axis of the primed frame. Observers in the unprimed frame measure the position of both ends of the stick at the same time t to them as the stick is moving along at speed V. (c) Using the Lorentz transformation, find the length $L \equiv (x_2 - x_1)$ of the stick in the unprimed frame, in terms of L_0 and the relative frame velocity V. Here x_2 and x_1 are the locations of each end of the stick, as measured in the unprimed frame. The fact that $L < L_0$ is an example of the fact that *moving lengths are contracted in their direction of motion.* This phenomenon is called the Lorentz contraction or the Lorentz–Fitzgerald contraction.

★   **Problem 2.2** *The invariance of transverse lengths.* A stick of length L_0 is placed at rest along the y' axis of the primed frame, extending from $y' = y'_1$ to $y' = y'_2$. Observers in the unprimed frame measure the position of both ends of the stick at the same time t to them as the stick is moving along at speed V. Using the Lorentz transformation of Eqs. (2.15), find the length $L \equiv (y_2 - y_1)$ of the stick in the unprimed frame, in terms of L_0 and the relative frame velocity V. Here y_2 and y_1 are the locations of each end of the stick, as measured in the unprimed frame. The fact that $L = L_0$ is an example of the fact that *moving transverse lengths are invariant under Lorentz transformations.*

★   **Problem 2.3** *The relativity of simultaneity.* Two clocks are placed at rest on the x' axis of the primed frame, clock A at $x' = 0$ and clock B at $x' = L_0$. They are therefore a distance L_0 apart in their mutual (primed) rest frame. Observers in the unprimed frame see both clocks moving at velocity V, B leading the way and A following it. Then at some particular time t, unprimed observers measure the readings of t'_A and t'_B of the two clocks. Show from the Lorentz transformation of Eqs. (2.15) that $t'_B < t'_A$, and that in fact $t'_B = t'_A - VL_0/c^2$. This is an example of the fact that *leading clocks lag*, i.e., that the clock leading the way reads a lesser time than the chasing clock. It also shows that simultaneity is not universal but relative. In nonrelativistic physics, if two events are simultaneous according to observers in one frame or reference, they are simultaneous in all frames. That is not true in relativity.

★ **Problem 2.4** A primed frame moves at $V = (3/5)c$ relative to an unprimed frame. Just as their origins pass, clocks at the origins of both frames read zero, and a flashbulb explodes at that point. Later, the flash is seen by observer A at rest in the primed frame, whose position is $x', y', z' = (3\,\mathrm{m}, 0, 0)$. (a) What does A's clock read when A sees the flash? (b) When A sees the flash, where is she located according to unprimed observers? (c) To unprimed observers, what do their own clocks read when A sees the flash? Use the Lorentz transformation of Eqs. (2.15).

★ **Problem 2.5** Synchronized clocks A and B are at rest in our frame of reference, a distance 5 light-minutes apart. Clock C passes A at speed $(12/13)c$ bound for B, when C, and also both A and B, read $t = 0$ in our frame. (a) What time does C read when it reaches B? (b) How far apart are A and B in C's frame? (c) In C's frame, when A passes C, what time does B read?

★ **Problem 2.6** Two spaceships are approaching one another. According to observers in our frame, (a) the left-hand ship moves to the right at $(4/5)c$ and the right-hand ship moves to the left at $(3/5)c$. How fast is the right-hand ship moving in the frame of the left-hand ship? (b) The left-hand ship moves to the right at speed $(1 - \epsilon)c$ and the right-hand ship moves to the left at $(1 - \epsilon)c$, where $0 < \epsilon < 1$. How fast is the right-hand ship moving in the frame of the left-hand ship? Show that this speed is less than c, no matter the value of ϵ within the range allowed.

★ **Problem 2.7** Astronaut A boards a spaceship leaving earth for the star Alpha Centauri, 4 light years from earth, while her friend B stays at home. The ship travels at speed $(4/5)c$, and upon arrival immediately turns around and travels back to earth at the same speed $(4/5)c$. (a) How much has A aged during the entire trip? (b) How much has B aged during the time A has been gone?

★★ **Problem 2.8** A*l* and Bert are identical twins. When Bert is 24 years old he travels to a distant planet at speed $(12/13)c$, turns around, and heads back at the same speed, arriving home at age 44. A*l* stays at home. (a) How old is A*l* when Bert returns? (b) How far away was the planet in A*l*'s frame? (c) Why can't Bert reasonably claim that from his point of view it was A*l* who was moving, so that A*l*'s clocks should be time dilated, making A*l* younger than Bert when they reunite?

★ **Problem 2.9** The Global Positioning System (GPS) features 24 earth satellites orbiting at altitude 20,200 km above earth's surface. Each satellite carries four highly precise atomic clocks; this precision is essential in allowing us to know our positions on the ground within a few meters or less. Special relativistic time-dilation effects, although tiny, must be taken into account. They are due to the speed v of the satellites relative to a clock at rest in some appropriate inertial frame. Let us take this reference clock to be a hypothetical clock at rest at the center of the earth. (To call such a clock inertial is only an approximation, because the earth has a small acceleration toward the sun and moon, which are themselves accelerating toward the center of our galaxy, etc., etc.) (a) Find the special relativistic time-dilation factor $\sqrt{1 - v^2/c^2}$ for clocks in a GPS satellite, expressed in the form $1 - \epsilon$, where ϵ is a very small number. (b) How much time would they lose in 1 year due to this effect? (There is a *second* relativistic effect on GPS clocks, as described in Chapter 10, due not to their velocity but to their altitude in earth's gravity. Given information: mass and mean radius of the earth 5.98×10^{24} kg and 6370 km; Newton's gravitational constant $G = 6.674 \times 10^{-11}$ m^3/(kg s^2).)

★ **Problem 2.10** Incoming high-energy cosmic-ray protons strike earth's upper atmosphere and collide with the nuclei of atmospheric atoms, producing a downward-directed shower of particles, including (among much else) the pions π^+, π^-, and π^0. The charged pions decay quickly into muons and neutrinos:

$$\pi^+ \rightarrow \mu^+ + \nu \quad \text{and} \quad \pi^- \rightarrow \mu^- + \nu.$$

The muons are themselves unstable, with a half-life of 1.52 μs in their rest frame, decaying into electrons or positrons and additional neutrinos. Nearly all muons are created at altitudes of about 15 km and more, and then those that have not yet decayed rain down upon the earth's surface. Consider muons with speeds $(0.995 \pm 0.001)c$, with their numbers measured on the ground and in a balloon-lofted experiment at altitude 12 km. (a) How far would such muons descend toward the ground in one half-life if there were no time dilation? (b) What fraction of these muons observed at 12 km would reach the ground? (c) Now take into account time dilation, in which the muon clocks run slow, extending their half-lives in the frame of the earth. What fraction of those observed at 12 km would make it to the ground? (Such experiments supported the fact of time dilation.)

★ **Problem 2.11** Suppose that in the distant future astronomers build a telescope so powerful they can see aliens on a planet that is 10 light years from earth. One day they observe the aliens board a spaceship and blast off toward earth. According to earth clocks, the ship and its alien crew arrive at the earth exactly 1 week later. Assuming the velocity of the ship was constant during almost the entire trip, find its velocity ratio $\beta = v/c$ in earth's frame, valid to three significant figures. (Note that $v/c < 1$.)

★★★ **Problem 2.12** A distant galactic nucleus ejects a jet of material at right angles (90°) to our line of sight. We know the distance of the galaxy from the redshift of its spectral lines, so we can calculate how far the jet has traveled in a given time using the very small but growing angle between the galactic nucleus and the jet as observed through our telescope. From this information we can find the velocity of the jet. Note that for such transverse motion it takes essentially the same time for light from the jet to reach us from the end of its journey as it does from its beginning, because it stays essentially the same distance from us throughout. But now suppose the jet is ejected at some angle θ relative to our line of sight, so the jet's transverse velocity component is $v_\perp = v \sin \theta$ and its velocity component toward us is $v_\parallel = v \cos \theta$. And because the jet is getting closer to us, the time it takes light to reach us from it becomes smaller and smaller. (a) In this case find an equation for the jet's *apparent* transverse velocity in the sky, defined as the transverse distance moved divided by the time interval as observed on earth, and show that this apparent velocity v_{app} can exceed the speed of light. (b) For a given actual velocity v, find the angle θ that *maximizes* v_{app}, and then find the magnitude of v_{app} in this case. (c) Evaluate such a maximal v_{app} for the case $v = 0.99c$. Such apparent superluminal velocities have often been observed by astronomers, even though no matter actually travels faster than light.

★ **Problem 2.13** A bullet train of rest length 500 m is chugging along a straight track at speed $(4/5)c$ when it enters a tunnel of length 400 m. Due to length contraction in the frame of the tunnel, the train apparently briefly fits inside the tunnel all at once. From the point of view of train passengers, however, it is the tunnel that is contracted, with a length of only 400 m × 3/5 = 240 m, so the 500-m train seemingly *cannot* fit inside all at once. The question is: Does the train fit inside the tunnel all at once, or not? Explain.

★★ **Problem 2.14** A carrot-slicing machine consists of eight parallel blades spaced 5 cm apart, held together in a framework that allows all the blades to descend at once upon an unsuspecting carrot laid out horizontally in the machine. The result is several carrot pieces of length 5 cm, plus random bits left over at each end. Now suppose that a carrot is made to move lengthwise at speed $(4/5)c$ into the machine just as the blades descend. The Lorentz contraction ensures that the carrot will be shorter in the machine frame than in its rest frame, so there will be fewer carrot pieces. Each of these non-end pieces will still have length 5 cm in the machine frame because that is the spacing of the blades, so it appears they must be *longer* than 5 cm when finally brought to rest. In fact, each should have rest length 5 cm/(3/5)c) = 8 1/3 cm. Now view the exact same procedure in the rest frame of the carrot. Then it is the slicing machine that moves at $(4/5)c$, so it contracts as a whole, and the distance between blades is Lorentz-contracted to $5 \text{ cm} \sqrt{1 - (4/5)^2}$ = 3 cm. That is, it seems that it produces carrot pieces 3 cm long in their rest frame. These conclusions (8 1/3 cm and 3 cm) cannot both be correct, since it is the same carrot that was involved in both sets of reasoning. Which is the correct answer (if either) and *why* is the other answer or answers wrong?

★★ **Problem 2.15** By differentiating the velocity transformation equations one can obtain transformation laws for acceleration. Find the acceleration transformations for the x component a_x, in terms of a_x, v_x, and V, the relative frame velocity.

★   **Problem 2.16** An electron moves at velocity $0.9c$. How fast must it move to double its momentum?

★   **Problem 2.17** An atomic nucleus starts at rest in the lab, and is then struck by two photons, one after the other, each with momentum p_γ in the same direction. The photons are absorbed in the nucleus. If the mass of the final (excited) nucleus is M^*, calculate its velocity.

★   **Problem 2.18** Two particles make a head-on collision, stick together, and stop dead. The first particle has mass m and speed $(24/25)c$, and the second has mass M and speed $(5/13)c$. Find M in terms of m.

★   **Problem 2.19** Spaceship A, moving away from the earth at velocity $(3/5)c$, is sending messages to spaceship B, which left the earth earlier at speed $(4/5)c$ in the same direction. The messages sent by A are contained in pulses sent by a laser on A, with the pulses separated by 100 fs in A's frame of reference. As B receives the pulses, what is the pulse separation according to the crew on B?

★   **Problem 2.20** An alien vessel is detected approaching earth at $(3/5)c$. An intercepting probe is sent from earth at speed $(4/5)c$ toward the vessel. As they approach one another the probe uses a pulsed laser to send a message to the oncoming aliens, where the time interval between pulses is 12 ps in the frame of the probe. What is the time interval between pulses as observed by the aliens?

★   **Problem 2.21** An organist on earth is playing Bach's Toccata and Fugue in D Minor, which is being broadcast by a powerful radio antenna. Travelers in a spaceship moving at speed $v = (3/5)c$ away from the earth are listening in. In what key do they hear the music?

★   **Problem 2.22** The Andromeda galaxy (also known to astronomers by catalogue number M31) is in our local group of galaxies, about 2.5 million light years from our own Milky Way (MW) galaxy (Figure 2.15). When using spectrometers to measure the wavelengths of light emitted by stars in M31, astronomers find the redshift to be $\Delta\lambda/\lambda = -0.001001$, where λ is the wavelength of the spectral line in the laboratory, $\Delta\lambda$ is the shift in wavelength, and the minus sign indicates that $\Delta\lambda < 0$, corresponding to a *blueshift*. (a) If one assumes this change in wavelength is due to the Doppler effect, how fast (km/s) is M31 approaching us? (b) If this velocity were also M31's velocity toward our galactic nucleus, and it did not change with time, how long would it take M31 to collide with the MW? [In fact, M31's velocity toward the MW nucleus is less than the result calculated in part (a), because the solar system is orbiting around our MW nucleus, with a component of velocity directed toward M31, so in fact M31 is moving only about 110 km/s toward our MW nucleus. We would also expect the M31/MW relative velocity of approach to increase with time due to their mutual gravitational attraction. Taking all this into account predicts they will collide in about 4 billion years.]

Fig. 2.15 The M31 (Andromeda galaxy) referred to in Problem 2.22. Image credit: NASA/JPL-Caltech, from the GALEX mission.

★★ **Problem 2.23** A proton moves in the x, y plane with velocity $v = (3/5)c$, at an angle of $45°$ to both the x and y axes. (a) Find all four components of the proton's four-vector velocity v^μ and evaluate the invariant square of its components $\eta_{\mu\nu}v^\mu v^\nu$. (b) Find all four components of the proton's four-vector velocity in a frame moving in the positive x direction at velocity $V = (4/5)c$. (c) Evaluate explicitly the invariant square of its components in this frame.

★★ **Problem 2.24** A particular pion π^+ decays in 26 ns in its own rest frame. Suppose a particle accelerator produces the pion with total energy $E = 100\ mc^2$, where m is its mass. (a) How far (in meters) will it travel before decaying? (b) A different pion has a kinetic energy equal to its mass energy. If it travels a distance D before decaying, find how long it lived in its own rest frame.

★ **Problem 2.25** A photon of total energy $E = 12,000$ MeV is absorbed by a nucleus of mass M_0, originally at rest. Afterwards, the excited nucleus has mass M and is moving at speed $(12/13)c$. Find its momentum in units MeV/c, and both M and M_0 in units MeV/c^2.

★ **Problem 2.26** A team plans to accelerate a probe of mass 2.0 kg away from the far side of the moon by a bank of lasers that push the probe with constant force F in the rest frame of the moon. What F would be required to accelerate the probe to velocity $0.9c$ in 1 week?

★   **Problem 2.27** Show that the momentum and velocity four-vectors are both time-like, and that the force four-vector is space-like.

★★   **Problem 2.28** A wave equation for light is

$$\frac{\partial^2 \phi}{\partial x^2} + \frac{\partial^2 \phi}{\partial y^2} + \frac{\partial^2 \phi}{\partial z^2} - \frac{1}{c^2}\frac{\partial^2 \phi}{\partial t^2} = 0,$$

where ϕ is a scalar potential. Show that the set of all linear transformations of the spacetime coordinates that permit this wave equation to be written as we did correspond to (i) four possible translations in space and time, (ii) three constant rotations of space, and (iii) three Lorentz transformations. Collectively, these are called the **Poincaré transformations** of spacetime.

★★   **Problem 2.29** *Two spaceships with string "paradox."* Consider two spaceships, both at rest in our inertial frame, a distance D apart, one behind the other. There is a light string of rest length D tied between them. Now the ships, both at the same time in our frame, begin to accelerate uniformly to the right, with the string still tied between them. The ships start at the same time and have the same acceleration, so the distance between them, and therefore the length of the string, must be constant in our frame. However, we know that a moving string *should* be Lorentz-contracted in its direction of motion, by the usual factor $\sqrt{1 - \beta^2}$. Therefore, does the "need" of the string to become shorter in our frame cause it to break eventually, or does the fact that its length remains the same in our frame mean that it will not break? Explain which is correct. *Hint*: The "proper length" of an accelerating object can be measured by observers in an inertial frame instantaneously comoving with the object, that is, in an inertial frame that at some moment is at rest relative to the object. This "paradox" was originally posed by E. Dewan and M. Beran in 1959 and later modified by J. S. Bell in 1987. As Bell describes in Chapter 9 of his book *Speakable and Unspeakable in Quantum Mechanics* (Cambridge University Press, 1987), some very good physicists have gotten the wrong answer, at least initially.

★★★ **Problem 2.30** (a) Prove that the time order of two events is the same in all inertial frames if and only if they can be connected by a signal traveling at or below speed c. (b) Suppose that in an unprimed inertial frame a particular signal from A to B can travel at velocity $v = 2c$. Then find a relative velocity V with a primed frame (where $|V| < c$) such that in the primed frame the same signal reaches B before it was sent by A.

★★   **Problem 2.31** In the text we derived the Doppler formulae for light. Using the same strategy, find the relativistic Dopper formulae for waves traveling at speed $v < c$. For example, the waves may be sound waves in some very stiff material whose sound speed is a few percent that of c.

★   **Problem 2.32** An algebraic expression is said to be **Lorentz covariant** if its form is the same in all inertial frames: the expression differs in two inertial frames $\mathcal{O}$ and $\mathcal{O}'$ only by putting prime marks on the coordinate labels. For example, $A^\mu \eta_{\mu\nu} B^\nu =$

K is a Lorentz-covariant expression, where A^μ and B^ν are four-vectors and K a constant. Under the Lorentz transformation, $A^\mu \eta_{\mu\nu} B^\nu = A^{\mu'} \Lambda^\mu_{\mu'} \eta_{\mu\nu} B^{\nu'} \Lambda^\nu_{\nu'} = A^{\mu'} \eta_{\mu'\nu'} B^{\nu'} = K$, where we used $\eta_{\mu'\nu'} = \Lambda^\mu_{\mu'} \eta_{\mu\nu} \Lambda^\nu_{\nu'}$. Because the indices come matched in pairs across a metric factor $\eta_{\mu\nu}$, the expression preserves its structural form. The quantity is also a **Lorentz scalar**: its *value* is unchanged under a Lorentz transformation. Which of the following quantities are Lorentz scalars, given that K is a constant and any quantity with a single superscript is a four-vector? (a) $KA^\mu \eta_{\mu\nu}$, (b) $C^\mu = D^\mu (A^\lambda \eta_{\lambda\nu} B^\nu)$, (c) $KA^\mu \eta_{\mu\nu} B^\lambda \eta_{\lambda\sigma} D^\nu F^\sigma$.

Problem 2.33 Consider a Lorentz-covariant expression that is *not* a Lorentz scalar, $C^\lambda = K^\lambda h(A^\mu \eta_{\mu\nu} B^\nu)$, where h is any function of the quantity in parentheses. Here, quantities with a single superscript are four-vectors. Under a Lorentz transformation, $A^\mu \eta_{\mu\nu} B^\nu$ is Lorentz covariant and is also a Lorentz scalar. Hence, its form and value are unchanged, which means that the function $h(A^\mu \eta_{\mu\nu} B^\nu)$ is unchanged in form or value as well. The quantity K^μ, however, is a four-vector; this means that it transforms as $K^\mu = \Lambda^\mu_{\mu'} K^{\mu'}$. The right-hand side of the equation for C^λ transforms as a four-vector as a whole, which implies that C^λ also transforms as a four-vector and observer $\mathcal{O}'$ can write $C^{\lambda'} = K^{\lambda'} h(A^{\mu'} \eta_{\mu'\nu'} B^{\nu'})$. This quantity is said to be a **Lorentz vector** (instead of a scalar), since it transforms as a four-vector: that is, its components change, but through the well-defined prescription for a four-vector. Which of the following quantities are Lorentz vectors, given that K is a Lorentz scalar and any quantity with a single subscript or superscript is a Lorentz vector? (a) $K\eta_{\mu\nu}$, (b) $C^\lambda = D^\mu A^\lambda \eta_{\mu\nu} B_\nu$, (c) $KA^\mu \eta_{\mu\nu} B^\lambda \eta_{\lambda\sigma} D^\nu F^\sigma$,

Problem 2.34 The concept of Lorentz covariance is important because it allows us to quickly determine the transformation properties of expressions under changes of inertial reference frames. The principle of relativity requires that all laws of physics are unchanged as seen by different inertial observers. Hence, we need to ensure that expressions reflecting statements of a law of physics are Lorentz covariant, *i.e.*, that they retain their structural form under Lorentz transformations. A useful application of this comes from the modified second law of dynamics:

$$f^\mu = \frac{dp^\mu}{d\tau}.$$

Forces that we insert on the left-hand side of this equation must be Lorentz-covariant expressions that transform as four-vectors. This ensures that observer $\mathcal{O}'$ can write simply

$$f^{\mu'} = \frac{dp^{\mu'}}{d\tau}.$$

For example, we could write $f^\mu = K^\mu$ with a constant four-vector K^μ. (a) Is a "relativistic spring law" $f^\mu = -(0, k\mathbf{r})$ for some constant k a Lorentz-covariant expression? (b) What about a modified spring law $f^\mu = -Kr^\mu = -k(ct, \mathbf{r})$? (c) What about Newtonian gravity $\mathbf{F} = -(k/r^3)\mathbf{r}$? Is such a force covariant?

★★ **Problem 2.35** Show that the most general Lorentz transformation can be written as a 4×4 matrix $\hat{\mathbf{\Lambda}}$ satisfying

$$\hat{\mathbf{\Lambda}}^{\mathrm{T}} \cdot \hat{\eta} \cdot \hat{\mathbf{\Lambda}} = \hat{\eta} \quad \text{and} \quad |\hat{\mathbf{\Lambda}}| = 1.$$

Since a Lorentz transformation is by definition a linear transformation of time and space that preserves the speed of light, you simply need to show that these two properties are necessary and sufficient for this. Note also that reflections get ruled out by the second condition by choice.

★★ **Problem 2.36** A π^0 meson with mass $m_\pi = 135.0$ MeV/c^2 is created in the upper atmosphere when a cosmic-ray proton collides with a nitrogen nucleus. The mean lifetime of π^0s is 8.4×10^{-17} s; they almost always decay into two photons. Suppose this particular pion has total energy $E = 500$ MeV and moves vertically downward toward the ground, and also that it decays in three mean lifetimes into two photons, one moving up and one moving down. (a) How far does the π^0 move relative to the ground from its creation until it decays? (b) Find the frequency of each final photon measured in the frame of the ground.

★ **Problem 2.37** A π^- meson with mass $m_\pi = 140.0$ MeV/c^2 is produced in a (p,p) collision in an accelerator. The pion subsequently decays into a muon and a muon-type antineutrino, in the reaction $\pi^- \rightarrow \mu^- + \bar{\nu}_\mu$. The antineutrino has a nonzero but very small mass, so in this calculation you can ignore it. The muon has a mass energy of 105.7 MeV/c^2. In the rest frame of the original pion, find (a) the total energy and three-momentum of the muon and (b) the total energy and three-momentum of the antineutrino.

★★ **Problem 2.38** The Higgs particle has a mass energy of 125 GeV/c^2. Once created it decays very quickly into various sets of particles: for example, about 60% of the time it decays into a $(b\bar{b})$ quark–antiquark pair. Such b quarks have a mass energy of about 4.2 GeV each. (The b quarks are also called "bottom" quarks, and the reaction is written $H \rightarrow b + \bar{b}$ or simply $H \rightarrow b\bar{b}$ for short.) Suppose a particular Higgs particle is moving at $v = (4/5)c$ in the lab and that it decays into a $(b\bar{b})$ pair with the b quark moving in the forward direction and the $\bar{b}$ quark moving in the backward direction. Find the energy and momentum in the lab frame of (a) the b quark and (b) the $\bar{b}$ quark.

★ **Problem 2.39** *Tachyons* are hypothetical (and so-far undetected) particles that always travel faster than light. (a) Show that all components of a tachyon's momentum four-vector are real if we assign the tachyon an imaginary mass, say $m = im_0$, where m_0 is real. (b) Then show that the invariant square of the momentum four-vector $\eta_{\mu\nu}p^\mu p^\nu$ is necessarily positive. Thus the world of particles could be separated into three regimes: (i) ordinary massive particles, with $\eta_{\mu\nu}p^\mu p^\nu < 0$; photons or other possible massless particles, with $\eta_{\mu\nu}p^\mu p^\nu = 0$; and tachyons, with $\eta_{\mu\nu}p^\mu p^\nu > 0$.

★   **Problem 2.40** *Pion photoproduction.* Positive π mesons can be created in the reaction
$\gamma + p \rightarrow n + \pi^+$, in which an incoming photon strikes a proton at rest, forming
a neutron and a π^+. (a) Find the threshold photon energy for this reaction given
the masses (in units MeV/c^2), $m_p = 938$; $m_n = 939.6$; and $m_{\pi^+} = 139.6$. (b) For
this photon energy, how fast is the CM frame moving relative to the lab frame,
expressed in the form V_{CM}/c? (c) What is the momentum of the initial photon in
the CM frame, expressed in units MeV/c?

★★   **Problem 2.41** Lambda (Λ^0) baryons can be created in high-energy $(p, \bar{p})$ collisions of
protons and antiprotons in the reaction $p + \bar{p} \rightarrow \Lambda^0 + k^+ + \bar{p}$, where k^+ is a positive
k meson. (a) Find the minimum (*i.e.*, threshold) energy required for the incident
antiproton if the target proton is at rest in the lab. The masses of the particles (in
MeV/c^2) are p or $\bar{p}$: 938.3; Λ^0: 1115.7; and k^+: 493.7. (b) Find the minimum energy
of each initial particle in a collider experiment, in which the total momentum is
zero. (c) Suppose that in the collider experiment the energy of each initial particle
is twice the minimum energy required. Find then how far the subsequent Λ^0 will
travel in the collider detector before it decays, assuming the Λ^0 lasts for a time 2.63
$\times 10^{-10}$ s (the mean lifetime of a Λ^0) in its own rest frame, and also assuming that
the final antiproton is at rest in the lab.

★   **Problem 2.42** Positive Sigma baryons Σ^+ can be created along with positive k
mesons k^+ in high-energy collisions of protons with protons, in the reactions
$p + p \rightarrow \Sigma^+ + k^+ + n$, where n is a neutron. (a) Find the minimum (*i.e.*, threshold)
energy required for the incident proton if the target proton is at rest in the lab. The
masses of the particles (in units MeV/c^2) are p: 938.3; Σ: 1189.4; k^+: 493.7; and
n: 939.6. (b) Find the minimum energy of each proton required in a (p, p) collider
experiment, in which the total momentum is zero. (c) Find the velocity of the CM
(*i.e.*, zero-momentum) frame in this experiment, relative to the lab frame in which
one initial proton is at rest, expressed as V_{CM}/c.

★   **Problem 2.43** Electrons (e^-) and antielectrons (e^+) (called *positrons*) each have mass
energy 0.511 MeV. A positron can be created in an (e^-, e^-) collision as long as
an electron is created along with it, thus conserving both electric charge and lepton
number. (Electrons have lepton number $+1$ and positrons have lepton number -1.)
(a) In a linear accelerator in which high-energy electrons are incident upon other
electrons at rest in the lab, what is the minimum required energy of each of the
incident electrons, in MeV? (b) If two beams of electrons are instead fired at one
another in a collider with equal but opposite momenta, what now is the minimum
energy of each electron required to create a positron, in MeV?

★   **Problem 2.44** (a) A photon of energy E_0 strikes a free electron at rest in the lab.
("Free" here means the electron is not bound inside an atom.) Is it possible for
the photon to be absorbed by the electron? If so, find the energy and momentum
of the final electron. If not, explain why not. (b) A free electron of energy E_0 is
moving in the lab. Is it possible for the electron to emit a photon, so that after the
emission there is a photon and an electron moving more slowly than before? If so,

find the final energy and the momentum of both the photon and the electron. If not, explain why not.

★ **Problem 2.45** The quantity $\lambda_C \equiv h/m_e c$ is called the "Compton wavelength" of the electron. (a) If a photon scatters off an electron at rest with scattering angle $\theta = 45°$, what is the photon's change of wavelength in terms of λ_C? (b) For what scattering angle is the change of wavelength a maximum, and what is the change of wavelength in that case?

★★ **Problem 2.46** Consider a relativistic elastic collision between two particles of equal mass, such as two protons. In the lab frame the target proton is at rest, and the incident proton has three-vector velocity v. For nonrelativistic equal-mass collisions the two protons emerge at right angles to one another, except for the special case where one of the protons moves straight ahead while the other is at rest. Is that also true for relativistic collisions? Prove that it is or that it is not. *Hint*: Draw before-and-after pictures of the three-vector velocities for each proton in the CM frame, and then transform to pictures in the lab frame.

★★★ **Problem 2.47** *A relativistic rocket.* In Chapter 1 we derived the differential equation of motion of a nonrelativistic rocket, by conserving both momentum and total mass over a short time interval Δt. That is, the momentum of the rocket at time t was set equal to the sum of the momenta of the rocket and bit of exhaust at time $t + \Delta t$, and similarly the mass of the rocket at t was set equal to the sum of the masses of the rocket and bit of exhaust at time $t + \Delta t$. We can find the equation of motion of a relativistic rocket in a similar way, except that the total mass is not conserved in this case; it is now the total momentum and the total energy that are conserved. At time t, let the rocket have mass m and velocity v; and at time $t + \Delta t$, let the rocket have mass $m + \Delta m$ (where $\Delta m < 0$) and velocity $v + \Delta v$, and let the bit of exhaust have mass ΔM and backward velocity $\bar{u}$. Note that $\Delta M \neq -\Delta m$ in relativistic physics. (a) Show from the velocity transformation that the velocity u of the bit of exhaust in the instantaneous rest frame of the rocket is given by

$$u = \frac{\bar{u} + v}{1 + \bar{u}v/c^2}.$$

(b) By conserving momentum show that, to first order in changes in m, v, and M:

$$\frac{(\Delta m)v}{\sqrt{1 - v^2/c^2}} + \frac{m\Delta v}{(1 - v^2/c^2)^{3/2}} = \frac{\Delta M\bar{u}}{\sqrt{1 - \bar{u}^2/c^2}}.$$

(c) Then conserve energy, again keeping no terms beyond those with first-order changes. Using both the conservation of momentum and conservation of energy expressions, show that the terms with ΔM can be eliminated. Then the results of problem (a) can be used to eliminate $\bar{u}$ in favor of u, and by dividing through by Δm and taking the limit $\Delta m \to 0$, show that

$$m\frac{dv}{dm} + u(1 - (v^2/c^2)) = 0,$$

which is the relativistic rocket differential equation of motion. Show also that this reduces to the equation for a classical rocket in the limit of small velocities.

★★ **Problem 2.48** By integrating the relativistic rocket differential equation of motion from the preceding problem, show that in terms of the ratio m/m_0, the relative rocket velocity v/c is given by

$$\frac{v}{c} = \frac{1 - (m/m_0))^{2u/c}}{1 + (m/m_0))^{2u/c}},$$

where m is the rocket mass at any time and m_0 is its mass at time $t = 0$ when the rocket starts from rest. We assume that the exhaust velocity $u = $ constant.

★★ **Problem 2.49** Consider the special case of a relativistic "photon" rocket in which the exhaust consists of photons only. The photons might be produced by an onboard laser or from the annihilation of particles and antiparticles carried with the rocket, for example. (a) From the result given in the preceding problem, how fast would the rocket be moving by the time it burned all its fuel, which was initially 90% of the rocket's mass? (b) Prove that for any given ratio of final to initial rocket mass, photon rockets are more efficient than rockets whose exhausts consist of massive particles, in that the final rocket velocity is greatest for photon rockets.

★★ **Problem 2.50** The captain of an interstellar photon-rocket spaceship wishes to maintain a constant acceleration a in the instantaneous rest frame of the ship, since that would provide a constant effective gravity for passengers. In that case, at what rate $|dm/dt|$ (as a function of time) should the ship's mass decrease with time?

★★ **Problem 2.51** *The transverse Doppler effect.* In Example 2.8 we derived the relativistic Doppler formulas for light sources that move either directly toward or away from the observer. Another possibility is that the source moves in a perpendicular direction, transverse to the observer's line of sight. In nonrelativistic physics there is no Doppler effect in this case. Show that if a light source is at rest at the origin of the primed frame, while moving at speed V in the x direction as seen by an unprimed observer, then if the momentum of the photons is purely in the y direction according to the observer, it follows that $\nu = \nu\prime\sqrt{1 - V^2/c^2}$. There is therefore a relativistic redshift in the case of transverse motion, related to the fact that the source's time is dilated in the observer's frame.

3 The Variational Principle

While Newton was still a student at Cambridge University, and before he had discovered his laws of particle motion, the French mathematician Pierre de Fermat proposed a startlingly different explanation of motion. Fermat's explanation was not for the motion of *particles*, however, but for *light rays*. In this chapter we explore Fermat's approach, and then go on to introduce techniques in variational calculus used to implement this approach, and to solve a number of interesting problems. We then show how Einstein's special relativity and the principle of equivalence help us demonstrate how variational calculus can be used to understand the motion of particles. All this is to set the stage for applying variational techniques to general mechanics problems in the following chapter.

3.1 Fermat's Principle

Imagine that a ray of light leaves a light source at point a and travels to some other given point b. Fermat proposed that out of all the infinite number of possible paths that the ray might take between the two points, it actually travels by the path of *least time*. For example, if there is nothing but vacuum between a and b, light traveling at constant speed c takes the path that minimizes the travel time, which of course is a straight line. Or suppose a piece of glass is inserted into part of an otherwise air-filled space between a and b. Given that light has speed $v = c/n$ in any medium with index of refraction n, the minimum-time path in this case is no longer a straight line: **Fermat's principle of least time** predicts that the ray will bend at the air–glass interface by an angle given by Snell's law, as shown in Figure 3.1.

More generally, a light ray might be traveling through a medium with index of refraction n that can be a continuous function of position, $n(x, y, z) \equiv n(\mathbf{r})$. In that case the time it takes the ray to travel an infinitesimal distance ds is

$$dt = \frac{\text{distance}}{\text{speed}} = \frac{ds}{c/n(\mathbf{r})}, \tag{3.1}$$

so the total time to travel from a to b by a particular path s is the integral

$$t = \int dt = \frac{1}{c} \int n(\mathbf{r}) \, ds. \tag{3.2}$$

The value of the integral depends upon the path chosen, so out of the infinite number of possible paths the ray might take, we are faced with the problem of finding the particular path for which the integral is a minimum. If we can find it, Fermat assures us that it is the path the light ray actually takes between *a* and *b*.

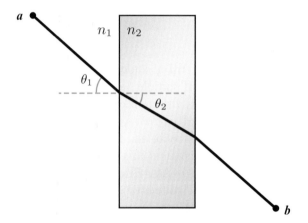

Fig. 3.1 Light traveling by the least-time path between *a* and *b*, in which it moves partly through air and partly through a piece of glass. At the interface the relationship between the angle θ_1 in air, with index of refraction n_1, and the angle θ_2 in glass, with index of refraction n_2, is $n_1\sin\theta_1 = n_2\sin\theta_2$, known as **Snell's law**. This phenomenon is readily verified by experiment.

Fermat's principle raises many questions, not least of which is: *How does the ray "know" that of all the paths it might take, it should pick out the least-time path?* In fact, a contemporary of Fermat named Claude Cierselier, who was an expert in optics, wrote

> ...*Fermat's principle can not be the cause, for otherwise we would be attributing knowledge to Nature: and here, by Nature, we understand only that order and lawfulness in the world, such as it is, which acts without foreknowledge, without choice, but by a necessary determination.*

In Chapter 5 we will elaborate on the profound reason *why* a light ray follows the minimum-time path. But in the meantime we focus on the general technique of finding the path that minimizes a given integral, such as the integral given in Eq. (3.2). The method is called the **calculus of variations**, or **functional calculus**, and that is the primary topic of this chapter.

3.2 The Calculus of Variations

The general methods of the calculus of variations were first worked out in the 1750s by the French mathematician Joseph-Louis Lagrange and the Swiss mathematician Leonard Euler, a century after Fermat proposed his principle. As an example of setting up these methods, return to the problem of finding the minimum-time path for a light ray traveling through a sheet of glass, as shown in Figure 3.1, except that now we allow the index of refraction in the glass to depend upon the coordinate x through the sheet, the horizontal axis in Figure 3.1. Let y be an axis perpendicular to x. Then what is the path $y(x)$ of the ray through the sheet?

The time to travel by any path is (using Eq. (3.2))

$$t = \frac{1}{c} \int ds \, n(x), \tag{3.3}$$

where ds is the distance between two infinitesimally nearby points along the path and c is the speed of light in vacuum. From the Pythagorean theorem we know that $ds = \sqrt{dx^2 + dy^2}$, so the time to travel by any path $y(x)$ through the sheet is

$$t = \frac{1}{c} \int n(x) \sqrt{dx^2 + dy^2} = \frac{1}{c} \int n(x) \sqrt{1 + \left(\frac{dy}{dx}\right)^2} \, dx$$

$$\equiv \frac{1}{c} \int n(x) \sqrt{1 + y'^2} \, dx. \tag{3.4}$$

In this case the integrand depends on *both* x and the path slope $y'(x)$. It is easy to imagine that the index of refraction n might also depend upon a transverse coordinate y, in which case the time for the ray to pass through the sheet of glass would be

$$t = \frac{1}{c} \int n(x, y) \sqrt{1 + y'^2} \, dx \equiv \int F(x, \, y(x), \, y'(x)) \, dx, \tag{3.5}$$

where the integrand $F(x, \, y(x), \, y'(x)) = (1/c) n(x, y) \sqrt{1 + y'^2}$ depends upon all three variables $x, y(x)$, and $y'(x)$. The calculus of variations shows us how to find the particular path $y(x)$ that minimizes this integral.

Example 3.1

Light Path Between Two Points in Glass

Consider first a very simple special case. Suppose the index of refraction n is *constant* throughout a sheet of glass, and that the endpoint of a light ray at $x = x_0, y = y_0$ is directly across the sheet from the beginning point at $x = 0, y = y_0$. Then the time to penetrate the sheet is

$$t = \frac{n}{c} \int_{x=0, y=y_0}^{x=x_0, y=y_0} \sqrt{1 + y'^2} \, dx. \tag{3.6}$$

By inspection, we can see that t is minimized if $y' = 0$ everywhere along the path, which is possible to arrange because the beginning point and endpoint are both at $y = y_0$. Therefore, according to Fermat's principle, the light path in this simple case is a straight line right through the sheet. No surprise here! ∎

More generally, Euler and Lagrange considered some arbitrary integral I of the form

$$I = \int F(x, y(x), y'(x))\, dx, \tag{3.7}$$

and the problem they wanted to solve was to find not only paths $y(x)$ that *minimize* I, but also paths that *maximize I*, or otherwise make *I stationary*. A "stationary" path is a particular path for which the integral I is nearly independent of slight variations in the path. We will make this definition precise in what follows. As we shall see, it is possible to have a stationary path that is neither a maximum nor a minimum.

How do we go about making I stationary? Let us revisit the more familiar problem of making an ordinary function stationary. Say we are given a function $f(x_1, x_2, \ldots) \equiv f(x_i)$ of a number of independent variables x_i with $i = 1, \ldots, N$, and we are asked to find the stationary points of this function. For the simpler case of a function with only two variables, we can visualize the problem as shown in Figure 3.2. We have a curved surface $f(x_1, x_2)$ over the x_1–x_2 plane, and we are looking for special points (x_1, x_2) where the surface is "locally horizontal." These can correspond to minima, maxima, or saddle points, as shown in the figure. Algebraically, we can phrase the general problem as follows. For every point $(x_1, x_2, \ldots)$, we move away by a small arbitrary distance δx_i:

$$x_i \rightarrow x_i + \delta x_i. \tag{3.8}$$

We then seek a special point $(x_1, x_2, \ldots)$ where this shift does not change the function $f(x_i)$ to *linear order* in the small shifts δx_i. This is what we mean by saying "locally horizontal." Employing a Taylor expansion, we can then write (see Appendix F)

$$f(x_i) \rightarrow f(x_i + \delta x_i) = f(x_i) + \frac{\partial f}{\partial x_j}\delta x_j + \frac{1}{2!}\frac{\partial^2 f}{\partial x_j x_k}\delta x_j \delta x_k + \cdots \tag{3.9}$$

Note that the j and k indices are repeated and hence summed over, using again the Einstein summation convention of Chapter 2. If the function is to remain constant up to linear order in δx_i, we then need N conditions

$$\frac{\partial f}{\partial x_j} = 0 \quad \text{with } j = 1, \ldots, N; \tag{3.10}$$

i.e., the slopes in all directions must vanish at the special point where the surface plateaus. This is because the δx_js are arbitrary and independent, and the only way for $(\partial f/\partial x_j)\delta x_j$ to vanish for an *arbitrary* δx_j is to have all the derivatives $\partial f/\partial x_j$ vanish.

The second-derivative terms involving $\partial^2 f/\partial x_j \partial x_k$ tell us how the surface curves away from the local plateau, whether the point is a minimum, maximum, or saddle point. Equation (3.10) typically yields a set of algebraic equations that can be

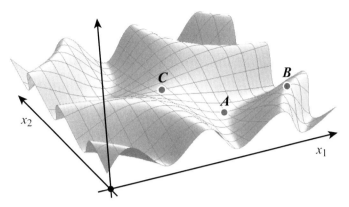

Fig. 3.2 A function of two variables $f(x_1, x_2)$ with a local minimum at point A, a local maximum at point B, and a saddle point at C.

solved for x_i, identifying the point(s) of interest. Formally, we write the condition to make a function $f(x)$ stationary as

$$\delta f \equiv f(x_i + \delta x_i) - f(x_i) = 0 \text{ to linear order in } \delta x_i$$

$$\Rightarrow \frac{\partial f}{\partial x_i} \delta x_i = 0 \Rightarrow \frac{\partial f}{\partial x_i} = 0. \tag{3.11}$$

Example 3.2 **Maximum of an Inverted Paraboloid**

Consider the function

$$f(x, y) = -a[(x - x_0)^2 + (y - y_0)^2], \tag{3.12}$$

where a, x_0, and y_0 are constants. What is the location of its maximum? We simply set both partial derivatives equal to zero, so that the function is stationary in each of the two directions:

$$\frac{\partial f}{\partial x} = -2a(x - x_0) = 0, \quad \frac{\partial f}{\partial y} = -2a(y - y_0) = 0 \tag{3.13}$$

and the maximum occurs at $x = x_0$, $y = y_0$. We could verify that this point is a maximum by taking second derivatives, but that is obvious here by inspecting the function.   ∎

The real problem now is to make stationary not just any arbitrary function, but the integral I as given by Eq. (3.7). The quantity I is different from a regular function as follows. A function $f(x_1, x_2, \ldots)$ takes as input a set of numbers $(x_1, x_2, \ldots)$, and gives back a number. The quantity I, however, takes as input *an entire function $y(x)$*, and gives back a single number. Take the function $f(x) = x^2$, for example: if $x = 3$, then $f(3) = 9$ – one number in, one number out. But to calculate a value for an integral $I = \int_a^b F(x, y(x), y'(x)) \, dx$ with (say) $F(x, y(x), y'(x)) = y(x)^2$ and $a = 0, b = 1$, we have to substitute into it not a single number, but an entire path $y(x)$. If, for example, $y(x) = 5x$ (hence with the boundary conditions $y(0) = 0$ and $y(1) = 5$), we would write

$$I = \int_0^1 (5x)^2 dx = \frac{25}{3}x^3 \Big|_0^1 = \frac{25}{3}. \tag{3.14}$$

The argument for I is then a path, an entire function $y(x)$. To make this explicit, we instead write I as

$$I[y(x)] = \int_{x_a}^{x_b} F(x, y(x), y'(x)) \, dx \tag{3.15}$$

with square brackets around the argument: I is not a function, but is called a **functional**, traditionally said to be a function of a function.

In general, a functional may take as argument several functions, not just one. But for now let us focus on the case of a functional depending on a single function. The question we want to address is then: how do we make such a functional stationary? This means we are looking for conditions that identify a set of paths $y(x)$ for which the functional $I[y(x)]$ is stationary or "locally horizontal." To do this, we can build upon the simpler example of making stationary a function. For any path $y(x)$, we look at a shifted path

$$y(x) \rightarrow y(x) + \delta y(x), \tag{3.16}$$

where $\delta y(x)$ is a function that is small everywhere, but is otherwise arbitrary. However, we require that at the endpoints of the integration in Eq. (3.15), the shifts vanish; *i.e.*, $\delta y(a) = \delta y(b) = 0$. This means that we do not perturb the boundary conditions on trial paths that are fed into $I[y(x)]$, because we only need to find the path that makes stationary the functional amongst the subset of all possible paths that *satisfy* the given fixed boundary conditions at the endpoints. We illustrate this in Figure 3.3. In this restricted set of trial paths, our functional extremization condition now looks very much like (3.11):

$$\delta I[y(x)] \equiv I[y(x) + \delta y(x)] - I[y(x)] = 0, \tag{3.17}$$

to linear order in $\delta y(x)$. We say: "the variation of the functional I is zero." For a function $f(x_1, x_2, \ldots)$, the condition amounted to setting all first derivatives of f to zero. Hence, we need to figure out how to differentiate a functional! Alternatively, we need to expand the functional $I[y(x) + \delta y(x)]$ in $\delta y(x)$ to linear order to identify its "first derivative."

Fortunately, we can deduce all the operations of functional calculus by thinking of a functional in the following way. Imagine that the input to the functional, the path $y(x)$, is evaluated only on a finite discrete set of points:

$$a < x < b \rightarrow x = a + n\epsilon \leq b, \tag{3.18}$$

for n a non-negative integer and ϵ small (see Figure 3.4). In the limit $\epsilon \rightarrow 0$ with the upper bound n going to ∞, we recover the original continuum problem.

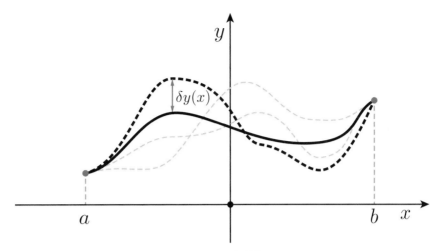

Fig. 3.3 A path $y(x)$ that can be used as input to the functional $I[f(x)]$. We look for that special path from which an arbitrary small displacement $\delta y(x)$ leaves the functional unchanged to linear order in $\delta y(x)$. Note that $\delta y(a) = \delta y(b) = 0$.

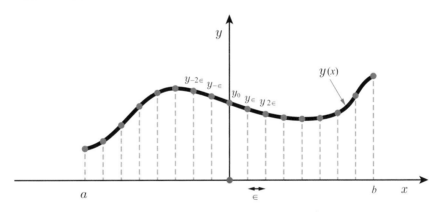

Fig. 3.4 The discretization of a smooth path. In functional calculus, the functional $y(x)$ can be treated as a collection of discrete points.

The functional is simply a function of a finite number of variables $y(a), y(a + \epsilon)$, $y(a + 2\epsilon), \ldots$. In the limit $\epsilon \to 0$, the set becomes infinitely dense. One can therefore view a functional as a function of an *infinite* number of variables. We can perform all needed operations on I in the discretized regime where I is treated as a function, and then take the $\epsilon \to 0$ limit at the end of the day.

Effectively, we may think of x in $y(x)$ as a discrete index y_x. We then have $I[y(x)] \to I(y_x)$, a function with a large but finite number of variables y_x, with $x \in \{a, \ldots, b\}$ a finite set. A functional then becomes a much more familiar animal: a function. The integral I may also depend upon $y'(x)$, which can be written in discretized form as $y'(x) \to (y_x - y_{x-\epsilon})/\epsilon$ by the definition of the derivative operation. We write it in shorthand as $y'(x) \to y_x'$, and the integration in Eq. (3.15) becomes a sum: $\int dx \rightarrow \sum_x \epsilon$. To summarize, we have a discretized form of our original functional:

$$I = \sum_x F(x, y_x, y_x') \, \epsilon. \tag{3.19}$$

We can now apply the shifts $y_x \rightarrow y_x + \delta y_x$, which also implies $y_x' \rightarrow y_x' + \delta y_x'$, where $\delta y_x' = (\delta y_x - (\delta y)_{x-\epsilon})/\epsilon = d(\delta y_x)/dx$. We then need the analogue of Eq. (3.11), or

$$\delta f = \frac{\partial f}{\partial x_i} \delta x_i = 0, \tag{3.20}$$

with $f \rightarrow I$ and $x_i \rightarrow y_x$. Starting from Eq. (3.19), we have

$$\delta I = \sum_x \left(\frac{\partial F}{\partial y_x} \delta y_x + \frac{\partial F}{\partial y_x'} \delta y_x' \right) \epsilon = 0. \tag{3.21}$$

In the $\epsilon \rightarrow 0$ limit, we retrieve the integral form

$$\delta I[y(x)] = \int_{x_a}^{x_b} \left(\frac{\partial F}{\partial y(x)} \delta y(x) + \frac{\partial F}{\partial y'(x)} \frac{d}{dx} (\delta y(x)) \right) dx = 0. \tag{3.22}$$

Integrating the second term by parts, we get

$$\int_{x_a}^{x_b} \frac{\partial F}{\partial y'(x)} \frac{d}{dx} (\delta y(x)) = \delta y(x) \frac{\partial F}{\partial y'(x)} \Big|_{x_a}^{x_b} - \int_{x_a}^{x_b} \delta y(x) \frac{d}{dx} \left(\frac{\partial F}{\partial y'(x)} \right) dx, \tag{3.23}$$

where the first term on the right vanishes because we have fixed the endpoints so that $\delta y(a) = \delta y(b) = 0$. Therefore, Eq. (3.22) becomes

$$\delta I[y(x)] = \int_{x_a}^{x_b} \left(\frac{\partial F}{\partial y(x)} - \frac{d}{dx} \left(\frac{\partial F}{\partial y'(x)} \right) \right) \delta y(x) \, dx = 0. \tag{3.24}$$

This integral might be zero because the integrand is zero for all x, or because there are positive and negative portions that cancel one another out. However, since *arbitrary* smooth deviation functions $\delta y(x)$ are permitted, the first alternative has to be the right one. For example, if $a < x_0 < b$ and the integrand happens to be positive from a to x_0 and negative from x_0 to b, so that by cancellation the overall integral is zero, the deviation function $\delta y(x)$ could be changed so that $\delta y(x) = 0$ from x_0 to b, which would force the integral to be positive. Therefore, the requirement that the integral vanishes for *arbitrary* smooth functions $\delta y(x)$ requires that

$$\frac{\partial F}{\partial y(x)} - \frac{d}{dx} \left(\frac{\partial F}{\partial y'(x)} \right) = 0, \tag{3.25}$$

which is known as **Euler's equation**. This equation was worked out by both Euler and Lagrange at around the same time, but we will call it simply "Euler's equation" because we reserve the term "Lagrange equations" for essentially the same equation when used in classical mechanics, as we shall see in Chapter 4.

Example 3.3 **The Straight Line**

Consider the problem of finding the shortest distance between two points on a two-dimensional plane. We need to minimize the expression

$$s = \int ds = \int \sqrt{dx^2 + dy^2} = \int_{x_a}^{x_b} \sqrt{1 + \left(\frac{dy}{dx}\right)^2}\, dx \equiv \int_{x_a}^{x_b} \sqrt{1 + y'^2}\, dx \qquad (3.26)$$

using Eq. (3.4) with $n(x) = $ constant. In that case, $F = \sqrt{1 + y'^2}$ in Eq. (3.7). Note that the integrand does not depend upon either x or $y(x)$ explicitly, so $\partial F/\partial y = 0$. Euler's equation (3.25) then becomes simply

$$\frac{d}{dx}\left(\frac{\partial F}{\partial y'}\right) = 0, \qquad (3.27)$$

so that

$$\frac{\partial F}{\partial y'} = \frac{y'}{\sqrt{1 + (y')^2}} = k, \qquad (3.28)$$

where k is a constant. Solving for y':

$$y' = \frac{\pm k}{\sqrt{1 - k^2}} \equiv m_1, \qquad (3.29)$$

which defines the constant m_1 in terms of the constant k. The integral of this equation is $y = m_1 x + m_2$, where m_2 is a constant of integration. That is, the shortest distance on a plane between two points is a straight line (!), where the slope m_1 and y-intercept m_2 may be found by requiring the line to pass through the endpoints $a = (x_a, y_a(x_a))$ and $b = (x_b, y_b(x_b))$.

Using the calculus of variations we have shown that among all smooth paths it is a straight line that makes the distance stationary. In this case stationary means minimum, because all nearby paths are longer. We showed earlier that minimizing the travel time of a light ray moving from a to b through a vacuum is equivalent to minimizing the distance traveled, so we have now also (no surprise) found that the minimum-time path for a light ray traveling in vacuum is a straight line.   ∎

Note the following two important features of Euler's equation (3.25).

- The derivatives with respect to y and y' are *partial*, but the derivative with respect to x is *total*. Suppose, for example, that $F(x, y(x), y'(x)) = xy(y')^2$. Then $\partial F/\partial y = (y')^2 x$ and $\partial F/\partial y' = 2xyy'$, so Euler's equation becomes

$$x(y')^2 - \frac{d}{dx}(2xyy') = x(y')^2 - [2yy' + 2x(y')^2 + 2xyy''] = 0. \qquad (3.30)$$

This is an ordinary differential equation whose solution $y(x)$ is the path we are looking for. That is, in the calculus of variations, *Euler's equation converts the problem of finding which path makes a particular integral stationary into a differential equation whose solution gives the path we want.*

- The variables x and y in Euler's equation do not have to represent Cartesian coordinates. The mathematics has no idea what x and y represent, as long as they are independent of one another. So if an integral I has the form of Eq. (3.15), but with x and y replaced by different symbols, the corresponding Euler's equation still holds. The total derivative occurring in the equation is always with respect to whatever variable of integration is chosen in the problem, which is called the independent variable. For example, if the integral to be made stationary has the form

$$I[q(t)] = \int F(t, q(t), q'(t))dt, \tag{3.31}$$

then the corresponding Euler equation is

$$\frac{\partial F}{\partial q(t)} - \frac{d}{dt}\left(\frac{\partial F}{\partial q'(t)}\right) = 0. \tag{3.32}$$

The variable t is therefore the *independent* variable, while $q(t)$ is referred to as the *dependent* variable, and $q'(t) \equiv dq/dt$.

3.3 Geodesics

The calculus of variations is best learned through examples. Let us proceed with a sequence of explicit cases. One application is to find **geodesics**, which are the stationary (usually shortest) paths between two points on a given surface.

Consider the problem of finding the shortest distance between two points on the surface of a sphere. We can use the polar angle θ and the azimuthal angle φ as the coordinates on a sphere (see Figure 3.5). If R is the radius of the sphere, an

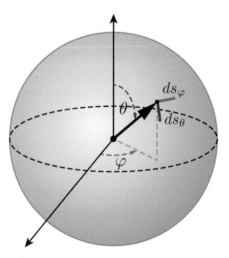

Fig. 3.5   The coordinates θ and φ on a sphere.

infinitesimal distance in the θ direction is $ds_\theta = R\,d\theta$ and an infinitesimal distance in the φ direction is $ds_\varphi = R\sin\theta d\varphi$. These two distances are perpendicular to one another, so the distance squared between any two nearby points is the sum of squares:

$$ds^2 = R^2 d\theta^2 + R^2 \sin^2\theta d\varphi^2. \tag{3.33}$$

There are two ways to write the total distance between two points, depending upon whether we use φ or θ as the variable of integration. If we use φ, then

$$s = R\int_a^b \sqrt{\theta'^2 + \sin^2\theta}\,d\varphi, \tag{3.34}$$

where $\theta' = d\theta/d\varphi$. The corresponding Euler equation is

$$\frac{\partial F}{\partial \theta} - \frac{d}{d\varphi}\frac{\partial F}{\partial \theta'} = 0, \tag{3.35}$$

where $F = \sqrt{\theta'^2 + \sin^2\theta}$. Alternatively, we can write

$$s = R\int_a^b \sqrt{1 + \sin^2\theta\,\varphi'^2}\,d\theta, \tag{3.36}$$

where $\varphi' = d\varphi/d\theta$ with the corresponding Euler equation

$$\frac{\partial F}{\partial \varphi} - \frac{d}{d\theta}\frac{\partial F}{\partial \varphi'} = 0, \tag{3.37}$$

and where in this case $F = \sqrt{1 + \sin^2\theta\,\varphi'^2}$. Both Euler equations are correct. Is one easier to use than the other?

Example 3.4 **Geodesics on a Sphere**

In the first alternative, Eq. (3.35) results in a second-order differential equation, since the first term $\partial F/\partial\theta \neq 0$ and by the time all the derivatives are taken the second term includes a second derivative θ''. The second alternative, Eq. (3.37), is much easier to use, because in that case $F = \sqrt{1 + \sin^2\theta\,\varphi'^2}$ is not an explicit function of φ, so the first term in Euler's equation vanishes. The quantity $\partial F/\partial\varphi'$ must therefore be constant in θ, since its total derivative is zero. This leaves us with only a first-order differential equation

$$\frac{\partial F}{\partial\varphi'} = \frac{\sin^2\theta\,\varphi'}{\sqrt{1 + \sin^2\theta\varphi'^2}} = k \tag{3.38}$$

for some constant k. This can be solved for φ' and rearranged to give

$$\varphi' = \pm\frac{k\csc^2\theta}{\sqrt{1 - k^2\csc^2\theta}}. \tag{3.39}$$

Then using the identity $\csc^2\theta = 1 + \cot^2\theta$ and substituting $q = \alpha\cot\theta$, where $\alpha = k/\sqrt{1-k^2}$, we have

$$\varphi = \alpha\int\frac{dq}{\sqrt{1-q^2}} = \alpha\sin^{-1}q + \beta,\qquad(3.40)$$

where $\alpha = \pm(\sqrt{1-k^2})/k$ and β is a constant of integration. Therefore the variables θ and φ are related by

$$\sin(\varphi - \beta) = q = \alpha\cot\theta.\qquad(3.41)$$

We can better understand the meaning of this result by multiplying through by $R\cos\theta$ and using the identity $\sin(\varphi - \beta) = \sin\varphi\cos\beta - \cos\varphi\sin\beta$, which gives

$$(\cos\beta)y - (\sin\beta)x = \alpha z,\qquad(3.42)$$

where $x = R\sin\theta\cos\varphi$, $y = R\sin\theta\sin\varphi$, and $z = R\cos\theta$, which are the Cartesian coordinates on the sphere. Equation (3.42) is the equation of a plane passing through the center of the sphere, which slices through the sphere in a **great circle**. So we have found that the solutions of Euler's equation are great-circle routes, as illustrated in Figure 3.6(a).

Unless one endpoint is at the antipode of the other, there is a shorter distance and a longer distance along the great circle that connects them. The shorter distance is a minimum path length under small deviations in path, as is well known by airline pilots. The larger distance is a stationary path that is neither a minimum nor a maximum under all small deviations in path. Paths that oscillate around this path are generally longer than the great-circle route, while some paths pulled to one side of the great-circle route are shorter. Both kinds are sketched in Figure 3.6(b). This behavior is typical of stationary paths that are neither absolute maxima nor absolute minima relative to all neighboring paths: some neighboring paths lead to smaller values and others lead to larger values of the integral I. In this case the set of all such paths represents a kind of saddle.

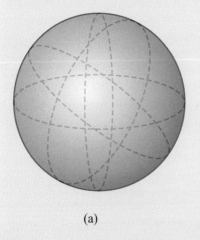

(a)

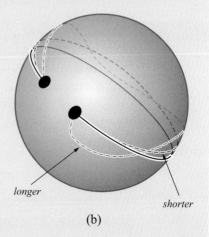

longer

shorter

(b)

Fig. 3.6 (a) Great circles on a sphere are geodesics. (b) Two paths nearby the longer of the two great-circle routes.   ∎

3.4 Brachistochrone

The **brachistochrone** ("shortest time") problem was invented and solved a half century *before* the work of Euler and Lagrange, and engaged some of the most creative people in the history of physics and mathematics. The problem is to find the shape of a track between two given points, such that a small block starting at rest at the upper point – and sliding without friction down along the track under the influence of gravity – arrives at the lower point in the shortest time. The two points *a* and *b*, and shapes of possible tracks between them, are illustrated in Figure 3.7.

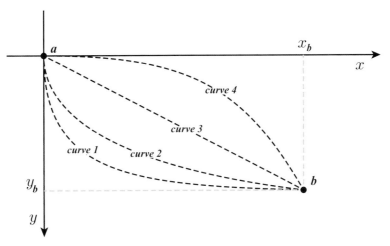

Fig. 3.7   Possible least-time paths for a sliding block.

We can guess the qualitative shape of the shortest-time track by physical reasoning. Of the four curves shown in Figure 3.7, it might seem that the straight line 3 is the shortest-time path, since it is the path of shortest distance. However, curve 2 has an advantage in that the block picks up speed more quickly, so that its greater average speed may more than make up for the greater distance it has to travel. Curve 1 permits the block to pick up speed still faster, but there is a risk that the slightly increased average speed might not outweigh the greater distance involved. There is no reason to choose curve 4, because a block will hardly get going in the first place and it also has to travel relatively far. A track whose shape is something like curve 2 should be the best choice.

To find the exact shape we choose coordinates as shown in Figure 3.7, with the origin at the release point, the positive y axis extending *downward*, and the final point designated by (x_b, y_b). The time to travel over a short distance is the distance divided by the speed, so the overall time is

$$t = \int \frac{ds}{v}, \tag{3.43}$$

where v is the varying speed of the block. The infinitesimal distance is again $ds = \sqrt{dx^2 + dy^2}$. Since v changes in general along the track, we need to express it in terms of the coordinates x and y to make sense of the integral. For this, we have energy conservation, which gives

$$E = \frac{1}{2}mv^2 + mg(-y) = 0, \tag{3.44}$$

since y and v are both zero initially. (We have used $-y$ in the potential energy because we are measuring y positive *downward*; *i.e.*, the potential $-mgy$ decreases for larger values of y.) For any given path, the time for the block to slide from beginning to end can be expressed either as

$$t = \int \frac{\sqrt{1 + y'^2}}{\sqrt{2gy}}\, dx, \tag{3.45}$$

where $y' = dy/dx$, or as

$$t = \int \frac{\sqrt{1 + x'^2}}{\sqrt{2gy}}\, dy, \tag{3.46}$$

where $x' = dx/dy$. The Euler equation for the latter expression is

$$\frac{\partial F}{\partial x} - \frac{d}{dy}\left(\frac{\partial F}{\partial x'}\right) = 0, \tag{3.47}$$

which is the right one to use, because F is not an explicit function of x, so the first term vanishes. Therefore

$$\frac{\partial F}{\partial x'} = \frac{1}{\sqrt{2gy}}\frac{x'}{\sqrt{1 + x'^2}} = k, \tag{3.48}$$

a constant. Solving for x':

$$x' = \frac{\pm k\sqrt{2gy}}{\sqrt{1 - 2k^2 gy}} \equiv \sqrt{\frac{y}{a - y}}, \tag{3.49}$$

where we have chosen the plus sign and defined $a \equiv 1/(2k^2 g)$. Integrating over y:

$$x = \int dx = \int dy \sqrt{\frac{y}{a - y}}, \tag{3.50}$$

which can be evaluated using the substitution

$$y = a \sin^2 \frac{\theta}{2} = \frac{a}{2}(1 - \cos\theta), \tag{3.51}$$

giving the result $x = (a/2)(\theta - \sin\theta)$, where we have chosen the constant of integration so that $x = 0$ when $y = 0$ (at $\theta = 0$), which is the release point. The resulting parametric equations

$$x = \frac{a}{2}(\theta - \sin\theta) \tag{3.52}$$

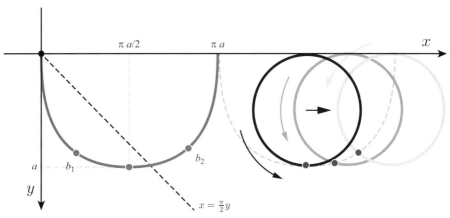

Fig. 3.8 A graph of a cycloid. If in darkness you watch a wheel rolling along a level surface, with a lighted bulb attached to a point on the outer rim of the wheel, the bulb will trace out the shape of a cycloid. In the diagram, the wheel is rolling along horizontally. For $x_b < (\pi/2)y_b$, the track may look like the segment from a to b_1; for $x_b > (\pi/2)y_b$, the segment from a to b_2 would be needed. Note that the size of the cycloids changes with a, which in turn depends on (x_b, y_b).

and

$$y = \frac{a}{2}(1 - \cos\theta) \tag{3.53}$$

are the equations of a *cycloid*, as shown in Figure 3.8. The quantities a and the final angle parameter θ_b can be determined from the coordinates (x_b, y_b) of the final point, although this ordinarily requires the solution of a transcendental equation. Only the first cycle of the cycloid is needed; if $x_b < (\pi/2)y_b$, the minimum-time path is a piece of the left half of the cycle, as shown in Figure 3.8 (the segment from point a to point b_1); if $x_b > (\pi/2)y_b$, the right half of the cycle is needed as well (the segment from a to b_2). That is, if $x_b > (\pi/2)y_b$, the sliding particle actually descends below y_b, and then comes up to meet y_b at the end. In either case the particle begins by falling vertically when it leaves the origin, to get the maximum possible initial acceleration. The cycloid has a vertical cusp at this point. The time required to fall to the final point can be found by returning to Eq. (3.46) and expressing x and y in terms of the parameter θ, according to Eqs. (3.52) and (3.53). The result is simply

$$t = \sqrt{\frac{a}{2g}} \int_0^{\theta_f} d\theta = \sqrt{\frac{a}{2g}}\, \theta_f. \tag{3.54}$$

In particular, if $(x_b, y_b) = (\pi a/2, a)$, so that a complete half-cycle of the cycloid is needed to connect the points, then $\theta_f = \pi$ and

$$t = \pi \sqrt{\frac{a}{2g}}. \tag{3.55}$$

This is the time it would take a particle to slide from the rim to the bottom of a smooth cycloidal bowl, where a is the depth of the bowl.[1]

Example 3.5

Fermat Again

We return to where we began the chapter, with Fermat's principle of stationary time, illustrated in Figure 3.9(a). Bringing to bear the calculus of variations, we can now find the path of a light ray in a medium like earth's atmosphere, where the index of refraction n is a continuous function of position. If a ray of light from a star descends through the atmosphere, it encounters an increasing density and an increasing index of refraction. We might therefore expect the ray to bend continuously, entering the atmosphere at some angle θ_a and reaching the ground at a steeper angle θ_b. For simplicity, take the earth to be essentially flat over the horizontal range of the ray and assume the index of refraction $n = n(y)$ only, where y is the vertical direction. The light travel time is then

$$t = \frac{1}{c} \int n(y) \sqrt{1 + y'^2}\, dx = \frac{1}{c} \int n(y) \sqrt{1 + x'^2}\, dy. \tag{3.56}$$

That is, we can use either x or y as the variable of integration, whichever is more convenient. Here it is more convenient to use y, because in that case Euler's equation is

$$\frac{\partial F}{\partial x} - \frac{d}{dy}\left(\frac{\partial F}{\partial x'}\right) = 0 \tag{3.57}$$

in which $\partial F / \partial x = 0$, so

$$\frac{d}{dy}\left(\frac{\partial F}{\partial x'}\right) = 0 \tag{3.58}$$

and

$$\frac{\partial F}{\partial x'} = \frac{n(y)\, x'}{\sqrt{1 + x'^2}} = k, \tag{3.59}$$

---

[1] A bit of history: on the afternoon of January 29, 1697, Sir Isaac Newton, who had left Cambridge the previous year to become Warden of the Mint in London, returned to his London home from a hard day at the Mint to find a letter from the Swiss mathematician Johann Bernoulli. The letter contained the unsolved brachistochrone problem, published the previous June. A challenge had gone forth to mathematicians to solve the problem, and they were given a time limit of 6 months to find the solution. Gottfried Wilhelm Leibniz, German mathematician and archrival of Newton for recognition as the original inventor of calculus, solved the problem but asked that the deadline be extended by an additional year so that everyone would have a chance to try it. Bernoulli agreed. Although presented as a general challenge, Bernoulli specifically sent the problem to Newton, who had not seen it before, to alert him to the problem and to try to stump him, thereby showing that he did not really understand calculus as well as the continental mathematicians. Newton's niece, Catherine Barton, was living with him in London at the time. She later testified that "Sr I. N. was in the midst of the hurry of the great recoinage and did not come home till four from the Tower very much tired, but did not sleep till he had solved it wch was by 4 in the morning." Newton sent off the solution that same morning to the Royal Society, and it was published anonymously in the February issue of *Philosophical Transactions*. Bernoulli had no doubt who was responsible, and wrote to a friend that it was "ex ungue Leonum" – "from the claws of the Lion." Aside from Newton, Leibniz, and Johann Bernoulli himself, the brachistochrone problem was solved by only two other mathematicians at that time, Bernoulli's older brother Jacob and the French mathematician de l'Hôpital. All of the solutions were ad hoc, involving algorithms suited to the particular problem, but not necessarily easily generalizable to a wider class of problems.

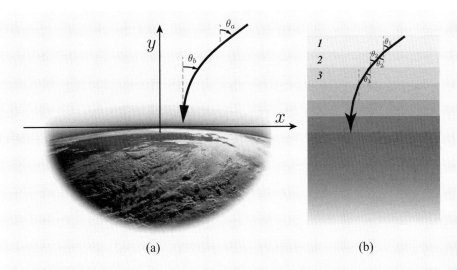

Fig. 3.9 (a) A light ray passing through a stack of atmospheric layers. (b) The same problem visualized as a sequence of adjacent slabs of air of different index of refraction.

a constant. The derivative $x' = dx/dy = \tan\theta(y)$, where $\theta(y)$ is the local angle of the ray relative to the vertical, so the quantity $x'/\sqrt{1+x'^2} = \sin\theta(y)$. Therefore

$$n(y)\sin\theta(y) = k, \tag{3.60}$$

a constant everywhere along the path. This result could also have been obtained immediately from Snell's law, by modeling the atmosphere as a large number of thin horizontal layers, where n is constant within each layer, but with n increasing slightly as one passes from one layer to the layer just beneath it. Snell's law is obeyed at each boundary: for example, $n_1\sin\theta_1 = n_2\sin\theta_2$ as shown in Figure 3.1. However, the angle θ_2 at which the ray enters layer 2 is the same angle at which the ray leaves layer 2 at the boundary with layer 3 (see Figure 3.9(b)). Therefore also $n_2\sin\theta_2 = n_3\sin\theta_3$, etc., so in the stack of layers it follows that $n(y)\sin\theta = $ constant. In the limit where the stack approaches an infinite number of layers of infinitesimal thickness, we get Eq. (3.60). Given a function $n(y)$, we can then find the specific path shape $y(x)$ from $\theta(y)$ (see the Problems section at the end of this chapter).

Note that the constancy of $n\sin\theta$ allows us to predict the ray angle θ_b at the ground without knowing the detailed index of refraction $n(y)$ or the path of the ray! If we know the indices of refraction at the top of the atmosphere n_a and at the ground n_b, and the angle at which the ray enters the atmosphere θ_a (from the true location of the star), we can find the angle at the ground θ_b – which is the angle at which a telescope would observe the star – as

$$n_a\sin\theta_a = n_b\sin\theta_b = \text{constant.} \tag{3.61}$$

∎

3.5 Several Dependent Variables

We have so far considered problems with one independent variable (such as x) and one dependent variable (such as $y(x)$). There are many additional problems that require two or more dependent variables. For example, to find the shortest-distance path between two given points in three-dimensional space, we would need both y and z as well as x to describe an arbitrary path. Consider the more general functional

$$I[t, y_i(x), y_i'(x)] = \int_{x_a}^{x_b} F(x,\ y_1(x), \ldots, y_N(x),\ y_1'(x), \ldots, y_N'(x))\, dx \qquad (3.62)$$

with $y_1(x), y_2(x), \ldots, y_N(x)$. We then have N dependent variables. The goal is to make I stationary under variations in *all* of the functions $y_i(x)$ with $i = 1, 2, \ldots, N$. In the preceding section, the single function $y(x)$ could be visualized as a path in the two-dimensional x, y space; in the more general case, the N functions $y_i(x)$ can be visualized as together defining a path in an $(N + 1)$-dimensional space, with axes $x, y_1, y_2, \ldots, y_N$. For example, the distance between the two points (x_a, y_a, z_a) and (x_b, y_b, z_b) in three dimensions is

$$s = \int ds = \int \sqrt{dx^2 + dy^2 + dz^2} = \int_{x_a}^{x_b} \sqrt{1 + y'^2 + z'^2}\, dx \qquad (3.63)$$

along a path described by $y(x)$ and $z(x)$, restricted to pass through the given endpoints. The three-dimensional path that minimizes s is a problem in the calculus of variations, and the integral is a simple case of the form written in Eq. (3.62). Analogues to the Euler equations can readily be found in the $(N + 1)$-dimensional case. Let the shift in the paths now be

$$y_i(x) \rightarrow y_i(x) + \delta y_i(x) \qquad (i = 1, \ldots, N). \qquad (3.64)$$

Therefore the functions $\delta y_i(x)$ describe the deviations of the arbitrary path $y_i(x)$. Looking back at Eq. (3.22), we note that the only difference is that we simply have more than one function on which I depends. We can then immediately extend Eq. (3.22) to

$$\delta I[y(x)] = \int_a^b \left(\frac{\partial F}{\partial y_i(x)} \delta y_i(x) + \frac{\partial F}{\partial y_i'(x)} \frac{d}{dx}(\delta y_i(x)) \right) dx = 0, \qquad (3.65)$$

where the index i is repeated and hence summed over. Applying the same technique of integration by parts for every i:

$$\int_a^b \frac{\partial F}{\partial y_i'(x)} \frac{d}{dx}(\delta y_i(x)) = \delta y_i(x) \frac{\partial F}{\partial y_i'(x)} \bigg|_a^b$$

$$- \int_a^b \delta y_i(x) \frac{d}{dx}\left(\frac{\partial F}{\partial y_i'(x)} \right) dx, \qquad (3.66)$$

we find again that the first term on the right vanishes because $\delta y_i(x_a) = \delta y_i(x_b) = 0$ by construction. Therefore, Eq. (3.65) becomes

$$\delta I[y(x)] = \int_a^b \left(\frac{\partial F}{\partial y_i(x)} - \frac{d}{dx}\left(\frac{\partial F}{\partial y_i'(x)} \right) \right) \delta y_i(x) dx = 0, \qquad (3.67)$$

from which we get N copies of the original Euler equation:

$$\frac{\partial F}{\partial y_i(x)} - \frac{d}{dx}\left(\frac{\partial F}{\partial y_i'(x)} \right) = 0 \quad \text{for } i = 1, \ldots, N. \qquad (3.68)$$

Note that, to deduce Eq. (3.68) from Eq. (3.67), we need to be sure that the y_is are independent variables so that the variations δy_i are also independent. We are now equipped to handle variational problems involving more than one dependent function.

Example 3.6

Geodesics in Three Dimensions

From Eq. (3.63), the setup for the problem of finding geodesics in three dimensions, we have $F = \sqrt{1 + y'^2 + z'^2}$, choosing x as the independent variable. Therefore, we use Eq. (3.68) with $N = 2$ and find

$$\frac{\partial F}{\partial y} - \frac{d}{dx}\frac{\partial F}{\partial y'} = 0 \quad \text{and} \quad \frac{\partial F}{\partial z} - \frac{d}{dx}\frac{\partial F}{\partial z'} = 0, \qquad (3.69)$$

which reduce to

$$\frac{\partial F}{\partial y'} = \frac{y'}{\sqrt{1 + y'^2 + z'^2}} = k_1 \quad \text{and} \quad \frac{\partial F}{\partial z'} = \frac{z'}{\sqrt{1 + y'^2 + z'^2}} = k_2, \qquad (3.70)$$

where k_1 and k_2 are constants. The equations can be decoupled by taking the sum of the squares of the two equations to show that the denominator of each equation is constant, or equivalently $y'^2 + z'^2 = $ constant, so from the equations it follows that y' and z' must themselves each be constants. Therefore the minimum-length path has constant slope in both the x–y and x–z planes, corresponding to a straight line, as expected. The constants can be determined by requiring the line to pass through the given endpoints. ∎

3.6 Mechanics from a Variational Principle

We now want to ask whether there is a *general* formulation of *mechanics* that is based entirely on a variational principle. This may be true, in part because both Newton's second law and variational principles can lead to second-order differential equations. Perhaps we can cast any classical mechanics problem in the form of a statement about finding the stationary paths of some functional.

Motivated by the examples already explored, a natural starting point is to extremize *travel time*. We start with the case of a free relativistic particle, and require the formalism to be Lorentz invariant from the outset. After all, the variational principle – if it is to lead to a law of physics – should look the same in all inertial frames. This immediately tells us to write the simple candidate functional

$$I = \int d\tau, \tag{3.71}$$

which is simply the proper time for a particle to travel between two fixed points in spacetime. We propose that extremizing this quantity leads to the trajectory of a free relativistic particle, equivalently described by

$$\frac{d}{dt}(\gamma m\mathbf{v}) = 0 \tag{3.72}$$

from Eq. (2.120).

Armed with the techniques developed in the previous sections, we can check whether this statement is correct. We write the functional in terms of the coordinate system of some inertial observer $\mathbb{O}$ using coordinates (ct, x, y, z):

$$I = \int d\tau = \int \frac{dt}{\gamma} = \int dt \sqrt{1 - \frac{\dot{x}^2 + \dot{y}^2 + \dot{z}^2}{c^2}}, \tag{3.73}$$

where we used the time-dilation relation $dt = \gamma \, d\tau$. We need to determine three functions $x(t)$, $y(t)$, and $z(t)$ that extremize the functional I whose independent variable is t. We can imagine that the endpoints of the trajectory are fixed, and so we have a familiar variational problem. We can then use Euler's Eqs. (3.68) with

$$F = \sqrt{1 - \frac{\dot{x}^2}{c^2} - \frac{\dot{y}^2}{c^2} - \frac{\dot{z}^2}{c^2}} \tag{3.74}$$

and $N = 3$. We have three equations

$$\frac{\partial F}{\partial x} - \frac{d}{dt}\frac{\partial F}{\partial \dot{x}} = 0, \quad \frac{\partial F}{\partial y} - \frac{d}{dt}\frac{\partial F}{\partial \dot{y}} = 0, \quad \frac{\partial F}{\partial z} - \frac{d}{dt}\frac{\partial F}{\partial \dot{z}} = 0, \tag{3.75}$$

from which it is straightforward to show that

$$\frac{d}{dt}(\gamma\dot{x}) = 0, \quad \frac{d}{dt}(\gamma\dot{y}) = 0, \quad \frac{d}{dt}(\gamma\dot{z}) = 0, \tag{3.76}$$

which are consistent with Eq. (3.72). This is already very promising: we can describe a free relativistic particle by extremizing the particle's proper time.

Let us next look at the low-velocity regime of our functional. We write Eq. (3.73) in an expanded form for $\beta = v/c \ll 1$:

$$I \simeq \int dt \left(1 - \frac{1}{2}\frac{v^2}{c^2} + \cdots\right). \tag{3.77}$$

The first term is a constant and so does not affect a variational principle: Euler's equations involve derivatives of F and hence constant terms in F may safely be dropped. The second term is quadratic in the velocity. We rewrite the functional as

$$I \to \int dt \left(\frac{1}{2}m\left(\dot{x}^2 + \dot{y}^2 + \dot{z}^2\right)\right). \tag{3.78}$$

In addition to dropping the constant shift term, we have also multiplied I from Eq. (3.77) by $-m c^2$ for convenience. This is a multiplication by a constant and hence, once again, does not affect the Euler equations (3.68). It makes things a little more suggestive, however: we are now extremizing the particle's nonrelativistic kinetic energy. If we now use Euler's equations (3.68) with $F = (1/2)mv^2$, we get the familiar three differential equations

$$\frac{d}{dt}(m\mathbf{v}) = 0, \tag{3.79}$$

as expected for a nonrelativistic particle. So far so good. *We have the expected results for free particles.* But how about problems that involve forces?

3.7 Motion in a Uniform Gravitational Field

Shortly after developing his special theory of relativity, Einstein saw a beautiful way to understand the effect of uniform gravitational forces, which he called the **principle of equivalence**. He later said that it was "the happiest thought of my life," because it was a wonderfully simple but powerful idea that became a crucial stepping stone to achieving his relativistic theory of gravity: **general relativity**.

The equivalence principle can be illustrated by experiments carried out in two spaceships, one accelerating uniformly in gravity-free empty space and one standing at rest in a uniform gravitational field, as shown in Figure 3.10. The acceleration a of the first ship is adjusted to be numerically equal, but opposite in direction, to the gravitational field g acting on the second ship. The equivalence

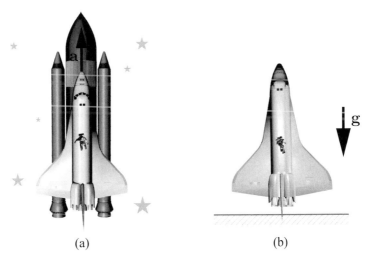

(a)   (b)

Fig. 3.10 Two spaceships, one accelerating in gravity-free space (a) and the other at rest on the ground (b). Neither observers in the accelerating ship nor those in the ship at rest on the ground can find out which ship they are in on the basis of any experiments carried out solely within their ship.

ship's bow

ship's stern

Fig. 3.11   A laser beam travels from the bow to the stern of the accelerating ship.

principle then claims that if observers in either one of the ships carry out any experiment whatever that is confined entirely within their own ship, the results cannot be used to determine which ship they are living in: the two situations are *equivalent*. This is a statement inspired by observation – dating back to Galileo's Pisa tower experiment equating inertial and **gravitational masses** – which Einstein then elevated to the stature of a principle of Nature.

We use the principle here to deduce two related effects of gravity that are not contained in Newton's theory: the gravitational frequency shift and the effect of gravity on the rate of clocks. We start by considering a particular thought experiment with light waves. An observer in the bow of the accelerating ship shines a laser beam at another observer in the stern of the ship, as shown in Figure 3.11. The experiment starts with the ship initially at rest. The laser emits monochromatic light of frequency ν_{em} in the rest frame of the laser. We assume that the distance traveled by the ship while the beam is traveling is very small compared with the length h of the ship, so that the time it takes for the beam to reach the stern is essentially $t = h/c$.

During this time the stern attains an additional small velocity $v = at = ah/c$. This velocity is small compared with the speed of light, so the ship suffers no appreciable length contraction.[2] The stern observer is moving toward the source, so will observe a *blueshift* due to the Doppler effect. The nonrelativistic Doppler formula is given by Eq. (2.133), approximated for small V as

$$\nu_{\text{ob}} = \nu_{\text{em}} \sqrt{\frac{1 + \frac{V}{c}}{1 - \frac{V}{c}}} \simeq \nu_{\text{em}} \left(1 + \frac{V}{c} \right) = \nu_{\text{em}} \left(1 + \frac{ah}{c^2} \right), \qquad (3.80)$$

and relates the observed frequency ν_{ob} to the emitted frequency ν_{em}.[3]

---

[2] More precisely, the ship's length contraction would scale as V^2/c^2. The physical effect we focus on arises from the Doppler shift, which is linear in V/c.

[3] We used the binomial expansion to derive this result (see Appendix F).

Now according to the equivalence principle, the same result will be observed in the ship at rest in a uniform gravitational field if we substitute the acceleration of gravity g for the rocket acceleration a. That is, if the observer at the top of the stationary ship shines light with emitted frequency ν_{em} toward the observer at the bottom, the bottom observer will see a blueshifted frequency

$$\nu_{\text{ob}} = \nu_{\text{em}}(1 + gy/c^2), \tag{3.81}$$

where now we have used the symbol y for the altitude of the upper clock above the lower clock. It is also true that the upper observer will see a redshift if he or she looks at a light beam sent off by the lower observer. In neither case can we blame the shift on Doppler, however, because neither observer is moving. Instead, the shift in this case is due to a difference in altitude of the two clocks at rest in a uniform gravitational field.

How can we *explain* the blueshift seen by the person in the stern, or bottom, of the stationary ship? If we think of the laser atoms that radiate light at the top as atomic clocks whose rate is indicated by the frequency of their emitted light, the observer at the bottom will be forced to conclude that these top clocks are running *fast* compared to similar clocks at the bottom of the ship. For suppose a clock at the top of the ship has a luminous second hand that emits light of frequency ν_{em}. In a time $t = 1$ s, the hand emits $t/\text{period} = t\nu_{\text{em}} = 1$ s $\cdot \nu_{\text{em}}$ wavelengths of light. The observer at the bottom must collect all these wavelengths, since none of them is created or destroyed in transmission. However, the frequency of the waves observed at the bottom is increased by the factor $(1 + gy/c^2)$, as shown in Eq. (3.81), which means that the observer at the bottom will collect all of these waves in less than 1 s according to her own clock:

$$t'\nu_{\text{ob}} = 1 \text{ s} \cdot \nu_{\text{em}} \Rightarrow t' = 1 \text{ s} \cdot \left(1 + \frac{gy}{c^2}\right)^{-1} < 1 \text{ s}. \tag{3.82}$$

That is, the second hand of the clock at the top appears to advance by 1 s in *less* than one second to the observer at the bottom, by the exact same factor. The observer at the top agrees with this judgment. This upper observer sees a redshift when looking at clocks at the bottom, so it is natural for a person at the top to believe that bottom clocks run slower than top clocks.

If such atomic clocks at high altitude run faster, it must be true that *all* clocks up high run faster, because they can be continuously compared with one another. And if all stationary clocks at high altitude run fast compared with all stationary clocks at lower altitude, we can conclude that time itself runs faster at higher altitude. That is, for time intervals Δt:

$$\Delta t_{\text{high}} = \Delta t_{\text{low}}(1 + gy/c^2). \tag{3.83}$$

This is the time difference for two clocks at rest, but at different altitudes, in a uniform gravitational field. Now let us expand the problem a bit by considering *two* clocks at ground level, clock A at the origin $(x_A, y_A) = (0,0)$ and the other, clock B, a distance x_B away at $(x, y) = (x_B, 0)$. The two clocks have been synchronized

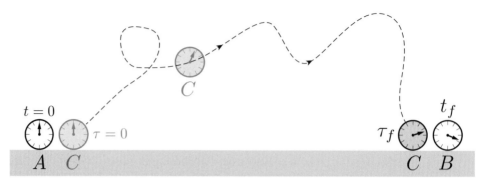

Fig. 3.12 Clock C travels by an arbitrary path between stationary ground clocks A and B. The left-hand side shows C leaving A when both clocks read zero. The right-hand side shows C arriving at B when B reads t_f and C reads τ_f. C *must* arrive at B when B's clock reads t_f, but C's reading τ_f will depend upon what path C takes in moving from A to B.

with one another, say by exploding a flashbulb halfway between them, and then observers right beside each clock setting their clocks to $t = 0$ when the flash arrives. Once synchronized, these two clocks remain synchronized because they remain in the same rest frame and at the same altitude.

Now we picture a third clock C that starts out at ground level with $(t, x, y) = (0, 0, 0)$, right beside ground clock A. It initially reads the same time $t = 0$ as ground clock A, and then moves by any path it likes, up and down, back and forth, but it must finally end up right beside ground clock B, when B has $(t, x, y) = (t_f, x_B, 0)$, where t_f stands for the final ground time when C arrives, as illustrated in Figure 3.12. The time interval for the trip according to the two ground clocks has therefore been $\Delta t = t_f$, since clock A initially read $t = 0$.

Clock C ends up moving the net horizontal distance $x_B - 0 = x_B$, but its time interval will most likely be different than t_f, both because it has moved with a nonzero velocity (and will therefore have run slow due to time dilation) and will likely have explored various altitudes during its trip (and so will have run fast due to the altitude effect on clocks as derived above). Therefore clock C ends up with $(t, x, y) = (\tau_f, x_B, 0)$, where τ_f is the final time it reads. Our goal is to find this final time τ_f read by clock C, which will depend upon what path it takes on its journey from A to B, and then find what particular path makes τ_f *extreme* for a given, fixed value of t_f, the reading on clock B when C arrives. Of course τ_f is just the total proper time clock C gains in following a particular path from the origin at A to the destination at B.

In an infinitesimal time dt according to either ground clock, clock C advances by the proper time

$$d\tau = dt(1 + gy/c^2)\sqrt{1 - v^2/c^2}, \tag{3.84}$$

with factors showing that it runs fast due to its altitude and slow due to its speed. For a nonrelativistic particle moving near earth's surface, both gh/c^2 and v^2/c^2 are very small, so

$$d\tau \cong dt(1 + gy/c^2)(1 - v^2/2c^2) \cong dt(1 + gy/c^2 - v^2/2c^2) \qquad (3.85)$$

using the binomial expansion $(1 + z)^n \simeq (1 + nz)$ to obtain the first expression and neglecting the product of two very small quantities to obtain the second expression. Therefore, as clocks A and B advance from time $t = 0$ to the later time t_f, with these approximations clock C advances by

$$\tau_f = \int_0^{t_f} dt(1 + gy/c^2 - v^2/2c^2). \qquad (3.86)$$

The Hafele–Keating Experiments

Both clock effects, the slowing due to velocity and the speeding up due to altitude, have been tested many times, in different situations and using different kinds of clocks. In 1971, J. C. Hafele and R. E. Keating tested both effects using human-made atomic clocks. They began by assembling a set of four atomic clocks which they carried with them on several consecutive commercial air flights traveling *eastward* around the world. As they traveled they would register the speed and altitude of the clocks as a function of time, using information provided by the various flight captains. They had assembled a comparable set of atomic clocks which remained in the laboratory at the U.S. Naval Observatory in Washington, D.C., which they could compare with the traveling clocks at the end of their trip. Of course, the stay-at-home clocks also moved throughout this time relative to an inertial frame, because the earth's surface steadily moved eastward due to the earth's rotation. For these two sets of clocks, the calculated net altitude (gravitational) effect, the net velocity effect, and the overall net effect were (in nanoseconds)

$$\begin{array}{ll} 144 \pm 14 & \text{altitude effect} \\ -184 \pm 18 & \text{velocity effect} \\ -40 \pm 23 & \text{net effect predicted.} \end{array} \qquad (3.87)$$

They could compare this net predicted effect with that of the stay-at-home clocks when they returned home. The observed difference in time was -59 ± 10 ns. So the predicted vs. observed time differences were the same within experimental error.

Then after a week of comparing clocks and recuperating, Hafele and Keating carried a comparable set of four atomic clocks *westward* around the world, again using regularly scheduled commercial flights. The data now looked as follows (in nanoseconds):

$$\begin{array}{ll} 179 \pm 18 & \text{altitude effect} \\ 96 \pm 10 & \text{velocity effect} \\ 275 \pm 21 & \text{net effect predicted.} \end{array} \qquad (3.88)$$

When they finally compared this net predicted effect with that of the stay-at-home clocks when they returned home, the observed difference in time was 273 ± 7 ns, in strong agreement with the calculated values for the traveling clocks. Note that their experiments tested both the equivalence principle and special relativistic time dilation.

A Reinterpretation

Now notice that if m is the mass of clock C, the overall time effect of Eq. (3.89) can also be written in the form

$$\tau_f = t_f - \frac{1}{mc^2} \int_0^{t_f} \left(\frac{1}{2}mv^2 - mgy \right) dt$$

$$= t_f - \frac{1}{mc^2} \int_0^{t_f} dt \, (T - U), \tag{3.89}$$

where we recognize that $T = (1/2)mv^2$ and $U = mgy$ are the kinetic and potential energies of clock C, if clocks A and B are at rest and have zero potential energy.

The value of τ_f depends not only upon the initial and final times 0 and t_f, but also upon the path the clock takes in getting from the beginning point to the endpoint. So looking at the problem of the three clocks in a uniform gravitational field, where clocks A and B are at rest on the ground and clock C has altitude $y(t)$ and moves with speed $v(t)$, we have shown that the proper time interval read by clock C, as it moves between the two given points while clocks A and B advance from time 0 to time t_f, is

$$\tau_f = t_f - \frac{1}{mc^2} \int_0^{t_f} dt \, (T - U), \tag{3.90}$$

where the integrand is now the *difference* between the kinetic and potential energies of the upper clock. Some of the possible paths of clock C are illustrated in the figure.

Let us now find that particular path of clock C which extremizes the proper time τ_f as it travels between the two given points, starting at fixed time $t = 0$ and ending at fixed time t_f according to clock B. We know how to do this. Extremizing τ here is the same as extremizing the functional

$$I \equiv \int_0^{t_f} \left(\frac{1}{2}mv^2 - mgy \right) dt, \tag{3.91}$$

with the integrand

$$F(x, \dot{x}, y, \dot{y}, z, \dot{z}) = \frac{1}{2}m(\dot{x}^2 + \dot{y}^2 + \dot{z}^2) - mgy, \tag{3.92}$$

since the t_f term is a constant. Euler's equations for the x, y, and z directions then give, respectively:

$$\ddot{x} = 0, \quad \ddot{y} = -g, \quad \text{and} \quad \ddot{z} = 0. \tag{3.93}$$

We immediately recognize these as the differential equations of motion resulting from Newton's second law $\mathbf{F} = m\mathbf{a}$ *for a particle in a uniform gravitational field.* The solutions of the equations are

$$x = v_{0x}t,$$

$$y = v_{0y}t - \frac{1}{2}gt^2,$$

$$z = 0, \tag{3.94}$$

where v_{0x} and v_{0y} are easily found by requiring that $(x, y) = (x_B, 0)$ at the final time t_f. Furthermore, if we use these equations of motion to find the shape $y(x)$ of the path, we find it is a parabola, as expected for trajectories in a uniform gravitational field.

Our goal of identifying a variational principle for the motion of a particle in a uniform gravitational field has been successful. Without ever using Newton's laws we found the correct equations of motion by using the calculus of variations. There remains the question: In finding the path that extremizes the proper time of the moving particle, as we have done, is the proper time in fact a *maximum, minimum,* or *saddle point* under local distortions in the path? This is left for a problem at the end of the chapter. There is also the question whether extremizing the proper time is still correct for *any* gravitational field, such as the inverse-square field around a spherically symmetric mass. That question we postpone until Chapter 10. Finally, the form of the functional we varied, as given in Eq. (3.91), namely the difference between the kinetic and potential energies of the particle, is highly suggestive, an idea worth exploring in the next section.

3.8 Arbitrary Potential Energies

We just reached the intriguing conclusion that the correct equations of motion for a nonrelativistic particle of mass m in a uniform gravitational field can be found by making stationary the functional

$$I = \int dt \left(\frac{1}{2}mv^2 - U \right) = \int dt\,(T - U), \tag{3.95}$$

where

$$T \equiv \frac{1}{2}mv^2 \tag{3.96}$$

is the particle's kinetic energy and

$$U = mgy \tag{3.97}$$

is its gravitational potential energy. It was the *difference* between the kinetic and gravitational potential energy that was needed in the integrand.

Can this approach be generalized? That is, suppose that a particle is subject to an *arbitrary* conservative force $\mathbf{F}$ for which a potential energy U can be defined, $\mathbf{F} = -\nabla U$. Does the form

$$I = \int dt \left(\frac{1}{2} m v^2 - U \right) = \int dt \, (T - U) \tag{3.98}$$

still work? Do we still get the correct $\mathbf{F} = m\mathbf{a}$ equations of motion?

Let us do a quick check, using Cartesian coordinates. Note that if $U = U(x,y,z)$ and $T = T(\dot{x},\dot{y},\dot{z})$, then the integrand in the variational problem, which we will denote by L, is

$$L(x,y,z,\dot{x},\dot{y},\dot{z}) \equiv T(\dot{x},\dot{y},\dot{z}) - U(x,y,z) = \frac{1}{2} m v^2 - U(x,y,z)$$

$$= \frac{1}{2} m \left(\dot{x}^2 + \dot{y}^2 + \dot{z}^2 \right) - U(x,y,z), \tag{3.99}$$

since $v^2 = \dot{x}^2 + \dot{y}^2 + \dot{z}^2$. Writing out the three associated Euler equations, we get the differential equations of motion

$$\frac{\partial L}{\partial x} - \frac{d}{dt} \frac{\partial L}{\partial \dot{x}} = 0, \qquad \frac{\partial L}{\partial y} - \frac{d}{dt} \frac{\partial L}{\partial \dot{y}} = 0, \qquad \frac{\partial L}{\partial z} - \frac{d}{dt} \frac{\partial L}{\partial \dot{z}} = 0, \tag{3.100}$$

where

$$\frac{\partial L}{\partial x} = -\frac{\partial U}{\partial x} = F_x \quad \text{and} \quad \frac{d}{dt} \frac{\partial L}{\partial \dot{x}} = \frac{d}{dt} m\dot{x} = m\ddot{x}, \quad \text{so} \quad F_x = m\ddot{x}, \tag{3.101}$$

with similar results in the y and z directions. That is, we have derived the three components of

$$-\nabla U = m\mathbf{a} \Rightarrow \mathbf{F} = m\mathbf{a} \tag{3.102}$$

as we hoped to find.

We have succeeded in showing that if all forces $\mathbf{F}$ in a problem are conservative, so they can be described by a potential energy function $U(x,y,z)$ with $\mathbf{F} = -\nabla U$, then Euler's equation with integrand $L \equiv T - U$ gives Newton's laws of motion. For *conservative* forces at least, and in Cartesian coordinates, Newton's laws are equivalent to a variational problem.

3.9 Summary

In this chapter we have shown that a variational principle – Fermat's principle of stationary time – can be used to find the paths of light rays. Such a variational principle seems totally unlike the approach of Newton to finding the paths of particles subject to forces. Yet we have shown that the associated calculus of variations or functional calculus allows us to convert the problem of making stationary a certain integral into a differential equation of motion. We applied these techniques to solve several interesting problems.

We then went on to show that the relativistic and nonrelativistic mechanics of a free particle can be understood from a variational principle, and extended that approach, using Einstein's principle of equivalence, to find the motion of nonrelativistic particles in uniform gravitational fields. The functional

$$I \equiv \int_{t_a}^{t_b} \left(\frac{1}{2}mv^2 - mgy \right) dt = \int_{t_a}^{t_b} (T - U) \, dt, \qquad (3.103)$$

where the integrand is the difference between the kinetic and gravitational potential energies of the particle, gives the correct differential equations of motion for a nonrelativistic particle.

Finally, we showed that an integrand of the form $L \equiv T - U$ gives the correct Newtonian equations of motion for *any* potential energy, *i.e.*, for mechanics problems in which all forces are conservative.

What is the meaning of this? Can the variational approach be generalized still further? Can we do something similar for *any* mechanics problem? One involving normal and tension forces? Or frictional forces? How about nonconservative forces in general, which do not have potentials? In short, can we always find the equations of motion of a particle through this program of extremizing an associated functional? These are questions for Chapter 4.

Problems

★ **Problem 3.1** Prove from Fermat's principle that the angles of incidence and reflection are equal for light bouncing off a mirror. Use neither algebra nor calculus in your proof! (*Hint*: The result was proven by Hero of Alexandria 2000 years ago.)

★ **Problem 3.2** An ideal converging lens focuses light from a point object onto a point image. Consider only rays that are straight lines except when crossing an air–glass boundary. Relative to the ray that passes straight through the center of the lens, do the other rays require more time, less time, or the same time to go from O to I? That is, in terms of Fermat's principle, is the central path a local minimum, a local maximum, or a stationary path that is neither a minimum nor a maximum?

★ **Problem 3.3** Light focuses onto a point I from a point O after reflecting off a surface that completely surrounds the two points. The shape of the surface is such that all rays leaving O (excepting the single ray which returns to O) reflect to I. (a) What is the shape of the surface? (b) Pick any one of the paths. Is it a path of minimum time, maximum time, or is it stationary but of neither minimum nor maximum time for all nearby paths?

★ **Problem 3.4** Consider the ray bouncing off the bottom of the surface in the preceding problem. Replace the surface at this point by an even more highly curved surface. The ray still bounces from O to I. Is the ray now a path of minimum time, maximum time, or is it stationary but of neither minimum nor maximum time? Compare with nearby paths that bounce once but are otherwise straight. Suppose the paths must bounce once but need not be segments of straight lines. What then?

★★   **Problem 3.5** When bouncing off a flat mirror, a light ray travels by a minimum-time path. (a) For what shape mirror would the paths of all bouncing light rays take equal times? (b) Is there a shape for which a bouncing ray would take a path of greatest time, relative to nearby paths?

★★   **Problem 3.6** A hypothetical object called a straight **cosmic string** (which may have been formed in the early universe and may persist today) makes the r, θ space around it conical. That is, set an infinite straight cosmic string along the z axis; the two-dimensional space perpendicular to this, measured by the polar coordinates r and θ, then has the geometry of a cone rather than a plane. Suppose there is a cosmic string between earth and a distant quasi-stellar object (QSO). What might we see when we look at this QSO? [Assume light travels in least-time paths (here also least-distance paths) relative to nearby paths.]

★★★   **Problem 3.7** Model earth's atmosphere as a spherical shell 100 km thick, with index of refraction $n_t = 1.00000$ at the top and $n_b = 1.00027$ at the bottom. Is a light ray's final angle φ_f relative to the normal at the ground greater or less than its initial angle φ_i relative to the normal at the top of the atmosphere? (Earth's radius is $R = 6400$ km.) Assume the ray strikes the upper atmosphere at a $45°$ angle.

★   **Problem 3.8** We seek to find the path $y(x)$ that minimizes the integral $I = \int f(x, y, y')dx$. Find Euler's equation for $y(x)$ for each of the following integrands f, and then find the solutions $y(x)$ of each of the resulting differential equations if the two endpoints are $(x, y) = (0, 1)$ and $(1, 3)$ in each case. (a) $f = ax + by + cy'^2$, (b) $f = ax^2 + by^2 + cy'^2$.

★★   **Problem 3.9** Find a differential equation obeyed by geodesics in a plane using polar coordinates r, θ. Integrate the equation and show that the solutions are straight lines.

★   **Problem 3.10** Find two first-order differential equations obeyed by geodesics in three-dimensional Euclidean space, using spherical coordinates r, θ, φ.

★★★   **Problem 3.11** Two-dimensional surfaces that can be made by rolling up a sheet of paper are called *developable* surfaces. Find the geodesic equations on the following developable surfaces and solve the equations. (a) A circular cylinder of radius R, using coordinates θ and z. (b) A circular cone of half-angle α (which is the angle between the cone and the axis of symmetry) using coordinates θ and ℓ, where ℓ is the distance of a point on the cone from the apex. *Hint*: Find the distance ds between nearby points on the surface in terms of $\ell, \alpha, d\theta$, and $d\ell$.

★★   **Problem 3.12** A torus can be defined by two radii: a large radius R running around the center of the torus, and a small radius r corresponding to a cross-sectional slice. Let R live in the x, y plane. Then if φ is an angle relative to the x axis and lying in the x, y plane, and θ is an angle within a cross-sectional slice, with $\theta = 0$ corresponding to the outermost radius of the torus $R + r$, then the Cartesian coordinates of points on the torus are

$$x = (R + r\cos\theta)\sin\varphi, \quad y = (R + r\cos\theta)\cos\varphi, \quad z = r\sin\theta.$$

(a) Find an expression for the distance ds between nearby points on the torus, using the angles φ and θ as coordinates. (b) Find a second-order differential equation for geodesics on the torus in terms of θ, θ', and θ'', where $\theta' = d\theta/d\varphi$, etc. (c) Show that paths with constant $\theta = 0$ or with constant $\theta = \pi$ are geodesics, but that a path with constant $\theta = \pi/2$ is *not* a geodesic.

Problem 3.13 Using Euler's equation for $y(x)$, prove that

$$\frac{\partial f}{\partial x} - \frac{d}{dx}\left(f - y'\frac{\partial f}{\partial y'}\right) = 0.$$

This equation provides an alternative method for solving problems in which the integrand f is not an explicit function of x, because in that case the quantity $f - y'\partial f/\partial y'$ is constant, which is only a first-order differential equation.

Problem 3.14 A line and two points not on the line are drawn in a plane. A smooth curve is drawn between the two points and then rotated about the given line. This is known as a surface of revolution. Find the shape of the curve that minimizes the area generated by the rotated curve. A lampshade manufacturer might use this result to minimize the material used to produce a lampshade of given upper and lower radii.

Problem 3.15 The time required for a particle to slide from the cusp of a cycloid to the bottom was shown in Section 3.4 to be $t = \pi\sqrt{a/2g}$. Show that if the particle starts from rest at any point *other* than the cusp, it will take this same length of time to reach the bottom. The cycloid is therefore also the solution of the *tautochrone*, or "equal-time" problem. *Hint*: The energy equation for the particle speed in terms of y written in Section 3.4 must be modified to take into account the new starting condition. [The tautochrone result was known to the author Herman Melville. In the chapter entitled "The try-works" in *Moby Dick*, the narrator Ishmael, on board the whaling ship Pequod, describes the great try-pots used for boiling whale blubber: "Sometimes they are polished with soapstone and sand, till they shine within like silver punchbowls. ... It was in the lefthand try-pot of the Pequod, with the soapstone diligently circling around me, that I was first indirectly struck by the remarkable fact, that in geometry all bodies gliding along the cycloid, my soapstone for example, will descend from any point in precisely the same time."]

Problem 3.16 Derive Snell's law from Fermat's principle.

Problem 3.17 A lifeguard is standing on the beach some distance from the shoreline, when he hears a swimmer calling for help. The swimmer is some distance offshore and also some lateral distance from the lifeguard. The lifeguard knows he can run twice as fast as he can swim. To minimize the time it takes to reach the swimmer, show that his path should consist of two line segments: relative to the shoreline, his running path should be at angle θ_1 and his swimming path should be at angle θ_2, where $\cos\theta_1 = 2\cos\theta_2$.

★   **Problem 3.18** Describe the geodesics on a right-circular cylinder. That is, given two arbitrary points on the surface of a cylinder, what is the shape of the path of minimum length between them, where the path is confined to the surface? *Hint:* A cylinder can be made by rolling up a sheet of paper.

★★   **Problem 3.19** A particle falls along a cycloidal path from the origin to the final point $(x,y) = (\pi a/2, a)$; the time required is $\pi\sqrt{a/2g}$, as shown in Section 3.4. How long would it take the particle to slide along a straight-line path between the same points? Express the time for the straight-line path in the form $t_{\text{straight}} = k t_{\text{cycloid}}$, and find the numerical factor k.

★   **Problem 3.20** A unique transport system is built between two stations 1 km apart on the surface of the moon. A tunnel in the shape of a full cycloid cycle is dug, and the tunnel is lined with a frictionless material. If mail is dropped into the tube at one station, how much later (in seconds) does it appear at the other station? How deep is the lowest point of the tunnel? (Gravity on the moon is about 1/6th that on earth.)

★   **Problem 3.21** A hollow glass tube is bent into the form of a slightly tilted rectangle, with rounded corners. Two small ball bearings are introduced into the tubes at the upper left, at the highest point of the rectangle. One rolls clockwise and the other counterclockwise down to the opposite corner at the bottom right, the lowest point of the rectangle. The balls are started out simultaneously from rest, and note that each ball must roll the same distance to reach the destination. The question is: Which ball reaches the lower corner first, or do they arrive simultaneously? Why?

★   **Problem 3.22** Assume earth's atmosphere is essentially flat, with index of refraction $n = 1$ at the top and $n = n(y)$ below, with y measured from the top, and the positive y direction downward. Suppose also that $n^2(y) = 1 + \alpha y$, where α is a positive constant. Find the light-ray trajectory $x(y)$ in this case.

★★   **Problem 3.23** Suppose that earth's atmosphere is as described in the preceding problem, except that $n^2(y) = 1 + \alpha y + \beta y^2$, where α and β are positive constants. Find the light-ray trajectory $x(y)$ in this case.

★★   **Problem 3.24** Consider earth's atmosphere to be spherically symmetric above the surface, with index of refraction $n = n(r)$, where r is measured from the center of the earth. Using polar coordinates r, θ to describe the trajectory of a light ray entering the atmosphere from high altitudes, (a) find a first-order differential equation in the variables r and θ that governs the ray trajectory; (b) show that $n(r) r \sin \varphi = $ constant along the ray, where φ is the angle between the ray and a radial line extending outward from the center of the earth. This is the analogue of the equation $n(y) \sin \theta = $ constant for a flat atmosphere.

★   **Problem 3.25** Using the result found in the preceding problem, and supposing that $n^2(r) = 1 + \alpha/r^2$ (where α is a constant), find the light-ray trajectory expressed either as $r(\theta)$ or $\theta(r)$.

Problem 3.26 (a) Show that the pressure difference between two points in an incompressible liquid of density ρ in static equilibrium is $\Delta P = \rho g s$, where s is the vertical separation between the two points and g is the local gravitational field. (b) The liquid is caused to flow through a horizontal pipe of varying cross-sectional area, so that its velocity depends upon position. In a particular section of pipe of length s, the pipe is narrowing, so that the fluid's acceleration has the constant value a. Find the pressure difference ΔP between one end of the section and the other, in terms of ρ and the change in the velocity squared (v^2) between the two ends of the section. Is the pressure larger or smaller at the narrower end of the section? (The result is an example of the *Bernoulli effect*.)

Problem 3.27 The surface of a paraboloid of revolution is defined by $z = a(x^2 + y^2)$, where a is a constant. Find the differential equation for a geodesic originating at a point $(x, y) = (x_0, 0)$ with slope $(dy/dx)_0 = 0$. Does the geodesic return to the same point?

Problem 3.28 According to Einstein's general theory of relativity, light rays are deflected as they pass by a massive object like the sun. The trajectory of a ray influenced by a central, spherically symmetric object of mass M lies in a plane with coordinates r and θ (so-called *Schwarzschild coordinates*); the trajectory must be a solution of the differential equation

$$\frac{d^2u}{d\theta^2} + u = \frac{3GM}{c^2}u^2,$$

where $u = 1/r$, G is Newton's gravitational constant, and c is the constant speed of light. (a) The right-hand side of this equation is ordinarily small. In fact, the ratio of the right-hand side to the second term on the left is $3GM/rc^2$. Find the numerical value of this ratio at the surface of the sun. The sun's mass is 2.0×10^{30} kg and its radius is 7×10^5 km. (b) If the right-hand side of the equation is neglected, show that the trajectory is a straight line. (c) The effects of the term on the right-hand side have been observed. It is known that light bends slightly as it passes by the sun and that the observed deflection agrees with the value calculated from the equation. Near a black hole, which may have a mass comparable to that of the sun but a much smaller radius, the right-hand side becomes very important, and there can be large deflections. In fact, show that there is a single radius at which the trajectory of light is a circle orbiting the black hole, and find the radius r of this circle.

Problem 3.29 A clock is thrown straight upward on an airless planet with uniform gravity g, and it falls back to the surface at a time t_f after it was thrown, according to clocks at rest on the ground. (a) Using the clock's motion as derived in Section 3.7, how much more time than t_f will have elapsed according to this moving clock, in terms of g, t_f, and c, the speed of light? (b) Now suppose that instead of the freely falling motion used in part (a), the moving clock has constant speed v_0 straight up for time $t_f/2$ according to ground clocks, and then moves straight down again at the same constant speed v_0 for another time interval $t_f/2$, according to ground clocks. How much more time than t_f will have elapsed according to this moving clock, in terms of v_0, g, c, and t_f? (c) Now find the value of v_0, keeping g and t_f fixed, which maximizes the final reading of the moving clock described in part (b). Then evaluate the final reading of this moving clock in terms of g, t_f, and c, and show that it is *less than* the final reading of the freely falling clock described in part (a). [This is a particular illustration of the fact that the path which *maximizes* the proper time is that of a *freely falling* clock, *i.e.*, a clock that moves according to Newton's laws. The reader could choose some alternative motion for a clock, and show again that as long as it returns to the beginning point at t_f according to ground clocks, its time will be less than that of the freely falling clock of part (a).]

★ **Problem 3.30** A skyscraper elevator comes equipped with two weighing scales: the first is a typical bathroom scale containing springs that compress when someone stands on it, and the second is the type often used in doctor's offices, where weights are adjusted to balance that of the patient. (a) A rider enters the elevator at the ground floor and stands on the first scale; it reads 150 lbs. Use the principle of equivalence to answer the following questions. (i) As the elevator accelerates upward, will the scale read less than, more than, or equal to 150 lbs? (ii) When the elevator reaches its maximum speed and continues rising at this speed, will the scale read less than, more than, or equal to 150 lbs? (iii) And as the elevator comes to rest at the top floor, what will it read? (b) The rider repeats the experiment, standing this time on the second scale. What will it read during each portion of the trip?

★ **Problem 3.31** A laser is aimed horizontally near the earth's surface, a distance y_0 above the ground; a pulse of light is then emitted. (a) How far will the pulse fall by the time it has traveled a distance L? (b) What is the value of L if the pulse falls by 0.1 nm, roughly the diameter of a hydrogen atom?

★ **Problem 3.32** Note that in the Hafele–Keating experiments the total error in the eastward and westward flights was comparable, ± 23 and ± 21 ns, respectively, but that the *percentage* error was much greater for the eastward flights. (a) What is the reason for that? What is the lesson one might draw for other experiments or theoretical calculations? (b) Note also that in the calculated differences between the traveling and stay-at-home clocks, the special-relativistic (velocity) effect is negative for the eastward flights and positive for the westward flights. Why was that?

★ **Problem 3.33** A hypothetical planet has an equatorial circumference of 40,000 km, a gravity $g = 10$ m/s^2, and completes one revolution every 24 h. Aircraft A circles eastward around the equator at constant altitude 10 km, while aircraft B circles westward around the equator, both at altitude 10 km, except for the brief takeoffs and landings, each requiring 40 h to make the trip from home base back to home base. Atomic clocks are carried on both planes and others are left at home. What are the calculated differences between the traveling and stay-at-home clocks due to (a) altitude effects, (b) velocity (special-relativistic) effects, and (c) the net predicted effect, both for clocks carried on A and on B?

4 Lagrangian Mechanics

A notable advance in mechanics took place nearly a century after Newton in the work of the French mathematician and physicist Joseph-Louis Lagrange (1736–1813). Lagrange invented no new fundamental physics, but rather looked at mechanics in a very different way, using variational techniques that gave fresh insight and provided powerful methods for finding differential equations of motion.

We already introduced variational principles in Chapter 3, and showed that they can give the equations of motion of nonrelativistic particles subject to arbitrary *conservative* forces. Although useful, that is not enough: now we need to see if other kinds of forces can be included in the variational approach. In this chapter we will find that forces of constraint can be included as well, which provides us with deep insights into mechanics and also enormous simplifications in problem solving. We will introduce Hamilton's principle and the Lagrangian, concepts that are so elegant we are encouraged to place them at the very heart of classical mechanics. We are further encouraged to do so in the following chapter, the capstone chapter to Part I of the book, where we show how they naturally emerge as we take the classical limit of the vastly more comprehensive theory of quantum mechanics.

4.1 Nonconservative Forces

In our initial exploration at the end of Chapter 3 we permitted only conservative forces, forces that can be derived from a potential energy function. Now what about *nonconservative* forces, like air resistance and sliding friction? These forces typically reduce the energy of a particle. For example, the force of friction $\mathbf{F}_f$ acting on a block sliding along a tabletop typically does negative work on the block, $dW = \mathbf{F}_f \cdot ds < 0$, since the direction of the frictional force is opposite to that of the block's displacement.

In the elegant techniques we will develop in this chapter, we will assume that no such nonconservative forces act on our system. This will not be a serious problem, however; nonconservative forces are typically *macroscopic* forces, caused by the interactions of our system with vast numbers of molecules in the air or on a surface; that is, nonconservative forces are typically not fundamental. It is *possible* to include such forces in the variational methods of this chapter, but it is seldom worth it. If friction is important in a particular problem, it is safest to revert to using $\mathbf{F} = m\mathbf{a}$ and forget the elegance. If we reject nonconservative forces, what is left?

There is another important type of force which is *neither* conservative nor nonconservative, as we shall now see.

4.2 Forces of Constraint and Generalized Coordinates

Suppose that a particular particle is free to move in all three dimensions, so three coordinates are needed to specify its position. The coordinates might be Cartesian (x, y, z), cylindrical (ρ, φ, z), spherical (r, θ, ϕ), as illustrated in Figure 4.1, or they might be any other complete set of three (not necessarily orthogonal) coordinates.[1]

A different particle may be less free: it might be constrained to move on a tabletop, or along a wire, or within the confines of a closed box, for example. The constraint is enforced by a *force of constraint*. It is a normal force F_N of the tabletop on a particle that keeps the particle from falling to the ground, and it is likewise the tension force F_T in a supporting wire that ensures that a pendulum bob moves in a circular arc as it swings back and forth. Such constraint forces are neither conservative nor nonconservative. They are not conservative, because their magnitudes are not functions of position only. For example, the tension in a pendulum wire can depend not only upon the angle of the pendulum relative to the vertical, but also upon how fast the bob is moving at that point, which depends on how fast the bob is pushed before it is released to swing on its own. Also, if (as is often the case) the force of constraint F_C is always perpendicular to the displacement of the particle δs, then F_C does no work, and so neither adds nor subtracts from the energy of the particle.

Sometimes the presence of a *constraint* means that fewer than three coordinates are required to specify the position of the particle. So if the particle is restricted to slide on the surface of a table, for example, only two coordinates are needed. Or if the particle is a bead sliding along a frictionless hoop, we need only a single coordinate, say the angle denoting the location of the bead along the hoop. In contrast, if the particle is confined to move within a closed three-dimensional box, the constraint does *not* reduce the number of coordinates required: we still need three coordinates to specify the position of the particle inside the box.

A constraint that reduces the number of coordinates needed to specify the position of a particle is called a **holonomic constraint**. The requirement that a particle move anywhere on a tabletop is a holonomic constraint, for example, because the minimum set of required coordinates is lowered from three to two, from (say) (x, y, z) to (x, y). The requirement that a bead move on a wire in the shape of a hoop is a holonomic constraint, because the minimum set of required

---

[1] Note that in spherical coordinates the radius r is the distance from the origin, while in cylindrical coordinates ρ is the distance from the vertical (z) axis.

coordinates is lowered from three to one, from (say) cylindrical coordinates (ρ, φ, z) to just φ. The requirement that a particle remains within a closed box is **nonholonomic**, because a requirement that $x_1 \leq x \leq x_2, y_1 \leq y \leq y_2, z_1 \leq z \leq z_2$ does not reduce the number of coordinates required to locate the particle.

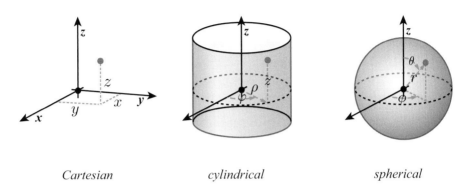

Cartesian   *cylindrical*   *spherical*

Fig. 4.1   Cartesian, cylindrical, and spherical coordinates.

For an unconstrained single particle, three coordinates are needed; or if there is a holonomic constraint, the number of coordinates is reduced to two or one. We call a *minimal* set of required independent coordinates **generalized coordinates** and denote them by q_k, where $k = 1$ up to 3. For each generalized coordinate there is a generalized velocity $\dot{q}_k = dq_k/dt$. For example, for a bead on a fixed hoop, one choice of generalized coordinates consists of the single coordinate $q_1 = \varphi$, the angle at which the bead is located along the hoop. The three cylindrical coordinates ρ, φ, and z are not generalized coordinates in this case: there are *two* constraint relations amongst the three coordinates which assure that the bead stays along the hoop – i.e., $r =$ constant and $z =$ constant. Note that a generalized velocity does not necessarily have the dimensions of length/time, just as a generalized coordinate does not necessarily have the dimensions of *length*. For example, the angle φ in cylindrical coordinates is dimensionless, and its generalized velocity $\dot{\varphi}$ has dimensions of inverse time.

4.3 Hamilton's Principle

The quantity

$$L = T - U \tag{4.1}$$

is called the **Lagrangian** of the particle. As we just saw at the end of Chapter 3, using the Lagrangian as the integrand in the variational problem gives us the correct equations of motion, at least in Cartesian coordinates, for *any* conservative force.

We now have an interesting proposal at hand: reformulate the equations of motion of nonrelativistic mechanics, $\mathbf{F} = d\mathbf{p}/dt$, in terms of a variational principle making stationary a certain functional. This has two benefits:

1. It is an interesting idea to think of dynamics as arising from making a certain physical quantity stationary; we will appreciate some of these aspects in due time, especially when we get to the chapters on the connections between classical and quantum mechanics.
2. This reformulation provides powerful computational tools that can allow us to solve complex mechanics problems with greater ease. The formalism also lends itself more transparently to computer algorithms.

Having chosen a set of generalized coordinates q_k for a particle, the integrand L in the variational problem, the **Lagrangian**, can be written

$$L = T - U = L(t, q_1, q_2, \ldots, \dot{q}_1, \dot{q}_2, \ldots) = L(t, q_k, \dot{q}_k), \qquad (4.2)$$

written in terms of the generalized coordinates, generalized velocities, and time.[2]

Given a mechanical system described by means of N dynamical generalized coordinates labeled $q_k(t)$, with $k = 1, 2, \ldots, N$, we define its **action** $S[q_k(t)]$ as the functional of the time integral over the Lagrangian L, from a starting time t_a to an ending time t_b:

$$S[q_k(t)] = \int_{t_a}^{t_b} dt\, L(t, q_1, q_2, \ldots, \dot{q}_1, \dot{q}_2, \ldots) \equiv \int_{t_a}^{t_b} dt\, L(t, q_k, \dot{q}_k). \qquad (4.3)$$

It is understood that the particle begins at some definite position $(q_1, q_2, \ldots)_a$ at time t_a and ends at some definite position $(q_1, q_2, \ldots)_b$ at time t_b. We then propose that, for trajectories $q_k(t)$ where S is *stationary – i.e.*, when

$$\delta S = \delta \int_{t_a}^{t_b} L(t, q_k, \dot{q}_k)\, dt = 0, \qquad (4.4)$$

the $q_k(t)$s satisfy the equations of motion for the system with the prescribed boundary conditions at t_a and t_b. This proposal was first enunciated by the Irish mathematician and physicist William Rowan Hamilton (1805–1865), and is called **Hamilton's principle**.[3] From Hamilton's principle and our discussion of the previous chapter, we deduce the N **Lagrange equations**

$$\frac{d}{dt}\frac{\partial L}{\partial \dot{q}_k} - \frac{\partial L}{\partial q_k} = 0 \qquad (k = 1, 2, \ldots, N). \qquad (4.5)$$

---

[2] Here, we are assuming that the Lagrangian does not involve dependence on higher derivatives of q_k, such as $\ddot{q}_k$. It can be shown that such terms lead to differential equations of the third or higher orders in time (see the Problems section at the end of this chapter). Our goal is to reproduce traditional Newtonian and relativistic mechanics involving differential equations that are no higher than second order.

[3] It is also sometimes called the **principle of least action** or the **principle of stationary action**. This can be confusing, however, because there is an older principle called the "principle of least action" that is quite different.

These then have to be the equations of motion of the system if Hamilton's principle is correct. Note that we *need* to express the Lagrangian in terms of generalized coordinates q_k since the variational principle assumes that the perturbed variables in the functional are *independent*: if we were to use instead all the variables describing the system, there could be constraint relations amongst them implying that the corresponding variations in Eq. (3.67) (writing $\delta y \to \delta q$) are *not* independent and the corresponding Eqs. (3.68) or (4.5) do not then follow.

Consider a general physical system involving only conservative forces and a number of particles – constrained or otherwise. We are proposing that we can describe the dynamics of this system fully through Hamilton's principle, using the Lagrangian $L = T - U$, the difference between the total kinetic energy and the total potential energy of the system – *written in terms of generalized coordinates*. For a single particle under the influence of a conservative force, and described with Cartesian coordinates, we have already shown that this is indeed possible (see Eqs. (3.101) through (4.1)). The question is whether we can extend this new technology to more general situations with several particles, constraints, and described with arbitrary coordinate systems. There are several issues we need to tackle in this process:

1. Does changing the coordinate system in which we express the kinetic and potential energies generate any obstacles to the formalism? The answer to this is "*no*," since the functional we extremize – which is the action – is a **scalar quantity**: its value does not change under coordinate transformations $q_k \to q_k'$

$$S = \int dt\, L(t, q_k, \dot{q}_k) = \int dt\, L(t, q_k', \dot{q}_k'). \tag{4.6}$$

The coordinate change simply relabels the stationary path of the functional; that is, the path at the extremum transforms as $q_k^{sol}(t) \to q_k'^{\,sol}(t)$, where $q_k'^{\,sol}(t)$ is the stationary path of S expressed in the new coordinates. Hence, we can safely perform coordinate transformations as long as we always write the Lagrangian as kinetic minus potential energy in our preferred coordinate system.

2. Constraints provide for relations between the variables describing a mechanical system, and so reduce their number to a minimal set of generalized coordinates. This was the premise of the variational principle: the generalized coordinates must be *independent*. Therefore, no constraint would interfere with the variational principle as long as we express $L = T - U$ in terms of the generalized coordinates. But constraints on the coordinates are due to forces in the system that restrict the dynamics. For example, the normal force pushes upward to make sure a block stays on the floor; likewise, the tension force in a rope constrains the motion of a pendulum bob. Can we be certain that these forces should not be included in the potential energy U that appears in L? To ensure that this is the case, we need to ascertain that such **constraint forces** do no work, and hence do

not have any net energetic contribution to U. This is not always easy to see. We will demonstrate the mechanism at work through examples, and then identify the general strategy.

3. Should we expect any obstacles to the formalism when we have more than one particle? Do we simply add the kinetic and potential energies of all the particles? With two or more particles, shouldn't we worry about Newton's third law? We will see soon that the Lagrangian formalism incorporates Newton's third law through symmetries of the Lagrangian and indeed can handle many-particle systems very well.

4. Where is Newton's first law in this formalism? This is indeed an important potential pitfall: *one should always write the kinetic energy and potential energy in $L = T - U$ as seen from the perspective of an inertial observer*. This is because our contact with mechanics is through the reproduction of Newton's second law, $\mathbf{F} = m\mathbf{a}$ – which is valid only in an inertial frame. Hence, the first law is hidden in the prescription $L = T - U$, which must be expressed from an inertial perspective.

The punchline of all this is that, for arbitrarily complicated systems with many constraints and involving many particles interacting with **conservative forces**, the Lagrangian formalism works, and is extremely powerful. Newton's second law follows from Hamilton's principle, the third law arises, as we will see, from symmetries of the Lagrangian, and the first law is respected by making sure that the Lagrangian is written from the perspective of an inertial observer. It is important to emphasize that the Lagrangian formalism does *not* introduce new physics. It is a mathematical reformulation of mechanics, both nonrelativistic and relativistic. What it does is give us powerful new technical tools to tackle problems with greater ease and less work; a deeper insight into the laws of physics and how Nature ticks; and, as we shall see later, how the classical world is linked to the quantum realm.

Example 4.1   **A Simple Pendulum**

An inertial observer sees that a small plumb bob of mass m is free to swing back and forth in a vertical x–z plane at the end of a string of length R. The position of the bob can be specified uniquely by its angle θ measured up from its equilibrium position at the bottom, so we choose θ as the generalized coordinate, as illustrated in Figure 4.2. The bob's kinetic energy is

$$T = \frac{1}{2}mv^2 = \frac{1}{2}m\left(\dot{x}^2 + \dot{y}^2 + \dot{z}^2\right) = \frac{1}{2}m\left(\dot{r}^2 + r^2\dot{\theta}^2 + \dot{z}^2\right) = \frac{1}{2}m\left(R^2\dot{\theta}^2\right). \tag{4.7}$$

We have introduced polar coordinates r and θ in the vertical x–z plane, and then implemented the constraint equations $r = R$ (with $\dot{r} = 0$) and $y = 0$ (with $\dot{y} = 0$). Its potential energy is $U = mgh = mgR(1 - \cos\theta)$, measuring the bob's height h up from its lowest point. The Lagrangian of the bob is therefore

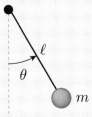

Fig. 4.2   The simple pendulum, with θ as the generalized coordinate.

$$L = T - U = \frac{1}{2}mR^2\dot{\theta}^2 - mgR(1 - \cos\theta). \qquad (4.8)$$

The constraint reduces the dynamics from three coordinates x, y, and z to only one, θ, a single *degree of freedom*. The single Euler equation in this case is

$$\frac{\partial L}{\partial \theta} - \frac{d}{dt}\frac{\partial L}{\partial \dot{\theta}} = -mgR\sin\theta - \frac{d}{dt}\left(mR^2\dot{\theta}\right) = 0, \qquad (4.9)$$

which reduces to the well-known "pendulum equation"

$$\ddot{\theta} + (g/R)\sin\theta = 0. \qquad (4.10)$$

Note that Eq. (4.9) (or (4.10)) is equivalent to $\tau = I\ddot{\theta}$, where the torque $\tau = -mgR\sin\theta$ is taken about the point of suspension (negative because it is opposite to the direction of increasing θ), and the moment of inertia of the bob is $I = mR^2$.

Note also two twists here. First, we switched from Cartesian to polar coordinates, which is no problem in the Lagrangian formalism, since the action is a scalar quantity. Second, we implemented constraints $r = R$, implying $\dot{r} = 0$, and $y = 0$, implying $\dot{y} = 0$. The former is responsible for holding the bob at fixed distance from the pivot, due to the **tension force** in the rope. The latter keeps the bob in the x–z plane. By implementing these constraints we reduced the problem from three to a single degree of freedom. There are two forces acting on the bob: gravity and the tension in the rope. The Lagrangian already accounts for the gravitational force through the gravitational potential energy. But what about the tension force responsible for the $r = R$ constraint? This tension in the rope does no work: it is always perpendicular to the motion of the bob, and hence the work contribution $\mathbf{T} \cdot d\mathbf{r} = 0$, where $\mathbf{T}$ is the tension force and $d\mathbf{r}$ is the displacement of the bob. Thus, our potential energy U – related to work done by a force – is simply the potential energy due to gravity alone, $U = mgh$. In general, whenever a contact force remains perpendicular to the displacement of the particle, it can safely be ignored in constructing the Lagrangian. Its impact on the problem comes instead through constraints that help identify a reduced set of independent generalized coordinates. By using a single degree of freedom in the Lagrangian, the general coordinate θ, we have taken care of the tension force as well – thus fully accounting for all the forces on the bob.   ∎

Example 4.2

A Bead Sliding on a Vertical Helix

A bead of mass m is slipped onto a *frictionless* wire wound in the shape of a helix of radius R, whose symmetry axis is oriented vertically in a uniform gravitational field, as shown in Figure 4.3. As always, we assume the description is from an inertial frame's perspective (unless explicitly stated otherwise). Using cylindrical coordinates ρ, φ, z, the base of the helix is located at $z = 0$, $\varphi = 0$, and the angle φ is related to the height z at any point by $\varphi = \alpha z$, where α is a constant with dimensions of inverse length. The gravitational potential energy of the bead is $U = mgz$, and its kinetic energy is $T = (1/2)mv^2 = (1/2)m[\dot\rho^2 + \rho^2\dot\varphi^2 + \dot z^2]$. However, the bead is constrained to move along the helix, so the bead's radius is constant at $\rho = R$, and (choosing z as the single generalized coordinate), $\dot\varphi = \alpha\dot z$. Therefore the kinetic energy of the bead is

$$T = \frac{1}{2}mv^2 = \frac{1}{2}m\left(\dot x^2 + \dot y^2 + \dot z^2\right) = \frac{1}{2}m\left(\dot\rho^2 + r^2\dot\varphi^2 + \dot z^2\right)$$

$$= (1/2)m[0 + \alpha^2R^2 + 1]\dot z^2, \tag{4.11}$$

where we switched to cylindrical coordinates first and then implemented the *two* constraints $\varphi = \alpha z$ and $\rho = R$. So the Lagrangian of the bead is

$$L = T - U = \frac{1}{2}m[1 + \alpha^2R^2]\dot z^2 - mgz \tag{4.12}$$

in terms of the single generalized coordinate z and its generalized velocity $\dot z$. Two constraints reduced the dynamics from three to only one degree of freedom. In Newtonian mechanics we need to take into account the normal force of the wire on the bead as one of the forces in $\mathbf{F} = m\mathbf{a}$; however, the normal force appears nowhere in the Lagrangian, because it does no work on the bead — it is always perpendicular to the bead's displacement. Once again, whenever a normal force is perpendicular to the displacement of a particle, we can safely ignore it in setting up the Lagrangian of the particle. In this simple case, with a single normal force, the simplification is not that impressive. However, in the more complicated scenarios we shall see later, the advantages of using the Lagrangian approach, in which the normal force never appears, will become more apparent. *Lagrange's method can save an enormous amount of effort.*

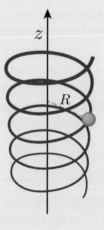

Fig. 4.3   A bead sliding on a vertically oriented helical wire.

Example 4.3 **Block on an Inclined Plane**

A block of mass m slides down a frictionless plane tilted at angle α to the horizontal, as shown in Figure 4.4. There is only a single degree of freedom here, say the distance X of the block from the bottom, measured along the incline. The gravitational potential energy is $mgh = mgX \sin \alpha$. Using X as the generalized coordinate, the velocity is $\dot{X}$, and the Lagrangian of the block is

$$L = T - U = \frac{1}{2}mv^2 - U = \frac{1}{2}m\dot{X}^2 - mgX \sin \alpha, \tag{4.13}$$

which depends explicitly upon the single coordinate (X) and the single velocity ($\dot{X}$). The Lagrange equation is

$$\frac{\partial L}{\partial X} - \frac{d}{dt}\frac{\partial L}{\partial \dot{X}} = -mg \sin \alpha - \frac{d}{dt}m\dot{X} = 0, \text{ or } -mg \sin \alpha = m\ddot{X}, \tag{4.14}$$

which is indeed the correct $\mathbf{F} = m\mathbf{a}$ equation for the block along the tilted plane.

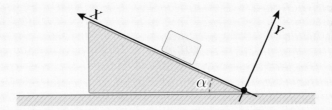

Fig. 4.4   Block sliding down an inclined plane.

In this problem, we judiciously chose our only degree of freedom as X, the distance along the inclined plane. If we had originally started with three coordinates, say X, Y, and Z, where Z is perpendicular to the page, we would then set $Z = 0 \Rightarrow \dot{Z} = 0$ to confine the motion to the plane of the paper; then, we would use the constraint $Y = 0 \Rightarrow \dot{Y} = 0$ since the block cannot sink into the inclined plane. We are then left with a single degree of freedom, the general coordinate X. The $Y = 0$ constraint is associated with the normal force that keeps the block on the incline. This normal force is always perpendicular to the displacement of the block along the incline and thus does no work. So by writing the Lagrangian in terms of a single generalized coordinate X we implicitly account for the normal force and only need to include the gravitational potential energy in the Lagrangian. ∎

4.4 Generalized Momenta and Cyclic Coordinates

In Cartesian coordinates the kinetic energy of a particle is $T = (1/2)m(\dot{x}^2 + \dot{y}^2 + \dot{z}^2)$, whose derivatives with respect to the velocity components are $\partial L/\partial \dot{x} = m\dot{x}$, etc., which are the components of momentum. So with generalized coordinates q_k, it is natural to define the **generalized momenta** p_k as

$$p_k \equiv \frac{\partial L}{\partial \dot{q}_k}. \tag{4.15}$$

In terms of p_k, the Lagrange equations become simply

$$\frac{dp_k}{dt} = \frac{\partial L}{\partial q_k}. \tag{4.16}$$

Now sometimes a particular coordinate q_l is *absent* from the Lagrangian. Its generalized velocity $\dot{q}_l$ is present, but not q_l itself. A missing coordinate is said to be a **cyclic coordinate** or an **ignorable coordinate**.[4] For any such coordinate the Lagrange equation (4.16) tells us that the time derivative of the corresponding generalized momentum is zero, so that particular generalized momentum is *conserved*:

$$\frac{dp_k}{dt} = \frac{\partial L}{\partial q_k} = 0. \tag{4.17}$$

One of the first things to notice about a Lagrangian is whether there are any cyclic coordinates, because any such coordinate leads to a conservation law that is also a first integral of motion. This means that the equation of motion for that coordinate is already half solved, in that it is only a first-order differential equation rather than the second-order differential equation one typically gets for a noncyclic coordinate.

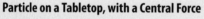

Example 4.4   **Particle on a Tabletop, with a Central Force**

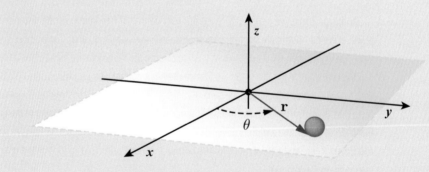

Fig. 4.5   Particle moving on a tabletop.

For a particle moving in two dimensions and subject to a central force, it is best to use polar coordinates (r, θ) about the origin, as shown in Figure 4.5. The kinetic energy of the particle is

$$T = \frac{1}{2}mv^2 = \frac{1}{2}m\,(\dot{x}^2 + \dot{y}^2 + \dot{z}^2) = \frac{1}{2}m\left[\dot{r}^2 + (r\dot{\theta})^2\right], \tag{4.18}$$

---

[4] Neither "cyclic" nor "ignorable" is a particularly appropriate or descriptive name for a coordinate absent from the Lagrangian, but they are nevertheless the conventional terms. In this book we will most often call any missing coordinate "cyclic."

where we used the constraint $\dot{z} = 0$ related to the normal force being applied by the table onto the particle vertically. This force does no work and hence does not appear explicitly in the Lagrangian.

We will assume here that any force acting on the particle is a central force, depending upon r alone, so the potential energy U of the particle also depends upon r alone. The Lagrangian is therefore

$$L = \frac{1}{2}m(\dot{r}^2 + r^2\dot{\theta}^2) - U(r). \tag{4.19}$$

Our two degrees of freedom are r and θ and they are our generalized coordinates. We note right away that in this case the coordinate θ is cyclic, so there must be a conserved quantity

$$p_\theta \equiv \frac{\partial L}{\partial \dot{\theta}} = mr^2\dot{\theta}, \tag{4.20}$$

which we recognize as the **angular momentum** of the particle. In Lagrange's approach, p_θ is conserved because θ is a cyclic coordinate; in Newtonian mechanics, p_θ is conserved because there is no torque on the particle, since we assumed that any force is a central force. The Lagrange equations

$$-\frac{d}{dt}\frac{\partial L}{\partial \dot{r}} = \frac{\partial L}{\partial r} \quad \text{and} \quad -\frac{d}{dt}\frac{\partial L}{\partial \dot{\theta}} = \frac{\partial L}{\partial \theta} \tag{4.21}$$

become

$$mr\dot{\theta}^2 + m\ddot{r} = -\frac{\partial U(r)}{\partial r} \quad \text{and} \quad -\frac{dp_\theta}{dt} = -\frac{d}{dt}\left(mr^2\dot{\theta}\right) = 0 \tag{4.22}$$

or

$$m(\ddot{r} - r\dot{\theta}^2) \equiv ma_r = F_r, \quad \text{and} \quad m(r\ddot{\theta} + 2\dot{r}\dot{\theta})ma_\theta = F_\theta = 0, \tag{4.23}$$

where the radial force is $F_r = -\partial U/\partial r$ and the radial and tangential accelerations are[a]

$$a_r = \ddot{r} - r\dot{\theta}^2 \quad \text{and} \quad a_\varphi = r\ddot{\theta} + 2\dot{r}\dot{\theta}. \tag{4.24}$$

In Example 1.6 we found (using $\mathbf{F} = m\mathbf{a}$) the equations of motion of a particle of mass m in two dimensions at the end of a spring of zero natural length and force constant k, with one end of the spring fixed. There we used Cartesian coordinates (x, y). Now we are equipped to write the equations of motion in polar coordinates instead. Equations (4.22) with $U = (1/2)kr^2$ give

$$m(\ddot{r} - r\dot{\theta}^2) = -kr \quad \text{and} \quad p_\theta = mr^2\dot{\theta} = \text{constant}. \tag{4.25}$$

That is, since the Lagrangian is independent of θ, we immediately get the first integral of motion $p_\theta = $ constant. Eliminating $\dot{\theta}$ between the two equations, we find the purely radial equation

$$\ddot{r} - \frac{(p_\theta)^2}{m^2r^3} + \omega_0^2 r = 0, \tag{4.26}$$

where $\omega_0 = \sqrt{k/m}$ is the natural frequency the spring–mass system would have if the mass were oscillating in one dimension (which in fact it would do if the angular momentum p_θ happened to be zero). Note that even though the motion is generally two-dimensional, Eq. (4.26) contains only $r(t)$. We can then find a second integral of motion because this equation has the form of a one-dimensional $ma = F$ equation with

$$m\ddot{r} = \frac{p_\theta^2}{mr^3} - m\omega_0^2 r \equiv -\frac{dU_{\text{eff}}}{dr} = F(r). \tag{4.27}$$

The *effective potential energy* U_{eff} can be found as

$$U_{\text{eff}}(r) = -\int^r F(r)\, dr = -\int^r \left(\frac{p_\theta^2}{mr^3} - m\omega_0^2 r \right) dr = \frac{p_\theta^2}{2mr^2} + \frac{1}{2}kr^2 \tag{4.28}$$

plus a constant of integration, which we might as well set to zero. Therefore, the second first integral of motion is given by energy conservation in this one-dimensional system:[b]

$$\frac{1}{2}m\dot{r}^2 + \frac{(p_\theta)^2}{2mr^2} + \frac{1}{2}kr^2 = \frac{1}{2}m\dot{r}^2 + U_{\text{eff}} = \text{constant} = E. \tag{4.29}$$

A sketch of U_{eff} is shown in Figure 4.6. Note that U_{eff} has a minimum, which is the location of an equilibrium point (the value of r for which $dU_{\text{eff}}(r)/dr = 0$ is of course also the radius for which $\ddot{r} = 0$; see Eq. (4.27)). If r remains at the minimum of U_{eff}, the mass is actually circling the origin. The motion about this point is stable because the potential energy is a minimum there, so for small displacements from equilibrium the particle oscillates back and forth about this equilibrium radius as it orbits the origin. In Section 4.9 we will calculate the frequency of these oscillations. For now, we can write a closed-integral form of the full solution of the problem by solving for $\dot{r}$ in Eq. (4.29) and integrating:

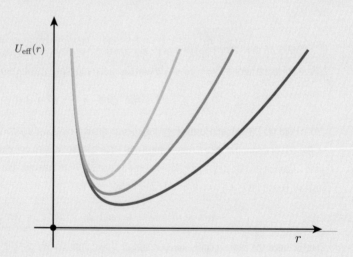

Fig. 4.6 The effective radial potential energy for a mass m moving with an effective potential energy $U_{\text{eff}} = (p^\theta)^2/2mr^2 + (1/2)kr^2$ for various values of p^θ, m, and k.

$$\int dt = \int \frac{dr}{\sqrt{2E/m - U_{\text{eff}}(r)}}. \tag{4.30}$$

This example demonstrates the use of coordinate transformations in the Lagrangian formalism. In this case, we see how useful it is to be able to change coordinates when solving problems in Lagrangian mechanics. ∎

[a] Note how easy it is to get the expressions for radial and tangential accelerations in polar coordinates using this method. They are often found in classical mechanics by differentiating the position vector $\mathbf{r} = r\hat{\mathbf{r}}$ twice with respect to time, which involves rather tricky derivatives of the unit vectors $\hat{\mathbf{r}}$ and $\hat{\boldsymbol{\theta}}$.

[b] See Example 1.9.

Example 4.5   **The Spherical Pendulum**

A ball of mass m swings on the end of an unstretchable string of length R in the presence of a uniform gravitational field g. This is often called the "spherical pendulum," because the ball moves as though it were sliding on the frictionless surface of a spherical bowl. We aim to find its equations of motion.

The ball has two degrees of freedom:

(i) It can move horizontally around a vertical axis passing through the point of support, corresponding to changes in its azimuthal angle ϕ. (On earth's surface this would correspond to a change in longitude.)

(ii) It can also move in the polar direction, as described by the angle θ. (On earth's surface this would correspond to a change in latitude.)

These angles are illustrated in Figure 4.7. In spherical coordinates, the velocity squared is

$$v^2 = \dot{r}^2 + r^2\dot{\theta}^2 + r^2\sin^2\theta\,\dot{\phi}^2 = R^2\dot{\theta}^2 + R^2\sin^2\theta\,\dot{\phi}^2 \tag{4.31}$$

using the constraint $r = R$. We are left with two degrees of freedom, with corresponding variables θ and ϕ. We know once again that the tension force is always perpendicular to the displacement in this problem, and so can be thrown away through the use of a constraint. Note that the velocities in the θ and ϕ directions are $v_\theta = R\dot{\theta}$ and $v_\phi = R\sin\theta\,\dot{\phi}$, which are perpendicular to one another; therefore

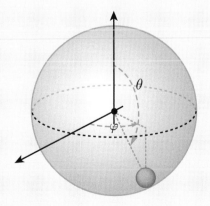

Fig. 4.7   Coordinates of a ball hanging on an unstretchable string.

$$T = \frac{1}{2}mv^2 = \frac{1}{2}mR^2(\dot{\theta}^2 + \sin^2\theta\,\dot{\phi}^2). \tag{4.32}$$

The altitude h of the ball, measured from its lowest possible point, is $h = R(1 - \cos\theta)$, so the potential energy can be written

$$U = mgh = mgR(1 - \cos\theta). \tag{4.33}$$

The Lagrangian is therefore

$$L = T - U = \frac{1}{2}mR^2(\dot{\theta}^2 + \sin^2\theta\,\dot{\phi}^2) - mgR(1 - \cos\theta). \tag{4.34}$$

The derivatives of L are

$$\frac{\partial L}{\partial \theta} = mR^2\sin\theta\cos\theta\,\dot{\phi}^2 - mgR\sin\theta, \quad \frac{\partial L}{\partial \dot{\theta}} = mR^2\dot{\theta}, \tag{4.35}$$

$$\frac{\partial L}{\partial \phi} = 0, \quad \frac{\partial L}{\partial \dot{\phi}} = mR^2\sin^2\theta\,\dot{\phi}, \tag{4.36}$$

and so the Lagrange equations are

$$\frac{\partial L}{\partial \theta} - \frac{d}{dt}\frac{\partial L}{\partial \dot{\theta}} = 0 \quad \text{and} \quad \frac{\partial L}{\partial \phi} - \frac{d}{dt}\frac{\partial L}{\partial \dot{\phi}} = 0. \tag{4.37}$$

Note that ϕ is cyclic, so the corresponding generalized momentum is

$$p_\phi = \frac{\partial L}{\partial \dot{\phi}} = mR^2\sin^2\theta\,\dot{\phi} = \text{constant}, \tag{4.38}$$

which is an immediate first integral of motion. We identify p_ϕ as the angular momentum about the vertical axis.

Now since

$$\frac{\partial L}{\partial \dot{\theta}} = mR^2\dot{\theta}, \tag{4.39}$$

the θ equation can be written

$$mR^2\ddot{\theta} = mR^2\sin\theta\cos\theta\,\dot{\phi}^2 - mgR\sin\theta, \tag{4.40}$$

so

$$\ddot{\theta} - \sin\theta\cos\theta\,\dot{\phi}^2 + \left(\frac{g}{R}\right)\sin\theta = 0. \tag{4.41}$$

We can eliminate the $\dot{\phi}^2$ term using $\dot{\phi} = p_\phi/(mR^2\sin^2\theta)$ from Eq. (4.38), to give

$$\ddot{\theta} - \left(\frac{p_\phi}{mR^2}\right)^2\frac{\cos\theta}{\sin^3\theta} + \left(\frac{g}{R}\right)\sin\theta = 0, \tag{4.42}$$

a second-order differential equation for the polar angle θ as a function of time.

Do we have to tackle this differential equation head-on? Not if we can find another integral of motion instead. We have already identified one first integral, the conservation of angular momentum (4.38) about the vertical axis. Another first integral is energy conservation

$$E = T + U = \frac{1}{2}mR^2(\dot{\theta}^2 + \sin^2\theta\,\dot{\phi}^2) + mgR(1 - \cos\theta), \tag{4.43}$$

valid because no work is being done on the ball aside from the work done by gravity, which is already accounted for in the potential energy. By combining the two conservation laws (4.38) and (4.43), we can eliminate $\dot{\phi}$ and write

$$E = \frac{1}{2}mR^2\dot{\theta}^2 + \frac{(p_\phi)^2}{2mR^2\sin^2\theta} + mgR(1 - \cos\theta) = \text{constant} \tag{4.44}$$

or

$$E = \frac{1}{2}mR^2\dot{\theta}^2 + U_{\text{eff}}, \tag{4.45}$$

where the "effective potential energy" is

$$U_{\text{eff}} = \frac{(p_\phi)^2}{2mR^2\sin^2\theta} + mgR(1 - \cos\theta). \tag{4.46}$$

This effective potential energy U_{eff} is sketched in Figure 4.8. The second term is the actual gravitational potential energy, while the first term is really a piece of the kinetic energy that has become a function of position only, thanks to angular momentum conservation. We can then solve for $\dot{\theta}$ in Eq. (4.45), separate the variables t and θ, and integrate to find

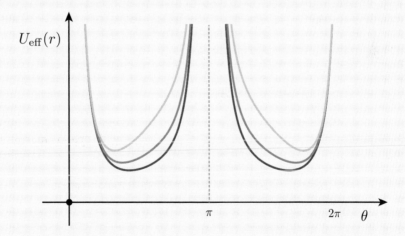

Fig. 4.8 A sketch of the effective potential energy U_{eff} for a spherical pendulum. A ball at the minimum of U_{eff} is circling the vertical axis passing through the point of suspension, at constant θ. The fact that there is a potential energy *minimum* at some angle θ_0 means that if disturbed from this value the ball will oscillate back and forth about θ_0 as it orbits the vertical axis.

$$t(\theta) = \sqrt{\frac{mR^2}{2}} \int_{\theta_0}^{\theta} \frac{d\theta}{\sqrt{(E - mgR) - (p_\phi)^2/(2mR^2 \sin^2 \theta) + mgR \cos \theta}}. \qquad (4.47)$$

In summary, the constraint reduced the number of degrees of freedom – from three to two in this case. The associated tension constraint force does no work, since it is perpendicular to the trajectory; we did not need to include its contribution in the Lagrangian. There were only two generalized coordinates, and we were able to solve the problem by finding two first integrals of motion.   ■

4.5 Systems of Particles

So far we have treated the motion of single particles only, described by at most three generalized coordinates and three generalized velocities. But often we want to find the motion of **systems of particles**, in which two or more particles may interact with one another, like two blocks on opposite ends of a spring, or several stars orbiting around one another. Can we still use the variational approach, by writing down a Lagrangian that contains the total kinetic energy and the total potential energy of the entire system?

Begin with a system of two particles, with masses m_1 and m_2, confined to move along a horizontal frictionless rail. Figure 4.9 shows a picture of the system, where we label the coordinates of the particles x_1 and x_2. We write an action for the system

$$S = \int dt \left(\frac{1}{2} m_1 \dot{x}_1^2 + \frac{1}{2} m_2 \dot{x}_2^2 - U(x_2 - x_1) \right) \qquad (4.48)$$

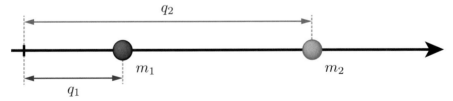

Fig. 4.9 Two interacting beads on a one-dimensional frictionless rail. The interaction between the particles depends only on the distance between them.

where, in addition to the usual kinetic energy terms, there is some unknown interaction between the particles described by a potential $U(x_2 - x_1)$. Note that we use the *total* kinetic energy, and we assume that the potential – hence the associated force law – depends only upon the distance between the particles. We then have two equations of motion with two generalized coordinates, x_1 and x_2, so that

$$\frac{d}{dt} \left(\frac{\partial L}{\partial \dot{x}_1} \right) - \frac{\partial L}{\partial x_1} = 0 \Rightarrow m_1 \ddot{x}_1 = -\frac{\partial U}{\partial x_1} \qquad (4.49)$$

and

$$\frac{d}{dt}\left(\frac{\partial L}{\partial \dot{x}_2}\right) - \frac{\partial L}{\partial x_2} = 0 \Rightarrow m_2 \ddot{x}_2 = -\frac{\partial U}{\partial x_2} = +\frac{\partial U}{\partial x_1}, \qquad (4.50)$$

where in the last step we used the fact that $U = U(x_2 - x_1)$. The equations are equivalent to Newton's second law applied to each of the two particles: kinetic energy is additive and each of its terms will generally give the $m\mathbf{a}$ part of Newton's second law for the corresponding particle. Hence, in multi-particle systems we need to consider the *total* kinetic energy T minus the *total* potential energy. Terms that mix the variables of different particles, such as $U(x_2 - x_1)$, will give the correct forces on the particles as well. In this case, we see that the action–reaction pair, $\partial U/\partial x_1 = -\partial U/\partial x_2$, comes out for *free*, and arises from the fact that the force law depends only on the distance between the particles! That is, Newton's third law is naturally incorporated in the formalism and originates from the fact that forces between two particles depend only upon the distance between the interacting entities, and not (say) their absolute positions.

In such a case the total momentum of the system must be conserved, which is easily verified simply by adding the two Eqs. (4.49) and (4.50), giving

$$\frac{d}{dt}(m_1 \dot{x}_1 + m_2 \dot{x}_2) = 0. \qquad (4.51)$$

These results suggest that there must be a more transparent set of generalized coordinates to use here, in which one of the new coordinates is cyclic, so that its generalized momentum will be conserved automatically. These new coordinates are the **center of mass** and **relative** coordinates

$$X \equiv \frac{m_1 x_1 + m_2 x_2}{M} \qquad \text{and} \qquad x \equiv x_2 - x_1, \qquad (4.52)$$

where $M = m_1 + m_2$ is the total mass of the system. Note that X and x are simply linear combinations of x_1 and x_2. Then in terms of X and x, it is straightforward to show that the Lagrangian of the system has the form

$$L = \frac{1}{2}M\dot{X}^2 + \frac{1}{2}\mu \dot{x}^2 - U(x), \qquad (4.53)$$

where $\mu \equiv m_1 m_2 / M$ is called the **reduced mass** of the system (note that μ is in fact smaller than either m_1 or m_2). Using this Lagrangian, it is obvious that the center of mass coordinate X is cyclic, so the corresponding momentum

$$P = \frac{\partial L}{\partial \dot{X}} = M\dot{X} \equiv m_1 \dot{x}_1 + m_2 \dot{x}_2 \qquad (4.54)$$

is conserved; this is simply the conservation of the total momentum. The setup easily extends to the case with an arbitrary number of particles – pairwise interacting with forces that depend on the distance between them. Action–reaction pairs would immediately arise realizing Newton's third law, and the sum of all the momenta would be conserved.

This problem is also an example of reducing a two-body problem to an equivalent one-body problem through a coordinate transformation. The motion of the center of mass of the system is trivial: the center of mass just drifts along at constant velocity. The interesting motion of the particles is their *relative* motion x, which behaves as though it were a single particle of mass μ and position $x(t)$ subject to the potential energy $U(x)$, with Lagrangian

$$L = \frac{1}{2}\mu\dot{x}^2 - U(x). \tag{4.55}$$

This problem exemplifies the fact that coordinate transformations of the Lagrangian can be a very powerful technique.

Example 4.6   **Pulleys Everywhere**

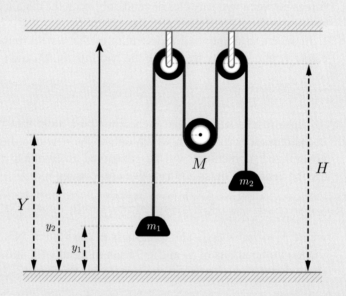

Fig. 4.10   A contraption of pulleys. We want to find the accelerations of all three weights. We assume that the three pulleys have negligible mass so they have negligible kinetic and potential energies.

Another classic set of mechanics problems involves pulleys, often lots of pulleys. Consider the setup shown in Figure 4.10. Two weights, with masses m_1 and m_2, hang on the outside of a three-pulley system, while a weight of mass M hangs on the middle pulley. We assume the pulleys and the connecting rope have negligible mass, so their kinetic and potential energies are also negligible. We will suppose for now that all three pulleys have the same radius R, but this will turn out to be of no importance. We want to find the accelerations of $m_1, m_2,$ and M. We construct a Cartesian coordinate system as shown in the figure, which remains fixed in the inertial frame of the ground. First of all, note that there are three massive objects moving in two dimensions, so we might think that we have six variables to track, x_1 and y_1 for weight m_1, x_2 and y_2 for weight m_2, and X and Y for weight M. We can then write the *total* kinetic energy

$$T = \frac{1}{2}m_1\left(\dot{x}_1^2 + \dot{y}_1^2\right) + \frac{1}{2}m_2\left(\dot{x}_2^2 + \dot{y}_2^2\right) + \frac{1}{2}M\left(\dot{X}^2 + \dot{Y}^2\right). \tag{4.56}$$

But that is obviously overkill. We know the dynamics is entirely vertical, so we can focus on y_1, y_2, and Y *only* and set $\dot{x}_1 = \dot{x}_2 = \dot{X} = 0$. But that is still too much. There are only *two* degrees of freedom in this problem! Just pick any two of y_1, y_2, or Y, and we can draw the figure uniquely, as long as we know the length of the rope. Another way of saying this is to write

$$\text{Length of rope} = (H - y_1) + 2\,(H - Y) + (H - y_2) + 3\pi R, \tag{4.57}$$

where H is the height of the upper-pulley centers, as shown in the figure. We can therefore write in general

$$y_1 + 2Y + y_2 = \text{constant}, \tag{4.58}$$

which can be used to eliminate one of the three variables. We choose to get rid of Y, writing

$$Y = -\frac{y_1 + y_2}{2} + \text{constant}, \tag{4.59}$$

which implies

$$\dot{Y} = -\frac{\dot{y}_1 + \dot{y}_2}{2}. \tag{4.60}$$

We can now write our kinetic energy in terms of two variables only, y_1 and y_2:

$$T = \frac{1}{2}m_1\dot{y}_1^2 + \frac{1}{2}m_2\dot{y}_2^2 + \frac{1}{2}M\left(\frac{\dot{y}_1 + \dot{y}_2}{2}\right)^2. \tag{4.61}$$

We next need the potential energy, which is entirely gravitational. We can write

$$\begin{aligned} U &= m_1 g\,y_1 + m_2 g\,y_2 + M g\,Y \\ &= m_1 g\,y_1 + m_2 g\,y_2 - M g\left(\frac{y_1 + y_2}{2}\right) + \text{constant}, \end{aligned} \tag{4.62}$$

where the zero of the potential was chosen at the ground, and we can also drop the additive constant term since it does not affect the equations of motion. In summary, we have a variational problem with the Lagrangian

$$\begin{aligned} L = T - U &= \frac{1}{2}m_1\dot{y}_1^2 + \frac{1}{2}m_2\dot{y}_2^2 + \frac{1}{2}M\left(\frac{\dot{y}_1 + \dot{y}_2}{2}\right)^2 \\ &\quad - m_1 g\,y_1 - m_2 g\,y_2 + M g\left(\frac{y_1 + y_2}{2}\right). \end{aligned} \tag{4.63}$$

There are two dependent variables y_1 and y_2, so we have two equations of motion:

$$\frac{d}{dt}\left(\frac{\partial L}{\partial \dot{y}_1}\right) - \frac{\partial L}{\partial y_1} = 0 \Rightarrow m_1\ddot{y}_1 + \frac{M}{4}\left(\ddot{y}_1 + \ddot{y}_2\right) = -m_1 g + \frac{M g}{2} \tag{4.64}$$

and

$$\frac{d}{dt}\left(\frac{\partial L}{\partial \dot{y}_2}\right) - \frac{\partial L}{\partial y_2} = 0 \Rightarrow m_2\ddot{y}_2 + \frac{M}{4}(\ddot{y}_1 + \ddot{y}_2) = -m_2 g + \frac{Mg}{2}. \tag{4.65}$$

We can now solve for $\ddot{y}_1$ and $\ddot{y}_2$:

$$\ddot{y}_1 = -g + \frac{4 m_2 g}{m_1 + m_2 + 4 m_1 m_2/M},$$

$$\ddot{y}_2 = -g + \frac{4 m_1 g}{m_1 + m_2 + 4 m_1 m_2/M}. \tag{4.66}$$

Note that these accelerations have magnitudes *less* than g, as we might expect intuitively. We can also find $\ddot{Y}$ from Eq. (4.60):

$$\ddot{Y} = -\frac{\ddot{y}_1 + \ddot{y}_2}{2} \Rightarrow \ddot{Y} = g - \frac{2(m_1 + m_2)g}{m_1 + m_2 + 4 m_1 m_2/M}. \tag{4.67}$$

The astute reader may rightfully wonder whether we have overlooked something in this treatment. We never encountered the tension force of the rope on each of the masses! Consider the two tension forces T_1 and T_2 at the end of this (or any) massless rope. If we wanted to account for such forces in a Lagrangian, we would need the associated energy, or **work**, they contribute to the system. Since the rope has zero mass, we know that $|T_1| = |T_2|$. The two tension forces, however, point in *opposite* directions. When one end of the rope moves by $\Delta s_1 > 0$ parallel to T_1, T_1 does work $W_1 = |T_1|\Delta s_1$. At the same time, the other end *must* move the same distance $\Delta s_2 = \Delta s_1$. However, at this other end, the tension force points opposite to the displacement, and the work is $W_2 = -|T_2|\Delta s_2 = -|T_1|\Delta s_1$. The *total* work is $W = W_1 + W_2 = 0$. Therefore, the tension forces along a massless rope will always contribute zero work, and hence cannot be associated with energy in the Lagrangian. The constraint given by Eq. (4.57), which eliminated one of the three variables, implements the physical condition that the rope has constant length. This is related to the existence of tension in the rope: if the tension were zero, there would be no rope and hence the constraint $\Delta s_1 = \Delta s_2$. Thus, the fact that there is no work associated with the tension is related to the use of the constraint to reduce the problem to two generalized coordinates. ∎

Example 4.7

A Block on a *Movable* Inclined Plane

Let us return to the classic problem of a block sliding down a frictionless inclined plane, as in Example 4.2, except that we will make things a bit more interesting: now the inclined plane itself is allowed to *move*. Figure 4.11(a) shows the system. A block of mass m rests on an inclined plane of mass M: both the block and the inclined plane are free to move without friction. The plane's angle is denoted by α. The problem is to find the acceleration of the block.

The observation deck is the ground, which is taken as an inertial reference frame. Note that the inclined plane is *not* an inertial reference frame, because it will accelerate to the left as the block slides down. We set up a convenient set of Cartesian coordinates, as shown in the figure. The origin is shifted to the top of the incline at zero time to make the geometry easier to analyze. We start by identifying the degrees of freedom. At first,

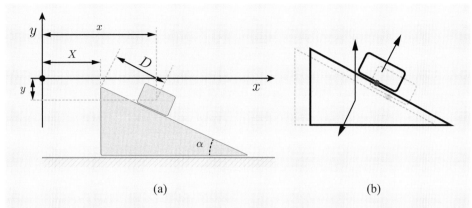

(a)   (b)

Fig. 4.11 A block slides along an inclined plane. Both block and inclined plane are free to move along frictionless contact surfaces. (a) The coordinate system used. (b) The normal contact forces at the block–inclined plane interface.

we can think of the block and inclined plane as moving in the two dimensions of the problem. The block's coordinates could be denoted by x and y, and the inclined plane's coordinates by X and Y. But we quickly realize that this would be overkill: if we specify X, and how far down the top of the incline the block is located (denoted by D in the figure), we can draw the figure uniquely. This is because the inclined plane cannot move vertically, either jumping off the ground or burrowing into it (that's one condition), so Y is unnecessary; and the horizontal position x or vertical position y of the block is determined by X and D (these correspond to the second condition that the block cannot sink through the incline). We then start with four coordinates, add two conditions or restrictions, and we are left with *two* degrees of freedom. The choice of the two remaining generalized coordinates is arbitrary, as long as the choice uniquely fixes the geometry. We will pick X and D, although we might have chosen X and x, for example.

Next, we need to write the Lagrangian. The starting point for this is the total kinetic energy of the system:

$$T = \frac{1}{2}m\left(\dot{x}^2 + \dot{y}^2\right) + \frac{1}{2}M\left(\dot{X}^2 + \dot{Y}^2\right) \tag{4.68}$$

in the inertial frame of the ground. Note that we *must begin by writing the kinetic energy in an inertial frame* before rewriting it in terms of the two degrees of freedom X and D alone. This requires a bit of geometry. Looking back at the figure, we can write the two constraint conditions as

$$Y = 0, \quad x = X + D\cos\alpha, \quad \text{and} \quad y = -D\sin\alpha. \tag{4.69}$$

This implies

$$\dot{Y} = 0, \quad \dot{x} = \dot{X} + \dot{D}\cos\alpha, \quad \text{and} \quad \dot{y} = -\dot{D}\sin\alpha. \tag{4.70}$$

We can now substitute these into Eq. (4.68) and get

$$T = \frac{1}{2}M\dot{X}^2 + \frac{1}{2}m\dot{X}^2 + \frac{1}{2}m\dot{D}^2 + m\dot{X}\dot{D}\cos\alpha. \tag{4.71}$$

Note that this result, in terms of the generalized coordinates and velocities, would have been very difficult to guess, especially the $\dot{X}\dot{D}$ term. Again, it is very important to start by writing the kinetic energy first in an inertial frame, and often important to use Cartesian coordinates in this initial expression to be confident that it has been done correctly.

We now need the potential energy of the system, which is entirely gravitational. The inclined plane's potential energy does not change. Since it is a constant, we need not add it to the Lagrangian: the Lagrange equations of motion involve partial derivatives of L and, hence, a constant term in L is irrelevant to the dynamics. In contrast, the block's potential energy *does* change. We can choose the zero of the potential at the origin of our coordinate system and write

$$U = mgy = -mgD \sin \alpha. \tag{4.72}$$

The Lagrangian then becomes

$$L = T - U = \frac{1}{2}M\dot{X}^2 + \frac{1}{2}m\dot{X}^2 + \frac{1}{2}m\dot{D}^2 + m\dot{X}\dot{D}\cos\alpha + mgD\sin\alpha. \tag{4.73}$$

We observe immediately that X is cyclic, so its corresponding momentum is conserved. Note that the total energy is also conserved. Therefore we can obtain the complete set of two first integrals of motion for our two degrees of freedom.

Nevertheless, to illustrate a different approach, we will tackle the full second-order differential equations of motion obtained directly from the Lagrange equations. Since we have two degrees of freedom X and D, we'll have two second-order equations. The equation for X is

$$\frac{d}{dt}\left(\frac{\partial L}{\partial \dot{X}}\right) - \frac{\partial L}{\partial X} = 0 \Rightarrow (m+M)\ddot{X} + m\ddot{D}\cos\alpha = 0, \tag{4.74}$$

and the equation for D is

$$\frac{d}{dt}\left(\frac{\partial L}{\partial \dot{D}}\right) - \frac{\partial L}{\partial D} = 0 \Rightarrow m\ddot{D} + m\ddot{X}\cos\alpha = mg\sin\alpha. \tag{4.75}$$

This is a system of two linear equations in two unknowns $\ddot{X}$ and $\ddot{D}$. Separating the two algebraically, we find

$$\ddot{X} = \frac{-mg \cos\alpha \sin\alpha}{M + m\sin^2\alpha} \quad \text{and} \quad \ddot{D} = \frac{(M+m)g \sin\alpha}{M + m\sin^2\alpha}. \tag{4.76}$$

Since we want the acceleration of the block in our inertial reference frame, we need to find $\ddot{x} \equiv a^x$ and $\ddot{y} \equiv a^y$. Differentiating Eq. (4.70) with respect to time, we find

$$a^x = \ddot{x} = \ddot{X} + \ddot{D}\cos\alpha \quad \text{and} \quad a^y = \ddot{y} = -\ddot{D}\sin\alpha. \tag{4.77}$$

Substituting our solution from Eqs. (4.76) into these, we have

$$a^x = \frac{Mg \sin\alpha \cos\alpha}{M + m\sin^2\alpha} \quad \text{and} \quad a^y = -\frac{(M+m)g \sin^2\alpha}{M + m\sin^2\alpha}. \tag{4.78}$$

It is always useful to look at various limiting cases to see if a result makes sense. For example, what if $\alpha = 0$, *i.e.*, what if the block moves on a horizontal plane? Both accelerations then vanish, as expected: if started at rest, both block and incline just stay put. Now what if the inclined plane is much heavier than the block, *i.e.*, $M \gg m$? We then have

$$a^x \simeq g \cos \alpha \sin \alpha \text{ and } a^y \simeq -g \sin^2 \alpha, \tag{4.79}$$

so that $a^y / a^x \simeq - \tan \alpha$, which is what we would expect if the inclined plane were not moving appreciably. Finally, if $M = 0$, then $a^x = 0$ and $a^y = -g$, so the block would fall straight down; this is exactly what would happen if there were no inclined plane at all.

The most impressive aspect of this example is the absence of any normal forces from our computations! With the traditional approach of problem solving, we would need to include several normal forces in the computation, shown in Figure 4.11(b): the normal force exerted by the inclined plane on the block, the normal force exerted by the ground on the inclined plane, *and* the normal force exerted by the block on the inclined plane as a reaction force. The role of these normal forces is to hold the inclined plane on the ground and to hold the block on the inclined plane. If we think of the contributions of the normal forces to the Lagrangian, we would want to include some potential energy terms for them. But potential energy is related to work done by forces. The normal force is often perpendicular to the direction of motion, and hence does no work, $\mathbf{N} \cdot \Delta \mathbf{r} = 0$, where N is a normal force and $\Delta \mathbf{r}$ is the displacement of a contact point. Therefore, there is no potential energy term to include in the Lagrangian to account for such normal forces. In our example, however, this is not entirely correct. While it is true for the normal force exerted by the ground onto the inclined plane, it is *not* true for the two normal forces acting between the block and the inclined plane. This is because the inclined plane is moving as well and the trajectory of the block is not parallel to the incline! However, there is another reason why this normal force is safely left out of the Lagrangian. These normal forces occur as an action–reaction pair; and the displacement of the interface between the block and the incline is the same for both forces, and hence the contributions to the total work or energy of the system from these two normal forces cancel. As we saw from the previous example, such force pairs do not appear in the Lagrangian. The computational step at the beginning, where we zeroed onto the degrees of freedom of the problem – going from X, Y, x, and y to X and D – is in a sense the process of accounting for the normal forces in the Lagrangian approach to the problem.

In Chapter 6, we will also learn of a way to force the inclusion of normal and tension forces in a Lagrangian even when we need not do so – for the purpose of finding the magnitude of a normal or tension force if so desired. For now, we are very happy to drop normal and tension forces from consideration. This can be a big simplification for problem solving: *fewer variables, fewer forces to consider, less work to do* (no pun intended). ∎

4.6 The Hamiltonian

We will now prove an enormously useful consequence of the Lagrange equations that will at the end look very familiar. First, take the *total* derivative of the Lagrangian $L(t, q_k, \dot{q}_k)$ with respect to time t. There are many ways in which L can change with time: it can change because of *explicit* dependence on t, but also because of *implicit* dependence on t through the time dependence of one or more of the coordinates $q_k(t)$ or velocities $\dot{q}_k(t)$. Therefore, from multivariable calculus and the chain rule, we have

$$\frac{dL(q_k, \dot{q}_k, t)}{dt} = \frac{\partial L}{\partial t} + \frac{\partial L}{\partial q_k}\dot{q}_k + \frac{\partial L}{\partial \dot{q}_k}\ddot{q}_k, \tag{4.80}$$

using the Einstein summation convention from Chapter 2. That is, since the index k is repeated in each of the last two terms, a sum over k is implied in each term; we have also used the fact that $d\dot{q}_k/dt \equiv \ddot{q}_k$. Now, take the time derivative of the quantity $\dot{q}_k(\partial L/\partial \dot{q}_k)$; again, with implied sums over k:

$$\frac{d}{dt}\left(\dot{q}_k\frac{\partial L}{\partial \dot{q}_k}\right) = \ddot{q}_k\frac{\partial L}{\partial \dot{q}_k} + \dot{q}_k\frac{d}{dt}\left(\frac{\partial L}{\partial \dot{q}_k}\right) = \ddot{q}_k\frac{\partial L}{\partial \dot{q}_k} + \dot{q}_k\frac{\partial L}{\partial q_k}, \tag{4.81}$$

using the product rule. We have also used the Lagrange equations to simplify the last term on the right. Note that this expression contains the same two summed terms that we found in Eq. (4.80). Therefore, subtracting Eq. (4.81) from Eq. (4.80) gives

$$\frac{\partial L}{\partial t} - \frac{d}{dt}\left(L - \dot{q}_k\frac{\partial L}{\partial \dot{q}_k}\right) = 0. \tag{4.82}$$

Now define the **Hamiltonian** H of a particle to be

$$H \equiv \dot{q}_k\, p_k - L, \tag{4.83}$$

where we have already defined the generalized momenta to be $p_i = \partial L/\partial \dot{q}_k$, and again a sum over k is implied. Equation (4.82) can be written

$$\frac{\partial L}{\partial t} = -\frac{dH}{dt}. \tag{4.84}$$

This result is particularly interesting if L is not an *explicit* function of time, *i.e.*, if $\partial L/\partial t = 0$. It shows that *if a Lagrangian L is not an explicit function of time, then the Hamiltonian H is conserved.*

What is the *meaning* of H? Suppose that the particle is free to move in three dimensions in a potential $U(x, y, z)$ without constraints, and that we are using Cartesian coordinates. Then $p_x = m\dot{x}$, $p_y = m\dot{y}$, etc., so $\sum_i \dot{q}_k p_k = m(\dot{x}^2 + \dot{y}^2 + \dot{z}^2)$. Therefore

$$H = m(\dot{x}^2 + \dot{y}^2 + \dot{z}^2) - \frac{1}{2}m(\dot{x}^2 + \dot{y}^2 + \dot{z}^2) + U(x, y, z)$$

$$= \frac{1}{2}m(\dot{x}^2 + \dot{y}^2 + \dot{z}^2) + U(x, y, z) = T + U = E, \tag{4.85}$$

which is simply the energy of the particle. Therefore, since this Lagrangian does not depend on time explicitly, this energy quantity is conserved, as expected.

Is H *always* equal to $E = T + U$? The answer is *no*, although very often it is. The precise conditions for which $H \neq E$ are worked out in Section 4.7.

Example 4.8   **Bead on a Rotating Parabolic Wire**

Suppose we bend a wire into the shape of a vertically oriented parabola defined in cylindrical coordinates by $z = \alpha\rho^2$, as illustrated in Figure 4.12: here z is the vertical coordinate, and ρ is the distance of a point on the wire from the vertical axis of symmetry. Using a synchronous motor, we can force the wire to spin at constant angular velocity ω about its symmetry axis. Then we slip a bead of mass m onto the wire and we want to determine its equation of motion – assuming that it slides without friction along the wire.

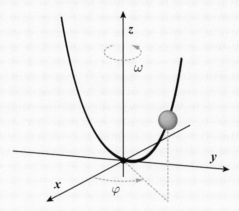

Fig. 4.12   A bead slides without friction on a vertically oriented parabolic wire forced to spin about its axis of symmetry.

We first have to choose generalized coordinate(s) for the bead. The bead moves in three dimensions, but because of the constraint confining the bead along the wire we need only a single generalized coordinate to specify the bead's position. For example, if we know the distance ρ of the bead from the vertical axis of symmetry, we also know its altitude z, $z = \alpha\rho^2$. And the bead also has no freedom to choose its azimuthal angle, because the synchronous motor turns the wire around at a constant rate. Given its angle φ_0 at time $t = 0$, its angle at other times is constrained to be $\varphi = \varphi_0 + \omega t$. Hence, we start with three coordinates and, using two constraint conditions, we end up with just one degree of freedom. It is convenient to choose the cylindrical coordinate ρ as the single generalized coordinate, although we could equally well choose the vertical coordinate z.

In cylindrical coordinates the square of the bead's velocity is the sum of squares of the velocities in the ρ, φ, and z directions:

$$v^2 = \dot{\rho}^2 + \rho^2\dot{\varphi}^2 + \dot{z}^2 = \dot{\rho}^2 + \rho^2\omega^2 + (2\alpha\rho\dot{\rho})^2, \tag{4.86}$$

using the constraints $\dot{\varphi} = \omega = $ constant and $z = \alpha\rho^2$ for the parabolic wire. The gravitational potential energy is $U = mgz = mg\alpha\rho^2$, so the Lagrangian becomes

$$L = T - U = \frac{1}{2}m[(1 + 4\alpha^2\rho^2)\dot{\rho}^2 + \rho^2\omega^2] - mg\alpha\rho^2. \tag{4.87}$$

We implemented two constraints, $\dot{\varphi} = \omega$ and $z = \alpha\rho^2$, and thus reduced the problem from three variables to one, the radius ρ. The contact force that keeps the bead on the wire is the normal force associated with these two constraints. We do *not* include its contribution to the Lagrangian, assuming that this normal force has no contribution to the energy of the system. That this is correct is far from obvious in this case, since this normal force *does* work! The normal force in this case is *not* perpendicular to the displacement. This work is associated with the energy input by the motor to keep the wire turning at constant rate ω. Let us proceed anyway, and revisit the issue at the end. The partial derivatives $\partial L/\partial\dot{\rho}$ and $\partial L/\partial\rho$ are easy to find, leading to the Lagrange equation

$$\frac{\partial L}{\partial \rho} - \frac{d}{dt}\frac{\partial L}{\partial \dot{\rho}} = m[4\alpha^2\rho\dot{\rho}^2 + \rho\omega^2 - 2g\alpha\rho] - m\frac{d}{dt}\left[(1 + 4\alpha^2\rho^2)\dot{\rho}\right] = 0, \qquad (4.88)$$

which results in a second-order differential equation of motion for ρ:

$$(1 + 4\alpha^2\rho^2)\ddot{\rho} + 4\alpha^2\rho\dot{\rho}^2 + (2g\alpha - \omega^2)\rho = 0. \qquad (4.89)$$

Are we stuck with having to solve this second-order differential equation? Are there no first integrals of motion? The coordinate ρ is not cyclic, so p_ρ is not conserved. However, note that L is *not an explicit function of time*, $\partial L/\partial t = 0$, so the Hamiltonian H *is* conserved according to Eq. (4.84). Conservation of H gives us a first integral of motion, so we are rescued: we do not have to solve the second-order differential equation (4.89) after all, given that we have one degree of freedom and one first integral of motion.

The generalized momentum is $p_\rho = \partial L/\partial\dot{\rho} = m(1 + r\alpha^2\rho^2)\dot{\rho}$, so the Hamiltonian is

$$H = \dot{\rho}p_\rho - L = m(1 + 4\alpha^2\rho^2)\dot{\rho}^2 - \frac{1}{2}m[(1 + 4\alpha^2\rho^2)\dot{\rho}^2 + \rho^2\omega^2] + mg\alpha\rho^2$$

$$= \frac{1}{2}m[(1 + 4\alpha^2\rho^2)\dot{\rho}^2 - \rho^2\omega^2] + mg\alpha\rho^2 = \text{constant}, \qquad (4.90)$$

which differs from the energy

$$E = T + U = \frac{1}{2}m[(1 + 4\alpha^2\rho^2)\dot{\rho}^2 + \rho^2\omega^2] + mg\alpha\rho^2 \qquad (4.91)$$

by $H - E = -m\rho^2\omega^2$. Note that this difference changes as ρ changes; that is because even though H is conserved, the energy E is not conserved, because the normal force of the wire does work on the bead.

Equation (4.90) is a first-order differential equation, which can be reduced to quadrature. Without going that far, we can understand a good deal about the motion just by using Eq. (4.90), and noting that it has a similar mathematical *form* to that of energy-conservation equations. That is, rewrite the equation as

$$H = \frac{1}{2}m(1 + 4\alpha^2\rho^2)\dot{\rho}^2 + U_{\text{eff}}, \qquad (4.92)$$

where

$$U_{\text{eff}} = \frac{1}{2}m(-\rho^2\omega^2) + mg\alpha\rho^2 = \frac{1}{2}m\rho^2(2g\alpha^2 - \omega^2). \qquad (4.93)$$

This effective potential is quadratic in ρ, with the interesting feature that its sign depends upon how the angular velocity ω of the wire is related to a critical angular velocity $\omega_{\text{crit}} \equiv (2g)^{1/2}\alpha$, as illustrated in

Figure 4.13. If $\omega < \omega_{\text{crit}}$, then U_{eff} rises with ρ, so the bead is stable at $\rho = 0$, the potential minimum. But if $\omega > \omega_{\text{crit}}$, then U_{eff} falls off with increasing ρ, so $\rho = 0$ is an unstable equilibrium point in that case: if the bead wanders even slightly from $\rho = 0$ at the bottom of the parabola, it will be thrown out indefinitely far. The stability is neutral if $\omega = \omega_{\text{crit}}$, meaning that if placed at rest at any point along the wire, the bead will stay at that point indefinitely; but if placed at any point and pushed outward it will keep moving outward, or if pushed inward it will keep moving inward.

This example shows that although the Hamiltonian function H is often equal to $E \equiv T + U$, this is not always so. Section 4.7 explains when and why they can differ. In any case, the Hamiltonian can be very useful, because it provides a first integral of motion if L is not explicitly time dependent.

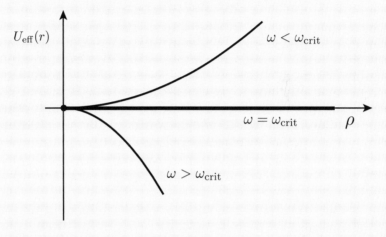

Fig. 4.13   The effective potential U_{eff} for the Hamiltonian of a bead on a rotating parabolic wire with $z = \alpha\rho^2$, depending upon whether the angular velocity ω is less than, greater than, or equal to $\omega_{\text{crit}} = \sqrt{2g}\,\alpha$.

Once again, the Lagrangian formalism avoids dealing with contact forces and accounts for them through constraints – simplifying the problem significantly. We leave it as an exercise for the reader to solve this same problem using traditional force body diagram methods so as to appreciate the power of the Lagrangian formalism.   ∎

4.7 When is $H \neq E$?

In Example 4.8 the Hamiltonian H was not equal to E. Why were they different, and why was H conserved while E was not?

The definition $H = \dot{q}_k \partial L / \partial \dot{q}_k - L$ (using the Einstein summation convention, implying a sum over k in the first term on the right, since k is repeated in that term) contains the Lagrangian $L = T - U$, where only the kinetic energy T depends upon the generalized velocities $\dot{q}_k$. Therefore:

$$H = \dot{q}_k \frac{\partial T}{\partial \dot{q}_k} + U - T. \tag{4.94}$$

Now let $\mathbf{r}(q_k, t)$ be the position vector of the particle from some arbitrary origin fixed in an inertial frame, in terms of the time and any or all of the generalized coordinates q_k. Then the velocity of the particle is

$$\mathbf{v} = \frac{d\mathbf{r}(q_k, t)}{dt} = \frac{\partial \mathbf{r}}{\partial t} + \frac{\partial \mathbf{r}}{\partial q_l} \dot{q}_l, \tag{4.95}$$

because $\mathbf{r}$ can change with time either from an explicit time dependence or because one or more of the generalized coordinates changes with time. Therefore its kinetic energy is $T = (1/2)mv^2 = (1/2)m\mathbf{v} \cdot \mathbf{v}$, which is

$$T = \frac{1}{2} \left[m \frac{\partial \mathbf{r}}{\partial t} \cdot \frac{\partial \mathbf{r}}{\partial t} + 2 \frac{\partial \mathbf{r}}{\partial t} \cdot \frac{\partial \mathbf{r}}{\partial q_l} \dot{q}_l + \frac{\partial \mathbf{r}}{\partial q_l} \dot{q}_l \cdot \frac{\partial \mathbf{r}}{\partial q_m} \dot{q}_m \right], \tag{4.96}$$

where we have used a different dummy summation index m in the final factor to distinguish it from the sum over l in the preceding factor. That is, by the Einstein summation convention the final term above is actually the product of two sums, one over l and one over m. Now we can take the partial derivative of T with respect to a particular one of the generalized velocities $\dot{q}_k$:

$$\frac{\partial T}{\partial \dot{q}_k} = \frac{1}{2} m \left[2 \frac{\partial \mathbf{r}}{\partial t} \cdot \frac{\partial \mathbf{r}}{\partial q_k} + 2 \frac{\partial \mathbf{r}}{\partial q_k} \cdot \frac{\partial \mathbf{r}}{\partial q_l} \dot{q}_l \right], \tag{4.97}$$

where there is a factor of two in the second term because $\dot{q}_k$ occurs in both summations in the last term of the expression for T. Therefore the sum

$$\dot{q}_k \frac{\partial T}{\partial \dot{q}_k} = m \left[\frac{\partial \mathbf{r}}{\partial t} \cdot \frac{\partial \mathbf{r}}{\partial q_k} \dot{q}_k + \frac{\partial \mathbf{r}}{\partial q_k} \dot{q}_k \cdot \frac{\partial \mathbf{r}}{\partial q_l} \dot{q}_l \right]$$

$$= 2T - m \frac{\partial \mathbf{r}}{\partial t} \cdot \left[\frac{\partial \mathbf{r}}{\partial t} + \frac{\partial \mathbf{r}}{\partial q_k} \dot{q}_k \right]$$

$$= 2T - m \frac{\partial \mathbf{r}}{\partial t} \cdot \frac{d\mathbf{r}}{dt}, \tag{4.98}$$

using Eqs. (4.96) and (4.97). The Hamiltonian H can therefore be written as

$$H = E - m \frac{d\mathbf{r}}{dt} \cdot \frac{\partial \mathbf{r}}{\partial t} = E - \mathbf{p} \cdot \frac{\partial \mathbf{r}}{\partial t}, \tag{4.99}$$

where $\mathbf{p}$ is the momentum of the particle in the chosen inertial frame. If the transformation $\mathbf{r} = \mathbf{r}(q_k, t)$ between the Cartesian coordinates $\mathbf{r} = (x, y, z)$ and the generalized coordinates q_k happens not to be an explicit function of time, i.e., if $\partial \mathbf{r}/\partial t = 0$, then the Hamiltonian is just $T + U$. This case occurs when there are no constraints or when any constraints are fixed in space. But if the constraints are moving, then the transformation $\mathbf{r} = \mathbf{r}(q_k, t)$ does generally depend upon time, and so in the likely case that there is a component of $\partial \mathbf{r}/\partial t$ in the direction of the particle's momentum $\mathbf{p}$, the Hamiltonian is not equal to $T + U$.

For the problem of the bead on a rotating parabolic wire, where the constraint is obviously moving, the position vector of the bead can be taken to be $\mathbf{r} = (x, y, z) = (r \cos \omega t, r \sin \omega t, \alpha r^2)$. In that case we found that $H = T + U - m\omega^2 r^2$, and it is easy to show that $m\omega^2 r^2 = \mathbf{p} \cdot \partial \mathbf{r}/\partial t$, as required by Eq. (4.99). It is clear that E is not conserved in this case because the rotating wire is continually doing work on the bead. The wire exerts a force F^θ in the tangential direction, which causes work to be done at the rate $dW/dt = F^\theta v^\theta = F^\theta r\omega$. From the elementary relationship $N^z = dL^z/dt$, with torque $N^z = F^\theta r$ and angular momentum $L^z = (\mathbf{r} \times \mathbf{p})^z = mr^2\omega$, it follows that

$$\frac{dW}{dt} = \omega \frac{d(mr^2\omega)}{dt} = m\omega^2 \frac{dr^2}{dt}, \tag{4.100}$$

so that the work done by the wire is $W = m\omega^2 r^2$ plus a constant of integration, which depends upon the initial location of the bead. Thus the energy $E = T + U$ of the bead increases by the work done upon it by the wire, so that E *minus* the work done must be conserved, and that difference $E - m\omega^2 r^2 = H$.

Note that:

1. H is conserved if the Lagrangian L is not an explicit function of time.
2. $H = E$ if the coordinate transformation $\mathbf{r} = \mathbf{r}(q_k, t)$ is not an explicit function of time.

Therefore it is possible to have $E = H$, with both E and H conserved, or neither conserved, and it is also possible to have $E \neq H$, with both conserved, neither conserved, or only one of the two conserved.

4.8 The Moral of Constraints

Let us summarize the steps we have used so far in setting up and preparing to solve Lagrange's equations.

1. Identify the degrees of freedom of each particle or object consistent with any constraints, and choose an appropriate set of generalized coordinates q_k for each.
2. Write the square of the velocity for each particle in terms of any convenient coordinates, usually Cartesian coordinates, in *some inertial reference frame*. Then re-express the kinetic energy in terms of the generalized coordinates q_k, the generalized velocities $\dot{q}_k$, and the time if needed; *i.e.*, $v^2 = v^2(q_k, \dot{q}_k, t)$. Then write the *total* kinetic energy T in terms of these variables.
3. Write the *total* potential energy in the form $U = U(q_k, t)$. Do *not* include any contributions from constraint forces.
4. Write the Lagrangian $L(t, q_1, q_2, \ldots, \dot{q}_1, \dot{q}_2, \ldots) = T - U$.
5. Identify any cyclic coordinates in L; that is, identify any coordinate q_l missing in the Lagrangian, even though its corresponding generalized velocity $\dot{q}_l$ is present. In this case the corresponding generalized momentum $p_l \equiv \partial L/\partial \dot{q}_l$ is *conserved*.

This gives a highly valued first integral of motion, *i.e.*, a differential equation that is first order rather than second order.

6. If L is not an explicit function of time, then the Hamiltonian $H \equiv \dot{q}_k\, p_k - L$ is conserved, providing another first integral of motion.

7. If there are more generalized coordinates in the problem than first integrals identified in the preceding steps, then one or more of the Lagrange equations of motion

$$\frac{\partial L}{\partial q_k} - \frac{d}{dt}\frac{\partial L}{\partial \dot{q}_k} = 0 \tag{4.101}$$

must be used as well, to obtain a complete set of differential equations. That is, if there are N generalized coordinates, we will generally need N mutually independent differential equations whose solutions will give the coordinates as functions of time. Some of these may be first-order equations, each corresponding to a conserved quantity, while others may be second-order equations coming from Eq. (4.101).

All these steps were justified: that the Lagrangian $L = T - U$ accounts for all conservative forces, that coordinate transformations at the Lagrangian level are justified and in fact very useful, and that constraints assure that the generalized coordinates are independent and a reduced set of degrees of freedom describes the dynamics fully. As one gets used to the steps outlined above, many stages of this algorithmic process become second nature and can be done mentally. With practice, one looks at a complex mechanical system, writes down the Lagrangian immediately on a single line, and in a few more lines, writes the equations of motion! To get there, however, one needs to practice, practice, practice.

Why can we avoid constraint forces in the Lagrangian? We saw example after example in which neglecting them works out fine. We see an emerging pattern. Constraint forces implement restrictions on the dynamics between *two objects in contact*. If both objects in question are part of the dynamical system (*i.e.*, they both contribute kinetic energy to the Lagrangian), these constraint forces must come in equal and opposite pairs. Since the contact point is the same, this always implies that such forces will not do work and hence need not appear in the Lagrangian. On the contrary, if only one of the two objects is part of the dynamical system, the other one must then have *prescribed* time evolution by definition: *i.e.*, the ground just sits there as a function of time, the pivot of the pendulum is fixed in position, a parabolic wire – on which a bead is sliding – is rotating at a given constant angular speed driven by a motor. In some of these cases, the constraint forces do no work because they are perpendicular to the displacement. But it is easy to see this statement in more generality: extend the Lagrangian to include the nondynamical system – the ground, the parabolic wire connected to a motor – by adding their *constant* kinetic energies to the Lagrangian. Then the constraint force becomes part of an internal action–reaction pair, which we know does not contribute to the Lagrangian! And the cost of adding the kinetic energy of the external agent to the Lagrangian is irrelevant: it is a constant shift to the Lagrangian since the relevant dynamics is, by definition, prescribed. This, we now see, is one of the central advantages of the Lagrangian formalism: *We can drop all constraint forces from the outset.*

4.9 Small Oscillations about Equilibrium

We often find that a system of interest is at or near a state of stable equilibrium. For example, a nearby chair rests on the floor, happily doing nothing. It is in a minimal energy configuration relative to nearby configurations. If we bump the chair it may wobble and slide a bit, and then quickly come to rest again in some new equilibrium state. When we bumped the chair, we added energy to the system, and the chair eventually dissipated this energy through friction and returned to a minimal-energy, motionless state.

In general, many if not most mechanical systems can be accorded an energy of the form

$$\text{Constant} \times \dot{q}_k^2 + U_{\text{eff}}(q_k) = E, \tag{4.102}$$

where the q_ks are the generalized coordinates and U_{eff} is an effective potential. We have seen this in several examples. For simplicity, imagine we have only one such coordinate, which we will call q. If the effective potential energy U_{eff} has an extremum at some particular point q_0, then that point is an equilibrium point of the motion, so if placed at rest at q_0 the particle will stay there. If q_0 happens to be a *minimum* of U_{eff}, as illustrated in Figure 4.14, q_0 is a *stable* equilbrium

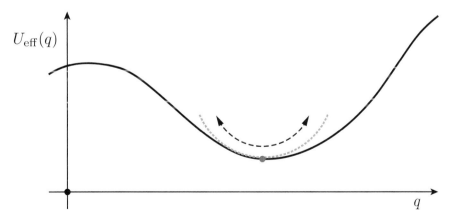

Fig. 4.14   An effective potential energy U_{eff} with a focus near a minimum. Such a point is a *stable* equilibrium point. The dotted parabola shows the leading approximation to the potential near its minimum. As the energy drains out, the system settles into its minimum, with the final moments being well approximated by harmonic oscillatory dynamics.

point, so that if the particle is displaced slightly from q_0 it will oscillate back and forth, never wandering far from that point. It is sometimes interesting to find the frequency of small oscillations about an equilibrium point. We can often do this by fitting the bottom of the effective potential energy curve to a parabola, which is the shape of the potential energy of a simple harmonic oscillator. That is, by the Taylor expansion, we write (see Appendix F))

$$U_{\text{eff}}(q) = U_{\text{eff}}(q_0) + \frac{dU_{\text{eff}}}{dq}\bigg|_{q_0} (q - q_0) + \frac{1}{2!}\frac{d^2 U_{\text{eff}}}{dq^2}\bigg|_{q_0} (q - q_0)^2 + \cdots \quad (4.103)$$

So if q_0 is the equilibrium point, by definition the second term vanishes, and the third term has the form $(1/2)k_{\text{eff}}(q-q_0)^2$, like that for a simple harmonic oscillator with center at q_0, where the effective force constant is given by $k_{\text{eff}} = U''_{\text{eff}}(q_0)$. The frequency of small oscillations is therefore

$$\omega = \sqrt{\frac{k_{\text{eff}}}{m}} = \sqrt{\frac{U''_{\text{eff}}(q_0)}{m}}. \quad (4.104)$$

Note that this explains the pervasiveness of the harmonic oscillator in Nature: systems will try to find their lowest-energy configurations by dissipating energy, so often find themselves near the minima of their effective potentials.[5] An example will demonstrate.

Example 4.9   **Particle on a Tabletop with a Central Spring Force**

In Example 4.4 we considered a particle moving on a frictionless tabletop subject to a central Hooke's-law spring force. There is an equilibrium radius for given energy and angular momentum for which the particle orbits in a circle of some radius r_0. We now want to find the oscillation frequency ω for the mass about the equilibrium radius if it were perturbed slightly from this circular orbit.

The effective potential in Example 4.4 was

$$U_{\text{eff}} = \frac{(p_\varphi)^2}{2mr^2} + \frac{1}{2}kr^2 \quad (4.105)$$

from Eq. (4.28); the first derivative of this potential is $U'_{\text{eff}}(r) = -(p_\varphi)^2/mr^3 + kr$. The equilibrium value of r is located where $U'_{\text{eff}}(r) = 0$; namely, where $r = r_0 = ((p_\varphi)^2/mk)^{1/4}$.

The second derivative of $U_{\text{eff}}(r)$ is

$$U''_{\text{eff}}(r) = \frac{3(p_\varphi)^2}{mr^4} + k, \quad (4.106)$$

so

$$U''_{\text{eff}}(r_0) = \frac{3(p_\varphi)^2}{m((p_\varphi)^2/mk)} + k = 3k + k = 4k = k_{\text{eff}} > 0. \quad (4.107)$$

---

[5] We focused for simplicity on a system with one degree of freedom. However, this conclusion generalizes to an effective potential depending on an arbitrary number of degrees of freedom. $U_{\text{eff}}(q_k)$ is then a multi-dimensional surface with generally a complex landscape of valleys and hills. We will revisit this scenario in Chapter 13.

From Eq. (4.104), the frequency of small oscillations about the equilibrium radius r_0 is therefore

$$\omega = \sqrt{\frac{U''_{\text{eff}}(r_0)}{m}} = \sqrt{\frac{4k}{m}} = 2\,\omega_0. \tag{4.108}$$

That is, for the mass orbiting the origin and subject to a central Hooke's-law spring force, the frequency of small oscillations about a circular orbit is just twice what it would be for the mass if it were oscillating back and forth about the center in one dimension.

We can also find the *shape* of the orbits, noting that the radial oscillations are small. The angular frequency of rotation is $\omega_{\text{rot}} = v_\phi/r_0 = (p_\varphi/mr_0)/r_0 = p_\varphi/(mr_0^2)$, where v_ϕ is the tangential component of velocity. But the equilibrium radius is $r_0 = ((p_\varphi)^2/mk)^{1/4}$, so the angular frequency of rotation is $\omega_{\text{rot}} = p_\varphi/(mr_0^2) = p_\varphi/[m\sqrt{(p_\varphi)^2/k}] = \sqrt{k/m} = \omega_0$, which is the same as the frequency of oscillation of the system as if it were oscillating in one dimension through the center. Therefore, the frequency of radial oscillations (4.108) is just twice the rotational frequency, so the orbits for small oscillations are *closed*: that is, the path retraces itself in every orbit, as shown in Figure 4.15. The small-oscillation path appears to be elliptical. In fact, it is *exactly* elliptical, even for large displacements from equilibrium, as we already saw in Chapter 1 using Cartesian coordinates.

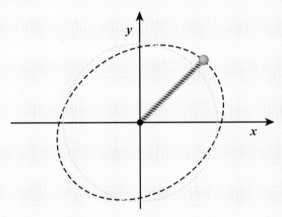

Fig. 4.15   The shape of the two-dimensional orbit of a particle subject to a central spring force, for small oscillations about the equilibrium radius. ∎

4.10 Recap

We began the chapter by describing a conservative mechanical system by N generalized coordinates $q_k(t)$, with $k = 1, 2, \ldots, N$, and then defining the Lagrangian

$$L(t, q_1, q_2, \ldots, \dot{q}_1, \dot{q}_2, \ldots) = T - U \tag{4.109}$$

as the difference between the kinetic and potential energies of the system, expressed in terms of the generalized coordinates q_k, generalized velocities $\dot{q}_k$, and time t. We then define the action $S[q_k(t)]$ of the system as the functional consisting of the time integral over the Lagrangian $L(t, q_1, q_2, \ldots, \dot{q}_1, \dot{q}_2, \ldots)$, from a starting time t_a to an ending time t_b:

$$S[q_k(t)] = \int_{t_a}^{t_b} dt\, L(t, q_1, q_2, \ldots, \dot{q}_1, \dot{q}_2, \ldots) \equiv \int_{t_a}^{t_b} dt\, L(t, q_k, \dot{q}_k), \qquad (4.110)$$

where it is understood that the particle or system of particles begins at some definite position $(q_1, q_2, \ldots)_a$ at time t_a and ends at some definite position $(q_1, q_2, \ldots)_b$ at time t_b.

Hamilton's principle then proposes that, for trajectories $q_k(t)$ where the action S is *stationary* – i.e., when

$$\delta S = \delta \int_{t_a}^{t_b} L(t, q_k, \dot{q}_k)\, dt = 0, \qquad (4.111)$$

the coordinates $q_k(t)$ satisfy the equations of motion for the system with the prescribed boundary conditions at t_a and t_b. That is, the variational principle $\delta S = 0$ gives the Lagrange equations

$$\frac{d}{dt}\frac{\partial L}{\partial \dot{q}_k} - \frac{\partial L}{\partial q_k} = 0 \qquad (k = 1, 2, \ldots, N), \qquad (4.112)$$

which are the second-order differential equations of motion for the system.

We also defined the Hamiltonian of the system to be

$$H \equiv \dot{q}_k\, p_k - L, \qquad (4.113)$$

where a sum over k is implied, and the generalized momenta p_k are defined as

$$p_k = \frac{\partial L}{\partial \dot{q}_k}. \qquad (4.114)$$

Then

$$\frac{\partial L}{\partial t} = -\frac{dH}{dt}, \qquad (4.115)$$

so if L is not an explicit function of time, it follows that the Hamiltonian is conserved. In this case, and also when a coordinate is cyclic, we are led to first integrals of motion – that is, first-order differential equations.

If the problem is simple, exact analytic solutions of the first- and second-order differential equations may be possible. If not, we can always solve the equations numerically on a computer. There is also a very common intermediate situation, when one or more of the equations is too difficult to solve exactly, but approximate analytic techniques can be used to find the answer as accurately as required. Finding a sufficiently accurate approximate analytic technique to solve a particular problem can be very entertaining, requiring as much creativity as setting up the physical problem in the first place. One of the pleasures of doing physics is to find a way, no holds barred, to solve a problem to the extent we need it solved, using back-of-the-envelope calculations, dimensional reasoning, approximate analytical techniques, numerical calculations, or whatever it takes to get the job done.

Note that we have confined ourselves in this chapter to *nonrelativistic* motion. Is it possible to generalize the action and the Lagrangian to provide the correct equations of motion for relativistic particles? The answer is a somewhat qualified "yes," but we will have to postpone a discussion of this interesting topic until Chapter 8, where we will find how to deal with relativistically correct electromagnetic potential energies.

This chapter has concentrated on a primary cornerstone of the modern subject of theoretical mechanics. It has presented a reformulation of mechanics by means of a variational principle, where the central quantity is the Lagrangian. All of dynamics is packaged into a single functional. Beyond the elegance of the approach, the Lagrangian formalism gives us two additional benefits. First, by eliminating the need to keep track of a myriad of forces, it helps us zero onto the relevant mechanics and the associated differential equations; thus, it provides us a powerful new technical tool for tackling complex mechanics problem. Second, as we shall see in the next chapter, the formalism presents a natural conceptual bridge from the classical to the quantum world. Historically, the Lagrangian and the related Hamiltonian formalisms (see Chapter 11) played key roles in the development of the subject of quantum mechanics and modern physics.

Problems

★ **Problem 4.1** In Example 4.3, we found the equation of motion of a block on an inclined plane, using the generalized coordinate X, the distance of the block from the bottom of the incline. Solve the equation for $X(t)$ in terms of an arbitrary initial position $X(0)$ and velocity $\dot{X}(0)$.

★ **Problem 4.2** A particle of mass m slides inside a smooth hemispherical bowl of radius R. Beginning with spherical coordinates r, θ, and φ to describe the dynamics, select generalized coordinates, write the Lagrangian, and find the differential equations of motion of the particle.

★★ **Problem 4.3** Example 4.2 featured a bead sliding on a vertically oriented helix of radius R. The angle θ about the symmetry axis was related to its vertical coordinate z on the wire by $\theta = \alpha z$. There is a uniform gravitational field g vertically downward. (a) Rewrite the Lagrangian and find the Lagrange equation, using θ as the generalized coordinate. (b) Are there any conserved dynamical quantities? (c) Write the simplest differential equation of motion of the bead, and go as far as you can to solve analytically for θ as a function of time.

★★ **Problem 4.4** One end of a wire is tied to a point A on the ceiling and the other end is tied to a point on a ring of radius R and negligible mass. The ring therefore hangs from the wire in a vertical plane and in a gravitational field g. A bead of mass m is threaded onto the ring so it can slide around the ring without friction. The lowest point on the ring is then tied to a second wire whose opposite end is attached to point B on the floor, where point B is directly beneath point A. The wires are then drawn taut. If the ring and attached wires are made to twist sideways through an angle φ away from equilibrium, a potential energy $(1/2)\kappa\varphi^2$ is set up in the wire. (a) Using angles θ and φ as generalized coordinates, where θ is the angle of the bead down from the top of the ring, find the kinetic and potential energies of the bead. (b) Find the equations of motion using Lagrange's equations. Assume that during the motion of the bead it remains entirely on one side of the ring, so it does not meet the wires at $\theta = 0$ and $\theta = \pi$.

★★   **Problem 4.5** A particle moves in a cylindrically symmetric potential $U(\rho, z)$. Use cylindrical coordinates ρ, φ, and z to parameterize the space.

   (a) Write the Lagrangian for an unconstrained particle of mass m (using cylindrical coordinates) in the presence of this potential.

   (b) Write the Lagrange equations of motion for ρ, φ, and z.

   (c) Identify any cyclic coordinates, and write a first integral corresponding to each.

★   **Problem 4.6** A particle of mass m slides inside a smooth paraboloid of revolution whose axis of symmetry z is vertical. The surface is defined by the equation $z = \alpha\rho^2$, where z and ρ are cylindrical coordinates, and α is a constant. There is a uniform gravitational field g. (a) Select two generalized coordinates for m. (b) Find T, U, and L. (c) Identify any ignorable coordinates, and any conserved quantities. (d) Show that there are two first integrals of motion, and find the corresponding equations.

★   **Problem 4.7** Repeat the preceding problem for a particle sliding inside a smooth cone defined by $z = \alpha r$.

★   **Problem 4.8** A spring pendulum features a pendulum bob of mass m attached to one end of a spring of force constant k and unstretched length R. The other end of the spring is attached to a fixed point on the ceiling. The pendulum is allowed to swing in a plane. Use r, the distance of the bob from the fixed point, and θ, the angle of the spring relative to the vertical, as generalized coordinates. (a) Find the kinetic and potential energies of the bob in terms of the generalized coordinates and velocities. (b) Find the Lagrangian. Are there any ignorable coordinates? (c) Are there any conserved quantities? (d) Find a complete set of equations of motion, including as many first integrals as possible.

★★★   **Problem 4.9** A pendulum is constructed from a bob of mass m on one end of a light string of length D. The other end of the string is attached to the top of a circular cylinder of radius R $(R < 2D/\pi)$. The string makes an angle θ with the vertical, as shown in Prob. 4.9, as the pendulum swings in the plane. There is a uniform gravity g directed downward. (a) Find the Lagrangian of the bob, using θ as the generalized

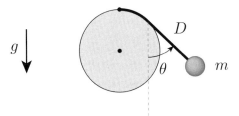

Prob. 4.9   A pendulum hanging from a cylinder.

coordinate. (b) Identify any first integrals of motion. (c) Find the frequency of small oscillations about the stable equilibrium point.

★   **Problem 4.10** A particle moves with a cylindrically symmetric potential energy $U = U(\rho, z)$, where ρ, φ, z are cylindrical coordinates. (a) Write the Lagrangian for an unconstrained particle of mass m in this case. (b) Are there any cyclic coordinates? If so, what symmetries do they correspond to, and what are the resulting constants of motion? (c) Write the Lagrange equation for each cyclic coordinate. (d) Find the Hamiltonian H. Is it conserved? (e) Find the total energy E. Is $E = H$? is E conserved? (f) Write the simplest (*i.e.*, lowest-order) complete set of differential equations of motion of the particle.

★★   **Problem 4.11** A plane pendulum is made with a plumb bob of mass m hanging on a Hooke's-law spring of negligible mass, force constant k, and unstretched length ℓ_0. The spring can stretch but is not allowed to bend. There is a uniform downward gravitational field g. (a) Select generalized coordinates for the bob, and find the Lagrangian in terms of them. (b) Write out the Lagrange equations of motion. (c) Are there any conserved quantities? If so, write down the corresponding conservation law(s). (d) If the bob is swinging back and forth, find the frequency of small oscillations in the general case where the spring can change its length while the bob is swinging back and forth.

★   **Problem 4.12** *Motion in a slowly changing uniform electric field.* A particle of mass m and charge q moves within a parallel-plate capacitor whose charge Q decays exponentially with time, $Q = Q_0 e^{-t/\tau}$, where τ is the time constant of the decay. Find the equations of motion of the particle. Ignore the effect of any magnetic field that may be generated from the changing eletric field.

★   **Problem 4.13** A particle of mass m travels between two points $x = 0$ and $x = x_1$ on earth's surface, leaving at time $t = 0$ and arriving at $t = t_1$. The gravitational field g is uniform. (a) Suppose m moves along the ground (keeping altitude $z = 0$) at steady speed. Find the total action S to go by this path. (b) Suppose instead that m moves along the least-action parabolic path. Show that the action in this case is

$$S = \frac{mx_1^2}{2t_1} - \frac{mg^2 t_1^3}{24}$$

and verify that it is less than the action for the straight-line path of part (a).

★★   **Problem 4.14** Suppose the particle of the preceding problem moves instead at constant speed along an isoceles triangular path between the beginning point and the endpoint, with the high point at height z_1 above the ground, at $x = x_1/2$ and $t = t_1/2$. (a) Find the action for this path. (b) Find the altitude z_1 corresponding to the least-action path among this class of constant-speed triangular paths. (c) Verify that the total action for this path is greater than that of the parabolic path of the preceding problem.

★★   **Problem 4.15** A plane pendulum consists of a light rod of length R supporting a plumb bob of mass m in a uniform gravitational field g. The point of support of the top end of the rod is forced to oscillate back and forth in the horizontal direction with $x = A \cos \omega t$. Using the angle θ of the bob from the vertical as the generalized coordinate, (a) find the Lagrangian of the plumb bob. (b) Are there any conserved dynamical quantities? (c) Find the simplest differential equation of motion of the bob.

★★   **Problem 4.16** Solve the preceding problem if instead of being forced to oscillate in the horizontal direction, the upper end of the rod is forced to oscillate in the vertical direction with $y = A \cos \omega t$.

★★   **Problem 4.17** A particle of mass m on a frictionless tabletop is attached to one end of a light string. The other end of the string is threaded through a small hole in the tabletop, and held by a person under the table. If given a sideways velocity v_0, the particle circles the hole with radius r_0. At time $t = 0$ the mass reaches an angle defined to be $\theta = 0$ on the tabletop, and the person under the table pulls on the string so that the length of the string above the table becomes $r(t) = r_0 - \alpha t$ for a period of time thereafter, where α is a constant. Using θ as the generalized coordinate of the particle, find its Lagrangian, identify any conserved quantities, find its simplest differential equation of motion, and get as far as you can using analytic means alone toward finding the solution $\theta(t)$ (or $t(\theta)$).

★   **Problem 4.18** A rod is bent in the middle by angle α. The bottom portion is kept vertical and the top portion is therefore oriented at angle α to the vertical. A bead of mass m is slipped onto the top portion and the bottom portion is forced by a motor to rotate at constant angular speed ω about the vertical axis. (a) Define a generalized coordinate for the bead and write down the Lagrangian. (b) Identify any conserved quantity or quantities and explain why it (or they) are conserved. (c) Find the generalized momentum of the bead and the Hamiltonian. (d) Are there any equilibrium points of the bead? If so, are they stable or unstable?

★★   **Problem 4.19** A wire is bent into the shape of a cycloid, defined by the parametric equations $x = A(\varphi + \sin \varphi)$ and $y = A(1 - \cos \varphi)$, where φ is the parameter $(-\pi < \varphi < \pi)$, and A is a constant. The wire is in a vertical plane, and is spun at constant angular velocity ω about a vertical axis through its center. A bead of mass m is slipped onto the wire. (a) Find the Lagrangian of the bead, using the parameter φ as the generalized coordinate. (b) Identify any first integral of motion of the bead. (c) Reduce the problem to quadrature. That is, show that the time t can be expressed as an integral over φ.

Problem 4.20 *Center of mass and relative coordinates.* Show that for two particles moving in one dimension, with coordinates x_1 and x_2, with a potential that depends only upon their separation $x_2 - x_1$, the Lagrangian

$$L = \frac{1}{2}m_1\dot{x}_1^2 + \frac{1}{2}m_2\dot{x}_2^2 - U(x_2 - x_1)$$

can be rewritten in the form

$$L = \frac{1}{2}M\dot{X}^2 + \frac{1}{2}\mu\dot{x}'^2 - U(x'),$$

where $M = m_1 + m_2$ is the total mass and $\mu = m_1 m_2/M$ is the "reduced mass" of the system, and $X = (m_1 x_1 + m_2 x_2)/M$ is the center of mass coordinate and $x' = x_2 - x_1$ is the relative coordinate.

Problem 4.21 Two blocks of equal mass m, connected by a Hooke's-law spring of unstretched length ℓ, are free to move in one dimension. Find the equations of motion of the system, using the relative and center of mass coordinates introduced in the preceding problem.

Problem 4.22 A small block of mass m and a weight of mass M are connected by a string of length D. The string has been threaded through a small hole in a tabletop, so the block can slide without friction on the tabletop, while the weight hangs vertically beneath the tabletop. We can let the hole be the origin of coordinates, and use polar coordinates r, θ for the block, where r is the block's distance from the hole, and z for the distance of the weight below the tabletop. (a) Using generalized coordinates r and θ, write down the Lagrangian of the system of block plus weight. (b) Write down a complete set of first integrals of the motion, explaining the physical meaning of each. (c) Show that the first integrals can be combined to give an equation of the form

$$E = \frac{1}{2}(M + m)\dot{r}^2 + U_{\text{eff}}(r)$$

and write out an expression for $U_{\text{eff}}(r)$. (d) Find the radius of a circular orbit of the block in terms of constants of motion. (e) Now suppose the block executes small oscillations about a circular orbit. What is the frequency of these oscillations? Is the resulting orbit of the block open or closed? That is, does the perturbed orbit of the block continually return to its former position or not?

Problem 4.23 Example 1.3 of Chapter 1 proposed that mined material on the moon might be projected off the moon's surface by a rotating boom that *slings* the material into space. Assume the boom rotates in a horizontal plane with constant angular velocity ω, and let r, the distance of the payload from the rotation axis at one end of the boom, be the single generalized coordinate. (a) Find the Lagrangian for a bucket of material of mass m that moves along the boom. (b) Find its equation of motion. (c) Solve it for $r(t)$, subject to the initial conditions $r = r_0$ and $\dot{r} = 0$ at $t = 0$. (d) If the boom has length R, find the radial and tangential components of the bucket's velocity, and its total speed, as it emerges from the end of the boom. (e) Find the power input $P = dE/dt$ into a bucket of mass m as a function of time. Is the power input larger at the beginning or end of the bucket's journey along the boom? (f) Find the torque exerted by the boom on the bucket, as a function of the position r of the bucket on the boom. There would be an equal but opposite torque back on the boom, caused by the bucket, which might break the boom. At what part of the bucket's journey would this torque most likely break the boom? (g) If $R = 100$ m and $r_0 = 1$ m, what must be the rotational period of the boom so that buckets will reach the moon's escape speed as they fly off the boom?

★★★   **Problem 4.24** Consider a vertical circular hoop of radius R rotating about its vertical symmetry axis with constant angular velocity Ω. A bead of mass m is threaded onto the hoop, so is free to move along the hoop. Let the angle θ of the bead be measured up from the bottom of the hoop.

(a) Write the Lagrangian in terms of the generalized coordinate θ. Are there any first integrals of motion?

(b) Show that there are *two* equilibrium angles of the bead for sufficiently small angular velocities Ω, but that there are *four* equilibrium angles if Ω is sufficiently large.

(c) For each of these equilibrium angles, find out whether that position of the bead is *stable* or *unstable*. That is, if the bead is displaced slightly from equilibrium, does it tend to move back toward the equilibrium angle, or does it depart farther and farther from it?

★   **Problem 4.25** In certain situations, it is possible to incorporate frictional effects in a simple way into a Lagrangian problem. As an example, consider the Lagrangian

$$L = e^{\gamma t} \left(\frac{1}{2} m\dot{q}^2 - \frac{1}{2} kq^2 \right).$$

(a) Find the equation of motion for the system.

(b) Do a coordinate change $s = e^{\gamma t/2} q$. Rewrite the dynamics in terms of s.

(c) How would you describe the system?

★★   **Problem 4.26** Consider a particle moving in three dimensions with Lagrangian $L = (1/2)m(\dot{x}^2 + \dot{y}^2 + \dot{z}^2) + a\dot{x} + b$, where a and b are constants. (a) Find the equations of motion and show that the particle moves in a straight line at constant speed, so that it must be a free particle. (b) The result of (a) shows that there must be another reference frame (x', y', z') such that the Lagrangian is just the usual free-particle Lagrangian $L' = (1/2)m(\dot{x}'^2 + \dot{y}'^2 + \dot{z}'^2)$. However, L' may also be allowed an additive constant, which cannot show up in Lagrange's equations. Find the Galilean transformation between (x, y, z) and (x', y', z') and find the velocity of the new primed frame in terms of a and b.

★ **Problem 4.27** Consider a Lagrangian $L' = L + df/dt$, where the Lagrangian is $L = L(q_k, \dot{q}_k, t)$ and the function $f = f(q_k, t)$. (a) Show that $L' = L'(q_k, \dot{q}_k, t)$, so that it depends upon the proper variables. Show that this would not generally be true if f were allowed to depend upon the $\dot{q}_k$ s. (b) Show that L' obeys Lagrange's equations if L does, by substituting L' into Lagrange's equations. Therefore, the equations of motion are the same using L' as using L, so the Lagrangian of a particle is not unique. (This problem requires care in taking total and partial derivatives!)

★ **Problem 4.28** Show that the function L' given in the preceding problem must obey Lagrange's equations if L does, directly from the principle of stationary action. Lagrange's equations do not have to be written down for this proof!

★★ **Problem 4.29** Consider the Lagrangian $L' = m\dot{x}\dot{y} - kxy$ for a particle free to move in two dimensions, where x and y are Cartesian coordinates, and m and k are constants. (a) Show that this Lagrangian gives the equations of motion appropriate for a two-dimensional simple harmonic oscillator. Therefore, as far as the motion of the particle is concerned, L' is equivalent to $L = (1/2)m(\dot{x}^2 + \dot{x}^2) - (1/2)k(x^2 + y^2)$. (b) Show that L' and L do *not* differ by the total time derivative of any function $f(x, y)$. Therefore, L' is not a member of the class of Lagrangians mentioned in the preceding problems, so there are even more Lagrangians describing a particle than suggested before.

★★ **Problem 4.30** Consider a Lagrangian that depends on second derivatives of the coordinates

$$L = L(q_k, \dot{q}_k, \ddot{q}_k, t).$$

Through the variational principle, find the resulting differential equations of motion.

★★ **Problem 4.31** A pendulum consists of a plumb bob of mass m on the end of a string that swings back and forth in a plane. The upper end of the string passes through a small hole in the ceiling, and the angle θ of the bob relative to the vertical changes with time as it swings back and forth. The string is pulled upward at constant rate through the hole, so the length R of the pendulum decreases at a constant rate, with $dR/dt = -\alpha$. (a) Find the Lagrangian of the bob, using θ as the generalized coordinate. (b) Find the Hamiltonian H. Is H equal to the energy E? Why or why not? (c) Is either H or E conserved? Why or why not?

★ **Problem 4.32** A spherical pendulum consists of a particle of mass m on the end of a string of length R. The position of the particle can be described by a polar angle θ and an azimuthal angle φ. The length of the string decreases at the rate $dR/dt = -f(t)$, where $f(t)$ is a positive function of time. (a) Find the Lagrangian of the particle, using θ and φ as generalized coordinates. (b) Find the Hamiltonian H. Is H equal to the energy? Why of why not? (c) Is either E or H conserved? Why or why not?

Problem 4.33 The Hamiltonian of a bead on a parabolic wire turning with constant angular velocity ω is

$$H = \frac{1}{2}m[(1 + 4\alpha^2 r^2)\dot{r}^2 - r^2\omega^2] + mg\alpha r^2,$$

where H is a constant. Reduce the problem to quadrature: that is, find an equation for the time t is terms of an integral over r.

Problem 4.34 A bead of mass m is placed on a vertically oriented circular hoop of radius R that is forced to rotate with constant angular velocity ω about a vertical axis through its center. (a) Using the polar angle θ measured up from the bottom as the single generalized coordinate, find the kinetic and potential energies of the bead. (Remember that the bead has motion due to the forced rotation of the hoop as well as motion due to changing θ.) (b) Find the bead's equation of motion using Lagrange's equation. (c) Is its energy conserved? Why or why not? (d) Find its Hamiltonian. Is H conserved? Why or why not? Is $E = H$? Why or why not? (e) Find the equilibrium angle θ_0 for the bead as a function of the hoop's angular velocity ω. Sketch a graph of θ_0 vs. ω. Notice that there is a "phase transition" at a certain critical velocity ω_{crit}. (f) Find the frequency of small oscillations of the bead about the equilibrium angle θ_0, as a function of ω.

Problem 4.35 A wire is bent into the shape of a quartic function $y = ax^4$ and oriented in a vertical plane, with x horizontal, y vertical, and a a positive constant. A bead of mass m is threaded onto the wire, and the wire is then forced to rotate with constant angular velocity Ω about the y axis. (a) Let x be the generalized coordinate for the bead and find its Lagrangian. (b) Is the bead's energy conserved? Why or why not? (c) Is the bead's angular momentum conserved about the vertical axis? Why or why not? (d) Find the bead's Hamiltonian. Is H conserved? Why or why not? (e) Is $E = H$? Why or why not? (f) Given $\Omega > 0$, are there any equilibrium positions of the bead? (g) If so, is each stable or unstable? For any stable equilibrium position, find the frequency ω of small oscillations about the equilibrium point, expressed as a multiple of Ω.

Problem 4.36 In Example 4.8 we analyzed the case of a bead on a rotating parabolic wire. The energy of the bead was not conserved, but the Hamiltonian was:

$$H = \frac{1}{2}m(1 + 4\alpha^2 r^2)\dot{r}^2 + \text{``}U_{\text{eff}}\text{''} = \text{constant},$$

where

$$U_{\text{eff}} = \frac{1}{2}mr^2(2g\alpha^2 - \omega^2).$$

There is an equilibrium point at $r = 0$ which is unstable if $\omega > \omega_0 \equiv \sqrt{2g}\,\alpha$, neutrally stable if $\omega = \omega_0$, and stable if $\omega < \omega_0$. Find the frequency of small oscillations about $r = 0$ if $\omega < \omega_0$.

★★ **Problem 4.37** A wire bent in the shape of a hyperbolic cosine function $y = a\cosh(x/x_0)$ is supported in a vertical plane, where x and y are the horizontal and vertical coordinates, respectively, and a and x_0 are positive constants. A bead of mass m is threaded onto the wire and is free to slide without friction along it, and is subject to uniform gravity g directed downward. (a) Find Lagrange's equations for the bead using x as the generalized coordinate, and (b) find the frequency of small oscillations of the bead about the lowest point of the wire.

★★ **Problem 4.38** The wire described in the preceding problem is now forced to rotate about its vertical axis of symmetry with constant angular velocity Ω. (a) Find Ω_c, the critical value of Ω for which the equilibrium point at $x = 0$ is no longer a *stable* equilibrium point, and find the values of x for which there is then a stable equilibrium point for the bead. (b) Find the frequency of small oscillations of the bead about each of any new equilibrium points.

★★★ **Problem 4.39** One point on a horizontal circular wire C of radius R is attached to a thin, vertical axle which turns at constant angular velocity Ω about the vertical axis, causing C to turn around with it. A bead of mass m moves without friction on C. (a) Show that relative to C the bead oscillates like a pendulum. (b) Find the frequency of small-amplitude oscillations of the bead in terms of any or all of R, Ω, and m.

★★ **Problem 4.40** A frictionless slide is constructed in the shape of a cycloid. The horizontal coordinate x and vertical coordinate y of the slide are given in parametric form by

$$x = A(\varphi + \sin\varphi) \quad y = A(1 - \cos\varphi),$$

where A is a constant. Here the y coordinate is positive *upward*. The slide is the portion of the cycloid with $= -\pi \le \varphi \le \pi$, with the bottom of the slide corresponding to $\varphi = 0$. There is a uniform gravitational field g in the negative y direction. (a) Find the Lagrangian of a small block of mass m moving along the slide, using φ as the generalized coordinate. (b) The block will oscillate back and forth near the bottom of the slide. Is its motion simple harmonic in the limit of small amplitudes? If not, explain why not; if so, find the angular frequency of oscillation ω in terms of any or all of A, m, and g.

★★★ **Problem 4.41** A mass M_1 is hung on an unstretchable string A, and the other end of string A is passed over a fixed, frictionless, nonrotating pulley P_1, as shown in Prob. 4.41. This other end of string A is then attached to the center of a second frictionless, nonrotating pulley P_2 of mass M_2, over which is passed a second nonstretchable string B, one end of which is attached to a hanging mass m_1 while the other end is attached to a hanging mass m_2. Let $X_1(t)$ be the length of string A beneath the center of pulley P_1; $X_2(t)$ be the length of string A beneath the center of P_1; $x_1(t)$ be the length of string B beneath the center of pulley P_2; and $x_2(t)$ be the length of string B beneath the center of pulley P_2. There is a uniform gravity g downward. (a) Find the total kinetic energy of the system, in terms of X_1, x_1, and the various masses. (b) Find the total potential energy of the system, measured

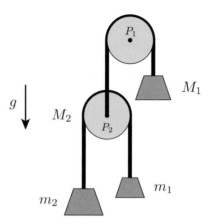

Three masses and two pulleys.

from the center of fixed pulley P_1. (c) Find the Lagrangian of the system. (d) Find the acceleration of mass M_1 in terms of given quantities.

★   **Problem 4.42** Consider a Lagrangian of the form

$$L = mc^2 \left(1 - \sqrt{1 - v^2/c^2}\right) - U(x, y, z).$$

Show that the resulting Lagrange equations give Newton's second law $\mathbf{F} = d\mathbf{p}/dt$ for a relativistic particle, if $F^i = -\partial U/\partial x^i$.

5

From Classical to Quantum and Back

We come now to the first of our "capstone chapters," in which we reflect upon topics from the prior four chapters in the light of contemporary physics. In these chapters we will illustrate how classical mechanics is related to modern physics, especially quantum mechanics and general relativity. Classical mechanics played a key role in developing modern physics in the first place, and in turn modern physics has given us deeper insights into the meaning and validity of classical mechanics.

In the first four chapters of this book we have reviewed Newtonian mechanics and Einstein's extension of it into the relativistic regime; we have also introduced variational calculus and used it to reformulate mechanics through *Hamilton's principle of stationary action*. However, classical mechanics, even extended into the realm of special relativity, has its limitations. It arises as a special case of the vastly more comprehensive theory of quantum mechanics. *Where* does classical mechanics fall short, and *why* is it limited?

The key to understanding this is Hamilton's principle. We will show how this principle comes about as a special case of the larger quantum theory. And so, since we can use Hamilton's principle to derive classical mechanics, we will reach a good understanding of when classical mechanics is valid and when we have to use the full apparatus of quantum theory. In this chapter we aim to set classical mechanics in context.

We begin with the behavior of waves in classical physics, and then show results of some critical experiments that upset traditional notions of light as waves and atoms as particles. We proceed to give a brief review of Richard Feynman's sum-over-paths formulation of quantum mechanics, which describes the actual behavior of light and atoms, and then show that Hamilton's principle emerges naturally in a certain limiting case.

5.1 Classical Waves

A string with uniform mass per unit length μ is held in a horizontal position under uniform tension T. What happens if we disturb the string? In particular, suppose that at time $t = 0$ we give the string some particular profile $y(x)$, and some velocity distribution $\partial y(x, t)/\partial t|_{t=0}$, where x is the horizontal coordinate and y is the transverse displacement (see Figure 5.1). Our goal is to find $y(x, t)$, the shape of the string at any later time in the regime, where y is small compared to the string's full length.

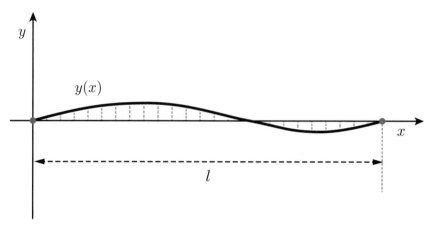

Fig. 5.1   A transverse small displacement of a string, with $y \ll l$.

Consider a very small slice of string of length Δx and mass $\Delta m = \mu \Delta x$, as shown in Figure 5.2. We ignore gravity, so the only forces acting on this piece are the string tensions to the right and to the left of it. If the string is displaced from equilibrium, the slice will generally be slightly curved. The two tension forces on the right and left therefore pull in slightly different directions, as shown in Figure 5.2, so the resulting unbalanced force causes Δm to accelerate. For small vertical displacements, the horizontal component of T remains essentially constant along the string, so Δm accelerates vertically, not horizontally.

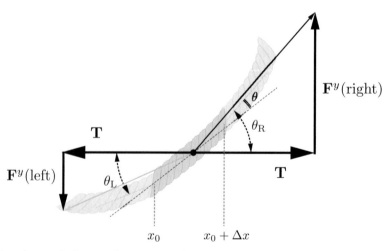

Fig. 5.2   A small slice of string; the horizontal components of the tension are labeled as T and must add to zero since the string does not move horizontally. The angles shown are exaggerated for clarity.

Let the left-hand end of Δm be located at x_0 and the right-hand end at $x_0 + \Delta x$. We can relate the slopes of the shape at x_0 and $x_0 + \Delta x$ to the angles θ_L and θ_R shown in the figure:

$$\tan \theta_L = \left.\frac{dy}{dx}\right|_{x_0} , \quad \tan \theta_R = \left.\frac{dy}{dx}\right|_{x_0 + \Delta x} . \tag{5.1}$$

For small displacement y, the angles θ_L and θ_R are small, so that

$$\theta_L \simeq \tan \theta_L, \ \theta_R \simeq \tan \theta_R. \tag{5.2}$$

Using Taylor series (see Appendix F)), the slopes of the string at the left and right are related by

$$\left.\frac{\partial y}{\partial x}\right|_{x_0 + \Delta x} = \left.\frac{\partial y}{\partial x}\right|_{x_0} + \left.\frac{\partial^2 y}{\partial x^2}\right|_{x_0} \Delta x + \cdots \tag{5.3}$$

If $y(x)$ is smooth and Δx is sufficiently small, we can neglect all but these first two terms. The vertical forces on the right-hand and left-hand sides of the slice Δm are

$$F^y(\text{left}) = T \tan \theta_L = T \left.\frac{\partial y}{\partial x}\right|_{x_0} \quad \text{and} \quad F^y(\text{right}) = -T \tan \theta_R = -T \left.\frac{\partial y}{\partial x}\right|_{x_0 + \Delta x} , \tag{5.4}$$

with the force at the left upward and the force at the right downward. The *net* vertical force on Δm is therefore

$$F^y{}_{\text{net}} = T \left[\left.\frac{\partial y}{\partial x}\right|_{x_0 + \Delta x} - \left.\frac{\partial y}{\partial x}\right|_{x_0} \right] = T \left.\frac{\partial^2 y}{\partial x^2}\right|_{x_0} \Delta x, \tag{5.5}$$

using Eq. (5.3). The vertical acceleration of Δm is $\partial^2 y/\partial t^2$, the second derivative of y with respect to *time*, keeping now the position x fixed. The left end of the slice of string moves up and down vertically, with position $y(x_0, t)$, velocity $\partial y/\partial t|_{x_0}$, and acceleration $\partial^2 y/\partial t^2|_{x_0}$. Newton's second law $F^y{}_{\text{net}} = \Delta m\, a$ therefore becomes

$$T \left.\frac{\partial^2 y}{\partial x^2}\right|_{x_0} \Delta x = \Delta m \left.\frac{\partial^2 y}{\partial t^2}\right|_{x_0} = \mu \Delta x \left.\frac{\partial^2 y}{\partial t^2}\right|_{x_0} , \tag{5.6}$$

so

$$\frac{\partial^2 y}{\partial x^2} - \left(\frac{\mu}{T}\right) \frac{\partial^2 y}{\partial t^2} = 0. \tag{5.7}$$

This is the **wave equation** of the string for small transverse displacements. It represents an infinite number of $F = ma$ equations, one for each value of x, for the infinite number of infinitesimal slices of the string.

For a single particle, an initial position and initial velocity determine the future position – given mass and forces – by solving ordinary differential equations. For a string, an initial shape $y(x, 0)$ and a velocity distribution $\partial y/\partial t|_{t=0}$ determine the future shape $y(x, t)$ – given the mass density μ and tension T – by solving a *partial* differential equation, which is the wave equation (5.7). It is most convenient to rewrite the wave equation as

$$\frac{\partial^2 y}{\partial x^2} - \frac{1}{v^2}\frac{\partial^2 y}{\partial t^2} = 0, \tag{5.8}$$

where v is a constant whose role will become apparent soon. In this case, we have

$$v = \sqrt{\frac{T}{\mu}}. \tag{5.9}$$

The wave equation is linear, which is a consequence of assuming that the displacements are small. As a result, it satisfies the **superposition principle**: two separate solutions of the equation can simply be added, and the result still satisfies the wave equation. A solution of the wave equation (5.7) consists of *any* two functions $f(u)$ and $g(u)$ of the combination $u = x \pm v t$:

$$y_{sol} = f(x + v t) + g(x - v t), \tag{5.10}$$

as can easily be verified by substituting this expression into Eq. (5.8). This implies that once $y(x,0)$ and $y'(x,0)$ are given, f and g are fixed. The wave equation simply tells us that the profiles of f and g evolve in time, without distortion, at speed v toward negative and positive x, respectively. An easy way to see this is to sketch the two functions at two instants in time (see Figure 5.3).

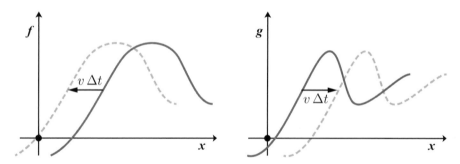

Fig. 5.3   The time evolution of the f and g profiles as dictated by the wave equation.

Note also that the wave equation is *real* in addition to being linear. This means that we can solve it with complex functions as well, with the real and imaginary parts as separate solutions: the equation splits into a real part and an imaginary part, each looking identical in form, applied to the real and imaginary parts of the complex solution.

Classical waves propagate also in fluids like air or water, characterized by macroscopic properties such as mass density and pressure. If a fluid is locally perturbed, sound waves can be set up in which both the local density and pressure oscillate, and the oscillations are propagated from the initial site throughout the material. In the case of small-amplitude waves, the density of the material has the form $\rho = \rho_0 + \Delta\rho$ and the pressure is $p = p_0 + \Delta p$, where ρ_0 and p_0 are the ambient density and pressure, and $\Delta\rho$ and Δp are the small perturbations that can propagate from place to place. We won't prove it here, but the disturbance $\Delta\rho$ obeys a similar wave equation to (5.8), but in three dimensions:

$$\nabla^2(\Delta\rho) - \left(\frac{\rho_0}{B}\right)\frac{\partial^2(\Delta\rho)}{\partial t^2} = 0, \tag{5.11}$$

where B is the **bulk modulus** of the material, a measure of the stiffness of the fluid defined by

$$B \equiv \frac{1}{\rho_0}\frac{dp}{d\rho}\bigg|_{\rho_0}. \tag{5.12}$$

The less the fractional change in density for a given pressure change, the greater the stiffness, and the greater the bulk modulus. The differential operator ∇^2 is the **Laplacian**, which in Cartesian coordinates takes the form

$$\nabla^2 = \frac{\partial^2}{\partial x^2} + \frac{\partial^2}{\partial y^2} + \frac{\partial^2}{\partial z^2}. \tag{5.13}$$

In the case of an **infinite plane wave** propagating in the x direction – that is, $\Delta\rho(x, y, z, t) = \Delta\rho(x, t)$ – the wave equation becomes

$$\frac{\partial^2(\Delta\rho)}{\partial x^2} - \left(\frac{\rho}{B}\right)\frac{\partial^2(\Delta\rho)}{\partial t^2} = 0, \tag{5.14}$$

which has the same form as the wave equation for a string (5.8) with speed

$$v = \sqrt{\frac{B}{\rho_0}}. \tag{5.15}$$

Another well-known classical wave is the *electromagnetic* wave, a hallmark of Maxwell's equations of electrodynamics. In vacuum the equations can be combined to produce a wave equation

$$\nabla^2\mathbf{E} - \frac{1}{c^2}\frac{\partial^2\mathbf{E}}{\partial t^2} = 0 \tag{5.16}$$

for the electric field $\mathbf{E}$, with a similar equation for the magnetic field $\mathbf{B}$. The electric and magnetic fields propagate together at the speed of light $v = c$, so Maxwell was able to achieve a grand synthesis of electricity, magnetism, and optics by showing that light waves are in fact electromagnetic waves. Again, the one-dimensional form of these equations, corresponding to an electromagnetic plane wave propagating in the x direction, has the same form as waves on a string (5.8). One difference here is that the electromagnetic wave equation is valid for *any* classical electromagnetic wave, whatever its amplitude. No linear approximation has to be made.

Among the infinite variety of solutions (5.10) to the one-dimensional wave equation – whether for waves on a string, sound waves, or light waves – are

sinusoidal traveling waves that propagate to the right or to the left. We write (see Figure 5.4)

$$y_{\text{sol}} = A_0 \sin \left[k \left(x \mp \frac{\omega}{k} t \right) - \varphi \right], \tag{5.17}$$

where A_0 is the **amplitude** of the wave, k is the **wave number**, ω is the **angular frequency**, and φ is the **phase angle**. Notice that these solutions are indeed a function of $x \pm vt$ as argued earlier in Eq. (5.10), with $v = \omega/k$. The upper (minus) sign corresponds to a wave traveling to the right, and the lower (plus) sign corresponds to a wave traveling to the left. The wave number is related to the wavelength λ by $k = 2\pi/\lambda$, and the angular frequency is related to the frequency ν (*i.e.*, cycles/s) by $\omega = 2\pi\nu$. The phase angle simply displaces the sinusoidal shape to the right (if φ is positive) or to the left (if φ is negative).

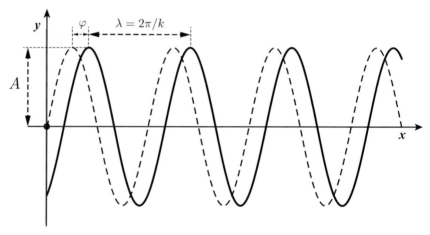

Fig. 5.4   A sinusoidal wave moving with speed $v = \omega/k$.

For waves on a string, we have the wave speed given by $v = \omega/k = \sqrt{T/\mu}$; for sound waves, we have $v = \omega/k = \sqrt{B/\rho}$; and for light waves, $v = \omega/k = c$, the speed of light.

The **intensity** I of a plane wave is proportional to the square of its amplitude A_0; that is

$$I = KA_0^2, \tag{5.18}$$

where K is a constant that depends upon the type of wave. Intensity is the energy/time passing through a unit area perpendicular to the wave velocity. For sound waves, higher intensity is higher volume; for light waves, higher intensity corresponds to brighter light.

The wave equation also has complex exponential traveling-wave solutions of the form

$$y(x, t) = A_0 e^{i(kx \mp \omega t - \varphi)}, \tag{5.19}$$

which is simple to verify by substitution, but is quite obvious from the fact that the real and imaginary parts of the complex exponential are given by Euler's formula

$$e^{i\theta} = \cos\theta + i\sin\theta. \tag{5.20}$$

So if we choose $\theta = kx \mp \omega t - \varphi$, then $y(x,t)$ is in fact the sum of two sinusoidal traveling waves with the same amplitude, frequency, and wave number, but differing in phase by $\pi/2$ (that is, the phase difference between the cosine and sine functions). Complex exponential solutions are often used in part because they are easier to work with mathematically (the derivative of an exponential is an exponential, for example). Then we can always take the real (or imaginary) part of the final result to get the physical result, which in classical physics corresponds to an observable quantity and must therefore be real. In quantum mechanics, as we will introduce in this chapter, the complex exponential form turns out to be the *natural* form to use.

The intensity of a complex wave is proportional to the product of the wave amplitude and the complex conjugate of the wave amplitude; this gives the real quantity

$$I = Ky(x,t)y^*(x,t) = K\left[A_0 e^{i(kx\mp\omega t-\varphi)}\right]\left[A_0^* e^{-i(kx\mp\omega t-\varphi)}\right] = K|A_0|^2 \tag{5.21}$$

as before.

Example 5.1 **Two-Slit Interference of Classical Waves**

When two or more traveling waves combine, we observe interference effects. Direct a plane sinusoidal wave from left to right at a double-slit system, for example, as shown in Figure 5.5. For concreteness, let us focus on the case of sound waves: the source would be a speaker. Only waves that pass through one of the slits make it through to the right-hand side. The resulting wave is then detected on a detecting plane, which is a screen or bank of detectors much farther along to the right – in this case an array of microphones. What will be observed by the detectors? Using the complex exponential form of the solution given by Eq. (5.19) at the position of the detector, we add the contributions from two slit sources by linear superposition:

$$\Delta\rho_T = \Delta\rho_1 + \Delta\rho_2 = A_0\left(e^{j(ks_1-\omega t-\varphi)} + e^{j(ks_2-\omega t-\varphi)}\right), \tag{5.22}$$

where s_1 is the distance of an observation point from slit 1 and s_2 is the distance from slit 2. Here we have used the fact that the wave number, frequency, and phase angle of each of the two contributions are the same (both slit waves originate from the same source on the left after all). We have also assumed that the amplitude A_0 of each wave as it reaches the detector is the same, which is an excellent approximation as long as the detecting plane is far away compared with the distance between the two slits (that is, we have $s_1 \sim s_2$).

The total wave at the right is then

$$\Delta\rho_T = A_0 e^{-i(\omega t+\phi)}\left(e^{jks_1} + e^{jks_2}\right). \tag{5.23}$$

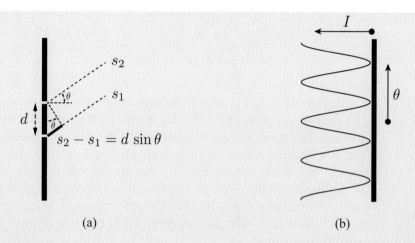

Fig. 5.5 Two paths for waves from slit system to detectors. We assume for simplicity that the detecting plane is very far from the slit system compared with the distance between the two slits, so the wave disturbances from each slit propagate essentially parallel to one another.

The intensity of the wave at the detection point becomes

$$I = K\Delta\rho_T^* \Delta\rho_T = K|A_0)|^2 \left(e^{-iks_1} + e^{-iks_2}\right)\left(e^{iks_1} + e^{iks_2}\right)$$
$$= 2K|A_0)|^2(1 + \cos(k(s_2 - s_1)))$$
$$= 4K|A_0)|^2 \cos^2(k(s_2 - s_1)/2), \tag{5.24}$$

using the identities $\cos\alpha = (e^{i\alpha} + e^{-i\alpha})/2$ and $\cos^2(\alpha/2) = (1/2)(1 + \cos\alpha)$. The difference $s_2 - s_1$ of the path lengths from the two slits to a point on the detecting plane is

$$s_2 - s_1 = d\sin\theta, \tag{5.25}$$

as shown in Figure 5.6(a), where θ is the angle between the two rays and the forward direction. The phase difference between the two waves is $\Phi \equiv k(s_2 - s_1) = (2\pi d/\lambda)\sin\theta$, so the intensity at an arbitrary angle θ, in terms of the intensity I_0 in the forward direction $\theta = 0$, is

$$I(\theta) = I_0 \cos^2(\Phi/2) \quad \text{where} \quad \Phi = \frac{2\pi d}{\lambda}\sin\theta, \tag{5.26}$$

as illustrated in Figure 5.6(b). There are alternating maxima and minima, with the maxima occurring at angles θ for which $n\lambda = d\sin\theta$, with $n = 0, \pm1, \pm2, \ldots$ These are points along the detecting plane where the sound volume is high – sandwiching points in between with zero or low volume. To discern this alternative pattern of intensity, we need to have $\lambda \leq d$ since $\sin \theta \leq 1$. The pattern predicted by Eq. (5.26) is indeed easily verified in the laboratory.

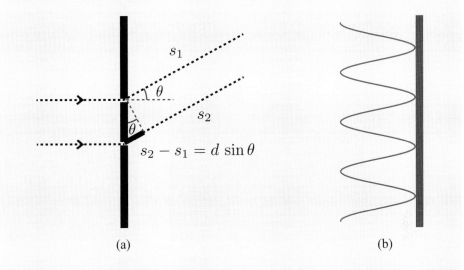

(a)   (b)

Fig. 5.6   (a) The relationship between $s_2 - s_1$, d, and θ. (b) The two-slit interference pattern.

The interference pattern from this double-slit example is typical for all classical waves. It is observed for light waves and sound waves alike, whenever two or more waves combine in the same region of space. As we shall now see, this wave-like behavior also helps us peek into the quantum nature of the world.

5.2 Two-Slit Experiments and Quantum Mechanics

According to Maxwell's equations light is an electromagnetic wave, so if we direct a beam of light at a double slit we should observe wave interference. But if we direct a beam of *atoms* at a double slit, classical mechanics teaches us that we should observe a bunch of atoms downstream of each slit, much like what would happen if we tossed ball bearings at a pair of slits. Atoms are particles, after all, so should not exhibit interference patterns. But what about actual experiments?

Light
Various light detectors can be used on the detecting plane, including photographic film, photomultipliers, charge-coupled devices, and others, depending upon the wavelength. If the wavelength of the light beam is smaller than the slit separation, a fairly bright light source is used, and fairly long exposures are made (the meaning of "fairly" here will soon become clear). The experimental intensities again show alternating maxima and minima, with maxima occurring where $n\lambda = d \sin \theta$.

But now crank the brightness of the light source way down, and observe what happens over short time intervals. Instead of seeing very low-intensity light spread over the detecting plane, as predicted by the interference formula (5.26), one finds that at first the light arrives at apparently random discrete locations. If the detector is a bank of photocells, for example, only certain cells will register the reception of light, while others (even at locations where the intensity should be a maximum) receive nothing at first. That is, light is seen to arrive in discrete lumps, or **photons** (see Figure 5.7(a)).

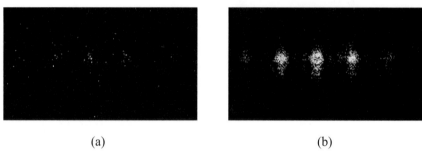

(a)   (b)

Fig. 5.7   (a) At very low-intensity light, individual photons appear to land on the screen randomly. (b) As the intensity is cranked up, the interference pattern emerges. Reprinted by permission from Ch. Kurtsiever, T. Pfau, and J. Mlynek, *Nature* **386**, 150 (1997).

The remarkable fact is that even though the photons arrive one at a time at the detectors, if we wait long enough, the large number of photons distribute themselves among the detectors *exactly as the interference formula predicts* (see Figure 5.7(b)). That is, in some sense light has both a particle nature (we observe single particles only in the detectors) and a wave nature (when huge numbers of photons have arrived at the detecting plane, the overall distribution shows the interference pattern predicted by wave theory). And if we close off one of the two slits, the pattern of photons shows no such interference.

By observing the number of photons arriving at the screen, and knowing the intensity, the wavelength λ, and hence the frequency $\nu = c/\lambda$ of the light, one finds that each photon must have an energy $E = h\nu$ and a momentum $p = E/c = h/\lambda$, where h is **Planck's constant**, $h = 6.627 \times 10^{-34}$ J $\cdot$ s. As we saw in Chapter 2, it was Albert Einstein in 1905 who first realized that light is not a continuous, Maxwellian wave after all, but consists of discrete photons, and that each photon is massless and has energy $E = h\nu$ and momentum $p = h/\lambda$.

The central puzzle is: If light consists of a stream of individual photons, so that in the case of two slits each photon presumably goes through one slit or the other slit and not both, how can they develop an interference pattern? How do photons "know" whether two slits are open or only one?

Atoms

Now project a beam of *helium* atoms at a pair of slits and observe their distribution on the detecting plane. A double-slit system has slits of width $a = 1$ μm and slit separation $d = 8$ μm. Each helium atom has mass $m = 6.68 \times 10^{-27}$ kg, and each can be detected by various counters as a discrete particle, where the detecting plane is a distance $D = 1.95$ m behind the slits. Our beam of helium atoms travels at speeds between 2.1 and 2.2 km/s.

The atoms *do* arrive at the screen in discrete lumps, as expected, but the distribution shows interference effects similar to what we observe with light! Figure 5.8 shows the actual results of this experiment.

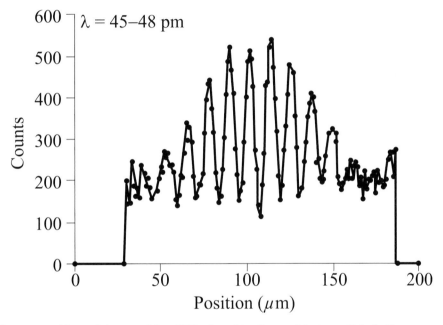

Fig. 5.8 Helium atoms with speeds between 2.1 and 2.2 km/s reaching the rear detectors, with both slits open. The detectors observe the arrival of individual atoms, *but the distribution shows a clear interference pattern as we would expect for waves!* We see how the interference pattern builds up one atom at a time. The first data set is taken after 5 min of counting, while the last is taken after 42 h of counting. Reprinted by permission from Ch. Kurtsiever, T. Pfau, and J. Mlynek; see their article in *Nature* **386**, 150 (1997).

The obvious question is: For helium atoms, as with photons, *what exactly is interfering?* The beam intensity can be turned so low that there is at most a single atom in flight at any given time, so atoms are not interfering with other atoms – yet the interference pattern eventually emerges. Each atom must be interfering with *itself* in some way. The interference distribution emerges only after many atoms have been detected. We can carry out similar experiments with atoms with different masses moving with different velocities. The results show that the wavelength λ deduced from a particular interference pattern on the screen is inversely proportional to both the atomic mass m and the velocity v of the atoms. That is:

$$\lambda = h/p, \tag{5.27}$$

where $p = mv$ is the momentum of the nonrelativistic atom and h again is Planck's constant. This is exactly the same relation between λ and p as for photons.[1] If one of the two slits is blocked off, so atoms can only penetrate one of the slits, the two-slit interference pattern goes away.

Even though an atom is detected at a specific spot much like a classical particle, the interference patterns show that an individual atom somehow "knows" whether there are two slits open or only one. How does it know that? If both slits are open, does it somehow probe both paths? *Does it in some sense take both paths?*

5.3 Feynman Sum-over-Paths

Thirty-one years ago, Dick Feynman told me about his "sum over histories" version of quantum mechanics. "The electron does anything it likes," he said. "It just goes in any direction at any speed, ... however it likes, and then you add up the amplitudes and it gives you the wave function." I said to him "You're crazy." But he wasn't. – Freeman Dyson, 1980.

According to the highly original American physicist Richard Feynman, the answer to the question posed at the end of the preceding section is *yes*! In his sum-over-paths formulation of quantum mechanics, atoms (or electrons or molecules or photons or ball bearings or anything...) *do* take all available paths between two points. If both slits are open in the experiments we have described, the particle goes through *both slits*. If one of the slits is closed, the particle goes through the remaining open slit.

How do we predict what will be observed in each case? According to quantum mechanics, there is no way we can tell where a particular photon or helium atom will go. This is not because our measuring devices do not yet have sufficient precision; it is because a particle does not *have* a definite position or momentum at any given time, and it does not travel by any single classical path. The best we can do is find the *probability* P that a particle will be observed at any particular location. Nature itself keeps track of only probabilities: this is the concept of **non deterministic reality**.

How do we find the probability distribution? Here are the rules of quantum mechanics relevant to this task.

---

[1] The wavelength $\lambda = h/p$ is called the **de Broglie wavelength**, because it was in his doctoral dissertation that the French physicist Louis de Broglie proposed that all particles have a wavelength $\lambda = h/p$. In the case of photons, we can increase the momentum by increasing the frequency of the light, since $p = h/\lambda = h\nu/c$. In the case of atoms, we can increase the momentum by increasing their velocity. And larger momentum leads to shorter de Broglie wavelength.

I. The probability P that a particle will be observed at a particular location is given by the absolute square of a total complex **probability amplitude** A_T to arrive there:

$$P = |A_T|^2 \equiv A_T^* A_T, \qquad (5.28)$$

where A_T^* is the complex conjugate of A_T.

II. The total probability amplitude for a particle to go from a to b is simply the sum of the probability amplitudes to go by every path available to it:

$$A_T = A_1 + A_2 + \cdots, \qquad (5.29)$$

where A_1 is the amplitude for path 1, A_2 is the amplitude for path 2, etc. We sum over all possible paths with fixed endpoints a and b. This sum includes all possible path trajectories between a and b with all possible momenta.[2]

III. The probability amplitude A to go from a source at a to a detector at b by some *particular* path is given by

$$A = A(t_0)e^{i\phi}, \qquad (5.30)$$

where $A(t_0)$ is the amplitude of the source when the particle leaves, and ϕ is a phase that depends upon the path. This phase is the Lorentz-invariant quantity

$$\phi = \frac{1}{\hbar} \int_a^b \eta_{\mu\nu} p^\mu dr^\nu = \frac{1}{\hbar} \int_a^b (\mathbf{p} \cdot d\mathbf{r} - E dt), \qquad (5.31)$$

where the coordinates at a are the initial position and time (x_0, y_0, z_0, t_0), and the coordinates at b are the final position and time (x_f, y_f, z_f, t_f), for a particular path.

These are the three simple rules for calculating the probability that a particle will be detected at time t.

Note from Rule III that if a particular path from a source at a to a detector at d is thought of as a sequence of path segments, for example, (1) $(a \rightarrow b)$, (2) $(b \rightarrow c)$, (3) $(c \rightarrow d)$, then the phase ϕ, being an integral over the entire path from a to d, is the sum of integrals for each segment of the path. That is, $\phi = \phi_1 + \phi_2 + \phi_3 + \cdots$, so the phase factor can be written

$$e^{i\phi} = e^{i(\phi_1 + \phi_2 + \phi_3 + \cdots)} = e^{i\phi_1} e^{i\phi_2} e^{i\phi_3} \cdots \qquad (5.32)$$

The amplitude to go by a particular path all the way from a source to a detector is therefore

$$A = A(t_0)e^{i\phi_1} e^{i\phi_2} e^{i\phi_3} \cdots, \qquad (5.33)$$

the amplitude at the source multiplied by the *product* of phase factors for each segment of the path.

---

[2] The sums over all paths and momenta consist of two *independent* sums. The state of the particle is prescribed at any given instant in time by specifying both its position and momentum independently. In the language we will introduce in Chapter 11, the sum is carried out in *phase space*.

Of all paths appearing in the sum (5.29), let us focus on a subset of interest – in an effort to simplify our job. It can be shown that going beyond this subset does not change the conclusion. We will consider only paths for which, at any instant in time, the momentum $\mathbf{p}$ is parallel to the displacement $d\mathbf{r}$, and the magnitudes of the momentum p and energy E are constant all along the path. Furthermore, we will fix the *energy* of the incoming particles. The probability amplitude that incoming particles with *definite* E and momentum p are detected at some definite position on the detection screen at time t – after going along a path of length s – then takes the form

$$A(t) = A_0 e^{\frac{i}{\hbar}(ps - Et)} = A_0 e^{iks - i\omega t}, \tag{5.34}$$

where A_0 is a constant. This expression can be obtained from Eq. (5.31), being careful to recast the amplitude from one involving particles that start at a definite position to one involving particles of definite energy. Here the magnitude of the wave number three-vector $\mathbf{k}$ is $k = 2\pi/\lambda = p/\hbar$ and $\omega = 2\pi\nu = E/\hbar$, where $\hbar \equiv h/2\pi$. Note that both relationships $p = h/\lambda$ and $E = h\nu$ are valid for massless photons as well as massive particles like helium atoms. This probability amplitude can be displayed as a two-dimensional vector called a **phasor** in the complex plane, as illustrated in Figure 5.9. The horizontal axis represents real numbers and the vertical axis imaginary numbers. Points not on either axis represent complex numbers with both real and imaginary parts. Placing the tail of the phasor at the origin, the length of the phasor is $|A_0|$ and its angle with the real axis is the total phase $ks - \omega t$. Adding probability amplitudes from different paths of different path lengths s translates then to adding phasors like two-dimensional vectors on a phasor diagram, as show in Figure 5.10.

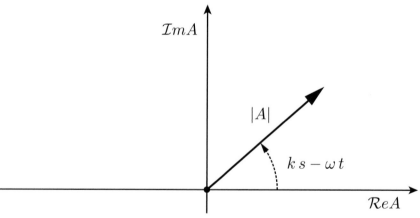

Fig. 5.9   A phasor $A_0 e^{i\phi}$ drawn in the complex plane. The real axis is horizontal and the imaginary axis is vertical. The absolute length of the phasor is $|A_0|$ and the angle between the phasor and the real axis is the phase $\phi = ks - \omega t$. Hence, as t changes, the phasor rotates.

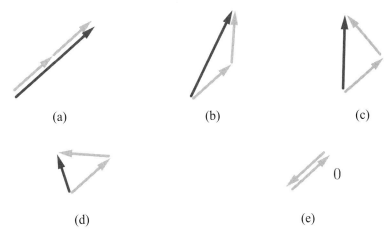

Fig. 5.10 The sum of two individual phasors with the same magnitudes $|A_0|$ but different phases. The result is a phasor that extends from the tail of the first to the tip of the second, as in vector addition. The difference in their angles in the complex plane is the difference in their phase angles. Shown are examples with phase differences equal to (a) zero, (b) $45°$, (c) $90°$, (d) $135°$, (e) $180°$.

The total probability amplitude for a monoenergetic beam of particles becomes, using Eqs. (5.29) and (5.34):

$$A_T = A_0 \left(e^{i(ks_1 - \omega t)} + e^{i(ks_2 - \omega t)} + \cdots \right) = A_0 e^{-i\omega t} \left(e^{iks_1} + e^{iks_2} + \cdots \right), \quad (5.35)$$

where $s_1, s_2, \ldots$ are the lengths of the various paths considered from the subset of the path we have focused on. These paths can zig-zag, go back and forth, in circles, any way they like as long as there are no physical barriers to prevent them.

5.4 Helium Atoms and the Two Slits, Revisited

We can now derive the probability distribution for particles passing through a double slit using the quantum rules. We will make a huge simplification for now, allowing particles to move along just two paths from the source to a detector, each path consisting of two straight-line segments joined at a slit, as illustrated in Figure 5.11. We assume also that both the source and the detector are far from the slit system, so the two paths from the source to the slits are essentially parallel to one another, and the two paths from the slits to the detector are also essentially parallel to one another.

The total probability amplitude is

$$A_T = A_0 e^{-i\omega t} \left(e^{iks_1} + e^{iks_2} \right), \quad (5.36)$$

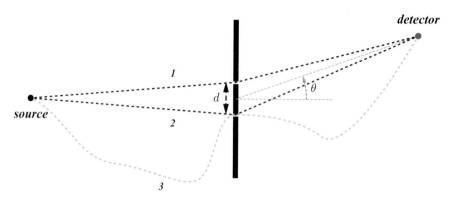

Fig. 5.11 Two paths from a source to a detector. Path 3 is possible but is not used in the text.

so the probability of observing a photon or atom at a particular detector is

$$P = A_T^* A_T = |A_0|^2 \left(e^{-iks_1} + e^{-iks_2}\right)\left(e^{iks_1} + e^{iks_2}\right)$$
$$= 2|A_0|^2(1 + \cos(k(s_2 - s_1)))$$
$$= 4|A_0|^2 \cos^2(k(s_2 - s_1)/2). \tag{5.37}$$

The probability that a particle is detected at arbitrary angle θ, in terms of the probability $P(0)$ of detecting it in the forward direction $\theta = 0$, is therefore

$$P(\theta) = P(0)\cos^2(\Phi/2) \quad \text{where} \quad \Phi = k(s_2 - s_1) = \frac{2\pi}{\lambda}(d\sin\theta), \tag{5.38}$$

using the same trig identities we used earlier for classical waves, where we found an *intensity* distribution $I(\theta) = I(0)\cos^2(\Phi/2)$. Naturally enough, if the probabilities of single-particle events obey the two-slit pattern, then if we collect a great many particles the intensity will have the same distribution. The formula agrees with the experimental results for photons or helium atoms as long as the de Broglie wavelengths are not extremely small compared with the slit separation d. That is, as long as we have $\lambda < d$ but not too small. The interference pattern observed is similar to that shown in Figure 5.8.

What do we see if the wavelength *is* extremely small? To understand this, remember that $\lambda = h/p$; hence, smaller wavelengths correlate with larger momenta. For a nonrelativistic particle we can achieve a large momentum $p = mv$ by, for example, increasing the mass: for instance, we can throw ball bearings at the two slits instead of helium atoms (assuming the ball bearings are smaller than the width of each slit of course). If we were to toss ball bearings at a double-slit system, we expect bunches of balls to accumulate downstream of each slit, with no interference pattern at all. It is true that some balls might nick the slit edges and be deflected to one side or the other, yet we would certainly see no interference pattern. For concreteness, consider an actual experiment with fast helium atoms. As before, each slit has width $a = 1\ \mu\text{m}$ and the two slits are separated by a distance $d = 8\ \mu\text{m}$.

Each atom has a mass $m = 6.68 \times 10^{-27}$ kg, and each can be detected by various counters as a discrete particle. For atoms with large velocities above 30 km/s, the results of actual experiments with both slits open are shown in Figure 5.12. They strike the screen with a bunch downstream of each slit, much like what we would find if we tossed ball bearings at a much larger slit system. This corresponds to the case where the de Broglie wavelength of the helium atoms $\lambda = h/mv \sim 10^{-11}$ m is much smaller than the width of each slit $a \sim 10^{-6}$ m (the high momentum is achieved by increasing the atoms' speed, not the mass of the particles as when considering ball bearings).

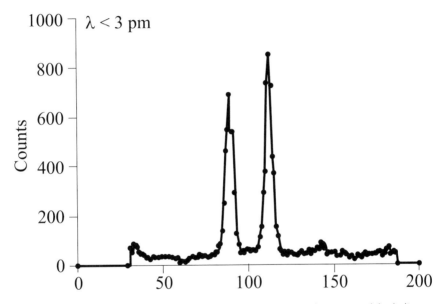

Fig. 5.12 High-velocity helium atoms, with speeds above 30 km/s, reaching the rear detectors, with both slits open. The detectors observe the arrival of individual atoms, and the distribution is what we would expect for classical particles. Reprinted by permission from Ch. Kurtsiever, T. Pfau, and J. Mlynek, *Nature* **386**, 150 (1997).

Whether we throw ball bearings, photons, or helium atoms, for sufficiently large momenta the interference pattern disappears. That is, any particle that penetrated a two-slit system would go through either one slit or the other, and for those going through the top slit it would make no difference whether the bottom slit is open or not, and for those going through the bottom slit it would make no difference whether the top slit is open or not. The distribution with both slits open is simply the sum of the distributions with only one slit open at a time.

So something else must be going on to explain why we do not see the two-slit interference pattern of Eq. (5.38) when the momenta are large. To see the resolution of this puzzle, we need to realize that the two-slit interference pattern is incomplete. Because each individual slit has a finite width a, there is an infinite number of nearby paths passing through each slit and we have ignored the effects of these

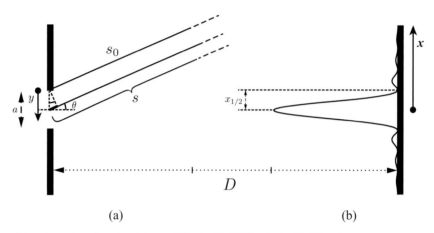

Fig. 5.13 (a) Path length as a function of position y within the slit. (b) The single-slit diffraction pattern.

additional paths. These have slightly different phases, so they interfere with one another significantly, especially when $\lambda < a$.

Consider a narrow slice of a single slit of width dy (see Figure 5.13). If we measure y up from the top of the slit, then $s = s_0 + y \sin \theta$, where s_0 is the distance of the top of the slit from the detector, as shown in Figure 5.13(a). The amplitude dA of all paths passing through the narrow slice of width dy will be proportional to dy, so[3]

$$dA = (C\,dy)\,e^{iks_0}\,e^{i(ky\sin\theta-\omega t)}, \qquad (5.39)$$

where C is a constant. The total amplitude to reach the detector, passing through a single slit of width a, is therefore

$$A_{\mathrm{T}} = C\,e^{i(ks_0-\omega t)} \int_0^a dy\, e^{iky\sin\theta} = C\,e^{i(ks_0-\omega t)}\left(\frac{e^{ika\sin\theta}-1}{ik\sin\theta}\right). \qquad (5.40)$$

The probability is equal to the absolute square of A_{T}:

$$P = A_{\mathrm{T}}^* A_{\mathrm{T}} = |C|^2 \left(\frac{e^{-iky\sin\theta}-1}{-ik\sin\theta}\right)\left(\frac{e^{iky\sin\theta}-1}{ik\sin\theta}\right)$$

$$= \frac{2|C|^2}{k^2\sin^2\theta}(1-\cos(ka\sin\theta)) = 2|C|^2 a^2\left(\frac{\sin^2\alpha}{\alpha^2}\right), \qquad (5.41)$$

where $\alpha \equiv (ka\sin\theta)/2$ and we have used the identity $\sin^2\alpha = (1/2)(1-\cos 2\alpha)$. Now $(\sin^2\alpha)/(\alpha^2) \to 1$ as $\alpha \to 0$, which is the maximum value this ratio can achieve. This distribution is called *single-slit diffraction*. The probability pattern in this case is

---

[3] To see this, note that if you double the size of dy, twice as many particles would go through dy – assuming uniform flux of particles across the single slit.

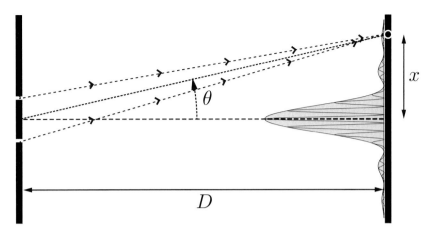

Fig. 5.14 The double slit with finite slit width a, with a screen at distance D. We can view the intensity on the screen as a function of the transverse distance x.

$$P = P_{\max}\left(\frac{\sin^2\alpha}{\alpha^2}\right) \quad \text{where} \quad \alpha \equiv \frac{ka\sin\theta}{2} = \frac{\pi a}{\lambda}\sin\theta \qquad (5.42)$$

and is depicted in Figure 5.13(b). The distribution has a maximum in the middle where $\alpha = 0$, and the first minimum at each side corresponds to $\alpha = \pm\pi$. The half-width of the central peak is therefore $\Delta\alpha = \pi$, which occurs at an angle θ_0 for which $\sin\theta_0 = \lambda/a$.

Let us start with λ small enough that the diffraction pattern has structure when the angles are small, *i.e.*, $\sin\theta \ll 1$ and so $\sin\theta \cong \theta$. Let D be the distance from the slit system to the detecting screen, and x be the distance on the screen from the midpoint of the wave pattern on the screen, as shown in Figure 5.14. Now $x/D = \tan\theta \cong \theta$ if $\theta \ll 1$, so the central peak of the single-slit diffraction pattern has a half-width on the screen of magnitude

$$\Delta x_{1/2} \simeq D\theta = \frac{D\lambda}{a}. \qquad (5.43)$$

Figure 5.15(a) illustrates the scenario. We see that the diffraction curves for the two slits essentially overlap. In this case the overall probability distribution is simply the product of the interference oscillation with the central diffraction envelope:

$$P = P_{\max}\cos^2\beta\left(\frac{\sin\alpha}{\alpha}\right)^2, \qquad (5.44)$$

where $\beta \equiv \pi d\sin\theta/\lambda$ and $\alpha \equiv \pi a\sin\theta/\lambda$. We have a familiar interference pattern from Figure 5.6 or 5.8 modulated by an envelope from the single-slit diffraction phenomenon. This is what we observe initially for photons or helium atoms – and were able to understand using Feynman's sum-over-paths formalism.

Now keep the same pair of slits and the same distance to the detecting plane, but decrease the beam wavelength by a factor of 20. That is, increase the momentum of the photons or helium atoms (in an extreme case, throw ball bearings!). The result is shown in Figure 5.15(b). The two peaks are now quite well separated. Clearly, as the wavelength is further reduced, the pattern becomes closer and closer to that corresponding to two bunches of particles formed downstream of each of the two slits. We see that, summing over all paths properly according to Feynman's sum-over-paths prescription, we observe the transition from quantum mechanics at longer de Broglie wavelengths to classical mechanics at shorter wavelengths.

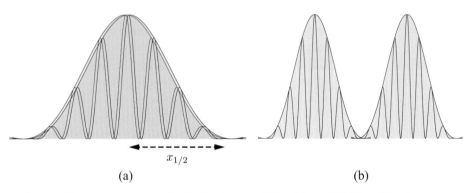

$x_{1/2}$

(a)   (b)

Fig. 5.15   Interference/diffraction patterns for a double slit with $a = d/4$ and $D = 1000d$. The diffraction curves serve as envelopes for the more rapidly oscillating interference pattern. (a) The pattern in the case $x_{1/2} = 10d$, where $x_{1/2}$ is the distance on the detecting plane between the center and the first minimum of the diffraction envelope. The diffraction curves of the two slits strongly overlap in this case, giving in effect a single diffraction envelope. (b) The pattern in the case $x_{1/2} = d/2$, showing that the two diffraction patterns have become separated, with the first minimum due to each slit at the same location in the center. This case corresponds to a wavelength smaller by a factor of 20 than the pattern shown in (a).

Feynman's sum-over-paths approach captures the full physics. The moral is that at larger wavelengths, we discern the interference and diffraction patterns that are the hallmarks of quantum mechanics, while at smaller wavelengths we recover our classical physical intuition. The subtle addition of phases of amplitudes accounts for the full picture.

Our conclusion from atomic-beam experiments is that helium atoms generally behave like *neither* ball bearings nor sound waves: they are neither classical particles nor classical waves, but retain some properties of each. They are detected as localized units like particles, but they show interference patterns like waves. In fact, we find that their particle and wave properties are related by $p = h/\lambda \equiv \hbar k$, where h is Planck's constant and $\hbar \equiv h/2\pi$.

5.5 The Emergence of the Classical Trajectory

Imagine the double slit of the previous section, but now without any barriers between the source and the detector (see Figure 5.16): in that case we need to consider an infinite number of paths between source and detector, with various path lengths $s(\alpha)$, ... Here α is a number that labels a path uniquely and $s(\alpha)$ is the length of the corresponding path (conveniently arranging the labeling so that paths with similar lengths have similar labels α). According to the quantum rules, the total probability amplitude for the particle to be detected at time t is

$$A_{\mathrm{T}} = A_0\, e^{-i\omega t} \int d\alpha\, e^{iks(\alpha)}, \tag{5.45}$$

where we have again assumed that all of the magnitudes A_0 (which may be complex numbers) are equal. Suppose that the path lengths (and therefore the phases) are all about the same for some particular set of paths. In that case the phasors associated with each term in the set add up to give a large total amplitude, as shown in Figure 5.17(a). If the path lengths are all quite different for another set of paths, then those phasors tend to cancel one another out, as shown in Figure 5.17(b).

Since we know that there is a shortest path between source and detector, that of the straight line, we know that the function $s(\alpha)$ has a minimum, say at $\alpha = \alpha_0$, corresponding to the straight-line path. It follows that

$$\left.\frac{ds}{d\alpha}\right|_{\alpha_0} = 0. \tag{5.46}$$

The Taylor series of $s(\alpha)$ about the point α_0 is (see Appendix F)

$$s(\alpha) = s(\alpha + 0) + \left.\frac{ds}{d\alpha}\right|_{\alpha_0}(\alpha - \alpha_0) + \frac{1}{2!}\left.\frac{d^2 s}{d\alpha^2}\right|_{\alpha_0}(\alpha - \alpha_0)^2 + \cdots \simeq s(\alpha_0),$$

$$\tag{5.47}$$

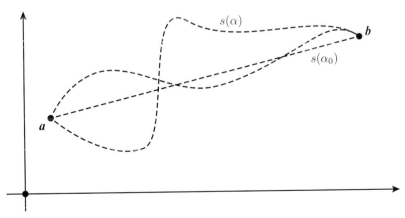

Fig. 5.16   A sampling of paths between source and detector with no barrier in between.

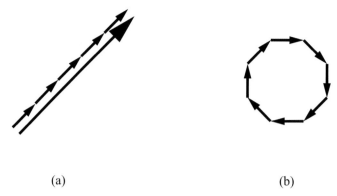

(a)   (b)

Fig. 5.17 The sum of a large number of phasors: (a) that are about the same; (b) that differ by constant amounts.

so the values of $s(\alpha)$ are the same to first order in $\Delta\alpha \equiv \alpha - \alpha_0$, since α_0 is at the minimum of the function. This means that if the set of paths is nearby the path of minimum length at α_0, then the phases will all be about the same, $\sim e^{iks(\alpha_0)}$, and the corresponding phasors will add up to give a large total as in Figure 5.17(a). Otherwise, the phases of paths very different from the minimum path will differ sufficiently from one another, so that the total phasor will be small, as in Figure 5.17(b). In the case of a free particle, the shortest path is a straight line, and the phase of nearby paths will be nearly the same, so the total probability amplitude will be large. That is, if a free particle travels from a to b by a straight-line path, the neighboring paths all have about the same length so their phases add up constructively. But if the particle travels by some arbitrary path, the neighboring paths differ more markedly in length from it, so the phases for these surrounding paths tend to cancel one another out. The sum in Eq. (5.29) (or the integral in Eq. (5.45)) is then dominated by the paths near the classical one, the straight line. While all paths are accessible, the shortest classical trajectory dominates the amplitude computation. This observation now justifies a previous assumption we made: in the double- and single-slit cases, when we considered only straight-line paths that kink at the slits, we were implicitly making use of this observation – these straight paths are expected to dominate the amplitude computation because of the phase cancellations for other paths.

Example 5.2 **A Class of Paths Near a Straight-Line Path**

Let s_{α_0} be the shortest distance between the source and the detector, corresponding to a straight-line path. We will sum the probability amplitudes for a certain subset of all paths near this path. These particular alternate paths consist of a straight line with a kink in the middle, where the kink is a distance $D = |n|D_0$ ($n = \pm 1, \pm 2, \ldots$) from the straight line, as shown in Figure 5.18. If the paths have length s_n, then by the Pythagorean theorem

$$(s_n/2)^2 = (s_0/2)^2 + (nD_0)^2. \tag{5.48}$$

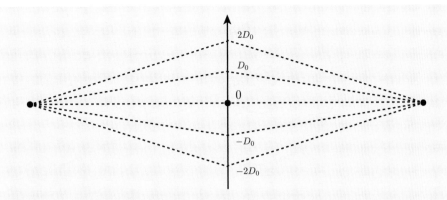

Fig. 5.18 A class of kinked paths midway between a source and a detector. The straight line is the shortest path, and the midpoint of the others is a distance $D = |n|D_0$ from the straight line, where $n = \pm 1, \pm 2, \ldots$

We will assume that $|n|D_0 \ll s_0$, so using the binomial approximation (see Appendix F):

$$s_n = s_0 \sqrt{1 + (2nD_0/s_0)^2} \cong s_0(1 + 2n^2 D_0^2/s_0^2). \tag{5.49}$$

Therefore the probability amplitude to go by a particular path of length s_n is proportional to

$$A_n \propto e^{iks} = e^{iks_0} e^{i\theta_n}, \tag{5.50}$$

where

$$\theta_n = \left(\frac{2kD_0^2}{s_0} \right) n^2 \quad |n| = 0, 1, 2, \ldots \tag{5.51}$$

Note that θ_n is the angle of the associated phasor A_n with respect to that of the straight-line path. As a particular case, suppose the phasor corresponding to the straight-line path is horizontal, and that $2kD_0^2/s_0 = \pi/200$. Then the angles θ_n are given in Table 5.1 for $n = 0, n = \pm 1, n = \pm 2, \ldots, n = \pm 25$. The sum of these phasors, all with the same length but in directions θ_n relative to the horizontal, will give the total phasor for these paths.

Table 5.1 The amplitude phase angles in terms of the integer n that characterizes the corresponding paths

n	θ_n	n	θ_n	n	θ_n	n	θ_n	n	θ_n
0	$0°$	± 6	$32.4°$	± 12	$129.6°$	± 18	$291.6°$	± 24	$518.4°$
± 1	$0.9°$	± 7	$44.1°$	± 13	$152.1°$	± 19	$324.9°$	± 25	$562.5°$
± 2	$3.6°$	± 8	$57.6°$	± 14	$176.4°$	± 20	$360.0°$	± 26	$608.4°$
± 3	$8.1°$	± 9	$72.9°$	± 15	$202.5°$	± 21	$396.9°$	± 27	$656.1°$
± 4	$14.4°$	± 10	$90.0°$	± 16	$230.4°$	± 22	$435.6°$	± 28	$705.6°$
± 5	$22.5°$	± 11	$108.9°$	± 17	$260.1°$	± 23	$476.1°$	± 29	$756.9°$

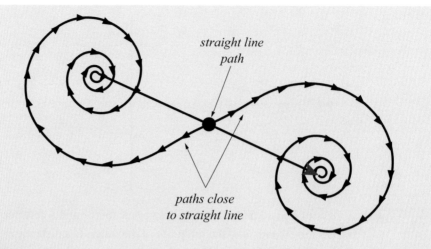

Fig. 5.19 Phasors up to $n = \pm 25$. The more distant paths wind up in spirals, contributing very little to the overall phasor sum.

The phasors are drawn in Figure 5.19. Those corresponding to $n = 0$ through $n = \pm 5$ are more or less aligned, so the paths neighboring the straight-line path are enhancing it. As $|n|$ increases the angles between successive phasors gradually increase, so the phasors begin to loop around, winding up in tighter and tighter spirals so they no longer make any important contribution to the total amplitude. The shape of these phasors is called a **Cornu spiral**. The sum of all the phasors up to $n = \pm 25$ is the long arrow shown. If we were to include additional phasors we would not change this sum very much. It is the straight-line path and its neighbors that contribute the most to the overall phasor, and therefore to the overall probability amplitude for the atom to go from the source to the detector. The classical path is the straight-line path in this case, but it is not the *only* path. ∎

How do we *know* that the classical path is not the only path that a particle takes? The best way to show the importance of the other paths is to block them off. If a highly collimated beam of particles travels from a source to a detecting screen, detectors will find that the particles are confined to a narrow region on the screen, in accord with the idea that the particles follow a straight-line path from source to screen. Therefore, if we were to introduce a narrow slit at a location directly between the source and the screen, it should make no difference, because according to classical ideas the particles are taking only that path anyway. But it *does* make a difference. With the narrow slit in place, the particles show a diffraction pattern on the screen as in Figure 5.13(b). Therefore without the screen, particles must be taking more than the straight-line path after all; in fact, according to quantum mechanics, they take all paths from source to screen that are not blocked off by some barrier.

Example 5.3

How Classical is the Path?

The phase of a free particle involves the product $ks = 2\pi s/\lambda$. If the wavelength is very small, so that $s/\lambda \gg 1$, then slight changes in s mean large changes in phase. The straight-line path between a and b is the minimum-distance path, so we already know that is the classical path. Neighboring paths have almost the same length, so they tend to add up in phase. But for a given path near the straight line, as the wavelength becomes smaller and smaller, the phase $2\pi s/\lambda$ changes more and more, so the corresponding phasors tend to spiral around and cancel out. That is, for very small wavelengths the set of mutually reinforcing paths becomes more and more constrained, closer and closer to the straight line; in the limit $s/\lambda \gg 1$, nonclassical paths become less and less important in the overall sum, so classical mechanics becomes a better and better approximation to the true situation. Take, for example, an electron moving at speed v. Classical motion is valid in the limit

$$\frac{s}{\lambda} = \frac{p\,s}{h} = \frac{mvs}{h} \gg 1. \tag{5.52}$$

An electron in a cathode-ray tube (such as an old-fashioned TV picture tube), with $s = 0.5$ m and $v = 10^8$ m/s, has the ratio

$$\frac{mvs}{h} = \frac{9.1 \times 10^{-31} \text{ kg} \cdot 10^8 \text{ m/s} \cdot 0.5 \text{ m}}{6.6 \times 10^{-34} \text{ J s}} \sim 10^{11}, \tag{5.53}$$

so such an electron moves on a classical path (in a TV tube we obviously want the electrons to move along a classical, deterministic path!).

But consider the electron in a hydrogen atom, where it has a typical speed of 10^6 m/s and a path length from one side of the atom to the other of order 10^{-10} m. The ratio in this case is

$$\frac{mvs}{h} = \frac{9.1 \times 10^{-31} \text{ kg} \cdot 10^6 \text{ m/s} \cdot 10^{-10} \text{ m}}{6.6 \times 10^{-34} \text{ J s}} \sim 0.1, \tag{5.54}$$

so the path in this case is not at all classical, but quantum-mechanically fuzzy. The electron in hydrogen does not move along a classical orbit anything like the motion of planets around the sun. However, the ratio for earth orbiting the sun is

$$\frac{mvs}{h} = \frac{6 \times 10^{24} \text{ kg} \cdot 3 \times 10^4 \text{ m/s} \cdot 1.5 \times 10^{11} \text{ m}}{6.6 \times 10^{-34} \text{ J s}} \sim 10^{73}, \tag{5.55}$$

which corresponds to supremely classical motion.

There are no sharp boundaries between classical and quantum behavior. All motion is really quantum mechanical, but classical behavior can in certain cases be an excellent approximation. A quick check on whether a system behaves classically amounts to the following: multiply the typical momentum and size scales in the problem; if the product is much larger than the Planck constant, the system can be well approximated classically. Otherwise, quantum mechanics is needed to describe it.[a]   ∎

---

[a] In the case of particles striking the double-slit setup, the condition of classicality was that the slit width a is much larger than the de Broglie wavelength, $\lambda \ll a$. From $p = h/\lambda$, this translates to $p\,a \gg h$ as expected. The key here is to identify the correct length scale, in this case the slit width a. A classical description through electromagnetic waves captures the quantum interference patterns as long as the intensity of light is high (or the number of photons is large).

5.6 Why Hamilton's Principle?

For free particles, we have already shown that quantum-mechanical amplitudes in Feynman's sum-over-paths add up *in phase* near the classical shortest path, so that in the limit of *small de Broglie wavelengths* one observes only the classical path. What we have *not* done so far is to show how *Hamilton's principle* comes about in general. This is the central question, because Lagrange's equations and all of the dynamical results that flow from them can be derived from making the action

$$S = \int_{t_0}^{t} L \, dt \tag{5.56}$$

stationary. Where does this principle come from? And how does a classical particle "know" that it should follow a stationary-action path?

To see the emergence of Hamilton's principle, we start from Eq. (5.31):

$$A = A(t_0) \, e^{\frac{i}{\hbar} \int_a^b \eta_{\mu\nu} p^\mu dr^\nu} = A(t_0) \, e^{\frac{i}{\hbar} \int_a^b (\mathbf{p} \cdot d\mathbf{s} - E dt)}, \tag{5.57}$$

where E and $\mathbf{p}$ are the particle's energy and momentum, and $d\mathbf{s}$ is the local infinitesimal distance vector tangent to its path. We remind the reader that this expression is for a *particular* path and one needs to sum over all possible paths by adding these phases to construct the final amplitude. Consider a small time step from some time t to $t + \Delta t$. We focus on the one-dimensional nonrelativistic case for simplicity: the generalized coordinate is q and the corresponding generalized momentum is p. The energy of the particle can then be written

$$E = K.E. + P.E. = \frac{p^2}{2m} + U(q), \tag{5.58}$$

where $U(q)$ is the potential energy. The corresponding amplitude ΔA then takes the form

$$\Delta A \propto e^{i\frac{\Delta t}{\hbar}\left(p\dot{q} - p^2/2m - U(q)\right)}, \tag{5.59}$$

where we have written $dq = \dot{q}\Delta t$ for small Δt. Since the quantum prescription involves summing over all paths between a and b, this involves integrating over all q *and* p at every time step. So we will encounter the integral

$$\int_{-\infty}^{\infty} dp \, e^{i\frac{\Delta t}{\hbar}\left(p\dot{q} - p^2/2m - U(q)\right)}. \tag{5.60}$$

Now

$$p\dot{q} - \frac{p^2}{2m} \equiv -\frac{(p - m\dot{q})^2}{2m} + \frac{m\dot{q}^2}{2}, \tag{5.61}$$

so we have to evaluate

$$\int_{-\infty}^{\infty} dp \, e^{i\frac{\Delta t}{\hbar}\left(-\frac{(p-m\dot{q})^2}{2m} + \frac{m\dot{q}^2}{2} - U(q)\right)} = e^{i\frac{\Delta t}{\hbar}\left(\frac{m\dot{q}^2}{2} - U(q)\right)} \int_{-\infty}^{\infty} dp \, e^{-i\frac{\Delta t}{\hbar}\frac{(p-m\dot{q})^2}{2m}}. \tag{5.62}$$

This "Gaussian integral" in p can be evaluated explicitly:

$$\int_{-\infty}^{\infty} dp\, e^{-i\frac{\Delta t}{\hbar}\frac{(p-m\dot{q})^2}{2m}} = \sqrt{\frac{2\pi m\hbar}{i\Delta t}}, \tag{5.63}$$

so in terms of proportionalities:

$$\Delta A \propto e^{i\frac{\Delta t}{\hbar}\left((1/2)m\dot{q}^2 - U(q)\right)}, \tag{5.64}$$

which should look familiar. It is simply

$$\Delta A \propto e^{i\frac{\Delta t}{\hbar}L}, \tag{5.65}$$

where L is the Lagrangian of the system. Multiplying such phases together from a to b to add the phasors from different small time steps then gives

$$\Delta A \propto e^{\frac{i}{\hbar}\int_a^b L dt} = e^{iS/\hbar}, \tag{5.66}$$

where now the action $S = \int_a^b L dt$ appears in the phase. To obtain the full amplitude, we still need to sum over all paths $q(t)$ – so far we only considered a sum over all possible momenta $p(t)$. Hence, the total amplitude has the form

$$A_{\mathrm{T}} = \text{constant} \times \sum_{\text{paths } q(t)} e^{iS/\hbar}. \tag{5.67}$$

The path about which all the amplitudes tend to add up is the path corresponding to motion near the extremum of the action $S = \int dt\, L$. This statement is nothing but Hamilton's principle from Section 4.3. Hence, we now understand more deeply why Hamilton's principle works: *it emerges naturally from the sum-over-paths, from the basic principles of quantum mechanics.*

5.7 The Jacobi Action

Now, having shown how to derive Hamilton's principle, the heart of classical mechanics, from sum-over-paths quantum mechanics, what more can we learn here? An answer is that we can also use the formalism of sum-over-paths to find the *shape* of paths taken by light rays and by nonrelativistic massive particles.

Light Rays

When light passes from one medium to another, say air to a refractive medium like glass, its frequency (*i.e.*, the energy/photon) remains constant, but the wavelength changes, because the velocity $v = \lambda\nu$ is less in glass than in air. If the medium has index of refraction n, where n may depend upon position, the speed of light in the medium is $v = c/n$, so with constant frequency the wavelength decreases by the factor $1/n$, and the wave number $k = 2\pi/\lambda$ increases by the factor n. So whereas in vacuum $\omega/k = c$ (or $k = \omega/c$), in a refractive medium $k = \omega/v = n\omega/c$. The quantum-mechanical amplitude for photons to travel by a particular path therefore becomes

$$z = z(t_0)e^{i\phi}, \tag{5.68}$$

where

$$\phi = \int_a^b k\,ds - \omega(t - t_0) = (\omega/c)\int_a^b n\,ds - \omega(t - t_0). \tag{5.69}$$

The corresponding phasors add up for paths near that path which extremizes the phase $\int_a^b n\,ds$. That is, they add up along the path that extremizes the optical path $\int_a^b n\,ds$. This is just Fermat's principle of stationary time, since the time for light to follow a particular path is

$$t = \int_a^b ds/v = \frac{1}{c}\int_a^b n\,ds. \tag{5.70}$$

According to Fermat, light "rays" take minimum-time paths between a and b. According to Feynman, photons take *all* paths between a and b, but it is mainly paths near the stationary path that contribute to the total amplitude.

Nonrelativistic Massive Particles

Up to now we have described the behavior of massive particles only when they are free, in which case their classical paths are straight lines. The probability amplitude for a path of length s in this case is

$$A = A_0 e^{(i/\hbar)(ps - Et)}, \tag{5.71}$$

where p and E are the momentum magnitude and energy, both constant along the path; and s is the path length.

For nonrelativistic particles encountering forces as they move from a to b, this expression cannot be correct, because even with a conserved energy $E = T + U$, the magnitude of the momentum $p = \sqrt{2mT} = \sqrt{2m(E - U(\mathbf{r}))}$ is not conserved because the potential energy generally depends upon position. The expression for the amplitude $A = A_0 e^{(i/\hbar)(ps - Et)}$ therefore no longer makes sense, because p keeps changing; it has to be replaced by its original form

$$A = A_0 e^{(i/\hbar)(\int_a^b p\,ds) - Et}, \tag{5.72}$$

where the varying momentum is integrated over the path between the source and the detector, as dictated by Eq. (5.31). The total probability amplitude of the particle beginning at point a and ending at point b at time t is therefore

$$A_{\mathrm{T}} = A_0 e^{-iEt/\hbar}\sum e^{(i/\hbar)\int_a^b p\,ds}, \tag{5.73}$$

summing over all paths between a and b.

For any particular path the integral is

$$\int_a^b p \, ds = \int_a^b \sqrt{2m(E - U(\mathbf{r}))} \, ds. \tag{5.74}$$

In the case of a free particle, the classical path is the path for which the product ps is minimized, which means (because p is fixed in this case) that s is minimized (which of course leads to a straight line). Now that potential energies have been included, it is the *integral* $\int_a^b \sqrt{E - U(\mathbf{r})} \, ds$ that has to be minimized.

In classical mechanics the integral

$$J = \int_a^b \sqrt{E - U(\mathbf{r})} \, ds \tag{5.75}$$

is called the **Jacobi action**, named for the German mathematician Carl Gustav Jacob Jacobi (1804–1851). Finding the path that makes J stationary is called the **Jacobi principle of stationary action**: this is indeed the classical path between a and b. Note that *time* does not appear in the Jacobi action, so making J stationary provides the path *shape*, but not the dynamical equations of motion.

Without quantum mechanics, Jacobi's principle seems quite mysterious. Why should a particle choose that path that minimizes (or maximizes, or otherwise makes stationary) the Jacobi action? Now we know why it works. According to Feynman's sum-over-paths, if the path s_0 happens to be the one that minimizes J, then nearby paths have nearly the same phase, so the corresponding phasors add up to give a large result. But paths surrounding some other, arbitrary path have more rapidly varying phases, so their phasors tend to cancel one another out. The integral $\int_a^b \sqrt{E - U(\mathbf{r})} \, ds$ is simply the integral $\int_a^b p \, ds$ in disguise.

Example 5.4

Path Shape for Particles in Uniform Gravity

A nonrelativistic particle of mass m moves in a uniform gravitational field g. What is the shape of its classical path?

The potential energy is $U = mgy$, so the Jacobi action is

$$J = \int_a^b \sqrt{E - mgy} \, ds = \int_a^b \sqrt{E - mgy} \sqrt{dx^2 + dy^2}. \tag{5.76}$$

We can choose either x or y as the independent variable: it is easier to choose y, because then the integral has the form

$$J = \int_a^b \sqrt{E - mgy} \sqrt{1 + (dx/dy)^2} dy = \int_a^b f(x, dx/dy, y) dy, \tag{5.77}$$

where x is missing from f, and so x is a cyclic coordinate. Therefore, from Euler's equation

$$\frac{\partial f}{\partial x} - \frac{d}{dy} \frac{\partial f}{\partial x'} = 0, \tag{5.78}$$

it follows that we have a first integral $\partial f / \partial x' = $ constant along the stationary path. That is:

$$\sqrt{E - mgy} \left(\frac{x'}{\sqrt{1 + x'^2}} \right) = \alpha, \text{ a constant.} \tag{5.79}$$

Solving for x':

$$x' \equiv \frac{dx}{dy} = \pm \frac{\alpha}{\sqrt{E - \alpha^2 - mgy}}. \tag{5.80}$$

Integrating both sides over y, choosing $x = 0$ and $y = y_0$ at the point where the path is horizontal (*i.e.*, where $x' = dx/dy = \infty$), and solving for $y(x)$ gives

$$y(x) = y_0 - \frac{mgx^2}{4(E - mgy_0)} \tag{5.81}$$

which is a parabola, as we knew it would be all along.

The classical path for a particle moving in a uniform gravitational field is a parabola, but according to quantum mechanics it is not the *only* path the particle takes. It actually takes all possible paths, but in the classical limit, where the de Broglie wavelength is very small compared with any physical length in the problem, the classical path is the only path for which the amplitudes of nearby paths all add up in phase. ∎

5.8 Summary

In summary, experiments with light and particle beams show the following features.

- Light, as photons, and massive particles as well, have both a particle and a wave nature. Their energy and momenta are related to their frequency and wavelength by the de Broglie relations

$$E = h\nu \quad \text{and} \quad p = h/\lambda, \tag{5.82}$$

where h is Planck's constant.
- Photons, electrons, atoms, and all other particles are detected as discrete lumps as though they were classical particles. If huge numbers of them are detected, they have an apparently continuous distribution.
- High-energy particles (photons, electrons, atoms, etc.) sent through a double-slit system appear to have the same bimodal distribution at the detecting plane as that predicted for classical particles.
- Lower-velocity particles (lower-frequency photons, slower electrons and atoms) show wave-like interference effects at the screen. Giving more speed to atoms makes the interference oscillations tighter, until they become so tight they cannot be separated, and they look just like the bimodal distribution predicted for classical particles.

- A particle takes all available paths from a source to a detector. The probability of reaching the detector is equal to the absolute square of the sum of amplitudes to travel by all possible paths. Near the classical path the amplitudes add up to give a strong total amplitude, especially if the de Broglie wavelength of the particle is small compared with the physical dimensions in an experiment.

- Fermat's principle for light and the Jacobi principle of stationary action for massive particles can be used to find particle path shapes. They follow from the quantum rules.

- Hamilton's principle, which we have used to find the dynamical equations of motion of massive particles, also follows from the quantum-mechanical sum-over-paths.

In Part I of this book we have encountered three strikingly different pictures of the nature of classical motion.

 (i) In the Newtonian picture, including the extension of this picture into the relativistic regime, a massive particle moves along a path determined by its initial position and velocity and the forces encountered along the way. The particle knows nothing of the future, where it is going or how long it will take to get there.

 (ii) In the picture accompanying Hamilton's principle, a massive particle moves along a single path that makes stationary the action S, a functional that depends upon the beginning points, endpoints, and potential energies encountered along the path. Rather than specifying the initial position and velocity, as in the Newtonian picture, with Hamilton's principle we specify the initial and final positions. A natural question for those with a Newtonian intuition is: How does the particle "know" to choose that path which makes stationary the action between given endpoints?

(iii) In the quantum-mechanical view, using the sum-over-paths formulation of Feynman, a particle, massive or massless, takes *all* paths between given initial and final points: that is how it "knows" the classical path. The probability that it will reach the final point at a certain time is given by the absolute square of a total complex probability amplitude, where the total amplitude is the sum of the amplitudes for all possible paths. The amplitudes reinforce one another along the path that makes stationary the action, especially if the de Broglie wavelength of the particle is small compared with any physical dimensions in the environment. Thus, Hamilton's principle for massive particles emerges naturally from quantum mechanics in the case of small-wavelength particles. The picture of motion here is more "Darwinian" than deterministic: every path is tried, but in the classical limit only the "fittest" (the stationary-action path) survives. The determinism of Newton is discovered to be only an illusion, emerging from the nondeterministic theory of quantum mechanics.

Problems

★★   **Problem 5.1** Maxwell's equations for the electric field **E** and magnetic field **B** are

$$\nabla \cdot \mathbf{E} = 4\pi\rho, \quad \nabla \cdot \mathbf{B} = 0, \quad \nabla \times \mathbf{E} = -\frac{1}{c}\frac{\partial \mathbf{B}}{\partial t}, \quad \nabla \times \mathbf{B} = \frac{4\pi}{c}\mathbf{J} + \frac{1}{c}\frac{\partial \mathbf{E}}{\partial t},$$

where ρ is the charge density, **J** is the current density, and ϵ_0 and μ_0 are (respectively) the permittivity and permeability of the vacuum, both constants. Derive the vacuum wave equations for **E** and for **B**, and relate ϵ_0 and μ_0 to the speed of electromagnetic waves c. *Hint*: You can use the vector identity $\nabla \times (\nabla \times \mathbf{A}) = \nabla(\nabla \cdot \mathbf{A}) - \nabla^2\mathbf{A}$ where **A** is any vector.

★   **Problem 5.2** Photons of wavelength 580 nm pass through a double-slit system, where the distance between the slits is $d = 0.16$ nm and the slit width is $a = 0.02$ nm. If the detecting screen is a distance $D = 60$ cm from the slits, what is the linear distance from the central maximum to the first minimum in the diffraction envelope?

★★   **Problem 5.3** Photons are projected through a double-slit system. (a) What must be the ratio d/a of the slit separation to slit width, so that there will be exactly nine interference maxima within the central diffraction envelope? (b) Is any change observed on the detecting screen if the photon wavelength is changed from λ_0 to $2\lambda_0$? If so, what? (c) If 10^4 photons are counted within the central interference maximum, about how many do you expect will be counted within the last interference maximum that fits within the central diffraction envelope?

★★   **Problem 5.4** A beam of monoenergetic photons is directed at a triple-slit system, where the distance between adjacent slits is d, and the photon wavelength is $\lambda = d/2$. Find the angles θ from the forward direction for which there are (a) interference maxima, (b) interference minima. (c) Then show that some maxima have the same maximum probability as the central peak, but that others have a smaller maximum. Find the ratio of the larger to the smaller maximum probabilities.

★   **Problem 5.5** A beam of 10-keV photons is directed at a double-slit system and the interference pattern is measured on the detecting plane. The wavelength of these photons is less than the slit separation. Then electrons are accelerated so their (nonrelativistic) kinetic energies are also 10 keV; these electrons are then directed at the same double-slit system, and their interference pattern is measured on the same detecting plane. If the distance between two adjacent photon interference maxima on the detecting screen is y_0, what is the distance between two adjacent *electron* interference maxima? (Note that the mass energy of an electron is 0.5 MeV.)

★   **Problem 5.6** Consider a grating composed of four very narrow slits each separated by a distance d. (a) What is the probability that a photon strikes a detector centered at the central maximum if the probability that a photon is counted by this detector with a single slit open is r? (b) What is the probability that a photon is counted at the first minimum of this four-slit grating if the bottom two slits are closed? (From *Quantum Physics* by John S. Townsend.)

Problem 5.7 Example 5.2 considered a set of kinked paths about a straight-line path. (a) Using the same set of alternative paths, suppose one considered the sum of phasors about the path with $n = 50$ instead of the sum about the $n = 0$ straight-line path. In particular, if one summed from $n = 25$ to $n = 75$, ± 25 about $n = 50$, how would the sum of phasors differ from the sum for paths about $n = 0$? What physical conclusion can you draw from this? (b) Now returning to the set of kinked paths about the straight-line $n = 0$ path, draw the phasor diagram if the wave number k of the particle were doubled (*i.e.*, if the de Broglie wavelength λ were halved). What can be concluded about the physical difference between this case and that used in Example 5.2?

Problem 5.8 Example 5.2 considers a particular class of paths near a straight-line path. A different class of paths consists of a set of *parabolas* of the form $y = n\alpha(1 - x/x_0)^2$ fit to the endpoints of the straight line at $(x, y) = (0, 0)$ and $(x, y) = (x_0, 0)$. Here α is a (small) constant, and $n = 0, \pm 1, \pm 2, \ldots$ Let $\alpha = 0.1x_0$, and draw a careful phasor diagram including enough integers n to see the Cornu spiral behavior and obtain a good estimate of the sum of all these phasors.

Problem 5.9 Judge whether or not the following situations are consistent with classical paths. (a) A nitrogen molecule moving with average kinetic energy $\langle 3/2 \rangle kT$ at room temperature $T = 300$ K (where k is Boltzmann's constant). (b) A typical hydrogen atom caught in a trap at temperature $T = 0.1$ K. (c) A typical electron in the center of the sun, at temperature $T = 15 \times 10^6$ K.

Problem 5.10 (a) What condition would have to be met so that the motion of a 135-g baseball would be inconsistent with a classical path? Is this a potentially feasible condition? (b) If we could adjust the value of Planck's constant, how large would it have to be so that the ball in a baseball game would fail to follow classical paths?

Problem 5.11 According to the Heisenberg indeterminacy principle $\Delta x \, \Delta p \geq \hbar$, the uncertainty in position of a particle multiplied by the uncertainty in its momentum must be greater than Planck's constant divided by 2π. The neutrons in a particular atomic nucleus are confined to be within a nucleus of diameter 2.0 fm (1 fm = 10^{-15} m). Can these neutrons be properly thought of as traveling along classical paths? Explain.

Problem 5.12 Show from the Newtonian equations $x = v_{0x}t$ and $y = y_0 - (1/2)gt^2$ for a particle moving in a uniform gravitational field g that the shape of its path is a parabola, given by

$$y = y_0 - \frac{mgx^2}{4(E - mgy_0)}, \tag{5.83}$$

the same result we found using the Jacobi principle of least action.

★★   **Problem 5.13** A particle of mass m can move in two dimensions under the influence of a repulsive spring-like force in the x direction, $F = +kx$. Find the shape of its classical path in the x, y plane using the Jacobi action.

★★   **Problem 5.14** An object of mass m can move in two dimensions in response to the simple harmonic oscillator potential $U = (1/2)kr^2$, where k is the force constant and r is the distance from the origin. Using the Jacobi action, find the shape of the orbits using polar coordinates r and θ; that is, find $r(\theta)$ for the orbit. Show that the shapes are ellipses and circles centered at the origin $r = 0$.

★★★   **Problem 5.15** A comet of mass m moves in two dimensions in response to the central gravitational potential $U = -k/r$, where k is a constant and r is the distance from the sun. Using the Jacobi action and polar coordinates (r, θ), find the possible shapes of the comet's orbit. Show that these are (a) a parabola, if the energy of the comet is $E = 0$; (b) a hyperbola, if $E > 0$; (c) an ellipse or a circle, if $E < 0$, where in each case $r = 0$ at one of the foci.

PART II

6

Constraints and Symmetries

An enormous advantage of using Lagrangian methods in mechanics is the simplifications that can occur when a system is constrained or if there are symmetries of some kind in the environment of the system. Constraints can be used to reduce the number of generalized coordinates so that solutions become more practicable. In this chapter we will illustrate this fact using the example of contact forces, and we will demonstrate the use of Lagrange multipliers to learn about the contact forces themselves.

Constraints are also typically associated with the breaking of symmetries. Lagrangian mechanics allows us to efficiently explore the relationship between symmetries in a physical situation and dynamical quantities that are conserved. These properties are nicely summarized in a theorem by the German mathematician Emmy Noether (1882–1935), and provide us with deep insight into the physics – in addition to helping us make important technical simplifications while solving problems.

We first discuss constraints and contact forces, and then symmetries and conservation laws.

6.1 Contact Forces

A square block of mass M rests on a horizontal floor, as shown in Figure 6.1. Newtonian mechanics tells us that the block experiences two forces: the downward pull of gravity Mg and the upward push of the floor, called the normal force N. A static scenario implies that $N = Mg$, so the forces sum to zero and there is no acceleration. Hence, the normal force adjusts its strength as needed to counteract the gravitational pull Mg. If the balance succeeds, the block stays put on the floor. If, however, the normal push of the floor is not enough because the block is too heavy, the floor would disintegrate and the block would fall through.

The normal force is an example of a **contact force**. It arises by virtue of the physical contact between two objects. Contact forces are always electromagnetic in origin. As two objects touch, the atoms at the contact interface push against one another through electromagnetic forces. Each of these tiny pushes may be negligible, but with 10^{23} atoms reinforcing the effect, we get a net, effective, *macroscopic* force we call the contact force. Hence, contact forces are not

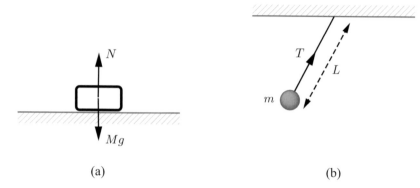

Fig. 6.1 (a) A block resting on the ground subject to the normal contact force that constrains it to horizontal motion. (b) A bob pendulum involves the tension contact force in the rope that constrains the bob from flying away.

fundamental: they are the sum of many complicated microscopic interactions. However, we can often characterize them by a simple, effective, phenomenological force "law." The normal force, the tension force, and the friction force are all examples. They really consist of intricate electromagnetic interactions between large numbers of constituent atoms.

Our square block rests on the floor. The net effect of the contact force is to *constrain* its vertical motion: the block can move only sideways and not up and down. The dynamics of the center of mass of the box is now reduced to two degrees of freedom from the original three. Similarly, the tension force in the rope of a bob pendulum implements a constraint that assures that the bob at the end of the rope does not fly way beyond a distance equal to the length of the rope. Many more such contact forces translate into statements of *constraints* on the degrees of freedom. In this section, we develop the technology of solving mechanics problems with various types of constraints.

Consider a mechanical system parameterized by N coordinates q_k, $k = 1, \ldots, N$. However, due to some constraint forces, there are P algebraic relations amongst these coordinates, given by

$$C_l(q_1, q_2, \ldots, q_N, t) = 0, \tag{6.1}$$

where $l = 1, \ldots, P$. This means we have effectively $N - P$ generalized coordinates or degrees of freedom, instead of the nominal N. In our problem of the block on the floor, for example, let us suppose that the floor is at $z = 0$. Then we can define $C = z$, so that the single algebraic equation $C = 0$ implies that $z = 0$. We started with $N = 3$ coordinates x, y, z; then we specified $P = 1$ constraint, the constraint that $z = 0$, where z is the vertical direction and the center of mass of the block rests at $z = 0$. We are then left with $N - P = 3 - 1 = 2$ degrees of freedom, x and y.

In tackling such situations with constraints, we have two choices. We could try to use the P relations (6.1) to eliminate P of the q_ks, and write the Lagrangian in terms of $N - P$ generalized coordinates. This is in a sense what we have been doing so far. For the example of the block, we would write

$$L = \frac{1}{2}m\left(\dot{x}^2 + \dot{y}^2 + \dot{z}^2\right) - mgz = \frac{1}{2}m\left(\dot{x}^2 + \dot{y}^2\right). \tag{6.2}$$

Alternatively, we may want to delay implementing the constraint. There are two good reasons for this. First, it may be difficult or inconvenient to eliminate P q_ks using the constraints (6.1). Second, we may be interested in finding the *constraint forces* underlying these constraint relations. For example, we may want to find the normal force on a block sliding along a curved rail; when this force vanishes, the block is losing contact with the rail, and we would be able to determine this critical point in the evolution.

Therefore we now focus on a method that delays the implementation of the constraints in a mechanical problem, instead dealing with all N q_ks in the problem. This requires a careful treatment, since the variational formalism assumes *independent* q_ks: the q_ks must not have relations amongst them such as those given by (6.1). Otherwise, in the process of extremizing the action we would get to the step

$$\delta I = \int \left[\frac{d}{dt}\left(-\frac{\partial L}{\partial \dot{q}_k}\right) + \frac{\partial L}{\partial q_k}\right]\delta q_k dt = 0, \tag{6.3}$$

and, since the δq_ks are not independent due to (6.1), we cannot conclude that the vanishing of the bracketed expression – that is the Lagrange equations of motion – is necessary individually for every k.

Instead, let us consider a *new* Lagrangian defined as

$$L' = L + \sum_{l=1}^{P} \lambda_l C_l, \tag{6.4}$$

where we have introduced P additional degrees of freedom labeled λ_l with $l = 1, \ldots, P$, each multiplying a related constraint equation from (6.1). For example, with the block on a horizontal floor, we would write

$$L' = \frac{1}{2}m\left(\dot{x}^2 + \dot{y}^2 + \dot{z}^2\right) - mgz + \lambda_1 z. \tag{6.5}$$

We now assume that the constraint equations (6.1) are *not* satisfied *a priori*. We then have $N + P$ degrees of freedom: N q_ks and P λ_ls. Correspondingly, we have $N + P$ equations of motion. For our example, we have $N + P = 3 + 1 = 4$ variables remaining: x, y, z, and λ_1.

The equations of motion for (6.4) are then

- P equations for the λ_ls:

$$\frac{d}{dt}\left(\frac{\partial L'}{\partial \dot{\lambda}_l}\right) - \frac{\partial L'}{\partial \lambda_l} = z = 0 \Rightarrow \frac{\partial L'}{\partial \lambda_l} = 0 \Rightarrow C_l = 0. \tag{6.6}$$

These are simply the original constraints (6.1)! But they now arise dynamically through the equations of motion and need not be implemented from the outset. The P parameters labeled λ_l are called **Lagrange multipliers**. For our simple example, we have a single Lagrange multiplier λ_1 with the equation of motion $z = 0$.

- N equations for the q_ks:

$$\frac{d}{dt}\left(\frac{\partial L'}{\partial \dot{q}_k}\right) - \frac{\partial L'}{\partial q_k} = 0. \tag{6.7}$$

In terms of the original L from (6.4), these look like

$$\frac{d}{dt}\left(\frac{\partial L}{\partial \dot{q}_k}\right) - \frac{\partial L}{\partial q_k} - \lambda_l \frac{\partial C_l}{\partial q_k} = 0. \tag{6.8}$$

Note the additional terms involving the λ_ls (l is repeated and hence summed over). We can rewrite these as

$$\frac{d}{dt}\left(\frac{\partial L}{\partial \dot{q}_k}\right) - \frac{\partial L}{\partial q_k} = \mathscr{F}_k \equiv \lambda_l \frac{\partial C_l}{\partial q_k}, \tag{6.9}$$

where we have defined the **generalized constraint forces** $\mathscr{F}_k$ that essentially enforce the constraints onto the q_k dynamics. For the block on the horizontal floor problem these become

$$m\ddot{x} = 0, \quad m\ddot{y} = 0, \quad m\ddot{z} = -mg + \lambda_1. \tag{6.10}$$

In this case it is obvious that λ_1 is the normal force. More generally, how do we relate the generalized constraint forces – and hence the λs – to the *actual* forces? For every object i in the problem located at position $\mathbf{r}_i$, denote the total constraint force acting on it by $\mathbf{F}_i$. We also know the relations $\mathbf{r}_i(q_k, t)$ that connect the position of every object to the generalized coordinates q_k. Using all this, we can relate the Lagrange multipliers to the constraint forces by noting that the right-hand side of the new Lagrange equations must be given by

$$\mathscr{F}_k = \lambda_l \frac{\partial C_l}{\partial q_k} = \frac{\partial(\lambda_l C_l)}{\partial q_k}. \tag{6.11}$$

Since $\lambda_l C_l$ is added to the Lagrangian, it should correspond to the energy or work associated with the constraint forces $\mathbf{F}_i^{\text{C}}$:

$$\lambda_l C_l = \int^{\mathbf{r}_i} \mathbf{F}_i^{\text{C}} \cdot d\mathbf{r}_i', \tag{6.12}$$

where i sums over the various entities tracked by the Lagrangian with corresponding position vectors $\mathbf{r}_i$. The prime on the measure $d\mathbf{r}_i'$ is to avoid a variable naming clash with the limit of the integral. Notice that if a force on a particle $\mathbf{F}_i^{\text{C}}$ were conservative and if we were to write it in terms of a potential energy $\mathbf{F}_i^{\text{C}} = -\nabla U_{\text{C}}$, this would correspond to adding $-U_C(\mathbf{r}_i)$ to the Lagrangian as expected. Since we have a map between these position vectors and the generalized coordinates $\mathbf{r}_i(q_k, t)$, we can write

$$\lambda_l C_l = \int^{\mathbf{r}_i} \mathbf{F}_i^{\mathbf{c}} \cdot d\mathbf{r}_i' = \int^{q_k} \mathbf{F}_i^{\mathbf{c}} \cdot \frac{\partial \mathbf{r}_i'}{\partial q_k'} dq_k', \qquad (6.13)$$

where we used the chain rule and the fact that $dt = 0$, since the expression added to the Lagrangian should be evaluated at a fixed time. We then have

$$\mathcal{F}_k = \lambda_l \frac{\partial C_l}{\partial q_k} = \frac{\partial (\lambda_l C_l)}{\partial q_k} = \mathbf{F}_i^{\mathbf{c}} \cdot \frac{\partial \mathbf{r}_i}{\partial q_k}. \qquad (6.14)$$

Once we determine the Lagrange multipliers λ_l, we can use this relation to read off constraint forces $\mathbf{F}_i^{\mathbf{c}}$. For the example at hand, we have

$$\lambda_1 = \mathbf{F}^{\mathbf{c}} \cdot \frac{\partial \mathbf{r}}{\partial z} = F_z^{\mathbf{c}} = N. \qquad (6.15)$$

Hence λ_1 is the normal force, as expected.

Before we apply this general and abstract treatment to particular examples, we note that the method of constraints can easily be generalized a bit further. Looking back at Eq. (6.3), we see that we only need the constraint to be in the form of a *variation*. That is, we only need to add the constraint in the *infinitesimal change* of the action or Lagrangian, not the Lagrangian itself as we did in Eq. (6.4). We write instead

$$\delta I = \int \left(\frac{d}{dt}\left(-\frac{\partial L}{\partial \dot{q}_k} \right) + \frac{\partial L}{\partial q_k} + \lambda_l a_{lk} \right) \delta q_k dt = 0, \qquad (6.16)$$

for some functions a_{lk} of q_k and t. That is, if the constraints on the generalized coordinates q_k can be written in the form

$$a_{lk}\delta q_k + a_{lt}\delta t = 0, \qquad (6.17)$$

where a_{lk} and a_{lt} are arbitrary functions of q_k and t, we can write the following set of $N + P$ equations of motion:

$$\frac{d}{dt}\left(\frac{\partial L}{\partial \dot{q}_k} \right) - \frac{\partial L}{\partial q_k} = \lambda_l a_{lk} = \mathcal{F}_k, \qquad (6.18)$$

$$a_{lk}\dot{q}_k + a_{lt} = 0. \qquad (6.19)$$

This is a useful generalization of the original formulation of the problem given by Eq. (6.6) and (6.9) because in a particular case perhaps

$$\frac{\partial a_{lk}}{\partial t} \neq \frac{\partial a_{lt}}{\partial q_k}. \qquad (6.20)$$

To see the further reach of this approach, note that if the constraints could be written as before in terms of P *algebraic* relations $C_l(t, q) = 0$, we could write

$$0 = dC_l = \frac{\partial C_l}{\partial q_k} dq_k + \frac{\partial C_l}{\partial t} dt = a_{lk}dq_k + a_{lt}dt, \qquad (6.21)$$

reading off the needed a_{lk} and a_{lt} as functions of q_k and t. However, in this special class of constraints, we *would* have

$$\frac{\partial a_{lk}}{\partial t} = \frac{\partial a_{lt}}{\partial q_k} \tag{6.22}$$

because of the commutativity of derivatives $\partial^2 C_l / \partial q_k \partial t = \partial^2 C_l / \partial t \partial q_k$. For the example at hand, we would write

$$dz = 0 \Rightarrow a_{1z} = 1, \quad a_{1t} = 0, \tag{6.23}$$

with Eq. (6.22) satisfied as expected. Algebraic constraints of the form $C_l(q, t) = 0$ are called **holonomic constraints**, already discussed briefly in Chapter 4. The type of constraints that cannot be expressed in this form but instead can be written in variational form (6.17) are **nonholonomic constraints**. In either case, Eqs. (6.18) and (6.19) give us a powerful recipe for investigating the problem of constraints.

In summary, when dealing with a mechanics problem involving constraints of the form (6.17), we may choose to delay the implementation of the constraints in an effort to extract any constraint forces acting on the system. To do so we would need to solve a set of $N + P$ differential equations given by (6.18) and (6.19). In addition to finding $q_k(t)$, this procedure also leads to P Lagrange multipliers that can be related to constraint forces through Eq. (6.14). Once again, the best way to learn this technology is through examples.

Example 6.1   **Rolling Down the Plane**

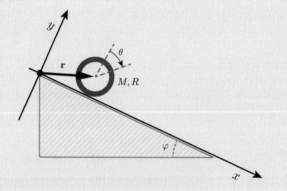

Fig. 6.2   A hoop rolling down an inclined plane without slipping.

Consider a hoop of radius R and mass M rolling down an inclined plane, as shown in Figure 6.2. To describe the hoop, we may prescribe three variables: a center of mass position in two dimensions $\mathbf{r}$, and a rotational angle θ:

$$\mathbf{r} = x\hat{\mathbf{x}} + y\hat{\mathbf{y}}, \quad \theta, \tag{6.24}$$

where the coordinate system is set up tilted, as shown in the figure. However, we know of two potential constraints. First, the hoop is prevented from falling through the incline because of the normal force. This contact force enforces the constraint

$$y = R \Rightarrow dy = 0. \tag{6.25}$$

Furthermore, if there is friction involved, the hoop may roll down without slipping. This rolling without slipping condition amounts to the constraint

$$dx = R \, d\theta. \tag{6.26}$$

At the end of the day, these two constraints suggest that the problem involves only $3 - 2 = 1$ degree of freedom, not three. The kinetic energy is

$$T = \frac{1}{2} M \left(\dot{x}^2 + \dot{y}^2 \right) + \frac{1}{2} M R^2 \dot{\theta}^2 \tag{6.27}$$

while the potential energy can be written as (dropping constant shifts)

$$U = -Mgx \sin \varphi, \tag{6.28}$$

where φ is the tilt of the incline. We may then proceed as usual by implementing the constraints (6.25) and (6.26) from the outset. We arrive at the Lagrangian with a single degree of freedom, which we may choose to be the x coordinate, so

$$L = T - U = M\dot{x}^2 + Mgx \sin \varphi. \tag{6.29}$$

The equation of motion then tells us the acceleration down the incline:

$$\ddot{x} = \frac{g}{2} \sin \varphi. \tag{6.30}$$

But what if we are interested in knowing the friction force, although we still do not care at all about the normal force? This means we will implement the normal force constraint given by (6.25) from the outset, eliminating the y coordinate, but we will delay implementing the frictional constraint given by (6.26). This leaves us with two of the original coordinates, x and θ, and a new third degree of freedom, λ_1, a Lagrange multiplier that we can use to find the friction force. We then need three differential equations. Looking back at (6.26) and mapping it onto the general form (6.19), we read off

$$a_{1\theta} = R, \quad a_{1x} = -1, \quad a_{1t} = 0. \tag{6.31}$$

We then write the Lagrangian in terms of x and θ only:

$$L = \frac{1}{2} m\dot{x}^2 + \frac{1}{2} M R^2 \dot{\theta}^2 + Mgx \sin \varphi \tag{6.32}$$

and use the modified equations of motion given by (6.18). This gives for the x direction

$$M\ddot{x} - Mg \sin \varphi = -\lambda_1 = \mathscr{F}_x \tag{6.33}$$

and for the θ direction

$$MR^2\ddot{\theta} = \lambda_1 R = \mathscr{F}_\theta. \tag{6.34}$$

We now have a set of three differential equations, given by (6.33), (6.34), and

$$R\dot{\theta} = \dot{x}, \tag{6.35}$$

which follows from the constraint (6.26). Solving this system of equations, we get

$$\ddot{x} = \frac{g}{2}\sin\varphi, \quad \lambda_1 = \frac{Mg}{2}\sin\varphi. \tag{6.36}$$

The novelty, of course, is the determination of λ_1, which we can now relate to the friction force through

$$\mathscr{F}_\theta = \lambda_1 R = \frac{MgR}{2}\sin\varphi = \mathbf{F}\cdot\frac{\partial\mathbf{r}}{\partial\theta} = \mathbf{F}\cdot\frac{\partial x}{\partial\theta}\hat{\mathbf{x}} = \mathbf{F}\cdot\frac{\partial(R\theta)}{\partial\theta}\hat{\mathbf{x}} = F_x R = \tau. \tag{6.37}$$

That is, the friction force $(Mg\sin\varphi)/2$ supplies a torque equal to $(MgR\sin\varphi)/2$. The method of Lagrangian multipliers has allowed us to selectively extract forces of constraint in a mechanical problem – without abandoning the elegant and powerful machinery of the Lagrangian formalism. ∎

Example 6.2   **Stacking Barrels**

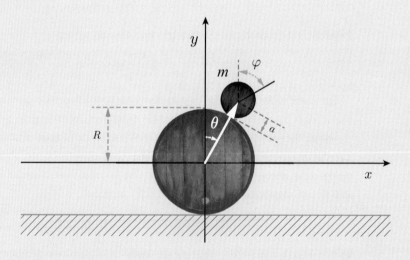

Fig. 6.3   Two barrels stacked, one on top of the other. The lower barrel is stationary, while the upper one rolls down without slipping.

Consider the problem of two cylindrical barrels, one on top of the other, as shown in Figure 6.3. The bottom barrel is fixed in position and orientation, but the top one, of mass m, is free to move. It starts rolling down from its initial position at the top, rolling without slipping due to friction between the barrels. The problem is to find the point along the lower barrel where the top barrel loses contact with it. That is, we need to find the moment when the normal force acting on the top barrel vanishes.

We are tracking the motion of the top barrel. Hence, we have *a priori* three variables to keep track of, two positions r and θ, and one rotational angle φ:

$$\mathbf{r} = r\hat{\mathbf{r}}, \quad \varphi, \tag{6.38}$$

where we use polar coordinates centered on the bottom barrel to track the position of the top barrel. A normal force acting on the top barrel enforces the constraint

$$r = R + a. \tag{6.39}$$

If the top barrel rolls without slipping, the friction force enforces an additional constraint

$$a d\varphi = R d\theta. \tag{6.40}$$

The full kinetic energy in terms of r, θ, φ is

$$T = \frac{1}{2}m \left(\dot{r}^2 + r^2\dot{\theta}^2 \right) + \frac{1}{2}ma^2\dot{\varphi}^2 \tag{6.41}$$

and the potential energy is

$$U = mgr\cos\theta. \tag{6.42}$$

Note that we included the rotational kinetic energy of the top, hoop-like barrel (whose endcaps have negligible mass), given by $\frac{1}{2}ma^2\dot{\varphi}^2$. Since we are only interested in the normal force, we implement the frictional constraint (6.40) from the outset to eliminate φ in favor of θ. We then get

$$L = T - U = \frac{1}{2}m \left(\dot{r}^2 + r^2\dot{\theta}^2 \right) + \frac{1}{2}mR^2\dot{\theta}^2 - mgr\cos\theta. \tag{6.43}$$

We do *not* implement the normal force constraint (6.39), which we now write in the canonical form (6.17):

$$dr = 0. \tag{6.44}$$

This means we will have three degrees of freedom: r, θ, and a Lagrange multiplier λ_1 associated with constraint (6.39). We can now read off the relevant coefficients of (6.19):

$$a_{1r} = 1, \quad a_{1\theta} = a_{1t} = 0. \tag{6.45}$$

The equations of motion follow from (6.18). For the r direction:

$$m\ddot{r} - mr\dot{\theta}^2 + mg\cos\theta = \lambda_1 = \mathscr{F}_r = \mathbf{F} \cdot \frac{\partial \mathbf{r}}{\partial r} = F_r = N \tag{6.46}$$

while for the θ direction:

$$mr^2\ddot{\theta} + mR^2\ddot{\theta} - mgr\sin\theta = 0. \tag{6.47}$$

The third and final equation follows from the constraint (6.39), which we now write as

$$\dot{r} = 0. \tag{6.48}$$

We may now proceed to solve this set of three differential equations. It is more elegant, however, to realize that energy is conserved in this problem (since all forces acting are conservative), so we can write

$$E = T + U = \frac{1}{2}m\left((R+a)^2\,\dot{\theta}^2\right) + \frac{1}{2}mR^2\dot{\theta}^2 + mg(R+a)\cos\theta \tag{6.49}$$

with $r = R + a$ from the constraint and where E is a constant. Arranging for the initial conditions $\theta(0) = 0$ and $\dot{\theta}(0) = 0$, we have $E = mg\,(R+a)$. This implies

$$\dot{\theta}^2 = \frac{2g\,(R+a)}{2\,R^2 + a^2 + 2Ra}\,(1 - \cos\theta)\,. \tag{6.50}$$

From Eq. (6.46), we have

$$-m\,(R+a)\,\dot{\theta}^2 + mg\,\cos\theta = \lambda_1 = N. \tag{6.51}$$

And setting $N = 0$ at the moment when the top barrel loses contact with the bottom one, we find

$$\dot{\theta}_c^2 = \frac{g\cos\theta_c}{R+a}, \tag{6.52}$$

where θ_c denotes the critical angle at which this condition is satisfied. Using Eq. (6.50), we find

$$\cos\theta_c = \frac{2}{3 + R^2/(R+a)^2}. \tag{6.53}$$

There are two interesting limiting cases. If the top barrel is tiny, we have $a/R \ll 1$, which leads to

$$\cos\theta_c = \frac{1}{2}. \tag{6.54}$$

In contrast, taking the opposite regime $a/R \gg 1$, one finds

$$\cos\theta_c = \frac{2}{3}. \tag{6.55}$$

∎

<table><tr><td>Example 6.3</td><td>

On the Rope

A classic problem is that of a bob pendulum consisting of a point mass m at the end of a rope of length l swinging in a plane (see Figure 6.4). We would like to determine the tension in the rope as a function of the angle θ. We start with two variables

$$\mathbf{r} = r\hat{\mathbf{r}} \tag{6.56}$$

using a polar coordinate system centered at the pivot, as shown in the figure. Hence, our variables are r and θ. However, we have a constraint enforced by the tension in the rope:

$$r = l \Rightarrow dr = 0 \Rightarrow a_{1r} = 1,\ a_{1t} = 0, \tag{6.57}$$

which we immediately use to read off the relevant coefficients for (6.19). The kinetic energy is
</td></tr></table>

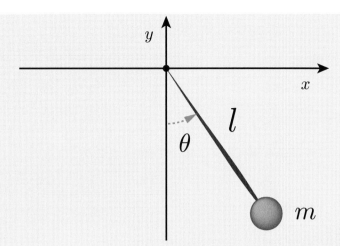

Fig. 6.4 A bob pendulum with one constraint given by the finite length of the rope.

$$T = \frac{1}{2}m\left(\dot{r}^2 + r^2\dot{\theta}^2\right) \tag{6.58}$$

while the potential energy is simply

$$U = -mgr\cos\theta. \tag{6.59}$$

The Lagrangian becomes

$$L = T - U = \frac{1}{2}m\left(\dot{r}^2 + r^2\dot{\theta}^2\right) + mgr\cos\theta, \tag{6.60}$$

where we keep track of both r and θ as independent degrees of freedom at the cost of introducing one Lagrange multiplier λ_1 associated with the constraint. The equation of motion for r comes from (6.18):

$$m\ddot{r} + mr\dot{\theta}^2 - mg\cos\theta = \lambda_1 = \mathbf{F} \cdot \frac{\partial \mathbf{r}}{\partial r} = F_r = T, \tag{6.61}$$

while that for θ is

$$\frac{d}{dt}\left(mr^2\dot{\theta}\right) + mgr\sin\theta = 0. \tag{6.62}$$

Our three degrees of freedom are associated with three differential equations, and the third comes of course from the constraint (6.57), which we now write as

$$\dot{r} = 0. \tag{6.63}$$

This is enough to determine all three variables of interest. Once again, however, it is easier to use energy conservation:

$$E = T + U = \frac{1}{2}ml^2\dot{\theta}^2 - mgl\cos\theta. \tag{6.64}$$

We start with initial conditions given by

$$\theta(0) = \theta_0, \quad \dot{\theta}(0) = 0 \Rightarrow E = -mgl \cos\theta_0. \tag{6.65}$$

We then have

$$\dot{\theta}^2 = \frac{2g}{l}\left(\cos\theta - \cos\theta_0\right). \tag{6.66}$$

This allows us to find the Lagrange multiplier λ_1 in terms of θ:

$$\lambda_1 = -mg\cos\theta + 2mg\left(\cos\theta - \cos\theta_0\right) = -mg\left(\cos\theta - 2\cos\theta_0\right) = T, \tag{6.67}$$

which is the tension in the rope.   ∎

Example 6.4

A Rubber Wheel on a Road

A wheel of radius R rolls on a two-dimensional surface without slipping. If the wheel cannot skid in any direction, we have an interesting problem containing constraints. Assuming that the wheel does not tip over, we can describe its state by specifying the coordinates x, y on the plane, the orientation angle θ measured from the x axis, and the roll angle φ – as shown in Figure 6.5. Because of the rolling without slipping condition, the roll angle can be related to x and y and hence can be eliminated from the outset:

$$R^2 d\varphi^2 = dx^2 + dy^2. \tag{6.68}$$

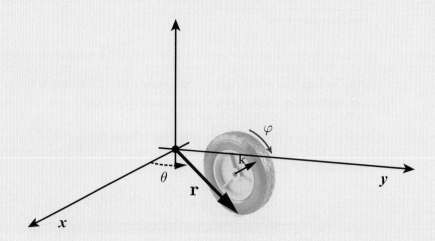

Fig. 6.5 The coordinates used to describe the rolling rubber wheel.

Furthermore, the wheel cannot move sideways, although it can make gradual turns. So there is indeed another constraint in the problem. How can we quantify it? If we write the vector perpendicular to the wheel as $\mathbf{k} = (\sin\theta, -\cos\theta)$, we want to say that the velocity of the wheel is always perpendicular to $\mathbf{k}$:

$$\mathbf{k} \cdot (\dot{x}, \dot{y}) = 0 \Rightarrow \dot{x}\sin\theta - \dot{y}\cos\theta = 0. \tag{6.69}$$

Hence, we start with four variables minus two constraints, and end up with two degrees of freedom. The orientation angle θ is a freedom we could actually take control of by *steering* the wheel by some external mechanical method. This would amount to imposing a third constraint. For example, if we externally lock the orientation so that $\dot{\theta} = 0$, Eq. (6.69) can be integrated and we have an algebraic relation between x and y. With such a holonomic constraint, the problem now involves only a single degree of freedom: θ is fixed to a constant, while x and y are related to each other algebraically; the wheel rolls in a straight line along a single direction.

Suppose however we steer the wheel in pursuit of a moving target! Say we aim the wheel toward a running person, with the wheel's orientation thus determined by the line joining its position and the position of the runner. This implies that θ is not constant, but instead is generally given by a function of x, y, and t: $\theta(x, y, t)$. In general, Eq. (6.69) becomes non-integrable, so we have a *nonholonomic* constraint. As seen in our general treatment earlier, however, we can still handle this problem since our new constraint is in the form (6.17). We write the Lagrangian of the wheel as the sum of its rotational and translational kinetic energies; and we then use Eqs. (6.18) and (6.19). We leave the treatment as a problem for the reader. ∎

6.2 Symmetries and Conservation Laws: A Preview

At a very fundamental level, physics is about identifying patterns in Nature. The field arguably begins with Tycho Brahe (1546–1601) – the first modern experimental physicist[1] – and Johannes Kepler (1571–1630) – the first modern theoretical physicist. In the sixteenth century, Brahe painstakingly accrued huge quantities of astronomical data about the location of planets and stars with unprecedented accuracy – using impressive observing instruments that he had designed and set up in his castle. Kepler pondered for years over Brahe's long tables until he could finally identify patterns underlying planetary dynamics, summarized in what we now call Kepler's three laws. Later on, Isaac Newton (1643–1727) referred to these achievements, among others, in his famous quote: "*If I have seen further it is only by standing on the shoulders of giants.*"

Ever since, physics has always been about identifying patterns in numbers, in measurements. And a pattern is simply an indication of a repeating rule, a constant attribute within seeming complexity, an underlying symmetry. In 1918, Emmy Noether published a seminal work that clarified the deep relations between symmetries and conserved quantities in Nature. In a sense, this work organizes physics into a clear diagram that gives us a bird's eye view of the myriad of branches of the field. Noether's theorem, as it is called, can change the way one thinks about physics in general. It is simple yet profound.

While one can study Noether's theorem in the context of Newton's formulation of mechanics, the methods in that case are quite cumbersome. Studying Noether's

---

[1] Or perhaps the first modern observational astronomer.

theorem using Lagrangian mechanics instead is an excellent demonstration of the power of the newer formalism.

In the following section we begin by recounting the connections between symmetries and conserved quantities already encountered in Chapter 4. We then use the variational principle to develop a statement of Noether's theorem, and go on to prove it and demonstrate its importance through examples.

6.3 Cyclic Coordinates and Generalized Momenta

Already in Chapter 4 we defined a *cyclic* coordinate q_k as a generalized coordinate *absent* from the Lagrangian. The corresponding generalized velocity $\dot{q}_k$ *is* present in L, but not q_k itself. For example, a particle free to move in three dimensions x, y, z with uniform gravity in the z direction has Lagrangian

$$L = \frac{1}{2}m(\dot{x}^2 + \dot{y}^2 + \dot{z}^2) - mgz, \tag{6.70}$$

so both x and y are cyclic, but z is not. Then from Lagrange's equations

$$\frac{d}{dt}\frac{\partial L}{\partial \dot{q}_k} - \frac{\partial L}{\partial q_k} = 0 \tag{6.71}$$

it follows that for each of the cyclic coordinates q_k the **generalized momentum**

$$p_k \equiv \frac{\partial L}{\partial \dot{q}_k} \tag{6.72}$$

is *conserved*, since $dp_k/dt = 0$. So in this example, both $p_x = m\dot{x}$ and $p_y = m\dot{y}$ are conserved, but p_z is not.

Now the interesting question is: *Why* is a particular coordinate cyclic, from a physical point of view? Why are x and y cyclic in our example, but z is not?

The answer is quite clear. The physical environment is invariant under displacements in the x and y directions, but not under vertical displacements. Changes in z mean that we get closer or farther from the ground, but changes in x or y make no difference whatever. Everything looks exactly the same if we displace ourselves horizontally. We say there is a *symmetry* under horizontal displacements, but not under vertical displacements. So through Lagrange's equations we see that the momenta p_x, p_y are conserved but p_z is not. Two symmetries, those in the x and y directions, have led to two conserved quantities.

If a generalized coordinate q_i is missing from the Lagrangian, that means there is a symmetry under changes in that coordinate: the physical environment, whether a potential energy or a constraint, is independent of that coordinate. And if the environment possesses the symmetry, the corresponding generalized momentum will be conserved. This is very important, not least because the equation of motion for that coordinate will be only first order rather than second order, so becomes much easier to solve.

Example 6.5

A Star Orbiting a Spheroidal Galaxy

A particular galaxy consists of an enormous sphere of stars somewhat squashed along one axis, so it becomes spheroidal in shape, as shown in Figure 6.6. A star at the outer fringes of the galaxy experiences the general gravitational pull of the galaxy. Are there any conserved quantities in the motion of this star?

We can answer this question by identifying any symmetries in the environment of the star. There are no translational symmetries; any finite translation takes one nearer or farther from the galaxy, so no component of linear momentum is conserved.

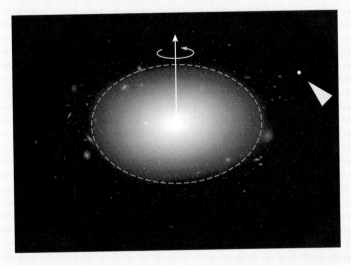

Fig. 6.6   An elliptical galaxy (NGC 1132) pulling on a star at the outer fringes. Image credit: Optical: NASA/ESA/STScI/M. West; X-ray: NASA/CXC/Penn State/G. Garmire.

There is, however, a symmetry under rotation about the squashed axis of the galaxy; if we imagine rotating about that axis, the shape of the galaxy, and therefore its gravitational field, will be unchanged. Therefore we expect that the *angular momentum* p_φ about this axis will be conserved. No other component of angular momentum will be conserved, because if we rotate about any other axis, the galaxy will look different. Therefore we expect conservation of angular momentum about a single axis only, and no conservation of linear momentum in any direction.

Mathematically, using cylindrical coordinates (ρ, φ, z), the kinetic and gravitational potential energies of the star are

$$T = \frac{1}{2}m(\dot{\rho}^2 + \rho^2\dot{\varphi}^2 + \dot{z}^2) \text{ and } U = U(\rho, z), \tag{6.73}$$

where the symmetric shape of the galaxy means that the gravitational potential depends on ρ and z, but not φ. Therefore the Lagrangian $L = T - U$ depends on ρ and z, but φ is cyclic. The corresponding generalized momentum $p_\varphi = \partial L/\partial\dot{\varphi}$, which is in fact the angular momentum about the z axis, is therefore conserved. ∎

Example 6.6 **A Charged Particle Moving Outside a Charged Rod**

An infinite straight dielectric rod is oriented in the z direction, and given a uniform electric charge per unit length. A point charge is free to move outside it, as shown in Figure 6.7. Are there any conserved quantities for the motion of the particle?

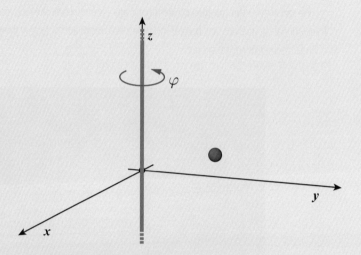

Fig. 6.7 A charged particle moving near a charged rod.

This time the environment of the particle has *two* symmetries: a symmetry corresponding to a rotation about the rod axis, and a symmetry corresponding to displacements in the z direction, along the rod axis. If φ is the angle about the rod in the x, y plane, then the generalized momentum p_φ, which is in fact the angular momentum of the particle about the z axis, is conserved, because of symmetry under rotation. The generalized momentum p_z is also conserved, because of symmetry under displacement in the z direction; this is the linear momentum of the particle in the z direction. The other components of linear momentum are not conserved, and the other components of angular momentum are also not conserved, because there is no symmetry under displacements in the x or y direction, or under rotations about any other axis.

Mathematically, the kinetic and potential energies in this case (using cylindrical coordinates) have the form

$$T = \frac{1}{2}m(\dot{\rho}^2 + \rho^2\dot{\varphi}^2 + \dot{z}^2) \quad \text{and} \quad U = U(\rho). \tag{6.74}$$

Therefore, when the Lagrangian $L = T - U$ is assembled, both φ and z are cyclic, so the corresponding generalized momenta p_φ, p_z are conserved, as we already knew from our more "physical" analysis. ∎

6.4 A Less Straightforward Example

Consider the simple mechanics problem we described in Section 4.5. Two particles, with masses m_1 and m_2, are constrained to move along a horizontal frictionless rail, as depicted in Figure 4.9. The action for the system is

$$S = \int dt \left(\frac{1}{2} m_1 \dot{q}_1^2 + \frac{1}{2} m_2 \dot{q}_2^2 - U(q_1 - q_2) \right), \tag{6.75}$$

with some interaction between the particles described by a potential $U(q_1 - q_2)$ that depends only on the distance between the particles.

Note that neither q_1 nor q_2 is cyclic, so it might seem that there are no symmetries in this problem. However, let us consider a simple *transformation* of the coordinates given by

$$q_1' = q_1 + C, \quad q_2' = q_2 + C, \tag{6.76}$$

where C is some arbitrary constant. So in this case, we have simply changed the origin of coordinates, which should have no effect on the physics. We then have

$$\dot{q}_1' = \dot{q}_1, \quad \dot{q}_2' = \dot{q}_2, \tag{6.77}$$

so the kinetic terms in the action are unchanged under this transformation. Furthermore, we also have

$$q_1' - q_2' = q_1 - q_2, \tag{6.78}$$

implying that the potential term is also unchanged. The action then preserves its overall structural form under the transformation

$$S \rightarrow \int dt \left(\frac{1}{2} m_1 (\dot{q}_1')^2 + \frac{1}{2} m_2 (\dot{q}_2')^2 - U(q_1' - q_2') \right). \tag{6.79}$$

This means that the equations of motion, written in the primed transformed coordinates, are identical to the ones written in the unprimed original coordinates. We can then say that the transformation given by (6.76) is a symmetry of our system. Physically, we are simply saying that – since the interaction between the particles depends only on the distance between them – a constant shift of both coordinates leaves the dynamics unaffected. Indeed, we showed in Chapter 4 that in this problem a simple linear transformation of coordinates to center-of-mass and relative coordinates produced a cyclic coordinate, the position of the center of mass. As a result, the momentum of the center of mass, that is the total momentum, is conserved.

It is also useful to consider an *infinitesimal* version of such a transformation. Assume that the constant C is small, $C \rightarrow \epsilon$; and write

$$q_k' - q_k = \epsilon \tag{6.80}$$

for $k = 1, 2$. We then say that the shift of the coordinates is a symmetry of our system. To make these ideas more useful, in the next section we extend this example by considering a general class of interesting transformations.

6.5 Infinitesimal Transformations

We consider two types of infinitesimal transformations: *direct and indirect.*

Direct Transformations

A **direct transformation** deforms the generalized coordinates of a system as in

$$\delta q_k(t) = q'_k(t) - q_k(t) \equiv \Delta q_k(t, q). \tag{6.81}$$

We use the notation Δ to label a direct transformation. Note that $\Delta q_k(t, q)$ is possibly a function of time and all of the generalized coordinates in the problem. In the example of Section 6.4, we used the special case $\Delta q_k(t, q) = $ constant. But it need not be so. Figure 6.8(a) depicts a direct transformation: it is an arbitrary, but *small* shift in the q_ks *at fixed time*. In the figure, a particular generalized coordinate q_k is shifted to slightly larger variable values for early times, and then to slightly smaller variable values for later times.

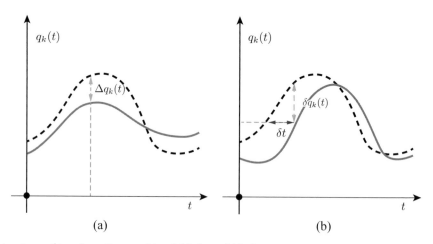

(a)   (b)

Fig. 6.8   The two types of transformations considered: (a) direct; (b) indirect.

Indirect Transformations

By contrast, an **indirect transformation** affects the generalized coordinates indirectly – through the transformation of the time coordinate

$$\delta t(t, q) \equiv t' - t. \tag{6.82}$$

Note again that the shift in time is assumed to be small, and this shift can itself depend on time. The small shifts in time then bring about small shifts in the generalized coordinates, in which case they have been affected *indirectly*:

$$q_k(t) = q_k(t' - \delta t) \simeq q_k(t') - \frac{dq_k}{dt'}\delta t \simeq q_k(t') - \dot{q}_k \delta t, \tag{6.83}$$

where we have used a Taylor expansion in δt to linear order only, since δt is small (see Appendix F). We also have used $dq_k/dt' = dq_k/dt$ since this term already multiplies a power of δt: to linear order in δt, $\delta t\, dq_k/dt' = \delta t\, dq_k/dt \equiv \delta t\, \dot{q}_k$. We then see that shifting time results in a shift in the generalized coordinates

$$\delta q_k = q_k(t') - q_k(t) = \dot{q}\, \delta t(t, q). \tag{6.84}$$

Compare with Eq. (6.81) and note that the coordinates are now evaluated at *different* times. Figure 6.8(b) shows how we can think of this effect graphically: a small change in time has caused a small change in the generalized coordinate $\delta q_k(t)$.

Combined Transformations

In general, we want to consider a transformation that may include *both* direct and indirect pieces. We write

$$\delta q_k = q_k'(t') - q_k(t) = q_k'(t') + \left(-q_k(t') + q_k(t')\right) - q_k(t)$$
$$= \left(q_k'(t') - q_k(t')\right) + \left(q_k(t') - q_k(t)\right) = \Delta q_k(t', q) + \dot{q}\, \delta t(t, q)$$
$$= \Delta q_k(t, q) + \dot{q}\, \delta t(t, q), \tag{6.85}$$

where in the last line we have equated t and t' since the first term is already linear in the small parameters. To specify a particular transformation, we could then provide a set of functions

$$\delta t(t, q) \quad \text{and} \quad \delta q_k(t, q), \tag{6.86}$$

from which Eq. (6.85) determines $\Delta q_k(t, q)$. For N degrees of freedom, that amounts to $N + 1$ functions of time and the q_ks. Let us look at a few examples.

Example 6.7 **Translations**

Consider a single particle in three dimensions, described by the three Cartesian coordinates $r^x = x, r^y = y$, and $r^z = z$. We also have the time coordinate $r^t = ct$. An infinitesimal constant spatial translation can be realized by

$$\delta r^i(t, x) = \epsilon^i, \quad \delta t(t, x) = 0 \Rightarrow \Delta r^i(t, x) = \delta\epsilon^i, \tag{6.87}$$

where $i = x, y, z$, and the ϵ^is are three small constants. A translation in *space* is then defined by

$$\{\delta t(t, x) = 0, \ \delta r^i(t, x) = \delta\epsilon^i\} \quad \textit{Translation in space.} \tag{6.88}$$

A constant translation in *time* would be given by

$$\delta r^i(t, x) = 0, \quad \delta t(t, x) = \delta\epsilon \Rightarrow \Delta r^i(t, x) = -\dot{r}^i\delta\epsilon \tag{6.89}$$

for constant ε. Notice that for a translation purely in time, we need to require that the *total* shifts in the r's — the $\delta r^i(t,x)$s — vanish. This then generates direct shifts, the Δr's, to compensate for the indirect effect on the spatial coordinates from the shifting of the time. A translation in time is then defined by

$$\{\delta t(t,x) = \delta\epsilon, \ \delta r^i(t,x) = 0\} \quad \textit{Translation in time.} \tag{6.90}$$

■

Example 6.8

Rotations

To describe constant rotations, we consider for simplicity a particle moving in two dimensions. We use the coordinates $r^x = x$ and $r^y = y$, and start by specifying

$$\delta t(t,x) = 0 \Rightarrow \delta r^i(t,x) = \Delta r^i(t,x), \tag{6.91}$$

with $i = x, y$. Next, we look at an arbitrary rotation angle θ using Eq. (2.21):

$$\begin{pmatrix} r^{x'} \\ r^{y'} \end{pmatrix} = \begin{pmatrix} \cos\theta & \sin\theta \\ -\sin\theta & \cos\theta \end{pmatrix} \begin{pmatrix} r^x \\ r^y \end{pmatrix}. \tag{6.92}$$

However, we need to focus on an infinitesimal version of this transformation: *i.e.*, we need to consider small angles $\delta\theta$. Using $\cos\delta\theta \sim 1$ and $\sin\delta\theta \sim \delta\theta$ to first order in $\delta\theta$, we then write[a]

$$\begin{pmatrix} r^{x'} \\ r^{y'} \end{pmatrix} = \begin{pmatrix} 1 & \delta\theta \\ -\delta\theta & 1 \end{pmatrix} \begin{pmatrix} r^x \\ r^y \end{pmatrix}. \tag{6.93}$$

This gives

$$\delta r^x(t,x) = r^{x'}(t) - r^x(t) = \delta\theta \, r^y(t) = \Delta r^x(t,x),$$

$$\delta r^y(t,x) = r^{y'}(t) - r^y(t) = -\delta\theta \, r^x(t) = \Delta r^y(t,x). \tag{6.94}$$

We now have a less trivial transformation. Rotations can be defined by

$$\{\delta t(t,x) = 0, \ \delta r^i(t,x) = \delta\theta \, \varepsilon^i_{\ j} r^j(t)\} \quad \textit{Two-dimensional spatial rotation}, \tag{6.95}$$

where j is summed over 1 and 2. We have also introduced a useful shorthand: $\varepsilon^i_{\ j}$ is the totally antisymmetric matrix in two dimensions, given by

$$\varepsilon^x_{\ x} = \varepsilon^y_{\ y} = 0, \ \ \varepsilon^x_{\ y} = -\varepsilon^y_{\ x} = 1, \tag{6.96}$$

which has allowed us to write the transformation in a more compact notation.

■

---

[a]Recall that the Taylor series expansions of $\cos\delta\theta$ and $\sin\delta\theta$ are $\cos\delta\theta = 1 - \delta\theta^2/2! + \delta\theta^4/4! - \ldots$ and $\sin\delta\theta = \delta\theta - \delta\theta^3/3! + \delta\theta^5/5! - \ldots$ (see Appendix F for more).

Example 6.9

Lorentz Transformations

To find the infinitesimal form of Lorentz transformations, we can start with the general transformation equations (2.15) and take the velocity parameter $\beta = v/c$ to be small. We need to be careful, however, to keep the leading-order terms in β in all expansions. Given our previous example, it is easier to map the problem onto a rotation with hyperbolic trig functions and the rapidity ξ, which (as shown in Chapter 2) is

$$
\begin{pmatrix} ct \\ x \\ y \\ z \end{pmatrix} = \begin{pmatrix} \cosh\xi & \sinh\xi & 0 & 0 \\ \sinh\xi & \cosh\xi & 0 & 0 \\ 0 & 0 & 1 & 0 \\ 0 & 0 & 0 & 1 \end{pmatrix} \begin{pmatrix} ct' \\ x' \\ y' \\ z' \end{pmatrix}, \tag{6.97}
$$

where $\gamma = \cosh\xi$, $\gamma\beta = \sinh\xi$. For simplicity, let us consider a particle in one dimension, with two relevant coordinates $r^t = ct$ and r^x. We take the rapidity $\xi \to \delta\xi$ to be small and use $\cosh\delta\xi \sim 1$ and $\sinh\delta\xi \sim \delta\xi$ to linear order in $\delta\xi$. Using the same steps as in the previous example, we find

$$
\delta r^t(t,x) = \delta\xi \, r^x, \quad \delta r^x(t,x) = \delta\xi \, r^t. \tag{6.98}
$$

Using Eq. (6.85), we then have

$$
\Delta r^x(t,x) = \delta r^x(t,x) - \dot{r}^x \frac{\delta r^t(t,x)}{c} = \delta\xi \, r^t - \frac{\delta\xi}{c} \dot{r}^x r^x. \tag{6.99}
$$

Note that

$$
\sinh\xi = \gamma\beta \Rightarrow \sinh\delta\xi \sim \delta\xi \sim \beta. \tag{6.100}
$$

The Lorentz transformations can then be defined by

$$
\{\delta r^t = \beta \, r^x, \ \delta r^x = \beta \, r^t, \ \delta r^y = \delta r^z = 0\} \quad \textit{Lorentz transformation}, \tag{6.101}
$$

where we added the effect on the transverse directions as well.

■

6.6 Symmetry

We define a **symmetry** to be a transformation that leaves the action unchanged in form. In the particular example of two interacting particles on a frictionless rail, it was simple to see that the transformation was indeed a symmetry. Now that we have a general class of transformations, we want to find a general condition that can be used to test whether a particular transformation – possibly a complicated one – is or is not a symmetry. We then need to look at how the action changes under a general transformation: for a symmetry, this change should vanish. We start with the usual form for the action:

$$
S = \int dt \, L(q, \dot{q}, t). \tag{6.102}
$$

Then we apply a general transformation given by $\Delta q_k(t, q)$ and $\delta t(t, q)$. It follows that

$$\delta S = \int \delta(dt)\, L + \int dt\, \delta(L),\qquad(6.103)$$

where we have used the Leibniz product rule $\delta(ab) = (\delta a)b + a(\delta b)$, since δ is an infinitesimal change. The first term is the change in the *measure* of the integrand:

$$\delta(dt) = dt\, \frac{\delta(dt)}{dt} = dt\, \frac{d}{dt}(\delta t),\qquad(6.104)$$

where in the last part we exchanged the order of δ and d, since they commute.[3] The second term has two parts

$$\delta(L) = \Delta(L) + \delta t\, \frac{dL}{dt}.\qquad(6.105)$$

The first part is the change in L resulting from its dependence on the q_ks and $\dot{q}_k$s. Hence, we labeled it as a direct change with a Δ. The second part is the change in L to linear order in δt due to the change in t. This comes from changes in the q_ks on which L depends, as well as changes in t directly, since t can make an explicit appearance in L. This is an identical situation to the linear expansion encountered for q_k in Eq. (6.85): there is a part from direct changes in the degrees of freedom, as well as a part from the transformation of time. Then, using multivariable calculus:

$$\Delta(L) = \frac{\partial L}{\partial q_k}\Delta q_k + \frac{\partial L}{\partial \dot{q}_k}\Delta(\dot{q}_k) = \frac{\partial L}{\partial q_k}\Delta q_k + \frac{\partial L}{\partial \dot{q}_k}\frac{d}{dt}(\Delta q_k).\qquad(6.106)$$

Note that in the last term we exchanged the orders of Δ and d/dt. We can now put everything together, so that

$$\delta S = \int dt\, \left(\frac{\partial L}{\partial q_k}\Delta q_k + \frac{\partial L}{\partial \dot{q}_k}\frac{d}{dt}(\Delta q_k) + \delta t\frac{dL}{dt} + L\frac{d}{dt}(\delta t)\right)$$

$$= \int dt\, \left(\frac{\partial L}{\partial q_k}\Delta q_k + \frac{\partial L}{\partial \dot{q}_k}\frac{d}{dt}(\Delta q_k) + \frac{d}{dt}(L\,\delta t)\right).\qquad(6.107)$$

Now given L, $\delta t(t, q)$, and $\delta q_k(t, q)$ (hence also $\Delta q_k(t, q)$), we can substitute these into Eq. (6.107) and check whether the expression vanishes. If it *does* vanish, we conclude that the given transformation $\{\delta t(t, q), \delta q_k(t, q)\}$ is a symmetry of our system. This shall be our notion of symmetry. A bit later, we will revisit this statement and generalize it further. For now, this is enough to move on to the heart of the topic, *Noether's theorem*.

---

[3] To see this, note that the commutativity of differentials follows from the general definition satisfied equally by δ and d; that is, a differential is the limit of a difference, as in $\lim_{\epsilon \to 0}(x_\epsilon - x)$.

6.7 Noether's Theorem

We start by simply stating the theorem:

Noether's theorem: *For every symmetry, i.e., for every set of transformations* $\{\delta t(t,q), \delta q_k(t,q)\}$ *that leave the action unchanged, there exists a quantity that is conserved under time evolution.*

A symmetry implies a conservation law. This is important for two reasons:

- First, a conservation law identifies a rule in the laws of physics. Virtually everything we have a name for in physics – mass, momentum, energy, charge, etc. – is tied by definition to a conservation law. Noether's theorem then states that fundamental physics is founded on the principle of identifying symmetries. If we want to know all the laws of physics, we need to ask: What are all the symmetries in Nature? From there, we find conservation laws and associated interesting conserved quantities. We can then study how these conservation laws can sometimes be violated. This leads us to equations that can predict the future. *It's all about symmetries.*
- Second, conservation laws have the form

$$\frac{d}{dt}\,(\text{something}) = 0 \Rightarrow \text{something} = \text{constant.} \qquad (6.108)$$

The "something" is typically a function of the degrees of freedom and the first derivatives of the degrees of freedom. The conservation statement then inherently leads to first-order differential equations. First-order differential equations are much more pleasant than second or higher-order equations. Thus, technically, finding the symmetries and corresponding conservation laws in a problem helps a great deal in solving and understanding the physical system.

The easiest way to understand Noether's theorem is to prove it, which is a surprisingly simple exercise.

Proof of the Theorem

The premise of the theorem is that we have a given symmetry $\{\delta t(t,q), \Delta q_k(t,q)\}$. This then implies, using Eq. (6.107), that

$$\delta S = 0 = \int dt \left(\frac{\partial L}{\partial q_k} \Delta q_k + \frac{\partial L}{\partial \dot{q}_k} \frac{d}{dt}\,(\Delta q_k) + \frac{d}{dt}(L\,\delta t) \right). \qquad (6.109)$$

We know that this equation is satisfied for *any* set of curves $q_k(t)$ by virtue of the assumption that $\{\delta t(t,q), \Delta q_k(t,q)\}$ constitutes a symmetry. And now comes the crucial step: what if the $q_k(t)$s satisfy the Lagrange equations of motion

$$\frac{d}{dt}\left(\frac{\partial L}{\partial \dot{q}_k} \right) = \frac{\partial L}{\partial q_k} \,? \qquad (6.110)$$

Of all possible curves $q_k(t)$, we pick the ones that satisfy the equations of motion. Given this additional statement, we can rearrange the terms in δS to give

$$0 = \int dt \, \frac{d}{dt} \left(\frac{\partial L}{\partial \dot{q}_k} \Delta q_k + L \, \delta t \right). \tag{6.111}$$

Since the integration interval is arbitrary, we then conclude that

$$\frac{d}{dt} \, (Q) = 0, \tag{6.112}$$

where

$$Q \equiv \frac{\partial L}{\partial \dot{q}_k} \Delta q_k + L \, \delta t. \tag{6.113}$$

Therefore we have a *conserved* quantity Q called the **Noether charge**.

Note a few important points:

- We used the equations of motion to prove the conservation law. However, we did *not* use the equations of motion to conclude that a particular transformation is a symmetry. The symmetry exists at the level of the action for any $q_k(t)$. The conservation law exists for trajectories that satisfy the equations of motion.
- The proof identifies explicitly the conserved quantity through Eq. (6.113). Knowing L, $\delta t(t, q)$, and $\delta q_k(t, q)$ (hence also $\Delta q_k(t, q)$ from Eq. (6.85)), this equation tells us immediately the conserved quantity associated with the given symmetry.

This proof also highlights a way to *generalize* the original definition of symmetry. All that was needed was to be able to write

$$\delta S = \int dt \frac{d}{dt} \, (K), \tag{6.114}$$

where K is some function that we can discover by using Eq. (6.107). If K turns out to be a constant, we would get $\delta S = 0$ trivially, and we are back to the situation just discussed. However, if K is *non*trivial, we get (using Lagrange's equations)

$$\delta S = \int dt \frac{d}{dt} \, (K) = \int dt \, \frac{d}{dt} \left(\frac{\partial L}{\partial \dot{q}_k} \Delta q_k + L \, \delta t \right), \tag{6.115}$$

from which it follows from the definitions that

$$\frac{d}{dt} \, (Q - K) = 0. \tag{6.116}$$

Therefore the conserved quantity is $Q - K$ rather than Q by itself. Since the interesting conceptual content of a symmetry is its associated conservation law, we want to turn the problem on its back: we want to *define* a symmetry through a conservation statement. So, here is a revised, more general statement:

$$\{\delta t(t, q), \delta q_k(t, q)\} \text{ is a symmetry if}$$

$$\delta S = \int dt \, \frac{dK}{dt} \text{ for some } K. \tag{6.117}$$

Noether's theorem then states that for every such symmetry, there is a conserved quantity given by $Q - K$.

To summarize, the general prescription is as follows:

1. Given a Lagrangian L and a candidate symmetry $\{\delta t(t, q), \delta q_k(t, q)\}$, use Eq. (6.107) to find δS. If $\delta S = \int dt \, dK/dt$ for some K we are to determine, $\{\delta t(t, q), \delta q_k(t, q)\}$ is indeed a symmetry.
2. If $\{\delta t(t, q), \delta q_k(t, q)\}$ is found to be a symmetry with some K, we can find an associated conserved quantity $Q - K$, with Q given by Eq. (6.113).

The best way to understand all of this is to look at a few examples.

Example 6.10 **Space Translations and Momentum**

We start with the spatial translation transformation in a fixed ith direction, as in our previous example (6.87):

$$\delta r^i(t, x) = \delta \epsilon^i, \quad \delta t(t, x) = 0 \Rightarrow \Delta r^i(t, x) = \delta \epsilon^i, \tag{6.118}$$

where $\delta \epsilon^i$ is a constant and $\delta r^k = \Delta r^k, = 0$ for $k \neq i$ (both i and k lie in the spatial directions). We next need a Lagrangian to test this transformation. Consider first a free nonrelativistic particle whose Lagrangian is

$$L = \frac{1}{2} m \, \dot{r}^k \dot{r}^k. \tag{6.119}$$

Substitute eqs. (6.118) and (6.119) into (6.107):

$$\delta S = \int dt \left(\frac{\partial L}{\partial q_k} \Delta q_k + \frac{\partial L}{\partial \dot{q}_k} \frac{d}{dt}(\Delta q_k) + \frac{d}{dt}(L \delta t) \right) \tag{6.120}$$

$$= \int dt \left(0 \cdot \Delta q_k + m \dot{r}^i \frac{d}{dt} \delta \epsilon^i + \frac{d}{dt}(0) \right) = 0 \text{ (no sum on } i\text{)},$$

since $\delta \epsilon^i$ is a constant. Therefore $K = \text{constant}$, so we have a symmetry. To find the associated conserved charge, we use Eq. (6.113) and find

$$Q^i = m \dot{r}^i \, \delta \epsilon^i \text{ (no sum on } i\text{)}. \tag{6.121}$$

We can repeat this for every $i = x, y, z$. We then have three charges for the three possible directions for translation. Any overall additive or multiplicative constant is arbitrary, since it does not affect the statement of conservation $\dot{Q}^i = 0$. Writing the conserved quantities as P^i instead, we can then state

$$P^i = m \dot{r}^i. \tag{6.122}$$

That is, *momentum* is the Noether charge associated with the symmetry of spatial translational invariance. If we have a physical system set up on a table and we notice that we can shift the table by any amount in any of the three spatial directions *without* affecting the dynamics of the system, we can conclude that there is a quantity – called momentum by definition – that remains constant in time.

We can then try to find the conditions under which this symmetry, and hence the conservation law, is violated. For example, we can add a simple harmonic oscillator potential to the Lagrangian:

$$L = \frac{1}{2}m\dot{r}^k\dot{r}^k - \frac{1}{2}k\,r^i r^i \quad \text{(no sum on } i\text{)}.$$ (6.123)

Using Eq. (6.107), we now get

$$\delta S = \int dt \left(-k\,r^i\delta\epsilon^i\right) \neq \int \frac{d}{dt}(K) \quad \text{(no sum on } i\text{)}.$$ (6.124)

Hence, momentum is no longer conserved in the ith direction and we can write

$$\dot{P}^i \neq 0 \Rightarrow \dot{P}^i \equiv F^i,$$ (6.125)

thus introducing the concept of **force**. We now see that the existence or non-existence of forces in Newton's second law arises from the existence or non-existence of a certain symmetry in Nature.

Newton's third law is also related to this idea: action–reaction pairs cancel each other, so that the total force on an isolated system is zero and hence the total momentum is conserved. To see this, look back at the two-particle system on a rail, described by the Lagrangian (6.75). Using once again Eq. (6.107) with $\delta t = 0$ and $\delta q_i = \delta\epsilon$, we get

$$\delta S = \int dt \left(-\frac{\partial U}{\partial q_1}\delta\epsilon + \frac{\partial U}{\partial q_2}\delta\epsilon\right) = 0$$ (6.126)

since

$$\frac{\partial}{\partial q_1}U(q_1 - q_2) = -\frac{\partial}{\partial q_2}U(q_1 - q_2).$$ (6.127)

The forces on each particle are $-\partial U/\partial q_1$ and $-\partial U/\partial q_2$, which are equal in magnitude but opposite in sign, since the potential energy has the form $U(q_1 - q_2)$ – note the relative minus sign between q_1 and q_2. These two forces form the action–reaction pair. The cancellation of forces arises because of the dependence of the potential and force on the distance $q_1 - q_2$ between the particles – which is what makes the problem translationally invariant as well. We now see that the third law is intimately tied to the statement of translational symmetry. The associated conserved quantity determined from Eq. (6.113) is

$$P_i = m_1\dot{q}_1 + m_2\dot{q}_2,$$ (6.128)

and so the Noether charge is now the *total* momentum of the system, and it is conserved. ∎

Example 6.11

Time Translation and the Hamiltonian

Next, let us consider *time* translational invariance. Due to its particular usefulness, we want to treat this example with greater generality. We focus on a system with an arbitrary number of degrees of freedom labeled by q_ks, with a general Lagrangian $L(q, \dot{q}, t)$. We propose the transformation

$$\delta t = \delta\epsilon, \quad \delta q_k = 0. \tag{6.129}$$

Therefore the degrees of freedom are left unchanged, but the time is shifted by a constant $\delta\epsilon$. This means that we need a direct shift

$$\delta q_k = 0 = \Delta q_k + \dot{q}_k \delta t = \Delta q_k + \delta\epsilon\,\dot{q}_k \Rightarrow \Delta q_k = -\delta\epsilon\,\dot{q}_k \tag{6.130}$$

to compensate for the indirect change in q_k induced by the shift in time. We then use Eq. (6.107) to find that the change in the action is

$$\delta S = \int dt \left(-\delta\epsilon\dot{q}_k \frac{\partial L}{\partial q_k} - \delta\epsilon\ddot{q}_k \frac{\partial L}{\partial \dot{q}_k} + \delta\epsilon \frac{dL}{dt} \right). \tag{6.131}$$

But we also know from multivariable calculus that

$$\frac{dL}{dt} = \frac{\partial L}{\partial t} + \frac{\partial L}{\partial q_k}\dot{q}_k + \frac{\partial L}{\partial \dot{q}_k}\ddot{q}_k, \tag{6.132}$$

so we can write simply

$$\delta S = \int dt\, \frac{\partial L}{\partial t} \delta\epsilon. \tag{6.133}$$

In general, since L depends on the q_ks as well as explicit ts, we need to consider the more restrictive condition for symmetry $\delta S = 0$, *i.e.*, we have $K = $ constant. This implies that we have time translational symmetry if

$$\frac{\partial L}{\partial t} = 0, \tag{6.134}$$

i.e., if the Lagrangian does not depend on time *explicitly*. If this is the case, we then have a conserved quantity given by Eq. (6.113):

$$Q = -\delta\epsilon\,\dot{q}_k \frac{\partial L}{\partial \dot{q}_k} + \delta\epsilon L. \tag{6.135}$$

Dropping the overall constant term $-\delta\epsilon$ and rearranging:

$$Q \to \dot{q}_k \frac{\partial L}{\partial \dot{q}_k} - L = H, \tag{6.136}$$

which we recognize as the **Hamiltonian** H of the system, already introduced in Chapter 4.

Consider, for example, the two-particle problem in one dimension described by the Lagrangian (4.48). We then find

$$H = \frac{1}{2}m_1\dot{q}_1^2 + \frac{1}{2}m_2\dot{q}_2^2 + U(q_1 - q_2), \tag{6.137}$$

which is obviously also the energy $E = T + U$ in this case. As shown in Section 4.7 of Chapter 4, the Hamiltonian H is equal to the energy E if the transformation $\mathbf{r} = \mathbf{r}(q_k, t)$, from the generalized coordinates to Cartesian coordinates, is not an explicit function of time. Therefore, if the results of a series of experiments do not depend upon when the experiments are performed, we expect that the Hamiltonian of the system will be conserved. In many circumstances $H = E$, so we often say that invariance under displacements in time indicates that the energy of the system is conserved. We can then look at dissipative effects involving loss of energy and learn about new physics through the nonconservation of energy.   ∎

Example 6.12

Rotations and Angular Momentum

Consider the problem of a nonrelativistic particle of mass m moving in two dimensions, in a plane labeled by $r^x = x$ and $r^y = y$. We add to the problem a central force with a Lagrangian of the form

$$L = \frac{1}{2} m \, \dot{r}^i \dot{r}^i - U(r^i r^i) \quad \text{summed on } i, \text{ with } i = x, y. \tag{6.138}$$

Note that the potential depends only on the distance $\sqrt{r^i r^i}$ of the particle from the origin of the coordinate system. Rotations are described by Eq. (6.95). We can then use Eq. (6.107) to test for rotational symmetry, where the change in the action is

$$\delta S = \int dt \left(\frac{\partial U}{\partial r^i} \Delta r^i + m \, \dot{r}^i \frac{d}{dt} (\Delta r^i) \right)$$
$$= \int dt \left(2 U' \delta\theta \, \varepsilon_{ij} r^i r^j + m \, \delta\theta \, \varepsilon_{ij} \dot{r}^i \dot{r}^j \right). \tag{6.139}$$

In the second line we wrote

$$\frac{\partial U}{\partial r^i} = \frac{\partial U}{\partial u} \frac{\partial u}{\partial r^i} = \frac{\partial U}{\partial u} (2 r^i) = 2 U' r^i, \tag{6.140}$$

where $u \equiv r^i r^i$, and where we have used the chain rule. We now want to show that $\delta S = 0$. Focus on the first term in Eq. (6.139):

$$2 U' \delta\theta \, \varepsilon_{ij} r^i r^j = 2 U' \delta\theta \, \varepsilon_{xy} r^x r^y + 2 U' \delta\theta \, \varepsilon_{yx} r^y r^x = 2 U' \delta\theta \, r^x r^y - 2 U' \delta\theta \, r^y r^x = 0. \tag{6.141}$$

Let us do this one more time, with more grace and elegance:

$$2 U' \delta\theta \, \varepsilon_{ij} r^i r^j = 2 U' \delta\theta \, \varepsilon_{ji} r^j r^i$$
$$= 2 U' \delta\theta \, \varepsilon_{ji} r^i r^j$$
$$= -2 U' \delta\theta \, \varepsilon_{ij} r^i r^j. \tag{6.142}$$

In the second line, we just relabeled the indices $i \rightarrow j$ and $j \rightarrow i$: since they are summed indices, it does not matter what they are called. In the third line, we used the fact that multiplication is commutative $r^j r^i = r^i r^j$. Finally, in the third line, we used the property $\varepsilon_{ij} = -\varepsilon_{ji}$ from Eq. (6.96). Hence, we have shown

$$2 U' \delta\theta \, \varepsilon_{ij} r^i r^j = -2 U' \delta\theta \, \varepsilon_{ij} r^i r^j \Rightarrow 2 U' \delta\theta \, \varepsilon_{ij} r^i r^j = 0. \tag{6.143}$$

The key idea is that ε_{ij} is antisymmetric in its indices while $r^i r^j$ is symmetric under the same indices. The sum of their product therefore cancels. The same is true for the second term in Eq. (6.139):

$$m\,\delta\theta\,\varepsilon_{ij}\dot{r}^i\dot{r}^j = -m\,\delta\theta\,\varepsilon_{ij}\dot{r}^i\dot{r}^j \Rightarrow m\,\delta\theta\,\varepsilon_{ij}\dot{r}^i\dot{r}^j = 0. \tag{6.144}$$

We will use this trick occasionally later on in other contexts. We thus have shown that our system is rotational symmetric:

$$\delta S = 0 \Rightarrow K = \text{constant}. \tag{6.145}$$

We can then determine the conserved quantity using Eq. (6.113):

$$Q = \frac{\partial L}{\partial \dot{r}^i}\Delta r^i = m\,\dot{r}^i\,\delta\theta\,\varepsilon_{ij}r^j. \tag{6.146}$$

Dropping a constant term $\delta\theta$, we write

$$l = m\,\varepsilon_{ij}\dot{r}^i r^j = m\,(\dot{r}^x r^y - \dot{r}^y r^x) = (\mathbf{r}\times m\mathbf{v})^z, \tag{6.147}$$

i.e., this is the z-component of the *angular momentum* of the particle. As expected, it is perpendicular to the plane of motion. Rotational symmetry implies conservation of angular momentum, and rotation about the z axis corresponds to angular momentum along the z axis.   ∎

Example 6.13   **Lorentz and Galilean Boosts**

How about a Lorentz transformation? Special relativity *requires* the Lorentz transformation to be a symmetry of any physical system: it is not a question of *whether* it is a symmetry of any particular system – *it better be*. We could then use Eq. (6.107), with Lorentz transformations, as a *condition* for sensible Lagrangians. Noether's theorem can be used to construct theories consistent with the required symmetries. In general, an experiment would identify a set of symmetries in a newly discovered system. Then the theorist's task is to build a Lagrangian that describes the system; and a good starting point would be to ensure that the Lagrangian has all the needed symmetries. We now see the power of Noether's theorem: it allows us to mold equations and theories to our symmetry needs.

Returning to Lorentz transformations, let us look at an explicit example and find the associated conserved charge. We consider a relativistic system, say a free relativistic particle with action

$$S = -mc^2 \int d\tau = -mc^2 \int dt\sqrt{1 - \frac{\dot{r}^i \dot{r}^i}{c^2}}. \tag{6.148}$$

A particular Lorentz transformation is given by Eq. (6.101), which we can then substitute into Eq. (6.107) to show that there is a symmetry. We will return to this case in the next example. For now, consider instead the limit of small speeds, *i.e.*, a nonrelativistic system with Galilean symmetry. We take a single free particle in one dimension with Lagrangian

$$L = \frac{1}{2}m\dot{x}^2. \tag{6.149}$$

The expected symmetry is Galilean, given by (1.1):

$$x = x' + Vt', \quad y = y', \quad z = z', \quad t = t'. \tag{6.150}$$

We can then write the infinitesimal version

$$\delta t = \delta y = \delta z = 0, \quad \delta x = -V t$$
$$\Rightarrow \Delta t = \Delta y = \Delta z = 0 \text{ and } \Delta x = -V t. \tag{6.151}$$

Note that we want to think of this transformation near $t \sim 0$, the instant in time when the two origins coincide, to keep it as a small deformation for any V. Using Eq. (6.107), we then find

$$\delta S = \int dt \, m \, V \, \dot{x}. \tag{6.152}$$

Before we become overly concerned by the fact that this quantity does not vanish, let us remember that all that is needed is $\delta S = \int dt \, dK/dt$ for *some* K. That is indeed the case:

$$\delta S = \int dt \, \frac{d}{dt} (m \, V x). \tag{6.153}$$

We then have

$$K = m \, V x + \text{constant}. \tag{6.154}$$

This system does have Galilean symmetry. We look at the associated Noether charge, using Eq. (6.113):

$$Q = \frac{\partial L}{\partial \dot{x}} \Delta x = m \, \dot{x} \, V t. \tag{6.155}$$

But this is not the conserved charge: the conserved quantity is

$$Q - K = m \, \dot{x} \, V t - m \, V x = \text{constant}. \tag{6.156}$$

Rewriting things, we have a simple first-order differential equation

$$\dot{x} t - x = \text{constant}. \tag{6.157}$$

Integrating this gives the expected linear trajectory $x(t) \propto t$. Unlike momentum, energy, and angular momentum, this conserved quantity $Q - K$ is not given its own glorified name. Since Galilean (or the more general Lorentz) symmetry is expected to be prevalent in all systems, this does not add any useful distinguishing physics to a problem. Perhaps if we were to discover a fundamental phenomenon that breaks Galilean/Lorentz symmetry, we could then revisit this conserved quantity and study its nonconservation. For now, this conserved charge is relegated to second-rate status.   ∎

Example 6.14   **Lorentz Invariance**

We now return to the case of Lorentz invariance. We saw in Chapter 2 that Lorentz transformations are a kind of rotation in spacetime – using hyperbolic trigonometric functions. Let us start by writing infinitesimal Lorentz transformations (6.101) in a more compact form along the lines of Eq. (6.95). We write

$$\delta r^\mu = \beta \varepsilon^{\mu\nu} r^\lambda \eta_{\nu\lambda} = \beta \varepsilon^{\mu\nu} r^\nu, \tag{6.158}$$

where μ and ν run over t and x only, and $\varepsilon_{\mu\nu}$ is nonzero only for

$$\varepsilon^{tx} = +1, \quad \varepsilon^{xt} = -1, \tag{6.159}$$

as for the case of rotations. Note that we need the metric factor $\eta_{\nu\lambda}$ to flip the sign of the time transformation to the correct form. This transformation corresponds to an infinitesimal boost in the $r^x = x$ direction. More generally, Lorentz boosts can be in any of three spatial directions. By symmetry, we can guess that the most general infinitesimal Lorentz boost *and* spatial rotation must have the form

$$\delta r^\mu = \omega^{\mu\nu} r^\lambda \eta_{\nu\lambda} \quad \text{\textit{Lorentz boosts and rotations}}, \tag{6.160}$$

where μ and ν now run over t, x, y, and z; the $\omega^{\mu\nu}$s are a set of boost or rotational angle parameters that satisfy the antisymmetry relation

$$\omega^{\mu\nu} = -\omega^{\nu\mu}. \tag{6.161}$$

For example, if all $\omega^{\mu\nu}$s are zero except for $\omega^{xy} = -\omega^{yx} = \delta\theta$, we get rotation about the z axis as in Eq. (6.95); in contrast, if all $\omega^{\mu\nu}$s are zero except for $\omega^{tx} = -\omega^{xt} = \beta$, we get Eq. (6.158). Correspondingly, $\omega^{yz} = -\omega^{zy}$ rotates about the x axis, $\omega^{xz} = -\omega^{zx}$ rotates about the y axis, $\omega^{ty} = -\omega^{yt}$ boosts in the y direction, and $\omega^{tz} = -\omega^{zt}$ boosts in the z direction. We then have a total of three rotations and three boosts as needed, packaged into one antisymmetric matrix $\omega^{\mu\nu}$.

Notice that we cannot have *general* Lorentz invariance without rotational invariance: otherwise, we can always break the Lorentz boost symmetry by simply realigning the axes by a rotation. To see where the requirement of Lorentz invariance leads us, consider the general change in the action given by Eq. (6.107):

$$\delta S = \int d\tau \left(\frac{\partial L}{\partial q_k} \Delta q_k + \frac{\partial L}{\partial \dot{q}_k} \frac{d}{d\tau} \left(\Delta q_k \right) + \frac{d}{d\tau} \left(L \, \delta\tau \right) \right), \tag{6.162}$$

where we have traded the t dependence for a dependence on proper time τ instead, expecting that this will make the analysis cleaner. This means that the action is

$$S = \int d\tau \, L, \tag{6.163}$$

and the dot in $\dot{q}_k$ denotes a derivative with respect to τ. We can rearrange Eq. (6.162) into

$$\delta S = \int d\tau \left(\frac{\partial L}{\partial q_k} \delta q_k + \frac{\partial L}{\partial \dot{q}_k} \frac{d}{d\tau} \left(\delta q_k \right) \right), \tag{6.164}$$

where we have used $\delta\tau = 0$ (since proper time is unchanged under Lorentz transformations and rotations) and hence $\delta q_k = \Delta q_k + \dot{q}_k \delta\tau = \Delta q_k$. Noting that $q_k \to r^\mu$, we can then write instead

$$\delta S = \int d\tau \left(\frac{\partial L}{\partial r^\mu} \delta r^\mu + \frac{\partial L}{\partial \dot{r}^\mu} \frac{d}{d\tau} \left(\delta r^\mu \right) \right). \tag{6.165}$$

We can now read off the condition of Lorentz invariance. For example, consider *any* function of $s \equiv r^\mu r^\nu \eta_{\mu\nu}$; that is, say L contains a piece $L \to L(r^\mu r^\nu \eta_{\mu\nu})$. We then get

$$\frac{\partial L}{\partial r^\mu} \delta r^\mu = \frac{\partial L}{\partial s} \frac{\partial s}{\partial r^\mu} \delta r^\mu = \frac{\partial L}{\partial s} 2 r^\nu \eta_{\mu\nu} \delta r^\mu = \frac{\partial L}{\partial s} 2 r^\nu \eta_{\mu\nu} \omega^{\mu\alpha} \eta_{\alpha\beta} r^\beta = 0. \tag{6.166}$$

The last statement follows because the product of xs is symmetric while the $\omega^{\mu\nu}$ matrix is antisymmetric: the argument is the same one we saw in Eq. (6.142). This mechanism of cancellation is so general that it will apply whenever the Lagrangian depends on mathematical objects that are properly Lorentz contracted: there are no "hanging" Lorentz indices, and pairs of Lorentz indices μ and ν ar always summed over through $\eta_{\mu\nu}$. We can then easily build Lagrangians L out of r^μs and $dr^\mu/d\tau$s, as long as we contract the indices properly. The result is guaranteed to be Lorentz invariant. The simplest case is that of a free relativistic particle, for which we have from Eq. (3.71) $L = 1$ – which obviously works. In Chapter 8, we will see the case of a charged relativistic particle where the Lorentz invariant form of L will be less trivial. ∎

Example 6.15

Sculpting Lagrangians from Symmetry

What if we were to *stipulate* a particular symmetry and ask for all possible Lagrangians that fit the mold? To be more specific, consider a one-dimensional system with a single degree of freedom, denoted by $q(t)$. We want to ask: *What are all possible theories that can describe this system with the conditions that they are Galilean invariant and invariant under time translations?* Time translational invariance is easy to handle: we need $\partial L/\partial t = 0$. The Galilean symmetry, however, is obtained from Eq. (6.151) in the previous example:

$$\delta t = 0 \text{ and } \delta x = \Delta x = -Vt. \tag{6.167}$$

Substituting this into Eq. (6.107), we get

$$\delta S = \int dt \left(-\frac{\partial L}{\partial x} V t - \frac{\partial L}{\partial \dot{x}} V \right) = \int dt \frac{d}{dt} K. \tag{6.168}$$

The question is to find the most general L that does the job for some K. This means we need

$$\frac{\partial L}{\partial x} t + \frac{\partial L}{\partial \dot{x}} = -\frac{d}{dt} K \equiv \frac{d}{dt} \widetilde{K}. \tag{6.169}$$

Note also that we are *not* allowed to use the equations of motion while imposing the desired symmetry! Using the chain rule with $\widetilde{K}(t, x, \dot{x}, \ddot{x}, \ldots)$, we can write

$$\frac{d}{dt} \widetilde{K} = \frac{\partial \widetilde{K}}{\partial t} + \frac{\partial \widetilde{K}}{\partial x} \dot{x} + \frac{\partial \widetilde{K}}{\partial \dot{x}} \ddot{x} + \cdots . \tag{6.170}$$

Comparing this to Eq. (6.169), we see that we need $\widetilde{K}(t, x)$ – a function of t and x only – since we know L is a function of t, x, and $\dot{x}$ only. Therefore we have

$$\frac{\partial L}{\partial x} t + \frac{\partial L}{\partial \dot{x}} = \frac{\partial \widetilde{K}}{\partial t} + \frac{\partial \widetilde{K}}{\partial x} \dot{x}. \tag{6.171}$$

We want a general form for L, yet $\widetilde{K}(t,x)$ is also arbitrary. Since the right-hand side is linear in $\dot{x}$, the left-hand side must be linear as well. This implies we need L to be a quadratic polynomial in $\dot{x}$:

$$L = f_1(x)\dot{x}^2 + f_2(x)\dot{x} + f_3(x), \qquad (6.172)$$

with three unknown functions $f_1(x)$, $f_2(x)$, and $f_3(x)$. Note that no time dependence is allowed because of time translational symmetry. Looking at the $\partial L / \partial x$ term, we can immediately see that we need $f_1(x) = C_1$, a constant independent of x: otherwise we generate a term quadratic in $\dot{x}$ that does not exist on the right-hand side of Eq. (6.171). Our Lagrangian now looks like

$$L = C_1\dot{x}^2 + f_2(x)\dot{x} + f_3(x). \qquad (6.173)$$

But the second term is irrelevant to the dynamics. This is because L will appear in the action integrated over time, and this second term can be integrated out:

$$\int dt\, f_2(x)\dot{x} = \int dt\, \frac{d}{dt}\left(F_2(x)\right) = F_2(x)\big|_{\text{boundaries}} \qquad (6.174)$$

for some function $F_2(x) = \int^x d\xi\, f_2(\xi)$. Hence the term does not depend on the shape of paths plugged into the action functional and cannot contribute to the statement of stationarity – otherwise known as the equation of motion. We are now left with

$$L \to C_1\dot{x}^2 + f_3(x). \qquad (6.175)$$

The condition (6.171) on L now looks like

$$\frac{\partial f_3(x)}{\partial x}t + 2C_1\dot{x} = \frac{\partial \widetilde{K}}{\partial t} + \frac{\partial \widetilde{K}}{\partial x}\dot{x}. \qquad (6.176)$$

Picking out the $\dot{x}$ dependences on either side, this implies

$$2C_1 = \frac{\partial \widetilde{K}}{\partial x}, \quad \frac{\partial f_3(x)}{\partial x}t = \frac{\partial \widetilde{K}}{\partial t}. \qquad (6.177)$$

Since we know that

$$\frac{\partial^2 \widetilde{K}}{\partial x \partial t} = \frac{\partial^2 \widetilde{K}}{\partial t \partial x}, \qquad (6.178)$$

differentiating the two equations in (6.177) leads to the condition

$$\frac{\partial^2 f_3(x)}{\partial x^2} = 0 \Rightarrow f_3(x) = C_2 x + C_3 \qquad (6.179)$$

for some constants C_2 and C_3. The Lagrangian is now

$$L = C_1\dot{x}^2 + C_2 x, \qquad (6.180)$$

where we set $C_3 = 0$ since a constant shift of L does not affect the equation of motion. We can now solve for $\widetilde{K}$ as well if we want to, using (6.177):

$$\widetilde{K} = 2C_1 x + \frac{C_2}{2}t^2 + \text{constant}. \qquad (6.181)$$

Now, let us focus on the important point, which is Eq. (6.180). We write the constants C_1 and C_2 as $C_1 = m/2$ and $C_2 = -m\,g$:

$$L = \frac{1}{2}m\dot{x}^2 - m\,g\,x. \tag{6.182}$$

That is, we have shown that the most general Galilean and time translation-invariant mechanics problem in one dimension necessarily looks like a particle in uniform gravity.[a] We were able to *derive* the canonical kinetic energy term and gravitational potential from a symmetry requirement. This is just a hint at the power of symmetries and conservation laws in physics. Indeed, all the known forces of Nature can be derived from first principles using symmetries![b]   ∎

[a] Or a uniform electrostatic potential or any other potential linear in x.

[b] See for example *Quantum Field Theory in a Nutshell* by A. Zee, and references therein.

6.8 Some Comments on Symmetries

Let us step back for a moment and comment on several additional issues about symmetries and conservation laws.

- If a system has N degrees of freedom, then the typical Lagrangian leads to N second-order differential equations (provided the Lagrangian depends on at most first derivatives of the variables). If we were lucky enough to solve these equations, we would parameterize the solution with $2N$ constants related to the boundary conditions. If our system has M symmetries, it would then have M conserved quantities. Each of the symmetries leads to a first-order differential equation, and hence a total of M constants of motion. In total, the conservation equations will give $2M$ constants to parameterize the solution: M from the constants of motion, and another M from integrating the first-order equations. These $2M$ constants would necessarily be related to the $2N$ constants mentioned earlier. What if we have $M = N$? The system is then said to be **integrable**. This means that all one needs to do is write the conservation equations and integrate them. We need not even consider any second-order differential equations to find the solution to the dynamics. In general, we will have $M \leq N$, and the closer M is to N, the easier it will be to solve the given physical problem. As soon as an experienced physicist sees a mechanics problem, he or she would first count the degrees of freedom, and then instinctively look for the symmetries and associated conserved charges. This immediately lays out a strategy for how to tackle the problem, based on how many symmetries one has versus the number of degrees of freedom.
- Noether's theorem is based on infinitesimal transformations: symmetries that can be built up from small incremental steps of deformations. There are other

symmetries in Nature that do not fit this prescription. For example, discrete symmetries are rather common. Reflection transformations, *e.g.*, time reflection $t \to -t$ or discrete rotations of a lattice, can be very important for understanding the physics of a problem. Noether's theorem does not apply to these. However, such symmetries are also often associated with conserved quantities. Sometimes, these are called **topological conservation laws**.

- Infinitesimal transformations can be catalogued rigorously in mathematics. A large and useful class of such transformations fall under the general topic of **group theory**, more particularly **Lie groups**. The Lie group catalogue (developed by the French mathematician Élie Cartan) is exhaustive. Many if not all of the entries in the catalogue are indeed realized in Nature in various physical systems.

6.9 Summary

In this chapter we presented a technique that allows us to determine constraint forces while working within the Lagrangian formalism. This involved the use of Lagrange multipliers – additional degrees of freedom that one adds to the action. This allows us to formulate a variational principle while being subject to constraints. This approach turns out to be very useful beyond mechanics, from thermodynamics to quantum field theory.

We also encountered a profound formulation of physics, Noether's theorem, that we can use to organize the subject in terms of a catalogue of symmetries in Nature and corresponding conservation laws. Once again, this topic extends beyond the realm of classical mechanics and pervades throughout our modern understanding of physics.

Problems

★★   **Problem 6.1** A particle of mass m slides inside a smooth hemispherical bowl of radius R. Use spherical coordinates r, θ, and ϕ to describe the dynamics. (a) Write the Lagrangian in terms of generalized coordinates and solve the dynamics. (b) Repeat the exercise using a Lagrange multiplier. What does the multiplier measure in this case?

★★   **Problem 6.2** A pendulum consisting of a ball at the end of a rope swings back and forth in a two-dimensional vertical plane, with the angle θ between the rope and the vertical evolving in time. The rope is pulled upward at a constant rate so that the length l of the pendulum's arm is decreasing according to $dl/dt = -\alpha \equiv$ constant. (a) Find the Lagrangian for the system with respect to the angle θ. (b) Write the corresponding equations of motion. (c) Repeat parts (a) and (b) using Lagrange multipliers.

★★ **Problem 6.3** A particle of mass m slides inside a smooth paraboloid of revolution whose surface is defined by $z = \alpha \rho^2$, where z and ρ are cylindrical coordinates. (a) Write the Lagrangian for the three-dimensional system using the method of Lagrange multipliers. (b) Find the equations of motion.

★★ **Problem 6.4** In certain situations, it is possible to incorporate frictional effects without introducing the dissipation function. As an example, consider the Lagrangian

$$L = e^{\gamma t}\left(\frac{1}{2}m\dot{q}^2 - \frac{1}{2}kq^2\right). \tag{6.183}$$

(a) Find the equation of motion for the system. (b) Make the coordinate change $s = e^{\gamma t/2}q$, and rewrite the dynamics in terms of s. (c) How would you describe the system?

★★ **Problem 6.5** A massive particle moves under the acceleration of gravity and without friction on the surface of an inverted cone of revolution with half angle α. (a) Find the Lagrangian in polar coordinates. (b) Provide a complete analysis of the trajectory problem. Do not integrate the final orbit equation, but explore circular orbits in detail.

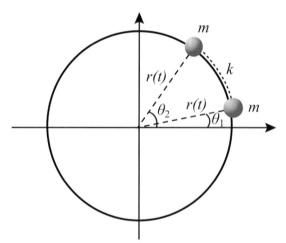

★★★ **Problem 6.6** A toy model for our expanding universe during the inflationary epoch consists of a circle of radius $r(t) = r_0 e^{\omega t}$ where we are confined to the one-dimensional world that is the circle. To probe the physics, imagine two point masses of identical mass m free to move on this circle without friction, connected by a spring of force constant k and relaxed length zero, as depicted in the figure above.

(a) Write the Lagrangian for the two-particle system in terms of the common radial coordinate r, and the two polar coordinates θ_1 and θ_2. Do *not* implement the radial constraint $r(t) = r_0 e^{\omega t}$ yet.

(b) Using a Lagrange multiplier for the radial constraint, write *four* equations describing the dynamics. In this process, show that

$$a_{1r} = 1, \quad a_{1t} = -\omega r. \tag{6.184}$$

(c) Consider the coordinate relabeling

$$\alpha \equiv \theta_1 + \theta_2, \quad \beta \equiv \theta_1 - \theta_2. \tag{6.185}$$

Show that the equations of motion of part (b) for the two angle variables θ_1 and θ_2 can be rewritten in decoupled form as

$$\ddot{\alpha} = C_1 \dot{\alpha}, \tag{6.186}$$

$$\ddot{\beta} = C_2 \dot{\beta} + C_3 \beta, \tag{6.187}$$

where C_1, C_2, and C_3 are constants that you will need to find.

(d) Identify a symmetry transformation $\{\delta t, \delta \alpha, \delta \beta\}$ for this system. Find the associated conserved quantity. What would you call this conserved quantity?

(e) Then find $\alpha(t)$ and $\beta(t)$ using Eqs. (6.186) and (6.187). Use the boundary conditions

$$\alpha(0) = \alpha_0, \quad \dot{\alpha}(0) = 0, \quad \beta(0) = 0, \quad \dot{\beta}(0) = C. \tag{6.188}$$

What is the effect of the expansion on the dynamics? Note: This conclusion is the same as in the more realistic three-dimensional cosmological scenario.

(f) Find the force on the particles exerted by the expansion of the universe. Write this as a function of $\alpha(t)$, $\beta(t)$, and $r(t)$; then show that its limiting form for later times in the evolution is given by

$$2 m \omega^2 r(t). \tag{6.189}$$

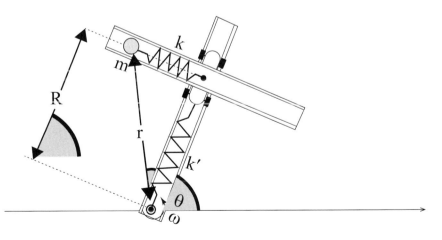

★★★   **Problem 6.7** The figure above shows a mass m connected to a spring of force constant k along a wooden track. The mass is restricted to move along this track without friction. The entire system is mounted on a toy wagon of zero mass resting on a track along a second frictionless beam. The wagon is connected by a spring of force constant k' to an axle about which the whole apparatus is spinning with constant angular speed ω. The figure is a top-down view, with gravity pointing into the page, and the rest length of each spring is zero. (a) First, write the Lagrangian of the system in terms of the four variables r, θ, R, and Θ shown in the figure, without implementing any constraints. (b) Identify two constraint equations. Implement the one, keeping the two tracks perpendicular to one another in the result of part (a) by eliminating R. Do *not* implement the constraint causing everything to spin at constant angular speed ω. (c) Introducing a Lagrange multiplier for the constraint having to do with the spin, write four differential equations describing the system. (d) Identify the force on the mass m due to the spin of the system, and find all conditions for which this force vanishes.

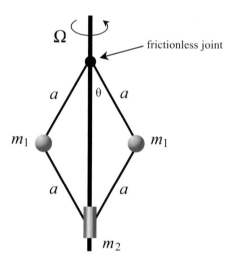

Problem 6.8 Consider the system shown in the figure above. The particle of mass m_2 moves on a vertical axis without friction and the entire system rotates about this axis with a constant angular speed Ω. The frictionless joint near the top assures that the three masses always lie in the same plane, and the rods of length a are all rigid. (*Hint:* Use the origin of your coordinate system at the upper frictionless joint.)

   (a) Find the equation of motion in terms of the single degree of freedom θ. (b) Using the method of Lagrange multipliers, find the torque on the masses m_1 due to the rotational motion. (c) Find a static solution in θ and identify the corresponding angle in terms of m_1, m_2, g, a, and Ω. Consider some limits/inequalities in your result and comment on whether they make sense. (d) Is the solution in part (c) stable? If so, what is the frequency of small oscillations about the configuration? (*Hint:* Use $\xi = \cos\theta$ and work on the Lagrangian instead of the equation of motion.)

Problem 6.9 Find the equations of motion for the example in the text of a wheel chasing a moving target using the nonholonomic constraint.

★★★   **Problem 6.10** Consider the example of the wheel from the example in the text, except that now we have no control over the wheel's steering – except of course at time zero. We start the wheel at some position on the plane, give it an initial roll ω_0 and an initial spin $\dot{\theta}_0$. Describe the trajectory, assuming that the wheel does not tip away from the vertical.

★★   **Problem 6.11** Consider a particle of mass m moving in two dimensions in the x–y plane, constrained to a rail-track whose shape is described by an arbitrary function $y = f(x)$. There is *no gravity* acting on the particle.

(a) Write the Lagrangian in terms of the x degree of freedom only.

(b) Consider some general transformation of the form

$$\delta x = g(x), \quad \delta t = 0, \tag{6.190}$$

where $g(x)$ is an arbitrary function of x. Assuming that this transformation is a symmetry of the system such that $\delta S = 0$, show that it implies the following differential equation relating $f(x)$ and $g(x)$:

$$\frac{g'}{g} = -\frac{1}{2(1+f'^2)}\frac{d}{dx}\left(f'^2\right), \tag{6.191}$$

where prime stands for derivative with respect to x (*not t*).

(c) Write a general expression for the associated conserved charge in terms of $f(x)$, $g(x)$, and $\dot{x}$.

(d) We will now specify a certain $g(x)$, and try to find the laws of physics obeying the prescribed symmetry; *i.e.*, for given $g(x)$, we want to find the shape of the rail-track $f(x)$. Let

$$g(x) = \frac{g_0}{\sqrt{x}}, \tag{6.192}$$

where g_0 is a constant. Find the corresponding $f(x)$ such that this $g(x)$ yields a symmetry. Sketch the shape of the rail-track. (*Hint*: $h(x) = f'^2$.)

★★   **Problem 6.12** One of the most important symmetries in Nature is that of *scale invariance*. This symmetry is very common (e.g., it arises whenever a substance undergoes a phase transition), fundamental (e.g., it is at the foundation of the concept of the *renormalization group*, for which a physics Nobel Prize was awarded in 1982), and entertaining (as you will now see in this problem).

Consider the action

$$S = \int dt \sqrt{h}\dot{q}^2 \tag{6.193}$$

of two degrees of freedom $h(t)$ and $q(t)$.

(a) Show that the following transformation (known as a scale transformation or dilatation)

$$\delta q = \alpha q, \quad \delta h = -2\alpha h, \quad \delta t = \alpha t \tag{6.194}$$

is a symmetry of this system.

(b) Find the resulting constant of motion.

Problem 6.13 A massive particle moves under the acceleration of gravity and without friction on the surface of an inverted cone of revolution with half angle α.

(a) Find the Lagrangian in polar coordinates.

(b) Provide a complete analysis of the trajectory problem. Use Noether charge when useful.

Problem 6.14 For the two-body central-force problem with a Newtonian potential, the effective two-dimensional orbit dynamics can be described by the Lagrangian

$$L = \frac{1}{2}\mu \left(\dot{r}^2 + r^2\dot{\phi}^2 \right) + \frac{k}{r} = \frac{1}{2}\mu \left(\dot{x}^2 + \dot{y}^2 \right) + \frac{k}{\sqrt{x^2 + y^2}}, \tag{6.195}$$

where $k > 0$ and we have chosen to use Cartesian coordinates.

(a) Show that the equations of motion become

$$\mu\ddot{x} = -k\frac{x}{(x^2 + y^2)^{3/2}}, \quad \mu\ddot{y} = -k\frac{y}{(x^2 + y^2)^{3/2}}. \tag{6.196}$$

(b) Consider the rotation

$$\delta x = \alpha y, \quad \delta y = -\alpha x, \quad \delta t = 0 \tag{6.197}$$

for small α. Show that this is a symmetry of the action.

Problem 6.15 In the previous problem show that the conserved Noether charge associated with the symmetry (6.197) is indeed the angular momentum $|\mathbf{r} \times \mu\mathbf{v}|$, which is naturally entirely in the z direction.

Problem 6.16 The two-body central-force problem we have been dealing with in the previous two problems also has another unexpected and amazing symmetry. Consider the transformation

$$\delta x = -\frac{\beta}{2}\mu y \dot{y}, \quad \delta y = \frac{\beta}{2}\mu \left(2x\dot{y} - y\dot{x} \right), \quad \delta t = 0 \tag{6.198}$$

for some constant β. This horrific velocity-dependent transformation is a symmetry if and only if the equations of motion (6.196) are satisfied – unlike other symmetries we've seen where the equations of motion need not be satisfied. It is said that it is an *on-shell symmetry*. Show that the change in the Lagrangian resulting from this transformation is given by

$$\delta L = \beta\mu k \frac{d}{dt}\left(\frac{x}{\sqrt{x^2 + y^2}} \right). \tag{6.199}$$

Therefore, it's a total derivative and generates a symmetry under our generalized definition of a symmetry. (*Hint:* You will need to use Eqs. (6.196) to get this result.)

★ **Problem 6.17** In the previous problem, show that the conserved charge associated with the symmetry is

$$Q \propto \mu^2 x\dot{y}^2 - \mu^2 y\dot{x}\dot{y} - \mu k\frac{x}{\sqrt{x^2 + y^2}}. \tag{6.200}$$

★★ **Problem 6.18** The hidden symmetry of the previous few problems is part of a twofold transformation – one of which is given by Eqs. (6.198), and another similar one that we have not shown. Together, they result in the conservation of a vector known as the Laplace–Runge–Lenz vector

$$\mathbf{A} = \mu\mathbf{v} \times (\mathbf{r} \times \mu\mathbf{v}) - \mu k\frac{\mathbf{r}}{r}. \tag{6.201}$$

Show that (6.200) is the x-component of this more general vector quantity. (*Hint:* You may find it useful to use the identity $\mathbf{a} \times (\mathbf{b} \times \mathbf{c}) = \mathbf{b}(\mathbf{a} \cdot \mathbf{c}) - \mathbf{c}(\mathbf{a} \cdot \mathbf{b})$.)

★★ **Problem 6.19** Show using Eq. (6.201) that

$$\frac{d\mathbf{A}}{dt} = 0. \tag{6.202}$$

Draw an elliptical orbit in the x–y plane and show on it the Laplace–Runge–Lenz vector $\mathbf{A}$. The existence of this conserved vector quantity is the reason why one can smoothly deform ellipses into a circle without changing the energy of the system. Mathematically, this additional hidden symmetry implies that the Newtonian problem is equivalent to a free particle on a three-dimensional sphere embedded in an abstract four-dimensional world. It is believed that this is a mathematical accident; no physical significance of this fourth dimension has yet been identified...

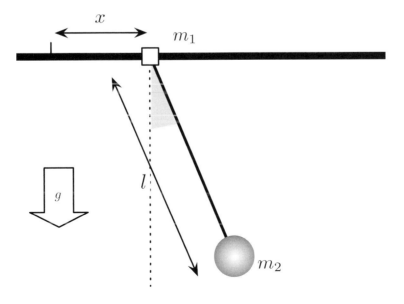

★★   **Problem 6.20** Consider a simple pendulum of mass m_2 and arm length l having its pivot on a point of support of mass m_1 that is free to move horizontally on a frictionless rail.

(a) Find the Lagrangian of the system in terms of the two degrees of freedom x and θ shown on the figure above. Do *not* assume small displacements. (b) Identify two symmetries and the two corresponding conservation laws. Write two first-order differential equations that describe the dynamics of the two degrees of freedom x and θ. Correspondingly, write a single nasty integral for $\theta(t)$.

7

Gravitation

In this chapter we describe motion caused by central forces, especially the orbits of planets, moons, and artificial satellites due to central gravitational forces. Historically, this is the most important testing ground of Newtonian mechanics. In fact, it is not clear how the science of mechanics would have developed if the earth had been covered with permanent clouds, obscuring the moon and planets from view. And Newton's laws of motion with central gravitational forces are still very much in use today, such as in designing spacecraft trajectories to other planets. Our treatment here of motion in central gravitational forces is followed in the next chapter with a look at motion due to electromagnetic forces, which can also be central in special cases, but are commonly much more varied, partly because they involve both electric and magnetic forces.

Throughout this chapter we focus on nonrelativistic regimes. The setting where large speeds are involved and gravitational forces are particularly large is the realm of general relativity – where Newtonian gravity fails to capture the correct physics. We explore such extreme scenarios in the capstone Chapter 10.

7.1 Central Forces

A **central force** on a particle is directed toward or away from a fixed point in three dimensions and is spherically symmetric about that point. In spherical coordinates (r, θ, ϕ) the corresponding potential energy is also spherically symmetric, $U = U(r)$ with no dependence on ϕ and θ.

For example, the sun, of mass m_1 (the source), exerts an attractive central force

$$\mathbf{F} = -G\frac{m_1 m_2}{r^2}\hat{\mathbf{r}} \tag{7.1}$$

on a planet of mass m_2 (the probe), where r is the distance between their centers and $\hat{\mathbf{r}}$ is a unit vector pointing away from the sun (see Figure 7.1). The corresponding gravitational potential energy is

$$U(r) = -\int F(r)\, dr = -G\frac{m_1 m_2}{r}, \tag{7.2}$$

with the choice $U(\infty) = 0$. Similarly, the spring-like central force from a fixed point (the source) on an attached (probe) mass is

$$\mathbf{F} = -k\mathbf{r} = -kr\,\hat{\mathbf{r}} \tag{7.3}$$

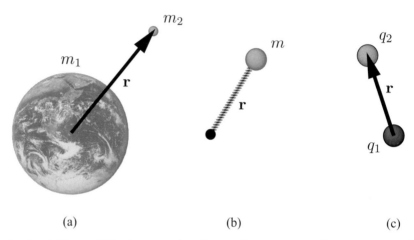

(a)   (b)   (c)

Examples of central forces: (a) Newtonian gravity pulling a probe mass m_2 toward a source mass m_1; (b) a three-dimensional spring fixed at one end; (c) a charge q_1 pulling on a charge q_2 through the electrostatic force.

and has a three-dimensional spring potential energy

$$U(r) = -\int F(r)\,dr = \frac{1}{2}kr^2, \tag{7.4}$$

with $U(0) = 0$. And the Coulomb force

$$\mathbf{F} = \frac{q_1 q_2}{4\pi\varepsilon_0 r^2}\hat{\mathbf{r}} \tag{7.5}$$

on a charge q_2 (the probe) due to a central charge q_1 (the source) has a Coulomb potential energy

$$U(r) = -\int F(r)\,dr = \frac{1}{4\pi\varepsilon_0}\frac{q_1 q_2}{r}, \tag{7.6}$$

with $U(\infty) = 0$. In all these cases, the force is along the direction of the line joining the centers of the source point and the probe object, and the potential energy is a function of the source–probe distance only.

The environment of a particle subject to a central force is invariant under rotations about any axis through the fixed point at the origin, so the angular momentum ℓ of the particle is conserved, as we saw in Chapter 4. Conservation of ℓ also follows from the fact that the torque $\boldsymbol{\tau} \equiv \mathbf{r} \times \mathbf{F} = 0$ due to a central force, if the fixed point is chosen as the origin of coordinates. The particle therefore moves in a *plane*,[1] because its position vector $\mathbf{r}$ is perpendicular to the fixed direction of $\boldsymbol{\ell} = \mathbf{r} \times \mathbf{p}$ (see Figure 7.2). Therefore, central-force problems are essentially two-dimensional.

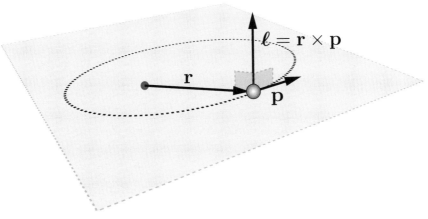

Fig. 7.2   Angular momentum conservation and the planar nature of central force orbits.

All this discussion assumes that the source of the central force is fixed in position: the sun, or the pivot of the spring, or the source charge q_1 are all at rest and lie at the origin of our coordinate system. What if the source object is also in motion? If it is accelerating, as is typically the case due to the reaction force exerted on it by the probe, the source then defines a non-inertial frame, so Newton's second law cannot be used in that source frame. Let us then proceed to tackle the more general situation, the so-called two-body problem involving two dynamical objects, both moving around, pulling on each other through a force that lies along the line that joins their centers.

7.2 The Two-Body Problem

We will now show that with the right choice of coordinates, the two-body problem is equivalent to a one-body central-force problem. If we can solve the one-body central-force problem, we can solve the two-body problem.

In the two-body problem there is a kinetic energy for each body and a mutual potential energy that depends only upon the distance between them. There are altogether six coordinates, three for the first body, $\mathbf{r}_1 = (x_1, y_1, z_1)$, and three for

---

[1] The plane in which a particle moves can also be defined by two vectors: (i) the radius vector to the particle from the force center and (ii) the initial velocity vector of the particle. Given these two vectors, as long as the central force remains the *only* force, the particle cannot move out of the plane defined by these two vectors. We are assuming that the two vectors are noncolinear; if $\mathbf{r}$ and $\mathbf{v}_0$ are parallel or antiparallel, the motion is obviously only one-dimensional, along a radial straight line.

the second, $\mathbf{r}_2 = (x_2, y_2, z_2)$, where all coordinates are measured from a fixed point in some inertial frame (see Figure 7.3). The alternative set of six coordinates used for the two-body problem are, first of all, three **center-of-mass** coordinates

$$\mathbf{R}_{cm} \equiv \frac{m_1 \mathbf{r}_1 + m_2 \mathbf{r}_2}{m_1 + m_2}, \tag{7.7}$$

already defined in Section 1.3: the CM vector extends from an arbitrary fixed point in some inertial frame to the center of mass of the bodies. There are also three **relative coordinates**

$$\mathbf{r} \equiv \mathbf{r}_2 - \mathbf{r}_1, \tag{7.8}$$

where the relative coordinate vector points from the first body to the second, and its length is the distance between them.

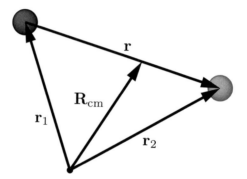

Fig. 7.3   The classical two-body problem in physics.

We can solve for $\mathbf{r}_1$ and $\mathbf{r}_2$ in terms of $\mathbf{R}_{cm}$ and $\mathbf{r}$:

$$\mathbf{r}_1 = \mathbf{R}_{cm} - \frac{m_2}{M}\mathbf{r} \quad \text{and} \quad \mathbf{r}_2 = \mathbf{R}_{cm} + \frac{m_1}{M}\mathbf{r}, \tag{7.9}$$

where $M = m_1 + m_2$ is the total mass of the system. The total kinetic energy of the two bodies, using the original coordinates for each, is[2]

$$T = \frac{1}{2}m_1 \dot{\mathbf{r}}_1^2 + \frac{1}{2}m_2 \dot{\mathbf{r}}_2^2, \tag{7.10}$$

which can be re-expressed in terms of the new generalized velocities $\dot{\mathbf{R}}_{cm}$ and $\dot{\mathbf{r}}$. The result is (see Problem 7.10)

$$T = \frac{1}{2}M\dot{\mathbf{R}}_{cm}^2 + \frac{1}{2}\mu\dot{\mathbf{r}}^2, \tag{7.11}$$

where

$$\mu = \frac{m_1 m_2}{M} \tag{7.12}$$

---

[2] Note that we adopt the linear algebra notation for a square of a vector $\mathbf{V}$: $\mathbf{V}^2 \equiv \mathbf{V} \cdot \mathbf{V} = |\mathbf{V}|^2 = V^2$.

is called the **reduced mass** of the two-body system (note that μ is less than either m_1 or m_2). The mutual potential energy is $U(r)$, a function of the distance r between the two bodies. Therefore the Lagrangian of the system can be written

$$L = T - U = \frac{1}{2}M\dot{\mathbf{R}}_{cm}^2 + \frac{1}{2}\mu\dot{\mathbf{r}}^2 - U(r) \tag{7.13}$$

in terms of $\mathbf{R}_{cm}$, $\mathbf{r}$, and their time derivatives. One of the advantages of the new coordinates is that the coordinates $\mathbf{R}_{cm} = (X_{cm}, Y_{cm}, Z_{cm})$ are cyclic, so the corresponding total momentum of the system $\mathbf{P} = M\dot{\mathbf{R}}_{cm}$ is conserved. That is, the center of mass of the two-body system drifts through space with constant momentum and constant velocity.

The remaining portion of the Lagrangian is

$$L \rightarrow \frac{1}{2}\mu\dot{\mathbf{r}}^2 - U(r) = \frac{1}{2}\mu(\dot{r}^2 + r^2\dot{\theta}^2 + r^2\sin^2\theta\,\dot{\phi}^2) - U(r), \tag{7.14}$$

which has the same form as that for a single particle of mass μ orbiting around a force center at the origin, written in polar coordinates. We already know that this problem is entirely two-dimensional, since the angular momentum vector is conserved. We can then choose our spherical coordinates so that the plane of the dynamics corresponds to $\theta = \pi/2$. This allows us to write a simpler Lagrangian with two degrees of freedom only:

$$L = \frac{1}{2}\mu(\dot{r}^2 + r^2\dot{\phi}^2) - U(r). \tag{7.15}$$

We can then immediately identify two constants of the motion.

(i) L is not an explicit function of time, so the Hamiltonian H is conserved, which in this case is also the sum of kinetic and potential energies:

$$E = \frac{1}{2}\mu(\dot{r}^2 + r^2\dot{\phi}^2) + U(r) = \text{constant}. \tag{7.16}$$

(ii) The angle ϕ is cyclic, so the corresponding generalized momentum p_ϕ, which we recognize as the angular momentum of the particle, is also conserved:

$$p_\phi \equiv \ell = \mu r^2\dot{\phi} = r(\mu r\dot{\phi}) = \text{constant}. \tag{7.17}$$

This is the magnitude of the conserved angular momentum vector $\boldsymbol{\ell} = \mathbf{r} \times \mathbf{p}$, written in our coordinate system, where $\mathbf{p} = \mu\mathbf{v}$.

With only two degrees of freedom remaining, represented by the coordinates r and ϕ, the two conservation laws of energy and angular momentum together form a complete set of first integrals of motion for a particle moving in response to a central force or in a two-body problem. This means the problem is integrable. We will proceed in the next section to solve for the motion explicitly in two different ways.

Before we find the solution, however, let us note an interesting attribute of such systems. Our original two-body problem collapsed into a two-dimensional one-body problem described by a position vector $\mathbf{r}$ pointing from the source m_1 to the probe m_2. This position vector traces out the *relative* motion of the probe about the source. Yet the source may be moving around and accelerating. Although it may appear that one is incorrectly formulating physics from the perspective of a potentially non-inertial frame – that of the source – this is not so. The elegance of the two-body central force problem arises in part from the fact that the information about the non-inertial aspect of the source's perspective is neatly tucked into a single parameter, μ: we are describing the relative motion of m_2 with respect to m_1 by tracing out the trajectory of a fictitious particle of mass $\mu = m_1 m_2 / (m_1 + m_2)$ about m_1. Our starting-point Lagrangian of the two-body problem was written from the perspective of a third entity, an inertial observer. Yet, after a sequence of coordinate transformations and simplifications, we have found that the problem is mathematically equivalent to describing the dynamics of an object of mass μ about the source mass m_1.

Note also that if we are in a regime where the source mass is much heavier than the probe, $m_1 \gg m_2$, we have $\mu \simeq m_2$. In such a scenario, the source mass m_1 is too heavy to be affected much by m_2's pull, so m_1 essentially stays put in an inertial frame, with m_2 orbiting around it. In this regime we recover the naive interpretation that one is tracing out the relative motion of a probe mass m_2 from the perspective of an inertial observer sitting with m_1.

7.3 The Effective Potential Energy

We start by analyzing the dynamics qualitatively, and in some generality, using the two conservation equations

$$E = \frac{1}{2}\mu(\dot{r}^2 + r^2\dot{\phi}^2) + U(r) \quad \text{and} \quad \ell = \mu r^2 \dot{\phi}. \tag{7.18}$$

We have a choice to make: we can use these two equations to eliminate either the time t or the angle ϕ. In this section we will be interested in using energy diagrams to see whether the trajectories of the probe are bound or unbound, and how long it takes the probe to move from one point to another. So we will first eliminate the angle ϕ between the two equations. In the next section we will eliminate t instead, which will allow us to find the orbital *shapes*.

From the angular momentum conservation equation we have $\dot{\phi} = \ell/mr^2$, so energy conservation gives

$$\frac{1}{2}\mu\dot{r}^2 + U_{\text{eff}}(r) = E, \tag{7.19}$$

where the **effective potential energy** is

$$U_{\text{eff}}(r) \equiv \frac{\ell^2}{2\mu r^2} + U(r). \tag{7.20}$$

Angular momentum conservation has allowed us to convert the *rotational* portion of the kinetic energy $(1/2)\mu r^2 \dot{\phi}^2$ into a term $\ell^2/2\mu r^2$ that depends on position alone, so it behaves just like a potential energy. Then the sum of this term and the "real" potential energy $U(r)$ (which is related to the central force $F(r)$ by $F(r) = -dU(r)/dr$) together form the effective potential energy. The extra term is often called the "centrifugal potential"

$$U_{\text{cent}}(r) \equiv \frac{\ell^2}{2\mu r^2} \qquad (7.21)$$

because its corresponding "force" $F_{\text{cent}} = -dU_{\text{cent}}/dr = +\ell^3/\mu r^3$ tends to push the orbiting particle away from the force center at the origin. By eliminating ϕ between the two conservation laws, they combine to form an equation that *looks* like a one-dimensional energy conservation law in the variable r. So as long as we add in the centrifugal potential energy, we can use all our experience with one-dimensional conservation-of-energy equations to understand the motion. In general, we can tell that if our $U_{\text{eff}}(r)$ has a minimum

$$U'_{\text{eff}}\big|_{r=R} = -\frac{\ell^2}{\mu r^3} + U'(r)\bigg|_{r=R} = 0, \qquad (7.22)$$

the system admits circular orbits at $r = R$. Such an orbit would be stable if $U''_{\text{eff}} > 0$, unstable if $U''_{\text{eff}} < 0$, and critically stable if $U''_{\text{eff}} = 0$:

$$U''_{\text{eff}}\big|_{r=R} = 3\frac{\ell^2}{\mu r^4} + U''(r)\bigg|_{r=R} \begin{cases} > 0 & \text{stable} \\ < 0 & \text{unstable} \\ = 0 & \text{critically stable} \end{cases}. \qquad (7.23)$$

We can also determine whether the system admits bounded noncircular orbits – where $r_{\text{min}} < r < r_{\text{max}}$ – or unbounded orbits – where r can extend all the way to infinity. Let us look at a couple of examples to see how the effective energy diagram method can be very useful.

7.3.1 Radial Motion for the Central-Spring Problem

The effective potential energy of a particle in a central-spring potential is

$$U_{\text{eff}}(r) = \frac{\ell^2}{2\mu r^2} + \frac{1}{2}kr^2, \qquad (7.24)$$

which is illustrated in Figure 7.4. At large radii the attractive spring force $F_{\text{spring}} = -dU(r)/dr = -kr$ dominates, but at small radii the centrifugal potential takes over, and the associated "centrifugal force," given by $F_{\text{cent}} = -dU_{\text{cent}}/dr = \ell^2/\mu r^3$, is positive for nonzero ℓ, and therefore outward, an inverse-cubed strongly repulsive force. To understand this effect, note that nonzero angular momentum necessitates that the particle comes in from infinity off-center – aimed not directly toward the origin $r = 0$: this leads to a minimum distance of approach as the particle "swings by" the center. From the perspective of the one-dimensional effective potential treatment, this is manifested by a "centrifugal force" or an **angular momentum barrier**.

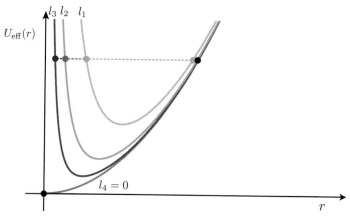

The effective potential for the central-spring potential for various angular momenta $l_1 > l_2 > l_3 > l_4 = 0$. The turning points are shown as colored discs.

We can also tell that this system admits only *bounded* orbits for $\ell \neq 0$: there is a minimum and maximum value of r for the dynamics. In this case, we will see that these bounded orbits are also *closed*. That is, after 2π's worth of evolution in ϕ, the probe again traces the same trajectory. To find the explicit shape of these trajectories – which will turn out to be ellipses – we will need to integrate our differential equations. We will come back to this in Section 7.4. For now, we can already answer interesting questions such as the time required for the probe to travel between two radii. Solving Eq. (7.19) (with $U_{\text{eff}} = \ell^2/2\mu r^2 + kr^2$) for $\dot{r}^2$ and taking the square root gives

$$\frac{dr}{dt} = \pm \sqrt{\frac{2}{\mu}\left(E - \frac{1}{2}kr^2 - \frac{\ell^2}{2\mu r^2}\right)}. \tag{7.25}$$

Separating variables and integrating:

$$t(r) = \pm \sqrt{\frac{\mu}{2}} \int_{r_0}^{r} \frac{r\,dr}{\sqrt{Er^2 - kr^4/2 - \ell^2/2\mu}}, \tag{7.26}$$

where we choose $t = 0$ at some particular radius r_0. We have reduced the problem to quadrature.

In fact, in this case the integral can be carried out analytically (see the Problems section at the end of this chapter), so we can find the time it takes the probe to move from any radius to any other radius.

Finally, for $\ell = 0$, the angular momentum barrier vanishes: this corresponds to a particle that is aimed directly toward the origin. The corresponding motion is then entirely radial (that is, it is truly one-dimensional) and the particle oscillates back and forth through the origin.

7.3.2 Radial Motion in Central Gravity

The effective potential energy of a particle in a central gravitational field is

$$U_{\text{eff}}(r) = \frac{\ell^2}{2mr^2} - \frac{GMm}{r}, \tag{7.27}$$

as illustrated in Figure 7.5. At large radii the inward gravitational force $F_{\text{grav}} = -dU(r)/dr = -GMm/r^2$ dominates, but at small radii the centrifugal potential takes over, and the associated "centrifugal force," given by $F_{\text{cent}} = -dU_{\text{cent}}/dr = \ell^2/mr^3$, is positive, and therefore outward, an inverse-cubed strongly repulsive force that pushes the planet away from the origin if it gets too close.

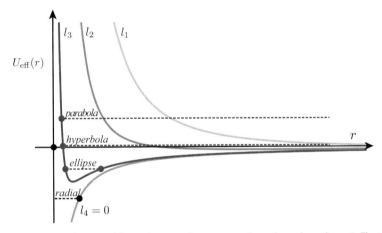

Fig. 7.5   The effective gravitational potential for various angular momenta $l_1 > l_2 > l_3 > l_4 = 0$. The turning points are shown as colored discs.

Two very different types of orbit are possible in this potential: **bound orbits** with energy $E < 0$, and **unbound orbits** with energy $E \geq 0$. Bound orbits do not escape to infinity. They include circular orbits with an energy E_{min} corresponding to the energy at the bottom of the potential well, where only one radius is possible, and there are orbits with $0 > E > E_{\text{min}}$, where the planet travels back and forth between inner and outer turning points while it is also rotating about the center. The minimum radius r_{min} is called the **periapse** for orbits around an arbitrary object, and specifically the **perihelion**, **perigee**, and **periastron** for orbits around the sun, the earth, and a star. The maximum radius r_{max} is called the **apoapse** in general, or specifically the **aphelion**, **apogee**, and **apastron**.

Unbound orbits are those with no outer turning point: these orbits extend out infinitely far. There are orbits with $E = 0$ that are just barely unbound: in this case the kinetic energy goes to zero in the limit as the orbiting particle travels infinitely far from the origin. And there are orbits with $E > 0$ where the particle still has nonzero kinetic energy as it escapes to infinity. In fact, we will see in the next section that orbits with energies $E = E_{\min}$ are circles, those with $E_{\min} < E < 0$ are ellipses, those with $E = 0$ are parabolas, and those with $E > 0$ are hyperbolas.

Now we can tackle the effective one-dimensional energy equation in (r, t) to try to obtain another first integral of motion. Our goal is to find $r(t)$ or $t(r)$, so we will know how far a planet, comet, or spacecraft moves radially in a given length of time, or how long it takes any one of them to travel between two given radii in its orbit.

Solving Eq. (7.19) (with $U_{\text{eff}} = \ell^2/2\mu r^2 - Gm_1m_2/r$) for $\dot{r}^2$ and taking the square root gives

$$\frac{dr}{dt} = \pm\sqrt{\frac{2}{\mu}\left(E + \frac{GM\mu}{r} - \frac{\ell^2}{2\mu r^2}\right)}, \tag{7.28}$$

using the fact that $m_1m_2 = M\mu$. Separating variables and integrating:

$$t(r) = \pm\sqrt{\frac{\mu}{2}}\int_{r_0}^{r}\frac{r\,dr}{\sqrt{Er^2 + GM\mu r - \ell^2/2\mu}}, \tag{7.29}$$

where we have chosen $t = 0$ at some particular radius r_0. We have reduced the problem to quadrature. In fact, the integral can also be carried out analytically, so we can calculate exactly how long it takes a planet or spacecraft to travel from one radius to another in its orbit (see Problem 7.16).

7.4 The Shape of Central-Force Orbits

Now we will find the *shape* of the orbits for the time evolution determined above. We will first eliminate the time t from the equations, leaving only r and ϕ. That is, we will find a single differential equation involving r and ϕ alone, which will give us a way to find the shape $r(\phi)$, the radius of the orbit as a function of the angle; or $\phi(r)$, the angle as a function of the radius.

Beginning with the first integrals

$$E = \frac{1}{2}m\dot{r}^2 + \frac{\ell^2}{2mr^2} + U(r) \quad \text{and} \quad \ell = mr^2\dot{\phi}, \tag{7.30}$$

we have two equations in the three variables, r, ϕ, and t. When finding the shape $r(\phi)$ we are not concerned with the time it takes to move from place to place, so we eliminate t between the two equations. Solving for dr/dt from the energy equation and dividing by $d\phi/dt$ from the angular momentum equation, we find that

$$\frac{dr}{d\phi} = \frac{dr/dt}{d\phi/dt} = \pm\sqrt{\frac{2m}{\ell^2}}r^2\sqrt{E - \ell^2/2mr^2 - U(r)}, \tag{7.31}$$

neatly eliminating t. Separating variables and integrating:

$$\phi = \int d\phi = \pm \frac{\ell}{\sqrt{2m}} \int^r \frac{dr/r^2}{\sqrt{E - \ell^2/2mr^2 - U(r)}}, \tag{7.32}$$

reducing the shape problem to quadrature. Further progress in finding $\phi(r)$ requires a choice of $U(r)$.

7.4.1 Central Spring-Force Orbits

A Hooke's-law spring force $\mathbf{F} = -k\mathbf{r}$ pulls on a particle of mass m toward the origin at $r = 0$. The force is central, so the particle moves in a plane with a potential energy $U = (1/2)kr^2$. What is the shape of its orbit? From Eq. (7.32):

$$\phi(r) = \pm \frac{\ell}{\sqrt{2m}} \int^r \frac{dr/r^2}{\sqrt{E - \ell^2/2mr^2 - (1/2)kr^2}}. \tag{7.33}$$

Multiplying top and bottom of the integrand by r and substituting $z = r^2$ gives

$$\phi(z) = \pm \frac{\ell}{2\sqrt{2m}} \int^z \frac{dz/z}{\sqrt{-\ell^2/2m + Ez - (k/2)z^2}}. \tag{7.34}$$

From integral tables online or in a book, we find that

$$\int^z \frac{dz/z}{\sqrt{a + bz + cz^2}} = \frac{1}{\sqrt{-a}} \sin^{-1}\left(\frac{bz + 2a}{z\sqrt{b^2 - 4ac}}\right), \tag{7.35}$$

where a, b, and c are constants, with $a < 0$. In our case $a = -\ell^2/2m, b = E$, and $c = -k/2$, so

$$\phi - \phi_0 = \pm \frac{\ell}{2\sqrt{2m}} \frac{1}{\sqrt{\ell^2/2m}} \sin^{-1}\left(\frac{bz + 2a}{z\sqrt{b^2 - 4ac}}\right)$$

$$= \pm \frac{1}{2} \sin^{-1}\left(\frac{Er^2 - \ell^2/m}{r^2\sqrt{E^2 - k\ell^2/m}}\right), \tag{7.36}$$

where ϕ_0 is a constant of integration. Multiplying by ± 2, taking the sine of each side, and solving for r^2 gives the orbital shape equation

$$r^2(\phi) = \frac{\ell^2/m}{E \mp (\sqrt{E^2 - k\ell^2/m})\sin 2(\phi - \phi_0)}. \tag{7.37}$$

Note that the orbit is closed (since $r^2(\phi + 2\pi) = r^2(\phi)$), and that it has a long axis (corresponding to an angle ϕ where the denominator is *small* because the second term subtracts from the first term) and a short axis (corresponding to an angle where the denominator is *large*, because the second term adds to the first term). In fact, the shape $r(\phi)$ is that of an **ellipse** with $r = 0$ at the *center* of the ellipse.[3]

The orbit is illustrated in Figure 7.6 for the case $\phi_0 = 0$ and with a minus sign in the denominator. The effect of changing the sign or using a nonzero ϕ_0 is simply to rotate the entire figure about its center, while keeping the "major" axis and the "minor" axis perpendicular to one another.

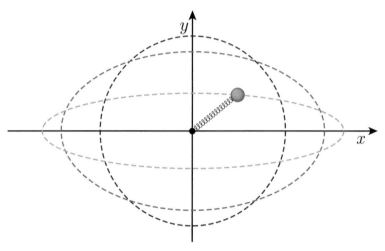

Fig. 7.6 Elliptical orbits due to a central spring force $\mathbf{F} = -k\mathbf{r}$.

7.4.2 The Shape of Gravitational Orbits

By far the most important orbital shapes are those for central gravitational forces. This is the problem that Johannes Kepler wrestled with in his self-described "War on Mars." Equipped with the observational data on the positions of Mars from Tycho Brahe, he tried one shape after another to see what would fit, beginning with a circle (which didn't work), various ovals (which didn't work), and finally an ellipse (which did). Now we can derive the shape by two different methods, by solving the integral of Eq. (7.32), and (surprisingly enough!) by *differentiating* Eq. (7.31).

By Direct Integration
For a central gravitational force the potential energy $U(r) = -GMm/r$, so the integral for $\phi(r)$ becomes

---

[3] A common way to express an ellipse in polar coordinates with $r = 0$ at the center is to orient the major axis horizontally and the minor axis vertically, which can be carried out by selecting the plus sign in the denominator and choosing $\phi_0 = \pi/4$. In this case the result can be written $r^2 = a^2b^2/(b^2 \cos^2 \phi + a^2 \sin^2 \phi)$, where a is the semi-major axis (half the major axis) and b is the semi-minor axis. In Cartesian coordinates $(x = r\cos\phi, y = r\sin\phi)$ this form is equivalent to the common ellipse equation $x^2/a^2 + y^2/b^2 = 1$.

$$\phi = \int d\phi = \pm \frac{\ell}{\sqrt{2m}} \int \frac{dr/r}{\sqrt{Er^2 + GMmr - \ell^2/2m}}, \tag{7.38}$$

which by coincidence is the same integral we encountered in Section 7.4.1 (using there the variable $z = r^2$ instead):

$$\int \frac{dr/r}{\sqrt{a + br + cr^2}} = \frac{1}{\sqrt{-a}} \sin^{-1}\left(\frac{br + 2a}{r\sqrt{b^2 - 4ac}}\right), \tag{7.39}$$

where now $a = -\ell^2/2m, b = GMm$, and $c = E$. Therefore

$$\phi - \phi_0 = \pm \sin^{-1}\left(\frac{GMm^2 - \ell^2}{\epsilon\, GMm^2 r}\right), \tag{7.40}$$

where ϕ_0 is a constant of integration and we have defined the **eccentricity**

$$\epsilon \equiv \sqrt{1 + \frac{2E\ell^2}{G^2M^2m^3}}. \tag{7.41}$$

We will soon see the geometrical meaning of ϵ. Taking the sine of $\phi - \phi_0$ and solving for r gives

$$r = \frac{\ell^2/GMm^2}{1 \pm \epsilon \sin(\phi - \phi_0)}. \tag{7.42}$$

By convention we choose the plus sign in the denominator together with $\phi_0 = \pi/2$, which in effect locates $\phi = 0$ at the point of closest approach to the center, the **periapse** of the ellipse. This choice changes the sine to a cosine, so

$$r = \frac{\ell^2/GMm^2}{1 + \epsilon \cos \phi}. \tag{7.43}$$

This equation gives the allowed shapes of orbits in a central gravitational field. Before identifying these shapes, we will derive the same result by a very different method that is often especially useful!

By Differentiation

Returning to Eq. (7.31) with $U(r) = -GMm/r$:

$$\frac{dr}{d\phi} = \pm \sqrt{\frac{2m}{\ell}} r^2 \sqrt{E - \frac{\ell^2}{2mr^2} + \frac{GMm}{r}}, \tag{7.44}$$

we will now *differentiate* it. The result turns out to be greatly simplified if we first introduce the *inverse radius* $u = 1/r$ as the coordinate. Then

$$\frac{dr}{d\phi} = \frac{d(1/u)}{d\phi} = -\frac{1}{u^2}\frac{du}{d\phi}. \tag{7.45}$$

Squaring this gives

$$\left(\frac{du}{d\phi}\right)^2 \equiv (u')^2 = \frac{2m}{\ell^2}\left(E - \frac{\ell^2 u^2}{2m} + (GMm)u\right). \tag{7.46}$$

Differentiating both sides with respect to ϕ:

$$2u'u'' = -2uu' + \frac{2GMm^2}{\ell^2}u'. \tag{7.47}$$

Then dividing out the common factor u', since (except for a circular orbit) it is generally nonzero, we find

$$u'' + u = \frac{GMm^2}{\ell^2}. \tag{7.48}$$

The most general solution of this *linear* second-order differential equation is the sum of the general solution of the homogeneous equation $u'' + u = 0$ and any particular solution of the full (inhomogeneous) equation. The general solution of the full equation can therefore be written in the form

$$u = A[1 + \epsilon \cos \phi], \tag{7.49}$$

where $A = GMm^2/\ell^2$. The shape of the orbit is therefore

$$r = \frac{1}{u} = \frac{\ell^2/GMm^2}{1 + \epsilon \cos \phi} = \frac{r_p(1 + \epsilon)}{1 + \epsilon \cos \phi}, \tag{7.50}$$

where r_p is the point of closest approach of the orbit to a fixed point called the **focus**; and we have again chosen $\phi = 0$ to be the point where r is a minimum, namely the radius

$$r_p \equiv \frac{\ell^2/GMm^2}{1 + \epsilon}. \tag{7.51}$$

Equation (7.50) is the same as Eq. (7.43), the result we found previously by direct integration, if, using Eq. (7.41), we identify

$$\epsilon = \sqrt{1 + \frac{2E\ell^2}{G^2M^2m^3}}. \tag{7.52}$$

Therefore, the orbit shapes are determined by two constants: either the energy E and the angular momentum l, or alternatively the eccentricity ϵ and the point of closest approach r_p. We also have relations (7.51) and (7.52) to map from one set of constants to the other. Even though this analysis has merely reproduced a result we already knew, the "trick" of substituting the inverse radius works for inverse-square forces, and will be a useful springboard later when we perturb elliptical orbits.

The shapes $r(\phi)$ given by Eq. (7.50) are known as **conic sections**, since they correspond to the possible intersections of a plane with a cone, as illustrated in Figure 7.7. There are only four possible shapes: (i) circles, (ii) ellipses, (iii) parabolas, and (iv) hyperbolas.

(i) For **circles**, the eccentricity $\epsilon = 0$, so the radius $r = r_p$, a constant independent of angle ϕ. The focus of the orbit is at the center of the circle.

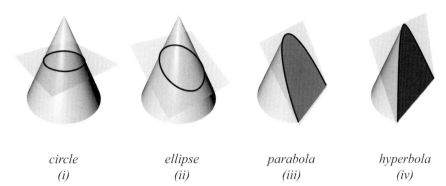

circle   ellipse   parabola   hyperbola
(i)   (ii)   (iii)   (iv)

Fig. 7.7   Conic sections: circles, ellipses, parabolas, and hyperbolas.

(ii) For **ellipses**, the eccentricity must obey $0 < \epsilon < 1$. Note from the shape equation that in this case, as with a circle, the denominator cannot go to zero, so the radius remains finite for all angles. There are two foci in this case, and r_p is the closest approach to the focus at the right in Figure 7.8, where the angle $\phi = 0$. Note that the force center at $r = 0$ is located at one of the foci of the ellipse for the gravitational force, unlike the ellipse for a central spring force of Section 7.4.1, where the force center was at the center of the ellipse.

The long axis of the ellipse is called the major axis, and half of this distance is the semi-major axis, denoted by the symbol a. The semi-minor axis, half of the shorter axis, is denoted by b. One can derive several properties of ellipses from Eq. (7.43) in this case:

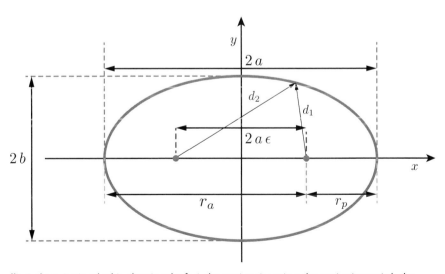

Fig. 7.8   An elliptical gravitational orbit, showing the foci, the semi-major axis a, the semi-minor axis b, the eccentricity ϵ, and the periapse and apoapse.

(a) The **periapse** and **apoapse** of the ellipse (the closest and farthest points of the orbit from the right-hand focus) are given, in terms of a and ϵ, by

$$r_p = a(1 - \epsilon) \quad \text{and} \quad r_a = a(1 + \epsilon), \tag{7.53}$$

respectively.

(b) The sum of the distances d_1 and d_2 from the two foci to a point on the ellipse is the same for all points on the ellipse.[4]

(c) The distance between the two foci is $2a\epsilon$, so the eccentricity of an ellipse is the ratio of this interfocal distance to the length of the major axis.

(d) The semi-minor and semi-major axes are related by

$$b = a\sqrt{1 - \epsilon^2}. \tag{7.54}$$

(e) The area of the ellipse is $A = \pi ab$.

(f) The shape of the orbit is entirely determined through two parameters that we can choose as the energy E and the angular momentum ℓ, or as the eccentricity ϵ and the periapse radius r_p, or even as the sizes of the semi-major and semi-minor axes a and b. Relations between these three pairs of constants (see Eqs. (7.51), (7.52), (7.53), and (7.54)) allow us to switch perspective as desired.

(iii) For **parabolas**, the eccentricity $\epsilon = 1$, so $r \to \infty$ as $\phi \to \pm\pi$, and the shape is as shown in Figure 7.9. One can show that every point on a parabola is equidistant from a focus and a line called the **directrix**, also shown on the figure.

(iv) For **hyperbolas**, the eccentricity $\epsilon > 1$, so $r \to \infty$ as $\cos\phi \to -1/\epsilon$. This corresponds to two angles, one between $\pi/2$ and π, and one between $-\pi/2$ and $-\pi$, as shown in Figure 7.9.

Example 7.1   **Orbital Geometry and Orbital Physics**

Let us relate the geometrical parameters of a gravitational orbit to the physical parameters, the energy E and angular momentum ℓ. The relationships follow from Eqs. (7.51), (7.52), (7.53), and (7.54). We first consider circles and ellipses, and then parabolas and hyperbolas.

For ellipses or circles, we may choose as geometrical parameters the eccentricity ϵ and semi-major axis length a (alternatively, we could choose a and the semi-minor axis length b). We have $a(1 - \epsilon^2) = \ell^2/GMm^2$, so the semi-major axis of an ellipse (or the radius of the circle) is related to the physical parameters by

$$a = \frac{\ell^2/GMm^2}{1 - \epsilon^2} = \frac{\ell^2/GMm^2}{1 - (1 + 2E\ell^2/G^2M^2m^3)} = -\frac{GMm}{2E}, \tag{7.55}$$

---

[4] Therefore the well-known property of an ellipse, that it can be drawn on a sheet of paper by sticking two straight pins into the paper some distance D apart, and dropping a loop of string over the pins, where the loop has a circumference greater than $2D$. Then sticking a pencil point into the loop as well, and keeping the loop taut, moving the pencil point around on the paper, the resulting drawn figure will be an ellipse.

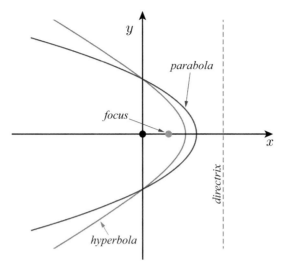

Fig. 7.9   Parabolic and hyperbolic orbits.

depending upon E but not ℓ. In summary, for ellipses and circles the geometrical parameters a, ϵ are related to the physical parameters by

$$a = -\frac{GMm}{2E} \quad \text{and} \quad \epsilon = \sqrt{1 + \frac{2E\ell^2}{G^2M^2m^3}}. \tag{7.56}$$

These can be inverted to give the physical parameters in terms of the geometrical parameters:

$$E = -\frac{GMm}{2a} \quad \text{and} \quad \ell = \sqrt{GMm^2a(1 - \epsilon^2)}. \tag{7.57}$$

Notice that $E < 0$, and the minimum energy, corresponding to a circular orbit, is given by

$$\epsilon = 0 \Rightarrow E_{\min} = -\frac{G^2M^2m^3}{2\,\ell^2}. \tag{7.58}$$

For parabolas and hyperbolas, we choose as geometrical parameters ϵ and r_p. We then have $r_p(1+\epsilon) = \ell^2/GMm^2$, where $\epsilon = 1$ for parabolas and $\epsilon > 1$ for hyperbolas. So the geometric parameters (r_p, ϵ) for these orbits are given in terms of the physical parameters E and ℓ by

$$r_p = \frac{\ell^2}{(1 + \epsilon)GMm^2} \quad \epsilon = \sqrt{1 + \frac{2E\ell^2}{G^2M^2m^3}}, \tag{7.59}$$

and inversely

$$E = \frac{GMm(\epsilon - 1)}{2r_p} \quad \ell = \sqrt{GMm^2r_p(1 + \epsilon)} \tag{7.60}$$

in terms of (r_p, ϵ). Note that for parabolas the eccentricity $\epsilon = 1$, so the energy $E = 0$. And for hyperbolas, we must have $E > 0$ since $\epsilon > 1$. ∎

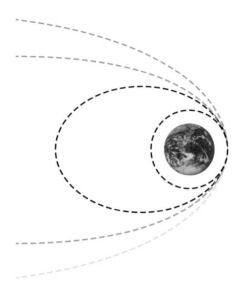

Fig. 7.10   The four types of gravitational orbits.

Finally, to summarize orbits in a purely central inverse-square gravitational field, there are four, and only four, types of orbits possible, as illustrated in Figure 7.10. There are circles ($\epsilon = 0$), ellipses ($0 < \epsilon < 1$), parabolas ($\epsilon = 1$), and hyperbolas ($\epsilon > 1$), with the gravitating object at one focus. Ellipses and circles are closed, bound orbits with negative total energy. Circles correspond to minimal-energy orbits for given angular momentum. Hyperbolas and parabolas are open, unbound orbits, which extend to infinity. Parabolic orbits have zero total energy, and hyperbolic orbits have positive total energy.

Circles (with $\epsilon = 0$) and parabolas (with $E = 0$) are so unique among the set of all solutions that mathematically one can say that they form "sets of measure zero," and physically one can say that they never occur in Nature. The orbits of planets, asteroids, and some comets are elliptical; other comets may move in hyperbolic orbits. There are no other orbit shapes for a central gravitational field. There are, for example, no "decaying" or "spiralling" purely gravitational orbits. There do exist straight-line paths falling directly toward or away from the central object, but these are really limiting cases of ellipses, parabolas, and hyperbolas. They correspond to motion with angular momentum $\ell = 0$, so the eccentricity $\epsilon = 1$. If the particle's energy is negative, it is the limiting case of an ellipse as $\epsilon \to 1$; if the energy is positive, it is the limiting case of a hyperbola as $\epsilon \to 1$; and if the energy is zero, it is a parabola with both $\epsilon = 1$ and $p_\phi = 0$.

7.5 Bertrand's Theorem

In the previous two sections we saw central potentials that admit bounded and unbounded orbits, and we found a way to calculate the orbital shapes. Bounded orbits are of particular interest, since they can potentially *close*, and we showed that the orbits are closed for both the central linear spring force and the central inverse-squared gravitational force. That is, after a certain finite number of revolutions, the probe starts tracing out its established trajectory – thus closing its orbit.

How general is this property of closure? What about the orbits due to other central forces? A beautiful and powerful result of mechanics is a theorem due to J. Bertrand, which states the following:

Bertrand's theorem: *The only central force potentials U(r) for which all bounded orbits are closed are the following:*

1. *The gravitational potential $U(r) \propto -1/r$.*
2. *The central-spring potential $U(r) \propto r^2$.*

The theorem asserts that, of all possible functional forms for a potential $U(r)$, only two kinds lead to the interesting situation in which all bounded orbits are closed! And these two potentials are just the ones we have treated in detail. The theorem is not very difficult to prove: we leave it to the Problems section at the end of this chapter.

So while it is interesting to find orbital shapes for other central forces, we know from this theorem that in such cases the probe will not generally return to the same point after completing one revolution.

7.6 Orbital Dynamics

As we saw already in Chapter 5, Kepler identified three rules that govern the dynamics of planets in the heavens:

1. *Planets move in elliptical orbits, with the sun at one focus.*
2. *Planetary orbits sweep out equal areas in equal times.*
3. *The periods squared of planetary orbits are proportional to their semi-major axes cubed.*

It took about a century to finally understand, through the work of Isaac Newton, the physical origins of these three laws. Armed with new powerful tools in mechanics, we have confirmed the first law of Kepler. To understand the second and third, we will need to do a bit more work.

7.6.1 Kepler's Second Law

There is an interesting consequence of angular momentum conservation for *arbitrary* central forces. Take a very thin slice of pie extending from the origin to the orbit of the particle, as shown in Figure 7.11. To a good approximation, becoming exact in the limit as the slice gets infinitely thin, the area of the slice is that of a triangle, $\Delta A = (1/2) \, (\text{base} \times \text{height}) = (1/2)r(r\Delta\phi) = (1/2)r^2\Delta\phi$.

If the particle moves through angle $\Delta\phi$ in time Δt, then $\Delta A/\Delta t = (1/2)r^2\Delta\phi/\Delta t$, so in the limit $\Delta t \to 0$:

$$\frac{dA}{dt} = \frac{1}{2}r^2\dot{\phi} = \frac{\mu r^2 \dot{\phi}}{2\mu} = \frac{\ell}{2\mu} = \text{constant}, \qquad (7.61)$$

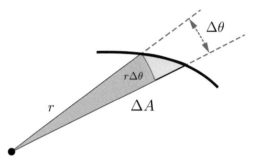

The area of a thin pie slice. The area of the light-shaded region comes with one additional power of $\Delta\theta$ compared to the area of the dark-shaded region – and hence can be neglected to leading order in small $\Delta\theta$.

since ℓ is constant and we have used Eq. (7.18). Therefore this *areal velocity*, the rate at which area is swept out by the orbit, remains constant as the particle moves. This in turn implies that the orbit *sweeps out equal areas in equal times*. Between t_1 and t_2, for example:

$$A = \int_{t_1}^{t_2}\left(\frac{dA}{dt}\right)dt = \int_{t_1}^{t_2}\left(\frac{\ell}{2\mu}\right)dt = \left(\frac{\ell}{2\mu}\right)(t_2 - t_1), \qquad (7.62)$$

which is the same as the area swept out between times t_3 and t_4 if $t_4 - t_3 = t_2 - t_1$.

We have therefore derived Kepler's second law. Kepler himself did not know *why* the law is true; the concepts of angular momentum and central forces had not yet been invented.[5] In the orbit of the earth around the sun, for example, the areas swept out in any 31-day month, say January, July, or October, must all be the same. To make the areas equal, in January, when the earth is closest to the sun, the pie slice must be fatter than in July, when the earth is farthest from the sun. Note that the tangential velocity $r\dot{\phi}$ must be greater in January to cover the greater distance in the same length of time, which is consistent with conservation of the angular momentum $\ell = mr^2\dot{\phi} = mr \times r\dot{\phi}$.

Although it was first discovered for orbiting planets, the equal-areas-in-equal-times law is also valid for particles moving in *any* central force, including asteroids, comets, and spacecraft around the sun; the moon and artificial satellites around the earth; and particles subject to a central attractive spring force or *any* central force.

---

[5] Kepler had effectively identified a symmetry in orbital dynamics through a conservation law, which we now understand through Noether's theorem of Chapter 6.

7.6.2 Kepler's Third Law

How *long* does it take planets to orbit the sun? And how long does it take the moon, and orbiting spacecraft or other earth satellite to orbit the earth?

From Eq. (7.62) in Section 7.6.1, the area traced out in time $t_2 - t_1$ is $A = (\ell/2m)(t_2 - t_1)$. The period of the orbit, which is the time to travel around the entire ellipse, is therefore

$$T = (2m/\ell)A = (2m/\ell)\pi ab = \frac{2m\pi a^2\sqrt{1 - \epsilon^2}}{\sqrt{GMm^2 a(1 - \epsilon^2)}}, \tag{7.63}$$

since the area of the ellipse is $A = \pi ab = a^2\sqrt{1 - \epsilon^2}$, and $\ell = \sqrt{GMm^2 a(1 - \epsilon^2)}$. This expression for the period simplifies to give

$$T = \frac{2\pi}{\sqrt{GM}}a^{3/2}. \tag{7.64}$$

It is interesting that the period depends upon the semi-major axis of the orbit, but *not* upon the eccentricity. Two orbits with the same semi-major axis have the same period, even though their eccentricities are different. And we thus arrive at Kepler's third law: the periods squared of planetary orbits are proportional to their semi-major axes cubed – that is, $T^2 \propto a^3$.

Example 7.2 **Halley's Comet**

Halley's Comet is named after the English astronomer, mathematician, and physicist Sir Edmund Halley (1656–1742), who was the first to determine that three comet sightings, separated from one another by about 76 years, were in fact visitations of the same object. The comet has been known as far back as 240 BC and probably longer, and was thought to be an omen when it appeared in 1066, the year of the Norman conquest at the Battle of Hastings. Mark Twain was born in 1835 at one of its appearances, and predicted (correctly) that he would die at its next appearance in 1910. It last passed through the earth's orbit in 1986 and will again in 2061.

From the comet's current period[a] of $T = 75.3$ years and observed perihelion distance $r_p = 0.586$ AU (which lies between the orbits of Mercury and Venus), we can calculate the orbit's (a) semi-major axis a, (b) aphelion distance r_a, and (c) eccentricity ϵ. (Note that 1 AU is the length of the semi-major axis of earth's orbit, 1 AU $= 1.5 \times 10^{11}$ m.)

(a) From Kepler's third law, which applies to comets in bound orbits as well as to all planets and asteroids, we can compare the period of Halley's Comet to the period of earth's orbit: $T_H/T_e = (a_H/a_e)^{3/2}$, so the semi-major axis has length

$$a_H = a_E(T_H/T_E)^{2/3} = 1\,\text{AU}\,(75.3\text{years}/1\text{year})^{2/3} = 17.8\,\text{AU}. \tag{7.65}$$

(b) The major axis therefore has length 2×17.8 AU $= 35.6$ AU, so the aphelion distance is at $r_a = 35.8$ AU $- r_p = 35.6$ AU $- 0.6$ AU $= 35.0$ AU from the sun. Halley's Comet retreats farther from the sun than the orbit of Neptune.

(c) The perihelion distance is $r_p = a(1 - \epsilon)$, so the eccentricity of the orbit is

$$\epsilon = 1 - r_p/a = 0.967. \tag{7.66}$$

The orbit is highly eccentric, as you would expect, since the aphelion is 36 times as far from the sun as the perihelion.

The orbit of Halley's Comet is inclined at about $18°$ to the ecliptic, *i.e.*, at about $18°$ to the plane of earth's orbit, as shown in Figure 7.12. It is also retrograde: the comet orbits the sun in the opposite direction from that of the planets, orbiting clockwise rather than counterclockwise looking down upon the solar system from above the sun's north pole.

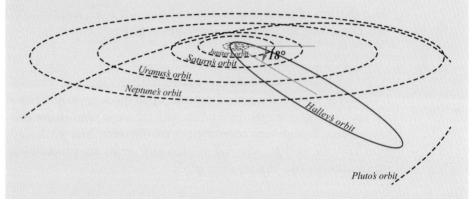

Fig. 7.12   The orbit of Halley's Comet.   ∎

[a] The period has varied considerably over the centuries, because the comet's orbit is easily influenced by the gravitational pull of the planets, especially Jupiter and Saturn.

7.6.3 Minimum-Energy Transfer Orbits

What is the best way to send a spacecraft to another planet? Depending upon what one means by "best," many routes are possible. But almost always the trajectory requiring the *least fuel* (assuming the spacecraft does not take advantage of "gravitational assists" from other planets along the way, which we will discuss later) is a so-called *minimum-energy transfer orbit* or "Hohmann" transfer orbit, which takes full advantage of earth's motion to help the spacecraft get off to a good start.

Typically the spacecraft is first lifted into low-earth orbit (LEO), where it circles the earth a few hundred kilometers above the surface. Then, at just the right time, the spacecraft is given a velocity boost "Delta v" that sends it away from the earth and into an orbit around the sun that reaches all the way to its destination. Once the spacecraft coasts far enough from earth that the sun's gravity dominates, the craft obeys all the central-force equations we have derived so far, including Kepler's laws. In particular, it coasts toward its destination in an elliptical orbit with the sun at one focus.

Suppose that in LEO the rocket engine boosts the spacecraft so that it ultimately attains a velocity v_∞ away from the earth. Then if the destination is Mars or one of the outer planets, it is clearly most efficient if the spacecraft is aimed so that this velocity v_∞ is in the *same* direction as the earth's velocity around the sun, because then the velocity of the spacecraft in the sun's frame will have its largest possible magnitude, $v_e + v_\infty$. The subsequent transfer orbit toward an outer planet is shown in Figure 7.13. The spacecraft's elliptical path is tangent to the earth's orbit at launch and tangent to the destination planet's orbit at arrival, just barely making it out to where we want it.

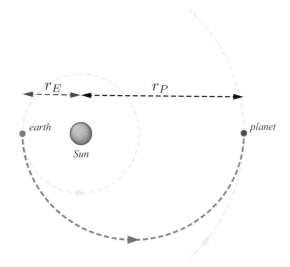

Fig. 7.13 A minimum-energy transfer orbit to an outer planet.

First we will find out how *long* it will take the spacecraft to reach its destination, which is easily found using Kepler's third law. Note that the major axis of the craft's orbit is $2a_C = r_e + r_P$, assuming the earth e and destination planet P move in nearly circular orbits with radii r_e and r_P, respectively. The semi-major axis of the transfer orbit is therefore

$$a_C = \frac{r_e + r_P}{2}. \tag{7.67}$$

From the third law, the period T_C of the craft's elliptical orbit obeys $(T_C/T_e)^2 = (a_C/r_e)^3$, in terms of the period T_e and radius r_e of the earth's orbit. The spacecraft travels through only half of this orbit on its way from earth to the planet, however, so the travel time is

$$T = \frac{T_C}{2} = \frac{1}{2}\left(\frac{r_e + r_P}{2r_e}\right)^{3/2} T_e, \tag{7.68}$$

which we can easily evaluate, since every quantity on the right is known.

Now we can outline the steps required for the spacecraft to reach Mars or an outer planet.

1. We first place the spacecraft in a circular "parking" orbit of radius r_0 around the earth. Ideally, the orbit will be in the same plane as that of the earth around the sun, and the rotation direction will also agree with the direction of earth's orbit. Using $\mathbf{F} = m\mathbf{a}$ in the radial direction centered on the earth, the speed v_0 of the spacecraft obeys

$$\frac{GM_e m}{r_0^2} = ma = \frac{mv_0^2}{r_0}, \tag{7.69}$$

so $v_0 = \sqrt{GM_e/r_0}$.

2. Then, at just the right moment, a rocket provides a boost Δv in the same direction as v_0, so the spacecraft now has an instantaneous velocity $v_0 + \Delta v$, allowing it to escape from the earth in the most efficient way. This will take the spacecraft from LEO into a *hyperbolic* orbit relative to the earth, since we want the craft to escape from the earth with energy to spare, as shown in Figure 7.14. Then, as the spacecraft travels far away, its potential energy $-GM_e m/r$ due to earth's gravity approaches zero, so its speed approaches v_∞, where, by energy conservation:

$$\frac{1}{2}mv_\infty^2 = \frac{1}{2}m(v_0 + \Delta v)^2 - \frac{GM_e m}{r_0}. \tag{7.70}$$

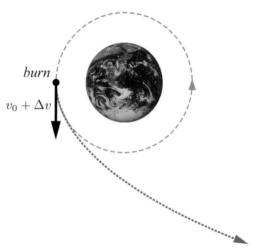

Fig. 7.14   Insertion from a parking orbit into the transfer orbit.

Solving for v_∞:

$$v_\infty = \sqrt{(v_0 + \Delta v)^2 - 2GM_e/r_0} = \sqrt{(v_0 + \Delta v)^2 - 2v_0^2}. \tag{7.71}$$

This is the speed of the spacecraft relative to the earth by the time it has essentially escaped earth's gravity, but before it has moved very far from earth's orbit around the sun.

3. Now, if we have provided the boost Δv at just the right time, when the spacecraft is moving in just the right direction, by the time the spacecraft has escaped from the earth its velocity v_∞ relative to the earth will be in the *same* direction as earth's velocity v_e around the sun, so the spacecraft's velocity in the sun's frame of reference will be as large as it can be for given v_∞:

$$v = v_\infty + v_e = \sqrt{(v_0 + \Delta v)^2 - 2v_0^2} + v_e. \qquad (7.72)$$

The earth has now been left far behind, so the spacecraft's trajectory from here on is determined by the sun's gravity alone. We have given it the largest speed v we can in the sun's frame for given boost Δv, to get it off to a good start.

4. The velocity v just calculated will be the speed of the spacecraft at the perihelion point of some elliptical Hohmann transfer orbit. What speed must this be for the transfer orbit to have the desired semi-major axis a? We can find out by equating the total energy (kinetic plus potential) of the spacecraft in orbit around the sun with the specific energy it has in an elliptical orbit with the appropriate semi-major axis a. That is:

$$E = T + U = \frac{1}{2}mv^2 - \frac{GMm}{r} = -\frac{GMm}{2a}, \qquad (7.73)$$

where m is the mass of the spacecraft, M is the mass of the sun, r is the initial distance of the spacecraft from the sun (which is the radius of earth's orbit), and a is the semi-major axis of the transfer orbit. Solving for v^2, we find

$$v^2 = GM\left(\frac{2}{r} - \frac{1}{a}\right), \qquad (7.74)$$

which is known as the *vis-viva* equation.[6] The quantities on the right are known, so we can calculate v, which is the sun-frame velocity the spacecraft must achieve.

Example 7.3 **A Voyage to Mars**

We will use this scenario to plan a trip to Mars by Hohmann transfer orbit. First, we can use Kepler's third law to find how long it will take for the spacecraft to arrive. The major axis of the spacecraft's orbit is $2a_C = r_e + r_M$, assuming the earth and Mars move in nearly circular orbits. The semi-major axis is therefore

$$a_C = \frac{r_e + r_M}{2} = \frac{1.50 + 2.28}{2} \times 10^8 \text{ km} = 1.89 \times 10^8 \text{ km}. \qquad (7.75)$$

The spacecraft travels through only half of this complete elliptical orbit on its way out to Mars, so the travel time is[a]

$$T = T_C/2 = \frac{1}{2}\left(\frac{1.89}{1.50}\right)^{3/2}(1 \text{ year}) = 258 \text{ days}. \qquad (7.76)$$

---

[6] *Vis-viva* means "living force," a term used by the German mathematician Gottfried Wilhelm Leibniz in a now-obsolete theory. The term survives only in orbital mechanics.

Now we will find the boost required in LEO to insert the spacecraft into the transfer orbit. We will first find the speed required of the spacecraft in the sun's frame just as it enters the Hohmann ellipse. From the *vis-viva* equation

$$v = \sqrt{GM \left(\frac{2}{r} - \frac{1}{a} \right)} = 32.7 \, \text{km/s}, \tag{7.77}$$

using $G = 6.67 \times 10^{-11} \, \text{m}^3/\text{kg s}^2$, $M = 1.99 \times 10^{30} \, \text{kg}$, $r = 1.50 \times 10^8 \, \text{km}$, and $a = 1.89 \times 10^8 \, \text{km}$. Compare this with the speed of the earth in its orbit around the sun,[b] $v_e = \sqrt{GM/r} = 29.7 \, \text{km/s}$.

Now suppose the spacecraft starts in a circular parking orbit around the earth, with radius $r_0 = 7000 \, \text{km}$, corresponding to an altitude above the surface of about 600 km. The speed of the spacecraft in this orbit is $v_0 = \sqrt{GM_e/r_0} = 7.5 \, \text{km/s}$. We then require that v_∞, the speed of the spacecraft relative to the earth after it has escaped from the earth, is $v_\infty = v - v_e = 32.7 \, \text{km/s} - 29.7 \, \text{km/s} = 3.0 \, \text{km/s}$. Solving finally for Δv in Eq. (7.72), we find that the required boost for this trip is

$$\Delta v = \sqrt{v_\infty^2 + 2v_0^2} - v_0$$

$$= \sqrt{(3.0 \, \text{km/s})^2 + 2(7.5 \, \text{km/s})^2} - 7.5 \, \text{km/s} = 3.5 \, \text{km/s}. \tag{7.78}$$

This boost of 3.5 km/s is modest compared with the boost needed to raise the spacecraft from earth's surface up to the parking orbit in the first place. Then, once the spacecraft reaches Mars, the rocket engine must provide an additional boost to insert the spacecraft into a circular orbit around Mars, or even to allow it to strike Mars's atmosphere at a relatively gentle speed. This is because the spacecraft, when it reaches the orbit of Mars in the Hohmann transfer orbit, will be moving considerably more slowly than Mars itself in the frame of the sun. Note that the Hohmann transfer orbit will definitely take the spacecraft out to Mars's orbit, but there are only limited launch windows; we have to time the trip just right so that Mars will actually be at that point in its orbit when the spacecraft arrives. ∎

[a] In his science fiction novel *Stranger in a Strange Land*, Robert Heinlein looks back on the first human journeys to Mars: "an interplanetary trip ... had to be made in free-fall orbits – from Terra to Mars, 258 Terran days, the same for return, plus 455 days waiting at Mars while the planets crawled back into positions for the return orbit."

[b] Earth's speed around the sun actually varies from 29.28 km/s at aphelion to 30.27 km/s at perihelion. It is not surprising that the spacecraft's speed of 32.7 km/s exceeds v_e; otherwise it could not escape outwards toward Mars against the sun's gravity.

Example 7.4

Gravitational Assists

There is no more useful and seemingly magical application of the Galilean velocity transformation of Chapter 1 than *gravitational assists*. Gravitational assists have been used to send spacecraft to destinations they could not otherwise reach because of limited rocket-fuel capabilities, including voyages to outer planets like Uranus and Neptune using gravitational assists from Jupiter and Saturn, and complicated successive visits to the satellites of Jupiter, gravitationally bouncing from one to another.

Suppose we want to send a heavy spacecraft to Saturn, but it has only enough room for fuel to make it to Jupiter. If the timing is just right and the planets are also aligned just right, it is possible to aim for Jupiter, causing the spacecraft to fly just *behind* Jupiter as it swings by that planet. Jupiter can pull on the spacecraft, turning its orbit to give it an increased velocity in the sun's frame of reference, sufficient to propel it out to Saturn.

The key words here are "in the sun's frame of reference," because in Jupiter's rest frame the trajectory of the spacecraft can be turned, but there can be no net change in speed before and after the encounter. When the spacecraft is still far enough from Jupiter such that Jupiter's gravitational potential energy can be neglected, the spacecraft has some initial speed v_0 in Jupiter's rest frame. As it approaches Jupiter, the spacecraft speeds up, the trajectory is bent, and the spacecraft then slows down again as it leaves Jupiter, once again approaching speed v_0. In Jupiter's own rest frame, Jupiter cannot cause a net increase in the spacecraft's speed because of energy conservation.

However, because of the deflection of the spacecraft, its speed *can* increase in the *sun's* rest frame, and this increased speed therefore gives the spacecraft a larger total energy in the sun's frame, perhaps enough to project it much farther out into the solar system.

Consider a special case to see how this works. Figure 7.15(a) shows a picture of a spacecraft's trajectory in the rest frame of Jupiter. The spacecraft is in a hyperbolic orbit about Jupiter, entering from below the picture and being turned by (we will suppose) a $90°$ angle by Jupiter. It enters with speed v_0 from below, and exits at the same speed v_0 toward the left. It has gained no energy in Jupiter's frame. Figure 7.15(b) shows the same trajectory drawn in the sun's frame of reference. In the sun's frame, Jupiter is moving toward the left with speed v_J, so the spacecraft's speed when it enters from beneath Jupiter (*i.e.*, as it travels away from the sun, which is much farther down in the figure) can be found by vector addition: it is $v_{\text{initial}} = \sqrt{v_0^2 + v_J^2}$, since v_0 and v_J are perpendicular to one another. However, the spacecraft's speed when it *leaves* Jupiter is $v_{\text{final}} = v_0 + v_J$, since in this case the vectors are parallel to one another. Obviously $v_{\text{final}} > v_{\text{initial}}$; the spacecraft has been sped up in the sun's frame of reference, so that it has more energy than before in that frame.

Clearly the trajectory must be tuned very carefully to get the right angle of flyby so that the spacecraft will be thrown in the correct direction and with the correct speed to reach its final destination.

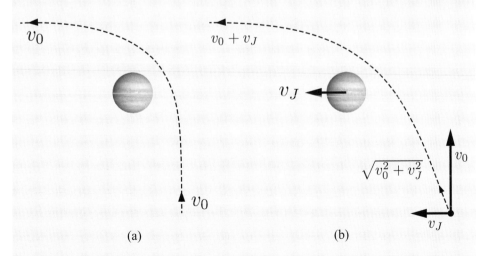

(a)   (b)

Fig. 7.15   A spacecraft flies by Jupiter, in the reference frames of (a) Jupiter and (b) the sun.■

7.7 The Virial Theorem in Astrophysics

We end this chapter with the virial theorem, which has wide applications through-out classical mechanics and statistical mechanics, although here we emphasize gravitational motion in astrophysical systems. Consider a collection of N point-like nonrelativistic particles, where the ith particle is at position $\mathbf{r}_i$, has momentum $\mathbf{p}_i$, and is subject to a net force $\mathbf{F}_i$. Then define a quantity $G \equiv \sum_i \mathbf{p}_i \cdot \mathbf{r}_i$, whose time derivative is

$$\frac{dG}{dt} = \sum_i \dot{\mathbf{p}}_i \cdot \mathbf{r}_i + \sum_i \mathbf{p}_i \cdot \dot{\mathbf{r}}_i = \sum_i \mathbf{F}_i \cdot \mathbf{r}_i + \sum_i m_i \dot{\mathbf{r}}_i \cdot \dot{\mathbf{r}}_i, \qquad (7.79)$$

where in the second equality we have used Newton's second law $\mathbf{F}_i = \dot{\mathbf{p}}_i$ and the fact that $\mathbf{p}_i = m\dot{\mathbf{r}}_i$ for nonrelativistic particles.

The final term on the right is $\sum_i m_i \dot{\mathbf{r}}_i \cdot \dot{\mathbf{r}}_i = \sum_i m_i \mathbf{v}_i \cdot \mathbf{v}_i = \sum_i m_i v_i^2 = 2T$, twice the total kinetic energy of the system. Therefore, so far we have found that

$$\frac{dG}{dt} = \sum_i \mathbf{F}_i \cdot \mathbf{r}_i + 2T. \qquad (7.80)$$

Our next step is to take the time average of each term in the equation, which we denote by brackets $< >$. Then, by definition of the time average over a total time τ, the time average of dG/dt is

$$\left\langle \frac{dG}{dt} \right\rangle \equiv \frac{1}{\tau} \int_0^\tau \frac{dG}{dt}\, dt = \frac{1}{\tau}[G(\tau) - G(0)] = \left\langle \sum_i \mathbf{F}_i \cdot \mathbf{r}_i \right\rangle + 2\langle T \rangle. \qquad (7.81)$$

Now if all motions in a problem are periodic with period τ, then $G(\tau) = G(0)$, so the left-hand side of the equation is zero. This is true for a single particle in a circular or elliptical orbit in a central gravitational or spring force, for example, if τ is the orbital period. Much more generally, suppose all motions are at least bounded, with an upper limit to G. Then, over a long period of time, *i.e.*, as τ becomes very large, the left-hand side of the equation becomes negligible. In that case

$$\langle T \rangle = -\frac{1}{2} \left\langle \sum_i \mathbf{F}_i \cdot \mathbf{r}_i \right\rangle \qquad (7.82)$$

for bounded systems in the limit of large times. This is the *virial theorem*.[7]

The theorem as written is valid for any number of nonrelativistic point particles, subject to arbitrary forces. Now let us temporarily specialize to a *single* particle subject to a conservative force, for which it is possible to write $\mathbf{F} = -\nabla U$, where

---

[7] The right-hand side of the equation is called the "virial of Clausius." The term "virial" derives from the Latin word *vis*, which has many meanings, including *force, strength, power,* and *energy,* among others.

U is the corresponding potential energy. Furthermore, suppose that the force is central, so $F = -dU/dr$. Then the virial theorem for that particle becomes

$$\langle T \rangle = -\frac{1}{2} \langle -\nabla U \cdot \mathbf{r} \rangle = -\frac{1}{2} \langle (-dU/dr)r \rangle. \tag{7.83}$$

With a potential energy of the form $U = \alpha r^{n+1}$, corresponding to a central power-law force $F = -dU/dr = -\alpha(n+1)r^n$, we have

$$\langle T \rangle = -\frac{1}{2} \langle \alpha r^n \cdot r \rangle = -\frac{1}{2} \langle \alpha r^{n+1} \rangle = \frac{n+1}{2} \langle U \rangle. \tag{7.84}$$

In particular, for a central gravitational force on an orbiting particle, like the earth orbiting the sun, we have $n = -2$, so $\langle T \rangle = -\frac{1}{2} \langle U \rangle$. It is easy to check that this is correct for the circular orbit of a planet around the sun, where time averages are unnecessary. It is not as easy to check for elliptical orbits, but the theorem still holds.

Example 7.5

Gravitating Systems of Particles

Now consider N particles that pull on one another with central gravitational forces, so all forces on a particle are due to other particles in the system. This might be a good approximation for the gravitational attractions of stars on one another in a globular cluster, for example, or for entire galaxies attracting one another in a cluster of galaxies like the Coma Cluster or the Virgo Cluster.

Let us start simply, by considering the case $N = 3$. The force of particle 2 on particle 1 is $\mathbf{F}_{12}$, and the force of particle 1 on particle 3 is $\mathbf{F}_{31}$, and so on. Therefore counting all six interactions:

$$\sum_{i=1}^{3} \mathbf{F}_i \cdot \mathbf{r}_i = \mathbf{F}_{12} \cdot \mathbf{r}_1 + \mathbf{F}_{13} \cdot \mathbf{r}_1 + \mathbf{F}_{21} \cdot \mathbf{r}_2 + \mathbf{F}_{23} \cdot \mathbf{r}_2 + \mathbf{F}_{31} \cdot \mathbf{r}_3 + \mathbf{F}_{32} \cdot \mathbf{r}_3. \tag{7.85}$$

However, by Newton's third law the force of particle 1 on particle 2 is equal but opposite to the force of particle 2 on particle 1, so for example $\mathbf{F}_{21} = -\mathbf{F}_{12}$. Therefore we can write more simply

$$\sum_{i=1}^{3} \mathbf{F}_i \cdot \mathbf{r}_i = \mathbf{F}_{12} \cdot (\mathbf{r}_1 - \mathbf{r}_2) + \mathbf{F}_{13} \cdot (\mathbf{r}_1 - \mathbf{r}_3) + \mathbf{F}_{23} \cdot (\mathbf{r}_2 - \mathbf{r}_3). \tag{7.86}$$

Now $(\mathbf{r}_1 - \mathbf{r}_2)$ is the vector that points directly from m_2 to m_1 and whose magnitude is the distance between these two particles, so

$$\mathbf{F}_{12} = -\left(\frac{dU_{12}}{dr}\right)\frac{(\mathbf{r}_1 - \mathbf{r}_2)}{|\mathbf{r}_1 - \mathbf{r}_2|}, \tag{7.87}$$

the product of the magnitude of the force $-dU_{12}/dr$ and a unit vector pointing from m_2 to m_1. Here $U_{12} = -Gm_1m_2/r_{12}$ is the mutual gravitational potential energy between the two stars or galaxies, where $r_{12} \equiv |\mathbf{r}_1 - \mathbf{r}_2|$. Therefore we find the surprisingly simple result

$$\mathbf{F}_{12} \cdot (\mathbf{r}_1 - \mathbf{r}_2) = -\left(\frac{dU_{12}}{dr}\right)r_{12} = -\frac{Gm_1m_2}{r_{12}^2}r_{12} = U_{12}. \tag{7.88}$$

Summing the terms, the virial theorem for three gravitating particles is simply

$$\langle T \rangle = -\frac{1}{2} \langle U_{12} + U_{13} + U_{23} \rangle = -\frac{1}{2} \langle U \rangle , \tag{7.89}$$

where U is the total gravitational potential energy of the three particles. *And by inspecting this derivation it is clear that the number of particles in the system makes no difference!* The virial theorem in the form just written should apply regardless of the number of particles, as long as T is their total kinetic energy and U is their total gravitational potential energy. If we have four particles, for example, we would have not three potential energies but six, corresponding to all possible pairs, but they would still sum to the total potential energy.

The first evidence for the existence of so-called "dark matter" in the universe was found by the Swiss astrophysicist Fritz Zwicky (1898–1974) in 1933, using the virial theorem in the context of the Coma Cluster of galaxies. He estimated the masses of galaxies in the cluster by their brightness, their velocities by their Doppler shifts,[a] and their distances apart by triangulation, knowing the distance of the cluster from us. In that way he estimated the average kinetic energies of the galaxies and their mutual gravitational potential energies as well. He found that the virial theorem was strongly violated, in that the apparent total kinetic energy was many times greater than the magnitude of the apparent potential energy, suggesting that the cluster should be flying apart, and could not be the more-or-less contained system it appeared to be. He reasoned, however, that if there were additional unseen mass within the cluster, then the cluster's total kinetic energy would scale linearly with mass, while its potential energy, which contains two factors of m, would scale quadratically. So a large amount of unseen matter in the cluster could bring the cluster into line with the virial theorem. This unseen matter is what we now call "dark matter," just as Zwicky did in 1933.   ∎

---

[a] The Doppler effect can give only the velocities parallel to our line of sight, say v_z, while the other components of velocity are unknown. He assumed random motions, however, after subtracting out the center-of-mass velocity of the cluster away from us due to the overall expansion of the universe. Therefore on average one expects $< v^2 > = < v_x^2 > + < v_y^2 > + < v_z^2 > = 3 < v_z^2 >$.

7.8 Summary

In this chapter we have treated in generality two-body problems involving a so-called central force law – a force that depends only on the distance between the two bodies and lies along the line joining them. This is a particularly important class of problems for two reasons. First, it encompasses the dynamics due to two of the most important forces of Nature: gravity and (as we shall see) the electrostatic force as well. Second, it involves nontrivial dynamical situations where the problem can be solved *exactly* through conservation laws. We will not be so lucky in most situations in mechanics.

The Newtonian theory of gravity has been enormously successful in predicting the motions of stars, planets, moons, comets, and spacecraft, even of entire galaxies of stars. It has been the most important testing ground of Newtonian mechanics itself. Nevertheless, in spite of its successes, the Newtonian theories of mechanics and gravitation fail the test of relativity, so must be replaced by a fully relativistic theory when speeds are close to that of light, and when we probe distances very close to large masses. A very brief introduction to that beautiful theory, Einstein's general theory of relativity, is presented in Chapter 10. Before we go there, however, we need to understand the motion of particles in electromagnetic fields and in non-inertial reference frames.

Problems

★   **Problem 7.1** Two satellites of equal mass are each in a circular orbit around the earth. The orbit of satellite A has radius r_A, and the orbit of satellite B has radius $r_B = 2r_A$. Find the ratio of their (a) speeds, (b) periods, (c) kinetic energies, (d) potential energies, (e) total energies.

★   **Problem 7.2** Halley's Comet passes through earth's orbit every 76 years. Make a close estimate of the maximum distance Halley's Comet gets from the sun.

★★   **Problem 7.3** Two astronauts are in the same circular orbit of radius R around the earth, 180° apart. Astronaut A has two cheese sandwiches, while Astronaut B has none. How can A throw a cheese sandwich to B? In terms of the astronaut's period of rotation about the earth, how long does it take the sandwich to arrive at B? What is the semi-major axis of the sandwich's orbit? (There are many solutions to this problem, assuming that A can throw the sandwich with arbitrary speeds.)

★   **Problem 7.4** Suppose that the gravitational force exerted by the sun on the planets were inverse r-squared, but not proportional to the planet masses. Would Kepler's third law still be valid in this case?

★   **Problem 7.5** Planets in a hypothetical solar system all move in circular orbits, and the ratio of the periods of any two orbits is equal to the ratio of their orbital radii *squared*. How does the central force depend on the distance from this sun?

★   **Problem 7.6** An astronaut is marooned in a powerless spaceship in circular orbit around the asteroid Vesta. The astronaut reasons that puncturing a small hole through the spaceship's outer surface into an internal water tank will lead to a jet action of escaping water vapor expanding into space. Which way should the jet be aimed so the spacecraft will most likely reach the surface of Vesta? Assume that the initial orbital radius is much larger than the radius of Vesta. (In Isaac Asimov's first published story *Marooned off Vesta*, the jet was oriented differently, but the ship reached the surface anyway.)

★   **Problem 7.7** A thrown baseball travels in a small piece or an elliptical orbit before it strikes the ground. What is the semi-major axis of the ellipse? (Neglect air resistance.)

★   **Problem 7.8** Assume that the period of elliptical orbits around the sun depends only upon G, M (the sun's mass), and a, the semi-major axis of the orbit. Prove Kepler's third law using dimensional arguments alone.

★   **Problem 7.9** A spy satellite designed to peer closely at a particular house every day at noon has a 24-h period, and a perigee of 100 km directly above the house. What is the altitude of the satellite at apogee? (Earth's radius is 6400 km.)

★★   **Problem 7.10** Show that the kinetic energy

$$K.E = \frac{1}{2}m_1 \dot{r}_1^2 + \frac{1}{2}m_2 \dot{r}_2^2$$

of a system of two particles can be written in terms of their center-of-mass velocity $\dot{\mathbf{R}}_{cm}$ and relative velocity $\dot{\mathbf{r}}$ as

$$K.E. = \frac{1}{2}M\dot{\mathbf{R}}_{cm}^2 + \frac{1}{2}\mu\dot{\mathbf{r}}^2,$$

where $M = m_1 + m_2$ is the total mass and $\mu = m_1 m_2/M$ is the reduced mass of the system.

★★   **Problem 7.11** Show that the shape $r(\varphi)$ for a central spring-force ellipse takes the standard form $r^2 = a^2 b^2/(b^2 \cos^2 \varphi + a^2 \sin^2 \varphi)$ if (in Eq. (7.37)) we use the plus sign in the denominator and choose $\varphi_0 = \pi/4$.

★   **Problem 7.12** Show that the period of a particle that moves in a circular orbit close to the surface of a sphere depends only upon G and the average density ρ of the sphere. Find what this period would be for *any* sphere having an average density equal to that of water. (The sphere consisting of the planet Jupiter nearly qualifies!)

★   **Problem 7.13** (a) Communication satellites are placed into geosynchronous orbits; that is, they typically orbit in earth's equatorial plane, with a period of 24 h. What is the radius of this orbit, and what is the altitude of the satellite above earth's surface? (b) A satellite is to be placed in a synchronous orbit around the planet Jupiter to study the famous "red spot." What is the altitude of this orbit above the "surface" of Jupiter? (The rotation period of Jupiter is 9.9 h, its mass is about 320 earth masses, and its radius is about 11 times that of earth.)

★   **Problem 7.14** The perihelion and aphelion of the asteroid Apollo are 0.964×10^8 km and 3.473×10^8 km from the sun, respectively. Apollo therefore swings in and out through earth's orbit. Find (a) the semi-major axis, (b) the period of Apollo's orbit in years, given earth's semi-major axis $a_E = 149.6 \times 10^6$ km. (Apollo is only one of many "Apollo asteroids" that cross earth's orbit. Some have struck the earth in the past, and others will strike it in the future unless we find a way to prevent it.)

★★ **Problem 7.15** The time it takes for a probe of mass μ to move from one radius to another under the influence of a central spring force was shown in the chapter to be

$$t(r) = \pm \sqrt{\frac{\mu}{2}} \int_{r_0}^{r} \frac{r \, dr}{\sqrt{Er^2 - kr^4/2 - \ell^2/2\mu}}, \tag{7.90}$$

where E is the energy, k is the spring constant, and ℓ is the angular momentum. Evaluate the integral in general, and find (in terms of given parameters) how long it takes the probe to go from the maximum to the minimum value of r.

★★ **Problem 7.16** (a) Evaluate the integral in Eq. (7.29) to find $t(r)$ for a particle moving in a central gravitational field. (b) From the results, derive the equation for the period $T = (2\pi/\sqrt{GM})a^{3/2}$ in terms of the semi-major axis a for particles moving in elliptical orbits around a central mass.

★★ **Problem 7.17** The sun moves at a speed $v_S = 220$ km/s in a circular orbit of radius $r_S = 30,000$ light years around the center of the Milky Way galaxy. The earth requires $T_E = 1$ year to orbit the sun, at a radius of 1.50×10^{11} m. (a) Using this information, find a formula for the total mass responsible for keeping the sun in its orbit, as a multiple of the sun's mass M_0, in terms also of the parameters v_S, r_S, T_E, and r_E. Note that G is not needed here! (b) Find this mass numerically.

★★ **Problem 7.18** The two stars in a double-star system circle one another gravitationally, with period P. If they are suddenly stopped in their orbits and allowed to fall together, show that they will collide after a time $P/4\sqrt{2}$.

★★ **Problem 7.19** A particle is subjected to an attractive central spring force $F = -kr$. Show, *using Cartesian coordinates*, that the particle moves in an elliptical orbit, with the force center at the *center* of the ellipse, rather than at one focus of the ellipse.

★★ **Problem 7.20** Use Eq. (7.32) to show that if the central force on a particle is $F = 0$, the particle moves in a straight line.

★★ **Problem 7.21** Find the central force law $F(r)$ for which a particle can move in a spiral orbit $r = k\theta^2$, where k is a constant.

★★ **Problem 7.22** Find two second integrals of motion for a particle of mass m in the case $F(r) = -k/r^3$, where k is a constant. Describe the shape of the trajectories, assuming that the angular momentum $\ell > \sqrt{km}$.

★★★ **Problem 7.23** A particle of mass m is subject to a central force $F(r) = -GMm/r^2 - k/r^3$, where k is a positive constant. That is, the particle experiences an inverse-cubed attractive force as well as a gravitational force. Show that if k is less than some limiting value, the motion is that of a precessing ellipse. What is this limiting value, in terms of m and the particle's angular momentum?

★★   **Problem 7.24** Find the allowed orbital shapes for a particle moving in a *repulsive* inverse-square central force. These shapes would apply to α-particles scattered by gold nuclei, for example, due to the repulsive Coulomb force between them.

★★   **Problem 7.25** A particle moves in the field of a central force for which the potential energy is $U(r) = kr^n$, where both k and n are constants, positive, negative, or zero. For what range of k and n can the particle move in a stable, circular orbit at some radius?

★★   **Problem 7.26** A particle of mass m and angular momentum ℓ moves in a central spring-like force field $F = -kr$. (a) Sketch the effective potential energy $U_{\mathrm{eff}}(r)$. (b) Find the radius r_0 of circular orbits. (c) Find the period of small oscillations about this orbit, if the particle is perturbed slightly from it. (d) Compare with the period of rotation of the particle about the center of force. Is the orbit closed or open for such small oscillations?

★★   **Problem 7.27** Find the period of small oscillations about a circular orbit for a planet of mass m and angular momentum ℓ around the sun. Compare with the period of the circular orbit itself. Is the orbit open or closed for such small oscillations?

★★   **Problem 7.28** (a) A binary star system consists of two stars of masses m_1 and m_2 orbiting about one another. Suppose that the orbits of the two stars are circles of radii r_1 and r_2, centered on their center of mass. Show that the period of the orbital motion is given by

$$T^2 = \frac{4\pi^2}{G(m_1 + m_2)}(r_1 + r_2)^3.$$

(b) The binary system Cygnus X-1 consists of two stars orbiting about their common center of mass with orbital period 5.6 days. One of the stars is a supergiant with mass 25 times that of the sun. The other star is believed to be a black hole with mass about 10 times the mass of the sun. From the information given, determine the distance between these stars, assuming that the orbits are circular.

★★   **Problem 7.29** A spacecraft is in a circular orbit of radius r about the earth. What is the minimum Δv the rocket engines must provide to allow the craft to escape from the earth, in terms of G, M_E, and r?

★★   **Problem 7.30** A spacecraft departs from the earth. Which takes less rocket fuel: to escape from the solar system or to fall into the sun?(Assume the spacecraft has already escaped from the earth, and do not include possible gravitational assists from other planets.)

★★   **Problem 7.31** After the engines of a 100 kg spacecraft have been shut down, the spacecraft is found to be a distance 10^7 m from the center of the earth, moving with a speed of 7000 m/s at an angle of $45°$ relative to a straight line from the earth to the spacecraft. (a) Calculate the total energy and angular momentum of the spacecraft. (b) Determine the semi-major axis and the eccentricity of the spacecraft's geocentric trajectory.

★★   **Problem 7.32** A 100 kg spacecraft is in circular orbit around the earth, with orbital radius 10^4 km and speed 6.32 km/s. It is desired to turn on the rocket engines to accelerate the spacecraft up to a speed so that it will escape the earth and coast out to Jupiter. Use a value of 1.5×10^8 km for the radius of earth's orbit, 7.8×10^8 km for Jupiter's orbital radius, and 30 km/s for the velocity of the earth. Determine (a) the semi-major axis of the Hohmann transfer orbit to Jupiter; (b) the travel time to Jupiter; (c) the heliocentric velocity of the spacecraft as it leaves the earth; (d) the minimum Δv required from the engines to inject the spacecraft into the transfer orbit.

★★   **Problem 7.33** The earth–sun L5 Lagrange point is a point of stable equilibrium that trails the earth in its heliocentric orbit by $60°$ as the earth (and spacecraft) orbit the sun. Some gravity-wave experimenters want to set up a gravity-wave experiment at this point. The simplest trajectory from earth puts the spacecraft on an elliptical orbit with a period slightly longer than 1 year, so that when the spacecraft returns to perihelion, the L5 point will be there. (a) Show that the period of this orbit is 14 months. (b) What is the semi-major axis of this elliptical orbit? (c) What is the perihelion speed of the spacecraft in this orbit? (d) When the spacecraft finally reaches the L5 point, how much velocity will it have to lose (using its engines) to settle into a circular heliocentric orbit at the L5 point?

★★★   **Problem 7.34** In *Stranger in a Strange Land*, Robert Heinlein claims that travelers to Mars spent 258 days on the journey out, the same for return, "plus 455 days waiting at Mars while the planets crawled back into positions for the return orbit." Show that travelers *would* have to wait about 455 days, if both earth–Mars journeys were by Hohmann transfer orbits.

★★   **Problem 7.35** A spacecraft approaches Mars at the end of its Hohmann transfer orbit. (a) What is its velocity in the sun's frame, before Mars's gravity has had an appreciable influence on it? (b) What Δv must be given to the spacecraft to insert it directly from the transfer orbit into a circular orbit of radius 6000 km around Mars?

★★   **Problem 7.36** A spacecraft parked in circular low-earth orbit 200 km above the ground is to travel out to a circular geosynchronous orbit, of period 24 h, where it will remain. (a) What initial Δv is required to insert the spacecraft into the transfer orbit? (b) What final Δv is required to enter the synchronous orbit from the transfer orbit?

★★   **Problem 7.37** A spacecraft is in a circular parking orbit 300 km above earth's surface. What is the transfer-orbit travel time out to the moon's orbit, and what are the two Δvs needed? Neglect the moon's gravity.

★★   **Problem 7.38** A spacecraft is sent from the earth to Jupiter by a Hohmann transfer orbit. (a) What is the semi-major axis of the transfer ellipse? (b) How long does it take the spacecraft to reach Jupiter? (c) If the spacecraft actually leaves from a circular parking orbit around the earth of radius 7000 km, find the rocket Δv required to insert the spacecraft into the transfer orbit.

★★★ **Problem 7.39** Find the Hohmann transfer-orbit time to Venus, and the Δvs needed to leave an earth parking orbit of radius 7000 km and later to enter a parking orbit around Venus, also of $r = 7000$ km. Sketch the journey, showing the orbit directions and the directions in which the rocket engine must be fired.

★ **Problem 7.40** Consider an astronaut standing on a weighing scale within a spacecraft. The scale by definition reads the normal force exerted by the scale on the astronaut (or, by Newton's third law, the force exerted on the scale by the astronaut). By the principle of equivalence, the astronaut can't tell whether the spacecraft is (a) sitting at rest on the ground in uniform gravity g, or (b) is in gravity-free space, with uniform acceleration a numerically equal to the gravity g in case (a). Show that in one case the measured weight will be proportional to the inertial mass of the astronaut, and in the other case proportional to the astronaut's gravitational mass. So if the principle of equivalence is valid, these two types of mass must have equal magnitudes.

★★★ **Problem 7.41** In **Bertrand's theorem**. Section 7.5, we stated a powerful theorem that asserts that the only potentials for which all bounded orbits are closed are $U_{\text{eff}} \propto r^2$ and $U_{\text{eff}} \propto r^{-1}$. To prove this theorem, let us proceed in steps. If a potential is to have bound orbits, the effective potential must have a minimum since a bound orbit is a dip in the effective potential. The minimum is at $r = R$ given by

$$U'(R) = \frac{\ell^2}{\mu R^3},\tag{7.91}$$

as shown in Eq. (7.22). This corresponds to a circular orbit which is stable if

$$U''(R) + \frac{3}{R}U'(R) > 0,\tag{7.92}$$

as shown in Eq. (7.23). Consider perturbing this circular orbit so that we now have an $r_{\min}$ and an $r_{\max}$ about $r = R$. Define the apsidal angle $\Delta\varphi$ as the angle between the point on the perturbed orbit at $r_{\min}$ and the point at $r_{\max}$. Assume $(R - r_{\min})/R \ll 1$ and $(r_{\max} - R)/R \ll 1$. Note that closed orbits require

$$\Delta\varphi = 2\pi\frac{m}{n}\tag{7.93}$$

for integer m and n and for all R.

- Show that

$$\Delta\varphi = \pi\sqrt{\frac{U'(R)}{3\,U'(R) + R\,U''(R)}}.\tag{7.94}$$

Notice that the argument under the square root is always positive by virtue of the stability of the original circular orbit.

- In general, any potential $U(r)$ can be expanded in terms of positive and negative powers of r, with the possibility of a logarithmic term

$$U(r) = \sum_{n=-\infty}^{\infty} \frac{a_n}{r^n} + a \ln r. \tag{7.95}$$

Show that, to have the apsidal angle independent of r, we must have $U(r) \propto r^{-\alpha}$ for $\alpha < 2$ and $\alpha \neq 0$, or $U(r) \propto \ln r$. Show that the value of $\Delta\varphi$ is then

$$\Delta\varphi = \frac{\pi}{\sqrt{2-\alpha}}, \tag{7.96}$$

where the logarithmic case corresponds to $\alpha = 0$ in this equation.

- Show that if $\lim_{r\to\infty} U(r) = \infty$, we must have $\lim_{E\to\infty} \Delta\varphi = \pi/2$. This corresponds to the case $\alpha < 0$. We then must have

$$\Delta\varphi = \frac{\pi}{\sqrt{2-\alpha}} = \frac{\pi}{2}, \tag{7.97}$$

or $\alpha = -2$, thus proving one of the two cases of the theorem.

- Show that for the case $0 \leq \alpha$, we can consider $\lim_{E\to 0} \Delta\varphi = \pi/(2-\alpha)$. This then implies

$$\Delta\varphi = \frac{\pi}{\sqrt{2-\alpha}} = \frac{\pi}{2-\alpha}, \tag{7.98}$$

which leaves only the possibility $\alpha = 1$, completing the proof of the theorem.

★★ **Problem 7.42** The luminous matter we observe in our Milky Way galaxy is only about 5% of the galaxy's total mass: the rest is called "dark matter," which seems to act upon all matter gravitationally but in no other way. As a rough approximation, we can therefore neglect luminous matter entirely as a source of gravity in understanding the dynamics of the galaxy. Along with many other stars, both much closer and much farther from the galactic center, we all circle about the center of the galaxy with about the same velocity 220 km/s. Our solar system in particular is 8.5 kiloparsecs (kpsc) from the galactic center (1 psc = 3.26 light years). (a) From this information, how must the dark-matter density ρ for this range of orbital radii depend upon r, the distance of an orbiting star from the galactic center? (b) The dark-matter density in the vicinity of the sun is thought to be $\rho_0 \simeq 0.3$ GeV/c^2 per cm^3. Assuming now that the radial dependence of density $\rho(r)$ found in part (a) is valid all the way to the center of the galaxy, what is the total mass of dark matter within the orbit of our sun as a multiple of one solar mass, where $M_{\text{sun}} = 2 \times 10^{30}$ kg? (c) Suppose several rogue stars are in highly noncircular orbits around the galactic center, perhaps as a result of collisions with one another. Which (if any) of Kepler's laws would then still be correct for these stars? Explain. (d) Consider a proposal that the radial dependence of dark-matter density as found in part (a) might still be valid for arbitrarily large distances from the center. Show that in fact this is *not* possible, and explain why.

★   **Problem 7.43** Within the solar system itself it is often thought that the density of unseen dark matter is quite uniform, with mass density $\rho_0 \simeq 0.3$ GeV/c^2 per cm^3 (the mass equivalent of about one proton per three cubic centimeters). The sun itself has mass $M_0 = 2 \times 10^{30}$ kg. (a) What fraction of a solar mass within the radius of earth's orbit might one expect in the form of dark matter? (The average radius of earth's orbit is 150×10^6 km.) (b) Would Kepler's second law still be valid for orbits of comets within the solar system? Explain. (c) Would Kepler's third law still be valid for the planets?

★   **Problem 7.44** Communications satellites are typically placed in orbits of radius r_{CS} circling the earth once per day. The 24 or so GPS (Global Positioning System) satellites are placed in one of six orbital planes, with each satellite circling the earth *twice* per day. (a) Find the radius of their orbits as a fraction of r_{CS}. (b) Low-earth orbit satellites typically have orbital periods of about 90 minutes. Find their radii as a fraction of r_{CS}.

★★★   **Problem 7.45** Trajectory specialists plan to send a spacecraft to Saturn requiring a gravitational assist by Jupiter. In Jupiter's rest frame the spacecraft's velocity will be turned $90°$ as it flies by, as illustrated in Figure 7.15(a). (a) If the nearest point on the spacecraft's path is a distance of $2R_J = 140,000$ km from the center of Jupiter, how fast (in km/s) is the spacecraft's speed in Jupiter's frame when it is at this nearest point? (b) In Jupiter's frame what is v_0 (as shown in the figure), the spacecraft's speed (in km/s) both long before and long after its encounter with Jupiter (but not so long before or after that its distance from the sun has changed appreciably)? (c) Note that long after the encounter, in the sun's frame the velocity of the spacecraft is $v_0 + v_J$ along the direction of Jupiter's motion around the sun, as illustrated in Figure 7.15(b). Is this velocity sufficient to take the spacecraft out to the orbit of Saturn? Explain. (Useful data: Jupiter has mass $M_J = 1.9 \times 10^{27}$ kg and radius $R_J = 70,000$ km. Its average orbital radius and velocity around the sun are about 780×10^6 km and 13 km/s, respectively. The average orbital radius of Saturn is 1.4×10^9 km. Newton's gravitational constant is $G = 6.67 \times 10^{-11}$ m^3/(kg s^2).)

★   **Problem 7.46** Show that the virial theorem is correct for a planet in circular orbit around the sun.

★★   **Problem 7.47** Show that the virial theorem is correct for a particle of mass m free to move in a plane, and attached to one end of a Hooke's-law spring exerting force $F = -kr$, if the particle is in (a) a circular orbit, (b) an elliptical orbit.

★★   **Problem 7.48** Suppose that in studying a particular globular cluster containing 10^5 stars, whose average mass is that of our sun, astronomers find that the total kinetic energy of the stars is 10 times that of the magnitude of their total potential energy. (a) Estimate the amount of dark matter in the cluster, expressed in solar masses. (b) Then assume there is no such thing as dark matter, but that the potential energy between two stars has the form $U(r) = -\alpha r^n$, where $n \neq -1$. Is there a value of n such that the virial theorem would be satisfied without dark matter?

★★ **Problem 7.49** An object named "Oumuamua" entered the inner solar system in 2017, coming closest to the sun on Sept 9, 2017, at a distance 0.255 AU from the sun, and with speed 87.7 km/s. (1 AU = 149, 598, 000 km is the average distance of the earth from the sun.) It was first observed through a telescope on Hawaii, hence it was given a Hawaiian name, which means "scout," or "first distant messenger." (a) Find the eccentricity of the orbit. What is the orbit's shape? (b) What was (and what will be) the limiting speed of the object when it is very far from the sun? Express the answer in meters/second and also in light years/million years. (c) If the object left a newly forming star system 1000 light years away from us, how long ago was the star system formed? (Although objects originating in the solar system can sometimes achieve eccentricities slightly greater than unity, due to gravitational-assist-type encounters with planets or other orbiting objects, the eccentricity of Oumuamua is too large to have been achieved in this way. Therefore astronomers believe that the object originated in another star system. There are probably many of these objects entering our solar system every year, but Oumuamua was the first we have observed.)

★★ **Problem 7.50** The cover of this book shows the paths of a number of stars orbiting a massive object named Sagittarius A-Star (Sgr A* for short) at the center of our Milky Way galaxy. One of these stars, called "S2," has an orbit whose period is 16.05 years, a semi-major axis of 970 AU, and a periastron distance of 120 AU. Assuming that S2 follows a Keplerian elliptical orbit, what is its orbital (a) eccentricity and (b) semi-minor axis? (c) Most importantly, according to these observations, what is the mass of Sgr A*, expressed as a multiple of the mass of the sun? (Note that 1 AU is the average distance of the earth from the sun.)

★★ **Problem 7.51** Using the observed characteristics of star S2's orbit as given in the preceding problem, and assuming it moves in a Keplerian elliptical orbit, find the speed of the star at periastron as a percentage of the speed of light. (b) If its orbit happened to be oriented so that at periastron S2 was moving directly towards the earth, by what factor would its light be blueshifted due to the Doppler effect?

8

Electromagnetism

While gravity was the first of the fundamental forces to be quantified and at least partially understood – beginning all the way back in the seventeenth century – it took an additional 200 years for physicists to unravel the secrets of a second fundamental force, the electromagnetic force. Ironically, it is the electromagnetic force that is by far the stronger of the two, and at least as prevalent in our daily lives. The fact that atoms and molecules stick together to form the matter we are made of, the contact forces we feel when we touch objects around us, and virtually all modern technological advances of the twentieth century, all these rely on the electromagnetic force. In this chapter, we introduce the subject within the Lagrangian formalism and demonstrate some familiar as well as unfamiliar aspects of this fascinating fundamental force of Nature.

8.1 Gravitation Revisited

Before we dive headfirst into the topic of motion caused by electromagnetic forces, which can be expressed in terms of electric and magnetic *fields*, we will begin by describing the somewhat simpler notion of gravitational forces proportional to a *gravitational* field.

In the preceding chapter we discussed Newtonian gravity as a force between two particles of masses m and M, separated by distance r, as given by

$$\mathbf{F} = -G\frac{Mm}{r^2}\hat{\mathbf{r}}. \tag{8.1}$$

It would seem that a "source" mass M somehow reaches out through space to pull on a "probe" mass m, in a kind of "action at a distance."[1] An alternative view, conceived in the nineteenth century, is that the source M first sets up a gravitational field $\mathbf{g}$ in the space surrounding it, where

$$\mathbf{g} = -G\frac{M}{r^2}\hat{\mathbf{r}}. \tag{8.2}$$

Then, if a point probe particle m is placed in this field, it experiences a force

$$\mathbf{F}_g = m\mathbf{g}. \tag{8.3}$$

---

[1] And vice versa. Which is the source and which the probe is arbitrary.

This force on m is obviously still given by Newton's inverse-square law of gravity, but there are a couple of advantages to introducing the field $\mathbf{g}$. First, we no longer have to swallow the idea of direct "action at a distance": rather, m responds only to the value of $\mathbf{g}$ at its own location.[2] Second, if the probe particle m is removed, and replaced by a *different* probe particle m', we do not have to start from scratch to calculate the force on it. The field at that point has not changed, so all we have to do to calculate the force is multiply $\mathbf{g}$ by m' instead of m.

If there is more than one source particle, the total gravitational field they produce is just the vector sum of the gravitational fields due to each source particle individually; that is, Newtonian gravitational fields add linearly, which is often called the *principle of superposition*. For example, we may wish to find the gravitational field due to an entire star. We conceptually divide up the star into small bits, compute the field due to each bit, and sum up all these small vector fields to account for the contribution from the entire star.

We can quantify the source mass distribution through a volume *mass density* function ρ_{M} that is not necessarily constant. In terms of ρ_{M}, the gravitational field obeys two differential equations, called the Newtonian gravitational field equations. They are

$$\nabla \cdot \mathbf{g} = -4\pi G \rho_{\mathrm{M}} \quad \text{and} \quad \nabla \times \mathbf{g} = 0. \tag{8.4}$$

The first of these states that the divergence of $\mathbf{g}$ is proportional to the mass density and is *negative*: the gravitational field points inward toward the source mass (i.e., it is *convergent* rather than *divergent*). The second equation states that the gravitational field has no curl, *i.e.*, it is irrotational: it dives directly into its sources, without added twists.

To see why this leads to the picture of gravity arising from (8.1), we can check that for $\rho_{\mathrm{M}} = M\delta^3(\mathbf{r})$, a source point mass M, one recovers the Newtonian force between two masses M and m (8.1). Here $\delta^3(\mathbf{r})$ is a three-dimensional **delta function**, which is infinite at $\mathbf{r} = 0$, zero elsewhere, and whose volume integral is unity. The total mass of the source mass distribution is then

$$\int \rho_{\mathrm{M}} dV = M \int \delta^3(\mathbf{r}) dV = M, \tag{8.5}$$

as expected. We then integrate the divergence equation from (8.4) over all space:

$$\int \nabla \cdot \mathbf{g}\, dV = -4\pi G \int \rho_{\mathrm{M}} dV. \tag{8.6}$$

Then we use Gauss's law while employing the spherical symmetry of the problem, as depicted in Figure 8.1. This gives

$$\int \nabla \cdot \mathbf{g}\, dV = \oint \mathbf{g} \cdot d\mathbf{A} = \oint (g\hat{\mathbf{r}}) \cdot (r^2 \sin\theta d\theta d\varphi\, \hat{\mathbf{r}}) = 4\pi r^2 g = -4\pi G M, \tag{8.7}$$

---

[2] Similarly, mass m sets up its own gravitational field $\mathbf{g}$ proportional to m in the space around itself, which exerts a force $m\mathbf{g}$ on M, which again is the same as Newton's law of gravity. The idea is that point particles do not feel the gravitational field they themselves produce, but only the field caused by other particles.

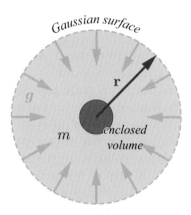

Fig. 8.1 A Gaussian surface probing the gravity around a point mass.

where we have used a spherical Gaussian surface centered at the source mass. This gives

$$\mathbf{g} = -\frac{GM}{r^2}\hat{\mathbf{r}} \quad \Rightarrow \quad \mathbf{F}_{\mathrm{g}} = M\mathbf{g} = -\frac{GMm}{r^2}\hat{\mathbf{r}}, \qquad (8.8)$$

as promised. Equations (8.3) and (8.4) are linear in $\mathbf{g}$ and ρ_{M}, so the gravitostatic field from (say) two point masses can be correctly built up from this point mass result by summing the fields from each individual point mass. And from this, we can expand to three, four, or infinitely many point masses. Hence, Eqs. (8.3) and (8.4) constitute an equivalent description of Newtonian gravity to that of (8.1) and the force superposition principle.

Note also that Eqs. (8.4) are differential equations that satisfy the existence and uniqueness theorem of differential equations: given a mass density ρ_{M} and appropriate boundary conditions for $\mathbf{g}$, there is a unique solution for $\mathbf{g}$. We then conclude that, armed with (8.1), we can in principle tackle any gravitostatic problem involving any complicated mass distribution. Let us see how our motivational problem of a probe inside a star could be handled using this general methodology.

Example 8.1 **Gravity Inside the Body of a Star**

Consider a spherical star of radius R, mass M, and constant volume density $\rho_0 = M/(4\pi R^3/3)$. We would like to find the force felt by a small probe of mass m that penetrates the star. Starting from Eq. (8.4), we integrate over a spherical Gaussian surface of radius $r < R$, as shown in Figure 8.2:

$$\int \nabla \cdot \mathbf{g}\, dV = -4\pi G \int \rho_0 dV = -4\pi G\, \rho_0 \left(\frac{4}{3}\pi r^3\right) = -4\pi G M \frac{r^3}{R^3}. \qquad (8.9)$$

That is, only the fraction of the mass *inside* the sphere of radius r contributes. The left-hand side is handled through **Gauss's law:**

$$\int \nabla \cdot \mathbf{g} \, dV = \oint \mathbf{g} \cdot d\mathbf{A} = 4\pi r^2 g. \tag{8.10}$$

We then have

$$\mathbf{g} = -GM\frac{r}{R^3}\hat{\mathbf{r}}. \tag{8.11}$$

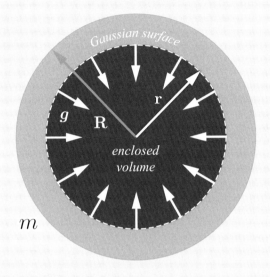

Fig. 8.2   A Gaussian surface probing the gravity inside a star of uniform volume mass density.

This gives a radial force on the probe of mass m:

$$\mathbf{F}_g = -GmM\frac{r}{R^3}\hat{\mathbf{r}}, \tag{8.12}$$

which increases with r until we reach the surface at $r = R$. Note that $\mathbf{F}_g = 0$ at the center of the star, and $\mathbf{F}_g = -GmM/R^2\hat{\mathbf{r}}$ at the star's surface.

Outside the star, the computation proceeds in a similar manner except that the Gaussian surface now encompasses the entire star mass. We then get the point particle result

$$\int \nabla \cdot \mathbf{g} \, dV = \oint \mathbf{g} \cdot d\mathbf{A} = 4\pi r^2 g = -4\pi Gm \Rightarrow \mathbf{g} = -\frac{GM}{r^2}\hat{\mathbf{r}}. \tag{8.13}$$

As far as gravity is concerned, the star looks like a point mass from the outside. Note that at the star surface $r = R$, the two gravitostatic fields (8.11) and (8.13) match up.   ∎

There is another enormously useful mathematical refinement in Newtonian gravitation theory that has important counterparts in electromagnetism, as we shall see in the next section. The gravitational field is irrotational, as expressed mathematically by $\nabla \times \mathbf{g} = 0$. Combining that with the universal vector identity

$\nabla \times \nabla\varphi = 0$, valid for *any* scalar field φ, allows us to express any gravitational field **g** in terms of a **gravitational scalar potential** φ by the equation

$$\mathbf{g} \equiv -\nabla\varphi_g. \tag{8.14}$$

That is, the gravitational field is the negative gradient of the gravitational scalar potential. Then the requirement that $\nabla \times \mathbf{g} = 0$ is automatically satisfied, and the other field equation $\nabla \cdot \mathbf{g} = -4\pi G \rho_{\mathrm{M}}$ can now be written in the form

$$\nabla^2\varphi_g = 4\pi G\rho_{\mathrm{M}}, \tag{8.15}$$

a differential equation known as **Poisson's equation**. Given a mass density ρ_{M}, we can try to solve Poisson's equation for the gravitational potential. If we are successful, the gravitational field is then given by $\mathbf{g} = -\nabla\varphi_g$ and the force on a probe m is $\mathbf{F} = m\mathbf{g}$.

This is the way in which gravitational fields are generally calculated for given mass densities. We may be able to solve Poisson's equation (a second-order partial differential equation) analytically (i.e., in terms of known functions), which usually requires that the mass distribution is quite symmetric. Or if we can't solve it analytically, in principle we can always solve it numerically with the help of a digital computer. Then from φ_g we can find **g**, and from **g** we can find the force on a probe mass m.

Before leaving Newtonian gravitational fields and potentials, we note in passing that the Newtonian gravitational field equations, either for **g** or φ_g, are not relativistically invariant, and so cannot be the correct fundamental law of gravitation, according to special relativity. This is seen most easily by noting that the spatial and time coordinates are not treated comparably in the theory. For example, the operator ∇^2 in Poisson's equation contains spatial derivatives but *no time derivatives*. So if we were to use a Lorentz transformation to change Poisson's equation in one (unprimed) frame to a new (primed) frame, we would gain some time derivatives in the new frame; therefore, the scalar potential in the new frame would not obey Poisson's equation. That field equation is therefore not covariant; it would look different in different inertial frames, so could not be a fundamental law of Nature. We return to the question of a relativistic theory of gravity in the capstone Chapter 10.

8.2 The Lorentz Force Law

In the mid-nineteenth century, Scottish physicist James C. Maxwell (1831–1879) combined the results of many experimental observations, having to do with currents, magnets, and even fuzzy cats, and formulated a set of equations describing a new force law, the **electromagnetic force**. Maxwell's equations can be written in Gaussian units as

$$\nabla \cdot \mathbf{E} = 4\pi \rho_Q, \quad \nabla \times \mathbf{E} = -\frac{1}{c}\frac{\partial \mathbf{B}}{\partial t},$$

$$\nabla \cdot \mathbf{B} = 0, \quad \nabla \times \mathbf{B} = \frac{4\pi}{c}\mathbf{J} + \frac{1}{c}\frac{\partial \mathbf{E}}{\partial t}. \tag{8.16}$$

Here $\mathbf{E}$ and $\mathbf{B}$ are the space time-dependent electric and magnetic vector fields. Through perfect vacuum they can relay a force between objects that carry an attribute called **electric charge**. The quantities ρ_Q and $\mathbf{J}$ are the charge density and current density of the charged stuff that is causing the corresponding $\mathbf{E}$ and $\mathbf{B}$ fields. For example, a single, isolated, and stationary point particle of electric charge Q located at a position $\mathbf{r}_0$ has charge and current densities

$$\rho_Q = Q\,\delta^3(\mathbf{r} - \mathbf{r}_0), \quad \mathbf{J} = 0, \tag{8.17}$$

where $\delta^3(\mathbf{r} - \mathbf{r}_0)$ is the three-dimensional delta function centered at $\mathbf{r} = \mathbf{r}_0$. The corresponding electric and magnetic fields are obtained from Maxwell's equations (8.16) as (with $\mathbf{R} \equiv \mathbf{r} - \mathbf{r}_0$)

$$\mathbf{E} = \frac{Q}{|\mathbf{R}|^2}\hat{\mathbf{R}}, \quad \mathbf{B} = 0, \tag{8.18}$$

known as the **Coulomb field**. This computation is very similar to the case of the gravitostatic field of a point mass encountered in the previous section.

Another classic example is that of a charge Q moving with constant velocity $\mathbf{v}$. At a position $\mathbf{R}$ away from the charge, one finds the **Biot–Savart** magnetic field

$$\mathbf{B} = \frac{Q\mathbf{v} \times \mathbf{R}}{c\,R^3}. \tag{8.19}$$

Given a probe particle of electric charge q in the presence of electric and magnetic fields – generated by other nearby charges described by ρ_Q and $\mathbf{J}$ – the force that the probe particle feels is given by

$$\mathbf{F}_{\text{em}} = q\mathbf{E} + \frac{q}{c}\mathbf{v} \times \mathbf{B}, \tag{8.20}$$

known as the **Lorentz force**. For example, taking the environment of the probe as consisting of the point charge Q of Eq. (8.17) and (8.18), the probe feels a force given by

$$\mathbf{F} = \frac{qQ}{|\mathbf{R}|^2}\hat{\mathbf{R}}, \tag{8.21}$$

where $\mathbf{R} \equiv \mathbf{r} - \mathbf{r}_0$ and the probe charge q is located at $\mathbf{r}$ while the source charge Q is located at $\mathbf{r}_0$ (see Figure 8.3).

This force law looks very familiar. Remember that the *gravitational* force experienced by a probe mass m located at $\mathbf{r}$ in the vicinity of a source mass M at $\mathbf{r}_0$ is given by

$$\mathbf{F} = -G\frac{mM}{|\mathbf{R}|^2}\hat{\mathbf{R}}, \tag{8.22}$$

so instead of the product $-GmM$ in Eq. (8.22), the strength of the electromagnetic force is proportional to the product of the charges Qq as in Eq. (8.21). The rest, the inverse-square distance law, is the same. This is not a coincidence: both forces have a geometrical origin and are tightly constrained by similar symmetries of Nature.

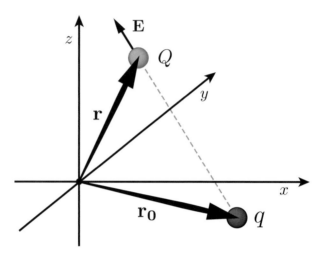

Fig. 8.3   The electrostatic Coulomb force between two charged particles.

Equations (8.16) consist of eight differential equations for the six field components $(E_x, E_y, E_z, B_x, B_y, B_z)$ tucked within $\mathbf{E}$ and $\mathbf{B}$ – sourced from some charge distribution described by ρ_Q and $\mathbf{J}$. An existence and uniqueness theorem in the theory of differential equations guarantees that, given ρ_Q and $\mathbf{J}$, Eqs. (8.16) always determine $\mathbf{E}$ and $\mathbf{B}$ uniquely. In turn, each of the charges making up ρ_Q and $\mathbf{J}$ experiences the Lorentz force (8.20) and evolves accordingly; which in turn influences the electric and magnetic fields via (8.16). Hence, we have a coupled set of differential equations for $\mathbf{E}$, $\mathbf{B}$, and the position of the charges – equations that have to be solved simultaneously, at least in principle.

In practice, this is a very hard problem. Fortunately, in many practical circumstances there is a clear separation of roles between the charges involved in the electromagnetic interactions. Some of the charges – called the *source* charges – have fixed and given dynamics and can be used to compute the electric and magnetic fields in a region of interest: that is, given ρ_Q and $\mathbf{J}$, we can use (8.16) to find $\mathbf{E}$ and $\mathbf{B}$. The remaining charges of interest are called *probe* charges. The electromagnetic fields they generate are negligible compared to those from the source charges, and their dynamics is described by the Lorentz force law (8.20) with given $\mathbf{E}$ and $\mathbf{B}$ *background* fields from the sources. This approximation scheme decouples the set of differential equations into two separate, more tractable, sets: (i) we first find the background $\mathbf{E}$ and $\mathbf{B}$ fields from ρ_Q and $\mathbf{J}$; (ii) then determine the trajectory of probes in the given $\mathbf{E}$ and $\mathbf{B}$ – ignoring the effects of the probe on these background fields. In this text, we focus on the second problem: the mechanics of probe charges in the background of *given* electric and magnetic fields.

The task at hand is then the following. Given some **E** and **B** fields, we want to study the dynamics of point charges using the Lagrangian formalism. That is, we want to incorporate the Lorentz force law (8.20) into a Lagrangian – which involves a variational principle.

We start by rewriting the background electric and magnetic fields in terms of new fields that make the underlying symmetries of Maxwell's equations more apparent. From (8.16), we know that the magnetic field is divergenceless. This implies that we can write

$$\mathbf{B} = \nabla \times \mathbf{A}, \tag{8.23}$$

due to the identity $\nabla \cdot (\nabla \times \mathbf{V}) = 0$ for *any* vector field **V** – thus trading the **B** field for a new vector field **A** called the **vector potential**. Using (8.16) also, we can therefore write

$$\nabla \times \left(\mathbf{E} + \frac{1}{c} \frac{\partial \mathbf{A}}{\partial t} \right) = 0, \tag{8.24}$$

which implies that

$$\mathbf{E} + \frac{1}{c} \frac{\partial \mathbf{A}}{\partial t} = -\nabla \phi \Rightarrow \mathbf{E} = -\nabla \phi - \frac{1}{c} \frac{\partial \mathbf{A}}{\partial t} \tag{8.25}$$

due to the general identity $\nabla \times \nabla f = 0$ for *any* function f – thus introducing a new *scalar* field ϕ which we call the **scalar potential**. Hence, we have traded the *six* fields in **E** and **B** for *four* fields in **A** and ϕ. The fact that we can do so is a reflection of a deep and foundational symmetry underlying the electromagnetic force law. Furthermore, given **E** and **B**, even **A** and ϕ are not unique! We can apply the following transformations to **A** and ϕ without changing **E** and **B** – and hence without affecting the force law and corresponding physics:

$$\mathbf{A} \to \mathbf{A} + \nabla f, \quad \phi \to \phi - \frac{1}{c} \frac{\partial f}{\partial t}, \tag{8.26}$$

for any arbitrary function $f(t, \mathbf{r})$. Thus, there is even less physical information in **A** and ϕ than it may seem. In fact, one can show that when the dust settles we are left with only *two* physical fields quantifying electrodynamics: from the original six fields in **E** and **B**, to the four in **A** and ϕ, now to only two. This remarkable symmetry of the theory is called **gauge symmetry**. Indeed, it is possible to *derive* classical electromagnetism – Maxwell's equations and the Lorentz force law – based solely on the principles of relativity and gauge symmetry.

Example 8.2

Potentials of a Point Charge

The electromagnetic fields from a point charge were given in Eq. (8.18). We can now find the scalar and vector potentials for a point charge – noting that the answer is not unique due to the gauge symmetry (8.26). Since $\mathbf{B} = 0$, we can choose

$$\mathbf{A} = 0 \tag{8.27}$$

satisfying Eq. (8.23). Note, however, that we could have chosen for $\mathbf{A}$ a vector field of the form $\mathbf{A} = \nabla f$ for an *arbitrary* function f because of the identity $\nabla \times \nabla f = 0$. The $\mathbf{A} = 0$ choice corresponds to the simplest case, for which $f = $ constant. To find the scalar potential ϕ we need to do more work. Starting from Eq. (8.25):

$$\mathbf{E} = -\nabla\phi - \frac{1}{c}\frac{\partial \mathbf{A}}{\partial t} = -\nabla\phi, \tag{8.28}$$

we can integrate both sides along *any* path between two points a and b:

$$\int_a^b \mathbf{E} \cdot d\mathbf{r} = -\int_a^b \nabla\phi \cdot d\mathbf{r} = -\int_a^b d\phi = -\phi(\mathbf{r}_b) + \phi(\mathbf{r}_a). \tag{8.29}$$

This then gives

$$\phi(R) = \frac{q}{R} \tag{8.30}$$

up to an arbitrary constant, using (8.18). The easiest way to see this is to use a simple radial path of integration. This potential is shown in Figure 8.4(a).

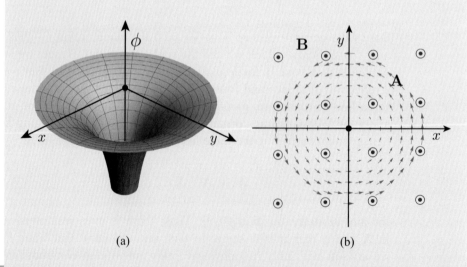

(a)   (b)

Fig. 8.4 (a) The electric potential ϕ from a point charge; only the dependence on x and y is shown. (b) The vector potential of a uniform magnetic field pointing out of the page. ∎

Example 8.3

Vector Potential for a Uniform Magnetic Field

Consider a region of space filled with a uniform magnetic field $\mathbf{B} = B_0\hat{\mathbf{z}}$, where B_0 is a constant, and arbitrary static electric field $\mathbf{E}$. We write $\mathbf{E} = -\nabla\phi$ from Eq. (8.25) and look for a *time-independent* vector potential $\mathbf{A}$ satisfying Eq. (8.23). Again, the choice is not unique. A particularly useful choice is

$$\mathbf{A} = \frac{1}{2}\mathbf{B} \times \mathbf{r}. \tag{8.31}$$

We can verify that the curl of this vector potential is indeed $\mathbf{B}$, as long as $\mathbf{B}$ is uniform. To do this, simply use the vector identity

$$\nabla \times (\mathbf{V}_1 \times \mathbf{V}_2) = \mathbf{V}_1(\nabla \cdot \mathbf{V}_2) - \mathbf{V}_2(\nabla \cdot \mathbf{V}_1) + (\mathbf{V}_2 \cdot \nabla)\mathbf{V}_1 - (\mathbf{V}_1 \cdot \nabla)\mathbf{V}_2. \tag{8.32}$$

For our case of a magnetic field pointing in the z direction, we then get

$$\mathbf{A} = -\frac{1}{2}By\hat{\mathbf{x}} + \frac{1}{2}Bx\hat{\mathbf{y}}. \tag{8.33}$$

This is a vector field circling about the z axis counterclockwise, lying in the x–y plane (see Figure 8.4(b)). ■

8.3 The Lagrangian for Electromagnetism

Maxwell wrote out his elegant equations based primarily on hundreds of experiments carried out by many physicists on the behavior of electric charges and currents, as well as the resulting concepts of electric and magnetic fields invented by Michael Faraday and others. Maxwell presumed his equations to be valid in some *inertial* frame of reference. Then one of his greatest achievements was to demonstrate that the equations predict waves of electric and magnetic fields that travel at the speed of light c, thus showing that light and electromagnetic waves are one and the same, resulting in a tremendous unification of electricity, magnetism, and optics in a single set of equations. This was the greatest single achievement of nineteenth-century physics.[3]

Following publication of Maxwell's equations in 1861, physicists entered a decades-long period of confusion about which inertial frame it was in which Maxwell's equations are valid. As we saw already in Chapter 2, it was not until Einstein published his theory of special relativity in 1905 that it was firmly established that Maxwell's equations are valid in *all* inertial frames, with the

---

[3] The reader is referred to any good textbook on electromagnetism for the derivation of light waves from Maxwell's equations; see, for example, *Classical Electrodynamics* by Roald K. Wangsness.

counterintuitive result that the speed of light must also be the same in all inertial frames.

The question we now want to address is the following: if an inertial observer $\mathcal{O}$ measures an electric field $\mathbf{E}$ and a magnetic field $\mathbf{B}$, what are the electric field $\mathbf{E}'$ and magnetic field $\mathbf{B}'$ as measured by an observer $\mathcal{O}'$ – moving as usual with constant velocity $\mathbf{V}$ along the common x or x' axis? We assume that the transformations relating these fields are linear in the fields, much like the Lorentz transformation of four-position or four-velocity; we also expect that they are linear in the relative velocity $\mathbf{V}$. We start with expressions of the form

$$\mathbf{E}' = \hat{\mathbf{a}}_1 \cdot \mathbf{E} + \hat{\mathbf{a}}_2 \cdot \mathbf{B}, \quad \mathbf{B}' = \hat{\mathbf{a}}_3 \cdot \mathbf{E} + \hat{\mathbf{a}}_4 \cdot \mathbf{B}, \tag{8.34}$$

where the $\hat{\mathbf{a}}_i$ are four 3×3 matrices whose components can depend on $\mathbf{V}$ at most linearly. We also require that the Lorentz force law (8.20) fits as the last three components of a four-force (2.119), as seen from Chapter 2. With all these conditions in place, one finds a unique solution for the $\hat{\mathbf{a}}_i$s. One can show that, given $\mathbf{E}$ and $\mathbf{B}$ as measured by an inertial observer $\mathcal{O}$, another inertial observer $\mathcal{O}'$, moving with respect to $\mathcal{O}$ with velocity $\mathbf{V}$, measures different electric and magnetic fields $\mathbf{E}'$ and $\mathbf{B}'$ given by

$$\mathbf{E}' = \gamma \left(\mathbf{E} + \frac{\mathbf{V}}{c} \times \mathbf{B} \right) + (1 - \gamma) \frac{\mathbf{E} \cdot \mathbf{V}}{V^2} \mathbf{V}, \tag{8.35}$$

$$\mathbf{B}' = \gamma \left(\mathbf{B} - \frac{\mathbf{V}}{c} \times \mathbf{E} \right) + (1 - \gamma) \frac{\mathbf{B} \cdot \mathbf{V}}{V^2} \mathbf{V}. \tag{8.36}$$

These rather complicated relations become greatly simplified when written in terms of the potentials ϕ, $\mathbf{A}$ and ϕ', $\mathbf{A}'$. Introduce a four-vector

$$A^\mu = (\phi, \mathbf{A}). \tag{8.37}$$

Then one can show that we have simply

$$A^{\mu'} = \Lambda^{\mu'}_\mu A^\mu, \tag{8.38}$$

where $\Lambda^{\mu'}_\mu$ is the usual Lorentz transformation matrix of Chapter 2. Therefore the information about the electromagnetic fields is now packaged in a four-vector field A^μ that transforms in a simple manner under Lorentz transformations.

We are now ready to develop a variational principle for the electromagnetic force law using Lorentz symmetry as a guiding principle. We start with the familiar relativistic action for a free point mass m:

$$S = -m c^2 \int d\tau = -m c^2 \int dt \sqrt{1 - \frac{v^2}{c^2}}. \tag{8.39}$$

Now we want to add a term to this action such that the particle experiences the Lorentz force law (8.20) as if it had a charge q in some background ϕ and $\mathbf{A}$ fields; and this combined action must be Lorentz invariant.[4] Looking back at the Lorentz force law (8.20), noting in particular that it is linear in the particle velocity and the background fields, there is only one Lorentz-invariant integral consistent with these statements and Lorentz symmetry:

$$\int A^\mu \eta_{\mu\nu} \frac{dx^\nu}{d\tau} d\tau \equiv \int A^\mu \eta_{\mu\nu} dx^\nu. \tag{8.40}$$

Adding an appropriate multiplicative constant, we get the full action

$$S = -m\,c^2 \int d\tau + \frac{q}{c} \int A^\mu \eta_{\mu\nu} dx^\nu. \tag{8.41}$$

We can rewrite the second term in some particular (unprimed) frame:

$$\frac{q}{c} \int A^\mu \eta_{\mu\nu} dx^\nu = \frac{q}{c} \int A^\mu \eta_{\mu\nu} \frac{dx^\nu}{dt} dt = \frac{q}{c} \int \left(-\phi\, c\, dt + \mathbf{A} \cdot \frac{d\mathbf{r}}{dt} dt \right)$$

$$= q \int dt \left(-\phi + \mathbf{A} \cdot \frac{\mathbf{v}}{c} \right), \tag{8.42}$$

so the full action becomes

$$S = -m\,c^2 \int dt\, \sqrt{1 - \frac{v^2}{c^2}} + q \int dt \left(-\phi + \mathbf{A} \cdot \frac{\mathbf{v}}{c} \right). \tag{8.43}$$

We leave it as an exercise for the reader to check that the resulting equations of motion from (8.41) reproduce the Lorentz force law (8.20):

$$\frac{d\mathbf{p}}{dt} = \frac{d}{dt} \left(\gamma_v m\, \mathbf{v} \right) = q \left(\mathbf{E} + \frac{\mathbf{v}}{c} \times \mathbf{B} \right) = \mathbf{F}_{\text{em}}. \tag{8.44}$$

We thus have a Lagrangian formulation of the electromagnetic force.

In the nonrelativistic limit, the action (8.43) becomes

$$L = \frac{1}{2} m \dot{r}^i \dot{r}^i - q\phi + \frac{q}{c} A^i \dot{r}^i, \tag{8.45}$$

with equations of motion

$$m\mathbf{a} = q\mathbf{E} + \frac{q}{c} \mathbf{v} \times \mathbf{B}, \tag{8.46}$$

after dropping terms quadratic in v/c while keeping linear terms. This set of nonrelativistic equations will be our focus in the next several examples.

---

[4] In fact, as mentioned earlier in the book, Lorentz invariance of the action we are about to write was the guiding principle for *developing* relativity and the associated Lorentz transformations. It may appear as a chicken-and-egg problem; in reality, one should think of the Lorentz symmetry and gauge symmetry as the fundamental requirements, with the physical consequences being relativity and electromagnetism.

8.4 The Two-Body Problem, Once Again

Consider now the familiar two-body problem once again for electromagnetic rather than gravitational interactions. Two point particles of masses m_1 and m_2, located at $\mathbf{r}_1$ and $\mathbf{r}_2$, respectively, carry electric charges q_1 and q_2. Let us first roughly estimate the relative importance of the various forces involved. Electromagnetic fields propagate with the speed of light in vacuum. This implies that if the particles are moving slowly compared to the speed of light, we may think of the electromagnetic fields cast about them as propagating essentially instantaneously – always reflecting their instantaneous positions. The electric field $\mathbf{E}$ from the point charge q_2 a distance $r = |\mathbf{r}_1 - \mathbf{r}_2|$ away generates a force on q_1 of the order $F_{\text{el}} = q_1 E \sim q_1 q_2 / r^2$. If q_2 is moving with a speed $v_2 \ll c$ while q_1 is moving with $v_1 \ll c$, q_1 experiences a magnetic force $F_{\text{m}} \sim q_1 v_1 B / c \sim q_1 q_2 v_1 v_2 / c^2 r^2$, where we used the Biot–Savart law from (8.19). Accelerating charges radiate electromagnetic energy which can add a level of complication. However, once again, this effect is much smaller in our nonrelativistic regime. Finally, the two masses interact gravitationally with a force of the order of $F_g \sim G m_1 m_2 / r^2$. Putting things together, we summarize

$$F_{\text{el}} \; : \; F_m \; : \; F_g \; \sim \; \frac{q_1 q_2}{r^2} \; : \; \frac{q_1 q_2 v_1 v_2}{c^2 r^2} \; : \; \frac{G m_1 m_2}{r^2}. \tag{8.47}$$

Since $v_1 v_2 \ll c^2$, we see that the magnetic force between the charges is less than the electric force by a factor v^2 / c^2. To compare the electrical and gravitational forces, consider the case of two electrons with mass $m \simeq 9 \times 10^{-28}$ g and charge $q \simeq 5 \times 10^{-10}$ esu. We get an estimate for $F_e \; : \; F_g \sim 1 : 10^{-43}$. Phrasing things gently, we need not care about gravitational forces! Gravity normally becomes relevant only when we are dealing with macroscopic quantities of *electrically neutral* matter.

The conclusion of all this is that, in the current nonrelativistic regime, we care only about the *electrostatic* force acting between the two point charges! So, using, (8.45) we can write the Lagrangian as

$$L = \frac{1}{2} m_1 \dot{\mathbf{r}}_1^2 + \frac{1}{2} m_2 \dot{\mathbf{r}}_2^2 - U(r), \tag{8.48}$$

where $U(r) = q_1 \phi(r)$ and $\phi(r)$ is the scalar potential due to source charge q_2 at the location of the probe charge q_1:

$$\phi = \frac{q_2}{|\mathbf{r}_1 - \mathbf{r}_2|} = \frac{q_2}{r}. \tag{8.49}$$

This gives the electric potential energy

$$U(r) = \frac{q_1 q_2}{r}. \tag{8.50}$$

We now have a familiar two-body problem with a central potential, so we can import the entire machinery developed in Chapter 7. We will not repeat the analysis, but instead step through the main benchmarks. We first factor away the trivial center-of-mass motion and write a Lagrangian for the interesting relative motion tracked by $\mathbf{r} \equiv \mathbf{r}_1 - \mathbf{r}_2$:

$$L = \frac{1}{2}\mu\dot{\mathbf{r}}^2 - \frac{q_1 q_2}{r}, \tag{8.51}$$

where $\mu = m_1 m_2 / (m_1 + m_2)$ is the reduced mass. This quickly leads to a one-dimensional problem in the radial direction, with fixed energy

$$E = \frac{1}{2}\mu\dot{r}^2 + U_{\text{eff}}(r) \tag{8.52}$$

and effective potential

$$U_{\text{eff}}(r) = \frac{l^2}{2\mu r^2} + \frac{q_1 q_2}{r}, \tag{8.53}$$

where l is the conserved angular momentum. All is very similar to the problem of two point masses interacting gravitationally, except for the following important observations:

- If $q_1 q_2 < 0$, *i.e.*, the charges are of opposite signs, the electric force between the point particles is *attractive*. Our entire analysis of orbits and trajectories from the gravitational analogue goes through with the simple substitution

$$-G m_1 m_2 \rightarrow q_1 q_2 \tag{8.54}$$

in all equations. We will then find closed orbits consisting of circles and ellipses, and open orbits consisting of parabolas and hyperbolas. For example, the radius of a stable circular orbit is given by[5]

$$r_p = \frac{|G m_1 m_2|}{2|E|} \rightarrow \frac{|q_1 q_2|}{2|E|}. \tag{8.55}$$

For a hydrogen atom with energy $|E| \simeq 13.6$ eV, we find $r_p \sim 10^{-10}$ m, a good estimate for the size of the ground state of the atom.
- If $q_1 q_2 > 0$, the electric force is *repulsive*. Unlike the gravitational force, the electromagnetic force then allows for repulsive effects. Let us look at this case more closely. When $q_1 q_2 > 0$, the formalism developed in Chapter 7 still goes through, except we need to be careful with certain signs. Figure 8.5 depicts the effective potential of the problem, noting that the dip in the potential has disappeared – signaling the absence of bound orbits.
The orbit trajectory becomes

$$r = \frac{\left(l^2 / G m_1 m_2 \right) (1/m_1)}{1 + \epsilon \cos\theta} \rightarrow \frac{\left(l^2 / q_1 q_2 m_1 \right)}{1 + \epsilon \cos\theta}, \tag{8.56}$$

---

[5] See Eq. (7.59) with $\epsilon = 0$.

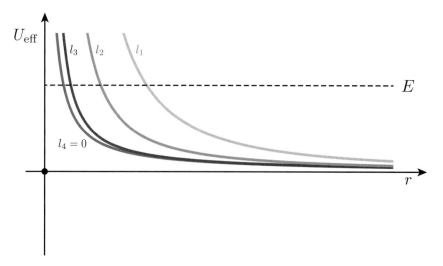

The effective Coulombic potential between two charges of the same sign for various angular momenta $l_1 > l_2 > l_3 > l_4 = 0$.

with eccentricity

$$\epsilon = \sqrt{1 + \frac{2El^2}{\left(G^2 m_1^2 m_2^2\right) m_1}} \rightarrow \sqrt{1 + \frac{2El^2}{\left(q_1^2 q_2^2\right) m_1}}. \tag{8.57}$$

However, since $q_1 q_2 > 0$, we see from Figure 8.5 that we necessarily have $E > 0$, and hence

$$\epsilon > 1. \tag{8.58}$$

This implies that now there are only hyperbolic trajectories. Obviously, with a repulsive force, we may not have bound orbits. The interesting physics problem becomes that of **particle scattering**.

8.5 Coulomb Scattering

Consider a point charge q_1 projected with some initial energy from infinity onto point charge q_2, as shown in Figure 8.6. We say that the repulsive electrostatic force from q_2 "scatters" the probe of charge q_1 at an angle that can be read off from Eq. (8.56) as

$$r \rightarrow \infty \Rightarrow \cos \theta_{1,2} \rightarrow -\frac{1}{\epsilon}, \tag{8.59}$$

as depicted in the figure.

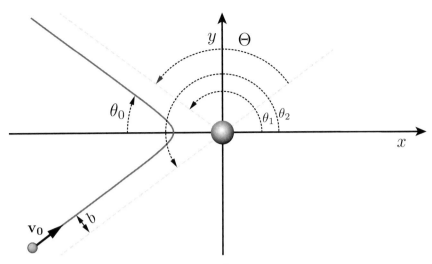

Hyperbolic trajectory of a probe scattering off a charged target.

The scattering angle Θ, as shown in the figure, is defined by

$$\theta_1 - \theta_2 = 2\theta_0 = \pi - \Theta, \tag{8.60}$$

which gives

$$\cos\Theta = \frac{2El^2/m_1 - q_1^2 q_2^2}{2El^2/m_1 + q_1^2 q_2^2}, \tag{8.61}$$

using Eq. (8.57). It is convenient to write the angular momentum l in terms of the so-called **impact parameter** b shown in the figure. Note that b would be the distance of closest approach between the probe and the target if there were no force between them. Looking at the initial configuration at $r \to \infty$, we have the angular momentum

$$l = m_1 v_0 b = b\sqrt{2\,m_1 E}, \tag{8.62}$$

where v_0 is the initial speed of the probe, related to the constant energy

$$E = \frac{1}{2}mv_0^2, \tag{8.63}$$

evaluated here at initial infinite separation. We then get

$$\cos\Theta = \frac{4\,b^2 E^2 - q_1^2 q_2^2}{4\,b^2 E^2 + q_1^2 q_2^2}. \tag{8.64}$$

Using the trigonometric identity

$$\cot^2\frac{\Theta}{2} = \frac{(1 + \cos\Theta)^2}{(1 - \cos\Theta)^2}, \tag{8.65}$$

we can simplify this expression further to

$$\cot \frac{\Theta}{2} = \frac{2\,bE}{q_1 q_2}.$$

(8.66)

This relation gives us the scattering deflection angle that a point charge q_1 would experience if projected from infinity with energy E and impact parameter b onto another point charge q_2. A scattering process such as this one is a powerful experimental probe into atomic structure and was instrumental in discovering the constituents of atoms. Put simply, one uses the electromagnetic force to poke into the electrically charged universe of the atom – by throwing charged particles at it.

While the initial energy E of the probe can be controlled, the impact parameter b is in practice impossible to measure on a per scattering atom basis. It is therefore useful to describe scattering processes through a quantity called the **scattering cross-section**. We focus on the change in the area within which a probe scatters in relation to a change in the impact parameter, as shown in Figure 8.7.

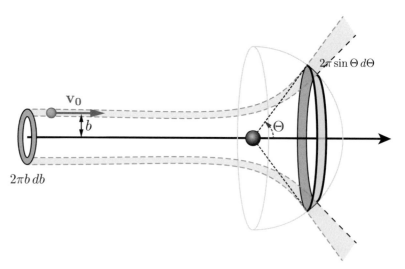

Fig. 8.7 Definition of the scattering cross-section in terms of change in impact area $2\pi b\,db$ and scattering area $2\pi \sin \Theta d\Theta$ on the unit sphere centered at the target.

The scattering cross-section $\sigma(\Theta)$ is defined as the change in initial impact area per change in scattering area on a unit sphere centered at the target:

$$\sigma(\Theta) \equiv \left| \frac{2\pi b\, db}{2\pi \sin \Theta d\Theta} \right| = \frac{b}{\sin \Theta} \left| \frac{db}{d\Theta} \right|.$$

(8.67)

We can then think of a stream of incident probe particles falling onto the target at various *unknown* impact parameters b with a *uniform* distribution in b. Then $\sigma(\Theta)$ is proportional to the probability of finding a scattered probe at an angle Θ on the unit sphere centered at the target:

$$\sigma(\Theta) \propto \text{Probability of scattering at angle } \Theta.$$

(8.68)

This probability can be measured by counting the number of particles scattered per incident particle – without needing to know the impact parameter b for every individual particle scattering. In the celebrated case of Coulomb scattering described in this example, one gets, using Eq. (8.66), the so-called **Rutherford scattering** cross-section[6]

$$\sigma(\Theta) = \frac{1}{4} \left(\frac{q_1^2 q_2^2}{2E} \right)^2 \csc^4 \frac{\Theta}{2}. \tag{8.69}$$

As shown in Figure 8.8, this probability is sharply peaked in the forward, $\Theta = 0$, direction.

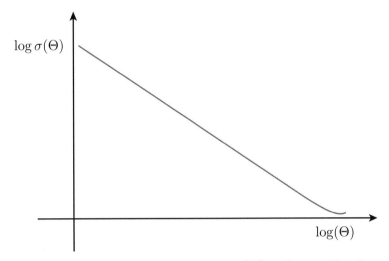

Fig. 8.8   The Rutherford scattering cross-section. The graph shows $\log \sigma(\Theta)$ as a function of $\log \Theta$.

It also depends strongly on the charges through a quartic power. Measuring the cross-section $\sigma(\Theta)$, and given the initial probe energy E, we can for example determine the charge of the target. A scattering process is effectively a way to looking into atoms using charges – much like looking into say neutral biological tissue using scattered light and a microscope.

Example 8.4   Snell Scattering

As an illustration of the concept of scattering cross-section, consider light scattered by a perfectly polished bead whose surface acts like a mirror. The bead's radius is R, as shown in Figure 8.9. We want to find the scattering cross-section of this bead as parallel light falls on it.

---

[6] It was the great twentieth-century experimental physicist Ernest Rutherford who derived this equation, and used it to show by experiment that there is a tiny, heavy, positively charged nucleus at the heart of every atom.

We start from Eq. (8.67). We then need to find $b(\Theta)$ using Snell's law. Looking at the figure, note that

$$b(\Theta) = R \sin\theta = R \sin\left(\frac{\pi}{2} - \frac{\Theta}{2}\right) = R \cos\left(\frac{\Theta}{2}\right). \tag{8.70}$$

Using this in Eq. (8.67), we get

$$\sigma(\Theta) = \frac{1}{4} b R \csc\frac{\Theta}{2} = \frac{1}{4}\frac{bR}{\sin\theta} = \frac{1}{4}R^2 = \frac{1}{4\pi}\left(\pi R^2\right). \tag{8.71}$$

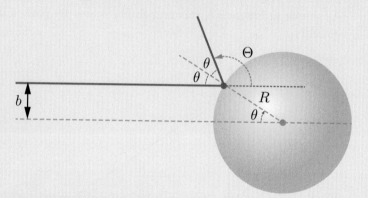

Scattering of light off a reflecting bead.

The *total* scattering cross-section is therefore

$$\sigma_T = \int_0^{2\pi} \int_0^{\pi} \sigma(\Theta) \sin\Theta d\Theta d\Phi = 4\pi \frac{1}{4\pi}\left(\pi R^2\right) = \pi R^2, \tag{8.72}$$

which is simply the cross-sectional area of the bead. This is because the scattering occurs on contact. As the interaction of the in-falling probe with the target becomes more long-ranged, the total scattering cross-section increases: the cross-section size seen by probes expands as the range of the interactions extends in space.

∎

8.6 Motion in a Uniform Magnetic Field

Now we will consider a nonrelativistic point particle of mass m moving in some given *static* magnetic field and with no electric fields present. The Lagrangian is then given by Eq. (8.45):

$$L = \frac{1}{2}m\left(\dot{x}^2 + \dot{y}^2 + \dot{z}^2\right) + \frac{q}{c}A^i \dot{r}^i. \tag{8.73}$$

If this background magnetic field **B** is uniform, say

$$\mathbf{B} = B\hat{\mathbf{z}}, \tag{8.74}$$

we can write a corresponding vector potential as

$$\mathbf{A} = -\frac{1}{2}By\hat{\mathbf{x}} + \frac{1}{2}Bx\hat{\mathbf{y}}, \tag{8.75}$$

as seen previously in Eq. (8.33). Using Eq. (8.45), the Lagrangian then becomes

$$L = \frac{1}{2}m\left(\dot{x}^2 + \dot{y}^2 + \dot{z}^2\right) - \frac{qB}{2c}\dot{x}y + \frac{qB}{2c}\dot{y}x, \tag{8.76}$$

from which we can find the equations of motion

$$m\ddot{x} = \frac{qB}{c}\dot{y}, \quad m\ddot{y} = -\frac{qB}{c}\dot{x}, \quad m\ddot{z} = 0. \tag{8.77}$$

The dynamics in the z direction is already decoupled. Setting the initial conditions

$$z(0) = 0, \quad \dot{z}(0) = \mathrm{V}^z, \tag{8.78}$$

we find

$$z(t) = \mathrm{V}^z t, \tag{8.79}$$

so the particle moves in the magnetic field direction as though it were a free particle. In the x and y directions, however, the motion is a bit more interesting. We can easily integrate the $\ddot{x}$ and $\ddot{y}$ equations once to find

$$\dot{x} = \omega_0\left(y - y_0\right), \quad \dot{y} = -\omega_0\left(x - x_0\right), \tag{8.80}$$

where

$$\omega_0 \equiv \frac{qB}{mc}, \tag{8.81}$$

while x_0 and y_0 are constants of integration whose role will soon become apparent. The new equations (8.80) suggest a change of variable

$$X \equiv x - x_0, \quad Y \equiv y - y_0, \tag{8.82}$$

to yield a somewhat simpler set of coupled equations

$$\dot{X} = \omega_0 Y, \quad \dot{Y} = -\omega_0 X. \tag{8.83}$$

These can be decoupled by differentiating with respect to time:

$$\ddot{X} = -\omega_0^2 X, \quad \ddot{Y} = -\omega_0^2 Y, \tag{8.84}$$

leading us to the familiar simple harmonic oscillator equations. We now see that the point charge circles in the x, y plane about the point (x_0, y_0), as shown in Figure 8.10(a). If $X > 0$, we have $\dot{Y} < 0$ with $\omega_0 > 0$. This implies that if $qB > 0$, the circling is in the clockwise direction in the x–y plane, as seen looking down the z axis.

Let us choose a particularly convenient set of initial conditions. First:

$$Y(0) = 0 \Rightarrow \dot{X}(0) = 0, \tag{8.85}$$

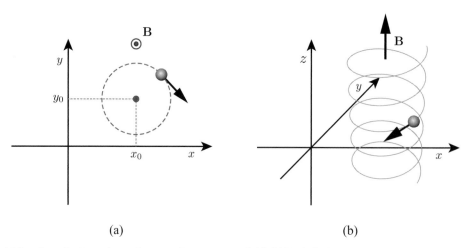

(a)   (b)

Fig. 8.10 (a) Top view of a charged particle in a uniform magnetic field. (b) The helical trajectory of the charged particle.

since $\dot{X} = \omega_0 Y$. Next:

$$X(0) = R \Rightarrow \dot{Y}(0) = V^y = -\omega_0 R \equiv C, \tag{8.86}$$

since $\dot{Y} = -\omega_0 X$, where we have denoted the radius of the circular trajectory as "R." We then find

$$X(t) = R\cos(\omega_0 t), \quad Y(t) = -R\sin(\omega_0 t). \tag{8.87}$$

In the original coordinates:

$$x(t) - x_0 = R\cos(\omega_0 t), \quad y(t) - y_0 = -R\sin(\omega_0 t). \tag{8.88}$$

As promised, this is a circle of radius R centered about (x_0, y_0), where

$$\left(\frac{x - x_0}{R}\right)^2 + \left(\frac{y - y_0}{R}\right)^2 = 1. \tag{8.89}$$

Superimposing the x, y motion onto the dynamics in the z direction, we get the celebrated helical trajectory depicted in Figure 8.10(b) of a charged particle in a uniform magnetic field.

The radius R and initial speed V^y are not independent. Using Eqs. (8.81) and (8.86), we can write

$$R = \frac{m c V^y}{q B}, \tag{8.90}$$

so it is clear that the radius of the circle is proportional to the initial speed in the x–y plane, and also that the larger the B field, the tighter the radius. We can use this setup to measure attributes of charged particles, such as their charge or speed, by measuring the radius of their circular trajectory in known uniform magnetic fields.

The phenomenon of charges circling in uniform magnetic fields was one of the first tools used by particle physicists to identify and measure properties of subatomic particles. Figure 8.11 shows a photograph from a **bubble chamber**.

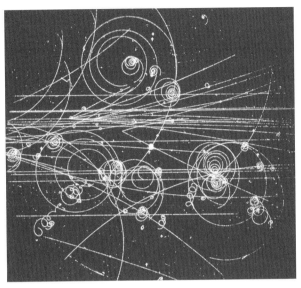

Fig. 8.11 The tracks of charged particles in a bubble chamber of superheated liquid hydrogen at CERN. The tightly coiled spirals are electrons or positrons (which coil in opposite directions); the straighter lines are more massive particles. Courtesy Science & Society Picture Library / SSPL / Getty Images.

Fast-moving charged particles enter a chamber immersed in a known external magnetic field. The box is filled with a superheated liquid,[7] such as liquid hydrogen. The incoming particles create a trail of bubbles as they pass through. The spirals shown in the figure represent actual trajectories of electrons and other more exotic subatomic particles![8] The spiral direction tells us the sign of the charge of the unknown particle. The radius of the spiral can be related to the particle's speed using Eq. (8.90). As the particle moves through the fluid in the box, it loses energy (and hence speed), and the radius of the spiral decreases. The device can be used to measure the charge-to-mass ratio q/m of many subatomic particles. This comparatively simple device was an important feature of the golden age of experimental particle physics. Unfortunately, bubble chambers cannot directly detect neutral particles like the π^0 meson or Λ hyperon. They can, however, detect electrically charged decay products such as the proton and negative pion in the

---

[7] "Superheated" means that the liquid's temperature is a bit above the boiling point. But unless there is a disturbance somewhere in the liquid, such as energetic electrons produced by the ionization of atoms in the liquid along the path of a relativistic charged particle passing through, there is no boiling, *i.e.*, no bubbles start to form.

[8] Some of the trajectories are from particles that enter from outside, while others are due to the most energetic electrons ejected from the bubble chamber's fluid atoms, such as hydrogen.

decay $\Lambda \rightarrow p + \pi^-$, and from the observed energies and momenta of the decay products learn a great deal about their neutral parent.[9]

Example 8.5 **Crossed Electric and Magnetic Fields**

Now we consider a nonrelativistic point charge q of mass m moving in the background of "crossed" uniform electric *and* magnetic fields, given by

$$\mathbf{E} = E_0 \hat{\mathbf{y}}, \quad \mathbf{B} = B_0 \hat{\mathbf{z}}. \tag{8.91}$$

That is, the fields are perpendicular to one another. If we now place a positively charged particle at rest in these fields and then let it go, where would you guess the particle would be found quite some time later? The actual result can be surprising, but useful in building up one's physical intuition. We will find the trajectory of the particle by first finding the vector and scalar potentials. From the discussion of a uniform magnetic field above, we can write

$$\mathbf{A} = \frac{1}{2} B_0 x \hat{\mathbf{y}} - \frac{1}{2} B_0 y \hat{\mathbf{x}}, \tag{8.92}$$

where we used Eq. (8.31). But we now also have a nontrivial electric static potential

$$\phi = - \int \mathbf{E} \cdot d\mathbf{s} = -E_0 y, \tag{8.93}$$

choosing the zero at $y = 0$. The Lagrangian follows from Eq. (8.45), and is given by

$$L = \frac{1}{2} m \left(\dot{x}^2 + \dot{y}^2 + \dot{z}^2 \right) + q E_0 y + \frac{q}{2c} B_0 x \dot{y} - \frac{q}{2c} B_0 y \dot{x}. \tag{8.94}$$

We next identify the canonical momenta

$$p^x = \frac{\partial L}{\partial \dot{x}} = m\dot{x} - \frac{q B_0}{2c} y, \quad p^y = \frac{\partial L}{\partial \dot{y}} = m\dot{y} + \frac{q B_0}{2c} x, \quad p^z = m\dot{z}. \tag{8.95}$$

Noting that the system is invariant under translations in the z direction, the momentum p_z must be conserved, so we can describe the evolution along the z axis as motion with constant speed:

$$\dot{p}^z = m\ddot{z} = 0. \tag{8.96}$$

Therefore we can focus on the two-dimensional evolution in the x–y plane by switching to an appropriate inertial frame moving in the z direction along with the particle.

Looking for other constants of motion, we note that the Lagrangian is not an explicit function of time, so the Hamiltonian

$$H = \dot{x}\frac{\partial L}{\partial \dot{x}} + \dot{y}\frac{\partial L}{\partial \dot{y}} - L = \frac{1}{2} m \left(\dot{x}^2 + \dot{y}^2 \right) - qE_0 y \tag{8.97}$$

must also be conserved, where we have already eliminated the z motion, as described above.

---

[9] Bubble chambers have long since been supplanted by other types of detectors, but none is better at showing off multiple particle tracks.

The *energy* of the particle is

$$E = K.E. + P.E. = \frac{1}{2}m\left(\dot{x}^2 + \dot{y}^2\right) + q\phi = \frac{1}{2}m\left(\dot{x}^2 + \dot{y}^2\right) - qE_0 y, \tag{8.98}$$

equal to the Hamiltonian in this case, with both conserved. Note that the magnetic field does not appear in the energy, consistent with the fact that *the magnetic force does no work*. We may have realized this from the form of the magnetic force law

$$\frac{dE_m}{dt} = \mathbf{F}_m \cdot \mathbf{v} = q\left(\frac{\mathbf{v}}{c} \times \mathbf{B}\right) \cdot \mathbf{v} = 0, \tag{8.99}$$

since the cross product of two vectors is perpendicular to both. The energy is a constant of motion and can be used to help find the trajectory of the particle.

The equations of motion in the *x* and *y* directions then take the simple form

$$m\dot{v}^x - \frac{q B_0}{c}v^y = 0, \quad m\dot{v}^y + \frac{q B_0}{c}v^x = q E_0, \tag{8.100}$$

and do very much reflect the presence of the magnetic field. We have written these equations in terms of the velocity instead of the position, since they are easier to decouple. We now take the time derivative of the first equation:

$$m\ddot{v}^x - \frac{q B_0}{c}\dot{v}^y = 0, \tag{8.101}$$

and then use the second equation to eliminate $\dot{v}_y$:

$$\ddot{v}^x + \frac{q^2 B_0^2}{m^2 c^2}v^x - \frac{q^2 E_0 B_0}{m^2 c} = 0. \tag{8.102}$$

This second-order linear equation is now fully decoupled and can easily be solved by using the form

$$v^x(t) = A_1 \cos\left(\omega_0 t + A_2\right) + C, \tag{8.103}$$

which is the sum of the general solution of the homogeneous equation (*i.e.*, the differential equation without the final term) and a particular solution of the full equation. Here A_1 and A_2 are integration constants, but ω_0 and C are constants determined by the differential equation. Substituting this expression into Eq. (8.102), we find that

$$\omega_0 = \frac{q B_0}{m c}, \quad C = \frac{E_0}{B_0}c. \tag{8.104}$$

Using Eq. (8.100), we can now find v_y:

$$v^y(t) = -A_1 \sin\left(\omega_0 t + A_2\right). \tag{8.105}$$

We have thus determined the velocity of the particle in terms of two constants of integration A_1 and A_2, which can be used to fix the initial velocity at (say) time $t = 0$. Let us in fact choose $v^x(0) = v^y(0) = 0$, so the particle starts at rest. Then we find

$$A_1 = -\frac{E_0}{B_0}c, \quad A_2 = 0; \tag{8.106}$$

or

$$v^x(t) = \frac{E_0}{B_0} c \left(1 - \cos \omega_0 t\right), \quad v^y(t) = \frac{E_0}{B_0} c \sin \omega_0 t. \tag{8.107}$$

Integrating the velocities (8.107) with respect to time gives $x(t)$ and $y(t)$:

$$x(t) = \frac{E_0}{B_0} \left(ct - \frac{c}{\omega_0} \sin \omega_0 t\right) \quad \text{and} \quad y(t) = \frac{E_0 c}{B_0 \omega_0} \left(1 - \cos \omega_0 t\right), \tag{8.108}$$

where we have chosen any constants of integration to make $x(0) = 0, y(0) = 0$.

This is an interesting result. Note in particular the term linear in time. To unravel the shape of the trajectory, we first note that the velocity vanishes periodically, with a period $P = 2\pi/\omega_0$, as seen from (8.107). Furthermore, the y motion is bounded with amplitude $E_0 c/(B_0 \omega_0)$, while the x motion is unbounded, growing on average linearly in time due to the term $x \rightarrow E_0 c\, t/B_0$. Figure 8.12(a) shows the trajectory. We see how both x and y oscillate with period P, with x advancing in discrete steps of $E_0 c\, P/B_0$. While the circling pattern about the magnetic field looks familiar, the drift in the x direction is somewhat counterintuitive, since the electric field points not in the x direction, but in the y direction.

We can understand why this is happening by breaking the motion into stages. A positive charge starts from rest at (say) $x = y = 0$, and then accelerates in the electric field direction (*i.e.*, the y direction) due to the force qE. But as the charge begins to move, the magnetic force takes hold, so the trajectory turns toward the positive x direction. Then as the magnetic field continues to curve the trajectory around, the charge begins to move back *against* the electric field direction (*i.e.*, in the negative y direction), and so is slowed and brought back to rest when $y = 0$. Then a new cycle begins: the electric field speeds up the charge and the previous motion repeats itself, and so on, and so on. Between each cycle there is a cusp at $y = 0$.

For arbitrary initial conditions the motion is the sum of a drift velocity $c\, E_0/B_0$ to the right and the type of circular motion previously described.

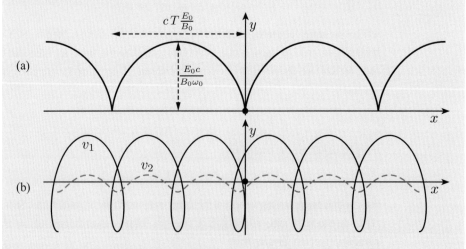

The trajectory of a particle in uniform crossed electric and magnetic fields: (a) for $v^x(0) = v^y(0) = 0$; (b) typical for other initial conditions. ■

Example 8.6

Mirror Mirror on the Wall

The previous two examples have taught us that probe charges spiral in a plane transverse to magnetic vector fields. But both examples involved *uniform* fields. An interesting new twist enters when a charge q moves in a *non*-uniform background field. Figure 8.13 shows a particularly interesting non-uniform magnetic field profile. The field has axial symmetry, so we adopt cylindrical coordinates ρ, φ, and z, as depicted in the figure. Because of this symmetry the magnetic field cannot depend on φ. Furthermore, the figure indicates that we want $B^\varphi = 0$. We then have two components to worry about, $B^\rho(\rho, z)$ and $B^z(\rho, z)$. But we know from Maxwell's equations that

$$\nabla \cdot \mathbf{B} = \frac{1}{\rho}\frac{\partial}{\partial\rho}(\rho B^\rho) + \frac{1}{\rho}\frac{\partial B^\varphi}{\partial\varphi} + \frac{\partial B^z}{\partial z} = 0, \tag{8.109}$$

where we have written the divergence in cylindrical coordinates.

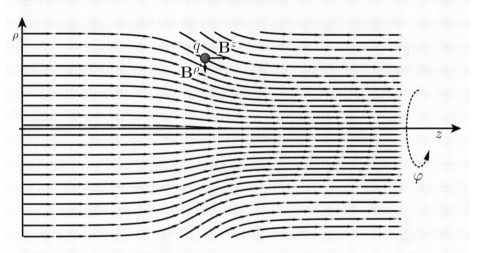

Fig. 8.13   A non-uniform magnetic field profile with axial symmetry.

In our case, since $B^\varphi = 0$, we have

$$\frac{\partial}{\partial\rho}(\rho B^\rho) = -\rho\frac{\partial B^z}{\partial z}. \tag{8.110}$$

Integrating this equation, we find

$$\rho B^\rho = -\int^\rho \rho'\frac{\partial B^z}{\partial z}d\rho' + f(z), \tag{8.111}$$

with some arbitrary function $f(z)$ arising from the ρ integration. To proceed further we make some simplifications. We consider a non-uniformity in B^z along the z direction only – we want $\partial B^z/\partial z \neq 0$ and B^z independent of ρ, with $\partial B^z/\partial\rho = 0$. Let us also arrange that $f(z) = 0$. These conditions can be achieved, at least approximately, by a careful arrangement of magnets. We then have

$$\rho B^\rho = -\frac{\rho^2}{2}\frac{\partial B^z}{\partial z} \Rightarrow B^\rho(\rho, z) = -\frac{\rho}{2}\frac{\partial B^z}{\partial z}. \tag{8.112}$$

Therefore if we want $\partial B^z / \partial z > 0$, for example, we must have a component of the magnetic field vector pointing radially inward – with a larger radial component the farther away we are from the z axis, as depicted in the figure. This configuration is known as a **magnetic mirror**.

We are now ready to tackle the problem of a point charge moving in this background magnetic field. We expect that spiraling about the z axis will be part of the dynamics; but there is also a new effect to be expected from the region where $\partial B^z / \partial z \neq 0$. To write the Lagrangian, we first need to find the potentials. With no electric field and a static magnetic field, we can write $\phi = 0$, using Eq. (8.25). We are then left with Eq. (8.23), written in cylindrical coordinates:

$$\left(\frac{1}{\rho}\frac{\partial A_z}{\partial \varphi} - \frac{\partial A_\varphi}{\partial z}\right)\hat{\boldsymbol{\rho}} + \left(\frac{\partial A_\rho}{\partial z} - \frac{\partial A_z}{\partial \rho}\right)\hat{\boldsymbol{\varphi}} + \left(\frac{1}{\rho}\frac{\partial}{\partial \rho}(\rho A_\varphi) - \frac{\partial A_\rho}{\partial \varphi}\right)\hat{\mathbf{z}}$$

$$= B^\rho(\rho,z)\hat{\boldsymbol{\rho}} + B^z(z)\hat{\mathbf{z}}. \tag{8.113}$$

The choice for **A** is not unique, so we need to make an educated guess. From the axial symmetry, we do not want **A** to have any dependence upon φ. Furthermore, using the example of uniform magnetic fields encountered in earlier examples, we expect that A_φ may play a central role as **A** whirls around the magnetic field. We therefore try

$$A_\rho(\rho,z) \neq 0, \quad A_\varphi(\rho,z) \neq 0, \quad A_z = 0. \tag{8.114}$$

Substituting these choices into Eq. (8.113), we get

$$\frac{\partial A_\rho}{\partial z} = 0, \quad \frac{\partial A_\varphi}{\partial z} = -B^\rho(\rho,z), \quad \frac{1}{\rho}\frac{\partial}{\partial \rho}(\rho A_\varphi) = B^z(z). \tag{8.115}$$

We can then immediately guess the solution

$$A_\varphi = \frac{\rho}{2}B^z(z), \quad A_\rho = 0, \tag{8.116}$$

which implies that

$$\frac{\partial A_\varphi}{\partial z} = \frac{\rho}{2}\frac{\partial B^z(z)}{\partial z} = -B^\rho(\rho,z), \tag{8.117}$$

as needed from (8.112), the no-magnetic-monopole condition encountered earlier. Hence, we have a good choice for a vector potential describing the desired magnetic field:

$$\mathbf{A} = \frac{\rho}{2}B^z(z)\hat{\boldsymbol{\varphi}}. \tag{8.118}$$

We can now write the Lagrangian of a probe charge of mass m and charge q using Eq. (8.45):

$$L = \frac{1}{2}m\left(\dot{\rho}^2 + \rho^2\dot{\varphi}^2 + \dot{z}^2\right) + \frac{q}{c}\frac{\rho^2}{2}\dot{\varphi}B^z(z), \tag{8.119}$$

expressed in cylindrical coordinates. The absences of φ and of time from this Lagrangian imply the conservation of canonical angular momentum and the Hamiltonian, respectively. We find that the canonical angular momentum is

$$p_\varphi = m\rho^2\dot{\varphi} + \frac{q}{2c}\rho^2 B^z(z) = \text{constant}, \tag{8.120}$$

and the Hamiltonian is

$$H = \frac{1}{2}m\left(\dot{\rho}^2 + \rho^2\dot{\varphi}^2 + \dot{z}^2\right) = \text{constant}, \tag{8.121}$$

which is also the kinetic energy in this case. Therefore, *the kinetic energy of the probe charge is conserved.* As discussed before, the magnetic force does no work: it can deflect and scramble the probe in complicated ways, but it cannot change its energy – in this case the probe's speed must remain constant. However, we have only two constants of motion for *three* equations of motion, so the overall problem is not integrable, and we have more work to do.

Next, we look at the equations of motion in the ρ and z direction. We find

$$m\ddot{\rho} = m\rho\dot{\varphi}^2 + \frac{q}{c}\rho\dot{\varphi}B^z(z) \tag{8.122}$$

and

$$m\ddot{z} = \frac{q}{2c}\rho^2\dot{\varphi}\frac{dB^z}{dz}. \tag{8.123}$$

We can use Eq. (8.120) to eliminate $\dot{\varphi}$ from these two equations. We first have

$$\dot{\varphi} = \frac{p_\varphi - m\rho^2\omega_0}{m\rho^2} = \frac{p_\varphi}{m\rho^2} - \omega_0, \tag{8.124}$$

where we have defined

$$\omega_0(z) \equiv \frac{q\,B^z(z)}{2\,m\,c}, \tag{8.125}$$

which is the natural circling angular frequency of the probe about the magnetic field, often called the *cyclotron frequency*, seen already in earlier examples. Note that this ω_0 is *z dependent*. The $\ddot{\rho}$ and $\ddot{z}$ equations then become

$$\ddot{\rho} = \frac{p_\varphi^2 - m^2\omega_0^2\rho^4}{m^2\rho^3} = \frac{\left(p_\varphi - m\rho^2\omega_0\right)\left(p_\varphi + m\rho^2\omega_0\right)}{m^2\rho^3} = \frac{\dot{\varphi}}{m\rho}\left(p_\varphi + m\rho^2\omega_0\right) \tag{8.126}$$

and

$$\ddot{z} = \frac{\omega_0}{m}\frac{1}{B^z}\frac{dB^z}{dz}\left(p_\varphi - m\rho^2\omega_0\right) = \rho^2\omega_0\dot{\varphi}\frac{1}{B^z}\frac{dB^z}{dz}. \tag{8.127}$$

Note that, for uniform magnetic field $dB^z/dz = 0$, choosing initial conditions such that $p_\varphi = -m\rho^2\omega_0$ leads to $\dot{\varphi} = -qB^z/mc$, $\ddot{\rho} = 0$, and $\ddot{z} = 0$ – i.e., the expected spiral motion about the axis of symmetry with radius ρ. With a non-uniform magnetic field, we can now see why this system is called a magnetic mirror. If we start the particle with $\dot{\varphi} < 0$ (for example by choosing $p_\varphi \sim -m\rho^2\omega_0$), as the particle travels from the region near $z = 0$ toward the magnetic funnel of Figure 8.14, the non-uniformity in the magnetic field enters through Eq. (8.127): this is a force in the negative z direction – assuming $dB^z/dz > 0$ and $\dot{\varphi} < 0$ – indicating that the probe is pushed *back* toward negative z. Meanwhile, Eq. (8.126) suggests that the probe is pushed toward $\rho = 0$, focusing along the z axis. The problem, is however, somewhat more intricate than this qualitative analysis may suggest: ω_0 is in fact a function of z in the region where the magnetic field is changing, and indeed it increases as the particle penetrates this region. Our set of equations of motion, Eqs. (8.124), (8.126), and (8.127), are tightly entangled.

To analyze the dynamics more quantitatively, it helps to make an approximation. We assume that the probe experiences the changes in the magnetic field very slowly compared to the timescale associated with its circling of the field lines. The latter faster time is determined by ω_0. Hence, at short timescales, the particle behaves much like a probe in a uniform magnetic field – spiraling around the field lines – with this magnetic field sampled from the vicinity of the probe. As the particle moves along the z direction, it samples larger and larger values of magnetic field, which increases ω_0 and hence the spinning frequency through Eq. (8.125). This is known as an **diabatic regime**: a slowly changing background parameter gradually shifting the evolution of a fast motion. We can arrange to be close to this regime by choosing

$$p_\varphi \simeq -m\,\rho^2\omega_0 \Rightarrow \dot{\varphi} \simeq -\frac{q\,B^z}{m\,c}, \tag{8.128}$$

which starts the particle in a circular trajectory around the symmetry axis z at the spin rate we already know from the case of a probe in a uniform magnetic field. We can then write the conserved Hamiltonian as

$$H \simeq \frac{1}{2}m\dot{\rho}^2 + \frac{1}{2}m\dot{z}^2 + \frac{1}{2}m\rho^2\dot{\varphi}\left(-\frac{q\,B^z}{m\,c}\right) = \frac{1}{2}m\dot{\rho}^2 + \frac{1}{2}m\dot{z}^2 + \mu B^z, \tag{8.129}$$

where in the last step we defined a new quantity called the **magnetic moment** of the circling probe:

$$\mu \equiv -\frac{1}{2}\frac{q}{c}\,\rho^2\dot{\varphi}. \tag{8.130}$$

The purpose of this juggling of terms is to show that μ is approximately conserved in the adiabatic regime. To see this, note that the Hamiltonian is conserved, so its time derivative is zero:

$$\frac{dH}{dt} = m\dot{\rho}\ddot{\rho} + m\dot{z}\ddot{z} + \dot{\mu}B^z + \mu\dot{B}_z = 0. \tag{8.131}$$

Looking back at Eq. (8.126), we see that $\ddot{\rho} \simeq 0$ in this regime since $p_\varphi \simeq -m\,\rho^2\omega_0$. We then have

$$\frac{dH}{dt} \simeq m\dot{z}\rho^2\omega_0\dot{\varphi}\frac{1}{B^z}\frac{dB^z}{dz} + \dot{\mu}B^z + \mu\frac{dB^z}{dz}\dot{z} = 0, \tag{8.132}$$

where we used Eq. (8.127). In the last term, we used the chain rule to write

$$\dot{B}_z = \frac{dB^z}{dz}\dot{z}. \tag{8.133}$$

Substituting for the magnetic moment in the first term of Eq. (8.132), we then get

$$\frac{d\mu}{dt} \simeq 0 \quad \textit{adiabatic regime}. \tag{8.134}$$

Since $\mu \sim \rho^2\dot{\varphi}$, this implies that the *traditional* angular momentum about the z axis is approximately conserved. As the particle samples stronger magnetic fields, it must spin faster by increasing $\dot{\varphi}$: as a result of this adiabatic conservation statement, we must then have the particle focus toward the z axis by decreasing ρ.

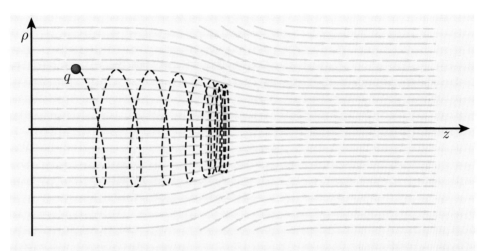

Fig. 8.14 A probe particle bouncing off a region of non-uniform magnetic field. This phenomenon is known as a magnetic mirror.

We can now easily determine the trajectory of the probe in the ρ–z plane. We take the time derivative of μ explicitly:

$$\frac{d\mu}{dt} \simeq 0 \Rightarrow \frac{d(\rho^2\dot\varphi)}{dt} \simeq 0 \Rightarrow 2\rho\dot\rho\dot\varphi + \rho^2\ddot\varphi \simeq 0. \tag{8.135}$$

Using Eq. (8.128), we get

$$2\rho\dot\rho\dot\varphi - 2\rho^2\dot\omega_0 \simeq 0. \tag{8.136}$$

Looking back at Eq. (8.125), we find

$$\dot\omega_0 = \frac{q\,\dot B_z}{2\,m\,c} = \frac{q\,\dot z}{2\,m\,c}\frac{dB^z}{dz}. \tag{8.137}$$

Finally, substituting this and Eq. (8.128) into Eq. (8.136), we get

$$\frac{\dot\rho}{\dot z} = -\frac{1}{2}\rho\frac{1}{B^z}\frac{dB^z}{dz} = \frac{B^\rho}{B^z}, \tag{8.138}$$

using Eq. (8.112). This implies that the probe tracks the magnetic field lines as they curve towards the z axis, all the while spinning faster and faster! That is, the number of field lines within an orbit remains constant in the adiabatic limit.

Using Eq. (8.127), we already argued that the probe is pushed away from the region of increasing magnetic field. In the Problems section at the end of this chapter, you will find the maximum extent to which the probe penetrates the region of dense magnetic fields, before bouncing back. This mirror effect can be very useful for trapping charged particles. If we arrange two such magnetic field profiles as in Figure 8.15, we have a

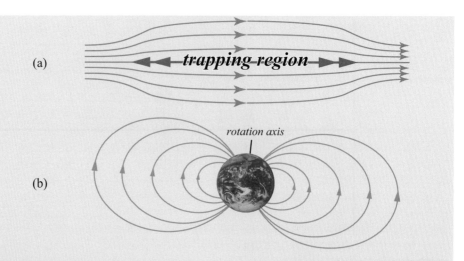

Fig. 8.15 (a) A magnetic bottle in the laboratory, used to trap charged particles. (b) Natural magnetic bottles formed as parts of earth's dipole magnetic field. They are an essential ingredient in producing aurora phenomena near both poles.

magnetic bottle in which charged particles can bounce back and forth. Indeed, the magnetic field lines of the earth create such a natural bottle, as shown in the figure. A plasma of space particles get trapped between the north and south pole as they bounce back and forth. At the poles these rapidly moving charged particles strike atoms in the atmosphere, raising atomic electrons to higher-energy states. When the electrons drop back to lower-energy states, photons are released – putting on a marvelous show of colors in the skies known as the aurora. ∎

Example 8.7   **Ion Trapping**

Imagine trapping a few ions – or even a single ion – in a small enclosure, poking it around, and observing the intricate physics within it as it reacts to external perturbations. This is something that physicists do regularly using the electromagnetic forces that rule the realm of atomic physics. While many such situations are most interesting because of the quantum physics they allow one to probe, the basic trapping mechanism can be understood using classical physics.

The task is to trap an ion of charge q using external electric and magnetic fields that we can tune arbitrarily. The simplest arrangement perhaps would consist of purely electrostatic fields that we might generate by some arrangement of charges placed far from the ion. This is however not the case, as clarified by the following theorem.

Earnshaw's theorem: *It is not possible to construct a stable stationary point for a probe charge using only electrostatic or only magnetostatic fields in vacuum.*

To see this for the case of electrostatic fields, consider a region of space where the ion probe is to sit and where we have some external electrostatic fields. There are no source charges in this region since these are far away from the trapping region, so we have $\nabla \cdot \mathbf{E} = 0$. Then using $\mathbf{E} = -\nabla \phi$, we get Laplace's equation for the electric potential ϕ:

$$\nabla^2 \phi = \frac{\partial^2 \phi}{\partial x^2} + \frac{\partial^2 \phi}{\partial y^2} + \frac{\partial^2 \phi}{\partial z^2} = 0. \tag{8.139}$$

The potential energy of an ion of charge q is then $U = q\phi$, which implies that $\nabla^2 U = 0$. For trapping the ion, we then need a minimum in this potential, which implies we need

$$\frac{\partial^2 U}{\partial x^2} > 0, \quad \frac{\partial^2 U}{\partial y^2} > 0, \quad \frac{\partial^2 U}{\partial z^2} > 0 \quad \text{while} \quad \frac{\partial^2 U}{\partial x^2} + \frac{\partial^2 U}{\partial y^2} + \frac{\partial^2 U}{\partial z^2} = 0, \tag{8.140}$$

which is obviously not possible! But a *saddle* surface – where say $\partial^2 U / \partial x^2 < 0$ while $\partial^2 U / \partial y^2 > 0$ and $\partial^2 U / \partial z^2 > 0$ – is the best one can do. However, even then the ion would quickly find a way down such a potential, running away to infinity along the x axis.

There are several ways we might circumvent this unfortunate situation. One is to consider time-varying electric fields. Imagine a saddle surface that is (say) spinning fast enough that every time the ion ventures a little down the potential, it is quickly pushed back into the middle. For example, one can construct what is known as a **Paul trap**. We leave this case to the Problems section at the end of this chapter. In this example, we discuss instead the so-called **Penning trap**, which involves both electrostatic *and* magnetostatic fields.

The idea of the Penning trap is to begin with a uniform magnetic field

$$\mathbf{B} = B\hat{\mathbf{z}} \tag{8.141}$$

which, as we now know, leads to a spiral trajectory of an ion, circling in the x–y plane with angular speed

$$\omega_0 = \frac{qB}{mc}. \tag{8.142}$$

Therefore such an ion seems to be confined in the x and y directions, but not at all in the z direction. To confine the ion in the z direction as well, we add an electrostatic field described by the electric potential

$$\phi = \frac{\phi_0}{D^2} \left(z^2 - \frac{x^2 + y^2}{2} \right), \tag{8.143}$$

where D is some length associated with the geometry of the system (see Figure 8.16). Note that this electric potential satisfies, as it must, the Laplace equation

$$\nabla^2 \phi = 0. \tag{8.144}$$

The nonrelativistic Lagrangian for the ion of charge q then becomes, using Eq. (8.45):

$$L = \frac{1}{2} m \left(\dot{x}^2 + \dot{y}^2 + \dot{z}^2 \right) - \frac{qB}{2c} \dot{x}y + \frac{qB}{2c} \dot{y}x - q \frac{\phi_0}{D^2} \left(z^2 - \frac{x^2 + y^2}{2} \right). \tag{8.145}$$

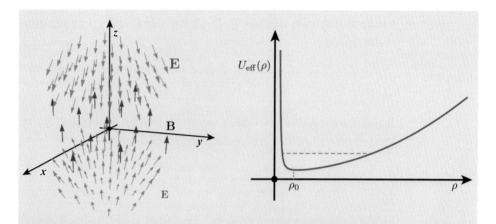

Fig. 8.16 The electric and magnetic fields for the Penning potential and the corresponding effective potential. At the minimum, we have a stable circular trajectory. In general, however, the radial extent will oscillate with frequency ω_0.

It is advantageous to switch to cylindrical coordinates ρ, φ, and z given the symmetries of the potential. We then get

$$L = \frac{1}{2}m \left(\dot{\rho}^2 + \rho^2\dot{\varphi}^2 + \dot{z}^2 \right) + \frac{qB}{2c}\rho^2\dot{\varphi} - q\frac{\phi_0}{D^2} \left(z^2 - \frac{\rho^2}{2} \right). \tag{8.146}$$

The dynamics in the z direction is then that of a simple harmonic oscillator

$$\ddot{z} = -\omega_z^2 z \tag{8.147}$$

with

$$\omega_z^2 = \frac{2q\phi_0}{mD^2}. \tag{8.148}$$

Note that we would need

$$q\,\phi_0 > 0 \tag{8.149}$$

to make sure the ion is trapped in the z direction. We can also write an "energy conservation" statement

$$\frac{1}{2}m\dot{z}^2 + \frac{1}{2}m\omega_z^2 z^2 = E_z = \text{constant}, \tag{8.150}$$

for some constant E_z. To see this, multiply Eq. (8.147) by $\dot{z}$ and integrate. For the φ equation of motion, one gets the angular momentum conservation law

$$m\rho^2\dot{\varphi} + \frac{qB}{2c}\rho^2 = p_\varphi = \text{constant}, \tag{8.151}$$

where p_φ denotes the angular momentum constant. Instead of looking at the ρ equation of motion, we realize that the Hamiltonian is also conserved, since $\partial L / \partial t = 0$. Therefore

$$H = \frac{1}{2}m \left(\dot\rho^2 + \rho^2 \dot\varphi^2 + \dot z^2 \right) + q \frac{\phi_0}{D^2} \left(z^2 - \frac{\rho^2}{2} \right) = \text{constant.} \tag{8.152}$$

This allows us to write an effective potential for a one-dimensional problem in the radial direction ρ – akin to the central force problem we have already seen:

$$\frac{1}{2}m\dot\rho^2 + U_{\text{eff}}(\rho) = H, \tag{8.153}$$

where

$$U_{\text{eff}}(\rho) = E_z - \frac{1}{2}I\,\omega_0\,c + \frac{p_\varphi^2}{2\,m\,\rho^2} + \frac{1}{8}m\rho^2 \left(\omega_0^2 - 2\omega_z^2 \right) \tag{8.154}$$

and where we also eliminated $\dot\varphi$ in favor of p_φ using Eq. (8.151). This potential is shown in Figure 8.16.

We identify a minimum at $\rho = \rho_0$:

$$\left. \frac{\partial U_{\text{eff}}}{\partial \rho} \right|_{\rho_0} = 0 \Rightarrow \rho_0^2 = \frac{2\,p_\varphi}{m} \left(\omega_0^2 - 2\omega_z^2 \right)^{-1/2}, \tag{8.155}$$

with the curvature near the minimum given by

$$\left. \frac{\partial^2 U_{\text{eff}}}{\partial \rho^2} \right|_{\rho_0} = m \left(\omega_0^2 - 2\,\omega_z^2 \right) > 0, \tag{8.156}$$

which is positive, since typically the oscillation frequency ω_z is much lower than ω_0:

$$\omega_z \ll \omega_0. \tag{8.157}$$

We may then write

$$\rho_0 \simeq \sqrt{\frac{2\,p_\varphi}{m\,\omega_0}}. \tag{8.158}$$

At this critical radius, we can find the angular speed $\dot\varphi$ using Eq. (8.151):

$$\dot\varphi \big|_{\rho_0} = -\frac{\omega_0}{2} + \frac{1}{2}\sqrt{\omega_0^2 - 2\,\omega_z^2} \simeq -\frac{\omega_z^2}{2\,\omega_0} \equiv -\omega_m, \tag{8.159}$$

where in the last step we used $\omega_z \ll \omega_0$. Therefore, $\omega_m \ll \omega_z \ll \omega_0$. The ion circles with radius ρ_0 in the x–y plane very slowly at frequency ω_m, while oscillating a little bit faster in the z direction at frequency ω_z. To see the role of the third frequency ω_0, we note that the general trajectory implied by the effective potential shown in Figure 8.16 involves also radial oscillation. The frequency of this oscillation is given by (8.156):

$$U_{\text{eff}}(\rho) \simeq U_{\text{eff}}(\rho_0) + \frac{1}{2}m \left(\omega_0^2 - 2\,\omega_z^2 \right) (\rho - \rho_0)^2, \tag{8.160}$$

which means that ρ oscillates with frequency $\sqrt{\omega_0^2 - 2\omega_z^2} \simeq \omega_0$. This is the third largest frequency in the problem, tuned by the strength of the external magnetic field. The combined motion is shown in Figure 8.17. It involves a slow circular trajectory in the x, y plane of large radius, on top of which is superimposed a slightly faster vertical oscillation in the z direction; and on top of these are superimposed fast epicycles with tight radii. This arrangement can in practice achieve ion trapping lasting for days. But eventually the configuration is unstable, and other considerations, such as the leaking of energy through electromagnetic radiation, invalidates the analysis. Electromagnetic trapping of charged particles is an extraordinarily difficult problem.

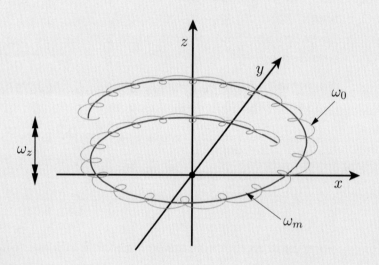

Fig. 8.17   The full trajectory of an ion in a Penning trap. A vertical oscillation along the z axis with frequency ω_z is superimposed onto a fast oscillation of frequency ω_0, while the particle traces a large circle with characteristic frequency ω_m. ∎

8.7 Relativistic Effects and the Electromagnetic Force

Our discussion of the dynamics of a probe charge in background electromagnetic fields focused on the nonrelativistic regime, where the probe's speed stays much smaller than the speed of light. While this is an adequate approximation in many situations, relativity and the speed of light are central to electromagnetism and there are many interesting situations where relativistic effects play a central role. In this section we revisit the problems of a probe in uniform electric and magnetic fields – presenting a full relativistic treatment to illustrate some of the new features due to relativity.

Example 8.8 | **Probe in a Uniform Magnetic Field**

We start with the example of a probe of charge q and mass m moving in the background of a uniform magnetic field **B**. We already know that the Lagrangian of a *free* relativistic particle leads to the equations of motion

$$\frac{d\mathbf{p}}{dt} = \frac{d}{dt}\left(\gamma\, m\, \mathbf{v}\right) = 0, \quad \frac{dE}{dt} = \frac{d}{dt}\left(\gamma\, m\, c^2\right) = 0. \tag{8.161}$$

When we add the term $q\mathbf{A}\cdot\mathbf{v}/c$ to the Lagrangian, we end up modifying the right-hand side of these equations to give

$$\frac{d}{dt}\left(\gamma\, m\, \mathbf{v}\right) = \frac{q}{c}\mathbf{v}\times\mathbf{B}, \quad \frac{dE}{dt} = \frac{d}{dt}\left(\gamma\, m\, c^2\right) = 0, \tag{8.162}$$

where **B** is a constant vector, and the electric field $\mathbf{E} = 0$. The conservation of relativistic energy $\gamma\, m\, c^2$ from the second equation tells us that the speed of the particle is *constant*. This then allows us to immediately write the first equation as

$$\gamma\, m\, \frac{d\mathbf{v}}{dt} = \frac{q}{c}\mathbf{v}\times\mathbf{B} \Rightarrow \frac{d\mathbf{v}}{dt} = \left(\frac{q\, c}{E}\right)\mathbf{v}\times\mathbf{B}. \tag{8.163}$$

The coefficient $q\, c/E$ is constant with constant energy E, so the trajectory of the probe is qualitatively the same as in the nonrelativistic case, in that it still spirals around the direction of **B**. The difference is that the angular frequency of rotation is now

$$\omega_0 = \frac{q\, c}{E}B. \tag{8.164}$$

In the slow speed limit, we have

$$E = \gamma\, m\, c^2 \simeq m\, c^2 + \frac{1}{2}m\, v^2 \simeq m\, c^2, \tag{8.165}$$

yielding the approximate angular frequency

$$\omega_0 \simeq \frac{q\, B}{m\, c}, \tag{8.166}$$

which is the expression we obtained earlier. As the speed of the probe approaches that of light, the denominator of Eq. (8.164) becomes larger than the nonrelativistic approximate counterpart. This implies that relativistic effects make the spiraling angular frequency smaller compared to the nonrelativistic estimate. In the limit $v \to c$, we have $E \to \infty$, implying $\omega_0 \to 0$: the probe does not spiral at all. This would then be the case for a hypothetical massless charged particle, but in fact there are no such particles.

The problem is more interesting than just portrayed. As is the case in all such situations, we ignored the electromagnetic fields due to the probe charge itself. But as a spiraling particle accelerates, it radiates energy by emitting electromagnetic waves and this leakage can become significant for a relativistic particle. The rate of energy loss for a probe undergoing circular motion can be shown to be

$$\frac{dE}{dt} = \frac{2}{3}\frac{q^2}{m^2 c^3}\gamma^2\left(\frac{d\mathbf{p}}{dt}\right)^2, \tag{8.167}$$

where we should note in particular the γ^2 factor. As the probe speed is increased, the power loss to electromagnetic radiation will quickly become sufficiently important that the electromagnetic fields from the probe cannot be ignored in determining its dynamics. Such effects are called **back-reaction effects** – the probe's own fields react back onto its dynamics. In this case, as energy is drained out of the probe faster and faster, we expect that the probe will slow down – its spiraling radius getting smaller and smaller. We can write a differential equation for the speed using Eq. (8.167):

$$\frac{d}{dt}\left(\gamma m c^2\right) = \frac{2}{3}\frac{q^4}{m^2 c^5}\gamma^2 \left(\mathbf{v} \times \mathbf{B}\right)^2 = \frac{2}{3}\frac{q^4}{m^2 c^5}\gamma^2 \left(v^2 B^2 - \mathbf{v} \cdot \mathbf{B}\right), \tag{8.168}$$

where we used the vector identity

$$(\mathbf{a} \times \mathbf{b}) \cdot (\mathbf{c} \times \mathbf{d}) = (\mathbf{a} \cdot \mathbf{c})(\mathbf{b} \cdot \mathbf{d}) - (\mathbf{a} \cdot \mathbf{d})(\mathbf{b} \cdot \mathbf{c}). \tag{8.169}$$

If we choose a scenario where **B** is oriented such that $\mathbf{v} \cdot \mathbf{B} = 0$, perpendicular to the plane of circular motion, we get

$$\frac{d\gamma}{dt} = \frac{2}{3}\frac{q^4 B^2}{m^3 c^7}\gamma^2 v^2 = \frac{v}{c^2}\gamma^3 \frac{dv}{dt}, \tag{8.170}$$

where in the last step we used the chain rule on $d\gamma/dt$. We then end up with a differential equation for the speed of the probe given by

$$\frac{dv}{dt} = \frac{2}{3}\frac{q^4 B^2}{m^3 c^5} v \sqrt{1 - \frac{v^2}{c^2}}. \tag{8.171}$$

One can find an exact solution to this equation and determine $v(t)$. We can see from this that the characteristic decay time for v is set by the combination $m^3 c^5/q^4 B^2$. However, this analysis is incomplete. In arriving at this result, we assumed that the first of Eqs. (8.162) holds. But the outgoing electromagnetic waves carry momentum and hence we may expect a modification of $d\mathbf{p}/dt$ as well. The full problem goes beyond the scope of this book. ∎

Example 8.9 **A Relativistic Probe in Crossed Uniform Electric and Magnetic Fields**

Now we will look back at the problem of a probe particle of mass m and charge q moving in the background of uniform electric *and* magnetic fields, **E** and **B**, which are perpendicular to one another, so $\mathbf{E} \cdot \mathbf{B} = 0$. We want to treat the full relativistic problem, where the probe's speed may not be much smaller than that of light. Along the same line of approach as in the previous example, we quickly arrive at the equations of motion

$$\frac{d\mathbf{p}}{dt} = q\mathbf{E} + \frac{q}{c}\mathbf{v} \times \mathbf{B}, \tag{8.172}$$

where the momentum $\mathbf{p} = \gamma m \mathbf{v}$. This is a complicated problem, but it can be simplified significantly through a physical trick. Knowing that these equations of motion are Lorentz invariant, we switch to a reference frame in which the background field is either entirely electric or entirely magnetic. Let us assume that $E < B$, and leave the opposite case as an exercise to the reader. Looking at the Lorentz transformations of the electric and magnetic fields, Eqs. (8.35) and (8.36), we see that if we choose a reference frame – call it $\mathcal{O}'$ – moving with velocity

$$\frac{\mathbf{V}}{c} = \frac{\mathbf{E} \times \mathbf{B}}{B^2}, \tag{8.173}$$

then the primed electric and magnetic fields become

$$\mathbf{E}' = 0, \quad \mathbf{B}' = \frac{1}{\gamma_V} \mathbf{B}, \tag{8.174}$$

where we have used the vector identity

$$\mathbf{a} \times (\mathbf{b} \times \mathbf{c}) = (\mathbf{a} \cdot \mathbf{c})\mathbf{b} - (\mathbf{a} \cdot \mathbf{b})\mathbf{c}. \tag{8.175}$$

Note that requiring $V < c$ implies that $E < B$, as needed. We have mapped the problem onto the previous example, a probe moving in a uniform *purely* magnetic background field. This is because the equations of motion of the probe in the new reference frame are simply

$$\frac{d\mathbf{p}'}{dt} = q\mathbf{E}' + \frac{q}{c}\mathbf{v}' \times \mathbf{B}' = \frac{q}{c}\mathbf{v}' \times \mathbf{B}', \tag{8.176}$$

because the equations of motion are Lorentz invariant and do not change structural form under change of inertial reference frame. From the perspective of this reference frame, the particle circles around the new $\mathbf{B}'$ field, which is a factor of γ smaller in magnitude than the original one. Switching back to the unprimed reference frame, we then superimpose on this circling motion a drift perpendicular to $\mathbf{E}$ and $\mathbf{B}$ given by the velocity $\mathbf{V}$. Hence, the relativistic problem is qualitatively very similar to the nonrelativistic version – except for a few factors of γ here and there! ∎

8.8 Summary

In this chapter we have extended the Lagrangian formalism to include the full effects of the electromagnetic force. In deriving the Lagrangian we once again learned to appreciate the role and power of symmetry in physics – as Lorentz symmetry closely guided us to the answer. We also encountered a new symmetry principle, gauge invariance, which plays a central role in physics. We will revisit the subject of gauge symmetry in the upcoming capstone Chapter 10.

Electromagnetism rules the technological world around us, and by tackling several real-life situations involving electromagnetic fields, we started to get a feel for the strange effects of electromagnetism: electric fields accelerate charges along their field lines, while the peculiar velocity-dependent magnetic force does no work, yet bends charges into helical trajectories.

Problems

★★ **Problem 8.1** Consider an infinite wire carrying a constant linear charge density λ_0. Write the Lagrangian of a probe charge Q in the vicinity, and find its trajectory.

★★ **Problem 8.2** Consider the oscillating Paul trap potential

$$U(z, \rho) = \frac{U_0 + U_1 \cos \Omega t}{\rho_0^2 + 2 z_0^2} \left(2 z^2 + \left(\rho_0^2 - \rho^2 \right) \right), \tag{8.177}$$

written in cylindrical coordinates. (a) Show that this potential satisfies Laplace's equation. (b) Consider a point particle of charge Q in this potential. Analyze the dynamics using a Lagrangian and show that the particle is trapped.

★ **Problem 8.3** Show that the Coulomb gauge $\nabla \cdot \mathbf{A} = 0$ is a consistent gauge condition.

★ **Problem 8.4** Find the residual gauge freedom in the Coulomb gauge.

★ **Problem 8.5** Show that the Lorentz gauge $\partial_\mu A_\nu \eta^{\mu\nu} = 0$ is a consistent gauge condition.

★ **Problem 8.6** Find the residual gauge freedom in the Lorentz gauge.

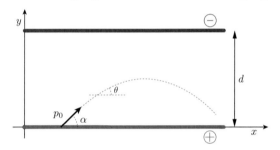

★★★ **Problem 8.7** An ultrarelativistic electron with $v \sim c$ and momentum p_0 enters a region between the two plates of a capacitor, as shown above. The plate separation is d and a voltage V is applied to the plates.

(a) Show that

$$\frac{d}{dt} (p_0 \cos \alpha \, \tan \theta) = -\frac{e V}{d},$$

where θ is the time-dependent angle the electron makes with the horizontal axis during its trek. (b) Write a differential equation for $y(x)$, assuming that

$$\frac{1}{c} \frac{d}{dt} \sim \frac{d}{dl}, \tag{8.178}$$

where $dl = \sqrt{dx^2 + dy^2}$. (c) Find the trajectory $y(x)$ by solving the differential equation from part (b).

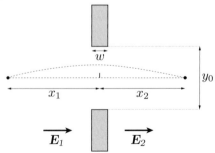

★★ **Problem 8.8** Charged particles are accelerated through a potential difference V_0 before falling onto a lens consisting of an aperture of height y_0 and thickness w, as shown above.

There is a uniform electric field $\mathbf{E}_1$ and $\mathbf{E}_2$ on the left and right of the aperture, respectively, as depicted. The figure also shows the trajectory of a charged particle of charge q emerging a distance x_1 from the aperture on the left and focusing a distance x_2 on the right. Assume $V_0 \gg E_1 x_1$ and $E_2 x_2$, and x_1 and $x_2 \gg y_0$. (a) Using $\nabla \cdot \mathbf{E} = 0$, show that inside the aperture we have

$$E^x \simeq (E_2 - E_1)w, \quad E^y \simeq -\frac{E_2 - E_1}{w}y.$$

(b) Show that

$$\frac{1}{x_1} + \frac{1}{x_2} \simeq \frac{E_2 - E_1}{2V_0}, \tag{8.179}$$

so that the aperture functions as a lens for charged particles.

★★ **Problem 8.9** A charged particle is circling a magnetic field that gradually increases in magnitude from B_1 to B_2 as the particle advances along the field lines. Show that the particle will be reflected if

$$v_{0\|} \leq v_{0\perp}\sqrt{\frac{B_2}{B_1} - 1}, \tag{8.180}$$

where $v_{0\|}$ and $v_{0\perp}$ are the components of the particle's velocity parallel and perpendicular to the magnetic field.

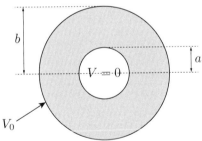

★★ **Problem 8.10** A coaxial cable has a grounded center and a voltage V_0 on the rim, as shown above.

A uniform magnetic field B_0 lies along the cylindrical axis of symmetry. Electrons propagate from the center to the rim. Find the mininum V_0 so that current can flow from the center to the rim.

★★★ **Problem 8.11** Neutrons have zero charge but carry a magnetic dipole moment μ. As a result, they are subject to a magnetic force given by $\mathbf{F} = (\mu \cdot \nabla)\mathbf{B}$. A beam of neutrons with $\mu = \mu_x \hat{\mathbf{x}}$ is moving along the z direction into a region of magnetic field $\mathbf{B}$. Find a simple form for $\mathbf{B}$ capable of *focusing* the beam and preventing it from dispersing. This means that small disturbances in the beam profile would not grow. *Hint*: For any vector field $\mathbf{V}$ satisfying $\nabla \cdot \mathbf{V} = 0$, we can write $\mathbf{V} = \nabla f$ for a function f satisfying $\nabla^2 f = 0$.

Problem 8.12 A magnetic monopole is a particle that casts out a radial magnetic field satisfying $\nabla \cdot \mathbf{B} = 4\pi q_m \delta(\mathbf{r})$, where q_m is the *magnetic charge* of the monopole. A nonrelativistic *electrically charged* particle of charge q is moving near a magnetic monopole of magnetic charge q_m. (a) Show that the magnetic field from the monopole takes the form

$$\mathbf{B} = \frac{q_m}{r^2}\hat{\mathbf{r}}.$$

(b) Write the equations of motion of the electrically charged particle assuming that the magnetic monopole remains stationary. (c) Find the constants of motion; in particular show that the so-called Fierz vector

$$\mathbf{Z} = \mathbf{r} \times \mathbf{p} - q\, q_m\, \frac{\mathbf{r}}{r} \tag{8.181}$$

remains constant.

Problem 8.13 Consider a charged *relativistic* particle of charge q and mass m moving in a cylindrically symmetric magnetic field with $B^\varphi = 0$. (a) Show that this general setup can be described with a vector potential that has one nonzero component $A^\varphi(\rho, z)$. (b) Write the equations of motion in cylindrical coordinates. (c) Consider circular orbits only and show that this implies that we need $B^\rho = 0$. Then find the form of B^z needed to achieve circular orbits.

Problem 8.14 For the previous problem, find the angular speed with which the particle spins about the magnetic field in terms of the radius of the circular orbit ρ and other constants in the problem.

Problem 8.15 A cyclotron is made of sheet metal in the form of an empty tuna-fish can, set on a table with a flat side down and then sliced from above through its center into two D-shaped pieces. The two "Dees" are then separated slightly so there is a small gap between them. A high-frequency alternating voltage is applied to the Dees, so they are always oppositely charged. At peak voltage there is therefore an electric field in the gap from the positive to the negative Dee that can accelerate charges across the gap. There is also a constant and uniform magnetic field applied vertically, i.e., perpendicular to the Dees, supplied by a large external electromagnet. Therefore after a charged particle has been accelerated across the gap it enters a Dee, where it follows a semicircular path due to the magnetic field and maintains constant speed because the electric field inside a Dee is negligible. By the time the charged particle has completed a semicircle it arrives back at the gap, but by now the charges on the two Dees have been reversed, so the particle is again accelerated in the gap, entering the previous Dee and then executing a larger semicircular path, this time because it is moving faster. As the particle moves faster and faster the semicircular paths increase in radius, so in effect the particle moves in a spiraling path until it reaches the outer edge of the machine, where by then it has achieved a very large kinetic energy due to the multiple accelerations it has received by repeatedly passing through the gap. It is then deflected out of the cyclotron where it causes a high-energy collision with other particles at rest in the lab. (a) Assuming that the charged particle is nonrelativistic, show that its kinetic energy by the time it reaches the outer radius R of the cyclotron is $T = q^2 B^2 R^2 / 2mc^2$, where q and m are the particle's charge and mass, B is the magnetic field, and R is the outer radius of the cyclotron. (b) If we want to accelerate protons to a kinetic energy of 16 MeV, what must be the applied magnetic field B (in Gauss) if the diameter of the cyclotron is 1.52 cm? Protons have mass energy $mc^2 = 938$ MeV and charge $q = 4.8 \times 10^{-10}$ esu. Note that 1 eV $= 1.602 \times 10^{-12}$ ergs $= 1.602 \times 10^{-19}$ Joules. In Gaussian units, B is measured in "Gauss" and in Standard International (SI) units, B is measured in "Teslas," where 1 Tesla $= 10^4$ Gauss.

Problem 8.16 A nice feature of the cyclotron described in the preceding problem is that the alternating current frequency applied to the "Dees" is a constant $\omega = qB/mc$ for nonrelativistic particles, regardless of their energy, so the circulating particles will arrive at the gaps at just the right time. No matter the radius at which a particle orbits, the time it takes to travel between two gap encounters is exactly the same. (a) Show that this is no longer true for relativistic particles. Find a new expression for ω in terms of q, B, m, c, and $\gamma \equiv (1 - \beta^2)^{-1/2} \equiv (1 - v^2/c^2)^{-1/2}$. (b) How might one design an "isochronous cyclotron," in which relativistic protons will still reach the gaps at the correct time, with the same constant-frequency alternating current applied to the Dees? (c) The TRIUMF isochronous cyclotron has a proton outer orbital radius of 7.9 m, where the protons have a kinetic energy of 510 MeV. What is the magnetic field strength at the outer orbit? (d) How fast are these protons moving, expressed as a fraction of the speed of light?

Problem 8.17 Several problems are encountered in trying to scale up cyclotrons to produce increasingly energetic protons. One of them is that the external magnets have to be made larger and larger, which is prohibitively expensive and ultimately becomes completely unfeasible. A newer generation of machines called synchrotrons were therefore invented in which protons can circulate at constant radius, so the magnets only need to cover a much smaller area. (a) In that case, how can relativistic protons be accelerated to higher and higher speeds if their orbital radius remains constant? (b) The Large Hadron Collider (LHC) of CERN (Organisation Européenne pour la Recherche Nucléaire) accelerates protons up to total energies as large as 7.0 TeV (1 TeV $= 10^3$ GeV $= 10^6$ MeV), or perhaps even larger. The circumference of the proton path is 27 km, lying in an underground tunnel near Geneva, partly in Switzerland and partly in France. What magnetic field B is required in this case?

Problem 8.18 Consider two inertial frames $\mathfrak{O}$ and $\mathfrak{O}'$, where $\mathfrak{O}'$ is moving with velocity $\mathbf{v}$ relative to $\mathfrak{O}$. We split all three-vectors into components parallel and perpendicular to the direction of the Lorentz boost, $\mathbf{v}$: for example, we have

$\mathbf{E} = \mathbf{E}_{\parallel} + \mathbf{E}_{\perp}$. Show that the Lorentz transformations of the electric and magnetic fields given in the text, Eqs. (8.35) and (8.36), can be written instead as

$$\mathbf{E}'_{\parallel} = \mathbf{E}_{\parallel}, \quad \mathbf{E}'_{\perp} = \gamma \left(\mathbf{E}_{\perp} + ((\mathbf{v}/c) \times \mathbf{B})_{\perp} \right),$$
$$\mathbf{B}'_{\parallel} = \mathbf{B}_{\parallel}, \quad \mathbf{B}'_{\perp} = \gamma \left(\mathbf{B}_{\perp} - (\mathbf{v}/c \times \mathbf{E})_{\perp} \right).$$

★★★ **Problem 8.19** We discovered in the text that the scalar and vector potentials are components of a four-vector $A^{\mu} = (\phi, \mathbf{A})$. In this problem, we will take as given the existence of this four-vector potential A^{μ} and, using the known Lorentz transformation of a four-vector and the relations of A^{μ} to $\mathbf{E}$ and $\mathbf{B}$, we want to *derive* the Lorentz transformations of $\mathbf{E}$ and $\mathbf{B}$. Consider two inertial frames $\mathfrak{O}$ and $\mathfrak{O}'$, where $\mathfrak{O}'$ is moving with velocity $\mathbf{v}$ relative to $\mathfrak{O}$. We split all three-vectors into components parallel and perpendicular to the direction of the Lorentz boost, $\mathbf{v}$: for example, we have $\mathbf{E} = \mathbf{E}_{\parallel} + \mathbf{E}_{\perp}$. Note that the gradient vector can also be decomposed as $\boldsymbol{\nabla} = \boldsymbol{\nabla}_{\parallel} + \boldsymbol{\nabla}_{\perp}$. (a) First show that $\mathbf{B}'_{\parallel} = \mathbf{B}_{\parallel}$. (b) Show next that $\boldsymbol{\nabla}'_{\parallel} = \gamma \left(\boldsymbol{\nabla}_{\parallel} + (\mathbf{v}/c^2)(\partial/\partial t) \right)$. (c) Finally, show that $\mathbf{B}'_{\perp} = \gamma \left(\mathbf{B}_{\perp} - ((\mathbf{v}/c) \times \mathbf{E})_{\perp} \right)$, as in the previous problem.

★★★ **Problem 8.20** In the previous problem, you derived the Lorentz transformations of the $\mathbf{B}$ field starting with the assumption that the scalar and vector potentials are components of a four-vector $A^{\mu} = (\phi, \mathbf{A})$. Using a similar approach, derive the Lorentz transformation of the electric field $\mathbf{E}$; show that you get $\mathbf{E}'_{\parallel} = \mathbf{E}_{\parallel}$ and $\mathbf{E}'_{\perp} = \gamma \left(\mathbf{E}_{\perp} + ((\mathbf{v}/c) \times \mathbf{B})_{\perp} \right)$. Note that this is a more involved computation than in the previous problem.

★★ **Problem 8.21** Using the Lorentz transformations of the $\mathbf{E}$ and $\mathbf{B}$ fields, show that $E^2 - B^2$ is a Lorentz invariant; that is, show that $E'^2 - B'^2 = E^2 - B^2$.

★★ **Problem 8.22** Using the Lorentz transformations of the $\mathbf{E}$ and $\mathbf{B}$ fields, show that $\mathbf{E} \cdot \mathbf{B}$ is a Lorentz invariant; that is, show that $\mathbf{E}' \cdot \mathbf{B}' = \mathbf{E} \cdot \mathbf{B}$.

★★ **Problem 8.23** Show that the action of a relativistic charged particle (8.43) is invariant under a gauge transformation.

★★★ **Problem 8.24** Using Noether's theorem, find the conserved quantity that results from the invariance of the action (8.43) under gauge transformation of the four-vector potential. For this, consider an infinitesimal but arbitrary gauge transformation.

★★ **Problem 8.25** Derive the equations of motion resulting from the action of a relativistic charged particle (8.43) and verify that you get the Lorentz force law.

★★★ **Problem 8.26** Show that Maxwell's equations given by (8.16) imply the *wave equations*

$$\nabla^2 \phi - \frac{1}{c^2} \frac{\partial^2 \phi}{\partial t^2} = -4\pi \rho_Q \tag{8.182}$$

and

$$\nabla^2 \mathbf{A} - \frac{1}{c^2} \frac{\partial^2 \mathbf{A}}{\partial t^2} = -\frac{4\pi}{c} \mathbf{J}. \tag{8.183}$$

To do this, you will need the vector identity $\nabla \times \nabla \times \mathbf{A} = \nabla(\nabla \cdot \mathbf{A}) - \nabla^2 \mathbf{A}$. You will also need to *fix* the gauge freedom using $\nabla \cdot \mathbf{A} + (1/c)\partial\phi/\partial t = 0$. The latter is allowed due to the gauge symmetry discussed in the text.

★ **Problem 8.27** Using the Lorentz transformation of the four-vector potential A^μ and the wave equations from the previous problem, deduce the Lorentz transformations of charge density ρ_Q and current density $\mathbf{J}$.

★★ **Problem 8.28** A relativistic particle with charge Q and mass M moves in the presence of a uniform electric field $\mathbf{E} = E_0\hat{z}$. The initial energy is K_0 and the momentum is p_0 in the $\hat{y}$ direction. Show that the trajectory in the y–z plane is described by

$$z = \frac{K_0 + Mc^2}{QE_0} \cosh\left(\frac{QE_0}{p_0 c} y\right). \tag{8.184}$$

★★ **Problem 8.29** A relativistic particle of charge Q and mass M is moving in uniform circular motion bound by a radial potential. We learned from Eq. (8.167) that the charge will lose energy to electromagnetic radiation. Assuming that this loss of energy is slow, we can describe the particle as gradually spiraling toward $r = 0$ while maintaining constant angular momentum. Apply this treatment to the ground state of the hydrogen atom, where the atomic radius is about 1 Å (10^{-10} m); estimate the time it takes for the electron to crash into the nucleus. Are you surprised? Why does this not happen?

★★ **Problem 8.30** A particle of charge Q and mass M moves through a region of uniform magnetic and gravitational fields described by constant field vectors $\mathbf{B}$ and $\mathbf{g}$, respectively. Show that the particle will have a drift velocity given by $Mc(\mathbf{g} \times \mathbf{B})/QB^2$.

★★ **Problem 8.31** A particle of charge Q and mass M starts at the origin of the coordinate system with initial speed v_0 in the $\hat{z}$ direction. There are uniform electric and magnetic fields E and B in the $\hat{x}$ direction. Find the location of the particle when it has reached one-half of its maximum value in z for the first time.

9

Accelerating Frames

As we saw in Chapter 1, Newton's laws are valid only for observers at rest in an *inertial* frame of reference. For example, Newton's second law $\mathbf{F} = m\mathbf{a}$ predicts that a body of mass m will move in a straight line at constant velocity ($\mathbf{a} = 0$) if there is no net force $\mathbf{F}$ on it as seen from the perspective of an inertial observer. But to an observer in a non-inertial frame, like an accelerating car or a rotating carnival ride, the same object will generally move in accelerated curved paths even when no forces act upon it.

The challenge is that we all live in non-inertial, *accelerating* reference frames, even if we are standing still on the ground. The earth rotates about its axis while also orbiting the sun; the sun orbits around the galactic center; our entire galaxy is accelerating toward the constellation Orion; and so on. So it seems we cannot use $\mathbf{F} = m\mathbf{a}$! How then can we do mechanics from the vantage point of actual, *non*-inertial frames?

In many tabletop situations, the effects of the non-inertial perspective are small and can be neglected. Yet even in these situations we often still need to quantify how small these effects are. Furthermore, learning how to study dynamics from the non-inertial vantage point turns out to be critical in understanding many other interesting phenomena, including the directions of large-scale ocean currents, the formation of weather patterns – including hurricanes and tornados, life inside rotating space colonies or accelerating spacecraft, and rendezvousing with orbiting space stations.

There is an infinity of ways a frame might accelerate relative to an inertial frame. Two stand out as particularly interesting and useful:

- linearly uniformly accelerating frames, and
- rotating frames.

9.1 Linearly Accelerating Frames

Observers living inside a uniformly accelerating spaceship experience an effective gravity attracting them toward the rear of the ship. If they hold up a ball and let it go, it "falls," or if they throw the ball, it also "falls," while tracing out a parabolic trajectory, as shown in Figure 9.1(a). Every experiment we try inside the ship proceeds just as though there were real gravity directed toward the rear of the ship. This of course is just Einstein's principle of equivalence, already introduced in Chapter 3.

There are two ways to find the motion of objects in a uniformly accelerating spaceship. We assume nonrelativistic motion throughout and therefore Galilean transformation rules apply.

Inertial Perspective

The first approach is to work out a particle's motion from the point of view of a hypothetical observer at rest in some inertial frame outside the ship (who *can* use Newton's laws) and then translate this motion into the accelerating ship itself. For example, take the case of the ball thrown sideways, from one cabin wall toward the opposite wall. In the frame of an external inertial observer the ball moves in a straight line at constant speed because there are no forces on it. Two pictures of the ball are shown in Figure 9.1(b): when the ball is launched, and when it reaches the opposite wall. It moves in a straight line at constant speed, and while doing so the spaceship accelerates upward with acceleration a_s from the inertial observer's point of view. In Lagrangian language, the inertial observer writes

$$L = T - U = \frac{1}{2}m\left(\dot{x}^2 + \dot{y}^2\right) \Rightarrow \ddot{y} = 0 \tag{9.1}$$

for the ball, and hence $y = y_0$ for all time. At a later time Δt when the ball strikes the other wall of the ship, the ship has moved upward a distance $(1/2)a_s\Delta t^2$ relative to the inertial observer, where a_s is the ship's acceleration. Therefore, if y_0 is the initial height of the ball from the cabin floor, *relative to the ship* the ball will subsequently have height $y_0 - (1/2)a_s\Delta t^2$. The ball strikes the opposite cabin wall lower down than the point from where it was thrown. Therefore the ball has "fallen" relative to the ship, and in fact its trajectory in the ship is parabolic. This behavior is exactly as though there were a uniform effective gravity $g_{\text{eff}} = a_s$ within the ship, numerically equal to the ship's acceleration, but directed toward the rear of the ship.

Non-inertial Perspective

The second way to do mechanics inside the uniformly accelerating ship is to find an equation analogous to $\mathbf{F} = m\mathbf{a}$ that is valid in the accelerating frame, and then solve the new equation to find the motion. The analogous equation in this case of uniform acceleration is easy to find. Suppose the accelerating frame of the ship is the primed frame and the external inertial frame is unprimed. The position vectors of a particle are therefore related by (see Figure 9.1(b))

$$\mathbf{r} = \mathbf{r}' + \frac{1}{2}\mathbf{a}_s t^2, \tag{9.2}$$

where $\mathbf{a}_s$ is the acceleration of the ship relative to the inertial, unprimed frame, and where we assume that the origins of the two frames coincide at time $t = 0$. By differentiating this equation, the velocities of the particle are related by $\mathbf{v} = \mathbf{v}' + \mathbf{a}_s t$ and the accelerations are related by $\mathbf{a} = \mathbf{a}' + \mathbf{a}_s$. That is, the acceleration of an

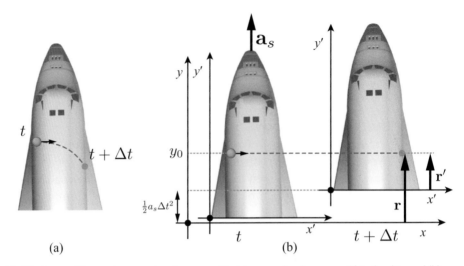

Fig. 9.1 A ball is thrown sideways in an accelerating spaceship (a) as seen by observers within the ship, and (b) as seen by a hypothetical inertial observer outside the ship. The position vectors $\mathbf{r}$ and $\mathbf{r}'$ are also shown.

object as seen in the unprimed inertial frame is the vector sum of two accelerations: the acceleration of the ship itself, plus the object's acceleration relative to the ship. Now Newton's second law $\mathbf{F} = m\mathbf{a}$ for an object of mass m is valid in the *inertial* frame, where $\mathbf{F}$ is the sum of the forces acting on the object; we now know that we can also write this in the form $\mathbf{F} = m\,\mathbf{a} = m(\mathbf{a}' + \mathbf{a}_{\mathrm{s}})$, or

$$\mathbf{F} - m\mathbf{a}_{\mathrm{s}} = m\mathbf{a}'. \tag{9.3}$$

Note that if we define $-m\mathbf{a}_{\mathrm{s}}$ to be a new "pseudoforce" $\mathbf{F}_{\mathrm{pseudo}} = -m\mathbf{a}_{\mathrm{s}}$, then in the accelerating primed frame

$$\mathbf{F}' = m\mathbf{a}', \tag{9.4}$$

where

$$\mathbf{F}' \equiv \mathbf{F} - m\mathbf{a}_{\mathrm{s}} = \mathbf{F} + \mathbf{F}_{\mathrm{pseudo}}, \tag{9.5}$$

the sum of the real forces $\mathbf{F}$ and the pseudoforce $-m\mathbf{a}_{\mathrm{s}}$. Therefore, in the example at hand with the accelerating rocket, we have $\mathbf{F} = 0$ but $\mathbf{F}' \neq 0$; the non-inertial observer wants to write Eq. (9.4) with $\mathbf{F}' = 0 - m\,\mathbf{a}_{\mathrm{s}} = m\mathbf{a}'$, which implies $\mathbf{a}' = -\mathbf{a}_{\mathrm{s}}$. Thus, the ball is seen by the non-inertial observer tracing out a parabolic trajectory with constant acceleration $-\mathbf{a}_{\mathrm{s}}$.

It is often convenient to define an **effective gravity**

$$\mathbf{g}_{\mathrm{eff}} \equiv -\mathbf{a}_{\mathrm{s}} \tag{9.6}$$

within the ship, so then the new pseudoforce is $m\mathbf{g}_{\mathrm{eff}}$. In going over to a uniform linearly accelerating frame we learn that we can use Newton's second law for an object, as long as we add an effective uniform gravitational force to all the real forces acting on an object as in Eq. (9.5). In the Lagrangian approach, this implies that we need to consider the addition of a pseudoforce potential to the Lagrangian. From the non-inertial perspective, we would need to write

$$L = T' - U', \tag{9.7}$$

where T' is the kinetic energy as measured in the non-inertial frame, and the potential U' is

$$U' = U + U_{\text{pseudo}}, \tag{9.8}$$

where U is any potential present in the inertial perspective, and U_{pseudo} is the potential energy from any pseudoforces. For the example at hand we can write

$$U_{\text{pseudo}} = mg_{\text{eff}}y', \tag{9.9}$$

which gives the new Lagrangian

$$L = \frac{1}{2}m\left((\dot{x}')^2 + (\dot{y}')^2\right) - mg_{\text{eff}}y', \tag{9.10}$$

written in terms of the primed coordinates within the spaceship.

Example 9.1   **Pendulum in an Accelerating Spaceship**

Passengers in a uniformly accelerating spaceship construct a simple pendulum, with a string of length R attached at its upper end to a fixed point in the ship, and a bob of mass m attached to its lower end, as shown in Figure 9.2. An easy way to find the equation of motion of the pendulum is to use Lagrange's equation in the accelerating frame, which we can do as long as we add the pseudoforce mg_{eff} to the bob in the form of an effective potential energy $U_{\text{eff}} = mg_{\text{eff}}y = mg_{\text{eff}}R(1 - \cos\theta')$ acting on the bob, where θ' is the angle of the string relative to the "vertical" in the spaceship frame, and y is measured up from the lowest point of the bob. The Lagrangian of the bob is therefore

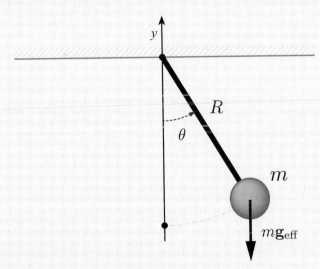

Fig. 9.2   A simple pendulum in an accelerating spaceship.

$$L = T' - U' = \frac{1}{2}mR^2 \left(\dot{\theta}'\right)^2 - mg_{\text{eff}}R(1 - \cos\theta') \tag{9.11}$$

in terms of the generalized coordinate θ'. The Lagrange equation

$$\frac{\partial L}{\partial \theta'} - \frac{d}{dt}\frac{\partial L}{\partial \dot{\theta}'} = -mg_{\text{eff}}R\sin\theta' - \frac{d}{dt}mR^2\dot{\theta}' = 0 \tag{9.12}$$

then gives the usual pendulum equation $\ddot{\theta}' + (g_{\text{eff}}/R)\sin\theta' = 0$, so an experiment done inside a uniformly accelerating frame without gravity gives the same results as the identical experiment done in an inertial frame containing a uniform gravitational field. ∎

9.2 Rotating Frames

Consider a number of space colonists living on the inside of a cylindrical rotating space colony, as illustrated in Figure 9.3. The colony is a long hollow tube of radius R rotating about its symmetry axis with angular velocity ω, far from any gravitating planet. Nevertheless, colonists find they are pressed against the inside of the rim, as though there were an *outward* gravitational force. They can look overhead and see other people living on other parts of the rim. If they toss a ball overhead, toward the symmetry axis, the ball rises and then falls back. If they throw it harder, however, it falls back, but to a point some distance away. Why do they experience an effective outward gravity, and why does a thrown ball behave as it does?

A rotating frame might be a frame attached to a merry-go-round turning at constant angular velocity or within a uniformly rotating space colony or placed on the rotating earth. As with linearly accelerating frames, there are two ways to deal with motion in rotating frames:

- Find the motion of an object as viewed from an external inertial frame in which $\mathbf{F} = m\mathbf{a}$ is valid, and then translate the results into the coordinates of the rotating frame.
- Reformulate $\mathbf{F} = m\mathbf{a} \rightarrow \mathbf{F}' = m\mathbf{a}'$ so that it is valid in the rotating primed coordinate system itself; then solve for the motion directly in the rotating frame, without ever referring back to an inertial frame.

Inertial Perspective

We can find the motion of a space colonist or ball from the point of view of a hypothetical external inertial observer who peers into the colony through a window on an endcap of the cylinder. This inertial observer is able to use $\mathbf{F} = m\mathbf{a}$, so has a way to predict the observed motions. The results can then be translated into the rotating frame of the colony. For example, the hypothetical external inertial observer looks at a colonist standing on the inside rim. As the colony turns, the

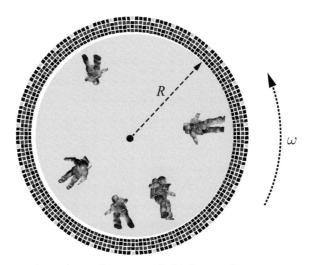

Fig. 9.3   Space colonists living on the inside rim of a rotating cylindrical space colony.

colonist moves in a circle of radius R at angular velocity ω, and is therefore accelerating toward the symmetry axis with centripetal acceleration $a_{\text{cent}} = \omega^2 R$. The inertial observer can use $\mathbf{F} = m\mathbf{a}$, so knows that there is some force causing this inward acceleration; the force is the normal force N of the rim on the colonist's feet: we then have $N = m\omega^2 R$. If the colonist happens to be standing on a scale, the reading of the scale is the normal force N, so the weight of the colonist is

$$N = m\omega^2 R = mg_{\text{eff}} \tag{9.13}$$

and the effective outward gravity at the colony rim is $g_{\text{eff}} = \omega^2 R$. When the colonist throws a ball, the ball is not in contact with anything, so the inertial observer will see that the ball goes in a straight line at constant speed. Let us look more quantitatively at a particular realization of this scenario as seen from the inertial perspective.

Example 9.2   **Throwing a Ball in a Rotating Space Colony**

A cylindrical space colony of radius R rotates with period P. A colonist standing on the rim throws a ball straight "up" toward the rotation axis with the particular speed $v = 2\pi R/P$. What is the path of the ball as seen by the colonist? How long does it take the ball to fall back to the rim? And how far along the rim must the colonist run, relative to the rim itself, to catch the ball?

Note that as seen by the inertial observer, the rim of the colony also moves with speed $v = 2\pi R/P$, the circumference divided by the period. If the colonist throws the ball toward the rotation axis with this same speed, then from the point of view of the outside inertial observer the ball moves at a $45°$ angle, as shown in Figure 9.4(a). In the inertial frame the ball subsequently moves in a straight line at constant velocity, since there is no force on it, so it intersects the rim one-quarter of the way around, as shown.

The ball's speed in this frame is

$$v = (v_x^2 + v_y^2)^{1/2} = \sqrt{2}\,(2\pi R/P), \tag{9.14}$$

so the time required for the ball to reach (*i.e.*, "fall to") the rim is

$$t = \frac{\text{distance}}{\text{speed}} = \frac{\sqrt{2}R}{\sqrt{2}(2\pi R/P)} = \frac{P}{2\pi}. \tag{9.15}$$

During this time the rim itself turns a distance

$$d = \text{speed} \times \text{time} = \frac{2\pi R}{P}\left(\frac{P}{2\pi}\right) = R, \tag{9.16}$$

so the ball strikes the rim a distance $(2\pi R/4 - R) = (\pi/2 - 1)R = 0.57R$ from the colonist. This is how far the colonist has to run to catch the ball, even though the ball was thrown vertically upward in the colony frame. The trajectory of the ball as seen by the colonist is as shown in Figure 9.4(b). Note that it rises vertically at the beginning, that its highest point is a distance $R/\sqrt{2}$ from the colony center, and that as it arrives at the rim it falls vertically (*i.e.*, perpendicular to the rim itself at that point), a distance $0.57R$ around the rim from where it started.

If a ball is thrown upward only a short distance – if its initial speed in the colony is only a small fraction of the colony's rim speed – the ball will behave more nearly like a ball thrown on earth, rising almost straight up and falling almost straight down with effective gravitational acceleration given by $-\omega^2 R$.

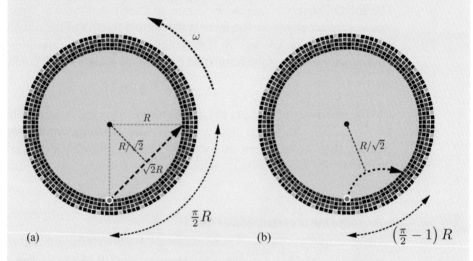

Fig. 9.4   Throwing a ball in a rotating space colony: (a) from the point of view of an external inertial observer; (b) from the point of view of a colonist.   ∎

Non-inertial Perspective

Is there an equation analogous to $\mathbf{F} = m\mathbf{a}$ that we can apply *directly* in a rotating frame? Much like the linearly accelerating case, the non-inertial perspective for rotating frames requires the addition of pseudoforces to Newton's second law. However, the rotating scenario is significantly more involved and benefits from a more general treatment than that explored by our particular example. We therefore relegate the discussion to a new section, treating the problem more generally – considering pseudoforces for both uniformly and non-uniformly rotating frames. After developing the formalism, we will come back to the space colony example – as seen from the perspective of the colonists.

9.3 Pseudoforces in Rotating Frames

We begin by finding the velocity $\mathbf{v}_{\text{in}}$ of a particle in an inertial frame in terms of its velocity $\mathbf{v}_{\text{rot}}$ in the rotating frame. This will allow us to write the Lagrangian directly from the perspective of the non-inertial rotating observer.

Suppose some vector $\mathbf{A}$ is *at rest* in the rotating frame. We take $\mathbf{A}$ to be perpendicular to the rotation axis, with its tail located at the rotation axis itself, as shown in Figure 9.5(a). Then if the rotating frame turns through angle $d\varphi$, the change in $\mathbf{A}$ is $d\mathbf{A}$, which is perpendicular to $\mathbf{A}$ in the limit of small $d\varphi$. The magnitude of $d\mathbf{A}$ is $dA = A d\varphi$. In vector cross-product form, we can write

$$dA = d\varphi \times \mathbf{A}, \tag{9.17}$$

where the vector $d\varphi$ points out of the page for counterclockwise rotation, according to the right-hand rule. Looking at Figure 9.5(b), we see that Eq. (9.17) actually holds whether $\mathbf{A}$ is perpendicular to the rotation axis or not, and whether its tail is at the rotation axis or not, since $|d\mathbf{A}| = |d\varphi \times \mathbf{A}| = |d\varphi||\mathbf{A}|\sin\theta$.

Now if $\mathbf{A}$ also changes in the rotating frame, say by the amount $d\mathbf{A}_{\text{rot}}$, we have to add $d\mathbf{A}_{\text{rot}}$ to the change due to frame rotation. That is:

$$d\mathbf{A}_{\text{in}} = d\mathbf{A}_{\text{rot}} + d\varphi \times \mathbf{A} \tag{9.18}$$

is the total change in the vector $\mathbf{A}$ observed from the inertial frame. Dividing by the small time interval dt during which the rotating frame turns by $d\varphi$, we find

$$\left.\frac{d\mathbf{A}}{dt}\right|_{\text{in}} = \left.\frac{d\mathbf{A}}{dt}\right|_{\text{rot}} + \omega \times \mathbf{A}, \tag{9.19}$$

where $\omega = d\varphi/dt$ is the instantaneous angular velocity of the rotating frame.[1] Equation (9.19) relates the time rate of change of *any* vector $\mathbf{A}$ between the inertial and rotating frames. Note also that ω need not be constant. We emphasize the generality of the transformation between inertial and rotating frames by simply writing the operator as

$$\left.\frac{d}{dt}\right|_{\text{in}} = \left.\frac{d}{dt}\right|_{\text{rot}} + \omega \times, \tag{9.20}$$

where the operator can operate on any vector $\mathbf{A}$.

---

[1] More formally, we have divided a small change $\Delta \mathbf{A}$ by a small change Δt and then taken the limit as $\Delta t \to 0$.

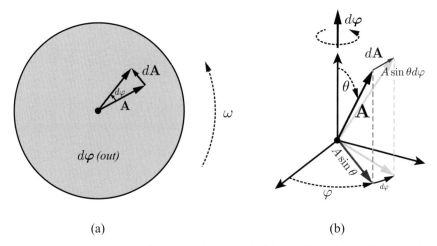

(a)   (b)

Fig. 9.5 A vector that is constant in a rotating frame changes in an inertial frame: (a) simple two-dimensional case; (b) general three-dimensional case.

In particular, let $\mathbf{A} = \mathbf{r}$, the position of the particle. Then $d\mathbf{r}/dt|_{\text{in}} = \mathbf{v}_{\text{in}}$ is the velocity of the particle measured in the inertial frame and $d\mathbf{r}/dt|_{\text{rot}} = \mathbf{v}_{\text{rot}}$ is its velocity in the rotating frame, so

$$\mathbf{v}_{\text{in}} = \mathbf{v}_{\text{rot}} + \boldsymbol{\omega} \times \mathbf{r}. \tag{9.21}$$

The second term on the right is the velocity the particle would have in the inertial frame if it were at rest in the rotating frame, a distance $\mathbf{r}$ from the axis.

We are now ready to write the Lagrangian from the perspective of the non-inertial rotating frame. We begin with the Lagrangian of a single particle of mass m, given by

$$L = \frac{1}{2}m\mathbf{v}^2 - U(\mathbf{r}), \tag{9.22}$$

with some arbitrary potential $U(\mathbf{r})$. Recall that the Lagrangian must be written in an *inertial* frame since the equations of motion we map onto, $\mathbf{F} = m\mathbf{a}$, constitute Newton's second law – written in an inertial frame. This means that the velocity $\mathbf{v}$ that appears in the Lagrangian is $\mathbf{v}_{\text{in}}$. We then have

$$L = \frac{1}{2}m\mathbf{v}_{\text{in}}^2 - U(\mathbf{r}) = \frac{1}{2}m\left(\mathbf{v}_{\text{rot}} + \boldsymbol{\omega} \times \mathbf{r}\right)^2 - U(\mathbf{r}). \tag{9.23}$$

The last form expresses the Lagrangian in terms of $\mathbf{v}_{\text{rot}}$. As such, it is most convenient to take $\boldsymbol{\omega}$ and $\mathbf{r}$ as vectors expressed in the rotating reference frame as well – even though in principle the scalar products that appear within L are invariant in this respect. We can now write

$$\frac{1}{2}m\left(\mathbf{v}_{\text{rot}}+\boldsymbol{\omega}\times\mathbf{r}\right)^{2}=\frac{1}{2}m\left(\mathbf{v}_{\text{rot}}\right)^{2}+m\mathbf{v}_{\text{rot}}\cdot\left(\boldsymbol{\omega}\times\mathbf{r}\right)+\frac{1}{2}m\left(\boldsymbol{\omega}\times\mathbf{r}\right)^{2}$$

$$=\frac{1}{2}m\mathbf{v}_{\text{rot}}^{2}+m\mathbf{v}_{\text{rot}}\cdot\left(\boldsymbol{\omega}\times\mathbf{r}\right)$$

$$+\frac{1}{2}m\omega^{2}r^{2}-\frac{1}{2}m\left(\boldsymbol{\omega}\cdot\mathbf{r}\right)^{2}, \tag{9.24}$$

where the last line was obtained by using the vector identity

$$\left(\mathbf{a}\times\mathbf{b}\right)\cdot\left(\mathbf{c}\times\mathbf{d}\right)=\left(\mathbf{a}\cdot\mathbf{c}\right)\left(\mathbf{b}\cdot\mathbf{d}\right)-\left(\mathbf{b}\cdot\mathbf{c}\right)\left(\mathbf{a}\cdot\mathbf{d}\right). \tag{9.25}$$

So we have found that in a rotating frame the Lagrangian can be written in terms of rotating coordinates as

$$L=\frac{1}{2}m\mathbf{v}_{\text{rot}}^{2}+m\mathbf{v}_{\text{rot}}\cdot\left(\boldsymbol{\omega}\times\mathbf{r}\right)+\frac{1}{2}m\omega^{2}r^{2}-\frac{1}{2}m\left(\boldsymbol{\omega}\cdot\mathbf{r}\right)^{2}-U(\mathbf{r}). \tag{9.26}$$

We can already see that from the perspective of the rotating frame, the kinetic energy is not simply $(1/2)m\mathbf{v}_{\text{rot}}^{2}$. To explore the effects of the additional terms, we need to look at the equations of motion in the rotating frame:

$$\frac{d}{dt}\left(\frac{\partial L}{\partial\dot{r}_{\text{rot}}^{i}}\right)=\frac{\partial L}{\partial r^{i}_{\text{rot}}}, \tag{9.27}$$

where all vectors in the Lagrangian are assumed written in terms of their components in the rotating frame (i is x, y, or z). We emphasize this by labeling the components of any vector in the rotating frame with the "rot" superscript or subscript. For the left-hand side, we find that

$$\frac{d}{dt}\left(\frac{\partial L}{\partial\dot{r}_{\text{rot}}^{i}}\right)=m\,\ddot{r}_{\text{rot}}^{i}+m\left(\boldsymbol{\omega}\times\mathbf{v}_{\text{rot}}\right)^{i}+m\left(\frac{d\boldsymbol{\omega}}{dt}\times\mathbf{r}\right)^{i}, \tag{9.28}$$

where we have used $\mathbf{v}_{\text{rot}}=d\mathbf{r}/dt$ and allowed for the possibility that $\boldsymbol{\omega}$ may vary with time. To compute the right-hand side of Eq. (9.27), it is convenient to rewrite the second term in the Lagrangian as

$$m\,\mathbf{v}_{\text{rot}}\cdot\left(\boldsymbol{\omega}\times\mathbf{r}\right)=m\,\mathbf{r}\cdot\left(\mathbf{v}_{\text{rot}}\times\boldsymbol{\omega}\right), \tag{9.29}$$

using the general vector identity $\mathbf{a}\cdot\left(\mathbf{b}\times\mathbf{c}\right)=\mathbf{b}\cdot\left(\mathbf{c}\times\mathbf{a}\right)=\mathbf{c}\cdot\left(\mathbf{a}\times\mathbf{b}\right)$. This helps in computing the spatial derivative of the Lagrangian more compactly:

$$\frac{\partial L}{\partial r^{i}_{\text{rot}}}=m\left(\mathbf{v}_{\text{rot}}\times\boldsymbol{\omega}\right)^{i}_{\text{rot}}+m\,\omega^{2}r^{i}_{\text{rot}}-m\left(\boldsymbol{\omega}\cdot\mathbf{r}\right)\omega^{i}-\frac{\partial U(\mathbf{r})}{\partial r^{i}_{\text{rot}}}. \tag{9.30}$$

Putting things together, we get the equation of motion in the rotating frame:

$$m\mathbf{a}_{\text{rot}}=m\,\omega^{2}\mathbf{r}_{\text{rot}}-m\left(\boldsymbol{\omega}\cdot\mathbf{r}\right)\boldsymbol{\omega}_{\text{rot}}-2\,m\left(\boldsymbol{\omega}\times\mathbf{v}_{\text{rot}}\right)_{\text{rot}} \tag{9.31}$$

$$-m\left(\frac{d\boldsymbol{\omega}}{dt}\times\mathbf{r}\right)_{\text{rot}}-\nabla_{\text{rot}}U(\mathbf{r}).$$

The first two terms on the right can be combined to give $-m\,\boldsymbol{\omega} \times (\boldsymbol{\omega} \times \mathbf{r})_{\text{rot}}$, using the vector identity $\mathbf{a} \times (\mathbf{b} \times \mathbf{c}) = (\mathbf{a} \cdot \mathbf{c})\,\mathbf{b} - (\mathbf{a} \cdot \mathbf{b})\,\mathbf{c}$. Then, *defining* the total force acting in the rotating frame as

$$\mathbf{F}_{\text{rot}} = m\mathbf{a}_{\text{rot}}, \tag{9.32}$$

it follows that

$$\mathbf{F}_{\text{rot}} = \mathbf{F}_{\text{in}} \underbrace{-m\,\boldsymbol{\omega} \times (\boldsymbol{\omega} \times \mathbf{r})_{\text{rot}}}_{\text{centrifugal}} \underbrace{-2\,m\,(\boldsymbol{\omega} \times \mathbf{v}_{\text{rot}})_{\text{rot}}}_{\text{Coriolis}} \underbrace{-m\,(\dot{\boldsymbol{\omega}} \times \mathbf{r})_{\text{rot}}}_{\text{Euler}}, \tag{9.33}$$

where $\mathbf{F}_{\text{in}} = -\nabla_{\text{rot}} U(\boldsymbol{r})$ is the sum of the real forces acting in the inertial frame. This implies that *in the rotating frame the particle is subject to three new pseudoforces in addition to real physical ones.* These are labeled the **centrifugal**, **Coriolis**, and **Euler** pseudoforces.

For notational simplicity, we will henceforth drop the "rot" labels on all the above expressions. We start by looking more closely at the centrifugal pseudoforce

$$\mathbf{F}_{\text{cen}} = -m\,\boldsymbol{\omega} \times (\boldsymbol{\omega} \times \mathbf{r}), \tag{9.34}$$

named so because in the rotating frame it pushes the particle *away* from the axis of rotation. Suppose for example the rotating frame turns counterclockwise, so the vector $\boldsymbol{\omega}$ points out of the page as in Figure 9.6. If $\mathbf{r}$ is measured directly out from the rotation axis, then $\mathbf{r}$ is perpendicular to $\boldsymbol{\omega}$, so $\boldsymbol{\omega} \times \mathbf{r}$ has magnitude ωr and points to the right as shown. Then since $\boldsymbol{\omega}$ is necessarily perpendicular to $\boldsymbol{\omega} \times \mathbf{r}$, the cross product $\boldsymbol{\omega} \times \boldsymbol{\omega} \times \mathbf{r}$ has magnitude $\omega^2 r$ and points toward the rotation axis, as shown in Figure 9.6(b). Finally, we have to multiply by $-m$, so $\mathbf{F}_{\text{cen}}$ has magnitude $m\omega^2 r$ and points *away* from the rotation axis. In vector form, since we have taken $\mathbf{r}$ to be perpendicular to the rotation axis, we can write

$$\mathbf{F}_{\text{cen}} = m\,\omega^2 \mathbf{r}_{\perp}, \tag{9.35}$$

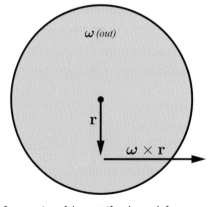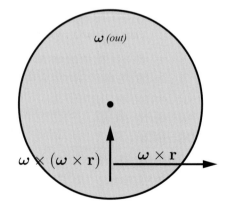

Fig. 9.6　Construction of the centrifugal pseudoforce.

where we have added a $\perp$ subscript to emphasize that this relation is only valid when $\mathbf{r} \to \mathbf{r}_\perp$ is perpendicular to $\boldsymbol{\omega}$. This is the familiar expression mv^2/r, since the tangential velocity of a particle at rest in the rotating frame is $v = r\omega$.

The second pseudoforce is the Coriolis pseudoforce[2]

$$\mathbf{F}_{\text{Cor}} = -2\,m\,\boldsymbol{\omega}\times\mathbf{v}_{\text{rot}}, \tag{9.36}$$

which depends on the velocity of the particle (but not its position) in the rotating frame, in contrast to the centrifugal pseudoforce that depends on the position of the particle but not its velocity. The Coriolis pseudoforce only acts when the particle is *moving* in the rotating frame, in any direction that is not parallel to $\boldsymbol{\omega}$. It tends to deflect the particle's path because $\boldsymbol{\omega}\times\mathbf{v}_{\text{rot}}$ is perpendicular to the direction of motion at any instant.

The third and final pseudoforce is known as the Euler term:

$$\mathbf{F} = -m\left(\frac{d\boldsymbol{\omega}}{dt}\right)\times\mathbf{r}, \tag{9.37}$$

which only acts if the rotating frame is speeding up or slowing down its rotation rate, or if the axis of rotation is changing direction. *In the rest of this chapter we will assume that our rotating frame turns at a steady rate in the same direction, so we will ignore this final term.*

In summary, in uniformly rotating reference frames we have to add two pseudoforces, Eqs. (9.34) and (9.36), to the real forces when using $\mathbf{F} = m\mathbf{a}$ in the rotating frame. Here $\mathbf{F}_{\text{cen}}$ depends on the particle's position but not its velocity in the rotating frame, and pushes the particle away from the axis of rotation. Also $\mathbf{F}_{\text{Cor}}$ depends on the particle's velocity in the rotating frame, but not its position, and pushes the particle in a direction perpendicular to its direction of motion, unless $\mathbf{v}_{\text{rot}}$ is parallel to $\boldsymbol{\omega}$, in which case the Coriolis pseudoforce vanishes.

Example 9.3

Rotating Space Colonies Revisited

Colonists in a rotating cylindrical space colony live on the inside rim of the colony. To outside inertial observers, a colonist standing on the rim travels in a circle of radius R, so accelerates toward the axis of rotation with $a = v^2/R = R\omega^2$. The force causing this acceleration is the *normal* force of the rim acting on the colonist's feet, $N = mR\omega^2$, which is in fact by definition the colonist's *weight*, since if the colonist were standing on a scale, that is what the scale would read. To observers in the non-inertial rotating frame of the colony, the normal force N is balanced by the pseudogravity $mg_{\text{eff}} = mR\omega^2$, i.e., the centrifugal pseudoforce (9.34), so in the rotating frame the net force is zero, and the colonist remains at rest.

Any object *moving* within the colony feels *both* the centrifugal *and* the Coriolis pseudoforce – as seen from the perspective of the colonists. Suppose that a colonist throws a ball tangent to the rim with speed $R\omega$ in the direction opposite to the rim's rotation direction. Then the ball is initially at rest in the inertial frame. Neglecting any air resistance, we know that it will remain at rest in the *inertial frame*, because there are no

---

[2] Named for the French engineer and mathematician Gustave-Gaspard Coriolis (1792–1843), who first understood and explained the effects of this pseudoforce.

real forces on it, as shown in Figure 9.7(a). From the colonist's perspective, there should be no radial motion; instead, the ball would appear to circle around parallel to the ground: the centrifugal pseudoforce $m\omega^2 R$ points outward and the Coriolis pseudoforce $-2m\omega \times \mathbf{v} = -2m\omega(\mathbf{R}\omega)$ points inward. These combine to give a net force $m\omega^2 R$ inward, just enough to "cause" the correct inward acceleration $\omega^2 R$ for circular motion.

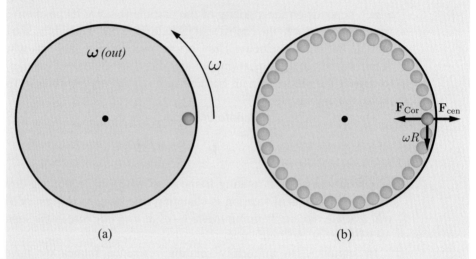

(a)   (b)

Fig. 9.7   Stroboscopic pictures of a ball thrown at speed $v = R\omega$ opposite to the direction of rotation: (a) as seen in an inertial frame; (b) as seen in the colony.   ∎

9.4 Pseudoforces on Earth

Every 24 h the earth makes a complete counterclockwise rotation as seen from a point above the north pole, in a frame in which the sun is at rest. That is, it takes 24 h for the sun to return to its highest point in the sky.[3] In the inertial frame in which the distant stars are at rest, this translates into an angular velocity of

$$\Omega = (2\pi/24 \text{ h} \times 3600 \text{ s/h})(366.5/365.5) = 7.292 \times 10^{-5}\,\text{s}^{-1}, \qquad (9.38)$$

where the factor $(366.5/365.5)$ changes solar time into sidereal (*i.e.*, star) time.

Sidereal time is the time it takes a distant star to return to its highest point in the sky. Because of the earth's orbit around the sun, its angular velocity is slightly greater in the inertial frame of the stars than in the rotating frame of the earth around the sun, as shown in Figure 9.8. The earth takes slightly *less* than 24 h to rotate once relative to the stars. In what follows we will say for short that earth rotates with a 24-h period, although it is really about 0.3% less than that in the inertial frame of

---

[3] This frame is not inertial because the sun moves with respect to the distant stars. The distant stars, however, do present an excellent inertial frame.

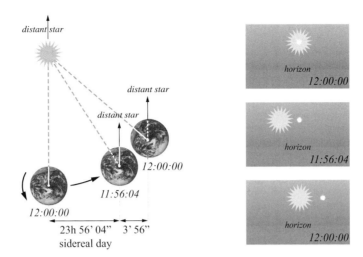

Fig. 9.8 The length of the day relative to the stars (sidereal time) is slightly longer than the length of the day relative to the sun.

the distant stars. Let us now look at the effects of this angular velocity from the perspective of the non-inertial frame that is the surface of the earth.

Centrifugal Pseudoforces

In earth's rotating frame, the centrifugal pseudoforce acts upon every particle within the body of the earth, causing the earth to bulge outward at the equator into an oblate spheroidal shape, as shown in Figure 9.9(a). All particles are subjected to a pseudoforce $m\Omega^2\rho$ outward from the axis of rotation as well as a gravitational force that is roughly toward the center. Here ρ is the distance from the rotation axis. This effect is largest at the equator, where $m\Omega^2\rho$ is about 0.3% of the acceleration of gravity; it is therefore a small but not necessarily negligible correction. If the earth were liquid, the shape of the surface in that case (called the **geoid**) would be everywhere perpendicular to the vector sum of gravity and the centrifugal pseudoforce (see Figure 9.9(b)), which together form an *effective* gravity, $\mathbf{g}_{\text{eff}} = \mathbf{g} - \Omega^2\rho$, where ρ is the vector distance from the rotation axis. There is no centrifugal force at the poles, so $\mathbf{g}_{\text{eff}}$ at that location is due entirely to gravity. At the equator, $\mathbf{g}_{\text{eff}}$ is about 0.5% less than at the poles, for two reasons: at the equator the centrifugal force is opposite to $\mathbf{g}$, so cancels some of it, and since earth bulges at the equator, an object placed there is farther from the center, and so experiences less real gravity. A plumb bob hung from a point above the surface will hang so that $\mathbf{g}_{\text{eff}}$ is parallel to the string. Therefore in the northern hemisphere the string points not toward earth's center, but to a point somewhat south of the center.

Coriolis Pseudoforces

Coriolis forces play an essential role in wind and weather patterns and ocean circulation on earth. They also deflect the trajectories of artillery shells, aircraft,

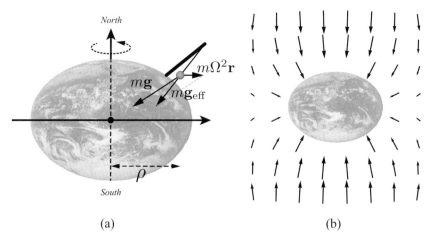

North

$m\Omega^2\mathbf{r}$

$m\mathbf{g}$

$m\mathbf{g}_{\text{eff}}$

ρ

South

(a)　　　　　　　　　　　　(b)

Fig. 9.9　(a) The earth bulges at the equator due to its rotation, which produces a centrifugal pseudoforce in the rotating frame. A plumb bob hanging near the surface experiences both gravitation and the centrifugal pseudoforce. The earth's bulge is shown greatly exaggerated. (b) The effective gravitational vector field around the earth from the perspective of the earth's rotating frame. The deformation of the pattern from that of a central force is shown greatly exaggerated.

and missiles. We concentrate here on effects within a localized region, so we can erect a local Cartesian system with the x, y plane on the surface (x to the *east* and y to the *north*, as in most maps), and z vertically upward, as shown in Figure 9.10(a). In these coordinates, the earth's angular velocity is

$$\mathbf{\Omega} = \Omega(\hat{\mathbf{y}}\cos\lambda + \hat{\mathbf{z}}\sin\lambda), \tag{9.39}$$

where λ is the latitude, which stays approximately constant for localized motion. A particle has velocity components $\mathbf{v} = (\dot{x}\hat{\mathbf{x}} + \dot{y}\hat{\mathbf{y}} + \dot{z}\hat{\mathbf{z}})$, so the Coriolis pseudoforce is

$$\begin{aligned}
\mathbf{F}_{\text{Cor}} &= -2m\mathbf{\Omega}\times\mathbf{v} \\
&= 2m\Omega\left[(\sin\lambda\,\dot{y} - \cos\lambda\,\dot{z})\hat{\mathbf{x}} - \sin\lambda\,\dot{x}\,\hat{\mathbf{y}} + \cos\lambda\,\dot{x}\,\hat{\mathbf{z}}\right]. \tag{9.40}
\end{aligned}$$

The most interesting special case is horizontal motion $\dot{z} = 0$ of the particle: the horizontal components of $\mathbf{F}_{\text{Cor}}$ are then

$$\mathbf{F}_{\text{Cor,horizontal}} = 2m\Omega\sin\lambda(\dot{y}\,\hat{\mathbf{x}} - \dot{x}\,\hat{\mathbf{y}}). \tag{9.41}$$

Note that in terms of the polar angle θ of the plane, as shown in Figure 9.10(b), the velocity of the particle is $\mathbf{v} = (\dot{x}, \dot{y}) = (v\cos\theta, v\sin\theta)$, so

$$\mathbf{F}_{\text{Cor,hor.}} = 2m\Omega\sin\lambda v(\sin\theta\,\hat{\mathbf{x}} - \cos\theta\,\hat{\mathbf{y}}) = -2m\Omega\sin\lambda v\,\hat{\boldsymbol{\theta}}, \tag{9.42}$$

where $\hat{\boldsymbol{\theta}}$ is a unit tangent vector, positive in the counterclockwise sense as shown. It follows that:

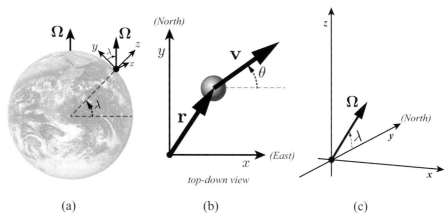

Fig. 9.10 (a) A set of three Cartesian coordinates placed on the earth. (b) The horizontal coordinates *x* and *y*.
(c) A three-dimensional view of the coordinate system setup on the ground, showing the angular
velocity vector of the earth inclined toward the north.

1. $\mathbf{F}_{\mathrm{Cor}}$ pushes moving objects toward the *right* of their direction of motion (*i.e.*,
   in the $-\hat{\theta}$ direction) in the northern hemisphere, where $\sin \lambda > 0$, and toward
   the *left* (in the $\hat{\theta}$ direction) in the southern hemisphere, where $\sin \lambda < 0$.
2. The magnitude of $\mathbf{F}_{\mathrm{Cor}}$ is independent of the direction of motion; the Coriolis
   pseudoforce pushes by the same amount whether the particle is headed north,
   south, east, or west.
3. $\mathbf{F}_{\mathrm{Cor}}$ is proportional to $\sin \lambda$, so it is largest at the poles and zero at the equator.

Example 9.4   **Coriolis Pseudoforces in Airflow**

Coriolis pseudoforces make an enormous impact on large-scale airflows. If the earth were not rotating, warm
air would rise near the equator and spread out at high altitudes toward both poles, where it would cool off
and sink again, and then flow at low altitudes back to the equator. Therefore one would expect predominantly
southward flow near the ground in the northern hemisphere, and northward flow in the southern hemisphere.
The Coriolis effect profoundly changes this pattern. As the high-altitude air warmed over the equator flows
northward in the northern hemisphere, it is deflected to the right, *i.e.*, eastward, by the Coriolis effect, and
part of it reaches only about $30°$ north latitude whereupon it sinks, and returns toward the equator flowing
northeast toward southwest. These flows are the well-known *trade winds*. In the southern hemisphere the
trade winds flow from southeast to northwest.

   At mid-latitudes one frequently encounters westerly winds, blowing from west to east in both hemi-
spheres. They include the well-known jet stream that speeds up airliners flying east and slows them down
flying west. Such moving air masses experience Coriolis forces tending to deflect them toward the equator.
However, they are often quite steady, retaining their flow for many hours at a time. The reason is that the
air is normally *warmer* toward the equator, so the resulting higher pressures can counteract the Coriolis
effect and the winds keep flowing eastward. This idealized pattern of steady eastward motion is called
geostrophic flow.

Air also tends to move from high-pressure to low-pressure regions in the atmosphere. In the absence of Coriolis effects, air would tend to flow straight from one to the other, perpendicular to the isobars, the surfaces of constant pressure. However, Coriolis forces tend to deflect inrushing winds to the right in the northern hemisphere, as shown in Figure 9.11, which makes the air whirl counterclockwise around the low-pressure region, as seen from above. This is the typical cyclonic pattern, characteristic of tornados and hurricanes in the northern hemisphere. In the southern hemisphere the circulation is clockwise, because the Coriolis forces then deflect inrushing winds to the left.

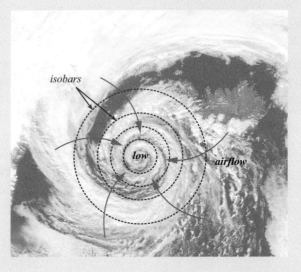

Fig. 9.11   Inflowing air develops a counterclockwise rotation in the northern hemisphere.   ∎

Coriolis effects are hard to observe in the laboratory (or a kitchen sink or bathroom tub), because they are easily overwhelmed by other effects. Water can swirl down a sink drain either clockwise or counterclockwise, depending upon the shape of the sink and the rotation of the water before the drain is opened. Supposedly, if one builds a symmetrical sink with negligible clockwise or counterclockwise bias, fills the sink with water and lets it stand for several hours so that it retains no memory of how it was filled, then the water will drain out with a counterclockwise swirl in the northern hemisphere and a clockwise swirl in the southern hemisphere. However, the story that the circulation of water emptying down a drain on board a ship suddenly changes from counterclockwise to clockwise as the ship steams south over the equator is an "old sailor's tale," not least because the Coriolis pseudoforce is zero at the equator and negligible nearby.

Example 9.5

Foucault's Pendulum

The French physicist Jean Leon Foucault (1819–1868) first publicly exhibited his famous pendulum in 1851 under the dome of the Pantheon in Paris. The 28-kg bob was supported by a wire 67 m long, resulting in an oscillation period of just over 16 s. The plane of oscillation of the pendulum slowly rotated in the clockwise sense as seen from above, completing a revolution in about 32 h.

If Foucault had mounted his pendulum at the north or south pole, the plane of oscillation would have completed a revolution in just 24 h, since that plane is fixed in an inertial frame, with the earth turning under the point of support once every 24 h, as shown in Figure 9.12. In the frame of the earth, the plane of oscillation rotates because of the Coriolis pseudoforce acting on the bob, continually deflecting it slightly to the right, and therefore in the clockwise sense in the northern hemisphere.

To analyze the problem, we arrange a non-inertial reference frame fixed with respect to the rotating earth at some arbitrary latitude λ, as shown in Figure 9.13(a). A close-up of the system is shown in Figure 9.13(b), along with the angular velocity vector $\boldsymbol{\Omega}$ and its components. In this coordinate system we have

$$\mathbf{r}_{\text{rot}} = (x, y, z)\,, \quad \mathbf{v}_{\text{rot}} = (\dot{x}, \dot{y}, \dot{z})\,, \quad \boldsymbol{\Omega} = (-\Omega\cos\lambda, 0, \Omega\sin\lambda)\,. \tag{9.43}$$

From Eq. (9.26), the Lagrangian is

$$
\begin{aligned}
L \simeq {}& \frac{1}{2}m\left(\dot{x}^2 + \dot{y}^2 + \dot{z}^2\right) - mgz \\
& - m\,\Omega\,y\dot{x}\,\sin\lambda + m\,\Omega\,x\dot{y}\,\sin\lambda \\
& + m\,\Omega\,z\dot{y}\,\cos\lambda - m\,\Omega\,y\dot{z}\,\cos\lambda,
\end{aligned}
\tag{9.44}
$$

where we have dropped all terms quadratic in Ω, since the rotation of the earth is a small effect and leading linear terms in Ω will be sufficient. For *small-angle* oscillations of the pendulum, we further expect that z will not change much compared with x and y; hence, we ignore terms involving two small quantities Ω and z or $\dot{z}$. We thus get a simpler approximate Lagrangian of the form

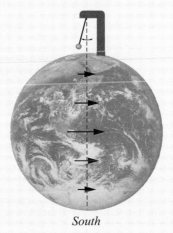

South

Fig. 9.12   Foucault's pendulum set up at the north pole.

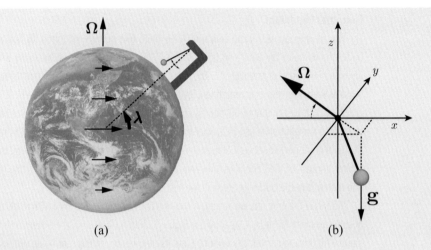

Fig. 9.13 (a) Foucault's pendulum set up at an arbitrary latitude λ. (b) The coordinate system in the rotating frame of the earth set up to analyze the trajectory of the pendulum.

$$L \simeq \frac{1}{2}m \left(\dot{x}^2 + \dot{y}^2 + \dot{z}^2\right) - mgz$$
$$- m\,\Omega\, y\,\dot{x}\,\sin\lambda + m\,\Omega\, x\,\dot{y}\,\sin\lambda. \tag{9.45}$$

There is one more ingredient to the dynamics: the constraint that the mass is attached to a fixed-length string. That is:

$$z = \sqrt{R^2 - x^2 - y^2} = R\sqrt{1 - \frac{x^2}{R^2} - \frac{y^2}{R^2}} \simeq R\left(1 - \frac{x^2}{2R^2} - \frac{y^2}{2R^2}\right), \tag{9.46}$$

for $x, y \ll R$, using the binomial approximation (see Appendix F). To leading order in the small parameters Ω and x/R, y/R we can focus only on the $x-y$ dynamics. We are then ready to write the relevant equations of motion using this Lagrangian:

$$\ddot{x} - (2\Omega \sin\lambda)\dot{y} + \omega_0^2 x = 0,$$
$$\ddot{y} + (2\Omega \sin\lambda)\dot{x} + \omega_0^2 y = 0, \tag{9.47}$$

where $\omega_0 = \sqrt{g/R}$ is the small-amplitude angular frequency of oscillation of the pendulum. We would like to solve these coupled linear differential equations for $x(t), y(t)$.

We will solve them first in a special case, and then in general. At the north pole, where $\lambda = \pi/2$, we already know the solution. The plane of oscillation is fixed in an inertial frame, as the earth turns with angular velocity Ω beneath it, so the plane of oscillation rotates *clockwise* with angular velocity Ω as seen by a person standing at the pole. Let the primed frame be the inertial frame and the unprimed frame be the frame of the rotating earth. Then from the usual rotation transformation, the relation between the primed and unprimed frames is

$$\left(\begin{array}{c} x \\ y \end{array} \right) = \left(\begin{array}{cc} \cos \Omega t & \sin \Omega t \\ -\sin \Omega t & \cos \Omega t \end{array} \right) \left(\begin{array}{c} x' \\ y' \end{array} \right). \tag{9.48}$$

The first and second time derivatives are

$$\left(\begin{array}{c} \dot{x} \\ \dot{y} \end{array} \right) = \left(\begin{array}{cc} \cos \Omega t & \sin \Omega t \\ -\sin \Omega t & \cos \Omega t \end{array} \right) \left(\begin{array}{c} \dot{x}' \\ \dot{y}' \end{array} \right) - \Omega \left(\begin{array}{cc} \sin \Omega t & -\cos \Omega t \\ \cos \Omega t & \sin \Omega t \end{array} \right) \left(\begin{array}{c} x' \\ y' \end{array} \right) \tag{9.49}$$

and

$$\left(\begin{array}{c} \ddot{x} \\ \ddot{y} \end{array} \right) = \left(\begin{array}{cc} \cos \Omega t & \sin \Omega t \\ -\sin \Omega t & \cos \Omega t \end{array} \right) \left(\begin{array}{c} \ddot{x}' \\ \ddot{y}' \end{array} \right) - 2\Omega \left(\begin{array}{cc} \sin \Omega t & -\cos \Omega t \\ \cos \Omega t & \sin \Omega t \end{array} \right) \left(\begin{array}{c} \dot{x}' \\ \dot{y}' \end{array} \right)$$
$$- \Omega^2 \left(\begin{array}{cc} \cos \Omega t & \sin \Omega t \\ -\sin \Omega t & \cos \Omega t \end{array} \right) \left(\begin{array}{c} x' \\ y' \end{array} \right). \tag{9.50}$$

Substituting these into the equations of motion (9.47) (using $\sin \lambda = \sin \pi/2 = 1$) yields the equations

$$\ddot{x}' \cos \Omega t + \ddot{y}' \sin \Omega t + \omega_0^2 (x' \cos \Omega t + y' \sin \Omega t) = 0, \tag{9.51}$$

$$-\ddot{x}' \sin \Omega t + \ddot{y}' \cos \Omega t + \omega_0^2 (-x' \sin \Omega t + y' \cos \Omega t) = 0, \tag{9.52}$$

where we have neglected terms containing Ω^2 compared with ω_0^2, since the angular frequency of rotation of the earth is much less than the angular frequency of oscillation of the pendulum.[a]

We can separate Eqs. (9.51) and (9.52) by multiplying (9.51) by $\cos \Omega t$ and (9.52) by $\sin \Omega t$, and subtracting the results. Then using the identity $\sin^2 \Omega t + \cos^2 \Omega t = 1$:

$$\ddot{x}' + \omega_0^2 x' = 0. \tag{9.53}$$

If instead we multiply Eq. (9.51) by $\sin \Omega t$ and (9.52) by $\cos \Omega t$ and *add* the results, then

$$\ddot{y}' + \omega_0^2 y' = 0. \tag{9.54}$$

These are simple-harmonic oscillator equations, showing that the pendulum moves in a fixed plane in the inertial (primed) frame. For example, if the initial values y_0' and $\dot{y}_0'$ are zero, Eq. (9.54) shows that $\ddot{y}_0''$ is also zero, so y' *remains* zero. The pendulum begins its motion purely in the x' direction, and it subsequently remains permanently in the x, z plane. And since this pendulum, which we have placed at the north pole, oscillates in a fixed plane in the primed frame, it rotates with angular velocity Ω relative to the earth. The rotation period is therefore $T = 2\pi/\Omega = 24$ h.

Now looking back at the equations of motion (9.47), we see that they are exactly the same at arbitrary latitude λ as they are at the north pole, *except* for Ω, which has to be replaced by $\Omega \sin \lambda$. So there is no need to do the calculations over again; the period of rotation of the plane of oscillation at *arbitrary* latitude must be

$$T = 2\pi/(\Omega \sin \lambda). \tag{9.55}$$

The latitude of the Pantheon in Paris is 48.85° north, so the rotation period of the first Foucault pendulum was $T = 24$ h$/(\sin 48.85°) = 24$ h$/0.753 = 31.9$ h.

A Foucault pendulum in the southern hemisphere rotates *counterclockwise* instead of clockwise. At both the north and south poles the period of rotation is 24 h, and at the equator it is $T = 24$ h$/\sin 0 = \infty$ (*i.e.*, the plane of oscillation doesn't change), since there is no Coriolis pseudoforce there at all.   ∎

[a] The periods are 24 h vs. (typically) 16 s, so $\Omega^2/\omega^2 \cong 3.4 \times 10^{-8}$.

9.5 Spacecraft Rendezvous and Docking

One of the least expected and delightful applications of the centrifugal and Coriolis pseudoforces arises in the problem of rendezvousing with orbiting satellites. For example, suppose an orbiting space station is in a circular orbit around the earth, and a spacecraft is in the same circular orbit some distance behind, as shown in Figure 9.14(a). The crew wants to rendezvous and dock with the station. In what direction should the pilot fire the spacecraft's thruster rocket so that the spacecraft will catch up with it?

Or an astronaut on a spacewalk finds herself stranded some distance from an orbiting space station with no tether and no fuel remaining in her portable rocket thruster, as shown in Figure 9.14(b). She is carrying a wrench, however, and decides that the only way to return to the station is to throw the wrench, giving herself a reactive impulse sufficient to get her home. In which direction, and with what speed, should she throw the wrench?

Or an astronaut tightening some bolts outside the space station accidentally bobbles a wrench, so it starts drifting away, as shown in Figure 9.14(c). What is the subsequent trajectory of the wrench? Is it gone forever?

These are related questions that can be answered most elegantly in the rotating frame in which the space station is at rest. Since the station is rotating around the earth, we must invoke the centrifugal and Coriolis pseudoforces as well as earth's gravity on the spacecraft, astronaut, and wrench.

Figure 9.15 shows the set of coordinates we will use. The object (spacecraft, astronaut, or wrench) is initially at distance r_0 from the center of the earth, and then moves to a different position $\mathbf{r}$, so that $\Delta\mathbf{r} = \mathbf{r} - \mathbf{r}_0$ is the vector displacement of the object. We will superimpose a set of Cartesian axes for the object, with x to the right, y upward, and z out of the page, such that $\mathbf{r}_0 = r_0\hat{\mathbf{y}}$. The object is initially in a circular orbit, so

$$\frac{GMm}{r_0^2} = ma = mr_0\omega^2. \tag{9.56}$$

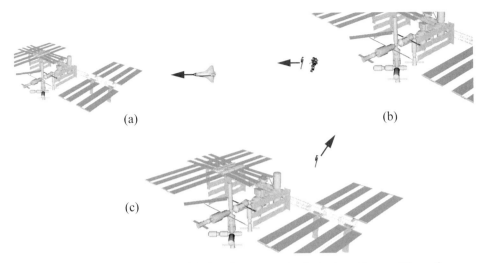

(a)

(b)

(c)

Fig. 9.14 (a) A spacecraft trying to rendezvous and dock with a space station in circular orbit around the earth. (b) A stranded astronaut trying to return to the space station by throwing a wrench. (c) An astronaut accidentally lets a wrench escape from the station. What is its subsequent trajectory?

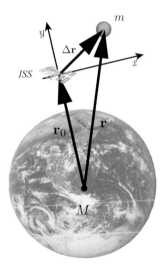

Fig. 9.15 Coordinates of the orbiting space station and object.

The angular velocity in its orbit is therefore

$$\omega = \sqrt{\frac{GM}{r_0^3}}. \tag{9.57}$$

Now consider both real forces and pseudoforces acting on the object in the rotating frame in which the space station is at rest. These are the real inward gravitational force, the outward centrifugal pseudoforce, and the Coriolis pseudoforce. Using unit vectors $\hat{\mathbf{x}}, \hat{\mathbf{y}},$ and $\hat{\mathbf{z}},$ we have

$$\boldsymbol{\omega} = \omega\hat{\mathbf{z}}, \ \Delta\mathbf{r} = x\hat{\mathbf{x}} + y\hat{\mathbf{y}}, \text{ and } \mathbf{r} = \mathbf{r}_0 + \Delta\mathbf{r} = x\hat{\mathbf{x}} + (r_0 + y)\hat{\mathbf{y}}, \qquad (9.58)$$

where we focus on the two-dimensional x–y plane: the dynamics in the z direction will be decoupled from the x–y dynamics and is relatively uninteresting (see the Problems section at the end of this chapter). Squaring the expression for $\mathbf{r}$, we also have

$$r^2 = x^2 + (r_0 + y)^2 = r_0^2 + 2r_0 y + (x^2 + y^2). \qquad (9.59)$$

We can now put our Lagrangian together using Eqs. (9.26), (9.58), and (9.59):

$$\begin{aligned}
L = &\frac{1}{2}m \left(\dot{x}^2 + \dot{y}^2\right) \\
&+ m \left(\dot{x}\hat{\mathbf{x}} + \dot{y}\hat{\mathbf{y}}\right) \cdot \left(\omega\hat{\mathbf{z}} \times (x\hat{\mathbf{x}} + (r_0 + y)\hat{\mathbf{y}})\right) \\
&+ \frac{1}{2}m\omega^2 \left(r_0^2 + 2r_0 y + (x^2 + y^2)\right) \\
&+ \frac{GMm}{\left(r_0^2 + 2r_0 y + (x^2 + y^2)\right)^{1/2}}.
\end{aligned} \qquad (9.60)$$

Evaluating the vector products, we get a simplified form

$$\begin{aligned}
L = &\frac{1}{2}m \left(\dot{x}^2 + \dot{y}^2\right) + m\omega \left(x\dot{y} - y\dot{x} - r_0\dot{x}\right) \\
&+ \frac{1}{2}m\omega^2 r_0^2 \left(1 + 2\frac{y}{r_0} + \frac{x^2}{r_0^2} + \frac{y^2}{r_0^2}\right) \\
&+ \frac{GMm}{r_0} \left(1 - \frac{y}{r_0} + \frac{y^2}{r_0^2} - \frac{x^2}{2r_0^2}\right),
\end{aligned} \qquad (9.61)$$

where we have used the binomial expansion $(1 + \varepsilon)^n \cong 1 + n\varepsilon$ to first order in any small quantity ε and consistently dropped terms of higher order than $(x/r_0)^2$ and $(y/r_0)^2$. That is, we are assuming that the object's distance from the space station is small compared with r_0, the distance of the space station from the center of the earth. We can simplify this further using Eq. (9.57), giving

$$L = \frac{1}{2}m \left(\dot{x}^2 + \dot{y}^2\right) + m\omega \left(x\dot{y} - y\dot{x} - r_0\dot{x}\right) + \frac{3}{2}m\omega^2 r_0^2 \left(1 + \frac{y^2}{r_0^2}\right). \qquad (9.62)$$

Lagrange's equations then provide the equations of motion

$$\ddot{x} = 2\omega\dot{y} \quad \text{and} \quad \ddot{y} = 3\omega^2 y - 2\omega\dot{x}, \qquad (9.63)$$

which are coupled linear second-order differential equations. One way to decouple them is to differentiate the second equation and substitute into it the first equation to eliminate the variable x. That is:

$$\frac{d^2(\dot{y})}{dt^2} = 3\omega^2\dot{y} - 2\omega\ddot{x} = 3\omega^2\dot{y} - 2\omega\left(2\omega\dot{y}\right) = -\omega^2(\dot{y}), \qquad (9.64)$$

which is the simple-harmonic oscillator equation in the variable $\dot{y}$! This equation in $\dot{y}$ has the usual general solution

$$\dot{y} = A' \cos(\omega t + \alpha), \tag{9.65}$$

where A' and α are constants. Then substituting this result into Eq. (9.63) above gives

$$\ddot{x} = 2\omega A' \cos(\omega t + \alpha), \tag{9.66}$$

which can be integrated once with respect to time to give

$$\dot{x} = 2A' \sin(\omega t + \alpha) + B', \tag{9.67}$$

where B' is a constant of integration. We can then integrate the equations for $\dot{y}$ and $\dot{x}$ one more time, to give the coordinates y and x as functions of time:

$$y = A \sin(\omega t + \alpha) + D,$$
$$x = -2A \cos(\omega t + \alpha) + B\omega t + C, \tag{9.68}$$

where $A \equiv A'/\omega$ and $B \equiv B'/\omega$. We began with two second-order differential equations, *i.e.*, the $\mathbf{F} = m\mathbf{a}$ equations in the x and y directions, so we should have *four* arbitrary constants in the solution. However, we seem to have *five*, A, B, C, D, and α. One of the five must therefore be superfluous, which we can track down by substituting our results back into the original differential equations as a consistency check. In particular, substituting our solutions back into Eq. (9.63) above, we must have

$$\ddot{y} = 3\omega^2 y - 2\omega\dot{x} \Rightarrow$$
$$-\omega^2 A \sin(\omega t + \alpha) = 3\omega^2 (A \sin(\omega t + \alpha) + D)$$
$$- 2\omega(2A\omega \sin(\omega t + \alpha) + B\omega), \tag{9.69}$$

which is true for *any* value of A, but only if $D = (2/3)B$. Therefore we can eliminate the constant D to give us the requisite number of arbitrary constants, A, B, C, and α, each of which can be determined from the initial values of x, y, $\dot{x}$, and $\dot{y}$ at $t = 0$.

We now choose once and for all the initial position to be located at $x(0) = y(0) = 0$, which allows us to fix B and C in terms of A and α. In fact, $B = -(3/2)A \sin \alpha$ and $C = 2A \cos \alpha$. The solutions for x, y, $\dot{x}$, and $\dot{y}$ then become

$$x/A = -2 \cos(\omega t + \alpha) - (3/2) \sin \alpha (\omega t) + 2 \cos \alpha,$$
$$y/A = \sin(\omega t + \alpha) - \sin \alpha,$$
$$\dot{x}/A = 2\omega \sin(\omega t + \alpha) - (3/2)\omega \sin \alpha,$$
$$\dot{y}/A = \omega \cos(\omega t + \alpha). \tag{9.70}$$

To also determine A and α, we would need to fix $\dot{x}(0)$ and $\dot{y}(0)$. Let us now explore the physical meaning of these equations by focusing on special cases.

Example 9.6

Rendezvous with a Space Station?

Suppose that a space station is in a circular orbit around the earth at altitude h, so that it has an orbital radius $r_0 = r_e + h$, an angular velocity $\omega = \sqrt{GM/r_0^3}$, and an orbital period $P = 2\pi/\omega$. A supply spacecraft is in the same circular orbit a distance Δs behind the station, as shown in Figure 9.16. The spacecraft pilot can clearly see the space station, and he intends to rendezvous and dock with it. Using the spacecraft thrusters, he gives the spacecraft a Δv boost directly toward the station, expecting to rendezvous with it in a time of about $\Delta s/\Delta v$.

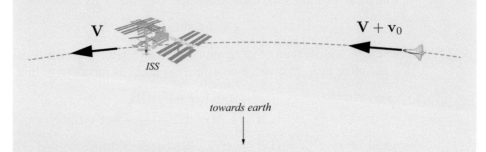

towards earth

Fig. 9.16   The spacecraft trajectory in the inertial frame at $t = 0$.

What in fact happens? Figure 9.17 depicts what happens from the perspective of the station. From the fourth equation of (9.70), the initial condition $\dot{y}(0) = 0$ means that the phase angle $\alpha = \pm\,\pi/2$. Then using the third equation of (9.70), the condition that $\dot{x}(0) < 0$ means that we must choose the *minus* sign, $\alpha = -\pi/2$. The four equations then become

$$x/A = -2\sin\omega t + (3/2)\omega t, \qquad\qquad y/A = 1 - \cos\omega t,$$

$$\dot{x}/A = -2\omega\cos\omega t + (3/2)\omega, \qquad\qquad \dot{y}/A = \omega\sin\omega t. \qquad (9.71)$$

Now we can find the constant A from the initial condition that $\dot{x}(0) = -\Delta v$. That is, $\dot{x}(0) = -A(2\omega - (3/2)\omega) = -A\omega/2 = -\Delta v$, so $A = \frac{2\Delta v}{\omega}$.

Using this result we can rewrite the four equations:

$$x = (2\Delta v/\omega)[-2\sin\omega t + (3/2)\omega t], \qquad\qquad y = (2\Delta v/\omega)[1 - \cos\omega t],$$

$$\dot{x} = 2\Delta v[-2\cos\omega t + (3/2)], \qquad\qquad \dot{y} = 2\Delta v\sin\omega t. \qquad (9.72)$$

We now have everything we need to plot the trajectory $y(x)$ of the spacecraft, as shown in Figure 9.17. Although the spacecraft initially moves toward the station, it soon veers away, moving off to larger and larger altitudes and then turning *backwards*, going *away* from the station. It keeps drifting backwards, and then eventually returns to the station's orbit, but at a position way behind the station itself.

How close does the spacecraft get to the station before veering off? We will plug in some typical numbers, setting $\dot{x} = 0$ for the spacecraft, at which point the spacecraft stops moving toward the station. So from the third equation above we have $\omega t = \cos^{-1}(3/4) = 0.723$. At that point

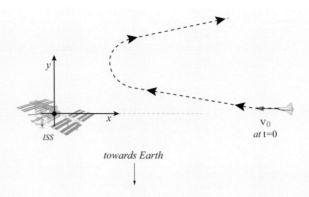

Fig. 9.17 Rendezvous with the space station? The bizarre trajectory, after starting off in the desired direction.

$x = (2\Delta v/\omega)[-2\sin(0.723) + (3/2)(0.723)] = -0.476\Delta v/\omega$ before the spacecraft turns around and heads away again from the station. If the altitude of the station is 280 km, its angular velocity is $\omega = \sqrt{GM/r_0^3} = 1.164 \times 10^{-3}\,\text{s}^{-1}$ and its orbital period is $P = 2\pi/\omega = 90$ min. Then if $\Delta v = 1.0$ m/s is the speed given to the spacecraft by its thrusters, the spacecraft will move a distance $\Delta x = 0.476(1.0)/(1.164 \times 10^{-3})$ m = 411 m ahead of its starting point in the orbit, in the frame of the space station. Therefore one might think that if the station is (say) only 400 m ahead of the spacecraft in their mutual orbits, they would indeed make a rendezvous. However, the spacecraft also moves *in the positive y direction* as it moves toward the station, increasing its altitude from the earth. In fact, when $\dot{x} = 0$, $y = (2\Delta v/\omega)(1 - 3/4) = (1/2)(1.0)/(1.164 \times 10^{-3})$ m = 430 m of additional altitude, and so it will likely miss the station.

   If the spacecraft does in fact miss the station, how long does it take it to return to its original orbital radius, with $y = 0$? That would correspond to $\omega t = 2\pi$, or time $t = 2\pi/\omega$, which is the orbital period P of the station.

   What is going on here? What could explain this bizarre trajectory? Let us look instead at the events from the point of view of the inertial frame in which the space station is in its circular orbit around the earth. Then we know from Kepler's first law that the trajectory of the spacecraft can only be an ellipse, and since the rocket boost gives an extra velocity to the spacecraft in the same direction it was already moving, the elliptical orbit will have its perigee at the point where the boost is made, and its apogee on the opposite side, as shown in Figure 9.18. Note that the spacecraft spends almost all of its time outside the orbit of the station, and that it returns once every orbital period. Furthermore, from Kepler's third law, since the semi-major axis of the spacecraft is somewhat larger than the radius of the station orbit, the period will be somewhat longer, in agreement with the fact that the spacecraft drifts farther and farther behind the station. So the apparently bizarre behavior of the trajectory in the station frame can be understood very well in the nonrotating, inertial frame. However, the counterintuitive nature of the motion in the rotating frame would make it difficult for a spacecraft pilot to rendezvous successfully, so help from radar and a computer would be useful even for an experienced pilot.

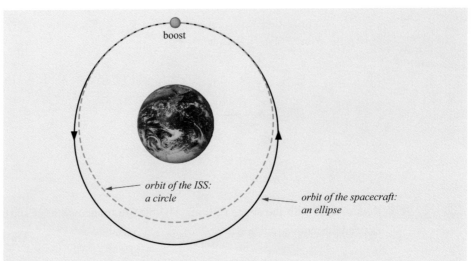

Fig. 9.18 Rendezvous with the space station? The initial boost.   ∎

Example 9.7   **Losing a Wrench?**

An astronaut tightening a bolt on the outside of the space station accidentally lets a wrench escape with an initial velocity 1.0 m/s directly away from the earth. Is it lost forever? We again assume that the station is in a 90-min circular orbit above the earth, at an altitude of 280 km and an angular velocity $\omega = 1.164 \times 10^{-3}$ s^{-1}.

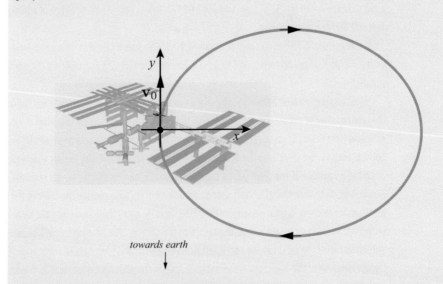

Fig. 9.19   Trajectory of a wrench in the rotating frame in which the space station is at rest. The wrench is thrown from the station vertically, away from the earth. It returns like a boomerang.

The initial velocity components of the wrench are $\dot{x}(0) = 0, \dot{y}(0) = 1.0$ m/s, so from the third and fourth equations of (9.70) we find that the phase angle $\alpha = 0$. Therefore Eqs. (9.70) become

$$x = 2A(1 - \cos\omega t), \quad y = A\sin\omega t,$$
$$v_x = 2A\omega\sin\omega t, \qquad v_y = A\omega\cos\omega t, \tag{9.73}$$

where $A\omega = \dot{y}(0) = 1.0$ m/s. We can now plot the trajectory using the equations for $x(t)$ and $y(t)$, with $A = (1.0\,\text{m/s})/\omega = 0.859$ km.

The result is shown in Figure 9.19. The wrench first moves away from the earth, then turns and lags behind the station, then turns toward the earth and reaches a maximum distance $4A = 3.436$ km behind the station. It then turns around and approaches the station again, ultimately returning from below like a boomerang. It returns when once again $x = 0, y = 0$, i.e., when $\omega t = 2\pi$, or $t = 2\pi/\omega = T = 90$ min. So to retrieve the lost wrench, the astronaut should get *beneath* the station and wait for it to return 90 min later. It is lost only temporarily.

Why does the wrench behave this way? It is easy to understand by considering what happens in the nonrotating frame in which the station is orbiting the earth. In this inertial frame, the wrench must move in the elliptical orbit shown in Figure 9.20. The station orbit is neither at the perigee or apogee of the ellipse. Notice that the wrench first travels outside the station's orbit, then crosses it again and travels inside the station's orbit until it returns to where it began, just as we saw in the rotating frame.

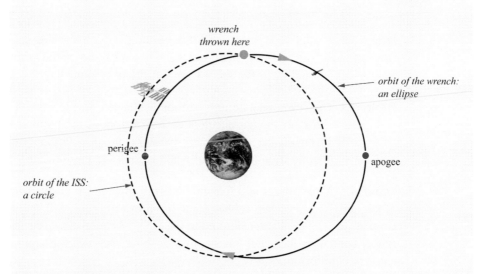

Fig. 9.20   Trajectory of the wrench in the nonrotating frame where the station is in circular orbit around the earth.   ∎

9.6 Summary

In this chapter we have developed the Lagrangian formalism (or equivalently New-ton's laws) from the perspective of certain types of non-inertial reference frames: from the viewpoint of a linearly accelerating frame with uniform acceleration, and from that of a rotating frame. The perspective of a rotating frame is of particular practical importance: all tabletop experiments that use an observer at rest on the ground of our planet fall into this category. We learned that the rotation of the earth results in subtle but detectable and sometimes extremely important effects arising from three pseudoforces, as seen from our non-inertial vantage point on the earth – the centrifugal, the Coriolis, and the Euler pseudoforces; and we explored in some detail the effects of the first two of these.

Generally, we can always add pseudoforces to the real forces in Newton's second law to use a non-inertial perspective that is often more convenient. But it is important to remember that there is no new physics here: all phenomena can be described through an inertial perspective using the unmodified Newton's laws *without* the addition of any pseudoforces. The addition of pseudoforces simply amounts to a mathematical transformation of coordinates from an inertial to a non-inertial one, which often simplifies the problem.

Problems

★★ **Problem 9.1** A satellite is in a polar orbit around the earth, passing successively over the north and south poles (see Figure 9.21). As we stand on the ground, what is the motion of the satellite as we see it from our rotating frame?

★ **Problem 9.2** The string on a helium balloon is attached inside a car at rest, as shown in Figure 9.22(a). If the car accelerates forward, does the balloon tilt forward or backward? If a is the car's acceleration and g is the gravitational field, what is the balloon's tilt angle from the vertical when it comes to equilibrium?

★ **Problem 9.3** A cork floats in a fishtank half full of water; it is attached to the bottom of the tank by a stretched rubber band, as shown in Figure 9.22(b). If the tank and contents are uniformly accelerated to the right, sketch the water surface, cork, and rubber band after the water has stopped sloshing back and forth and the system has come to equilibrium.

★ **Problem 9.4** In the rotating frame of the earth, stars appear to orbit in circles, with a period of 24 h. Show that the centrifugal and Coriolis pseudoforces acting together provide the net force needed in the rotating frame to cause a star to orbit as described.

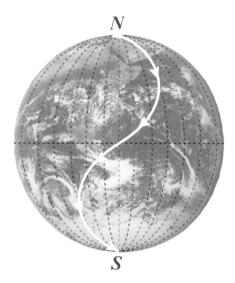

Fig. 9.21 Path of a satellite in a polar orbit, shown in the rotating frame in which earth is at rest. Dashed lines represent the equator and longitude lines.

(a) (b)

Fig. 9.22 (a) A balloon in a car. (b) A cork in a fishtank.

★ **Problem 9.5** A cylindrical space colony rotates about its symmetry axis with period 62.8 s. If the effective gravity felt by colonists standing on the inner rim is one earth "gee," what is the radius of the colony? What then is the percentage difference in the effective gravity acting upon the head and on the feet of a 2-m-tall colonist?

★ **Problem 9.6** (a) A uniformly rotating merry-go-round spins clockwise as seen from above. A rider stands at the rotation axis, and then slowly walks radially outward toward the rim. How will the centrifugal and Coriolis pseudoforces affect her? How will she have to lean to keep from falling over? (b) Now suppose the merry-go-round is spinning up, starting from rest and spinning faster and faster in the clockwise sense. The rider again starts at the center and walks outward. What new effect will the rider notice, and how will that affect how she has to lean to avoid falling?

★   **Problem 9.7** In *Rendezvous with Rama* by Arthur C. Clarke, observers inside a cylindrical spaceship view a waterfall, which originates at one of the endcaps at a point halfway between the rotation axis and rim, and then "falls" to the rim. The spaceship is rotating clockwise about its symmetry axis as seen in an external inertial frame, looking along the entire spaceship axis at the waterfall in the distance. Explain why the water does not fall straight as seen by people within the ship. Which way does it bend?

★★   **Problem 9.8** A train runs around the inside of the outer rim of a cylindrical space colony of radius R and angular velocity ω along its symmetry axis. How would the effective gravity on the passengers depend upon the train's speed v relative to the rim: (a) if the train travels *in* the rotation direction of the rim; (b) if it travels in the *opposite* direction? (c) If you were designing a train system, which way would you make the train run?

★   **Problem 9.9** Why don't we notice Coriolis effects when we walk, drive cars, or throw baseballs? In contrast, why may Coriolis effects be significant for long-range artillery or moving air masses?

★★   **Problem 9.10** In 1914 there was a World War I naval battle between British and German battle cruisers near the Falkland Islands, at 52° south latitude (i.e., $\lambda = -52°$). Guns on the British ships fired 12-inch shells at German ships up to about 15 km distant. The great majority of the shells missed their targets, due to the constant rolling of the ships, defensive maneuvers by the Germans, and perhaps other factors. After the battle (which the British won), another possible reason was offered: Coriolis deflections. The story goes that the British were used to battles in the northern hemisphere, where projectiles deflect toward the right, and aimed their guns incorrectly for the Falklands battle, where projectiles deflect toward the left. There seems to be some controversy over whether or not Coriolis effects were important in the battle. The purpose of this problem is to estimate their magnitude.

   The British guns reportedly had a muzzle velocity of 823 m/s. (a) For a target 15 km away, and pretending there was no air resistance, what must have been the elevation angle (the angle up from the horizontal) of the guns? Note that the guns were apparently limited to elevation angles of 15° or less. (b) By about how much would the shells have missed their target due to the sideways Coriolis effect? (Note that *if* the British used gun-aiming tables appropriate for 52° *north* latitude, which is appropriate for the North Atlantic, then the miss distance of the shells would have been about twice as much as the southern hemisphere deflection alone.)

★★★   **Problem 9.11** In World War I the German army set up an enormous cannon (which they called the "Paris gun") to fire shells at Paris, 120 km away from the cannon situated at a point NNE of Paris. The muzzle velocity was 1640 m/s. (a) Neglecting both air resistance and the Coriolis effect, find two solutions for the elevation angle of the gun, assuming the altitudes of the launching and target points were the same. (b) With these same assumptions, for each of the two possible elevation angles, how long would the shell have taken to reach its target, and what maximum altitude would it have achieved? Compare your results with the actual flight time 182 s and maximum altitude 42 km. What do you think is the primary reason for the large discrepancies? (c) With the same assumptions as in part (a), calculate the Coriolis deflection of the shell aimed at Paris for the larger elevation-angle solution. Be sure to include the Coriolis deflections due to both the horizontal and vertical components of the shell's velocity. This result might give at least a very rough estimate for the actual Coriolis deflection. For simplicity, assume the shell began traveling due south. Is its deflection toward the east or toward the west?

★★ **Problem 9.12** A satellite in low-earth orbit with a 90-min period passes over the north pole, headed south along the $0°$ line of longitude passing through Greenwich, England. (a) What is its longitude when it reaches the latitude of Greenwich ($\lambda = 50°$)? (b) When it reaches the equator, what angle does its trajectory make with the equator as seen by an observer on earth? (c) How close does it come to passing over the south pole?

★ **Problem 9.13** (a) Find the centrifugal acceleration of a particle on the earth's surface at the equator, due to earth's rotation, as a fraction of the gravitational field g at that point. (b) Do the same for the centrifugal acceleration due to the motion of earth around the sun. Note that this acceleration is small compared with that due to the axial rotation.

★★ **Problem 9.14** Suppose we flatten and smooth out the ice at the south pole, and place a hockey puck at rest on the ice exactly at the pole. We then give it a small velocity, initially along longitude $0°$. Pretend that there is no friction between the puck and the ice, and that there is no air resistance either. (a) If it reaches a final point $90°$ longitude west when it is 100.0 m from the pole, what was its initial speed? (b) At this final point, what is its speed relative to the ice? (c) What force or pseudoforce is responsible for the increased speed?

★★ **Problem 9.15** A merry-go-round has a 5-m radius and rotates with a 10-s period. If one "gee" is the gravitational force/mass experienced by a person standing still on the earth, how many gees are felt by a person walking from the center toward the rim of the merry-go-round at velocity 1 m/s in a straight line, due to the Coriolis pseudoforce? At what sideways angle is the person likely to lean while walking? How many gees are felt when the person is standing 3 m from the center, due to the centrifugal pseudoforce? At what angle is the person likely to lean backwards at this point?

★★ **Problem 9.16** A ball is dropped from height h by someone standing still on the earth's equator. (a) Does it fall to the east or west of a point just beneath the position from which it was dropped? (b) When it strikes the ground, how far is the ball from the point originally directly beneath it, in terms of g, h, and ω, earth's angular velocity? (Pretend there is no air resistance.) (c) Explain the direction found in (a) using conservation of the ball's angular momentum in the inertial frame in which earth rotates toward the east.

★★   **Problem 9.17** A ball at a point on the earth with latitude λ is thrown vertically upward
to a small altitude h. (a) Does the ball fall to the east or west of its starting point?
(b) Show that the ball strikes the ground a distance $(4/3)\Omega g \cos \lambda (2\,h/g)^{3/2}$ from
its starting point. (c) Explain qualiltatively the ball's path using conservation of
angular momentum in the inertial frame in which the earth rotates eastward.

★   **Problem 9.18** Show that the usual formula $P \; = \; 2\pi \sqrt{R/g}$ for the period of
small-amplitude oscillations of a pendulum of length R becomes instead
$P = 2\pi \sqrt{R/g}(\sqrt{m_I/m_G})$ if the inertial and gravitational masses of the pendulum
bob differ. (Newton himself built pendulums with plumb bobs made of different
materials. He would swing two of them side by side, both with the same length R,
to see if he could detect a difference in period apart from experimental errors. He
could not. Nevertheless, it is interesting that he conceived of the possibility they
might be different.)

★★   **Problem 9.19** If an artillery shell is fired a short distance from a point on earth's
surface at latitude λ, with speed v_0 and an angle of inclination α to the horizontal,
show that (pretending there is no air resistance) its lateral deflection when it strikes
the ground is

$$d = (4v_0^3/g^2)\omega \sin \lambda(\sin^2 \alpha \cos \alpha),$$

where ω is the earth's angular velocity.

★★   **Problem 9.20** An artillery shell is projected due north from a point at latitude λ at an
angle of 45° to the horizontal, and aimed at a target whose distance is D, where D is
small compared with earth's radius. (a) Show that due to the Coriolis pseudoforce,
and neglecting air resistance, the shell will miss its target by a distance

$$d = \left(\frac{2D^3}{g}\right)^{1/2} \omega \left(\sin \lambda - \frac{1}{3}\cos \lambda\right),$$

where ω is earth's angular velocity. (b) Evaluate this distance for $\lambda = 30°$ and $D =$
50 km. (c) What is the physical reason for the deviation to the east near the north
pole, but to the west both on the equator and near the south pole?

★   **Problem 9.21** In a rotating cylindrical space colony of radius R and angular velocity
ω about its axis of rotation, a ball of mass m is thrown with speed $v = \omega R/2$ from
a point halfway between the rotation axis and rim, in a direction exactly opposite
to the rotation direction, as seen by colonists. (a) State the nature of its subsequent
path in the inertial frame in which the cylinder is seen to rotate. (b) Sketch the ball's
path in the rotating frame of the colony, and show that this path is predicted by the
pseudoforces acting upon it.

★   **Problem 9.22** A ball is released from rest in the frame of a rotating cylindrical
space colony, at a point halfway between the rotation axis and rim. (a) Sketch
the subsequent path of the ball as seen by an inertial observer who sees the colony
rotating counterclockwise with angular velocity ω. (b) If the colonist is directly

"beneath" the ball when it is released (*i.e.*, at the rim at a point along a line connecting the rotation axis and release point), how far must the colonist run to catch the ball, in terms of the colony radius R? Sketch the path of the ball as seen by the colonist.

★ **Problem 9.23** A cylindrical space colony of radius R rotates with angular velocity ω about its symmetry axis. A colonist standing on the rim throws a ball straight "up" (*i.e.*, aimed at the rotation axis) with speed $v = R\omega$ from the colonist's point of view. (a) Sketch the subsequent path of the ball as seen by an inertial observer to whom the colony is rotating counterclockwise. *Hint*: First find the initial velocity of the ball in the inertial frame. (b) Sketch the ball's path as seen in the colony frame. (c) How far around the rim must the colonist run to catch the ball?

★★ **Problem 9.24** For the film *2001 Space Odyssey,* director Stanley Kubrick had a giant centrifuge constructed, of diameter 11.6 m. On the movie set, motors rotated the centrifuge about a horizontal axis, like a Ferris wheel. This was the home for fictional astronauts on their long journey to the planet Jupiter, providing artificial gravity throughout the trip.

(a) In one scene, astronaut Dr. Frank Poole is seen jogging all the way around the circumference of the centrifuge, requiring about 25 s to do so. What was the rotational period of the centrifuge on the movie set, and how fast was he jogging?

(b) In the movie, it appears that Poole is jogging in a gravity approximately the same as on earth. (No surprise!) Assuming that $g_{\text{effective}} = g_{\text{earth}}$ while standing at rest on the centrifuge rim, and that the centrifuge was actually rotating on the spaceship en route to Jupiter, what would the rotational period of the centrifuge have to be?

(c) Suppose the movie astronaut was 1.9 m tall. By what percentage less would the artificial gravity be on his head than on his feet, just standing on the centrifuge rim?

(d) Would it make any difference if the fictional astronaut were jogging *in* the direction of rotation or *opposite* to it? If so, what would be the effect of jogging in the two directions?

★ **Problem 9.25** (a) Prove that there is no work done by the Coriolis pseudoforce acting on a particle moving in a rotating frame. (b) If the Coriolis pseudoforce were the *only* force acting on a particle, what could you conclude about the particle's speed in the rotating frame?

★★ **Problem 9.26** Show that the formula $d\mathbf{A} = d\boldsymbol{\theta} \times \mathbf{A}$ for the change in an inertial frame of a vector $\mathbf{A}$ that is stationary in a rotating frame is still valid when $\mathbf{A}$ is not perpendicular to $\boldsymbol{\Omega}$ and when the tail of $\mathbf{A}$ is not situated at the rotation axis.

★ **Problem 9.27** A well-known actor and television crew, filming a travel documentary, were driving south in Africa when they were approached by a local citizen as they neared the equator. He offered (for a fee) to demonstrate the change in swirl direction of water in a hand basin. They all walked a few minutes north of the equator, he filled the basin with water, removed a plug in a hole in the middle of the basin, and sure enough, the water swirled counterclockwise as the water drained out. They then walked a few minutes south of the equator, repeated the experiment, and the water swirled clockwise this time as the water rushed out. It would have been interesting to learn how much the local man made demonstrating his skill over and over for equator-crossing tourists! How might the local man have produced the result?

★★   **Problem 9.28** Prevailing winds in middle latitudes of the northern hemisphere are westerly, blowing from west to east at typical speed v. The tendency of the Coriolis pseudoforce to deflect the path southward is typically balanced by a horizontal pressure gradient that keeps the air flowing eastward. The pressure gradient is related through the ideal gas law to the north–south temperature gradient, with warmer air in the south. Find an expression for the wind speed in terms of the temperature gradient $\Delta T/\Delta x$, the latitude λ, the earth's angular velocity Ω, and any necessary gas constants. Estimate the typical temperature gradient and find the typical flow velocity. Are the results reasonable? Would you expect prevailing winds in comparable southern latitudes to be westerly or easterly?

★★   **Problem 9.29** The Gulf Stream flows northward off the Florida coast, so tends to be deflected eastward. This causes the water level to rise on the eastern side, since the more stationary Atlantic waters cannot easily be moved aside. The higher waters on the eastern side provide the higher pressures needed to counteract the Coriolis force, so the stream is relatively undeflected. Looking northward, the stream looks as shown in Figure 9.23, with a greatly exaggerated eastern rise. Using a thin vertical slice of water and balancing the pressure and Coriolis forces upon it on the left and right, find an expression for the slope dy/dx of the surface in terms of the earth's angular velocity ω, the latitude λ, the acceleration of gravity g, and the stream velocity v. The westernmost islands of the Bahamas are only about 80 km from the east coast of Florida. Between them the Gulf Stream flows somewhat in excess of 1 m/s, and the sea level is about 0.5 m higher at the Bahamas. Are these measurements consistent with your results?

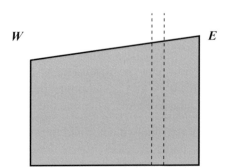

Fig. 9.23   Tilt of the northward-flowing Gulf Stream surface, looking north.

★★★ **Problem 9.30** A spherical asteroid of radius R and uniform density ρ rotates with angular velocity Ω about an axis through its center. Visiting astronauts drill a smooth hole from one point on the equator clear through the asteroid's center to a point on the opposite side. (a) If an astronaut falls into one end of the hole, how long does it take her to reach the opposite side? (b) During the trip she is pressed against the side of the hole with force N. Find N in terms of W, her weight on the asteroid's surface at one of the poles, as a function of her distance r from the center of the asteroid. Can N exceed W?

★ **Problem 9.31** (a) Find the radius of a toroidal space colony spinning once every 2 min, if the effective gravity for colonists living within the torus is 10 m/s². (b) Arriving tourists dock at the central hub, and are then transported to the torus by an elevator running through a "spoke" of the wheel. The elevator runs at constant speed, except for brief periods of acceleration at the beginning and end, and requires 1 min for the journey. Plot quantitatively the centrifugal and Coriolis accelerations of a rider as a function of time while the elevator is running at constant speed, as a fraction of 10 m/s². Would the Coriolis acceleration be noticeable?

★★ **Problem 9.32** A cylindrical space station rotating with angular velocity Ω contains an atmosphere with molecular weight M and temperature T. Show that if p_0 is the atmospheric pressure at the rotation axis, the pressure at radius ρ is $p = p_0 \exp(M\Omega^2 \rho^2 / 2RT)$, where R is the ideal gas constant. If the station has a radius 100 m, an effective rim gravity 10 m/s², and an oxygen atmosphere at temperature 300 K, what is the ratio of the rim pressure to that at the rotation axis? Would this difference be important to inhabitants who travel from the rim to the axis?

Problem 9.33 An astronaut is stranded in space above the earth, in the same orbit as a space station, but 200 m behind it. Both are circling 280 km above earth's surface in a 90-min orbit. The astronaut and spacesuit together have a mass of 100 kg. (a) In what direction, and with what speed, can the astronaut throw a 1-kg wrench so that the recoil will allow the astronaut to reach the vehicle and safety? (b) How long will it take the astronaut to arrive? (c) Sketch the trajectory of the astronaut and of the wrench after the throw, in the rest frame of the station, and then sketch the trajectory of each in the nonrotating, inertial frame in which the station orbits the earth.

★★★ **Problem 9.34** Only the centrifugal and Coriolis pseudoforces act upon a particular projectile moving within a rotating cylindrical space colony of radius R. (a) Find the differential equations of motion of the projectile in the rotating frame, using Cartesian coordinates centered on the rotation axis. (b) Decouple and solve the differential equations, to find expressions for $x(t)$ and $y(t)$ in terms of four arbitrary constants of integration. (c) Evaluate the constants in terms of the initial conditions $x_0, y_0, v_{x_0}, v_{y_0}$. (d) A colonist on the rim at $(x_0, y_0) = (R, 0)$ throws a ball toward the rotation axis with velocity $(v_{x_0}, v_{y_0}) = (-v_0, 0)$. Find a general expression for the time at which the ball returns to the rim, and show that in the limit as v_0 becomes small, the time agrees with what you would expect in a uniform gravitational field $g = R\Omega^2$.

★★   Problem 9.35 A uniform electric charge density ρ fills a very long stationary cylinder. (a) Show from Gauss's law $\oint \mathbf{E} \cdot d\mathbf{S} = 4\pi q_{\text{in}}$ that the electric field within the cylinder is $\mathbf{E} = 2\pi\rho\mathbf{r}$, where $\mathbf{r}$ is the radius vector out from the symmetry axis. Here q_{in} is the charge within an appropriate Gaussian surface. (b) A uniform magnetic field $\mathbf{B} = B_0\hat{z}$ is created in the same region of space. Including the effects both $\mathbf{E}$ and $\mathbf{B}$, find an expression for the force exerted on a test charge q placed within the cylinder. (c) Show that if the test charge has the proper charge/mass ratio q/m, there exists a *rotating* frame in which the charge is bound to move just as in part (b) with *no* electromagnetic fields at all. Find this ratio q/m, and the angular velocity ω of the rotating frame.

★★   Problem 9.36 Show that for the problem of a spacecraft rendezvous and docking discussed in the text, the dynamics in the z direction decouples from that in the x–y plane. What is the equation of motion of the z coordinate?

★★★   Problem 9.37 *Space visits for everyone?* An alternative way to visit space has been proposed: a space station of mass m is tethered to one end of a long cable and the other end of the cable is attached to a point on the earth's equator, a distance R from the center of the earth. In the rotating frame of the earth, three forces act on the station: the centrifugal pseudoforce $mr\omega^2$ outward, and earth's gravity GMm/r^2 and the cable tension T inward, where r is the distance of the station from the center of the earth. As one goes to larger radii the centrifugal pseudoforce grows while gravity decreases, so there must be a radius r_0 where these forces balance, so that the station will remain in place. Then one could ride an elevator up the cable and get some spectacular views and experience zero gees without using any rocket fuel. (a) Assuming $T = 0$, find the distance r_0 from the earth's center to the station where the forces balance, in terms of G, M, and ω, the angular velocity of the earth's rotation. (b) Of course the cable will require some tension $T(r)$ if an elevator is to travel up and down along it. This might be achieved by placing the space station at a somewhat greater distance $r_0 + \Delta r_0$ from the earth's center, requiring a positive downward tension force for it to stay in place. Let the cable have uniform mass per unit length λ. Then show that the tension $T(r)$ obeys the equation

$$\frac{dT}{dr} = \lambda\left(\frac{GM}{r^2} - r\omega^2\right). \tag{9.74}$$

(c) At what radius is the tension in the cable a maximum? (d) Find the tension T_s in the cable just where it is attached to the space station. Assume here that $\Delta r_0 \ll r_0$, and so keep only first-order terms in Δr_0. Express your answer in terms of m, ω, and Δr. (e) Find a general expression for the tension $T(r)$ anywhere along the cable, in terms of $T_s, \lambda, \omega, r_0, G, M$, and r. (f) In particular, what is the cable tension at $r = r_0$? (g) At what radius is the cable tension a minimum? (h) Find the minimum value of Δr_0 required, in terms of other given parameters, so that the cable will never have a negative tension anywhere along its length, because that would cause it to buckle.

10

From Black Holes to Random Forces

We have come now to the second capstone chapter in the book, in which we extend some of the classical mechanics from the preceding four chapters into the context of more recent developments in physics. We begin with gravitation, including some of the ideas that led Einstein to go way beyond Newton's nonrelativistic theory to find a fully *relativistic* theory of gravitation. After years of strenuous efforts, his work finally culminated in his stunningly original and greatest single achievement, the **general theory of relativity**. He was able to predict three effects that could be measured in the solar system, which he used to check his theory. We will cover all three of these. Then we will introduce so-called "magnetic gravity," which contains the leading terms in general relativity in a form much like Maxwell's equations for electromagnetism. Next, we delve just a bit deeper into gauge symmetry in Maxwell's theory, partly because it deepens our understanding of electromagnetism but also because gauge symmetry has played such a large role in physical theories over the past many decades. Finally we introduce stochastic forces, which are not fundamental forces but the result of huge numbers of small collisions.

10.1 Beyond Newtonian Gravity

By 1905 it was already clear to Albert Einstein that his special theory of relativity was consistent with Maxwell's electromagnetism, but *not* with Newton's gravitation. The fundamental problem is that while the equations of motion $\mathbf{F} = -(GMm/r^2)\hat{\mathbf{r}} = m\mathbf{a}$ for a probe particle m in central gravity are invariant under the **Galilean** transformations[1] of Chapter 1, they are *not* invariant under the relativistic **Lorentz** transformations of Chapter 2. So if the equations are true in one inertial frame of reference they cannot be true in another – using the correct Lorentz transformations – and so could not be a fundamental law of physics according to the principle of relativity.

There were several clues that gradually led him to the correct theory. One was the apparently trivial fact that according to $\mathbf{F} = -(GMm/r^2)\hat{\mathbf{r}} = m\mathbf{a}$, the mass

---

[1] As just one indication of this, the distance r between M and m is not invariant under a Lorentz boost. As another, Newton's law of gravity seems to imply instantaneous communication between M and m, which is not possible in a relativistic world. See examples in Chapter 6 for more details on how to explore the transformations and symmetries of Lagrangians.

m cancels out on both sides: all masses m have the same acceleration in a given gravitational field according to Newton's theory and experiments as well. This is not as trivial as it seems, however, because the two ms have very different meanings. The m in GMm/r^2 is called the **gravitational mass**; it is the property of a particle that causes it to be attracted by another particle. The m in ma is called the **inertial mass**; it is the property of a particle that makes it sluggish, resistant to acceleration. The fact that these two kinds of mass appear to be the same is consistent with Newton's theory, but not explained by it. Einstein wanted a *natural* explanation.

As discussed at the end of Chapter 3, this thinking helped to generate in Einstein his "happiest thought," the **principle of equivalence**. The equivalence of gravitational and inertial masses is an immediate consequence of the equivalence of (i) a uniformly accelerating frame without gravity and (ii) an inertial reference frame containing a uniform gravitational field (see the Problems section at the end of this chapter for more). Einstein therefore saw a deep connection between *gravitation* and *accelerating reference frames*.

A second clue is the type of geometry needed within accelerating frames, as shown in the following **thought experiment**.

| Example 10.1 | **A Thought Experiment** |

A large horizontal turntable rotates with angular velocity ω. A reference frame rotating with the turntable is an accelerating frame, because every point on it is accelerating toward the center with $r\omega^2 = a$. A colony of ants living on the turntable is equipped with meter sticks to make measurements (see Figure 10.1). A *second* colony of ants lives on a *non*rotating horizontal glass sheet suspended above the turntable; these ants are also equipped with meter sticks, and can make distance measurements on the glass while they are watching beneath them the rotating ants making similar measurements directly on the turntable itself.

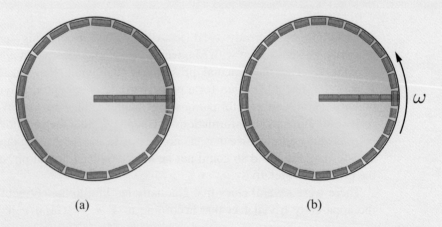

(a)   (b)

| Fig. 10.1 | An ant colony measures the radius and circumference of its rotating home: (a) the measuring sticks of the inertial ants from their perspective; (b) the measuring sticks of the rotating ants as seen from the perspective of the inertial ants. |

Both ant colonies can measure the radius and circumference of the turntable, as shown in Figure 10.1. The *inertial* ants on the glass sheet, looking down on the turntable beneath them, lay out a straight line of sticks from the turntable's center to its rim, and so find that the radius of the turntable (as measured on the glass sheet) is R_0. They also lay meter sticks end-to-end on the glass sheet around the rim of the turntable (as they see it through the glass), and find that the circumference of the turntable (according to their measurements) is $C_0 = 2\pi R_0$, verifying that Euclidean geometry is valid in their inertial frame.

The ants living on the turntable make similar measurements. Laying meter sticks along a radial line, they find that the radius is R. Laying sticks end-to-end around the rim, they find that the circumference of the turntable is C. Meanwhile the inertial ants, watching the rotating ants beneath them, find that the rotating-ant meter sticks laid out radially have no Lorentz contraction relative to their own inertial sticks, because the rotating sticks at each instant move *sideways* rather than *lengthwise*. So the inertial ants see that the rotating ants require exactly the same number of radial sticks as the inertial ants do themselves; in other words, both sets of ants measure the same turntable radius, $R = R_0$.

However, the meter sticks laid out around the turntable *rim* by the rotating ants are moving with speed $v = R_0\omega$ in the direction of their lengths, and so will be Lorentz-contracted as observed in the inertial frame. More of these meter sticks will be needed by the rotating ants to go around the rim than is required by the inertial ants. Therefore it must be that the circumference C measured by the rotating ants is *greater* than the circumference C_0 measured by the inertial ants. Since the measured radius is the same, this means that in the accelerating frame $C > 2\pi R$. The logical deduction that $C > 2\pi R$ in the accelerating frame means that the geometry actually measured in the rotating frame is **non-Euclidean**, since the measurements are in conflict with Euclidean geometry.   ∎

So this thought experiment shows that there appears to be a connection between *accelerating frames* and **non-Euclidean geometry**. Now Euclidean geometry is the geometry on a *plane*, while non-Euclidean geometries are the geometries on *curved surfaces*. Draw a circle on the curved surface of the earth, for example, such as a constant-latitude line in the northern hemisphere (see Figure 10.2). Then the north pole is at the center of the circle. The radius of the circle is a line on the sphere extending from the center to the circle itself; in the case of the earth, this is a line of constant longitude.

Then it is easy to show that the circumference and radius obey $C < 2\pi R$. The geometry on a two-dimensional curved space like the surface of the earth is non-Euclidean. The opposite relationship holds if a circle is drawn on a saddle, with the center of the circle in the middle of the saddle; in that case one can show that $C > 2\pi R$, so that the geometry on a curved saddle is also non-Euclidean.

All this suggests a question: Are *gravitational fields* therefore associated with *curved spaces*? Let us emphasize how special is the gravitational force in this aspect. If we were to write Newton's second law for a probe of gravitational *and* inertial mass m near a larger mass M, we would get

$$m\,\mathbf{a} = -G\frac{Mm}{r^2}\hat{\mathbf{r}}, \tag{10.1}$$

as we already know. Because the *ms* on both sides of this equation are the same, they cancel, and we find

$$\mathbf{a} = -G\frac{M}{r^2}\hat{\mathbf{r}}. \tag{10.2}$$

That is, all objects fall in gravity with the *same* acceleration, which depends only on the source mass M and its location. In the absence of air resistance, elephants and feathers fall with the same gravitational acceleration. Contrast this with the Coulombic force

$$m\,\mathbf{a} = \frac{Qq}{r^2}\hat{\mathbf{r}}, \tag{10.3}$$

where the probe has mass m and charge q, and the source has charge Q. We see that m does not cancel out, and the acceleration of a probe under the influence of the electrostatic force depends on the probe's attributes: its mass *and* charge. This dependence of a probe's acceleration on its mass is generic to all force laws except gravity! The gravitational acceleration is very special in that it has a universal character – independent of the attributes of the object it acts upon. Hence, gravity lends itself to be tucked into the fabric of space itself: all probes gravitate in the same way, and thus perhaps we can think of gravity as an attribute of curved space itself!

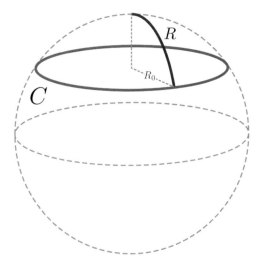

Fig. 10.2 Non-Euclidean geometry: the radius R and circumference C of a circle drawn on a sphere. Note that $C < 2\pi R$ since $R_0 < R$ and $C = 2\pi R_0$.

The next thought experiment is an illustration of how this distortion of space due to gravity can affect time as well.

Example 10.2 **Another Thought Experiment**

Two clocks A and B are at rest in a uniform gravitational field g, with A on the ground and B at altitude h directly above. At time $t = t_0$ on A, A sends a light signal up to B, arriving at B at time t_0' according to B, as shown in Figure 10.3. Later, when A reads t_1, A sends a second light signal to B, which arrives at t_1' according to B. The time interval on A is $\Delta t_A = t_1 - t_0$, and the time interval on B is $\Delta t_B = t_1' - t_0'$, which is greater than Δt_A, because from the principle of equivalence presented at the end of Chapter 7, high-altitude clocks run fast compared with low-altitude clocks. Now notice that the light signals together with the two clock worldlines form a parallelogram in spacetime. Two of the sides are the parallel vertical time-like world lines of the clocks in the figure, and the other two are the slanted null lines, corresponding to light signals. In Euclidean geometry, opposite sides of a parallelogram have the same length. But in the spacetime parallelogram of Figure 10.3, the two parallel vertical lines, which are the clock worldlines, have time-like "lengths" (measured by clock readings) $c\Delta t_A$ and $c\Delta t_B$, which are not equal! Therefore when gravity is added to spacetime, the geometry becomes non-Euclidean, and hence space*time* has in some sense become *curved*. If there were no gravity, the high and low clocks would tick at the same rate, so the two clock worldlines would have the same time-like length, and the parallelogram would obey the rules of Euclidean geometry, corresponding to a *flat*, Euclidean spacetime. There is no notion of global time even within a fixed frame whenever gravity is around. It seems then that *gravity can curve spacetime.*

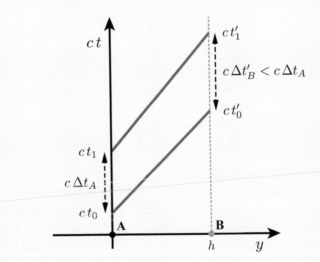

Fig. 10.3 Successive light rays sent to a clock at altitude h from a clock on the ground. Here time is measured on the vertical axis, and altitude on the horizontal axis. ∎

As he learned from the work of Minkowski, Einstein knew that physics takes place in the arena of four-dimensional spacetime. From the clues that (i) gravity is related to accelerating reference frames, (ii) accelerating reference frames are related to non-Euclidean geometries, and (iii) non-Euclidean geometries are realized on curved spaces, he became convinced that gravity is an effect of curved four-dimensional spacetime.

Einstein's quest to see how gravity is related to geometry ultimately led him to the famous **field equations** of 1915. These equations showed how spacetime geometry was affected by the particles and energy within it, and how curved spacetime bends the trajectories of particles and energy. General relativity has been summarized in a nutshell by the American physicist John Archibald Wheeler: *"Matter tells space how to curve; space tells matter how to move."*

10.2 The Schwarzschild Geometry

A presentation of Einstein's field equations and a thorough discussion of their solutions would take us very far afield! However, many interesting features of the equations are contained in their most famous solution, the solution for the curved spacetime in the vacuum surrounding a central spherically symmetric mass M like the sun.[2] The solution takes the form of a spacetime **metric** analogous to the Minkowski metric of special relativity corresponding to flat spacetime. In spherical coordinates (r, θ, ϕ), the *flat* Minkowski metric is

$$ds^2 = -c^2 dt^2 + dr^2 + r^2(d\theta^2 + \sin^2\theta d\phi^2), \tag{10.4}$$

which, for example, we can use to measure physical distances and times. For example, to compute the physical distance ℓ between two points, we compute $\ell \equiv \int ds$ along a line joining the two points, measuring both simultaneously, with $dt = 0$.

Consider a probe of mass m in the vicinity of a spherically symmetric nonrotating star of mass M. The curved geometry surrounding such a mass was discovered in 1915, shortly after Einstein published his theory, by the German physicist and astronomer Karl Schwarzschild. Remarkably enough, Schwarzschild was a soldier in the German army at the time, fighting in World War I. His metric was the first nontrivial (i.e., non-flat spacetime) exact solution found of the field equations. He published his solution later that same year, and died the following year of a disease he contracted on the Russian front. The **Schwarzschild metric** is

$$ds^2 = -\left(1 - \frac{2\mathcal{M}}{r}\right)c^2 dt^2 + \left(1 - \frac{2\mathcal{M}}{r}\right)^{-1} dr^2 + r^2(d\theta^2 + \sin^2\theta d\phi^2), \tag{10.5}$$

where $\mathcal{M} \equiv GM/c^2$. Note that as $M \to 0$ the spacetime becomes the flat Minkowski metric (10.4). In the Schwarzschild metric, the star sits at the origin

---

[2] Of course the sun is not quite spherically symmetric due to its rotation. But it is close enough for our applications here.

of the coordinates ($r = 0$) and the metric described the curved spacetime only for $r > R$, where R is the radius of the star.

Schwarzschild's metric contains the same spherical coordinates r, θ, ϕ and time t as in Minkowski's flat-space metric, so the meanings of the coordinates seem obvious. However, in reality their meanings are more subtle. In a general-relativistic metric, as we have already mentioned, *physical* distances (those measured by real meter sticks) correspond to integrals over space-like intervals, $s = \int ds$ – space-like means that $ds^2 > 0$ in the metric; while physical times (those read by real clocks) correspond to integrals over time-like intervals, $s = \int ds$ – where $ds^2 \equiv -c^2 d\tau^2 < 0$. Here $d\tau$ is the infinitesimal time interval read locally by a clock tracing an infinitesimal time-like path in spacetime, and $\tau = \int d\tau$ is the total interval read by a clock carried along some finite path.

Take for example two points in the Schwarzschild geometry that are separated purely radially, with identical values of $\theta \equiv \theta_0$ and $\phi \equiv \phi_0$. What is the physical distance between them at some fixed time t_0? We can label the two points $(t_0, r_1, \theta_0, \phi_0)$ and $(t_0, r_2, \theta_0, \phi_0)$ where we will assume that $r_2 > r_1$ as shown in Figure 10.4. Then for a radial distance between the two points at the fixed time t_0, we have $dt = 0$, $d\theta = 0$, and $d\phi = 0$, so the spacetime interval ds obeys

$$ds^2 = \left(1 - \frac{2\mathcal{M}}{r}\right)^{-1} dr^2. \qquad (10.6)$$

Therefore the actual physical distance between the two points is *not* $r_2 - r_1$, but

$$\Delta s \equiv \int_{r_1}^{r_2} ds = \int_{r_1}^{r_2} \frac{dr}{\sqrt{1 - 2\mathcal{M}/r}}. \qquad (10.7)$$

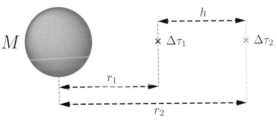

Fig. 10.4 Two spacetime points in the Schwarzschild geometry at Schwarzschild coordinate time $t = t_0$. Later on, we place two clocks at r_1 and r_2: the proper time intervals $\Delta\tau_1$ and $\Delta\tau_2$ for the two clocks are also shown.

Note that if $\mathcal{M} = 0$ the spacetime reduces to flat Minkowski spacetime and the physical distance between the two points really is $r_2 - r_1$, as we might expect. It is also true that if both points have values of $r \gg 2\mathcal{M}$, then the spacetime becomes Minkowskian, and the physical distance between the two points approaches $r_2 - r_1$. But if $\mathcal{M} \neq 0$ and r_1 is *not* large compared with $\mathcal{M}$, the spacetime is curved and the physical distance between the two points is *greater* than $r_2 - r_1$. That is, the radii r_2 and r_1 are merely convenient coordinate labels and do not necessarily represent measured physical distances. Note that we must assume that $r > 2\mathcal{M}$ everywhere along the path; otherwise the integrand becomes imaginary and we are in real trouble (we return later to the question of what happens if $r \leq 2\mathcal{M}$. In fact, spacetime points with $r = 2\mathcal{M} = 2GM/c^2$ turn out to have a profound physical significance, as we shall see.)

Now suppose instead that we have two spacetime points at the same *spatial* locations (r_0, θ_0, ϕ_0), but taking place at two different values of the time coordinate, at t_1 and t_2. That is, the points might represent two successive events in the history of a particle at rest. What is the physical time interval (the time interval read by a clock) between these two events? Using $dr = d\theta = d\phi = 0$ in Eq. (10.5), we have

$$ds^2 = -c^2 d\tau^2 = -c^2 \left(1 - \frac{2\,\mathcal{M}}{r_0} \right) dt^2, \tag{10.8}$$

so

$$\Delta\tau \equiv \tau_2 - \tau_1 = \int_{t_1}^{t_2} \sqrt{1 - \frac{2\mathcal{M}}{r_0}}\, dt = \sqrt{1 - \frac{2\mathcal{M}}{r_0}}\, \Delta t, \tag{10.9}$$

where $\Delta t \equiv t_2 - t_1$. Again, if $\mathcal{M} = 0$ or if $r_0 \to \infty$, the spacetime is flat and the physical time interval $\tau_2 - \tau_1$ reduces to $t_2 - t_1$. Now we understand the physical meaning of the coordinate time t: it is the reading of clocks at rest far from the central mass, *i.e.*, with value of r_0 such that $r_0 \to \infty$ (or more realistically $r_0 \gg \mathcal{M}$). Clocks at rest at smaller values of r run at the slower rate

$$\Delta\tau = \sqrt{1 - \frac{2\mathcal{M}}{r_0}}\, \Delta t < \Delta t. \tag{10.10}$$

In fact, as long as $r_0 > 2\mathcal{M} = 2GM/c^2$, where M is the central mass, *the smaller the radius at which a clock sits, the slower it runs*. Again we get in serious trouble if we try to imagine a clock at rest at some radius $r \leq 2\mathcal{M}$.

Example 10.3   **Radial Light and Gravitational Redshifts**

Picture two clocks at rest at different distances from some central mass. The first clock sits at distance r_1 from the center, and the second clock sits at the larger distance r_2, as shown in Figure 10.4 above. We ourselves are also at rest next to the second clock, and can watch the first clock through a telescope. As we watch the first clock, does it appear to run at the same rate as our own clock at r_2, or does it run fast or slow relative to our own clock?

First we need to look at the propagation of a radial light beam from r_1 to r_2. As in special relativity, light moves along null trajectories in spacetime, so that in the Schwarzschild geometry

$$ds^2 = 0 = -c^2 \left(1 - \frac{2\mathcal{M}}{r} \right) dt^2 + \left(1 - \frac{2\mathcal{M}}{r} \right)^{-1} dr^2 \tag{10.11}$$

for radially directed light. Solving for dr/dt:

$$\frac{dr}{dt} = \pm c \left(1 - \frac{2\mathcal{M}}{r} \right), \tag{10.12}$$

where we will now choose the plus sign since the light signals are traveling outward. It follows by integration that the coordinate time required for light to travel outward from r_1 to r_2 is

$$t_2 - t_1 \equiv \int_{t_1}^{t_2} dt = \int_{r_1}^{r_2} \frac{dr}{1 - 2\mathcal{M}/r}. \tag{10.13}$$

The integral can be solved exactly (see the Problems section at the end of this chapter), but for now note that if two successive light signals are sent from r_1 to r_2, *the propagation time is the same for each*. So if light with wavelength λ_1 is emitted at r_1, one crest at t_1 and the next crest at $t_1 + \Delta t_1$, then the two crests arrive at r_2 at times t_2 and $t_2 + \Delta t_2$, where $\Delta t_2 = \Delta t_1$. The coordinate time differences are the same before and after.

However, as we have seen, local rest clocks do not generally read coordinate times. If the initial time interval between two wave crests at r_1 is Δt_1, then the rest clock at r_1 reads a time interval

$$\Delta \tau_1 = \sqrt{1 - \frac{2\mathcal{M}}{r_1}} \, \Delta t_1, \tag{10.14}$$

and when the two wave crests arrive at r_2 the local rest clock there reads a time interval

$$\Delta \tau_2 = \sqrt{1 - \frac{2\mathcal{M}}{r_2}} \, \Delta t_2. \tag{10.15}$$

In our case $\Delta t_2 = \Delta t_1$, so the ratio of local clock rates is

$$\frac{\Delta \tau_2}{\Delta \tau_1} = \sqrt{\frac{1 - 2\mathcal{M}/r_2}{1 - 2\mathcal{M}/r_1}}. \tag{10.16}$$

Now the wavelength of the light beam is $\lambda_1 = c\Delta\tau_1$ when emitted at r_1, and $\lambda_2 = c\Delta\tau_2$ when received at r_2, so the ratio of the received and emitted wavelengths is the same as that for the time intervals:[a]

$$\frac{\lambda_2}{\lambda_1} = \sqrt{\frac{1 - 2\mathcal{M}/r_2}{1 - 2\mathcal{M}/r_1}}. \tag{10.17}$$

This ratio is greater than one, since $r_2 > r_1$, so the light has undergone a *redshift* as it travels outward. This is called the **gravitational redshift** and has been observed many times in light from our sun and other stars. The shift is especially large for the light from white-dwarf stars, where the observed shift of well-known spectral lines is used to measure the ratio of mass to radius for these stars. The gravitational redshift was first measured on earth using gamma-ray photons emitted by atomic nuclei.

Note that the gravitational redshift is *infinite* for light emitted by an object for which $2\mathcal{M}/r \equiv 2GM/rc^2 = 1$. We will revisit this question.

It is also clear from the analysis that if an observer at rest at some smaller radius r_1 views an emitter at some larger radius r_2, that observer will see a gravitational *blueshift*. As a rule of thumb, "falling" light gets bluer, and "rising" light gets redder.   ■

> [a]We have assumed here that light moves at speed c according to measurements made using local clocks and meter sticks. This can be shown from the results we have already derived (see the Problems section at the end of this chapter).

Example 10.4 **Comparison with the Equivalence Principle**

Let us compare two clocks at rest, one at r_1 and the other at larger radius r_2, in the weak-gravity limit, as shown in Figure 10.4 previously. The exact ratio of their rates (with the same coordinate time interval Δt for each – measured by the observer at $r \to \infty$ is

$$\frac{\Delta \tau_2}{\Delta \tau_1} = \frac{\sqrt{1 - 2\mathcal{M}/r_2}}{\sqrt{1 - 2\mathcal{M}/r_1}}. \tag{10.18}$$

In particular, for rest clocks near the earth (or anywhere in the solar system, for that matter), their distances from any central mass easily satisfy $r \ll 2\mathcal{M}$, so we can use the binomial expansion for small $2\,\mathcal{M}/r$ yielding (see Appendix F)

$$\frac{\Delta \tau_2}{\Delta \tau_1} = \left(1 - \frac{2\mathcal{M}}{r_2}\right)^{1/2} \left(1 - \frac{2\mathcal{M}}{r_1}\right)^{-1/2} \tag{10.19}$$

$$\simeq \left(1 - \frac{\mathcal{M}}{r_2}\right)\left(1 + \frac{\mathcal{M}}{r_1}\right) \simeq \left[1 + \mathcal{M}\left(\frac{1}{r_1} - \frac{1}{r_2}\right)\right],$$

where in the last equality we neglected the product of two small terms. Now suppose in addition that $r_2 = r_1 + h$, where $h \ll r_1$. Then

$$\frac{1}{r_2} = \frac{1}{r_1 + h} = \frac{1}{r_1}\left(1 + \frac{h}{r_1}\right)^{-1} = \frac{1}{r_1}\left(1 - \frac{h}{r_1}\right), \tag{10.20}$$

so finally

$$\frac{\Delta \tau_2}{\Delta \tau_1} = 1 + \frac{\mathcal{M}h}{r_1^2} = 1 + \left(\frac{GM}{r_1^2 c^2}\right)h = 1 + \frac{gh}{c^2}, \tag{10.21}$$

where $g = GM/r_1^2$ is the Newtonian gravitational field at r_1. This is exactly the result we found previously from the principle of equivalence: clocks at higher altitude run fast compared with clocks at lower altitude, by a factor $(1 + gh/c^2)$, where g is the local gravitational field, and h is the altitude difference between the two clocks. That is, the principle of equivalence is built into general relativity. If experiments showed that the principle of equivalence is false, then general relativity must be false as well. That has not happened.   ■

10.3 Geodesics in the Schwarzschild Spacetime

We now ask how a probe particle of mass $m \ll M$ would move in the vicinity of the source mass M responsible for the Schwarzschild metric (10.5). The answer to this question is a natural extension of how we do it in special relativity: particles move along trajectories which extremize proper time. Since $ds^2 = -c^2 d\tau^2$ along a time-like worldline in both special and general relativity, we can extremize either $\int ds$ or $\int d\tau$. Hence, according to general relativity, a massive particle subject to nothing but gravity must move on *time-like geodesics* in four-dimensional spacetime – by extremizing once again the functional

$$S = c \int d\tau. \tag{10.22}$$

From spherical symmetry we know that the geodesics will lie in a plane, which (as usual) we take to be the equatorial plane $\theta = \pi/2$, which leaves the two degrees of freedom r and ϕ. Therefore we seek to find those paths that make stationary the functional

$$S = \int \sqrt{\left(1 - \frac{2\mathcal{M}}{r}\right) c^2 dt^2 - \left(1 - \frac{2\mathcal{M}}{r}\right)^{-1} dr^2 - r^2 d\phi^2}$$

$$= \int d\tau \sqrt{\left(1 - \frac{2\mathcal{M}}{r}\right) c^2 \dot{t}^2 - \left(1 - \frac{2\mathcal{M}}{r}\right)^{-1} \dot{r}^2 - r^2 \dot{\phi}^2}, \tag{10.23}$$

where $\dot{t} = dt/d\tau$, $\dot{r} = dr/d\tau$, and $\dot{\phi} = d\phi/d\tau$. Note that while in nonrelativistic physics the time t is an independent variable, and not a coordinate, in relativistic physics the time $t(\tau)$ has become one of the coordinates, and the independent variable is the proper time τ read by a clock carried along with the moving particle. This calculus of variations problem is now a familiar one: the Lagrangian is the square-root integrand in Eq. (10.23), and τ replaces t as the independent variable.

   Before we proceed, there is a way to simplify things before making further computations. Note first of all that (since $ds^2 = -c^2 d\tau^2$)

$$\left(1 - \frac{2\mathcal{M}}{r}\right) c^2 \dot{t}^2 - \left(1 - \frac{2\mathcal{M}}{r}\right)^{-1} \dot{r}^2 - r^2 \dot{\phi}^2 = c^2, \tag{10.24}$$

a constant along the worldline of the particle. This fact can be used to help show that making stationary the integral in Eq. (10.23) is equivalent to making stationary the same integral, with the same integrand but with the square root removed (see the Problems section at the end of this chapter). So our action becomes

$$S \to \int d\tau \left[\left(1 - \frac{2\mathcal{M}}{r}\right) c^2 \dot{t}^2 - \left(1 - \frac{2\mathcal{M}}{r}\right)^{-1} \dot{r}^2 - r^2 \dot{\phi}^2\right], \tag{10.25}$$

where the effective Lagrangian is the expression in brackets. Of our three coordinates, t and ϕ are cyclic, so the corresponding generalized momenta are conserved, giving us two first integrals of motion

$$p_t = \frac{\partial L}{\partial \dot{t}} = -2c^2 \left(1 - \frac{2\mathcal{M}}{r} \right) \dot{t} \equiv -2c^2 \mathscr{E},$$

$$p_\phi = \frac{\partial L}{\partial \dot{\phi}} = 2r^2 \dot{\phi} \equiv 2\mathscr{L}, \tag{10.26}$$

where $\mathscr{E}$ and $\mathscr{L}$ are constants. For the third first integral we see that L is not an explicit function of the independent variable τ, so the Hamiltonian is conserved. However, it turns out that this is equivalent to Eq. (10.24), already in the form of a first integral of motion.

We can now eliminate $\dot{t}$ and $\dot{\phi}$ in Eq. (10.24) using Eq. (10.26), giving

$$\dot{r}^2 - \frac{2\mathcal{M}c^2}{r} + \frac{\mathscr{L}^2}{r^2} \left(1 - \frac{2\mathcal{M}}{r} \right) = c^2(\mathscr{E}^2 - 1). \tag{10.27}$$

This will look more familiar if we divide by two, multiply by the mass m of the orbiting particle, recall that $\mathcal{M} \equiv GM/c^2$, and write $\mathscr{L} \equiv \ell/m$:

$$\frac{1}{2}m\dot{r}^2 - \frac{GMm}{r} + \frac{\ell^2}{2mr^2} \left(1 - \frac{2GM}{rc^2} \right) = \frac{mc^2}{2}(\mathscr{E}^2 - 1) \equiv E, \tag{10.28}$$

which has the familiar mathematical form of a one-dimensional conservation of energy equation (!) in the radial direction

$$E = \frac{1}{2}m\dot{r}^2 + U_{\text{eff}}(r), \tag{10.29}$$

where the effective potential is

$$U_{\text{eff}}(r) \equiv -\frac{GMm}{r} + \frac{\ell^2}{2mr^2} \left(1 - \frac{2GM}{rc^2} \right). \tag{10.30}$$

The first two terms of $U_{\text{eff}}(r)$, which are by far the largest in familiar circumstances like the solar system, are exactly the same as the corresponding effective potential for Newtonian gravity (7.27). If this were not true, the theory would be dead in the water, because we know that Newtonian gravitation is extremely accurate, at least within the solar system. Note then that Einstein's gravity predictions for our probe's orbit can be approximated by those of Newtonian gravity when we can drop the last term in the effective potential (10.30), or when

$$\frac{GM}{rc^2} \ll 1 \; ; \tag{10.31}$$

that is, for small source masses or large distances from the source, which means in either case "weak gravity." In general, however, Einstein gravity can become important in other regimes as well, ones involving even weak gravity under the right circumstances.

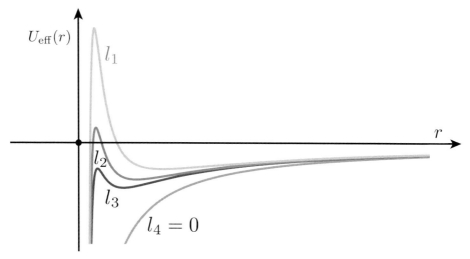

Fig. 10.5 Effective potential for the Schwarzschild geometry for various angular momenta $l_1 > l_2 > l_3 > l_4 = 0$.

In Newtonian mechanics, recall that the first term $-GMm/r$ in Eq. (10.30) is the gravitational potential and the second term $\ell^2/2mr^2$ is the angular momentum barrier. The *third* new term is obviously relativistic, since it involves the speed of light. It has the effect of diminishing the angular momentum barrier for small r, and can make the centrifugal term *attractive* rather than *repulsive*, as shown in Figure 10.5. This effect cannot be seen for the sun or most stars: coordinate distances r where the last term of (10.30) induces important dynamical effects lie within the body of the star – where the Schwarzschild metric cannot be used to describe the curvature of spacetime. The relativistic term can however have a small but observable effect on the inner planets, as we will show in the next example.

Example 10.5 **The Precession of Mercury's Perihelion**

By the end of the nineteenth century astronomers knew there was a problem with the orbit of Mercury. In the sun's frame, the perihelion of Mercury's orbit does not keep returning to the same spot. The perihelion slowly *precesses*, so that each time Mercury orbits the sun the perihelion occurs slightly later than it did on the previous revolution. The main reason for this is that the other planets pull slightly on Mercury, so the force it experiences is not purely central. Very accurate methods were worked out to calculate the total precession of Mercury's perihelion caused by the other planets, and although the calculations explained most of the precession, Mercury actually precesses by 43 seconds of arc per century *more* than the calculations predicted. Einstein was aware of this discrepancy when he worked on his general theory, and was intensely curious whether the effects of relativity might explain the 43 seconds/century drift.

We begin with the conservation equations (10.26) and (10.29):

$$E = \frac{1}{2}m\dot{r}^2 - \frac{GMm}{r} + \frac{\ell^2}{2mr^2}\left(1 - \frac{2GM}{rc^2}\right) \quad \text{and} \quad \mathscr{L} = \frac{\ell}{m} = r^2\dot{\phi}. \tag{10.32}$$

Using the chain rule, and again defining the inverse radius $u = 1/r$ as coordinate, we have

$$\dot{r} \equiv \frac{dr}{d\tau} = \frac{dr}{du}\frac{du}{d\phi}\frac{d\phi}{d\tau} = -\frac{1}{u^2}u'\frac{\ell}{mr^2} = -\frac{\ell}{m}u', \tag{10.33}$$

where $u' \equiv du/d\phi$. Substituting this result into Eq. (10.32) gives

$$E = \frac{\ell^2}{2m}(u'^2 + u^2) - GMmu - \frac{GM\ell^2}{mc^2}u^3. \tag{10.34}$$

Then differentiating with respect to ϕ, we get a second-order differential equation for the orbital shape $u(\phi)$:

$$u'' + u = \frac{GMm^2}{\ell^2} + \frac{3GM}{c^2}u^2, \tag{10.35}$$

which is the same equation we found for the nonrelativistic Kepler problem except for the second term on the right — which makes the equation nonlinear. We don't have to solve the equation exactly, however, because the new term is very small. We can solve it to sufficient accuracy using what is called **first-order perturbation theory**.

In first-order perturbation theory we assume a solution of the form $u = u_0 + \delta u_1$, where (in our example) u_0 is the known exact solution of the linear equation without the new relativistic term, u_1 is a correction due to the relativistic term, and δ is a small dimensionless parameter which we are free to choose later. Our goal is to find u_1 and see whether the corrected solution leads to a precession of Mercury's orbit. Substituting $u = u_0 + \delta u_1$ into Eq. (10.35), we find that

$$u_0'' + u_0 + \delta(u_1'' + u_1) = \frac{GMm^2}{\ell^2} + \frac{3GM}{c^2}(u_0 + \delta u_1)^2. \tag{10.36}$$

The function u_0 obeys the nonrelativistic Kepler equation

$$u_0'' + u_0 = \frac{GMm^2}{\ell^2}, \tag{10.37}$$

with solution

$$u_0 = A(1 + \epsilon \cos\phi), \tag{10.38}$$

where ϵ is the eccentricity of the classical elliptical orbit and $A \equiv GMm^2/\ell^2$, as shown in Chapter 7, where ℓ is Mercury's angular momentum. We have chosen a constant of integration so that $\phi = 0$ at perihelion (minimum r, maximum u.) The part left over is

$$\delta(u_1'' + u_1) = \frac{3GM}{c^2}(u_0 + \delta u_1)^2 = \frac{3GM}{c^2}[A(1 + \epsilon \cos\phi) + \delta u_1]^2. \tag{10.39}$$

The quantity GM/c^2 has dimensions of length, and A has dimensions of inverse length, so the ratio GMA/c^2 is dimensionless. The ratio is also very small if we choose M to be the sun's mass and the angular momentum

ℓ to be that for Mercury's orbit. Therefore this ratio is both dimensionless and small, so it makes sense to choose

$$\delta \equiv \frac{GMA}{c^2}, \tag{10.40}$$

leaving us with

$$u_1'' + u_1 = (3A)(1 + \epsilon \cos \phi)^2, \tag{10.41}$$

where we have neglected additional δ and δ^2 terms on the right to keep everything in the differential equation only through first-order terms in δ. In fact, it would be inconsistent to keep additional terms on the right, because we have kept only terms through first order in δ on the left. Expanding what remains on the right:

$$u_1'' + u_1 = 3A(1 + \epsilon \cos \phi)^2 = 3A \left[1 + 2\epsilon \cos \phi + \frac{\epsilon^2}{2}(1 + \cos 2\phi) \right], \tag{10.42}$$

where we have used the identity $\cos^2 \phi = (1/2)(1 + \cos 2\phi)$. This gives us three linearly independent terms on the right:

$$u_1'' + u_1 = 3A \left[\left(1 + \frac{\epsilon^2}{2} \right) + 2\epsilon \cos \phi + \frac{\epsilon^2}{2} \cos 2\phi \right]. \tag{10.43}$$

Note that this equation is *linear*, and that its general solution is the sum of the solution of the homogeneous equation (with zero on the right) and a particular solution of the full equation. We do not need the solution of the homogeneous equation, however, because it is the same as that for the u_0 equation, so contributes nothing new. And (because of the linearity of the equation) the particular solution of the full equation is just the sum of the particular solutions due to each of the three terms on the right, taken one at a time. That is, $u_1 = u_1^{(1)} + u_1^{(2)} + u_1^{(3)}$, where[a]

$$u_1'' + u_1 = 3A(1 + \epsilon^2/2) \quad \text{with solution } u_1^{(1)} = 3A(1 + \epsilon^2/2),$$
$$u_1'' + u_1 = 3A(\epsilon^2/2) \cos 2\phi \quad \text{with solution } u_1^{(3)} = -(\epsilon^2 A/2) \cos 2\phi,$$
$$u_1'' + u_1 = 6A\epsilon \cos \phi \qquad \text{with solution } u_1^{(2)} = 3\epsilon A\phi \sin \phi. \tag{10.44}$$

Altogether, the new contributions in first order add up to

$$u_1 = 3A \left[1 + \frac{\epsilon^2}{2} - \frac{\epsilon^2}{6} \cos 2\phi + \epsilon\phi \sin \phi \right]. \tag{10.45}$$

The only term here that can cause a perihelion precession is the $\phi \sin \phi$ term, the so-called **secular term**, since it is the only term that does not return to where it began after a complete revolution, *i.e.*, as $\phi \rightarrow \phi + 2\pi$. The other terms can cause a slight change in orbital shape, but not a precession. So including the secular term together with the zeroth-order terms, we write

$$u \simeq u_0 + \delta u_1^{(2)} = A(1 + \epsilon \cos \phi) + 3A\delta(\epsilon \phi \sin \phi) = A(1 + \epsilon \cos \phi) + \frac{3GM}{c^2}A^2 \epsilon \phi \sin \phi,$$
(10.46)

which is all we need for the purpose of tracing the precession to leading order. The perihelion corresponds to the minimum value of r, or the maximum value of u, so at perihelion

$$\frac{du}{d\phi} = 0 = -A\epsilon \sin \phi + \frac{3GM}{c^2}A^2 \epsilon (\sin \phi + \phi \cos \phi),$$
(10.47)

which has a solution at $\phi = 0$, but *not* at $\phi = 2\pi$. So we look for a solution at $\phi = 2\pi + \Delta$ for some small Δ, where Δ represents the angle of precession. For small Δ, $\sin(2\pi + \Delta) = \sin \Delta \cong \Delta$ and $\cos(2\pi + \Delta) \cong 1$ to first order in Δ. Therefore at the end of one revolution, Eq. (10.47) gives

$$0 = -A\epsilon\Delta + \frac{3GM}{c^2}A^2 \epsilon [\Delta + (2\pi + \Delta)].$$
(10.48)

However, the Δs in the square-bracket term are small compared with 2π, so for consistency they must be neglected. Then from Chapter 7 we find that

$$A = GMm^2/\ell^2 = \frac{1}{r_p(1 + \epsilon)} = \frac{1}{a(1 - \epsilon^2)}.$$
(10.49)

Here $r_p = a(1 - \epsilon)$, where r_p is the perihelion and a is the semi-major axis of the elliptical orbit. So the precession per revolution is then

$$\Delta = \frac{6\pi GMA}{c^2} = \frac{6\pi GM}{c^2 a(1 - \epsilon^2)}.$$
(10.50)

The data for Mercury's orbit is $a = 5.8 \times 10^{10}$ m, $\epsilon = 0.2056$, and $M = M_{sun} = 2.0 \times 10^{30}$ kg. The result is

$$\Delta = 5.04 \times 10^{-7} \text{ radians/revolution.}$$
(10.51)

We can convert this result to seconds of arc/century, using the facts that Mercury orbits the sun every 88 days and that there are $60 \times 60 = 3600$ seconds of arc in one degree:

$$\delta = 5.04 \times 10^{-7} \frac{\text{rad}}{\text{rev}} \left(\frac{360 \text{ deg}}{2\pi \text{ rad}} \right) \left(\frac{3600 \text{ s}}{\text{deg}} \right) \left(\frac{1 \text{ rev}}{88 \text{ days}} \right) \left(\frac{365 \text{ days}}{\text{year}} \right) \left(\frac{100 \text{ years}}{\text{century}} \right)$$

$$= 43 \frac{\text{seconds of arc}}{\text{century}} \,!$$
(10.52)

After the extraordinary efforts and frequent frustrations leading up to his discovery of general relativity, here was Einstein's payoff. He had successfully explained a well-known and long-standing conundrum. Later he wrote to a friend: "For a few days, I was beside myself with joyous excitement." And in the words of his scientific biographer Abraham Pais: "This discovery was, I believe, by far the strongest emotional experience in Einstein's scientific life, perhaps in all his life. Nature had spoken to him. He had to be right." ∎

---

[a] See the Problems section at the end of this chapter.

Example 10.6

Bending of Light by the Sun

Light is massless, so is not affected by Newtonian gravity: a light beam moving past the sun should not be bent according to classical physics. Using general relativity, however, the beam moves through a curved spacetime, which may cause a deflection. To find out, we begin again with the Schwarzschild metric:

$$ds^2 = -(1 - 2\mathcal{M}/r)c^2 dt^2 + (1 - 2\mathcal{M}/r)^{-1} dr^2 + r^2(d\theta^2 + \sin^2\theta d\phi^2), \tag{10.53}$$

where now $ds^2 = 0$, since (just as in the special relativity of Chapter 2) massless particles follow null paths in spacetime. From symmetry considerations, a massless particle must still move in a plane defined by its initial velocity vector and the center of the central mass. Therefore we can again take the effective Lagrangian of the particle to be

$$L = (1 - 2\mathcal{M}/r)c^2 \dot{t}^2 - (1 - 2\mathcal{M}/r)^{-1}\dot{r}^2 - r^2\dot{\phi}^2, \tag{10.54}$$

but with two modifications. First, since the trajectory is null, from the metric we must set the effective Lagrangian to be $L = 0$. Second, no longer can we take $\dot{t} \equiv dt/d\tau, \dot{r} \equiv dr/d\tau$, etc., because the proper time interval is $d\tau = 0$ for massless particles. However, we can let $\dot{t} \equiv dt/d\lambda, \dot{r} \equiv dr/d\lambda$, etc., where λ is some other parameter along the path, called an "affine" parameter.

Using this Lagrangian we can find three first integrals of motion. Two of them follow from the fact that both t and ϕ are cyclic coordinates:

$$\dot{t} = \frac{\alpha}{1 - 2\mathcal{M}/r}, \qquad \dot{\phi} = \frac{\ell}{r^2}, \tag{10.55}$$

where α and ℓ are constants. The third comes from setting $L = 0$ and eliminating $\dot{t}$ and $\dot{\phi}$ using the first two, i.e.,

$$\dot{r}^2 + \frac{\ell^2}{r^2}(1 - 2\mathcal{M}/r) = \alpha^2. \tag{10.56}$$

As with massive particles, we now define a new independent variable $u \equiv 1/r$; it is then straightforward to show that $\dot{r} = -\ell u'$, where $u' \equiv du/d\phi$. Therefore

$$u'^2 + u^2(1 - 2\mathcal{M}u) = \alpha^2/\ell^2. \tag{10.57}$$

Differentiating with respect to ϕ:

$$u'' + u = 3\mathcal{M}u^2. \tag{10.58}$$

If $\mathcal{M} = 0$ the spacetime is flat; in that case $u'' + u = 0$ with general solution $u = A\cos(\phi - \phi_0)$, where A and ϕ_0 are the two constants of integration. That is:

$$r\cos(\phi - \phi_0) = 1/A, \tag{10.59}$$

the equation of a straight line, as expected. In particular, if we choose $\phi_0 = 0$ the light beam moves in the y direction and its point of closest approach to the origin is at $\phi = 0$, with $r = 1/A$ at that point.

Now allow $\mathcal{M} \neq 0$. Returning to the full equation, we try the perturbation expansion

$$u = u_0 + \delta u_1, \tag{10.60}$$

where u_0 is the straight-line path we have already found for $\mathcal{M} = 0$, u_1 is a new function representing a correction to the straight line, and δ is some small dimensionless number which we have the freedom to select later on. That is, we assume that the new solution u includes only a small deviation from a straight-line path. Substituting into the differential equation, we have

$$u_0'' + u_0 + \delta(u_1'' + u_1) = 3\mathcal{M}(u_0 + \delta u_1)^2. \tag{10.61}$$

We have made no approximations so far. Now by definition $u_0'' + u_0 = 0$, and for simplicity we choose to define

$$\delta = \frac{3\mathcal{M}}{r_0} \equiv \frac{3GM}{r_0 c^2}, \tag{10.62}$$

where M is the sun's mass and r_0 is the distance of closest approach of the beam to the center of the sun, which can be no smaller than the radius R of the sun itself. It is easy to verify that $\delta \ll 1$, and that it is dimensionless. Therefore to first order in δ (neglecting terms in δ^2, etc.), we have

$$u_1'' + u_1 = r_0 u_0^2 = r_0 A^2 \cos^2 \phi = (1/2r_0)(1 + \cos 2\phi), \tag{10.63}$$

where δ has canceled out on both sides and we have used the trig identity $\cos^2 \phi = (1/2)(1 + \cos 2\phi)$. This is an inhomogeneous linear differential equation, whose solution is the sum of the general solution of the homogeneous equation $u_1'' + u_1 = 0$ and a particular solution of the full equation. We can ignore the general solution of the homogeneous equation, because that gives us nothing new: it is the straight-line path we have already included in the equation for u_0. It is straightforward to show that a particular solution of the full equation is

$$u_1 = (1/2r_0)[1 - (1/3)\cos 2\phi]. \tag{10.64}$$

Therefore to first order in the small quantity δ, the light-beam path is given by

$$u = u_0 + \delta u_1 = (1/r_0)\cos\phi + \frac{3GM}{r_0 c^2} \frac{1}{2r_0}[1 - (1/3)\cos 2\phi]. \tag{10.65}$$

We are interested to see if this corresponds to a deflection as the beam passes by the sun. As the beam approaches from far away, the initial radius is effectively infinite, so the initial value $u = 0$. The same is true as the beam departs far from the sun. With the zeroth-order term alone, u vanishes if $\phi = \pm\pi/2$. So we assume that with the full solution, $u = 0$ when $\phi = \pm(\pi/2 + \Delta)$, where Δ is a some very small angle. Therefore (canceling a $1/r_0$)

$$0 = \cos(\pi/2 + \Delta) + \left(\frac{3GM}{2r_0 c^2}\right)[1 - (1/3)\cos 2(\pi/2 + \Delta)]. \tag{10.66}$$

Now

$$\cos(\pi/2 + \Delta) = (\cos \pi/2)\cos\Delta - (\sin \pi/2)\sin\Delta = -\sin\Delta \cong -\Delta \tag{10.67}$$

and

$$\cos 2(\pi/2 + \Delta) = \cos \pi \cos 2\Delta - \sin \pi \sin 2\Delta = -\cos 2\Delta \cong -1, \tag{10.68}$$

to first order in the small angle Δ, so we are left with

$$\Delta = \left(\frac{2GM}{r_0 c^2} \right). \tag{10.69}$$

Clearly Δ is the deflection of the beam as it approaches the near-point from infinity, and also the deflection in the beam as it recedes to infinity, so the total deflection of the beam as it passes the sun is

$$2\Delta = \left(\frac{4GM}{r_0 c^2} \right). \tag{10.70}$$

If the closest approach of the beam is the sun's "radius" R itself, the bending is

$$2\Delta = \left(\frac{4GM_S}{Rc^2} \right) = 1.75 \text{ seconds of arc.} \tag{10.71}$$

This small bending was predicted by Einstein in 1915. The only way to detect it at that time was to wait for a total solar eclipse, so that the apparent position of stars as their light is bent by the sun could be seen and compared with their true positions in the sky. It was also necessary to wait until the end of the "Great War," so that expeditions to total solar eclipse sites could be undertaken. This was carried out by teams of British astronomers in 1919, and Einstein's prediction was verified. As newspapers picked up the story, Einstein became world famous, and for the rest of his life and beyond he was an icon of science. Since that time much more accurate measurements have been made, which agree with Einstein with a very high degree of confidence.   ∎

10.4 The Event Horizon and Black Holes

In Schwarzschild's metric we have already noticed that the particular radius

$$R_0 = 2\mathcal{M} \equiv 2GM/c^2, \tag{10.72}$$

where M is the central spherically symmetric mass, has special properties. At radius R_0, two of the metric coefficients are either zero or infinity. If we could place a clock at rest at R_0, it would appear from our equations that the clock would not advance in time at all! An outward-directed light beam would have zero velocity when emitted at R_0, whether time is measured by local rest clocks or clocks at infinity. Also, outward-directed light from R_0 would suffer an infinite redshift.

Should we be concerned about these strange features? We only need to worry about them if the Schwarzschild metric is *valid* at R_0, which means that if we calculate R_0 from the value of the central mass, this particular radius is situated in the vacuum *outside* the mass. That is, the Schwarzschild geometry is a solution of Einstein's equations in vacuum, so if the calculated value of R_0 lies inside the central mass, *i.e.*, at a radius where there is mass present, then the Schwarzschild geometry is invalid there, so we don't need to worry about these strange results. In fact, R_0 for the earth is just under 1 cm and R_0 for the sun is about 3 km, far inside the actual radii of those bodies.

However, it turns out that in the gravitational collapse of stars or galactic nuclei, it *is* possible for massive objects to collapse to radii comparable to R_0. For example, highly massive stars reach a point in their evolution when the entire star collapses and the outer layers then explode outwards to form a supernova, while the core of the star may continue to collapse to R_0 and beyond. The centers of galaxies, including our own, can be a collecting point for millions or billions of stars, as well as gas and dust, which can also collapse to radii comparable to R_0.

Therefore it is necessary to make a careful analysis of the physics of light and particles near R_0. Some features are still controversial, while others seem to be well established.

The *event horizon* of a spherical black hole is located at the coordinate radius R_0. This means that if we are in the vacuum at some radius $R > R_0$ we can observe events (through a telescope, for example) which take place at all points within our radius, up until $r \to R_0$. In fact, we are unable to see events that take place at any radius $r \le R_0$, which is why R_0 is called the event horizon. Any point at or within R_0 is "over the horizon" as far as we are concerned.

Also, since light cannot escape from any radius $r \le R_0$, we see nothing there, which is why a massive spherically symmetric body whose radius has collapsed to $r \le R_0$ is called a "black hole." Derivations of the properties of event horizons and black holes require analyses far beyond the scope of this book, so we leave it to the reader to pursue these fascinating topics.

Example 10.7

Representative Black Holes

Some black holes are the result of the gravitational collapse of a star. A giant star burns up its nuclear fuel relatively quickly, and ultimately, when its internal heat is no longer sufficient to support it against gravity, it collapses. The collapse reheats the star to such enormous temperatures that unburned nuclear fuels are ignited explosively, heating the outer parts in reactions that blow them outward, burning so brightly that the star is called a supernova. Meanwhile the inner core continues to collapse, and if it is heavy enough it becomes a black hole. A famous example is Cygnus X1, about 6000 light years away from us, first observed in the constellation Cygnus by an X-ray telescope. It is a black hole that is one member of a double-star system; the other member is an "ordinary" supergiant. Material from the ordinary star is gradually pulled away by the gravitational attraction of the black hole, and forms an "accretion disk" that orbits the black hole, a bit like Saturn's rings orbiting Saturn. Unlike Saturn's rings, however, there is enormous friction within the accretion disk, due to high-velocity collisions between particles in the disk. As a result, the disk is heated to millions of degrees, hot enough that it emits the observed X-rays. Orbital analyses estimate that the mass of the black hole is about 14.8 solar masses. This means that the radius of its event horizon is about 44 km, scaled linearly up from the 3 km event-horizon radius of a single solar mass.

More recently, *pairs* of black holes (called "black hole binaries") have been detected from the **gravitational waves** they emit as they orbit around one another. The first detection was on September 14, 2015: the two black holes had masses of approximately 29 solar masses and 36 solar masses, and were situated about 1.3 billion light years away from us. They were detected by two LIGO (Laser Interferometer Gravitational-Wave Observatory) instruments, located in Livingston, Louisiana and Hanford, Washington. Since then other detections of black-hole binaries have been made in this way. Gravitational waves are ripples in the curvature of spacetime, which propagate outwards from the source at the speed of light. As they are emitted by the rotating black-hole binary, energy is lost from the system, so the black-hole orbits decay into orbits of smaller and smaller radii and higher and higher rotational frequencies. Ultimately the two event horizons touch one another, so that they merge into a single event horizon surrounding a single black hole. In the first event observed, about three solar masses of energy were radiated outward in the gravitational waves, leaving a single black hole of about $29 + 36 - 3 = 62$ solar masses. The entire observation lasted about 0.2 s: during this time the rotational frequency (and the frequency of the emitted gravitational waves) rose from frequencies of a few cycles/second up to about 250 cycles/s. Translating these gravitational wave frequencies into sounds, the signal could be heard as a kind of "chirp," equivalent to what one would hear by running one's finger from the left end of a piano keyboard up to "middle C." As LIGO continues to be tuned and improved, it is expected that something like one or two events will be observed per month, most originating very far away in the universe.

Finally, there are also black holes in the nuclei of many galaxies, including our own. These are not the remnant of single supernova explosions, but may have formed by gravitational collapse of especially dense regions or a huge number of supernova explosions quite early in the history of the universe. They continue to grow as more matter falls into them, while other matter condenses into stars that orbit these central black holes to form the surrounding galaxies. The nucleus of our own Milky Way galaxy harbors a black hole of approximately 4.5 million solar masses. It is located in the constellation Sagittarius, and is named Sagittarius A*. Scaling up from a single solar mass, the event horizon of this black hole has a radius of about 13.5 million km: that is, the event horizon is less than one-tenth the distance of the earth from the sun, within which there are 4.5 million solar masses. We observe the system by the light surrounding it, including occasional flares, caused by the chaotic infall of dust, gas, and stars which are gradually increasing the mass of the black hole. Dust clouds, gas clouds, and stars orbit the galactic nucleus. Our solar system is approximately 26,000 light years from the nucleus, in an orbit that takes some 225–250 million years to complete. So in the 4.5 billion years since our solar system was formed, we have completed approximately 18 orbits around the black hole at the center of our galaxy.   ∎

10.5 Magnetic Gravity

It is possible to capture the leading relativistic corrections to Newtonian gravity using an analogy with electromagnetism. This is because a similar symmetry principle – gauge symmetry – underlies both force laws. We start with Maxwell's equations from Chapter 8:

$$\nabla \cdot \mathbf{E} = 4\pi\rho, \quad \nabla \times \mathbf{E} = -\frac{1}{c}\frac{\partial \mathbf{B}}{\partial t},$$

$$\nabla \cdot \mathbf{B} = 0, \qquad \nabla \times \mathbf{B} = \frac{4\pi}{c}\mathbf{J} + \frac{1}{c}\frac{\partial \mathbf{E}}{\partial t}, \tag{10.73}$$

where $\mathbf{E}$ and $\mathbf{B}$ are the electric and magnetic fields, and ρ and $\mathbf{J}$ are the charge and current densities. Given $\mathbf{E}$ and $\mathbf{B}$, the Lorentz force on a probe of charge q moving with velocity $\mathbf{v}$ is

$$\mathbf{F}_{em} = q\mathbf{E} + \frac{q}{c}\mathbf{v} \times \mathbf{B}. \tag{10.74}$$

We already encountered the Coulomb fields that solve Maxwell's equations for a point charge at rest:

$$\mathbf{E} = \frac{Q}{r^2}\hat{\mathbf{r}}, \quad \mathbf{B} = 0, \tag{10.75}$$

and noted the similarity between the corresponding electrostatic force law and Newtonian gravity:

$$\mathbf{F}_{em} = \frac{qQ}{r^2}\hat{\mathbf{r}}, \quad \mathbf{F}_{g} = -G\frac{mM}{r^2}\hat{\mathbf{r}}. \tag{10.76}$$

We are now ready to go beyond Newtonian gravity as described by the usual gravitational field $\mathbf{g}$ (see discussion in Section 8.1). Inspired by the full form of Eqs. (10.73) and (10.74), we propose an additional field, a **magnetic gravitational vector field b**, such that

$$\nabla \cdot \mathbf{g} = -4\pi G\,\rho_{\mathrm{m}}, \quad \nabla \times \mathbf{g} = -\frac{1}{c}\frac{\partial \mathbf{b}}{\partial t},$$

$$\nabla \cdot \mathbf{b} = 0, \qquad \nabla \times \mathbf{b} = -\frac{4\pi G}{c}\mathbf{J}_{\mathrm{m}} + \frac{1}{c}\frac{\partial \mathbf{g}}{\partial t}, \tag{10.77}$$

where ρ_{m} is the volume mass density of some source mass distribution, and $\mathbf{J}_{\mathrm{m}}$ is the mass current density. The force law on a probe of mass m and velocity $\mathbf{v}$ then turns out to be

$$\mathbf{F}_{g} = m\mathbf{g} + 4\frac{m}{c}\mathbf{v} \times \mathbf{b}, \tag{10.78}$$

which means that we now have a velocity-dependent gravitational force arising from the motion of mass! This expression has the same form as the Lorentz force on an electric charge, except for the substitution of mass for charge *and the factor of 4 in the "magnetic" term*. Notice that this "magnetic" term is also always multiplied by a factor of v/c, so it is very small at nonrelativistic speeds. This modification of Newtonian gravity is indeed correct to linear order in v/c. It can be derived from general relativity, even though the derivation is a subtle one with regards to truncating the corrections at the v/c linear level. This modified force law can be used, for example, to understand a gravitational gyroscope precession effect which has been verified experimentally.

From our experience with electromagnetism, we know how to add to the Lagrangian of a probe so that we reproduce this extended gravitational force law. Namely, we write

$$L \rightarrow L + m \int dr^\mu a^\nu \eta_{\mu\nu}, \tag{10.79}$$

where a^ν is a new gravitational four-vector potential field defined as

$$a^\mu = (a_0, \mathbf{a}) \tag{10.80}$$

with

$$\mathbf{g} = -\nabla a_0 - \frac{1}{c}\frac{\partial \mathbf{a}}{\partial t}, \quad \mathbf{b} = \nabla \times \mathbf{a} \tag{10.81}$$

as expected.

Example 10.8

Black Strings

Picture an infinite straight string of constant linear mass density λ_0 moving along its length at speed V, as shown in Figure 10.6. That is, a mass current $\lambda_0 V$ – i.e., mass per unit time – flows along the extent of the wire. It may represent a model for a so-called "black string" – a theorized configuration of mass which is infinite, straight, and with constant mass density, but no tension. We want to find the force on a spaceship of mass M and velocity $\mathbf{v}$ that has ventured nearby. Given the cylindrical symmetry of the mass configuration, to find $\mathbf{g}$ we can integrate the first of Eqs. (10.77) over the volume of a cylinder centered on the string of (say) radius ρ and height h:

$$\int \nabla \cdot \mathbf{g} \, d\text{Vol} = -4\pi G \int \rho_m \, dV \Rightarrow \oint \mathbf{g} \cdot d\mathbf{A} = -4\pi G \lambda_0 h, \tag{10.82}$$

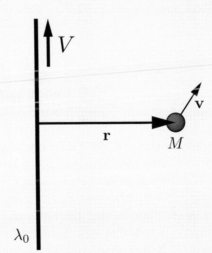

Fig. 10.6 An infinite linear mass distribution moves upward with speed V while a probe of mass M ventures nearby.

where we have used Gauss's theorem on the left-hand side of the equation, and the resulting surface integral is over the lateral surface area of the cylinder. This gives

$$2\pi \rho h g = -4\pi G \lambda_0 h \Rightarrow \mathbf{g} = -2G \frac{\lambda_0}{\rho} \hat{\rho}. \tag{10.83}$$

To find the magnetic gravitational vector field $\mathbf{b}$, we integrate the last of Eqs. (10.77) over a disk centered on the string of radius ρ through which the mass current protrudes:

$$\int (\nabla \times \mathbf{b}) \cdot d\mathbf{A} = -\frac{4\pi G}{c} \int \mathbf{J}_m \cdot d\mathbf{A} \Rightarrow \oint \mathbf{b} \cdot d\mathbf{l} = -\frac{4\pi G}{c} \lambda_0 V, \tag{10.84}$$

where we employed Stokes's theorem on the left-hand side, and the resulting line integral is over a circle around the black string. The right-hand side is simply the mass current, mass per unit time, passing through the disk. Then

$$2\pi \rho b = -\frac{4\pi G}{c} \lambda_0 V \Rightarrow \mathbf{b} = -\frac{2G}{c} \frac{\lambda_0 V}{\rho} \hat{\varphi}. \tag{10.85}$$

We can now compute the force on the probe using Eq. (10.78). We find

$$\mathbf{F}_g = m\mathbf{g} + \frac{4m}{c} \mathbf{v} \times \mathbf{b} = -2G \frac{m\lambda_0}{\rho} \hat{\rho} - \frac{8G m \lambda_0 V}{c^2} \frac{}{\rho} \mathbf{v} \times \hat{\varphi}. \tag{10.86}$$

The surprising result is that the second term, arising from the magnetic gravity effects, can generate a *repulsive* force, since we can flip the sign of V at will by changing the direction of mass flow in the black string. However, this is a small effect compared to the more familiar attractive first term. The reason for this is that the ratio of the second to the first term scales as Vv/c^2: we need to move the string matter and the probe very fast to generate a repulsive magnetic gravitational force large enough to compete with the omnipresent attractive effect of the first term. Indeed, as speeds approach c, the electromagnetic gravity framework we developed in these last two examples breaks down and the full theory of general relativity is needed to reach correct conclusions. ∎

10.6 Gauge Symmetry

The central symmetry of Maxwell's equations, gauge symmetry, is one of the most profound principles underlying the laws of Nature. Indeed, positing the symmetry as a starting point is enough information to *derive* Maxwell's equations and the Lorentz force law. While this beautiful derivation is simple and elegant, it requires either quantum mechanics or classical field theory to appreciate. However, it is worthwhile noting that such gauge symmetry principles in fact underlly all known forces of Nature. As a brief exposition of the workings of this symmetry principle in electromagnetism, we demonstrate here the process of "gauge fixing" – a technique that can be very handy in practical problem solving in the classical dynamics of probe charges.

The gauge symmetry of electromagnetism was already described in Chapter 8:

$$\phi \to \phi - \frac{1}{c}\frac{\partial f}{\partial t} \quad \text{and} \quad \mathbf{A} \to \mathbf{A} + \nabla f, \tag{10.87}$$

where ϕ and $\mathbf{A}$ are the scalar and vector potentials and $f(t, \mathbf{r})$ is any scalar function of time and space. This gauge transformation is a *symmetry* because it has no effect on the electric and magnetic fields

$$\mathbf{E} = -\nabla\phi - \frac{1}{c}\frac{\partial \mathbf{A}}{\partial t}, \qquad \mathbf{B} = \nabla \times \mathbf{A}, \tag{10.88}$$

and so it also has no effect on any observed physics.

For given electric and magnetic fields, the transformation provides us with the freedom to choose a variety of scalar and vector potentials. Given this freedom, it is customary to *fix the gauge* so as to make the manipulation of the potentials more convenient in a particular situation. For example, we may choose the so-called **static gauge**

$$\phi = 0 \quad \textit{static gauge}. \tag{10.89}$$

We can see that this is always possible: imagine you begin with some ϕ and $\mathbf{A}$ such that $\phi \neq 0$. Then apply a gauge transformation of ϕ and $\mathbf{A}$, which we know does not change the electromagnetic fields and the associated physics, such that

$$\phi' = \phi - \frac{1}{c}\frac{\partial f}{\partial t} = 0. \tag{10.90}$$

That is, we must find a function f such that this equation is satisfied. For any ϕ, this equation indeed has a solution f. This is a rather strange **gauge choice** since it sets the electric scalar potential to zero. But this is entirely legal. Note that it does *not* imply that the electric field is zero, since we still have

$$\mathbf{E} = -\frac{1}{c}\frac{\partial \mathbf{A}}{\partial t} \tag{10.91}$$

in this gauge choice – but the scalar potential term in the Lagrangian (8.43) is absent.

Another interesting aspect of gauge fixing is that, typically, the process does not necessarily fix *all* of the gauge freedom. In the case of the static gauge, we can still apply a gauge transformation f_0 such that

$$\phi' = 0 \to \phi'' = 0 = \phi' - \frac{1}{c}\frac{\partial f_0}{\partial t} = -\frac{1}{c}\frac{\partial f_0}{\partial t} \tag{10.92}$$

without changing the gauge condition that the electric potential is zero. We see from this expression that this is possible if

$$\frac{\partial f_0}{\partial t} = 0, \tag{10.93}$$

that is, if the gauge transformation function f_0 is time independent. Therefore some of the original freedom of the gauge symmetry remains even after gauge fixing. This is known as **residual gauge freedom**, for obvious reasons.

Another very common gauge choice is the **Coulomb gauge**

$$\nabla \cdot \mathbf{A} = 0, \tag{10.94}$$

and there is also a Lorentz-invariant version known as the **Lorentz gauge**

$$\frac{\partial A_\mu}{\partial x_\nu} \eta_{\mu\nu} = 0. \tag{10.95}$$

In this manner, depending on the details of the problem at hand, one can make the gauge choice for which the Lagrangian of a probe particle becomes most easily analyzed.

Example 10.9

Maxwell's Equations Using the Lorentz Gauge

Maxwell's equations are given at the beginning of Section 10.5 in terms of the electric and magnetic fields $\mathbf{E}$ and $\mathbf{B}$. They may be rewritten in terms of the scalar and vector potentials, using $\mathbf{E} = -\nabla\phi - \frac{1}{c}\partial\mathbf{A}/\partial t$ and $\mathbf{B} = \nabla \times \mathbf{A}$. The result is that two of the four Maxwell equations are satisfied automatically, and the other two are

$$-\nabla^2\phi - \frac{1}{c}\frac{\partial}{\partial t}\nabla \cdot \mathbf{A} = 4\pi\rho,$$

$$-\nabla^2\mathbf{A} - \frac{1}{c^2}\frac{\partial^2\mathbf{A}}{\partial t^2} = -\frac{4\pi}{c}\mathbf{J} + \nabla\left(\nabla \cdot \mathbf{A} + \frac{1}{c}\frac{\partial\phi}{\partial t}\right). \tag{10.96}$$

Each of the two equations involves both ϕ and $\mathbf{A}$, which makes them tricky to solve. But we can make a gauge choice to simplify the equations. Notice that if we use the Lorentz gauge condition

$$\frac{\partial\phi}{\partial ct} + \nabla \cdot \mathbf{A} = 0, \tag{10.97}$$

we can eliminate $\mathbf{A}$ in favor of ϕ in the first equation and ϕ in favor of $\mathbf{A}$ in the second equation, resulting in

$$\nabla^2\phi - \frac{1}{c^2}\frac{\partial^2\phi}{\partial t^2} = -4\pi\rho,$$

$$\nabla^2\mathbf{A} - \frac{1}{c^2}\frac{\partial^2\mathbf{A}}{\partial t^2} = -\frac{4\pi}{c}\mathbf{J}. \tag{10.98}$$

The two equations have now been decoupled, with one equation in ϕ and the other in $\mathbf{A}$. We recognize each as a wave equation, with sources consisting of the charge density ρ for the ϕ equation and the current density $\mathbf{J}$ for the $\mathbf{A}$ equation.

Even more elegantly, using the four-vector potential $A = (\phi, \mathbf{A})$ and defining a current density four-vector $J = (\rho c, \mathbf{J})$, the equations can be combined into a single wave equation (with four components)

$$\nabla^2 A = -(4\pi/c)J. \tag{10.99}$$

The use of the Lorentz gauge has allowed us to write the four Maxwell equations, in the fields **E** and **B** and sources ρ and **J**, as a single second-order differential equation in the four-vector potential with the current density four-vector as source.   ∎

10.7 Stochastic Forces

Consider a particle of dust floating in the air, tracing an irregular trajectory as it is bumped around by the air molecules around it. Or consider a bacterium swimming against the random forces of water molecules around it; or a minuscule spring with randomly fluctuating spring constant describing the forces between two nanoparticles. Such random forces abound in physics and can significantly affect the dynamics of tiny particles. Fundamentally, these are not new force laws: they are indeed mostly electromagnetic in origin. However, they are characterized by a level of randomness that requires us to treat them differently.

Randomness, or the lack of determinism in physics, arises from two possible sources:

- Quantum mechanics portrays a probabilistic picture of the world. Every state of a system is allowed, and all that Nature keeps track of is the likelihood of one state or another. Hence, in quantum mechanics, degrees of freedom fluctuate and we talk about the average values of measurements. Such fluctuations are called **quantum fluctuations**. They are important to the dynamics whenever the characteristic scale of the action – that is, the characteristic energy scale times the characteristic timescale – is of order $\hbar$.

- In the context of deterministic classical physics or probabilistic quantum mechanics, a high-level complexity in the dynamics of a system can arise from the involvement of a very large number of particles that are interacting through complex nonlinear force laws. The 10^{25} molecules of air may each individually be well described by classical deterministic trajectories, yet such a description is in practice impossible and in essence undesirable. The complex interactions between the molecules results in *ergodic dynamics*: an evolution that explores the various possible configurations of the whole system in a pseudorandom (often chaotic) pattern. The setup is then best quantified by average values of observables and statistical fluctuations. Such fluctuations then originate from complexity – whether within classical or quantum mechanics. Their size is determined by a macroscopic parameter known as *temperature*: the higher the temperature of the complex system, the larger the statistical fluctuations. Fluctuations of this type are referred to as **thermal fluctuations**. If a probe particle is tracked as it interacts with such an ergodic system – *i.e.*, a dust particle in air – the probe's dynamics is significantly affected by the background complexity if the energy of the probe is of order $k_B T$, where k_B is called "Boltzmann's constant" and T is the temperature of the background.

Whether quantum or thermal in origin, the effect of fluctuations on the *classical* dynamics of a probe or particle can be described in the same language. Let us focus on the example of a dust particle floating in air – subject to thermal fluctuations. The central idea is that there are two timescales in the problem. The background system – in this case the air molecules – forms a large **thermal reservoir** whose dynamics is not affected much by the dust particle; its dynamics is associated with a very short timescale compared to the timescale of evolution of the dust particle. In contrast, the latter can be treated classically, but is then subject to randomly and quickly fluctuating forces from the background molecules. We are interested in describing the trajectory of the dust particle in such a setting.

Let us say we have a dust particle of mass m – moving in one dimension for simplicity – evolving according to the equation of motion

$$m\ddot{x} = m\dot{v} = -\alpha v + \sigma f(t), \tag{10.100}$$

where αv is an effective frictional force arising from the particle's interaction with the background fluid molecules and $f(t)$ is a random force of the same origin changing in time very quickly compared to the evolution of the x coordinate. This problem is known as the **Ornstein–Uhlenbeck process**; a simpler form of it without the αv frictional term is the celebrated **Brownian motion** problem of Einstein. In fact, through a more fundamental statistic treatment, one can *derive* the αv term from the effect of the background fluctuations. Looking at this equation, we are immediately led to a couple of mathematical puzzles. First, if $f(t)$ is randomly fluctuating, we need to clarify what exactly we mean by randomness. Second, we expect $x(t)$ and $v(t)$ to have zigzagging profiles, as shown in Figure 10.7, which would make them very different from the nice differentiable smooth functions we are used to in the differential equations normally encountered

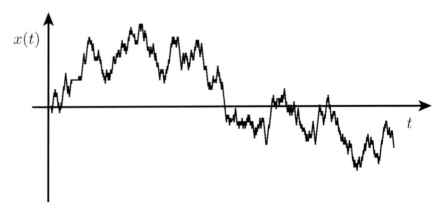

Fig. 10.7   A stochastic evolution of the position of a particle subject to random forces.

in mechanics. In fact, we may suspect that $x(t)$ and $v(t)$ are not strictly speaking *functions* in the usual sense. Let us tackle each of these issues separately.

If $f(t)$ is a randomly fluctuating variable, we can talk about its statistical moments to describe it. For example, the average $\overline{f(t)}$ would be a well-defined quantity. We may write for example

$$\overline{f(t)} = 0 \tag{10.101}$$

for a force fluctuating about a zero value. Here, by an average we mean a *time average* over timescales much longer than the short timescale associated with the fluctuations. Another quantity that quantifies the randomness of this force is

$$\overline{f(t_0)(f(t_0 + t))} = C(t). \tag{10.102}$$

This quantity measures the correlation of fluctuations over time. Expecting that the system is time translationally invariant on timescales of interest, we have written the right-hand side as a function of t only, and not t_0. Typically, $C(t)$ is an exponentially decaying function, implying that as we look at fluctuations at larger separations in time t, they quickly appear more and more uncorrelated. The Fourier transform of $C(t)$ is known as the **spectral density** of the force

$$S(\omega) = \frac{1}{2\pi} \int_{-\infty}^{\infty} e^{-i\omega t} C(t) dt. \tag{10.103}$$

A particularly interesting random force law arises when $S(\omega)$ is a constant independent of ω. We then say that $f(t)$ describes **white noise**: its spectral density is frequency independent. From Eq. (10.103), this implies we must have

$$C(t) = \delta(t). \tag{10.104}$$

That is, the force fluctuations are totally uncorrelated over time, or the timescale of fluctuations is essentially zero – much smaller than any other timescale in the problem. This is an idealization that is strictly speaking unrealistic, but it turns out to be a good approximation in many situations.

Now that we have a rigorous definition of the randomness of $f(t)$, we need to tackle the issue that $x(t)$ and $v(t)$ are not well-behaved "functions." We have the equation

$$\dot{v} = -\alpha v + \sigma f(t), \tag{10.105}$$

where $f(t)$ is meant to be a random white noise force – which is then expected to lead to a jagged trajectory for the probe. Instead of this equation, we propose to write

$$\frac{dV}{dt} = -\alpha V + \sigma f(t) \tag{10.106}$$

with the following caveat: the capitalized variable $V(t)$ represents many velocities of the probe *in an ensemble of many realizations of the evolution*. $V(t)$ is *not* a function, but a placeholder for many possible values of $V(t)$ realized when the experiment is run many times – all measured at time t. Such an object is called a **stochastic process**, and Eq. (10.106) is known as a **stochastic differential equation**. It describes the evolution of an ensemble of systems under the influence of a random force law.

Much like regular differential equations, one can develop mathematical machinery to manipulate and solve such stochastic equations. All such techniques can be derived by going back to the premise that a capitalized variable represents a measurement in an ensemble with a given probability distribution inherited from the random force $f(t)$. Then basic statistical methods can be used to compute statistical moments of (for example) $V(t)$.

An important theorem of stochastic differential equations can help us solve Eq. (10.106).

Theorem: *The stochastic differential equation*

$$\frac{dZ(t)}{dt} = a(t)Z(t) + b(t)f(t) + c(t), \tag{10.107}$$

where $f(t)$ is a random noise, and $Z(t)$ is a stochastic process with initial condition

$$Z(0) = C, \tag{10.108}$$

where C is a process with a given probability distribution, is solved by

$$Z(t) = \varphi(t)\left(C + \int_0^t \frac{c(s)}{\varphi(s)}ds + \int_0^t \frac{b(s)}{\varphi(s)}f(s)ds\right) \tag{10.109}$$

with

$$\varphi(t) = e^{\int_0^t a(s)ds}. \tag{10.110}$$

This is a powerful theorem that can be used to solve many stochastic differential equations. The proof of the theorem is beyond the scope of this book. For now, we want to use it to solve our physics problem described by Eq. (10.106).

Mapping Eq. (10.106) onto Eq. (10.107), we identify

$$Z(t) \to V(t), \quad a(t) \to -\alpha, \quad b(t) \to \sigma, \quad c(t) \to 0. \tag{10.111}$$

We then have

$$\varphi(t) = e^{-\alpha t}. \tag{10.112}$$

And we get the full solution

$$V(t) = e^{-\alpha t}C + \sigma \int_0^t e^{-\alpha(t-s)}f(s)ds. \tag{10.113}$$

We can now use the statistical properties of the white-noise force given by Eqs. (10.101), (10.102), and (10.104) to compute the statistical properties of the probe's velocity process $V(t)$. For example, we immediately get

$$\overline{V(t)} = e^{-\alpha t}\overline{C} \tag{10.114}$$

and

$$\overline{V(s)V(t)} = e^{-2\alpha t}\overline{C^2} + \sigma^2 \int_0^t \int_0^s e^{-\alpha(s-s')}e^{-\alpha(t-s'')}\overline{f(s')f(s'')}ds'ds''$$

$$= e^{-2\alpha t}\overline{C^2} + \frac{\sigma^2}{2\alpha}e^{-\alpha(t+s)}\left(e^{2\alpha\,\min(t,s)} - 1\right). \tag{10.115}$$

For concreteness, let us specify a specific boundary condition: let the initial velocity C represent a process with probability distribution

$$\mathrm{Prob}(C) = \delta(c), \tag{10.116}$$

implying we start with zero initial velocity with no statistical spread at all. We then have

$$\overline{C} = 0, \quad \overline{C^2} = 0. \tag{10.117}$$

This then implies that

$$\overline{V(t)} = 0, \quad \overline{V(t)^2} = \frac{\sigma^2}{2\alpha}\left(1 - e^{-2\alpha t}\right). \tag{10.118}$$

We can compute higher moments of $V(t)$ to check that $V(t)$ has a Gaussian probability profile, and hence is entirely quantified by these first two statistical moments (see corresponding problem in the Problems section at the end of this chapter). From this result, we see that as $t \to \infty$, we have

$$\overline{V(t)^2} \to \frac{\sigma^2}{2\alpha} \tag{10.119}$$

exponentially in time. Hence, the initial zero velocity of the probe gets "fuzzed out" along a Gaussian distribution which, at large times, has a standard deviation of $\sigma/\sqrt{2\alpha}$. Note that σ tunes the strength of the random forces, while α tunes the strength of the frictional forces. The effect of the random forces is then to spread the velocity of the probe from zero to a fixed range; and this range is larger for smaller frictional forces!

From these results, we can also determine the statistical properties of the position of the probe. We have

$$X(t) = X(0) + \int_0^t V(s)ds. \tag{10.120}$$

Using Eqs. (10.115) and (10.120), we can easily find $\overline{X(t)}$ and $\overline{X(s)X(t)}$. We leave this problem as an exercise for the reader and quote the result: for $t \to \infty$, one finds that the position of the probe is given by

$$\overline{X(t)^2} \to \frac{\sigma^2}{\alpha^2}t, \tag{10.121}$$

with $\overline{X(t)} = 0$. The probability distribution for $X(t)$ is Gaussian with variance $\overline{X^2} - \overline{X}^2$ given by (10.121). This is because it is a sum of stochastic variables that are Gaussian, as implied by Eq. (10.120) (see Problem 10.21). This is the celebrated **random walk** result of statistical physics. The probe evolves in time like a random uncorrelated sequence of steps: if you divide t into N snapshots (say each of fixed small duration $\delta t = t/N$), at each snapshot the probe goes one step left or one step right with equal probability – resulting in a spread in position scaling as $\sqrt{N} \propto \sqrt{t}$. This conclusion can easily be generalized to a dust particle evolving in three dimensions.

Example 10.10

Entropic Force

As we just argued, the evolution of a dust particle floating around in air proceeds as a random walk, starting at say $x = 0$ and, as shown by (10.121), expanding its average distance from the origin as

$$\Delta x(t) = \frac{\sigma}{\alpha}\sqrt{t}. \tag{10.122}$$

This process is known as **diffusion** – with the diffusion constant D defined as

$$2D \equiv \frac{\sigma^2}{\alpha^2}. \tag{10.123}$$

It is as if there is a force pulling on the dust particle away from the origin. This "force" is unlike any other we have seen so far: it has a statistical origin, and we call it an **entropic force**. The dust particle is more likely to be found further away from the starting point as time progresses because it has access to more possible configurations at larger distances from the origin than at smaller ones. To see this, note that the particle's position in the x direction has a Gaussian probability profile as alluded to earlier:[a]

$$P(x)dx = \frac{1}{\Delta x(t)\sqrt{2\pi}}e^{-x^2/2\Delta x(t)^2}dx, \tag{10.124}$$

which is the probability of finding the dust particle between x and $x + dx$. As time passes, $\Delta x(t)$ becomes larger and hence the probability profile spreads. If we were to think of a collection of dust particles evolving in this way, the number of dust particles per unit length would be proportional to $P(x)$. With time, the dust particles are spreading out towards a uniform equilibrium configuration. This means the average spreading space available to each dust particle scales as $1/P(x)$ (i.e., the inverse of the density). The number of configurations available to each dust particle then increases as $1/P(x)$, as a function of x. We define the **entropy** of any statistical system as

$$S \equiv k_B \ln[\text{number of available statistical configurations}], \tag{10.125}$$

where k_B is Boltzmann's constant. Hence, for the problem at hand, we can say that the entropy of a single dust particle is given by

$$S(x) = -k_B \ln P(x) + \text{constants} \simeq k_B \frac{x^2}{2\Delta x^2} + S_0, \tag{10.126}$$

where S_0 is an x-independent constant that will be irrelevant for our upcoming discussion.

As a general statement of statistical physics, an isolated system of fixed energy is most likely to be found in a configuration that *maximizes* its entropy. An entropic force on a particle arises from this principle and can be defined as

$$\mathbf{F} = T\nabla S, \tag{10.127}$$

where T is the temperature of the whole system, and the entropy S of the particle depends on its position. For example, equilibrium is reached when S becomes uniform (independent of the position of the particle) and thus entropic forces that drive the system toward statistically more likely configurations vanish. In our example, we have an entropic force acting on the dust particle given by

$$F = T\frac{dS}{dx} = \frac{k_B T}{\Delta x^2}x. \tag{10.128}$$

This is a force that points away from the origin and results in the spreading or diffusion of dust particles into the surrounding space. While the entropic force is not a force in the traditional sense of the term, its effect on statistical dynamics is nevertheless very much force-like. ∎

[a]We focus on the one-dimensional problem. In three dimensions, we have three copies of this probability distribution and the argument is replicated three times.

10.8 Summary

In this capstone chapter we have sampled a selection of advanced topics that put in perspective our discussions in the previous four chapters. We learned about the corrections to Newtonian gravity as they arise in the context of a new revolutionary formulation of gravitation – one that involves the warping of space and time by mass and energy. We also elaborated on a fundamental symmetry of electromagnetism, the so-called gauge symmetry. Indeed, gauge symmetries underlie all known forces of Nature, including the force of gravity. Finally, we also touched upon another class of forces of a very different nature, entropic forces that drive complex systems toward statistically more likely configurations. There are two other fundamental forces that we have not explored in this capstone chapter: the weak and strong forces. Both of these forces become strong and relevant only when quantum mechanics is also necessary to capture the correct physics. Hence, neither of these cases can be usefully explored within the realm of classical mechanics.

Problems

★★ **Problem 10.1** Verify the particular solutions given of the inhomogeneous first-order equations for the perihelion precession, as given in Eq. (10.34).

★★ **Problem 10.2** The metric of flat, Minkowski spacetime in Cartesian coordinates is $ds^2 = -c^2 dt^2 + dx^2 + dy^2 + dz^2$. Show that the geodesics of particles in this spacetime correspond to motion in straight lines at constant speed.

★★   **Problem 10.3** The geodesic problem in the Schwarzschild geometry is to make stationary the functional $S = \int I \, d\tau$, where

$$I = \sqrt{(1 - 2\mathcal{M}/r)c^2\dot{t}^2 - (1 - 2\mathcal{M}/r)^{-1}\dot{r}^2 - r^2\dot{\varphi}^2},$$

with $\dot{t} = dt/d\tau$, etc. Use this integrand in the Euler–Lagrange equations to show that one obtains exactly the same differential equations of motion in the end if the square root is removed; *i.e.*, if we instead make stationary the functional $S' = \int I^2 \, d\tau$. You may use the fact that $(1 - 2\mathcal{M}/r)c^2\dot{t}^2 - (1 - 2\mathcal{M}/r)^{-1}\dot{r}^2 - r^2 d\dot{\varphi}^2 = c^2$, a constant along the particle path. The result is important because it is often much easier to use I^2 in the integrand rather than I.

★★   **Problem 10.4** Show that there are no stable circular orbits of a particle in the Schwarzschild geometry with radius less than $6GM/c^2$.

★★   **Problem 10.5** Show from the effective potential corresponding to the Schwarzschild metric that if U_{eff} can be used for arbitrarily small radii, there are actually *two* radii at which a particle can be in a circular orbit. The outer radius corresponds to the usual stable, circular orbit such as a planet would have around the sun. Find the radius of the inner circular orbit, and show that it is unstable, so that if the orbiting particle deviates slightly outward from this radius it will keep moving outward, and if it deviates slightly inward it will keep moving inward.

★★   **Problem 10.6** Kepler's second law for classical orbits states that planets sweep out equal areas in equal times. Is that still true in Schwarzschild spacetime, assuming orbital radii $r > 2GM/c^2$? (a) First suppose that "time" here means the coordinate time t in Schwarzschild coordinates. (b) Then suppose instead that "time" means the proper time τ of the planets themselves.

★   **Problem 10.7** Earth's orbit has a semi-major axis $a = 1.496 \times 10^8$ km and eccentricity $\epsilon = 0.017$. Find the general relativistic precession of the earth's perihelion in seconds of arc per century.

★★   **Problem 10.8** Sometimes more than one coordinate system can usefully describe the same spacetime geometry. This is true in particular for the Schwarzschild geometry surrounding a spherically symmetric mass M. The usual Schwarzschild metric is

$$ds^2 = -(1 - 2\mathcal{M}/r)c^2 dt^2 + (1 - 2\mathcal{M}/r)^{-1} dr^2 + r^2 d\Omega^2, \qquad (10.129)$$

where $\mathcal{M} \equiv GM/c^2$ and $d\Omega^2 \equiv (d\theta^2 + \sin^2\theta d\varphi^2)$. The so-called "isotropic" metric, describing exactly the same spacetime, has the form

$$ds^2 = -(1 - 2\mathcal{M}/r)c^2 dt^2 + e^{2u}(d\bar{r}^2 + \bar{r}^2 d\Omega^2), \qquad (10.130)$$

with the same dt^2 term, while the other terms contain a new radial coordinate $\bar{r}$ instead of r, and where $u = u(\bar{r})$. (a) Find $\bar{r}$ in terms of r and $\mathcal{M}$, choosing a constant of integration so that $\bar{r} \to r$ as $r \to \infty$. (b) What is an advantage of using the isotropic metric?

★ **Problem 10.9** The geometry on the surface of a sphere is non-Euclidean, so the circumference C and radius R of a circle drawn on the sphere do not obey $C = 2\pi R$, where for example the circumference is a constant-latitude path and the radius is drawn on the sphere down from the north pole along a constant-longitude path. Suppose we measure latitude by the angle, measured from the center of the sphere, between the north pole and the constant-latitude path. (a) If the angle is $90°$, what is the coefficient α in $C = \alpha R$? (b) If in effect *pi* were 3.00000 instead of 3.14156..., what would be the angle in that case? (c) The feature that $C < 2\pi R$ is a property of a positively curved surface. In Euclidean geometry, given a line and a point exterior to the line, there is one and only one line through the given point that is parallel to the given line, parallel meaning that the two lines never meet. What is the analogue statement for a positively curved surface?

★★ **Problem 10.10** Before the age of relativity, some people calculated that light would be deflected by the sun in a classical model in which light consists of particles of tiny mass m moving at speed c, pulled by the sun's Newtonian gravity. Find in that case the approximate deflection of a light beam in terms of any or all of m, c, G, M, the sun's mass, and R, the distance of closest approach of the light beam from the sun's center. Compare your result with the actual deflection of light in the Schwarzschild spacetime as derived in the chapter.

★ **Problem 10.11** (a) Find the escape velocity $dr/d\tau$ of a particle of mass m starting from rest at radius $r_0 = 4GM/c^2$ in a Schwarzschild spacetime of mass M, where τ is read on the particle's own clock. (b) Then find the escape velocity $dr/d\tau$ of the particle, starting at rest from the same point, where now τ is read on a clock that remains at rest at r_0.

★★ **Problem 10.12** Tachyons are hypothetical particles (never observed, at least so far) that always travel faster than light. Therefore in general relativistic spacetimes they would follow space-like (rather than time-like or null) geodesics. Prove that the deflection of such a particle in passing by the sun would be less than that for light.

★ **Problem 10.13** Consider two concentric coplanar circles in the Schwarzschild metric surrounding the sun, with measured circumferences C_1 and C_2. In terms of C_1 and C_2, find an expression for (a) the radial coordinate distance Δr between them, (b) the radial measured distance between them.

★★★ **Problem 10.14** The Robertson–Walker metrics

$$ds^2 = -c^2 dt^2 + a(t)^2 \left[\frac{dr^2}{1 - k(r/R)^2} + r^2 d\theta^2 + r^2 \sin^2\theta d\varphi^2 \right] \qquad (10.131)$$

are applicable to universes that are both spatially homogenous and isotropic. That is, they have no preferred positions or directions. The spacetimes also feature a universal time t and a constant R with dimensions of length. There are three possible choices for the constant k: $k = 1, 0,$ or -1, which correspond to three-dimensional spatial geometries that have constant positive curvature ($k = +1$), constant negative curvature ($k = -1$), or are flat ($k = 0$). Here $a(t)$ is called the "scale factor" of the universe; if $a(t)$ grows with time distant galaxies become farther apart, or if $a(t)$ shrinks distant galaxies come closer together. The function $a(t)$ can be found using Einstein's field equations of general relativity, given the kind of matter, radiation, or other quantities that live in the universe. The result is the "Friedmann" equations, in which the universe is filled with a material having uniform mass density ρ and uniform pressure p; we can also add "dark energy" as represented by the constant Λ. There are then two independent equations for $a(t)$:

$$(1) \quad \frac{\dot{a}^2}{a^2} + \frac{kc^2}{a^2} = \frac{8}{3}\pi G\rho + \frac{\Lambda}{3}c^2 \tag{10.132}$$

and

$$(2) \quad \frac{2\ddot{a}}{a} + \frac{\dot{a}^2}{a^2} + \frac{kc^2}{a^2} = \frac{8\pi Gp}{c^2} + \Lambda c^2, \tag{10.133}$$

where overdots represent time derivatives and G is Newton's gravitational constant. (a) First suppose the universe is spatially flat, with $k = 0$, and that the cosmological constant Λ is also zero. Also suppose the energy density consists entirely of mass density ρ, which decreases as the universe expands so that $\rho a(t)^3 = \rho_0 a_0^3$, where ρ_0 and a_0 are the current mass density and scale factor. In that case solve the Friedmann equations to find $a(t)$ in terms of t, G, ρ_0, and a_0. It is thought that this is a good approximation to the situation for our universe in most of its history so far. It is called the "matter-dominated" period. (b) Repeat part (a) except suppose the energy density consists entirely of photons in thermal equilibrium, in which case the energy density obeys $\rho a(t)^4 = \rho_0 a_0^4$. This situation is thought to be a good approximation for our universe for a hundred thousand years or so early on, and is called the "radiation-dominated" period. (c) Finally, repeat part (a) for the case $k = 0, \rho = 0, p = 0$, but $\Lambda = $ constant > 0. This may be a good approximation to our universe for a brief time after the big bang began; it is called the "inflationary" period for reasons that will be apparent from the solution. It may also be a good approximation for our universe in the distant future. (d) At the current time the universe seems to be behaving as though it were driven by both dust-like matter and the cosmological constant Λ. Sketch a graph of $a(t)$ vs. t extending from times long ago to times in the distant future, showing what happens as the universe gradually transitions from one form of dominance to the other. Figure 10.8 shows a picture of a narrow patch of the sky taken by the Hubble telescope: we see thousands of galaxies moving away from us due to the expansion of the universe.

★★   **Problem 10.15** Inspired by Eqs. (10.77), write gravitational field vectors describing a gravitational wave of angular frequency ω propagating in vacuum in the positive z direction, specifying both the "electric" and "magnetic" field vectors. Assume the "electric" gravitational field amplitude is given by g_0.

Fig. 10.8 This picture was taken by the Hubble telescope over the course of 10 consecutive days. The image contains about 10,000 galaxies located at a distance of about 13 billion light years. All the galaxies in the picture are moving away from us because of the expansion of the universe. Credits: R. Williams (STScI), the Hubble Deep Field Team and NASA Horizon Telescope Collaboration.

★★ **Problem 10.16** Show that there is exactly one radius at which a light beam can move in a circular orbit around a spherical black hole, and find this radius. Then show that the orbit is unstable, by showing that a tangential beam beginning at a slightly larger radius will spiral outward and never return, and a tangential beam beginning at a slightly smaller radius will spiral into the black hole. Figure 10.9 shows the first picture ever taken of a black hole, that of the supermassive black hole at the center of the M87 galaxy.

★★ **Problem 10.17** Write the Lagrangian of a charged particle in terms of potentials in the case where we use the static gauge condition, and show that it appears to be different from the Lagrangian in the absence of any gauge fixing. Then show that even though the Lagrangian is different, the Lagrange equations of motion are the same.

★★ **Problem 10.18** Show that for a Gaussian probability distribution

$$p(x) = \frac{e^{-\frac{(x-x_0)^2}{2\,a^2}}}{\sqrt{2\,\pi\,a^2}},$$

all the moments are given by

$$\langle (x-x_0)^n \rangle = 1 \times 3 \times 5 \times (n-1) \times a^n \tag{10.134}$$

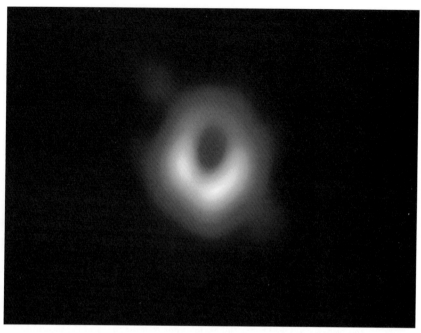

Fig. 10.9 The supermassive black hole at the center of the M87 galaxy. This is the first image of a black hole ever captured. The black hole is 6.5 million times more massive than our sun and is 53 million light years away. Image credit: Event Horizon Telescope Collaboration.

for even n, and are zero otherwise. Hence the Gaussian distribution is entirely characterized by its mean x_0 and deviation a.

★★ **Problem 10.19** (a) Show that for any probability distribution, if we compute the generating function $Z(\beta) \equiv \langle e^{\beta X} \rangle$ for arbitrary β and X being the stochatic variable, we can compute all moments using

$$\langle X^n \rangle = \lim_{\beta \to 0} \left(\frac{d}{d\beta} \right)^n Z(\beta).$$

(b) Use the generating function to compute the moments of a Gaussian stochastic variable. (c) Use the generating function to compute the moments of a stochastic variable with uniform probability distribution: $p(x) = 1/(2a)$ for $(x_0 - a) \le x \le (x_0 + a)$, and $p(x) = 0$ otherwise.

★★★ **Problem 10.20** For the stochastic equation studied in the text, show that

$$\overline{X(t)^2} = \frac{\sigma^2}{\alpha^2} \left(t + \frac{1}{2\alpha} e^{-2\alpha t} \right) \to \frac{\sigma^2}{\alpha^2} t. \tag{10.135}$$

★★ **Problem 10.21** Using the generating function Z introduced in an earlier problem, show that: (a) If X is a stochastic variable with a Gaussian distribution with mean x_0 and variance σ^2, then $a + bX$ is a stochastic variable with a Gaussian distribution with mean $a + bx_0$ and variance $b^2\sigma^2$. (b) If X and Y are Gaussian stochastic variables with means x_0 and y_0 respectively, and variances σ_X^2 and σ_Y^2, then $X+Y$ is a Gaussian stochastic variable with mean $x_0 + y_0$ and variance $\sigma_X^2 + \sigma_Y^2$.

★★★ **Problem 10.22** Show that if the initial condition C of the linear stochastic differential equation introduced in the text has a Gaussian distribution, so does the solution of the stochastic differential equation. *Hint*: You might want to compute generating functions as in previous problems.

★★ **Problem 10.23** Show that in the case of a particle executing a random walk as described by the statistical moments of its position computed in the text, the probability function $p(x, t)$ satisfies the so-called **diffusion equation**

$$\frac{\partial p}{\partial t} = D \frac{\partial^2 p}{\partial x^2},$$

where the constant D is called the diffusion coefficient.

★ **Problem 10.24** From statistical mechanics, for each degree of freedom q of a free system in thermal equilibrium at temperature T, the corresponding thermal fluctuations of $\dot{q}$ are given by

$$\left\langle \frac{1}{2} m \dot{q}^2 \right\rangle = \frac{kT}{2}. \tag{10.136}$$

Here m is the mass associated with the kinetic energy expression written in terms of q. Mapping this setup on the random Brownian motion dynamics elaborated in the text, find a relation between T, σ, and α; that is, a relation between temperature, random force strength, and friction. This is a form of the **fluctuation–dissipation theorem**.

★ **Problem 10.25** A team of researchers has long tracked the path of a star named S2 that orbits the supermassive black hole Sagittarius A^* at the center of our Milky Way galaxy. (The orbit is one of those shown on the cover of this book.) The orbital period of S2 is 16.05 years, its semi-major axis is 970 AU, where 1 AU is the average distance of the earth from the sun, and its orbital eccentricity is 0.88. The mass of the central black hole is approximately 4 million solar masses, where 1 solar mass is 2×10^{30} kg. The researchers also found that the highly elliptical orbit of S2 precesses by approximately 12 min of arc per revolution. Is this value consistent with the predictions of general relativity for a Schwarzschild spacetime?

★★ **Problem 10.26** Suppose that the orbit of star S2, as described in the preceding problem, lies in a plane that is perpendicular to our line of sight. Then, at periastron, when S2 is a distance 120 AU from the central black hole, there will be both a transverse Doppler effect and a gravitational redshift for light from S2 to earth. (a) Find the sum of these frequency shifts. (The result has been confirmed by observations. The transverse Doppler effect is described in Problem 2.4.) (b) Calculate the Schwarzschild radius (the Event Horizon radius) of the central Sgr A^* black hole, and compare it with the periastron distance of S2. (The actual Event Horizon radius differs somewhat from the Schwarzschild radius because Sgr A^* is undoubtedly rotating.)

PART III

11

Hamiltonian Formulation

In Chapter 3 we showed how mechanics can be reformulated in the language of a variational principle and Lagrangians, powerful technologies that have helped us unravel Newtonian dynamics more transparently and efficiently. These deep dives into mathematical physics also inspired the development of quantum mechanics many years later.

The Lagrangian $L(q_i, \dot{q}_i, t)$ of a system generally depends upon a set of generalized coordinates q_i, the corresponding generalized velocities $\dot{q}_i$, and the time t. The coordinates q_i live in what is called a *configuration space*, the space described by all of the q_i. The Lagrangian obeys the Lagrange equations

$$\frac{d}{dt} \frac{\partial L}{\partial \dot{q}_i} - \frac{\partial L}{\partial q_i} = 0, \tag{11.1}$$

second-order differential equations, one for each of the coordinates. We also introduced the generalized momentum $p_i \equiv \partial L / \partial \dot{q}_i$, and noted that if the corresponding coordinate q_i is missing from the Lagrangian (*i.e.*, is "cyclic" or "ignorable") then the generalized momentum p_i is conserved, which follows directly from the corresponding Lagrange equation. The p_i then allowed us to define the Hamiltonian

$$H = \dot{q}_k p_k - L(q_i, \dot{q}_i, t), \tag{11.2}$$

where, using the summation convention, $\dot{q}_k p_k \equiv \dot{q}_1 p_1 + \dot{q}_2 p_2 + \cdots + \dot{q}_n p_n$ for a system with n generalized coordinates. Often H turns out to be the total energy E of the system, but not always. So it is natural to ask the question: "Of what use is H itself, whether it is the energy or not?"

One reason why H can be important was presented in Chapter 3, based upon the result that

$$\frac{dH}{dt} = -\frac{\partial L}{\partial t}. \tag{11.3}$$

The Hamiltonian H is obviously *conserved* if L is not an explicit function of time. This is certainly important, giving us a conservation law in some cases where we might not otherwise suspect one. *But this result only scratches the surface of the meaning and usefulness of the Hamiltonian.*

In fact, we will show in this chapter that there is a quite different framework – known as the *Hamiltonian formalism* – that describes the same fundamental physics as Newtonian mechanics or the Lagrange method. However, just as we found with the Lagrange method, the Hamiltonian description of mechanics gives us a new perspective that opens up a deeper understanding of mechanics, is sometimes advantageous in problem solving, and has also played a crucial role in the emergence of quantum mechanics. Therefore, our goal in this chapter is to develop the Hamiltonian formalism, to explore examples that elucidate the advantages and disadvantages of this new approach, and to develop the powerful related formalisms of canonical transformations, Poisson brackets, and Liouville's theorem.

As a first step along this path we take the seemingly innocuous step of rewriting the Hamiltonian *not* as a function of the q_is, $\dot{q}_i$s, and t, the generalized coordinates, velocities, and time, but as a function of the q_is, p_is, and t, the generalized coordinates, *momenta*, and time. In effect we want to eliminate the velocities $\dot{q}_i$ in favor of the momenta p_i in the Hamiltonian. This might seem to be a simple task, but how exactly do we do it?

11.1 Legendre Transformations

The way to make the transformation is to use *Legendre transforms*.[1] A Legendre transform has many uses in physics. In addition to mechanics, it plays a particularly prominent role in thermodynamics and statistical mechanics. We begin with a general statement of the problem we wish to address.

Consider a function of two independent variables $A(x, y)$ – presumably of some physical importance – whose derivative

$$z = \frac{\partial A(x, y)}{\partial y} \tag{11.4}$$

is a measurable quantity that may be more interesting than the original independent variable y itself. For example, $A(x, y)$ may be a Lagrangian $L(q, \dot{q})$ and

$$p_k = \frac{\partial L}{\partial \dot{q}_k} \tag{11.5}$$

the momentum of a particle, a quantity that is conserved under certain circumstances, and therefore may be more interesting than the generalized velocity $\dot{q}_k$ we began with. We now want to eliminate the less important independent variable y in $A(x, y)$ in favor of the more important independent variable z:

$$y \to z. \tag{11.6}$$

Note that the independent variable x is just along for the ride here: we will not mention it again until the end of this section.

---

[1] Adrien-Marie Legendre (1752–1833) was a French mathematician and physicist who made a number of important contributions to applied mathematics and mathematical physics.

We may be tempted to accomplish our goal of eliminating y as follows: start with Eq. (11.4), and invert it to get $y(x,z)$; then substitute the result into $A(x, y(x,z)) \equiv B(x,z)$, thus eliminating y and retrieving a function of x and z alone. However, this naive approach throws away some of the information within $A(x,y)$! That is, unfortunately, $B(x,z)$ does *not* contain all the information in the original $A(x,y)$. To see this, consider an explicit example. Let

$$A(x,y) = x^2 + (y-a)^2 \tag{11.7}$$

where a is some constant. Then

$$z = \frac{\partial A}{\partial y} = 2\,(y-a) \Rightarrow y(x,z) = \frac{z}{2} + a. \tag{11.8}$$

Finally, eliminating y, we find that

$$A(x, y(x,z)) = x^2 + \frac{z^2}{4} \equiv B(x,z), \tag{11.9}$$

so we have eliminated y in favor of z. However, we have lost the constant a! We would, for example, get the *same* $B(x,z)$ for two different functions $A(x,y)$ with different constants for a. Thus, the naive substitution $y \rightarrow z$ loses information present in the original function $A(x,y)$. If $A(x,y)$ were a Lagrangian, for example, we could lose part of the dynamical information if we attempted to describe things with the transformed functional. We need instead a transformation that preserves *all* the information in the original function or functional.

The reason why the naive substitution does not work is simple. Our new independent variable $z = \partial A/\partial y$ is a *slope* of $A(x,y)$. Knowing the slope of a function everywhere does *not* determine the function itself: we can still shift the function around while maintaining the same slopes, as illustrated in Figure 11.1(a). To delineate the shape of $A(x,y)$, we need *both* its slopes and relevant intercepts of the tangent lines that envelop $A(x,y)$, as depicted in Figure 11.1(b).

Let us denote the intercepts of such straight lines by $B(x,z)$, one for each slope[2] z. At every y there is a slope z, as well as a corresponding intercept $B(x,z)$. It is now easy to see that given $A(x,y)$ we can find $B(x,z)$, and vice versa: geometrically, we can see that given $A(x,y)$ we can determine the envelope of straight lines, and given the envelope of straight lines we can reconstruct the shape of $A(x,y)$.

Algebraically, we can capture these statements by writing the negative of the intercept of each straight line in Figure 11.2 as

$$B(x,z) = zy - A(x,y), \tag{11.10}$$

where $y(x,z)$ is viewed as a function of x and z by using

$$z = \frac{\partial A(x,y)}{\partial y} \tag{11.11}$$

---

[2] Note that if $A(x,y)$ is not monotonic in y, we may get a multiple-valued function in z for $B(x,z)$.

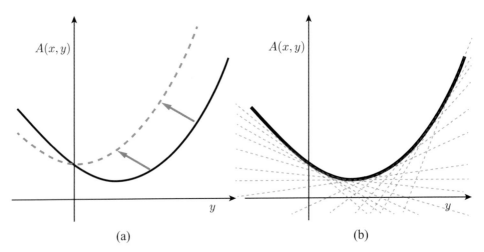

Fig. 11.1 (a) Two functions $A(x, y)$, differing by a shift, whose naive transformation through $y \rightarrow z$ led to the same transformed function $B(x, z)$. (b) The envelope of $A(x, y)$ consisting of slopes *and* intercepts completely describes the shape of $A(x, y)$.

to solve for $y(x, z)$. The vertical coordinate is $A(x, y)$, the slope is $z = \partial A / \partial y$, the horizontal coordinate is y, and the *negative* intercept is B. Therefore the equation of the straight line is $A(x, y) = z\,y + (-B(x, z))$. All the information in $A(x, y)$ can hence be found in a catalogue of the slopes and intercepts of all the straight lines tangent to the curve $A(x, y)$.

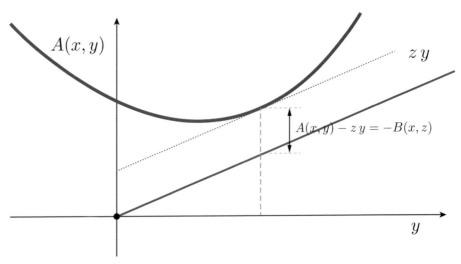

Fig. 11.2 The Legendre transformation of $A(x, y)$ as $B(x, z)$.

This is the approach of Legendre to eliminating the variable y in favor of the new variable z. As argued above, $B(x, z)$ contains *all* the original information in the function $A(x, y)$. In short, instead of the naive substitution $A(x, y(x, z)) \rightarrow B(x, z)$

we started out with, we need to consider $z\,y(x,z) - A(x,y(x,z)) \to B(x,z)$. $B(x,z)$ is then known as the **Legendre transformation** of $A(x,y)$.

Let us summarize the process. We start with a given function $A(x,y)$ and then replace y with

$$z = \frac{\partial A(x,y)}{\partial y}. \tag{11.12}$$

We invert the latter equation to get $y(x,z)$. Then we write the new Legendre transform of $A(x,y)$ as

$$B(x,z) = z\,y(x,z) - A(x,y(x,z)). \tag{11.13}$$

This is actually easy to remember and use. When making a Legendre transform from a function $A(x,y)$ to a new function $B(x,z)$, we set the new function $B(x,z)$ equal to the product of the independent variable y we intend to eliminate and the new independent variable z that replaces it, and subtract from this product the original function $A(x,y)$. Note that this procedure looks very similar to how we defined H back in Chapter 3, in that we began with $L(q_i,\dot{q}_i,t)$ and defined $H = \dot{q}_k p_k - L(q_i,\dot{q}_i,t)$. There are two differences, however. (i) In the case $L \to H$ there is an additional passive variable t. This is no problem, however, because t is just along for the ride, and is unaffected by the transformation. (ii) In the case $L \to H$ we have N products $\dot{q}_k p_k$ summed on $k = 1, 2, \ldots, N$ instead of the single product zy, where N is the number of degrees of freedom in the system. So in the $L \to H$ case we are carrying out multiple Legendre transforms, one for each product in the sum. The final result is a Hamiltonian $H(q_i,p_i,t)$, a function of the generalized coordinates, momenta, and time only, without generalized velocities $\dot{q}_i$.

Now to return to the general case. We will show how easy it is to invert a Legendre transformation, as for example converting a Hamiltonian back to a Lagrangian. We note that

$$dB = zdy + ydz - \left(\frac{\partial A}{\partial x}\right)dx - \left(\frac{\partial A}{\partial y}\right)dy. \tag{11.14}$$

Using the chain rule, we can also write

$$dB = \left(\frac{\partial B}{\partial x}\right)dx + \left(\frac{\partial B}{\partial z}\right)dz, \tag{11.15}$$

so from (1.9), and Eqs. (1.11) and (1.12), we get

$$y = \frac{\partial B}{\partial z} \tag{11.16}$$

and

$$-\frac{\partial A}{\partial x} = \frac{\partial B}{\partial x}. \tag{11.17}$$

Hence, the process of the inverse Legendre transform goes as follows. Given $B(x, z)$, the inverse Legendre transform replaces z with y by starting from (11.16)

$$y = \frac{\partial B}{\partial z},\qquad(11.18)$$

inverting it to get $z(x, y)$, and substituting in Eq. (11.13), or

$$A(x, y) = z(x, y)\, y - B(x, z(x, y))\qquad(11.19)$$

to retrieve $A(x, y)$. The variables y and z are called the **active variables**, while x is called the **passive** or **spectator** variable of the transform. All along, we also have the relation between the derivative of passive variables given by Eq. (11.17).

Example 11.1

A Simple Legendre Transform

Let us compute properly the Legendre transform of the function we already encountered:

$$A(x, y) = x^2 + (y - a)^2.\qquad(11.20)$$

Start with the derivative, which is to become our new independent variable:

$$z = \frac{\partial A(x, y)}{\partial y} = 2\,(y - a).\qquad(11.21)$$

Solving for y:

$$y(x, z) = \frac{z}{2} + a.\qquad(11.22)$$

The Legendre transform of $A(x, y)$ is then

$$B(x, z) = z\, y(x, z) - A(x, y(x, z)) = z\left(\frac{z}{2} + a\right) - \left(x^2 + \frac{z^2}{4}\right)$$

$$= \frac{1}{4}(z + 2a)^2 - x^2 - a^2,\qquad(11.23)$$

demonstrating that we have now kept track of the original a dependence in the transform $B(x, z)$. We can also verify Eq. (11.17):

$$\frac{\partial B}{\partial x} = -2x = -\frac{\partial A}{\partial x}.\qquad(11.24)$$

∎

11.2 Hamilton's Equations

We can now derive the celebrated Hamilton equations. We have already found a way to write the Hamiltonian $H = H(q_i, p_i, t)$ in terms of the generalized coordinates, momenta, and time, without generalized velocities. In Hamiltonian theory the qs and ps are called the *canonical coordinates* and *canonical momenta* of the system. For example, consider a one-dimensional simple harmonic oscillator with the Lagrangian

$$L = T - U = \frac{1}{2}m\dot{x}^2 - \frac{1}{2}kx^2. \tag{11.25}$$

The corresponding momentum is $p_x = \partial L/\partial\dot{x} = m\dot{x}$, the linear momentum of the particle. Then the Hamiltonian is

$$H = \dot{q}_k p_k - L = \dot{x}p_x - \left(\frac{1}{2}m\dot{x}^2 - \frac{1}{2}kx^2\right)$$

$$= m\dot{x}^2 - \left(\frac{1}{2}m\dot{x}^2 - \frac{1}{2}kx^2\right) = \frac{1}{2}m\dot{x}^2 + \frac{1}{2}kx^2, \tag{11.26}$$

a correct result, but not yet in canonical form. We need to eliminate $\dot{x}$, which we can do simply by using $\dot{x} = p^x/m$, giving finally

$$H = \frac{(p^x)^2}{2m} + \frac{1}{2}kx^2 \tag{11.27}$$

in terms of the canonical coordinate x and the corresponding canonical momentum p_x. *No velocities remain in H.*

Now we are prepared to derive Hamilton's equations. From our original definition $H = \dot{q}_k p_k - L(q_i, \dot{q}_i, t)$ (remember we are summing on k) we can take the differential of H to get

$$dH = \dot{q}_k dp_k + p_k d\dot{q}_k - \frac{\partial L}{\partial q_k}dq_k - \frac{\partial L}{\partial \dot{q}_k}\dot{q}_k - \frac{\partial L}{\partial t}dt. \tag{11.28}$$

The second and fourth terms on the right of this equation cancel one another because by definition $p_i = \partial L/\partial\dot{q}_i$ for each coordinate q_i. Furthermore, the Lagrange equation for q_i is

$$\frac{\partial L}{\partial q_i} = \frac{d}{dt}\frac{\partial L}{\partial \dot{q}_i} = \frac{d}{dt}p_i = \dot{p}_i, \tag{11.29}$$

so the third term on the right above can be rewritten in terms of $\dot{p}_i$, giving altogether

$$dH = \dot{q}_k dp_k - \dot{p}_k dq_k - \frac{\partial L}{\partial t}dt. \tag{11.30}$$

Now there is a quite different way to write dH, using the fact that we have sworn to express it in terms of the canonical variables alone, $H = H(q_i, p_i, t)$. Taking the differential in this case gives, from multivariable calculus:

$$dH(q_i, p_i, t) = \frac{\partial H}{\partial q_k}dq_k + \frac{\partial H}{\partial p_k}dp_k + \frac{\partial H}{\partial t}dt. \tag{11.31}$$

We now have two expressions for dH, which we can compare with one another.

Note that since the coordinates, momenta, and time are all independent variables, the coefficients of $dq_i, dp_i,$ and dt must be separately equal between the two expressions for H. Therefore

$$\frac{\partial H}{\partial q_i} = -\dot{p}_i, \qquad \frac{\partial H}{\partial p_i} = \dot{q}_i, \qquad \text{and} \qquad \frac{\partial H}{\partial t} = -\frac{\partial L}{\partial t}. \qquad (11.32)$$

The first two of these equations are **Hamilton's equations of motion**. Note that whereas Lagrange's approach provides us with N second-order differential equations, one for each degree of freedom, Hamilton's approach provides us with $2N$ *first-order* differential equations, two for each degree of freedom, along with a transformation of the passive time derivatives, $\partial H/\partial t = -\partial L/\partial t$.

Let us summarize this interesting transformation from second to first-order differential equations. Given a Lagrangian with N degrees of freedom, we transform it to a Hamiltonian

$$L(q_k, \dot{q}_k, t) \rightarrow H(q_k, p_k, t) \qquad (11.33)$$

with $2N$ independent degrees of freedom: the q_ks and the p_ks. To do this, we write

$$p_k = \frac{\partial L}{\partial \dot{q}_k}, \qquad (11.34)$$

which gives the functions $p_k(q, \dot{q}, t)$. We invert these functions to get $\dot{q}_k(q, p, t)$ and substitute in

$$H(q, p, t) = p_k \dot{q}_k - L. \qquad (11.35)$$

The dynamics is now tracked by the variables $q_k(t)$ and $p_k(t)$. This $2N$-dimensional space is called **phase space**. The time evolution is described in phase space through $2N$ *first*-order differential equations

$$\dot{q}_k = \frac{\partial H}{\partial p_k}, \quad \dot{p}_k = -\frac{\partial H}{\partial q_k}. \qquad (11.36)$$

Hamilton's equations (11.36) consist of twice as many equations as the Lagrange equations, but they are always first order rather than second order, and therefore have certain advantages.

Let us emphasize once again that in order to use Hamilton's equations, it is essential to write the Hamiltonian function in terms of the generalized coordinates q_k and their canonical momenta p_k! *There must be no generalized velocities $\dot{q}_k$ remaining in H!* It is a common mistake to write the Hamiltonian by its definition $H = \sum_i p_i \dot{q}_i - L(q_k, \dot{q}_k, t)$ and forget to eliminate the $\dot{q}_k$ in favor of the p_k and q_k before using Hamilton's equation $\dot{p}_k = \partial H/\partial q_k$. Doing this will give an incorrect equation of motion.

In terms of analytic problem solving, Hamilton's equations add to our arsenal of techniques. However, the real advantages of Hamilton's equations are not primarily in analytic problem solving, but in the following:

(i) They give insight into understanding motion, particularly in phase space. The Hamiltonian framework gives us a qualitative bird's-eye perspective of dynamics without solving any differential equations at all.

(ii) They are more immediately appropriate for numerical solutions, a very important advantage since relatively few problems in mechanics can be solved exactly in terms of established functions. That is, they are more easily implemented in computer algorithms than second-order differential equations, resulting in more stable numerical solutions of complex systems.

(iii) They provide a natural bridge from classical to quantum mechanics, a bridge that was exploited in very different ways by two originators of quantum mechanics, the Austrian-born physicist Erwin Schrödinger (1887–1961) and the German physicist Werner Heisenberg (1901–1976).

11.3 Phase Space

Figure 11.3 depicts a two-dimensional cross-section of a phase space. A point in phase space is a complete description of the system at an instant in time. Any such point may be viewed as a complete specification of the initial conditions at time zero, and we evolve from this point along the $2N$-dimensional vector field

$$\{\dot{q}_1, \dot{p}_1, \dot{q}_2, \dot{p}_2, \ldots, \dot{q}_N, \dot{p}_N\} = \left\{\frac{\partial H}{\partial p_1}, -\frac{\partial H}{\partial q_1}, \frac{\partial H}{\partial p_2}, -\frac{\partial H}{\partial q_2}, \ldots, \frac{\partial H}{\partial p_N}, -\frac{\partial H}{\partial q_N}\right\},$$

(11.37)

as shown in the figure.

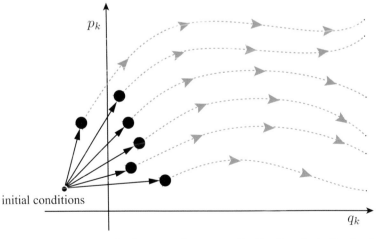

Fig. 11.3 The two-dimensional cross-section of a phase space for a system. The flow lines depict Hamiltonian time evolution.

The initial state of the system then traces a smooth line in phase space as it evolves into the future. Notice that this evolution is *almost* a gradient flow – that is a flow along the gradient of a function, H. It is so *except* for a minus sign in half of the terms of Eq. (11.37). This additional "twist" lies at the heart of dynamics. We will revisit it at the end of this part of the book, along with the insight it gives us into quantum mechanics.

Example 11.2

The Simple Harmonic Oscillator

Again, let us consider a particle of mass m moving in one dimension under the influence of a spring with force constant k. As we have already shown, in the essential canonical form the Hamiltonian is

$$H(x, p) = \frac{p^2}{2m} + \frac{1}{2}kx^2,$$ (11.38)

which is also the energy in this case. *Note that the generalized velocity $\dot{x}$ is (correctly) absent from the Hamiltonian.* The equations of motion are now first order:

$$\dot{x} = \frac{\partial H}{\partial p} = \frac{p}{m}, \quad \dot{p} = -\frac{\partial H}{\partial x} = -kx.$$ (11.39)

While already very simple, we can nevertheless implement this dynamics on a computer. More interestingly, we can now visualize the evolution in an interesting way, as shown in Figure 11.4. From Eq. (11.35), we know the Hamiltonian is conserved:

$$\frac{dH}{dt} = -\frac{\partial L}{\partial t} = 0.$$ (11.40)

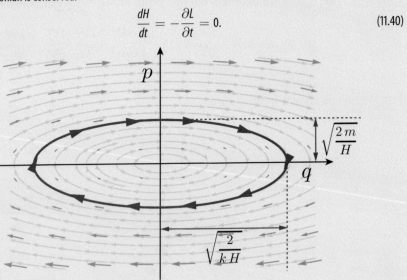

Fig. 11.4   The phase space of the one-dimensional simple harmonic oscillator.

Hence, the trajectories in phase space are contours of constant H. From Eq. (11.38), we see that these are ellipses with semi-major and semi-minor axes as shown in the figure. Note also the direction of flow in phase space: pick any point on an ellipse as an initial condition, and from the sign of p deduce the direction of flow as shown.

An interesting aspect of this picture is that we are able to get a quick bird's-eye view of the dynamics in the space of all initial conditions. In this simple example, there are no interesting regions of the phase space that result in qualitatively different evolutionary patterns. However, in a more complex system, a quick look at the phase space can immediately identify interesting basins of initial conditions, as we shall see. Nevertheless, the phase-space picture already constitutes a quick proof that all time developments of the simple harmonic oscillator are necessarily *closed* and *bounded*. This means we expect that there is a period after which the time evolution repeats itself; and also that the particle can never fly off to infinity. These statements are nontrivial, particularly for more complex systems. Note also that drawing the phase space does not involve solving any differential equations: it is simply the task of drawing contours of the algebraic expression given by H.

Now what if we want to solve our first-order differential equations (11.39) *analytically*? While their first-order nature is welcome, the two equations are in fact coupled. To decouple them, we unfortunately need to take a time derivative of the first equation of (11.39):

$$\ddot{x} = \frac{\dot{p}}{m} = -\frac{k}{m}x, \tag{11.41}$$

bringing us back to the second-order differential equation that is the simple harmonic oscillator equation. In this case, the Hamiltonian picture did not help us solve the time evolution beyond what the Lagrangian formalism can do more easily. From that perspective it seems like a waste of time. But even in this simple case, we learned about the geometry of the space of initial conditions through phase space, and as we shall see, we developed a framework particularly suited for a numerical, computer-based solution of the dynamics.

As the systems of interest get more and more complicated, we will see more and more benefits from analyzing it with the Hamiltonian formalism. We can think of the Hamiltonian picture as one of several different ways of analyzing a system, each having advantages and disadvantages, and together these methods make up a powerful arsenal of tools that help us understand complex dynamics.   ∎

Example 11.3   **A Bead on a Parabolic Wire**

Consider a bead of mass m constrained to move along a vertically oriented parabolic wire in the presence of a uniform gravitational field g, as shown in Figure 11.5. We write the Lagrangian as

$$L = \frac{1}{2}m\left(\dot{x}^2 + \dot{y}^2\right) - mgy, \tag{11.42}$$

with the constraint

$$y = \frac{x^2}{2}, \tag{11.43}$$

which is the shape of the wire. Implementing the constraint, we get a Lagrangian with a single degree of freedom:

$$L = \frac{m}{2}\left(1 + x^2\right)\dot{x}^2 - \frac{mg}{2}x^2. \tag{11.44}$$

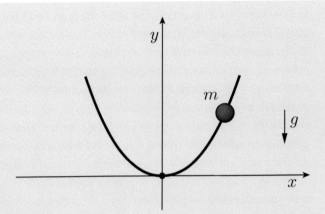

Fig. 11.5 A bead constrained to a parabola.

We expect a two-dimensional phase space, say parameterized by $\{x, p^x\}$, where[a]

$$p^x = \frac{\partial L}{\partial \dot{x}} = m\left(1 + x^2\right)\dot{x} \Rightarrow \dot{x} = \frac{p^x}{m\left(1 + x^2\right)}. \tag{11.45}$$

We can then write the Hamiltonian in terms of x and p_x by eliminating $\dot{x}$:

$$H = p^x\dot{x} - L = \frac{\left(p^x\right)^2}{2m\left(1 + x^2\right)} + \frac{mg}{2}x^2. \tag{11.46}$$

Once again, since $\partial L/\partial t = 0$, the Hamiltonian is conserved, as implied by Eq. (11.35). Thus, trajectories in phase space follow contours of constant H. Figure 11.6 shows a plot of the contours of H in phase space. We see a much richer structure of initial conditions than the case of the simple harmonic oscillator of the previous example. In particular, we see the stable point at $x = p_x = 0$. And we note that the system spends a great deal of time at the turning points x_{min} and x_{max}. Indeed, we can easily perform statistics on the figure to determine the fraction of time the particle is near the turning points, find the maximum and even minimum momentum, and determine various qualitative and quantitative aspects of the dynamics – all without solving the equations of motion. We also note, as expected intuitively, that the orbits are all bounded and closed.

The two first-order equations of motion are given by (Eqs. (11.36))

$$\dot{x} = \frac{p^x}{m\left(1 + x^2\right)} \tag{11.47}$$

and

$$\dot{p}^x = -mgx + \frac{x(p^x)^2}{m\left(1 + x^2\right)^2}. \tag{11.48}$$

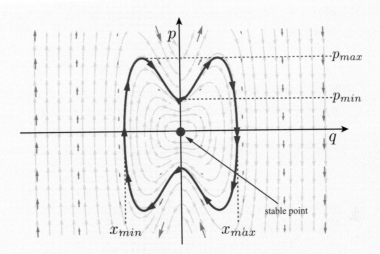

Fig. 11.6 The phase space of the one-dimensional particle on a parabola.

Once again, these are coupled first-order differential equations. Attempting to decouple them leads generically to second-order equations, the Lagrangian equations of motion. However, it is particularly easy to code these first-order differential equations into a Runge–Kutta computer algorithm. ∎

[a] Strictly speaking, the canonical momentum naturally lives in the so-called "cotangent" space and its spacetime components should be denoted by subscripts, not superscripts (which is reserved for vectors). We will not make this distinction here and we write canonical momenta as vectors with superscripts. With a mostly positive metric convention, all expressions remain correct.

Example 11.4

A Charged Particle in a Uniform Magnetic Field

A particle of mass m and charge q moves around in three dimensions in the background of given electric and magnetic fields. The Lagrangian was developed in Chapter 8 and is given by

$$L = \frac{1}{2}m\left(\dot{x}^2 + \dot{y}^2 + \dot{z}^2\right) - q\phi + \frac{q}{c}\dot{x}A^x + \frac{q}{c}\dot{y}A^y + \frac{q}{c}\dot{z}A^z, \tag{11.49}$$

where $\phi(x, y, z, t)$ and $\mathbf{A}(x, y, z, t)$ are, respectively, the electric potential and the vector potential. To transform to the Hamiltonian picture, we write the canonical momenta (see Eq. (11.34))

$$p^x = \frac{\partial L}{\partial \dot{x}} = m\dot{x} + \frac{q}{c}A^x,$$

$$p^y = \frac{\partial L}{\partial \dot{y}} = m\dot{y} + \frac{q}{c}A^y,$$

$$p^z = \frac{\partial L}{\partial \dot{z}} = m\dot{z} + \frac{q}{c}A^z, \tag{11.50}$$

or more compactly

$$\mathbf{p} = m\mathbf{v} + \frac{q}{c}\mathbf{A} \Rightarrow \mathbf{v} = \frac{\mathbf{p}}{m} - \frac{q}{mc}\mathbf{A}, \tag{11.51}$$

where in the last step we solved for $\mathbf{v}$ in preparation for eliminating the $\mathbf{v}$ dependence in the Hamiltonian (11.35):

$$H = \mathbf{v} \cdot \mathbf{p} - L = \frac{1}{2m}\left(\mathbf{p} - \frac{q}{c}\mathbf{A}\right)^2 + q\phi. \tag{11.52}$$

In the Hamiltonian picture, the effect of the electromagnetic fields is then simply the shifting of the momenta $\mathbf{p} \rightarrow \mathbf{p} - (q/c)\mathbf{A}$ and the addition of the electric potential energy $q\phi$.

As we saw in Chapter 8, the magnetic field is given in terms of the vector potential by $\mathbf{B} = \nabla \times \mathbf{A}$. For a uniform magnetic field in the z direction, $\mathbf{B} = B_0\hat{\mathbf{z}}$, we can write a vector potential (see Eq. (8.31))

$$\mathbf{A} = -\frac{1}{2}B_0 y\,\hat{\mathbf{x}} + \frac{1}{2}B_0 x\,\hat{\mathbf{y}}, \tag{11.53}$$

which is given in the Coulomb gauge choice. Substituting this into Eq. (11.52) and noting that since $\partial L/\partial t = 0$ the Hamiltonian is conserved, we have

$$H = \frac{1}{2m}\left(p^x - \frac{q\,y\,B_0}{2c}\right)^2 + \frac{1}{2m}\left(p^y + \frac{q\,x\,B_0}{2c}\right)^2 + \frac{1}{2m}(p^z)^2, \tag{11.54}$$

describing the constant Hamiltonian contours in phase space. As expected, the dynamics in the z direction is that of a free particle. In phase space, if we focus on the $x-p^x$ cross-section, for example, we have off-center ellipses as shown in Figure 11.7. In the (x, p^x) coordinates, the center is located at $(-2\,c\,p^y/q\,B_0, q\,y\,B_0/2\,c)$ and the radii are $(\sqrt{8\,m\,c^2 H/q^2 B_0^2}, \sqrt{2\,m\,H})$. In the $x-y$ plane, as we know, the particle would be circling around. More interestingly, we have shifted *circles* of radius

$$R = \sqrt{\frac{8\,m\,c^2}{q^2 B_0^2}\left(H - \frac{(p^z)^2}{2\,m}\right)}. \tag{11.55}$$

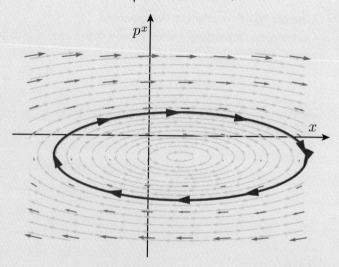

Fig. 11.7 The flow lines in the $x-p_x$ cross-section of phase space for a charged particle in a uniform magnetic field. ∎

11.4 Canonical Transformations

From the perspective of the Hamiltonian formalism, dynamics plays out in *phase space*. Figure 11.8(a) shows a two-dimensional cross-section of a phase space. The flow lines depict the time evolution of the system with various initial conditions.

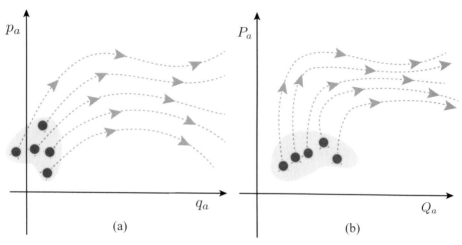

Fig. 11.8 (a) The flow lines in a given phase space. (b) The same flow lines as described by transforming coordinates and momenta.

Based on our experience, we know that coordinate transformations can be very useful when tackling problems in physics. What if we were to apply a coordinate transformation directly in phase space:

$$q_k \to Q_k(q,p,t), \quad p_k \to P_k(q,p,t) \; ? \tag{11.56}$$

To specify the full coordinate transformation in phase space, it then seems we would need $2N$ functions. Such a general coordinate transformation would also deform and distort the time-evolution flow pattern as illustrated in Figure 11.8(b). In general, the new flow lines may not be *Hamiltonian*: by that we mean that the elegant attribute of time evolution in phase space as one given by the twisted gradients (*i.e.*, Eq. (11.37)) of a function called the Hamiltonian may not persist in $\{Q,P\}$ space:

$$\dot{Q} \neq \frac{\partial f}{\partial P}, \quad \dot{P} \neq -\frac{\partial f}{\partial Q} \tag{11.57}$$

for any arbitrary function $f(Q,P)$. *The more interesting transformations in phase space are obviously those that preserve this structure of Hamiltonian dynamics*, because then the new canonical variables Q, P would be just as viable as the original variables q, p. We therefore look for a subset of all possible transformations that preserve Hamiltonian flow; that is, through which we obtain a new Hamiltonian

$$H(q,p) \rightarrow \tilde{H}(Q,P) \tag{11.58}$$

such that

$$\dot{q}_k = \frac{\partial H}{\partial p_k}, \quad \dot{p}_k = -\frac{\partial H}{\partial q_k} \tag{11.59}$$

implies

$$\dot{Q}_k = \frac{\partial \tilde{H}}{\partial P_k}, \quad \dot{P}_k = -\frac{\partial \tilde{H}}{\partial Q_k}. \tag{11.60}$$

This is a nontrivial condition on the allowed transformations. We henceforth refer to such transformations as **canonical transformations**.

Let us find the general attributes of such canonical transformations. The structure of the Hamiltonian equations (11.59) and (11.60) proceeds from the variational principle: the equations of motion are at the extremum of a single functional called the action. If the Hamiltonian flow is to be preserved by a canonical transformation of phase space, and the equations of motion in the old and new variables are to be describing the same physical situation, we must require that the actions in the old and new coordinates remain unchanged:

$$S[q, \dot{q}, t] = S[Q, \dot{Q}, t]. \tag{11.61}$$

The action is the time integral of the Lagrangian, $S = \int L \, dt$, so this implies that

$$L(q, \dot{q}, t) = \tilde{L}(Q, \dot{Q}, t) + \frac{dF}{dt}, \tag{11.62}$$

where the difference between the two Lagrangians can be a total time derivative of an arbitrary function F. That is, as long as that function vanishes at early and late times, the integral

$$\int_{t_0}^{t_1} dt \, \frac{dF}{dt} = F(t_1) - F(t_0) = 0. \tag{11.63}$$

In that case the time integral of dF/dt is inconsequential to the equations of motion. Now since $L(q, \dot{q}) = \dot{q}_k p_k - H$ and $\tilde{L}(Q, \dot{Q}, t) = \dot{Q}_k P_k - \tilde{H}$, we have

$$\dot{q}_k p_k - H = \dot{Q}_k P_k - \tilde{H} + \frac{dF(q, p, Q, P, t)}{dt}. \tag{11.64}$$

Note that so far we have assumed that q, p, Q, P, and t are all independent of one another. Therefore the chain rule of multivariable calculus tells us that

$$\frac{dF}{dt} = \frac{\partial F}{\partial t} + \frac{\partial F}{\partial q_k} \dot{q}_k + \frac{\partial F}{\partial p_k} \dot{p}_k + \frac{\partial F}{\partial Q_k} \dot{Q}_k + \frac{\partial F}{\partial P_k} \dot{P}_k. \tag{11.65}$$

There are huge numbers of functions F we might use here. One particular class of functions we might choose has the form

$$F \rightarrow F_1(q, Q, t), \tag{11.66}$$

a function of q, Q, and t *only*, so that we take q, Q, and t to be independent of one another. Then, comparing both sides of Eq. (11.64), we must have

$$p_k = \frac{\partial F_1(q, Q, t)}{\partial q_k}, \qquad P_k = -\frac{\partial F_1(q, Q, t)}{\partial Q_k}, \quad \text{and} \quad H = \widetilde{H} - \frac{\partial F_1}{\partial t}. \qquad (11.67)$$

To obtain the desired transformations $Q_k(q, p, t)$ and $P_k(q, p, t)$, we need to invert the first equation to get $Q_k(q, p, t)$; and use this in the second equation to get $P_k(q, p, t)$. Finally, we can use the third equation to solve for the desired new Hamiltonian $\widetilde{H}$. We will shortly show an example of this procedure.

We have therefore found that we can generate a canonical transformation from $\{q_k, p_k\}$ to $\{Q_k, P_k\}$ using any function of the form $F_1(q, Q, t)$ and its derivatives. The function $F_1(q, Q, t)$ is called the **generator** of this canonical transformation.

Example 11.5

Transforming the Simple Harmonic Oscillator

Consider once again the simple harmonic oscillator, with a particle of mass m free to move in one dimension, connected to a spring with Hamiltonian

$$H = \frac{p^2}{2m} + \frac{1}{2}m\omega^2 q^2, \qquad (11.68)$$

where $\omega = \sqrt{k/m}$ is the natural angular frequency and k is the spring constant. The phase space is two-dimensional, parameterized by $\{q, p\}$. Let us apply the canonical transformation using the particular generator

$$F_1(q, Q, t) = qQ. \qquad (11.69)$$

From Eqs. (11.67), we immediately get

$$p = Q \quad \text{and} \quad P = -q, \qquad (11.70)$$

which can be inverted to give $Q(q, p) = p$ and $P(q, p) = -q$. The transformation exchanges position and momenta – except for the ubiquitous minus sign that is the hallmark of Hamiltonian dynamics. The new Hamiltonian is then

$$\widetilde{H} = \frac{p^2}{2m} + \frac{1}{2}m\omega^2 q^2 + 0 = \frac{Q^2}{2m} + \frac{1}{2}m\omega^2 P^2, \qquad (11.71)$$

and the new equations of motion are

$$\dot{Q} = \frac{\partial \widetilde{H}}{\partial P} = m\omega^2 P \quad \text{and} \quad \dot{P} = -\frac{\partial \widetilde{H}}{\partial Q} = -\frac{Q}{m}. \qquad (11.72)$$

This $\{Q, P\}$ system is physically equivalent to the $\{q, p\}$ system. We may say that, as far as dynamics is concerned, coordinates and momenta can be mixed and even exchanged. They seem to be different facets of the same overall physical information, somewhat like the mixing of energy and momentum when changing inertial perspectives in relativistic dynamics.   ∎

How about a canonical transformation that is simply the identity transformation

$$p_k = P_k, \qquad q_k = Q_k, \tag{11.73}$$

and where F_1 is not a function of time? Using Eqs. (11.67), we find

$$p_k = \frac{\partial F_1}{\partial q_k} = P_k \Rightarrow F_1 = P_k q_k + f(Q) \tag{11.74}$$

and so, using this result:

$$P_k = -\frac{\partial F_1}{\partial Q_k} = -\frac{\partial f}{\partial Q_k} \Rightarrow f(Q) = -Q_k P_k \ ! \tag{11.75}$$

These results are *inconsistent* with our initial assumption that F_1 is a function of q_k, Q_k, and t only. Hence, it seems that $F_1(q, Q, t)$ cannot generate the simplest of all transformations, the identity transformation! Obviously, the identity transformation is canonical, hence we must have missed something in going from Eqs. (11.64) and (11.65) to (11.66).

Let us go back to Eqs. (11.64) and (11.65). *The generator $F_1(q, Q, t)$ is not the only class of generators we can use to solve these equations.* We start by writing

$$\dot{Q}_k P_k = -Q_k \dot{P}_k + \frac{d}{dt}(Q_k P_k) \tag{11.76}$$

using the product rule. Substituting this in Eq. (11.64), we get

$$\dot{q}_k p_k - H = -Q_k \dot{P}_k - \tilde{H} + \frac{d}{dt}(Q_k P_k + F_1). \tag{11.77}$$

Writing a new generator

$$F_2 = F_1(q, Q, t) + Q_k P_k, \tag{11.78}$$

it is now straightforward to show that a generator of the form $F_2(q, P, t)$ has the correct structure to satisfy Eq. (11.77) if

$$p_k = \frac{\partial F_2}{\partial q_k}, \qquad Q_k = \frac{\partial F_2}{\partial P_k}, \tag{11.79}$$

and

$$H = \tilde{H} - \frac{\partial F_2}{\partial t}. \tag{11.80}$$

That is, we again compare both sides of Eq. (11.81) using (11.83) and (11.84). Unlike $F_1(q, Q, t)$, however, the function $F_2(q, P, t)$ *does* include the identity transformation. Consider

$$F_2(q, P, t) = q_k P_k \tag{11.81}$$

where k is summed over. Using Eqs. (11.79), we get

$$p_k = P_k, \qquad Q_k = q_k, \tag{11.82}$$

which is the sought-for identity transformation.

This treatment also makes it clear that we can have two additional classes of generators of canonical transformations: a generator $F_3(p, Q, t)$ and a generator $F_4(p, P, t)$.

To find $F_3(p, Q, t)$, start by writing

$$\dot{q}_k p_k = -q_k \dot{p}_k + \frac{d}{dt}(q_k p_k)$$

(11.83)

and substitute in Eq. (11.64). We then need

$$F \rightarrow F_3(p, Q, t)$$

(11.84)

with

$$q_k = -\frac{\partial F_3}{\partial p_k}, \qquad P_k = -\frac{\partial F_3}{\partial Q_k},$$

(11.85)

and

$$H = \widetilde{H} - \frac{\partial F_3}{\partial t}.$$

(11.86)

To find $F_4(p, P, t)$ use both tricks (11.76) and (11.83) in (11.64). We then need

$$F \rightarrow F_4(p, P, t)$$

(11.87)

with

$$q_k = -\frac{\partial F_4}{\partial p_k}, \qquad Q_k = \frac{\partial F_4}{\partial P_k},$$

(11.88)

and

$$H = \widetilde{H} - \frac{\partial F_4}{\partial t}.$$

(11.89)

This concludes the list of all possible canonical transformations. They are described by four classes of generators $F_1(q, Q, t)$, $F_2(q, P, t)$, $F_3(p, Q, t)$, and $F_4(p, P, t)$. To summarize, the transformations are found from these generators as follows:

$$
\begin{aligned}
&\text{For } F_1(q, Q, t): \quad p_k = \partial F_1/\partial q_k, \quad P_k = -\partial F_1/\partial Q_k. \\
&\text{For } F_2(q, P, t): \quad p_k = \partial F_2/\partial q_k, \quad Q_k = \partial F_2/\partial P_k. \\
&\text{For } F_3(p, Q, t): \quad q_k = -\partial F_3/\partial p_k, \quad P_k = -\partial F_3/\partial Q_k. \\
&\text{For } F_4(p, P, t): \quad q_k = -\partial F_4/\partial p_k, \quad Q_k = \partial F_4/\partial P_k.
\end{aligned}
$$

(11.90)

And as always we have

$$H = \widetilde{H} - \frac{\partial F}{\partial t}.$$

(11.91)

Notice the pattern in these equations: a coordinate and a momentum are paired in each statement, i.e., p_k with q_k and P_k with Q_k in $F_1(q, Q, t)$; and the rest are obtained by exchanging within these pairs – along with a flip of a sign for every exchange. The four generators $F_1(q, Q, t)$, $F_2(q, P, t)$, $F_3(p, Q, t)$, and $F_4(p, P, t)$ are related to each other by the so-called Legendre transformations:

$$F_2(q, P, t) = F_1(q, Q, t) + Q_k P_k,$$

$$F_3(p, Q, t) = F_1(q, Q, t) - q_k p_k,$$

$$F_4(p, P, t) = F_1(q, Q, t) + P_k Q_k - q_k p_k. \tag{11.92}$$

Example 11.6 Identities

A particularly simply class of canonical transformations are the so-called "identities." One can easily check the following base transformations:

$$\text{For } F_1(q, Q, t) = q_k Q_k \Rightarrow p_k = Q_k, \quad P_k = -q_k. \tag{11.93}$$

$$\text{For } F_2(q, P, t) = q_k P_k \Rightarrow p_k = P_k, \quad Q_k = q_k. \tag{11.94}$$

$$\text{For } F_3(p, Q, t) = p_k Q_k \Rightarrow q_k = -Q_k, \quad P_k = -p_k. \tag{11.95}$$

$$\text{For } F_4(p, P, t) = p_k P_k \Rightarrow q_k = -P_k, \quad Q_k = p_k. \tag{11.96}$$

We see that the simplest nontrivial transformation for $F_1(q, Q, t)$ is the exchange of coordinates and momenta (with a minus sign twist); for $F_2(q, P, t)$, it is the usual identity transformation; for $F_3(p, Q, t)$, it is a reflection of both coordinates and momenta; and finally for $F_4(p, P, t)$, it is again an exchange of coordinates and momenta. ∎

Example 11.7 Infinitesimal Transformations and the Hamiltonian

Infinitesimal transformations – those that are *almost* the identity – are often useful in physics as the building blocks of larger transformations. Let us consider the class of infinitesimal canonical transformations

$$F_2 = q_k P_k + \epsilon G(q, P, t), \tag{11.97}$$

where ϵ is taken as small, and $G(q, P, t)$ is an unknown function. From Eq. (11.90), we get

$$p_k = P_k + \epsilon \frac{\partial G}{\partial q_k}, \quad Q_k = q_k + \epsilon \frac{\partial G}{\partial P_k}. \tag{11.98}$$

These transformations may look familiar. To see why, let us pretend we chose $G(q, P, t)$ such that $P_k = p_k(t + \delta t)$ and $Q_k = q_k(t + \delta t)$ and $\epsilon = \delta t$. That is, we transform the q_ks and p_ks to their values a very short time later. We then get

$$\dot{P}_k = -\frac{\partial G(q, P, t)}{\partial q_k}, \quad \dot{q}_k = \frac{\partial G(q, P, t)}{\partial P_k}. \tag{11.99}$$

Since the Q_ks and P_ks differ from the q_ks and p_ks by an amount of order ϵ, to linear order in ϵ we can write these equations as

$$\dot{p}_k = -\frac{\partial G(q,p,t)}{\partial q_k}, \quad \dot{q}_k = \frac{\partial G(q,p,t)}{\partial p_k}. \tag{11.100}$$

If we then identify

$$H(q,p,t) \rightarrow G(q,P,t) \tag{11.101}$$

we notice that these are simply the Hamilton equations of motion! Put differently, Hamiltonian evolution is a canonical transformation with the infinitesimal generator of the transformation being the Hamiltonian function itself! We can then view time evolution as a canonical transformation to coordinates an instant into the future at every time step.   ∎

Example 11.8 **Point Transformations**

Another interesting example is to transform the configuration-space coordinates q alone. These are called "point transformations," and they may or may not be canonical transformations as well. Consider the case of a two-dimensional phase space and a coordinate transformation

$$Q = f(q). \tag{11.102}$$

For this to be also a *canonical* transformation, we need to transform the associated momenta in a specific way. We may use

$$F_2 = f(q)P \Rightarrow p = \frac{\partial f}{\partial q}P. \tag{11.103}$$

Hence, as long as we transform the momenta as

$$p = \frac{\partial f}{\partial q}P, \tag{11.104}$$

we are guaranteed that the point transformation in this case is also canonical.   ∎

11.5 Poisson Brackets

Identifying the Hamiltonian as a generator of infinitesimal canonical transformations suggests that the interesting structure of phase space lies in general canonical transformations. We can view this in analogy to Lorentz symmetry in coordinate space. We learned in Chapter 2 that the invariance of the metric or line element

$$ds^2 = -c^2 dt^2 + dx^2 + dy^2 + dz^2 \tag{11.105}$$

under coordinate transformations plays a central role in defining Lorentz transformations, and the invariance of the laws of the physics under these transformations was a founding principle of the laws of mechanics. In the Hamiltonian picture, our playground is phase space instead of configuration (*i.e.*, coordinate) space, and canonical transformations play an equally central role in prescribing dynamics in phase space, as we have just learned. What is then the invariant object in phase space – the analogue of the "metric" – which is left invariant under canonical transformations?

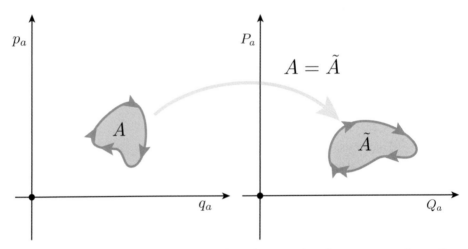

Fig. 11.9 The transformation of phase space under a canonical transformation. Area elements may get distorted in shape, but the area of each element must remain unchanged.

Consider a 2N-dimensional phase space parameterized by $\{q_k, p_k\}$. A canonical transformation generated by $F_1(q_a, Q_a)$ relabels phase space with $\{Q_k, P_k\}$. Here, q_a is one of the many q_ks, and Q_a is one of the Q_ks. Focus on this particular q_a, p_a plane as shown in Figure 11.9. The line integral over a closed path of dF_1 must vanish because the path is closed; that is

$$\oint dF_1 = 0. \tag{11.106}$$

But we can also write

$$\oint dF_1 = \oint \left(\frac{\partial F_1}{\partial q_a} dq_a + \frac{\partial F_1}{\partial Q_a} dQ_a \right)$$

$$= \oint (p_a dq_a - P_a dQ_a) = 0 \quad \text{(no sum over } a) \tag{11.107}$$

using Eq. (11.90). This implies

$$\oint p_a dq_a = \oint P_a dQ_a \Rightarrow \int dp_a dq_a = \int dP_a dQ_a \quad \text{(no sum over } a), \tag{11.108}$$

where in the last step we have used Stokes's theorem

$$\oint \psi \mathbf{dl} = \int \hat{\mathbf{n}} \times \nabla \psi dA \tag{11.109}$$

relating a line integral over a closed path in two dimensions to the area integral over the area enclosed by the path for any ψ. This implies that

$$\oint \psi(p)dq = \int \frac{d\psi}{dp} dp dq. \tag{11.110}$$

To see this, write the infinitesimal path vector in phase space as $\mathbf{dl} = dq\,\hat{\mathbf{q}} + dp\,\hat{\mathbf{p}}$, where $\hat{\mathbf{q}}$ and $\hat{\mathbf{p}}$ are unit vectors in the direction of increasing q and p, respectively, and note that the orientation of the path is correlated with the orientation of the area normal $\hat{\mathbf{n}} = \hat{\mathbf{p}} \times \hat{\mathbf{q}}$.

Since the closed path is arbitrary, and hence the enclosed area is arbitrary, Eq. (11.108) implies that canonical transformations preserve area in phase space (see Figure 11.9). Equivalently, the *measure* in phase space must be invariant:

$$dp_a dq_a = dP_a dQ_a = \text{Det}\left[\frac{\partial(P_a, Q_a)}{\partial(p_a, q_a)}\right] dp_a dq_a \quad \text{(no sum over } a) \tag{11.111}$$

where on the right we have written the Jacobian of the corresponding canonical transformation

$$\text{Jacobian} = \text{Det}\left[\frac{\partial(P_a, Q_a)}{\partial(p_a, q_a)}\right] = 1 \quad \text{(no sum over } a) \tag{11.112}$$

which must equal unity if canonical transformations are to preserve phase space area. We remind the reader that the transformation matrix $\partial(A, B)/\partial(x, y)$ with two functions $A(x, y)$ and $B(x, y)$ is defined as

$$\frac{\partial(A, B)}{\partial(x, y)} \equiv \begin{pmatrix} \partial A/\partial x & \partial A/\partial y \\ \partial B/\partial x & \partial B/\partial y \end{pmatrix}. \tag{11.113}$$

Now we introduce a new notation, the so-called **Poisson bracket**:[3]

$$\{A, B\}_{x,y} \equiv \text{Det}\left[\frac{\partial(A, B)}{\partial(x, y)}\right]. \tag{11.114}$$

This allows us to write Eq. (11.112) as

$$\{P_a, Q_a\}_{q_a, p_a} = 1 \quad \text{no sum over } a. \tag{11.115}$$

We can extend our argument to different combinations of q_a, p_a pairs in the full $2N$-dimensional phase space (see the Problems section at the end of this chapter), and we conclude that canonical transformations preserve phase space "volume":

$$dq_1 dp_1 dq_2 dp_2 \cdots dq_k dp_k = dQ_1 dP_1 dQ_2 dP_2 \cdots dQ_k dP_k. \tag{11.116}$$

---

[3] Named for the French mathematician and physicist Simeon Denis Poisson (1781–1840).

We then write the Poisson bracket of two functions $A(q,p)$ and $B(q,p)$ in $2N$-dimensional phase space as

$$\{A,B\}_{q,p} \equiv \frac{\partial A}{\partial q_k}\frac{\partial B}{\partial p_k} - \frac{\partial A}{\partial p_k}\frac{\partial B}{\partial q_k}, \tag{11.117}$$

where we have now extended the definition to $2N$ dimensions by summing over all two-dimensional subspaces labeled by q_k, p_k (since the index k is repeated in this expression). One can now show that canonical transformations preserve this generalized Poisson bracket. That is

$$\{A,B\}_{q,p} = \{A,B\}_{Q,P}. \tag{11.118}$$

Therefore we have found that the equivalent of the metric invariant of Lorentz transformations in phase space is the phase space integration measure (11.116), *or equivalently the differential operator from* (11.117). Reversing the statement, it is also possible to show that all phase space transformations that preserve the measure (11.116), or equivalently the Poisson bracket (11.117), are canonical transformations (see the Problems section at the end of this chapter). For example, we can use the preservation of Poisson brackets as a test of the canonicality of a transformation.

Example 11.9 **Position and Momenta**

Using (11.117), one can easily show that the Poisson brackets of the ps and qs are

$$\{p_a, q_b\}_{q,p} = \frac{\partial p_a}{\partial q_k}\frac{\partial q_b}{\partial p_k} - \frac{\partial p_a}{\partial p_k}\frac{\partial q_b}{\partial q_k} = \delta_{ab}, \tag{11.119}$$

where δ_{ab} is the Kronecker delta. That is, p_a and q_b "Poisson commute" for $a \neq b$; if $a = b$, the result is unity. We can similarly see that

$$\{q_a, q_b\}_{q,p} = \{p_a, p_b\}_{q,p} = 0. \tag{11.120}$$

Given a candidate canonical transformation $Q_k(q,p,t)$ and $P_k(q,p,t)$, we can *test* for canonicality by verifying that

$$\{P_a, Q_b\}_{q,p} = \delta_{ab}, \quad \{Q_a, Q_b\}_{q,p} = \{P_a, P_b\}_{q,p} = 0, \tag{11.121}$$

since

$$\{P_a, Q_b\}_{q,p} = \{p_a, q_b\}_{q,p},$$
$$\{Q_a, Q_b\}_{q,p} = \{q_a, q_b\}_{q,p}, \tag{11.122}$$
$$\{P_a, P_b\}_{q,p} = \{p_a, p_b\}_{q,p}$$

for canonical transformations.   ∎

Example 11.10 **The Simple Harmonic Oscillator Yet Again**

Consider the simple harmonic oscillator from Eq. (11.68), with Hamiltonian

$$H = \frac{p^2}{2m} + \frac{1}{2}m\omega^2 q^2. \tag{11.123}$$

The structure of the Hamiltonian as a sum of squares suggests a transformation of the form

$$q(Q,P) \propto \sqrt{\frac{2}{m\omega^2}} \sin Q, \quad p(Q,P) \propto \sqrt{2m}\cos Q, \tag{11.124}$$

since the identity $\cos^2 Q + \sin^2 Q = 1$ would simplify the new Hamiltonian $\widetilde{H}$. Let us try to transform to a new Hamiltonian that looks like

$$\widetilde{H} = \omega P. \tag{11.125}$$

This would be interesting, since the equations of motion would then imply

$$\dot{Q} = \frac{\partial \widetilde{H}}{\partial P} = \omega, \quad \dot{P} = -\frac{\partial \widetilde{H}}{\partial Q} = 0, \tag{11.126}$$

which can immediately be solved, yielding

$$Q(t) = \omega t + Q_0, \quad P(t) = P_0. \tag{11.127}$$

To achieve this transformation, write

$$q(Q,P) = \sqrt{\frac{2P}{m\omega}} \sin Q, \quad p(Q,P) = \sqrt{2m\omega P}\cos Q \tag{11.128}$$

according to Eq. (11.123). But is this a canonical transformation? If not, we would *not* have the evolution equations given by (11.126). To test for canonicality, we check the Poisson bracket

$$\{q,p\}_{q,p} = 1 = \{q,p\}_{Q,P} = \left\{ \sqrt{\frac{2P}{m\omega}} \sin Q, \sqrt{2m\omega P}\cos Q \right\}_{Q,P}. \tag{11.129}$$

Using the definition (11.117), we indeed verify, after some algebra, that this holds. Similarly, we can show that $\{q,q\}_{Q,P} = \{p,p\}_{Q,P} = 0$. With these three statements, we conclude that the transformation (11.128) is indeed canonical. Substituting the solution (11.127) in Eqs. (11.128), we find the solution in the original variables

$$q(Q,P) = \sqrt{\frac{2P_0}{m\omega}} \sin\omega t + Q_0, \quad p(Q,P) = \sqrt{2m\omega P_0}\cos\omega t + Q_0, \tag{11.130}$$

which by now have become very familiar. We have thus demonstrated a new strategy for tackling a dynamical system: first attempt to find/guess a canonical transformation to simplify the Hamiltonian; then verify the canonicality using the Poisson bracket. Sometimes guessing at a strategic canonical transformation turns out to be easier than tackling the original Hamiltonian in its full glory. ∎

In the Problems section at the end of this chapter, we explore some of the most important properties of Poisson brackets. In particular, one can show that the following identities follow from definition (11.117):

1. Anticommutativity

$$\{A, B\}_{q,p} = -\{B, A\}_{q,p}. \tag{11.131}$$

2. Distributivity

$$\{A, bB + cC\}_{q,p} = b\{A, B\}_{q,p} + c\{A, C\}_{q,p}, \tag{11.132}$$

   where b and c are constants.

3. A modified notion of *associativity*

$$\{AB, C\}_{q,p} = \{A, C\}_{q,p} B + A\{B, C\}_{q,p}. \tag{11.133}$$

4. The **Jacobi identity**

$$\left\{A, \{B, C\}_{q,p}\right\}_{q,p} + \left\{B, \{C, A\}_{q,p}\right\}_{q,p} + \left\{C, \{A, B\}_{q,p}\right\}_{q,p} = 0. \tag{11.134}$$

These properties of the Poisson bracket are so central to its role in mechanics that they can be used to *define* it: a bilinear operation that takes two functions of canonical coordinates and momenta, that is, the Poisson bracket, and gives a third in such a manner that it can be naturally used to describe time evolution in phase space. In the final chapter we will also revisit the central role of the Poisson bracket in transitioning from classical to quantum mechanics.

11.6 Poisson Brackets and Noether's Theorem

We can also relate Poisson brackets to Noether's theorem as introduced in Chapter 6. First of all, the Poisson bracket can be used to write Hamilton's equations of motion as

$$\dot{q} = \{q, H\}_{q,p}, \quad \dot{p} = \{p, H\}_{q,p}, \tag{11.135}$$

as can easily be verified using (11.117). Note that *this makes explicit* the fact that canonical transformations do not change the structural form of Hamilton's equations. More generally, we can write for any function $A(q, p, t)$

$$\frac{dA}{dt} = \frac{\partial A}{\partial t} + \frac{\partial A}{\partial q_k}\dot{q}_k + \frac{\partial A}{\partial p_k}\dot{p}_k = \frac{\partial A}{\partial t} + \{A, H\}_{q,p} \tag{11.136}$$

using the chain rule and Eqs. (11.135). If a function $A(q, p, t)$ is conserved, we then have

$$\frac{\partial A}{\partial t} + \{A, H\}_{q,p} = 0. \tag{11.137}$$

The most common case is when $A(q, p, t) = A(q, p)$ does not depend on time explicitly; we then have the condition for the conservation of A as

$$\{A, H\}_{q,p} = 0. \tag{11.138}$$

We say that, if the function A "Poisson commutes" with the Hamiltonian, it is a conserved quantity. Using this, it is possible to show that, if $A(q, p)$ and $B(q, p)$ are conserved, so is their Poisson bracket $\{A(q, p), B(q, p)\}$ (see the Problems section at the end of this chapter). A more general statement also exists that applies to the case where A and B depend explicitly on time. This means that if you have two conserved quantities, you might be able to construct a third one, and then a fourth one, and then a fifth, etc. by Poisson bracketing them with each other. This can be a very powerful technique, specially if the system is exactly integrable. Unfortunately, in many cases this strategy does not lead to new independent conserved quantities.

Finally, the Poisson bracket also plays a natural role in canonical transformations. For an infinitesimal transformation

$$F_2 = q_k P_k + \epsilon G(q, P, t), \tag{11.139}$$

we can write

$$\delta A = \frac{\partial A}{\partial q_k} \delta q_k + \frac{\partial A}{\partial p_k} \delta p_k = \frac{\partial A}{\partial q_k} \epsilon \frac{\partial G}{\partial p_k} - \frac{\partial A}{\partial p_k} \epsilon \frac{\partial G}{\partial q_k}$$

$$= \epsilon \{A, G\}_{q,p}. \tag{11.140}$$

In this expression, we have used the fact that $\partial G / \partial P_k \simeq \partial G / \partial p_k$, taking $G(q, P, t)$ to be the same as $G(q, p, t)$: this is because the difference between p and P is linear in ϵ and the derivatives of G appearing in the expression above are multiplied by ϵ already. This means that, to linear order in ϵ, we can take p and P to be equivalent in this expression. While the generator $G(q, P, t)$ is strictly a function of P and not p, in talking about such transformations of a function $A(q, p, t)$, one calls $G(q, p, t)$ the "generator" of the transformation, bearing in mind that this is a correct statement only in expressions already expanded to linear order in ϵ. We can then figure out the transformation of a function $A(q, p, t)$ under the action of a generator G by computing the Poisson bracket of A and G.

Example 11.11 **Translations, Once Again**

Consider a two-dimensional phase space with a single q and a single p as the coordinates. We want to find the transformation of a function $A(q, p, t)$ under a transformation generated by $G = p$. We have

$$\delta A = \epsilon \{A, p\} = \frac{\partial A}{\partial q}. \tag{11.141}$$

Now, notice that if we were to infinitesimally *translate* the canonical coordinate as in $q \to q + \epsilon$, we would have

$$\delta A = A(q + \epsilon) - A(q) = \epsilon \frac{\partial A}{\partial q}, \tag{11.142}$$

to leading order in ϵ. Comparing this to Eq. (11.141), we see that the generator $G = p$ generates *translations* in q. In a similar way, we can identify the generators of various useful transformations such as rotations. Notice in particular that Eqs. (11.135) imply the generator of time translations is the Hamiltonian. ∎

An interesting thing happens if we take $A(q, p, t)$, the subject of a transformation, as the Hamiltonian of the physical system. We then have

$$\delta H = \epsilon\{H, G\}. \tag{11.143}$$

If the transformation generated by G is a symmetry of the Hamiltonian, we must then have

$$\delta H = 0 = \epsilon\{H, G\}, \tag{11.144}$$

which implies that the bracket of H and G vanishes. As discussed earlier, this then implies that G is a constant of motion. That is, the generator of a transformation that leaves the Hamiltonian unchanged is nothing but the conserved quantity associated with the symmetry it generates! For every transformation that is a symmetry, we then have a conserved quantity – which we just identified as the generator of the transformation. *This is Noether's theorem in Hamiltonian language.* Looking at the previous example, where we found that $G = p$ generates translations, we see that for a Hamiltonian that is translationally invariant we expect that the canonical momentum p would be conserved.

Example 11.12   The Power of the Bracket

The Poisson bracket, as a bi-linear operator that takes two inputs and satisfies the four properties listed above, carries within it a lot of the structure that underlies infinitesimal transformations, and in particular time translations in phase space. To demonstrate this, imagine we are given a Hamiltonian of the form

$$H = \frac{p^2}{2m} + U(q), \tag{11.145}$$

with the equations of motion (11.135) expressed in terms of Poisson brackets. Beyond this, imagine we only know the four properties of the bracket – and not for example its particular form given by (11.117); and we are also given the "algebra" satisfied by the canonical coordinate and momentum

$$\{q, p\} = 1, \quad \{q, q\} = \{p, p\} = 0, \quad \{q, C\} = \{p, C\} = 0, \tag{11.146}$$

where C is a constant independent of q and p. From Eqs. (11.135), we can write

$$\dot{q} = \{q, H\} = \left\{ q, \frac{p^2}{2m} + U(q) \right\} = \left\{ q, \frac{p^2}{2m} \right\} + \{q, U(q)\}, \tag{11.147}$$

where we used the distributivity property of the bracket. For the first term, we have

$$\left\{q, \frac{p^2}{2m}\right\} = \frac{1}{2m}\{q, pp\} = -\frac{1}{2m}\{pp, q\}$$

$$= -\frac{1}{2m}\{p, q\}\, p - \frac{1}{2m}p\,\{p, q\} = -\frac{p}{m}\{p, q\}, \tag{11.148}$$

where we used the anticommutativity and special associativity properties of the bracket. For the second term in (11.147), we get

$$\{q, U(q)\} = \left\{q, \sum_{n=0}^{\infty} \frac{1}{n!}U_0^{(n)} q^n\right\}, \tag{11.149}$$

where we have used a Taylor expansion of $U(q)$ around $q = 0$, and $U_0^{(n)}$ is the nth derivative of $U(q)$ at $q = 0$ (see Appendix F). Using again the properties of the bracket, we can simplify this to

$$\{q, U(q)\} = \{q, U_0\} + \{q, q\} \times \text{messy expression} = \{q, U_0\}, \tag{11.150}$$

where in the last step we used the anticommutativity of the bracket to deduce that $\{q, q\} = 0$. We then get the expected relation

$$\dot{q} = \frac{p}{m} \tag{11.151}$$

using $\{q, p\} = 1$. We can similarly show that

$$\dot{p} = -\frac{\partial U}{\partial q} \tag{11.152}$$

as needed. Therefore, classical dynamics is captured by: (1) the four properties of the Poisson bracket; (2) the algebra of the canonical coordinates and momenta; and (3) the form of the Hamiltonian in terms of canonical coordinates and momenta. This observation will become very important when we discuss transitioning from classical to quantum mechanics in Chapter 15. ∎

11.7 Liouville's Theorem

Consider a set of initial conditions in phase space whose time evolution we wish to trace, as illustrated in Figure 11.10.

Let ΔN denote the number of such initial conditions, and ΔV the volume of phase space they occupy initially. The density of such initial conditions is then

$$\rho(q, p, t) = \frac{\Delta N}{\Delta V}. \tag{11.153}$$

As time evolves, ΔN remains unchanged, since any state of the system cannot suddenly disappear as we evolve forward in time. Furthermore, ΔV must remain unchanged by the Hamiltonian evolution: Hamiltonian evolution is a canonical

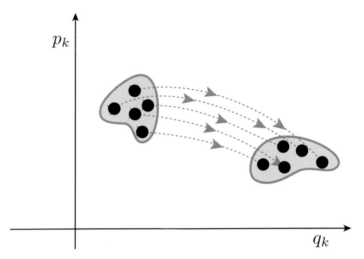

Fig. 11.10 A depiction of Liouville's theorem: the density of states of a system evolves in phase space in such as way that its total time derivative is zero.

transformation, and canonical transformations preserve phase space volume! The shape of the volume element may get twisted and compressed as shown in the figure, but the volume itself remains unchanged. This implies

$$\frac{d\rho}{dt} = 0 = \frac{\partial \rho}{\partial t} + \{\rho, H\}_{q,p} \Rightarrow \frac{\partial \rho}{\partial t} = -\{\rho, H\}_{q,p} \qquad (11.154)$$

using Eq. (11.136). This is known as **Liouville's theorem:**[4] *The density of states in phase space remains constant in time.* This powerful theorem plays an important role in statistical mechanics and fluid dynamics. It also packages within it the seeds of quantization, as we shall see at the end of the book. We leave further exploration of Liouville's theorem to the problems and to Chapter 15. In particular, there are problems at the end which illustrate that Liouville's theorem for particle points in the phase plane maintain constant density as time goes on. The number of points within a given area (or volume) in the phase plane remains constant, and so does the area or volume. Therefore the density of points remains constant as well.

11.8 Summary

We arrived at several important results in this chapter. Perhaps the most important are the Hamiltonian equations and the insights we can glean from exploring phase space. The Lagrange equations, and now the Hamilton's equations as well, offer profound insights into classical mechanics, even though no new physics has been introduced. That is, the Lagrange equations, which are typically second-order

---

[4] Named for the French mathematician Joseph Liouville (1809–1882).

ordinary differential equations, are built upon Newton's laws of motion so do not contain new physics. And now the Hamiltonian equations, which are *first-order* differential equations, are similarly built upon Newton's laws, so do not contain any fundamentally different physics. However, each approach has given us fresh insights, while also adding to our arsenal of techniques for solving specific problems.

Problems

★   **Problem 11.1** Find the Legendre transform $B(x, z)$ of the function $A(x, y) = x^4 - (y + a)^4$, and verify that $-\partial A/\partial x = \partial B/\partial x$.

★   **Problem 11.2** In thermodynamics, the enthalpy H (no relation to the Hamiltonian H) is a function of the entropy S and pressure P such that $\partial H/\partial S = T$ and $\partial H/\partial P = V$, so that

$$dH = TdS + VdP,$$

where T is the temperature and V the volume. The enthalpy is particularly useful in isentropic and isobaric processes, because if the process is isentropic or isobaric, one of the two terms on the right vanishes. But suppose we wanted to deal with isothermal and isobaric processes, by constructing a function of T and P alone. Define such a function, in terms of H, T, and S, using a Legendre transformation. (The defined function G is called the Gibbs free energy.)

★   **Problem 11.3** In thermodynamics, for a system such as an enclosed gas, the internal energy $U(S, V)$ can be expressed in terms of the independent variables of entropy S and volume V, such that $dU = TdS - PdV$, where T is the temperature and P the pressure. Suppose we want to find a related function in which the volume is to be eliminated in favor of the pressure, using a Legendre transformation. (a) Which is the passive variable, and which are the active variables? (b) Find an expression for the new function in terms of U, P, and V. (The result is the enthalpy H or its negative, where the enthalpy H is unrelated to the Hamiltonian H.)

★★ **Problem 11.4** The energy of a relativistic free particle is the Hamiltonian $H = \sqrt{p^2c^2 + m^2c^4}$ in terms of the particle's momentum and mass. (a) Using one of Hamilton's equations in one dimension, find the particle's velocity v in terms of its momentum and mass. (b) Invert the result to find the momentum p in terms of the velocity and the mass. (c) Then find the free-particle Lagrangian for a relativistic particle using the Legendre transform

$$L(v) = pv - H.$$

(d) Beginning with the same Hamiltonian, generalize parts (a), (b), and (c) to a relativistic particle free to move in three dimensions.

★   **Problem 11.5** The Lagrangian for a particular system is

$$L = \dot{x}^2 + a\dot{y} + b\dot{x}\dot{z},$$

where a and b are constants. Find the Hamiltonian, identify any conserved quantities, and write out Hamilton's equations of motion for the system.

★   **Problem 11.6** A system with two degrees of freedom has the Lagrangian

$$L = \dot{q}_1^2 + \alpha\dot{q}_1\dot{q}_2 + \beta\dot{q}_2^2/2,$$

where α and β are constants. Find the Hamiltonian, identify any conserved quantities, and write out Hamilton's equations of motion.

★   **Problem 11.7** Write the Hamiltonian and find Hamilton's equations of motion for a simple pendulum of length ℓ and mass m. Sketch the constant-H contours in the θ, p_θ phase plane.

★★   **Problem 11.8** (a) Write the Hamiltonian for a spherical pendulum of length ℓ and mass m, using the polar angle θ and azimuthal angle φ as generalized coordinates. (b) Then write out Hamilton's equations of motion, and identify two first integrals of motion. (c) Find a first-order differential equation of motion involving θ alone and its first time derivative. (d) Sketch contours of constant H in the θ, p_θ phase plane, and use it to identify the types of motion one expects.

★   **Problem 11.9** A Hamiltonian with one degree of freedom has the form

$$H = \frac{p^2}{2m} + \frac{kq^2}{2} - 2aq^3 \sin \alpha t,$$

where m, k, a, and α are constants. Find the Lagrangian corresponding to this Hamiltonian. Write out both Hamilton's equations and Lagrange's equations, and show directly that they are equivalent.

★★   **Problem 11.10** A particle of mass m slides on the inside of a frictionless vertically oriented cone of semi-vertical angle α. (a) Find the Hamiltonian H of the particle, using generalized coordinates r, the distance of the particle from the vertex of the cone, and φ, the azimuthal angle. (b) Write down two first integrals of motion, and identify their physical meaning. (c) Show that a stable circular (constant-r) orbit is possible, and find its value of r for given angular momentum p_φ. (d) Find the frequency of small oscillations ω_{osc} about this circular motion, and compare it with the frequency of rotation ω_{circle}. (e) Is there a value of the tilt angle α for which the two frequencies are equal? What is the physical significance of the equality?

★   **Problem 11.11** A particle of mass m is attracted to the origin by a force of magnitude k/r^2. Using plane polar coordinates, find the Hamiltonian and Hamilton's equations of motion. Sketch constant-H contours in the (r, p_r) phase plane.

★★   **Problem 11.12** A double pendulum consists of two strings of equal length ℓ and two bobs of equal mass m. The upper string is attached to the ceiling, while the lower end is attached to the first bob. One end of the lower string is attached to the first bob, while the other end is attached to the second bob. Using generalized coordinates θ_1 (the angle of the upper string relative to the vertical) and θ_2 (the angle of the lower string relative to the vertical), find (a) the Lagrangian of the system (*Hint*: It can be tricky to find the kinetic energy of the lower bob in terms of the angles and their time derivatives. Use Cartesian coordinates initially; then convert these to generalized coordinates), (b) the canonical momenta, (c) the Hamiltonian in terms of the angles and their first derivatives. Are there any constants of the motion? If so, what are they, and why are they constants? (Note that to go on and find the motion of the system using Hamilton's equations, one must first write $H(\theta_1, \theta_2, p_{\theta_1}, p_{\theta_1})$, without $\dot{\theta}_1$ and $\dot{\theta}_2$. This step, and the next step of solving the equations, involves a lot of algebra. This illustrates the fact that in somewhat complicated problems one could long since have written out Lagrange's equations and solved them, by the time one has even written out the Hamiltonian in canonical form.)

Problem 11.13 A double Atwood's machine consists of two massless pulleys, each of radius R, some massless string, and three weights, with masses m_1, m_2, and m_3. The axis of pulley 1 is supported by a strut from the ceiling. A piece of string of length ℓ_1 is slung over the pulley, and one end of the string is attached to weight m_1 while the other end is attached to the axis of pulley 2. A second string of length ℓ_2 is slung over pulley 2; one end is attached to m_2 and the other to m_3. The strings are inextendible, but otherwise the weights and pulley 2 are free to move vertically. Let x be the distance of m_1 below the axis of pulley 1, and y be the distance of m_2 below the axis of pulley 2. (a) Find the Lagrangian $L(x, y, \dot{x}, \dot{y})$. (b) Find the canonical momenta p_x and p_y, in terms of $\dot{x}$ and $\dot{y}$. (c) Find the Hamiltonian of the system in terms of $x, y, \dot{x}$, and $\dot{y}$. (Note that to go on and find the motion of the system using Hamilton's equations, one must first write $H(x, y, p_x, p_y)$, without $\dot{x}$ and $\dot{y}$. This step, and the next step of solving the equations, involves a lot of algebra. This illustrates the fact that in somewhat complicated problems one could long since have written out Lagrange's equations and solved them, by the time one has even written out the Hamiltonian in canonical form.)

Problem 11.14 A massless unstretchable string is slung over a massless pulley. A weight of mass $2m$ is attached to one end of the string and a weight of mass m is attached to the other end. One end of a spring of force constant k is attached beneath m, and a second weight of mass m is hung on the spring. Using the distance x of the weight $2m$ beneath the pulley and the stretch y of the spring as generalized coordinates, find the Hamiltonian of the system. (a) Show that one of the two coordinates is ignorable (*i.e.*, cyclic). To what symmetry does this correspond? (b) If the system is released from rest with $y(0) = 0$, find $x(t)$ and $y(t)$.

Problem 11.15 (a) A particle is free to move only in the x direction, subject to the potential energy $U = U_0 e^{-\alpha x^2}$, where α and U_0 are positive constants. Sketch constant-Hamiltonian curves in a phase diagram, including values of H with $H <$

U_0, $H = U_0$, and $H > U_0$. (b) Repeat part (a) if $U_0 < 0$ and $\alpha > 0$, for values of H including those with $0 > H > U_0$, $H = 0$, and $H > 0$.

Problem 11.16 A cyclic coordinate q_k is a coordinate absent from the Lagrangian (even though $\dot{q}_k$ is present in L). (a) Show that a cyclic coordinate is likewise absent from the Hamiltonian. (b) Show from the Hamiltonian formalism that the momentum p_k canonical to a cyclic coordinate q_k is conserved, so $p_k = \alpha = $ constant. Therefore one can ignore both q_k and p_k in the Hamiltonian. This led E. J. Routh to suggest a procedure for dealing with problems having cyclic coordinates. He carried out a transformation from the $q, \dot{q}$ basis to the q, p basis only for the cyclic coordinates, finding their equations of motion in the Hamiltonian form, and then used Lagrange's equations for the noncyclic coordinates. Denote the cyclic coordinates by $q_{s+1}, \ldots, q_n$, then define the *Routhian* as

$$R(q_1, \ldots, q_n ; \dot{q}_1, \ldots, \dot{q}_s ; p_{s+1}, \ldots, p_n ; t) = \sum_{i=s+1}^{n} p_i \dot{q}_1 - L.$$

Show (using R rather than H) that one obtains Hamilton-type equations for the $n - s$ cyclic coordinates while (using R rather than L) one obtains Lagrange-type equations for the noncyclic coordinates. The Hamilton-type equations are trivial, showing that the momenta canonical to the cyclic coordinates are constants of the motion. In this procedure one can in effect "ignore" the cyclic coordinates, so "cyclic" coordinates are also "ignorable" coordinates.

Problem 11.17 Show that the Poisson bracket of two constants of the motion is itself a constant of the motion, even when the constants depend explicitly on time.

Problem 11.18 Prove the anticommutativity and distributivity of Poisson brackets by showing that (a) $\{A, B\}_{q,p} = -\{B, A\}_{q,p}$ and (b) $\{A, B + C\}_{q,p} = \{A, B\}_{q,p} + \{A, C\}_{q,p}$.

Problem 11.19 Show that Hamilton's equations of motion can be written in terms of Poisson brackets as

$$\dot{q} = \{q, H\}_{q,p}, \quad \dot{p} = \{p, H\}_{q,p}.$$

Problem 11.20 A Hamiltonian has the form

$$H = q_1 p_1 - q_2 p_2 + a q_1^2 - b q_2^2,$$

where a and b are constants. (a) Using the method of Poisson brackets, show that

$$f_1 \equiv q_1 q_2 \quad \text{and} \quad f_2 \equiv \frac{1}{q_1}(p_2 + b q_2)$$

are constants of the motion. (b) Then show that $\{f_1, f_2\}$ is also a constant of the motion. (c) Is H itself constant? Check by finding q_1, q_2, p_1, p_2 as explicit functions of time.

★★★   **Problem 11.21** Show, using the Poisson bracket formalism, that the *Laplace–Runge–Lenz vector*

$$\mathbf{A} \equiv \mathbf{p} \times \mathbf{L} - \frac{mk\mathbf{r}}{r}$$

is a constant of the motion for the Kepler problem of a particle moving in the central inverse-square force field $F = -k/r^2$. Here $\mathbf{p}$ is the particle's momentum, and $\mathbf{L}$ is its angular momentum. *Hint*: Write $L_k = \varepsilon_{kln}x_l p_n$ and you might need to use the identity $\varepsilon_{ijk}\varepsilon_{ilm} = \delta_{jl}\delta_{km} - \delta_{jm}\delta_{kl}$.

★★   **Problem 11.22** A beam of protons with a circular cross-section of radius r_0 moves within a linear accelerator oriented in the x direction. Suppose that the transverse momentum components (p_y, p_z) of the beam are distributed uniformly in momentum space, in a circle of radius p_0. If a magnetic lens system at the end of the accelerator focuses the beam into a small circular spot of radius r_1, find, using Liouville's theorem, the corresponding distribution of the beam in momentum space. Here what may be a desirable focusing of the beam in position space has the often unfortunate consequence of broadening the momentum distribution.

★★   **Problem 11.23** A large number of particles, each of mass m, move in response to a uniform gravitational field g in the negative z direction. At time $t = 0$, they are all located within the corners of a rectangle in (z, p_z) phase space, whose positions are: (1) $z = z_0, p_z = p_0$; (2) $z = z_0 + \Delta z, p_z = p_0$; (3) $z = z_0, p_z = p_0 + \Delta p$; and (4) $z = z_0 + \Delta z, p_z = p_0 + \Delta p$. By direct computation, find the area in phase space enclosed by these particles at times (a) $t = 0, (b)\ t = m\Delta z/p_0$, and (c) $t = 2m\Delta z/p_0$. Also show the shape of the region in phase space for cases (b) and (c).

★★   **Problem 11.24** In an electron microscope, electrons scattered from an object of height z_0 are focused by a lens at distance D_0 from the object and form an image of height z_1 at a distance D_1 behind the lens. The aperture of the lens is A. Show by direct calculation that the area in the (z, p_z) phase plane occupied by electrons leaving the object (and destined to pass through the lens) is the same as the phase area occupied by electrons arriving at the image. Assume that $z_0 \ll D_0$ and $z_1 \ll D_1$ (from *Mechanics*, 3rd edn, by Keith R. Symon).

★   **Problem 11.25** Show directly that the transformation

$$Q = \ln\left(\frac{1}{q}\sin p\right), \quad P = q\cot p$$

is canonical.

★★   **Problem 11.26** Show that if the Hamiltonian and some quantity Q are both constants of the motion, then the nth partial derivative of Q with respect to time must also be a constant of the motion.

★　　　**Problem 11.27**　Prove the Jacobi identity for Poisson brackets

$$\left\{A,\{B,C\}_{q,p}\right\}_{q,p} + \left\{B,\{C,A\}_{q,p}\right\}_{q,p} + \left\{C,\{A,B\}_{q,p}\right\}_{q,p} = 0.$$

★　　　**Problem 11.28**　(a) Find the Hamiltonian for a projectile of mass m moving in a uniform gravitational field g, using coordinates x, y. (b) Then find Hamilton's equations of motion and solve them.

★★　　**Problem 11.29**　(a) Find the Hamiltonian for a projectile of mass m moving in a force field with potential energy $U(\rho, \varphi, z)$, where ρ, φ, z are cylindrical coordinates. (b) Find Hamilton's equations of motion. (c) Solve them as far as possible if $U = U(\rho)$ alone.

★　　　**Problem 11.30**　Consider a particle of mass m with relativistic Hamiltonian $H = \sqrt{p^2 c^2 + m^2 c^4} + U(x, y, z)$, where U is its relativistic potential energy. Find the particle's equations of motion.

★　　　**Problem 11.31**　We found Hamilton's equations by starting with the Lagrangian $L(q_i, \dot{q}_i, t)$ and using a Legendre transformation to define the Hamiltonian $H(q_i, p_i, t)$. Now starting with the Hamiltonian and Hamilton's equations, use a reverse Legendre transformation to define L, and show that one obtains the Lagrange equations.

★　　　**Problem 11.32**　Suppose that for some situations the coordinates p, q are canonical. Show that the transformed coordinates $P = \frac{1}{2}(p^2 + q^2), Q = \tan^{-1}(q/p)$ are also canonical.

★★　　**Problem 11.33**　Prove that if one makes two successive canonical transformations, the result is also canonical.

★　　　**Problem 11.34**　Prove that the Poisson bracket is invariant under a canonical transformation.

★★　　**Problem 11.35**　A plane pendulum consists of a rod of length R and negligible mass supporting a plumb bob of mass m that swings back and forth in a uniform gravitational field g. The point of support at the top end of the rod is forced to oscillate vertically up and down with $y = A \cos \omega t$. Using the angle θ of the rod from the vertical as the coordinate, (a) find the Lagrangian of the bob. (b) Find the Hamiltonian H. Is $H = E$, the energy? Is either one or both conserved? (c) Write out Hamilton's equations of motion.

★★　　**Problem 11.36**　A plane pendulum consists of a string supporting a plumb bob of mass m free to swing in a vertical plane and free to swing subject to uniform gravity g. The upper end of the string is threaded through a hole in the ceiling and steadily pulled upward, so the length of the string beneath the point in the ceiling is $\ell(t) = \ell_0 - \alpha t$, where α is a positive constant. (a) Find the Lagrangian of the plumb bob. (b) Find its Hamiltonian H. Is $H = E$, the energy of the bob? (c) Write out Hamilton's equations of motion. (d) Solve them assuming $l(t)$ is changing slowly and the angle of the pendulum remains small.

★★ **Problem 11.37** At time $t = 0$ a large number of particles, each of mass m, is strung out along the x axis from $x = 0$ to $x = \Delta x$, with momenta p_x varying from $p = p_0$ to $p = p_0 + \Delta p$. No forces act on the particles and they do not collide. (a) Show that the points representing these particles fill a rectangle in the x, p_x phase plane, and sketch it – identifying the four points at the corners of the rectangle with their positions and momenta. (b) Sketch the locations of the same particles in the phase plane some time t_1 later, where $t_1 > mx_0/p_0$. (c) What then is the shape of the area on the phase plane occupied by all of these particles? (d) Prove that the area of the occupied region at t_1 is the same as it was at $t = 0$. (Note that if the number of points and the area are both unchanged, then the average density of points is also unchanged, in accord with the Liouville theorem.)

★★★ **Problem 11.38** Any spherically symmetric function of the canonical coordinate and momentum of a particle can depend only on r^2, p^2, and $\mathbf{r} \cdot \mathbf{p}$. Show that the Poisson bracket of any such function f with a component of the particle's angular momentum is zero. In particular, show that $\{L_z, f\} = 0$, where $L_z = (\mathbf{r} \times \mathbf{p})_z$.

★★ **Problem 11.39** Write the Hamiltonian of a free particle of mass m in a reference frame that is rotating uniformly with angular velocity $\boldsymbol{\omega}$ with respect to an inertial frame.

12

Rigid-Body Dynamics

Watching a shoe tumble erratically as it flies through mid-air may be entertaining, but – to anyone without a background in rigid-body dynamics – it can look quite troubling. There is no net torque acting on the shoe, yet the rotational motion looks and is rather complicated. However, with the powerful tools provided by the Lagrangian formalism we are well equipped to tackle this subject, and go beyond it to more complicated examples of rotational motion. We start with a definition of a rigid body, and then proceed to introduce the Euler angles that can be used to describe the orientation of an object in three-dimensional space. With this scaffolding established, we can go on to describe torque-free dynamics, and then full rotational evolution with nonzero torque. For simplicity, throughout this chapter we restrict our discussion to nonrelativistic dynamics.

A **rigid body** is an object in which the distance between any two of its constituent bits remains fixed. No bodies can be completely rigid in Nature, however, because that would require instantaneous communication between all of their parts. If a ball hits one end of a bat, the other end of the bat cannot react immediately. In fact, we even know from special relativity that what might be "immediate" in one frame of reference cannot be "immediate" in others, so perfectly rigid bodies are impossible. Nevertheless, the concept of rigid bodies can greatly simplify our description of how objects behave. A metal rod, a glass hoop, a frisbee, all qualify as rigid bodies as long as we can ignore their tendency to bend slightly or even break apart. Up to now, we have represented extended objects as point bits located at their center of mass. In this chapter we will probe more deeply and describe the dynamics of the orientation of the object in three-dimensional space as it rotates, tumbles, or precesses.

12.1 Rotation About a Fixed Axis

We begin for simplicity with the case of rigid-body rotation about a fixed axis, such as the rotation of a bicycle wheel about its axle with the bicycle held in place (Figure 12.1). If the wheel is spinning about such a fixed axis with angular velocity ω, then the velocity of any small bit of the wheel at some instant is $\mathbf{v} = \omega \times \mathbf{r}$, where $\mathbf{r}$ is the vector extending from the axis to the bit. The angular momentum of the individual bit of mass m, as defined in Chapter 1, is therefore

$$\boldsymbol{\ell} \equiv \mathbf{r} \times \mathbf{p} = \mathbf{r} \times m\mathbf{v} = m\mathbf{r} \times (\omega \times \mathbf{r}) = m[\omega(\mathbf{r} \cdot \mathbf{r}) - \mathbf{r}(\mathbf{r} \cdot \omega)], \qquad (12.1)$$

where in the last step we used the vector identity $\mathbf{A} \times (\mathbf{B} \times \mathbf{C}) = \mathbf{B}(\mathbf{A} \cdot \mathbf{C}) - \mathbf{C}(\mathbf{A} \cdot \mathbf{B})$. Now suppose we orient the axis of rotation along the z direction so that $\boldsymbol{\omega} = \omega\hat{\mathbf{z}}$ and sum over all bits in the rotating body. Then the total angular momentum of the body is

$$
\begin{aligned}
\mathbf{L} \equiv \sum_i \boldsymbol{\ell}_i &= \sum_i m_i[(\omega\hat{\mathbf{z}})(x_i^2 + y_i^2 + z_i^2) - \mathbf{r_i}(z_i\omega)] \\
&= \sum_i m_i\omega[(x_i^2 + y_i^2 + z_i^2)\hat{\mathbf{z}} - (x_i\hat{\mathbf{x}} + y_i\hat{\mathbf{y}} + z_i\hat{\mathbf{z}})(z_i)] \\
&= \sum_i m_i(x_i^2 + y_i^2)\omega\,\hat{\mathbf{z}} + \sum_i m_i(-x_iz_i)\omega\,\hat{\mathbf{x}} + \sum_i m_i(-y_iz_i)\omega\,\hat{\mathbf{y}},
\end{aligned}
$$

$$(12.2)$$

where we label each bit by a discrete index i and sum over all bits.

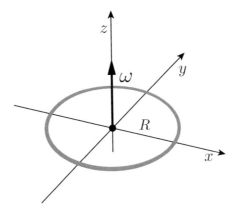

Fig. 12.1   Fixed-axis rotation of a bicycle wheel.

Notice that even though the angular velocity $\boldsymbol{\omega}$ is in the z direction, the angular momentum of the body generally has components in all three directions. In other words, *the angular momentum is not necessarily parallel to the angular velocity.* In this case, however, the components of $\mathbf{L}$ in the $\hat{\mathbf{x}}$ and $\hat{\mathbf{y}}$ directions *are indeed* zero and the angular velocity is parallel to the angular momentum. This is because we chose our coordinate system in a manner that makes the symmetry of the wheel manifest. In particular, notice that the shape of the wheel remains unchanged if we were to reflect in the x or y direction, $x \to -x$ or $y \to -y$. This implies that, if we take a bit of mass located at a certain value of x and z in our coordinate system, it can be paired up with an equal bit of mass located at $-x$ and z. Such pairs will count the same in the sum above except for a minus sign. In other words, these two bits will cancel each other out in the sum

$$
\sum_i m_i(-x_iz_i)\omega = \sum_i m_i(-y_iz_i)\omega = 0. \tag{12.3}
$$

So, given the setup, the angular momentum is parallel to the angular velocity, and is given simply by

$$\mathbf{L} = I\boldsymbol{\omega}, \tag{12.4}$$

where the *moment of inertia* of the object is defined as

$$I \equiv \sum_i m_i(x_i^2 + y_i^2) \equiv \sum_i m_i(r_\perp^2)_i, \tag{12.5}$$

where $r_\perp^2$ is the square of the distance of a bit from the axis of rotation. If the object is an infinitely thin wheel whose entire mass is located at the rim $r = R$, the moment of inertia becomes $I_{\text{ring}} = MR^2$, where M is its mass and R its radius. If a portion of the mass is located at smaller radii, however, as it is for a real wheel, the moment of inertia will be less than MR^2.

We can also find the kinetic energy T of a rigid body rotating about the fixed axis. The speed of the ith bit as it circles the axis of rotation is $v_i = (r_\perp)_i \omega$, so its kinetic energy of rotation is $T_i = (1/2)m_i v_i^2 = (1/2)m_i(r_\perp)_i^2\omega^2$. The total kinetic energy of the rigid body is therefore given by the familiar result

$$T = \sum_i T_i = \frac{1}{2}\sum_i m_i(r_\perp)_i^2\omega^2 = \frac{1}{2}I\omega^2. \tag{12.6}$$

Finally, it is also straightforward to show that the time rate of change of the angular momentum is caused by the net torque $\mathbf{N}$ on the object, as given by the equation $\mathbf{N} = d\mathbf{L}/dt = I d\omega/dt$. So overall, for objects that are symmetric about a fixed axis of rotation, we find the familiar results $\mathbf{L} = I\boldsymbol{\omega}$ for the angular momentum, $T = \frac{1}{2}I\omega^2$ for the kinetic energy, and $\mathbf{N} = d\mathbf{L}/dt = I d\omega/dt$ for the net torque on the object.

12.2 Euler's Theorem

So far we have described only the very simple case of rotation of a rigid body about a fixed axis. We will now consider rigid-body rotation in any circumstance whatever. Figure 12.2 shows an arbitrarily shaped rigid body. The first task is to describe the body's position and orientation in space. Pick any point within it and imagine that we tag it with a colored dot; the choice of this colored dot within the body of the object is completely arbitrary. As the object tumbles around and moves from place to place, we can describe the trajectory of the colored dot as a point particle. But that is not enough to describe the state of the rigid body. In addition, we need to describe its orientation about the colored dot. Hence, if we provide, for any instant in time, the location of the colored dot *and* the orientation of the object about the colored dot, we can reconstruct the state of the object at that instant. We may then say that the state of a rigid body can be described by a combination of a translation and a rotation: a translation of a fixed point together with a rotation of the object about this fixed point. Hence, we define the term **tagged point**: a fixed point in the rigid body of our own choosing with respect to which the motion is decomposed into translational and rotational parts.

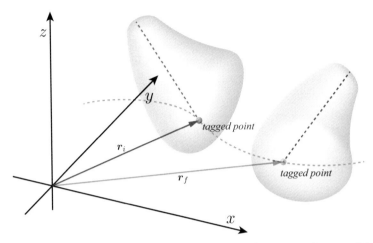

Fig. 12.2 The evolution of a rigid body tracked by the trajectory traced by an arbitrary tagged point and the changing orientation of the rigid body about this point.

That such a decomposition is always possible is the statement of a theorem by the Swiss mathematician Leonard Euler: *All spatial transformations that leave distances unchanged must be a combination of a translation and a rotation.*

While any point within a rigid body is satisfactory as a tagged point, there are two especially *convenient* choices. If the rigid body has a fixed pivot – that is, if it is like a pendulum swinging from a pivot – then a judicious choice for the tagged point is the pivot point itself, as shown in Figure 12.3(a).

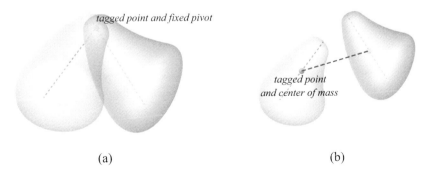

(a)   (b)

Fig. 12.3 (a) A natural choice for a tagged point for a pivoted rigid body is the pivot point itself; this is particularly useful when the pivot is at rest in an inertial reference frame. (b) When no appropriate pivot exists, the center of mass is a natural choice for the tagged point.

In contrast, if the object does not have an obvious pivot, the best choice for a tagged point is the center of mass (CM) (see Figure 12.3(b))

$$\mathbf{R}_{cm} = \sum_i \frac{\Delta m_i \mathbf{r}_i}{M} = \frac{1}{M} \int \mathbf{r} \, dm = \frac{1}{M} \int \rho \mathbf{r} \, dV, \qquad (12.7)$$

where the vector $\mathbf{r}$ is the distance of a point of mass dm from the CM. That is, we have first divided the rigid body into small bits labeled by the index i located at positions $\mathbf{r}_i$ and the total mass is $M = \sum_i \Delta m_i$. The sum is over all bits of the rigid body, which in the continuum limit we can write as an integral of bits of mass $dm = \rho \, dV$, where ρ is the volume mass density of the object. Note that the rigid body might not have constant density; indeed, it realistically has a variable density $\rho(\mathbf{r})$.

Now, how do we describe the orientation of the object about the tagged point?

12.3 Rotation Matrices and the Body Frame

To describe the orientation of a rigid body, we will need to set up two different coordinate systems. Figure 12.4 shows the **lab frame** coordinates – labeled x, y, z – which typically is an inertial frame from which we watch a rigid body tumble around and fly through space. The prime coordinate system – labeled x', y', z' – is conventionally chosen to be fixed in the rigid body and is called the **body frame**. That is, when the body tumbles around, so do the x', y', and z' axes; if we were sitting in the body frame and used the x', y', and z' axes to describe our observations, we would see the rigid body motionless while the rest of the world tumbles around us.

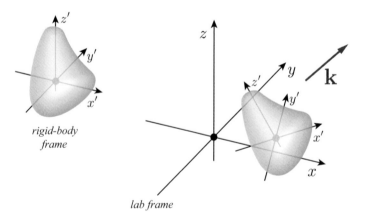

rigid-body
frame

lab frame

Fig. 12.4 The lab and body frame setups used to describe the orientation of a rigid body in three-dimensional space. Also shown is a *fixed* vector **k** which will have different components (k^x, k^y, k^z) and $(k^{x'}, k^{y'}, k^{z'})$ in the lab and body frame coordinate systems respectively – reflecting the orientation difference between the two frames.

We henceforth will always label the body frame with primed coordinates, while the lab frame will be labeled with unprimed coordinates. Note also that a vector, like a position or velocity vector, will generally have very different components when written in the lab versus body frame coordinates. For example, if we were to use a position vector to pinpoint a fixed location in the rigid body relative to its center of mass, this vector would tumble around with the body as seen from the lab frame. So, its components in the lab frame would be time dependent. But this same vector would appear fixed from the perspective of the body frame; hence, its components with respect to the primed coordinate system would be time independent.

To describe the orientation of a rigid body, we need to describe the orientation of its body frame relative to the lab frame. This will involve rotations, and we need *three* angles to describe a general rotation in three dimensions. Think of the orientation of an airplane as shown in Figure 12.5(a), often characterized by its *pitch, roll,* and *yaw*. We see that three independent numbers are needed to fully describe the plane's orientation, or that of any other rigid body. Another way to see this is to realize that the orientation of an object can be fully described by specifying the direction of its body frame z' axis, and a single angle of rotation about this z' axis (see Figure 12.5(b)). Identifying the aim of the z' axis requires two numbers – say the two angles in spherical coordinates that define a ray parallel to it, plus the single angle of rotation about the axis. Again we have three angles in total.

Consider a rigid body oriented in a certain arbitrary way in space at a snapshot in time, as shown in Figure 12.4. One way we can quantify the rigid body's orientation is by relating the components of some *fixed* vector between the lab and body coordinate systems. Let the rotation be described by a 3×3 matrix $\hat{\mathcal{R}}$, then we write the components of a fixed vector $\mathbf{k}$ as

$$\mathrm{k}^{i'} = \hat{\mathcal{R}}^{i'}_{\ j}\mathrm{k}^j. \tag{12.8}$$

Here, k^j are the components of the vector $\mathbf{k} = (\mathrm{k}^x, \mathrm{k}^y, \mathrm{k}^z)$ in the lab coordinate system, and $\mathrm{k}^{i'} = (\mathrm{k}^{z'}, \mathrm{k}^{y'}, \mathrm{k}^{z'})$ are the components of the *same* vector in the rotated body coordinate system – the *primed* coordinates. We are also using the Einstein summation convention and are therefore summing over the j index. Note that the vector $\mathbf{k}$ is a fixed vector; however, it has two separate sets of components that are different in general, one in the lab frame – denoted by unprimed labels $(\mathrm{k}^x, \mathrm{k}^y, \mathrm{k}^z)$, and one in the body frame – denoted by primed labels $(\mathrm{k}^{z'}, \mathrm{k}^{y'}, \mathrm{k}^{z'})$. Knowing the relation between these components through Eq. (12.8) can be used to tell us how the body coordinate system is oriented relative to the lab frame, and thus we learn about the orientation of the rigid body. For example, consider the rotation matrix

$$\hat{\mathcal{R}}^{i'}_{\ j} \rightarrow \hat{\mathcal{R}}_z = \begin{pmatrix} \cos\alpha_z & \sin\alpha_z & 0 \\ -\sin\alpha_z & \cos\alpha_z & 0 \\ 0 & 0 & 1 \end{pmatrix}, \tag{12.9}$$

with j representing a column index and i' representing a row index. If this rotation is to relate the components of a fixed vector between the lab frame and the body frame of a rigid body, it implies that the body frame, and correspondingly the rigid body, is to be viewed as rotated by an angle α_z about the lab frame's z axis.

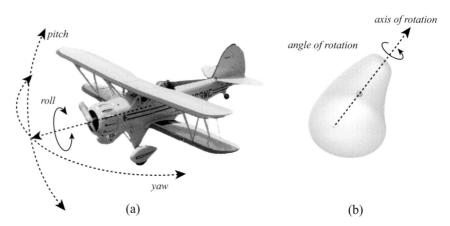

(a)   (b)

Fig. 12.5 (a) The three orientational degrees of freedom of an airplane. (b) An alternative quantification of orientation through the specification of body frame axis and an angle of rotation about this axis.

More generally, the *defining property* of a rotation is that it does not change the length of vectors. That is, we need to ensure that

$$\mathrm{k}^{i'}\mathrm{k}^{i'} = \mathrm{k}^{i}\mathrm{k}^{i}, \tag{12.10}$$

which implies that

$$\mathrm{k}^{i'}\mathrm{k}^{i'} = \hat{\mathcal{R}}^{i'}_{\ j}\mathrm{k}^{j}\hat{\mathcal{R}}^{i'}_{\ k}\mathrm{k}^{k} = \mathrm{k}^{j}\hat{\mathcal{R}}^{i'}_{\ j}\hat{\mathcal{R}}^{i'}_{\ k}\mathrm{k}^{k}, \tag{12.11}$$

where we have rearranged the terms since they represent *components* of matrices and vectors, *i.e.*, numbers, that are commutative under multiplication. We must then have

$$\hat{\mathcal{R}}^{i'}_{\ j}\hat{\mathcal{R}}^{i'}_{\ k} = \delta_{jk}, \tag{12.12}$$

where δ_{jk} is Kronecker's delta. Note however that the first factor corresponds to the components of the *transpose* of the $\hat{\mathcal{R}}$ matrix

$$\hat{\mathcal{R}}^{i'}_{\ j} = (\hat{\mathcal{R}}^{T})_{j}^{\ i'}, \tag{12.13}$$

where T stands for matrix transposition, where column and row indices are exchanged on the right-hand side. We then have

$$\hat{\mathcal{R}}^{i'}_{\ j}\hat{\mathcal{R}}^{i'}_{\ k} = (\hat{\mathcal{R}}^{T})_{j}^{\ i'}\hat{\mathcal{R}}^{i'}_{\ k} = (\hat{\mathcal{R}}^{T}\cdot\hat{\mathcal{R}})_{jk} = \delta_{jk}, \tag{12.14}$$

noting that the sum over i' amounts to matrix multiplication. We then need

$$\hat{\mathcal{R}}^{T}\cdot\hat{\mathcal{R}} = \mathbf{1}\,; \tag{12.15}$$

that is, the matrix $\hat{\mathcal{R}}$ must be both *orthogonal* and *normalized*, *i.e.*, *orthonormal*. We note however that $\hat{\mathcal{R}}$ may also implement a reflection of the coordinates while preserving distances and the statement of orthogonality. A reflection can be written in matrix form as

$$\hat{\mathcal{R}} = \begin{pmatrix} -1 & 0 & 0 \\ 0 & -1 & 0 \\ 0 & 0 & -1 \end{pmatrix}. \tag{12.16}$$

So we need to separate reflections from rotations. To do this, we define a rotation matrix as one satisfying *both* the conditions

$$\hat{\mathcal{R}}^T \cdot \hat{\mathcal{R}} = \mathbf{1} \quad \text{and} \quad \text{Det } \hat{\mathcal{R}} = +1. \tag{12.17}$$

The condition on the determinant rules out reflections. Equation (12.17) *defines* a rotation in general.

We can then ask the mathematical question: What are all 3×3 matrices $\hat{\mathcal{R}}$ that satisfy the conditions (12.17)? This exercise leads to the following conclusion: any such matrix is parameterized by three independent angular parameters. For example, a rotation of the coordinates by an angle α_z about the z axis is given by Eq. (12.9). Similarly, rotations about the x and y axes are

$$\hat{\mathcal{R}}_x = \begin{pmatrix} 1 & 0 & 0 \\ 0 & \cos\alpha_x & \sin\alpha_x \\ 0 & -\sin\alpha_x & \cos\alpha_x \end{pmatrix}, \quad \hat{\mathcal{R}}_y = \begin{pmatrix} \cos\alpha_y & 0 & -\sin\alpha_y \\ 0 & 1 & 0 \\ \sin\alpha_y & 0 & \cos\alpha_y \end{pmatrix},$$
$$\tag{12.18}$$

with angles α_x about the x axis and angle α_y about the y axis, respectively. An arbitrary rotation is a product of three such matrices with three independent parameters α_x, α_y, and α_z. As required, a product of orthogonal matrices is orthogonal, and the determinant of a product of matrices with unit determinants has unit determinant. It can be shown that any rotation matrix can be written in this product form. Note that rotation matrices are in general noncommuting; that is, for two rotations $\hat{\mathcal{R}}_1$ and $\hat{\mathcal{R}}_2$, we have

$$\hat{\mathcal{R}}_1 \cdot \hat{\mathcal{R}}_2 \neq \hat{\mathcal{R}}_2 \cdot \hat{\mathcal{R}}_1. \tag{12.19}$$

It is easy to show, for example, that the order of rotations makes a difference when twisting a book successively about two perpendicular axes.

Example 12.1   **Rotations in Higher Dimensions**

Consider D-dimensional space, where $D \geq 2$. What would a rotation be in such a space of arbitrary dimensions? The defining property for rotations still comes from requiring linear transformations that preserve distances. Therefore, we can still use Eq. (12.17) as a condition on a $D \times D$ rotation matrix $\hat{\mathcal{R}}$. In the commonplace world of three dimensions, we could have three rotations: one about each of the x, y, and z axes. But looking at the form of the rotation matrix, we see that the orthogonality condition is satisfied by a cosine–sine mixing of two directions at a time. For example, for rotation about the z axis, we mix the x and y axes using the familiar trigonometric functions, and orthogonality follows from $\cos^2 + \sin^2 = 1$. In higher dimensions, we can still satisfy the orthogonality condition by mixing any two of the D directions of the coordinates with the same cosine and sine pattern. That is, a general rotation in D dimensions is a product of a number of rotations, each mixing two of the D-space dimensions. The number of different rotations in D dimensions is then simply the number of independent ways we can pair up D axes. We then have

$$\text{Number of independent angles in } D \text{ dimensions} = \frac{D(D-1)}{2}. \tag{12.20}$$

For example, for $D = 2$ we have only a single angle, as we know. For $D = 3$ we have three angles. But for $D = 9$, we would have $9 \times 8/2 = 36$ different independent rotations – and not nine as we might have guessed! ∎

We now have a plan for quantifying the orientation of a rigid body. We will use rotation matrices that relate the components of fixed vectors between the lab and the body frames. But first we must settle on a convention that fixes the particular sequence of rotation matrices that we want to use.

12.4 The Euler Angles

An arbitrary orientation of a rigid body is represented by a 3×3 rotation matrix that tells us how a fixed vector's components in the lab frame are related to the same vector's components in the body frame. As discussed in the previous section, such a rotation matrix is decomposable into a product of three independent rotations. Therefore the orientation state of the object is described by three angular parameters. This decomposition is not unique. For example, we might write

$$\hat{\mathcal{R}}(\alpha_x, \alpha_y, \alpha_z) = \hat{\mathcal{R}}_x(\alpha_x) \cdot \hat{\mathcal{R}}_y(\alpha_y) \cdot \hat{\mathcal{R}}_z(\alpha_z) \tag{12.21}$$

or even

$$\hat{\mathcal{R}}(\alpha'_x, \alpha'_y, \alpha'_z) = \hat{\mathcal{R}}_y(\alpha'_y) \cdot \hat{\mathcal{R}}_x(\alpha'_x) \cdot \hat{\mathcal{R}}_z(\alpha'_z), \tag{12.22}$$

with a different set of angles α'_x, α'_y, and α'_z – yet yielding the same rotation. So we need to establish a convention. If your name happens to be Euler, whose name comes up frequently in this chapter,[1] then the convention you introduce has a greater chance of permanence. In fact, Euler made a choice that has stayed with

---

[1] Leonhard Euler was easily the most prolific mathematician of all time. He was known by contemporaries as "Analysis Incarnate." A later biographer wrote that "Euler calculated without apparent effort, as men breathe, or as eagles sustain themselves in the wind." Toward the end of his life he became totally blind, but continued to dictate mathematical papers to his wife while his grandchildren played on his knee.

us ever since. So along with everyone else, we adhere to Euler's convention and the corresponding set of angles known as the **Euler angles**.

According to Euler, a general rotation is defined through three rotations

$$\hat{\mathcal{R}}(\varphi, \theta, \psi) = \hat{\mathcal{R}}_3(\psi) \cdot \hat{\mathcal{R}}_2(\theta) \cdot \hat{\mathcal{R}}_1(\varphi) \tag{12.23}$$

where we define, in order:

- First, rotate about the z axis by φ:

$$\hat{\mathcal{R}}_1(\varphi) = \begin{pmatrix} \cos\varphi & \sin\varphi & 0 \\ -\sin\varphi & \cos\varphi & 0 \\ 0 & 0 & 1 \end{pmatrix}. \tag{12.24}$$

- Second, rotate about the *new x* axis by θ:

$$\hat{\mathcal{R}}_2(\theta) = \begin{pmatrix} 1 & 0 & 0 \\ 0 & \cos\theta & \sin\theta \\ 0 & -\sin\theta & \cos\theta \end{pmatrix}. \tag{12.25}$$

- Third, rotate about the *newest z* axis by ψ:

$$\hat{\mathcal{R}}_3(\psi) = \begin{pmatrix} \cos\psi & \sin\psi & 0 \\ -\sin\psi & \cos\psi & 0 \\ 0 & 0 & 1 \end{pmatrix}. \tag{12.26}$$

Figure 12.6 depicts three angles φ, θ, and ψ, known as the *Euler angles*. The coordinate axes of the body frame are *fixed* within the rigid body and are labeled by primed coordinates x', y', and z'; the orientation of these axes can then be described with respect to the lab frame coordinates x, y, and z through the three Euler angles. As the figure illustrates, this is a three-step prescription that takes us from x–y–z to x'–y'–z'. The corresponding rotation matrix (12.23) relates the components of any vector between the lab and body reference frames. The lab frame is also sometimes referred to as the "lab axes" or the "space axes."

Let us then put this machinery to use. Let $\mathbf{R}$ be the position vector of the tagged point in the rigid body measured from the origin of the lab frame, as shown in Figure 12.7.

If we pick any other point a in the rigid body we may denote its position vector by $\mathbf{r}_a$, measured from the origin of the lab frame. As the body moves and tumbles around, both $\mathbf{R}$ and $\mathbf{r}_a$ will in general change in both direction and magnitude. Now consider the position of this second point with respect to the tagged point: we call this vector $\boldsymbol{r}_a$ pointing from the tagged point. We have

$$\mathbf{r}_a = \mathbf{R} + \boldsymbol{r}_a. \tag{12.27}$$

Now an interesting observation: the components of $\boldsymbol{r}_a$ in the *body frame*, denoted as $(r_a^{x'}, r_a^{y'}, r_a^{z'})$, are guaranteed to be time independent since $\boldsymbol{r}_a$ is fixed with respect to the body frame. This same vector in the *lab frame*, now with components denoted as (r_a^x, r_a^y, r_a^z), is in contrast generally time dependent, tracking how the

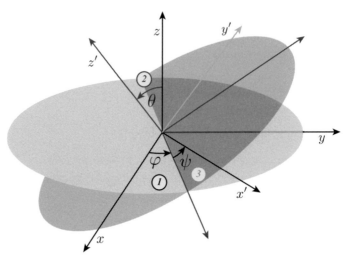

Fig. 12.6 The Euler angles describing the orientation of a rigid body.

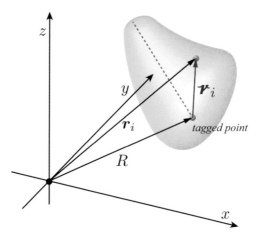

Fig. 12.7 The decomposition of the position vector of a bit of the rigid body in terms of the position of the tagged point **R** and the position of the bit with respect to the tagged point $\boldsymbol{r}_a{}'$.

body tumbles around in space. Using the Euler angles, we can write a relation between the components of $\boldsymbol{r}_a$ as

$$
\begin{pmatrix} r_a^{x'} \\ r_a^{y'} \\ r_a^{z'} \end{pmatrix} = \hat{\mathcal{R}}(\varphi, \theta, \psi) \cdot \begin{pmatrix} r_a^{x} \\ r_a^{y} \\ r_a^{z} \end{pmatrix}
$$

$$
\Rightarrow \begin{pmatrix} r_a^{x} \\ r_a^{y} \\ r_a^{z} \end{pmatrix} = \left[\hat{\mathcal{R}}(\varphi, \theta, \psi) \right]^{T} \cdot \begin{pmatrix} r_a^{x'} \\ r_a^{y'} \\ r_a^{z'} \end{pmatrix}, \qquad (12.28)
$$

where we transpose the rotation matrix to invert it, given that it is an orthogonal matrix. Note that

$$\left[\hat{\mathscr{R}}(\varphi,\theta,\psi)\right]^T = \left[\hat{\mathscr{R}}_1(\varphi)\right]^T . \left[\hat{\mathscr{R}}_2(\theta)\right]^T . \left[\hat{\mathscr{R}}_3(\psi)\right]^T$$

$$= \hat{\mathscr{R}}_1(-\varphi).\hat{\mathscr{R}}_2(-\theta).\hat{\mathscr{R}}_3(-\psi). \tag{12.29}$$

Given that the $r_a^{i'}$s are always constant in time, we see that the time evolution of the r_a^i – which describes how the bits in the rigid body are moving in the lab frame and hence how the rigid body is tumbling around – must be compensated by the time evolution of the Euler angles φ, θ, and ψ. Hence, knowing $\varphi(t)$, $\theta(t)$, and $\psi(t)$ can be used to deduce the exact orientation of the rigid body in three-dimensional space as a function of time. Our ultimate goal is then to find equations of motion that can be solved for these Euler angles as a function of time.

The usefulness of having vectors that are fixed in the body frame leads us to introduce a notational convenience. *A scripted font ($\boldsymbol{r}$ versus $\mathbf{r}$) will always signify that the vector is fixed with respect to the rigid body*: as the rigid body moves and tumbles around, $\boldsymbol{r}$ remains unchanged as seen from the perspective of an observer at rest and tumbling with the rigid body. Hence, whenever we have components of scripted vectors that are primed, as in $r^{i'}$, we may assume they are constant in time.

Equations (12.28) are central relations for quantifying the orientation of a rigid body. If we learn how the Euler angles are evolving in time, we can construct the time-dependent rotation matrix $\hat{\mathscr{R}}(\varphi,\theta,\psi)$; then, using any fixed vector in the body, we can deduce from Eq. (12.28) how the components of this vector evolve in the lab frame – thus tracking the orientation of the rigid body from the lab's perspective.[2]

12.5 Infinitesimal Rotations

Now that we know how to quantify the time-evolving orientation of a rigid body, we next consider how to quantify the rate at which a rigid body might be spinning about an arbitrary axis. We start by decomposing the rigid body's motion into a sum of a translation and a rotation. Let the position of the tagged point be denoted by $\mathbf{R}$. In a small time increment Δt, the tagged point translates by a small amount $\Delta\mathbf{R}$. More interestingly, how can we describe the small angular rotation of the object during this time interval? At the given instant in time, the rotation of the object can be specified by prescribing an axis of rotation as shown in Figure 12.8, and a

---

[2] Note that throughout our discussion the rotation matrices are used to rotate the coordinate system between lab and body frames – as opposed to rotating vectors directly. This type of rotation is known as **passive rotation** – where the rotations are applied to the *coordinate system* and as a result the components of fixed vectors change. This is to be contrasted with **active rotation** – where the rotation is applied to the vector directly, not to the coordinate system, and hence the components of the vector change because the vector itself is rotated. To avoid confusion, we have used and will continue to use only passive rotations throughout.

small rotation angle about it, $\Delta\alpha$. The axis gives a direction in three-dimensional space, and the angle a scalar number. We can represent this set of three numbers by a vector ω: the direction of ω is along the instantaneous axis of rotation, and its magnitude is defined as $\Delta\alpha/\Delta t$. Naturally, this vector is called the **angular velocity** vector. As the system evolves, the direction of the axis of rotation and the rate of rotation can change; therefore, the vector ω generally evolves in both direction and magnitude.

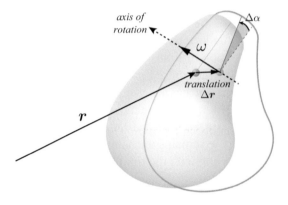

Fig. 12.8 An infinitesimal evolution of a rigid body. The full motion of the rigid body is the sum of a small displacement in the position of the tagged point plus an infinitesimal rotation about an axis through the tagged point. The rotation portion can be described by an instantaneous axis of rotation and a rate of spin $\omega = \Delta\alpha/\Delta t$ about this axis. ω is the angular velocity vector and is aligned along the axis of rotation.

Consider an arbitrary point within the rigid body whose position is denoted by $\boldsymbol{r}$ relative to a tagged point. We then have

$$\mathbf{r} = \mathbf{R} + \boldsymbol{r}, \tag{12.30}$$

where $\mathbf{r}$ is the position of the point relative to the lab frame. Using Euler's theorem, the instantaneous axis of rotation goes through the tagged point as shown in the figure. We can then write

$$\frac{d\mathbf{r}}{dt} = \frac{d\mathbf{R}}{dt} + \frac{d\boldsymbol{r}}{dt}. \tag{12.31}$$

The $\mathbf{V} \equiv d\mathbf{R}/dt$ part is the velocity of the tagged point. The more interesting quantity is the second part that tells us about the instantaneous rotation of the rigid body. As shown in Chapter 9, we have

$$\mathbf{v} = \frac{d\boldsymbol{r}}{dt} = \boldsymbol{\omega} \times \boldsymbol{r} \tag{12.32}$$

from the lab perspective. The direction of $\mathbf{v}$ is conveniently obtained by the right-hand rule of the cross product.

Since the velocity $d\boldsymbol{r}/dt$ of every point in the rigid body can be decomposed, as any velocity vector, into a vector Galilean sum of component velocities, the angular velocity $\boldsymbol{\omega}$ can also be decomposed into component angular velocities. For example, we can write

$$\boldsymbol{\omega} = \boldsymbol{\omega}_1 + \boldsymbol{\omega}_2, \qquad (12.33)$$

where $\boldsymbol{\omega}_1$ and $\boldsymbol{\omega}_2$ describe a decomposition of the full angular velocity into two others with different axes and magnitudes, as shown in Figure 12.9. The velocity of the corresponding point in the rigid body is then

$$\boldsymbol{\omega}_1 \times \boldsymbol{r} + \boldsymbol{\omega}_2 \times \boldsymbol{r} = \boldsymbol{v}_1 + \boldsymbol{v}_2. \qquad (12.34)$$

Addition is a commutative operation, so infinitesimal rotations are commutative. However, once a finite rotation is built from many infinitesimal ones, such large rotations no longer commute. Angular velocity describes a *small* change in angle about an axis, however, per a *small* interval in time, so for such small rotations the order of rotation doesn't matter. That is why we can simply add angular velocity vectors to obtain a total rate of spin.

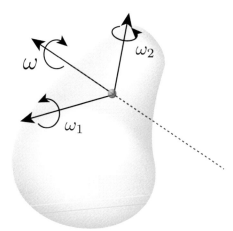

Fig. 12.9   Combining angular velocities through vector addition. This is a commutative operation for infinitesimal rotations, but is not commutativity for consecutive large rotations.

Now we will use the Euler-angle formalism to find the components of the angular velocity vector.

Example 12.2   **Angular Velocity Transformation**

Consider the angular velocity vector $\boldsymbol{\omega}$ that describes the instantaneous rate of rotation of a rigid body. Remember that given $\boldsymbol{\omega}$ at a snapshot in time, we know that the object of interest is spinning about the direction of this vector at a rate ω. Hence, if we can find out the time evolution of the components of the angular velocity vector from the lab perspective, we can reconstruct the rotational dynamics of the rigid body.

As discussed above, we label the components of $\boldsymbol{\omega}$ in the laboratory frame by $(\omega^x, \omega^y, \omega^z)$; and those in the rigid body frame by $(\omega^{x'}, \omega^{y'}, \omega^{z'})$. And we now know how to relate these given $\varphi(t)$, $\theta(t)$, and $\psi(t)$. Note that the components of $\boldsymbol{\omega}$ in either the body frame or the lab frame are not necessarily constant in general. Depending how complicated is the motion of the rigid body, the angular velocity will evolve in time in complicated ways in both perspectives. Since the orientation of the object is tracked by the Euler angles, it should however be possible to write the angular velocity vector in terms of $\dot{\varphi}$, $\dot{\theta}$, and $\dot{\psi}$, the rates of change of the Euler angles.

To do this, we divide $\boldsymbol{\omega}$ into three parts, as shown in Figure 12.10

$$\boldsymbol{\omega} = \boldsymbol{\omega}^{(I)} + \boldsymbol{\omega}^{(II)} + \boldsymbol{\omega}^{(III)}, \tag{12.35}$$

where (I), (II), and (III) denote various rotation axes. The total is a sum over spins aligned with particular axes. We can read off these individual spin rates from the figure as

$$\omega^{(I)} = \dot{\varphi}, \;\; \omega^{(II)} = \dot{\theta}, \;\; \omega^{(III)} = \dot{\psi}, \tag{12.36}$$

each aligned as shown. We can now write the components of $\boldsymbol{\omega}^{(I)}$, $\boldsymbol{\omega}^{(II)}$, and $\boldsymbol{\omega}^{(III)}$ in the *body frame*, and transform back to the laboratory frame. For example, we have

$$\left(\begin{array}{c} \omega^{(I)x'} \\ \omega^{(I)y'} \\ \omega^{(I)z'} \end{array} \right) = \hat{\mathcal{R}}(\varphi, \theta, \psi) \cdot \left(\begin{array}{c} 0 \\ 0 \\ \dot{\varphi} \end{array} \right) = \left(\begin{array}{c} \dot{\varphi} \sin\theta \sin\psi \\ \dot{\varphi} \sin\theta \cos\psi \\ \dot{\varphi} \cos\theta \end{array} \right). \tag{12.37}$$

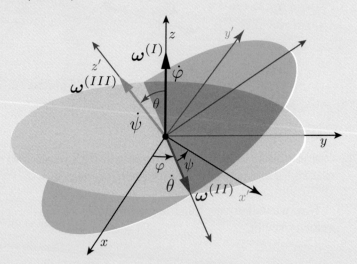

Fig. 12.10 The decomposition of an angular velocity vector into its Euler components.

For $\boldsymbol{\omega}^{(II)}$ we need to be more careful, since its components need to be rotated back to the laboratory only partially. We have

$$
\begin{pmatrix} \omega^{(II)x'} \\ \omega^{(II)y'} \\ \omega^{(II)z'} \end{pmatrix} = \hat{\mathcal{R}}_3(\psi) \cdot \hat{\mathcal{R}}_2(\theta) \cdot \begin{pmatrix} \dot{\theta} \\ 0 \\ 0 \end{pmatrix} = \begin{pmatrix} \dot{\theta}\cos\psi \\ -\dot{\theta}\sin\psi \\ 0 \end{pmatrix}. \tag{12.38}
$$

Finally, for $\omega^{(III)}$, we have

$$
\begin{pmatrix} \omega^{(III)x'} \\ \omega^{(III)y'} \\ \omega^{(III)z'} \end{pmatrix} = \hat{\mathcal{R}}_3(\psi) \cdot \begin{pmatrix} 0 \\ 0 \\ \dot{\psi} \end{pmatrix} = \begin{pmatrix} 0 \\ 0 \\ \dot{\psi} \end{pmatrix}. \tag{12.39}
$$

The total angular velocity vector is a sum of the three component vectors; in the body frame, this becomes

$$
\begin{aligned}
\boldsymbol{\omega} &= (\omega^{x'}, \omega^{y'}, \omega^{z'}) \\
&= \left(\dot{\varphi}\sin\theta\sin\psi + \dot{\theta}\cos\psi, \dot{\varphi}\sin\theta\cos\psi - \dot{\theta}\sin\psi, \dot{\varphi}\cos\theta + \dot{\psi} \right). \tag{12.40}
\end{aligned}
$$

It is also very useful to write the angular velocity vector in the laboratory frame coordinates. To do this, we can use the relation (12.28). After some algebra, one finds that

$$
\begin{aligned}
\boldsymbol{\omega} &= (\omega^x, \omega^y, \omega^z) \\
&= (\dot{\psi}\sin\theta\sin\varphi + \dot{\theta}\cos\varphi, -\dot{\psi}\sin\theta\cos\varphi + \dot{\theta}\sin\varphi, \dot{\psi}\cos\theta + \dot{\varphi}). \tag{12.41}
\end{aligned}
$$

Given $\varphi(t)$, $\theta(t)$, and $\psi(t)$, we can then determine the angular velocity vector at any instant in time in the laboratory and body frames. ∎

12.6 Angular Momentum

Having specified the orientation of a rigid body, and written its spin in terms of Euler angles, we will now find an expression for its total angular momentum $\mathbf{L}_{\text{tot}}$ as measured in the inertial frame of the laboratory. Summing over the angular momenta of all bits in the body:

$$
\mathbf{L}_{\text{tot}} = \sum_i \boldsymbol{\ell}_i = \sum_i (\mathbf{r_i} \times \mathbf{p_i}), \tag{12.42}
$$

where the position vector of the ith bit is $\mathbf{r}_i = \mathbf{R} + \boldsymbol{r}_i$. Here, $\mathbf{R}$ is the position vector from some origin in the lab frame to the tagged point in the body, and $\boldsymbol{r}_i$ is the position of the ith bit measured from the tagged point. The momentum $\mathbf{p_i}$ of the ith bit is

$$
\mathbf{p}_i = m_i \mathbf{v}_i = m_i \left(\mathbf{V} + \dot{\boldsymbol{r}}_i \right) = m_i \left(\mathbf{V} + \boldsymbol{\omega} \times \boldsymbol{r}_i \right), \tag{12.43}
$$

in which $\mathbf{V}$ is the velocity of the tagged point in the lab frame and ω is the instantaneous angular velocity of the rigid body. Altogether:

$$\mathbf{L}_{\text{tot}} = \sum_i \mathbf{r}_i \times (m_i \mathbf{v}_i) = \sum_i (\mathbf{R} + \boldsymbol{r}_i) \times (\mathbf{V} + \omega \times \boldsymbol{r}_i)\,\boldsymbol{r}_i m_i$$

$$= \mathbf{R} \times (M\mathbf{V}) + \mathbf{R} \times (\omega \times M\boldsymbol{\mathscr{R}}_{\text{cm}}) + M\boldsymbol{\mathscr{R}}_{\text{cm}} \times \mathbf{V} + \sum_i m_i \boldsymbol{r}_i \times (\omega \times \boldsymbol{r}_i)$$

$$\equiv \mathbf{L}_1 + \mathbf{L}_2 + \mathbf{L}_3 + \mathbf{L}_4, \tag{12.44}$$

where $\boldsymbol{\mathscr{R}}_{\text{cm}}$ is the position vector *measured from the tagged point* to the center of mass:

$$\sum m_i \boldsymbol{r}_i = M\boldsymbol{\mathscr{R}}_{\text{cm}}. \tag{12.45}$$

(Note the difference between this expression and the sum $\sum m_i \mathbf{r}_i = M\mathbf{R}_{\text{cm}}$, where $\mathbf{R}_{\text{cm}}$ extends from the origin in the laboratory frame to the center of mass of the body.) The scripted $\boldsymbol{\mathscr{R}}_{\text{cm}}$ is a vector *fixed* in the body frame; its components in the body frame must be constant in time. Now consider the two most common situations, each of which will turn out to simplify our life a great deal.

(1) First, suppose the tagged point is the *center of mass* of the rigid body, so $\mathbf{R} = \mathbf{R}_{\text{cm}}$, which implies that $\boldsymbol{\mathscr{R}}_{\text{cm}} = 0$. Therefore in this case we have $\mathbf{L}_2 = \mathbf{L}_3 = 0$. We then also have $\mathbf{V} = \mathbf{V}_{\text{cm}}$, so

$$\mathbf{L}_1 = \mathbf{R}_{\text{cm}} \times (M\mathbf{V}_{\text{cm}}), \tag{12.46}$$

which is often called the *orbital angular momentum*. This term is independent of any rotation of the rigid body itself; it would be the same if the entire rotating object were condensed to a single point mass located at the center of mass.

(2) Alternatively, suppose the tagged point is a *fixed pivot point*, so then $\mathbf{R} = $ constant, resulting in $\mathbf{V} = 0$. In that case $\mathbf{L}_1 = \mathbf{L}_3 = 0$. It is also often convenient to choose the origin of coordinates to be at the pivot point itself, in which case $\mathbf{R} = 0$, which eliminates $\mathbf{L}_2$. This leaves only the purely rotational term $\mathbf{L}_4$ as the total angular momentum of a rotating body with a fixed pivot point. Note however that since the coordinates used to calculate $\mathbf{L}_4$ are not the same about the CM as they are about some other fixed pivot point, $\mathbf{L}_4$ will be different in the two cases.

In summary, if our tagged point is the center of mass of the rigid body, there are two kinds of angular momentum, the **orbital angular momentum** $\mathbf{L}_1 = \mathbf{R}_{\text{cm}} \times (M\mathbf{V}_{\text{cm}})$ and the **spin angular momentum** $\mathbf{L}_4$ about the center of mass. If our tagged point is instead a *fixed* pivot point which we choose as the origin of coordinates, there is only one kind of angular momentum, the spin angular momentum $\mathbf{L}_4$ about the pivot point.

Now we have to understand the spin angular momentum, which we have written as $\mathbf{L}_4$. We have

$$\mathbf{L}_4 = \sum_i m_i \boldsymbol{r}_i \times (\boldsymbol{\omega} \times \boldsymbol{r}_i)$$

$$= \sum_i m_i [\boldsymbol{\omega}\left(r_i^2\right) - (\boldsymbol{\omega} \cdot \boldsymbol{r}_i)\,\boldsymbol{r}_i]$$

$$= \int dm\,[\boldsymbol{\omega}\left(r^2\right) - (\boldsymbol{\omega} \cdot \boldsymbol{r})\,\boldsymbol{r}], \qquad (12.47)$$

where in the second step we used the vector identity $\mathbf{A} \times (\mathbf{B} \times \mathbf{C}) = \mathbf{B}(\mathbf{A} \cdot \mathbf{C}) - \mathbf{C}(\mathbf{A} \cdot \mathbf{B})$ and in the last step we took the continuum limit $\sum_i m_i \to \int dm$ for the convenience of eliminating subscripts. Depending upon the shape of the rigid body, we could let $dm = \rho dV$, σdS, or λdl, where ρ, σ, λ, are, respectively, the volume, surface, and linear mass densities of the rigid body.

Now we can write out the components of the spin angular momentum $\mathbf{L}_4$, which we will now call simply "$\mathbf{L}$," assuming we have already accounted for any orbital angular momentum. We then write

$$\mathbf{L} = \int dm\,[\boldsymbol{\omega}\left(r^2\right) - (\boldsymbol{\omega} \cdot \boldsymbol{r})\,\boldsymbol{r}]. \qquad (12.48)$$

This means that if the tagged point is the center of mass, the total angular momentum is

$$\mathbf{L}_{\text{tot}} = \mathbf{R}_{\text{cm}} \times (M\mathbf{V}_{\text{cm}}) + \mathbf{L}, \qquad (12.49)$$

while if there is instead a fixed pivot point, we have simply

$$\mathbf{L}_{\text{tot}} = \mathbf{L} \qquad (12.50)$$

with no orbital angular momentum term.

Now let us look at this spin angular momentum more closely. From Eq. (12.48), the x component of $\mathbf{L}$ in the lab frame, for example, is

$$L^x = \int dm[\omega_x(x^2 + y^2 + z^2) - x(\omega_x x + \omega_y y + \omega_z z)]$$

$$= \left[\int dm(y^2 + z^2)\right]\omega_x + \left[\int dm(-xy)\right]\omega_y + \left[\int dm(-xz)\right]\omega_z, \qquad (12.51)$$

in which we have separated out the quantities that depend upon the shape, size, and density distributions of the rigid body as given by the integrals shown, from the angular velocity components ω^x, ω^y, and ω^z in the lab frame. Similar expressions can be written for L^y and L^z, so altogether

$$L^x = I_{xx}\omega^x + I_{xy}\omega^y + I_{xz}\omega^z,$$

$$L^y = I_{yx}\omega^x + I_{yy}\omega^y + I_{yz}\omega^z,$$

$$L^z = I_{zx}\omega^x + I_{xy}\omega^y + I_{zz}\omega^z, \qquad (12.52)$$

which can most clearly be written in the form of a matrix equation:

$$\hat{\mathbf{L}} = \hat{\mathbf{I}} \cdot \hat{\omega},$$ (12.53)

where the vectors $\hat{\mathbf{L}}$ and $\hat{\omega}$ are column matrices each consisting of three Cartesian components, and $\hat{\mathbf{I}}$ is written as a 3×3 square symmetric matrix called the **moment of inertia matrix**. That is:

$$\begin{pmatrix} L^x \\ L^y \\ L^z \end{pmatrix} = \begin{pmatrix} I_{xx} & I_{xy} & I_{xz} \\ I_{yx} & I_{yy} & I_{yz} \\ I_{zx} & I_{zy} & I_{zz} \end{pmatrix} \begin{pmatrix} \omega^x \\ \omega^y \\ \omega^z \end{pmatrix},$$ (12.54)

where the diagonal elements of $\hat{\mathbf{I}}$ are

$$I_{xx} = \int dm \left(y^2 + z^2 \right), \quad I_{yy} = \int dm \left(x^2 + z^2 \right), \quad I_{zz} = \int dm \left(x^2 + y^2 \right),$$ (12.55)

and the necessary off-diagonal elements are (since $\hat{\mathbf{I}}$ is symmetric)

$$I_{xy} = \int dm \left(-xy \right), \quad I_{xz} = \int dm \left(-xz \right), \quad I_{yz} = \int dm \left(-yz \right),$$ (12.56)

or more compactly

$$I_{ab} = \int dm \left(r^2 \delta_{ab} - r^a r^b \right).$$ (12.57)

Note that all these expressions are written in the lab frame, as suggested by the unprimed indices on L, ω, and I. We can write similar expressions in the body frame by simply "priming" all indices. Equation (12.53) is the general expression that can be decomposed in component form in either coordinate system. In essence, this equation factors away the angular velocities from the expression for angular momentum. The rest, denoted as a "moment of inertia matrix" $\hat{\mathbf{I}}$, depends only on the way mass is distributed within the rigid body. Note also that, by inspection of Eq. (12.57), we can see that the moment of inertia matrix is necessarily symmetric: exchanging the row and column indices a and b does not change the expression.

Now let us find $\hat{\mathbf{I}}$ in a familiar case, the case of a thin hoop.

Example 12.3 **A Hoop**

Consider a hoop of radius R, mass M, and uniform mass density, which can model the bicycle wheel we described at the beginning of the chapter. We want to compute the moment of inertia matrix for this rigid body in the lab frame. First, we need to choose a tagged point and a coordinate system. One possibility is shown on the left of Figure 12.11, where the tagged point is at the center of the hoop, and the coordinate axes are aligned as shown. Given that this is an object extended along one dimension with some constant *linear* mass density λ, we start with a line integral obtained from Eq. (12.57):

$$I_{ab} = \int \lambda \left(r^2 \delta_{ab} - r^a r^b \right) dl, \tag{12.58}$$

where we used $dm = \lambda \, dl$, and the integral circles around the hoop with differential path length $dl = R \, d\theta$, with the linear mass density given by $\lambda = M/(2\pi R)$.

The coordinates of the bits in the hoop are then traced by the angle θ:

$$r^x = R\cos\theta, \quad r^y = R\sin\theta, \quad r^z = 0, \tag{12.59}$$

which give

$$I_{xx} = \frac{M}{2\pi R} \int_0^{2\pi} \left(R^2 - R^2\cos^2\theta \right) R \, d\theta = \frac{MR^2}{2\pi}(2\pi - \pi) = \frac{MR^2}{2}, \tag{12.60}$$

$$I_{yy} = \frac{M}{2\pi R} \int_0^{2\pi} \left(R^2 - R^2\sin^2\theta \right) R \, d\theta = \frac{MR^2}{2\pi}(2\pi - \pi) = \frac{MR^2}{2}, \tag{12.61}$$

$$I_{zz} = \frac{M}{2\pi R} \int_0^{2\pi} \left(R^2 - 0 \right) R \, d\theta = MR^2 \tag{12.62}$$

for the diagonal components.

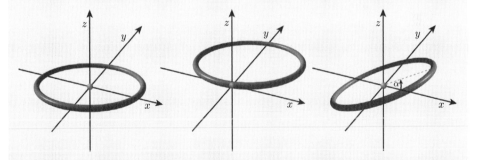

Fig. 12.11 The computation of the moment of inertia of a hoop. The tagged point at the center; the tagged point at the rim; the hoop inclined by an angle α off the x axis.

For the off-diagonal components, since the matrix is symmetric we only need to compute *half* of them:

$$I_{xy} = I_{yx} = \frac{M}{2\pi R} \int_0^{2\pi} \left(R^2\cos\theta\sin\theta \right) R \, d\theta = 0, \tag{12.63}$$

$$I_{xz} = I_{zx} = \frac{M}{2\pi R} \int_0^{2\pi} (0) \, R \, d\theta = 0, \tag{12.64}$$

$$I_{yz} = I_{zy} = \frac{M}{2\pi R} \int_0^{2\pi} (0) \, R \, d\theta = 0. \tag{12.65}$$

The moment of inertia matrix of the hoop about its center is therefore

$$\hat{\mathbf{I}} = \begin{pmatrix} MR^2/2 & 0 & 0 \\ 0 & MR^2/2 & 0 \\ 0 & 0 & MR^2 \end{pmatrix}. \tag{12.66}$$

Notice how the symmetry of the hoop comes in to cancel the off-diagonal matrix elements – for the same reason we encountered earlier when computing the angular momentum of the wheel and noticed that only one sum survives in the result. In that case, we had fixed axis rotation, *i.e.*, $\boldsymbol{\omega} = \omega\,\hat{z}$, and the only relevant part of the moment of inertia was labeled I, corresponding to I_{zz} in the current more general treatment.

Another possible arrangement for the tagged point is shown in the middle diagram of Figure 12.11, with the tagged point now located on the rim. Since the origin of the body axes is shifted, we need to recompute the moment of inertia matrix. We still begin with

$$I_{ab} = \int \lambda \left(r^2 \delta_{ab} - r^a r^b \right) \, dl, \tag{12.67}$$

but now $\boldsymbol{r}$ points from the rim to any arbitrary bit of the hoop. We can write

$$\boldsymbol{r} = (r^x, r^y, r^z) = (R\cos\theta, R\sin\theta + R, 0), \tag{12.68}$$

which gives

$$r^2 = 2R^2 \left(1 + \sin\theta \right), \tag{12.69}$$

and we still have $dl = Rd\theta$. Using similar calculations, we still find that

$$I_{xz} = I_{yz} = I_{xy} = 0, \tag{12.70}$$

and we also still have

$$I_{yy} = \frac{1}{2}MR^2. \tag{12.71}$$

However, $\hat{\mathbf{I}}_{xx}$ and $\hat{\mathbf{I}}_{zz}$ are different from before:

$$I_{xx} = \frac{3}{2}MR^2, \quad I_{zz} = 2MR^2. \tag{12.72}$$

Intuitively, since the hoop is shifted along the y axis, more mass is displaced away from the x and z axes, so that I_{xx} and I_{zz} have increased.

To appreciate the role of the off-diagonal terms of the moment of inertia matrix, consider the scenario on the right of Figure 12.11 – where the hoop is centered at the origin but it is tilted at an arbitrary angle α from the x axis. Once again, our starting point is Eq. (12.67), but now we have

$$\boldsymbol{r} = (r^x, r^y, r^z) = (R\cos\alpha\cos\theta, R\sin\theta, -R\sin\alpha\cos\theta). \tag{12.73}$$

The easiest way to see this is to realize that the new axes are related to the old ones by a rotation about the y axis by angle α. We still have

$$r^2 = R^2 \tag{12.74}$$

just as in the first scenario, and we still have $dl = R\,d\theta$, so we still find

$$I_{xy} = I_{zy} = 0, \tag{12.75}$$

but we now have

$$I_{zx} = \frac{1}{2}MR^2 \sin\alpha \, \cos\alpha. \tag{12.76}$$

Notice that this component vanishes whenever the hoop is aligned with the axes, when $\alpha = 0, \pi/2, 3\pi/2, 2\pi$. For the diagonal components we find

$$I_{xx} = \frac{1}{2}MR^2(2 - \cos^2\alpha), \quad I_{yy} = \frac{1}{2}MR^2, \quad I_{zz} = \frac{1}{2}MR^2(2 - \sin^2\alpha). \tag{12.77}$$

Once again, I_{yy} is unchanged. In all three cases, there is no change in the mass distribution as we look along the y axis. However, both I_{xx} and I_{zz} do change. In particular, note how these two components of the matrix exchange roles between $\alpha = 0$ and $\alpha = \pi/2$, as the hoop lies in the x–y plane first then in the z–y plane. Note also that if the hoop were rotating about the y axis with respect to this coordinate system, we would have a changing angle $\alpha(t)$, which implies that the moment of inertia matrix would have time-dependent components!   ■

From the previous example, we notice a couple of potential complications in studying rigid body dynamics:

- If we were to compute the moment of inertia matrix in the lab frame, and if the rigid body were to tumble around as we would expect and perhaps hope for, the moment of inertia matrix would depend upon time as the mass distribution itself varies with time. However, if we were to compute the moment of inertia matrix in the *body* frame instead, the matrix would be constant in time, since the mass distribution would then be fixed. Therefore it normally seems best to compute the moment of inertia matrix in the body frame. This implies, however, that we would need to write the component form of Eq. (12.53) in primed coordinates, and so deal with components of angular velocity and angular momentum expressed in the body frame. We would then need to use the rotation matrix (12.23) to transform these components back to the lab frame. The cost of writing the moment of inertia matrix in the body frame is therefore the occasional need to transform between lab and body frames. Nevertheless, the benefits of this approach normally outweigh the costs.
- Note that if the coordinate system is not neatly aligned with the symmetry axes, as in the final part of the example of the hoop, there are nonzero off-diagonal elements in the moment of inertia matrix. As we shall see in the next section, if we can find a coordinate system whose axes are aligned with the natural symmetry axes of the rigid body, the moment of inertia matrix will be diagonal.

Therefore, the lessons we draw from the previous examples are twofold: first, write the moment of inertia matrix in the body frame instead of the lab frame; second, choose the orientation of the body frame axes so that the moment of inertia matrix expressed in the body frame is diagonal.

12.7 Principal Axes

Our goal in this section is to elaborate on the most computationally beneficial strategy for choosing the body frame of a rigid body. Our goal is to write the moment of inertia matrix in this strategically chosen body frame so that most of its entries are zero, and, by virtue of being expressed in a body frame, the matrix is constant in time.

First, we want to fix the location of the origin of the body coordinate system. The natural choice is the tagged point with respect to which the motion of the rigid body was decomposed into translational and rotational parts. We are not done, however. Within these criteria we can still orient our axes in infinitely many ways. All these configurations are related to each other by rotations, which change the time-independent moment of inertia matrix components by

$$\hat{\mathbf{I}} \rightarrow \hat{\mathscr{A}} \cdot \hat{\mathbf{I}} \cdot \hat{\mathscr{A}}^T, \tag{12.78}$$

where $\hat{\mathscr{A}}$ is a rotation matrix that reorients the body frame. We write it as $\hat{\mathscr{A}}$ instead of $\hat{\mathscr{R}}$ to avoid confusion with the Euler rotation matrix that connects the body frame to the lab frame. The matrix $\hat{\mathscr{A}}$ transforms one choice of body frame to another. Since $\hat{\mathbf{I}}$ is a real symmetric matrix, we can use this freedom to orient our axes so that the moment of inertia matrix is *diagonal*: we can always find an orthogonal transformation $\hat{\mathscr{A}}$ – a rotation – that diagonalizes any real symmetric matrix. This typically corresponds to aligning the axes with the symmetry axes of the shape of the rigid body. So in practice we do not need to find a sometimes hideous transformation to do the job if our rigid body has enough symmetries that allow us to quickly guess at the appropriate orientation of the axes. This choice of axes for our coordinate system is referred to as the choice of **the principal axes**. Note that, given that the moment of inertia matrix is diagonal in the principal axes frame, we must have in Eq. (12.57)

$$\int dm \, r^{a'} \, r^{b'} = 0 \quad \text{for } a' \neq b', \text{ in the principal axis frame.} \tag{12.79}$$

So from now on we will always compute the moment of inertia matrix in the principal axes coordinate system of the body frame – with origin fixed at the tagged point in the rigid body. We then write

$$\hat{\mathbf{I}} = \begin{pmatrix} I_{x'x'} & 0 & 0 \\ 0 & I_{y'y'} & 0 \\ 0 & 0 & I_{z'z'} \end{pmatrix} = \begin{pmatrix} I_1 & 0 & 0 \\ 0 & I_2 & 0 \\ 0 & 0 & I_3 \end{pmatrix}, \tag{12.80}$$

where we will write $x' \to 1$, $y' \to 2$, and $z' \to 3$ to emphasize that these are components in a special body coordinate system which diagonalizes the moment of inertia matrix. We then need to compute only three moments of inertia, one along each of the three principal axes. Therefore, the physical content of the moment of inertia matrix is quantified by three independent numbers out of the six possible independent ones for a general 3×3 symmetric matrix.

In summary, the components of the spin angular momentum expressed in the body principal axes frame, obtained from Eq. (12.53), become

$$L^{x'} = I_{x'x'}\omega^{x'} \Rightarrow L_1 = I_1\,\omega_1, \tag{12.81}$$

$$L^{y'} = I_{y'y'}\omega^{y'} \Rightarrow L_2 = I_2\,\omega_2, \tag{12.82}$$

$$L^{z'} = I_{z'z'}\omega^{z'} \Rightarrow L_3 = I_3\,\omega_3, \tag{12.83}$$

where L_1, L_2, L_3 and ω_1, ω_2, ω_3 are components of the angular momentum and angular velocity in the body principal axes frame.

Example 12.4 **Symmetry and Principal Axes**

The off-diagonal components of the moment of inertia matrix come from

$$\int dm\, r^{a'}\, r^{b'} \tag{12.84}$$

in Eq. (12.57) and so arise from any *asymmetric* mass distribution about the axis passing through the origin and perpendicular to the a' and b' direction. Therefore, to diagonalize the inertia matrix one needs to align the principal axes along the symmetry axes of the rigid body. Figure 12.12 shows several rigid bodies with corresponding principal axes. Notice that the axes always orient along the symmetry axes of the object. In the last example in the figure, however, given the asymmetries in the shape of the object, it is not obvious how to orient the principal axes. So sometimes we need to rely on rigorous algebraic methods and diagonalize a matrix directly. This is an approach of last resort, when we really do need to find the principal axes but can't identify them by inspection.

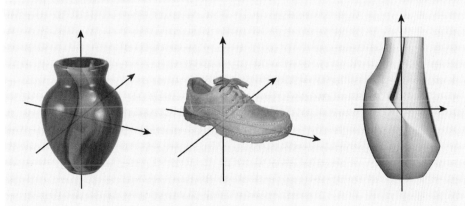

Fig. 12.12   The principal axes of various rigid bodies.   ∎

The techniques involved in diagonalizing the moment of inertia matrix illustrate how to solve an *eigenvalue problem*, a type of problem very common in both classical and quantum physics. From linear algebra, diagonalizing the matrix $\hat{\mathbf{I}}$ amounts to solving the following determinant equation for the unknown $\mathcal{I}$:

$$\begin{vmatrix} I_{x'x'} - \mathcal{I} & I_{x'y'} & I_{x'z'} \\ I_{y'x'} & I_{y'y'} - \mathcal{I} & I_{y'z'} \\ I_{z'x'} & I_{z'y'} & I_{z'z'} - \mathcal{I} \end{vmatrix} = 0. \tag{12.85}$$

This is a cubic equation in $\mathcal{I}$; its three solutions are the three principal moments of inertia. These three solutions are called the *eigenvalues* of the problem, and the corresponding eigenvectors tell us the direction of the set of axes, *i.e.*, the principal axis frame. Let us look at an example.

Example 12.5 **Diagonalizing the Moment of Inertia Matrix**

A particular rigid body of mass M and radius R has a moment of inertia matrix

$$\hat{\mathbf{I}} = \begin{pmatrix} 3/4 & 0 & 1/4 \\ 0 & 1/2 & 0 \\ 1/4 & 0 & 3/4 \end{pmatrix} MR^2 \tag{12.86}$$

for some set of body axes x', y', z'. We want to find the principal moments of inertia and the directions of the principal axes relative to the given set of axes. Note that the moment of inertia matrix is symmetric, so it can be diagonalized. We then write

$$\begin{vmatrix} 3/4 - \mathcal{I} & 0 & 1/4 \\ 0 & 1/2 - \mathcal{I} & 0 \\ 1/4 & 0 & 3/4 - \mathcal{I} \end{vmatrix} = 0, \tag{12.87}$$

where we have divided out the MR^2 term for convenience; we should however keep this in mind so that, at the end, we multiply the solution for $\mathcal{I}$ we obtain by MR^2. Expanding the determinant, we have

$$(1/2 - \mathcal{I}) \begin{vmatrix} 3/4 - \mathcal{I} & 1/4 \\ 1/4 & 3/4 - \mathcal{I} \end{vmatrix} = (1/2 - \mathcal{I}) \left[(3/4 - \mathcal{I})^2 - 1/16 \right]$$

$$= (1/2 - \mathcal{I})(1 - \mathcal{I})(1/2 - \mathcal{I}), \tag{12.88}$$

so that the three principal moments of inertia are $I_1 = (1/2)MR^2$, $I_2 = (1/2)MR^2$, and $I_3 = MR^2$, restoring the MR^2 factor in $\mathcal{I} = 1/2$ and $\mathcal{I} = 1$. These are in fact the principal moments of inertia of a hoop, as encountered earlier! Hence, this rigid body is indeed a hoop of mass M and radius R. We can also find the principal axes in terms of the initial axes for any one of the principal moments, by substituting a principal moment of inertia $\mathcal{I}$ back into the three algebraic equations

$$\begin{pmatrix} 3/4 - \mathcal{I} & 0 & 1/4 \\ 0 & 1/2 - \mathcal{I} & 0 \\ 1/4 & 0 & 3/4 - \mathcal{I} \end{pmatrix} \begin{pmatrix} a^{x'} \\ a^{y'} \\ a^{z'} \end{pmatrix} = 0, \tag{12.89}$$

and finding the ratios of $a^{x'}$, $a^{y'}$, and $a^{z'}$. If we select $\mathscr{I} = 1$ for example, then the algebraic equations give $-a^{x'}/4 + a^{z'}/4 = 0$ and $a^{y'} = 0$. That is, the principal axis corresponding to MR^2 contains an equal part of the initial x' and z' axes and contains no part of y. In other words, to find the principal 3 axis we rotate the initial coordinate system about the y' axis by $45°$.   ∎

Evaluating the integrals involved in moments of inertia can be easy or complicated. However, there are a couple of theorems that often help enormously. We will first state and prove each theorem, and then apply them to the moments of inertia of a hoop.

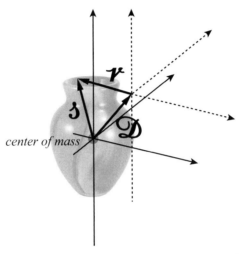

Fig. 12.13 The effect of translating the origin of the principal axes from the center of mass of a rigid body to another point, while keeping the axes parallel.

The parallel axis theorem: Suppose I_i^{cm} (with $i = 1, 2,$ or 3) are known moments of inertia for a rigid body about some *principal axes* that pass through the *center of mass* of the body at the origin $\mathscr{O}_{\mathrm{cm}}$, and let $\hat{\mathbf{I}}_{a'b'}$ be entries in the moment of inertia matrix about axes that are parallel to the principal axes but instead have origin at a point $\mathscr{O}$ that is displaced from $\mathscr{O}_{\mathrm{cm}}$ by a vector $\mathscr{D}$, as shown in Figure 12.13. We want to find $\hat{\mathbf{I}}_{a'b'}$ from the I_a's.

Let $\mathfrak{s}$ be the position vector of a bit of mass measured from the center of mass $\mathscr{O}_{\mathrm{cm}}$ of the rigid body as shown in the figure. We write

$$\mathfrak{s} = \boldsymbol{r} + \mathscr{D}, \tag{12.90}$$

where $\boldsymbol{r}$ is the position of this same bit as measured from the new origin $\mathscr{O}$. Notice how all these vectors are in scripted font as they represent vectors fixed in the rigid body frame. Substituting (12.90) into the expression for moment of inertia from (12.57), we get

$$I_{a'b'} = \int dm \left((\delta^2 + \mathscr{D}^2 - 2\,\delta \cdot \mathscr{D})\delta_{a'b'} - \delta^{a'}\delta^{b'} - \mathscr{D}^{a'}\mathscr{D}^{b'}\delta^{a'}\mathscr{D}^{b'} + \mathscr{D}^{a'}\delta^{b'} \right).$$

$$(12.91)$$

Note that we must have

$$I_i^{\text{cm}} = \int dm \left(\delta^2 - \delta^i \delta^i \right),$$

$$(12.92)$$

where no sum over the repeated index i is implied. We also know from Eq. (12.79) that

$$\int dm\,\delta^{a'}\delta^{b'} = 0 \quad \text{for } a' \neq b'.$$

$$(12.93)$$

This cancels the fourth term in Eq. (12.91) when $a' \neq b'$. Furthermore, since δ has its origin at the center of mass, we have

$$\int dm\,\delta \propto \delta_{\text{cm}} = 0.$$

$$(12.94)$$

This cancels all terms in Eq. (12.91) that are linear in $\delta^{a'}$. These three relations simplify Eq. (12.91). We can now state the theorem: Let I_i^{cm} be the known moment of inertia for a rigid body about some *principal axes* that pass through the *center of mass* of the body at the origin $\mathbb{O}_{\text{cm}}$, and let $I_{a'b'}$ be the moment of inertia matrix for the body about axes that are parallel to the first axes but whose origin is located at $\mathbb{O}$. Also let the vector distance from $\mathbb{O}_{\text{cm}}$ to $\mathbb{O}$ be given by the displacement vector $\mathscr{D}$. Then the diagonal components of the new (parallel) moment of inertia matrix are

$$I_{a'a'} = I_{a'}^{\text{cm}} + M\mathscr{D}^2 - M\mathscr{D}^{a'}\mathscr{D}^{a'},$$

$$(12.95)$$

with no sum on the index a', and for the off-diagonal components we have

$$I_{a'b'} = -M\mathscr{D}^{a'}\mathscr{D}^{b'} \quad \text{for } a' \neq b'.$$

$$(12.96)$$

In arriving at these two relations, we used Eqs. (12.92), (12.93), and (12.94) in (12.91). These two relations, Eq. (12.95) and Eq. (12.96), are known as the **parallel axis theorem**. The theorem allows us to quickly compute the new components of the moment of inertia tensor from the ones with respect to the center of mass principal axes frame. Notice in particular that, if we were to shift the axes *along* one of the parallel axes, Eq. (12.96) would always vanish since two components of $\mathscr{D}$ would necessarily vanish; this implies that translating along a principal axis leaves the moment of inertia in diagonal form. Hence, we have an infinite family of principal axes frames, all related to one another by translations along their axes.

Example 12.6 | **Translating the Hoop**

Let us apply the parallel axis theorem to compute the moment of inertia matrix for the hoop about its rim, as carried out earlier by brute force (the case corresponding to the middle of Figure 12.11). For the diagonal components we have

$$I_{x'x'} = I_1 + MR^2 - 0 = \frac{3}{2}MR^2, \tag{12.97}$$

$$I_{y'y'} = I_2 + MR^2 - MR^2 = \frac{1}{2}MR^2, \tag{12.98}$$

$$I_{z'z'} = I_3 + MR^2 - 0 = 2MR^2. \tag{12.99}$$

And since the translation is along the y axis, the off-diagonal components remain zero. These results agree with those of Eqs. (12.71) and (12.72). ∎

The perpendicular axis theorem for plane lamina: Plane lamina are idealized rigid bodies confined to a plane, which we take to be the (x', y') plane. Then since $z' = 0$ for plane lamina, their principal moment of inertia elements are

$$I_{x'x'} = \int dm \left(y'^2 \right), \quad I_{y'y'} = \int dm \left(x'^2 \right), \quad I_{z'z'} = \int dm \left(x'^2 + y'^2 \right), \tag{12.100}$$

while the off-diagonal elements are zero, since we assumed principal axes. Now it is obvious from these equations that

$$I_{x'x'} + I_{y'y'} = I_{z'z'}. \tag{12.101}$$

This is the perpendicular axis theorem for plane lamina, where the lamina is in the (x', y') plane. It is good to keep the perpendicular axis theorem in mind, because it can save a lot of time and effort if the rigid body is sufficiently symmetric. Also, many three-dimensional objects are sums over plane lamina, so the theorem is more useful than one might think at first. See the Problems section at the end of this chapter for more.

Example 12.7 | **Perpendicular Axis Theorem and the Hoop**

We can apply the perpendicular axis theorem to our example of the hoop, where the x' and y' axes are in the plane of the hoop and where z' is the axis of symmetry perpendicular to the hoop. We took the hoop to be infinitely thin, so it qualifies as a plane lamina. The moment of inertia in the z' direction is $I_{z'z'} = I_3 = MR^2$, so by the perpendicular axis theorem the sum $I_1 + I_2 = I_3 = MR^2$. But it is clear from symmetry that $I_1 = I_2$, so it follows that $I_1 = I_2 = (1/2)MR^2$, just as we found earlier by direct calculation. ∎

12.8 Torque

The net torque on a body is the cause of changes in its angular momentum, just as the net force on the body is the cause of changes in its linear momentum. Let us remind ourselves about how torques originate. We start by writing Newton's second law for every bit making up a rigid body:

$$\mathbf{F}_i = m_i \mathbf{a}_i = m_i \frac{d\mathbf{v}_i}{dt}. \tag{12.102}$$

We then cross both sides of this equation by the position vector of each bit $\mathbf{r}_i$ and sum over i:

$$\sum_i \mathbf{r}_i \times \mathbf{F}_i = \frac{d}{dt} \sum_i m_i \mathbf{r}_i \times \mathbf{v}_i = \frac{d\mathbf{L}_{\text{tot}}}{dt}, \tag{12.103}$$

where we have explicitly indicated that the angular momentum in this expression is the *total* angular momentum, which is a sum of spin angular momentum *plus* orbital angular momentum, as described in Eqs. (12.49) and (12.49). On the left-hand side the quantity is then defined as the **torque**:

$$\mathbf{N}_{\text{tot}} \equiv \sum_i \mathbf{r}_i \times \mathbf{F}_i. \tag{12.104}$$

Typically, in this sum, internal forces between the bits making up the rigid body cancel pairwise due to Newton's third law. So, we may write instead

$$\mathbf{N}_{\text{tot}} = \sum_i \mathbf{r}_i \times \mathbf{F}_i^{\text{ext}}, \tag{12.105}$$

where the $\mathbf{F}_i^{\text{ext}}$s account for *external* forces only. We then have a rotational analogue of Newton's second law useful for studying rotational dynamics:

$$\mathbf{N}_{\text{tot}} = \frac{d\mathbf{L}_{\text{tot}}}{dt}. \tag{12.106}$$

As for angular momentum, we can divide up the torque by splitting the dynamics of the rigid body into a translational and a rotational part. Substituting $\mathbf{r}_i = \boldsymbol{r}_i + \mathbf{R}$ into the definition of torque (12.105), we have

$$\mathbf{N}_{\text{tot}} = \sum_i \boldsymbol{r}_i \times \mathbf{F}_i^{\text{ext}} + \mathbf{R} \times \mathbf{F}_i^{\text{ext}} = \mathbf{N} + \mathbf{R} \times \mathbf{F}, \tag{12.107}$$

where the second term is the torque acting on the rigid body as a whole, which involves the sum of all external forces. In the Lagrangian formalism the notion of torque is often not directly encountered unless one is interested in unraveling constraints through Lagrange multipliers. However, Eq. (12.106) can be a useful parallel approach to understanding the dynamics of the rigid body.

12.9 Kinetic Energy

When finding the Lagrangian for a rigid body, we first need to write its kinetic and potential energies. We start by focusing on the kinetic energy. We set up our laboratory coordinate system as shown in Figure 12.7 and divide up the rigid body into bits of small mass Δm_i. We denote the tagged point's location by the position vector $\mathbf{R}$, so

$$\mathbf{r}_i = \mathbf{R} + \boldsymbol{r}_i \tag{12.108}$$

for every bit of the rigid body. The velocity of a bit is then given by

$$\mathbf{v}_i = \dot{\mathbf{R}} + \boldsymbol{\omega} \times \boldsymbol{r}_i = \mathbf{V} + \boldsymbol{\omega} \times \boldsymbol{r}_i \tag{12.109}$$

using Eq. (12.32), where $\mathbf{V}$ is the translational velocity of the tagged point. The kinetic energy of the rigid body is simply the sum of the kinetic energies of all the bits. That is:

$$T_{\text{tot}} = \frac{1}{2} \sum_i \Delta m_i \mathbf{v}_i^2 = \frac{1}{2} \sum_i \Delta m_i \left(\mathbf{V} + \boldsymbol{\omega} \times \boldsymbol{r}_i \right) \cdot \left(\mathbf{V} + \boldsymbol{\omega} \times \boldsymbol{r}_i \right). \tag{12.110}$$

Note that this is written from an inertial perspective. Expanding this expression, we identify three different parts

$$T_{\text{tot}} = \frac{1}{2} M \mathbf{V}^2 + \sum_i m_i \mathbf{V} \cdot \left(\boldsymbol{\omega} \times \boldsymbol{r}_i \right) + \frac{1}{2} \sum_i m_i \left(\boldsymbol{\omega} \times \boldsymbol{r}_i \right) \cdot \left(\boldsymbol{\omega} \times \boldsymbol{r}_i \right)$$

$$= T_1 + T_2 + T_3, \tag{12.111}$$

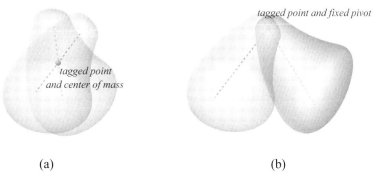

tagged point and fixed pivot

tagged point and center of mass

(a)   (b)

Fig. 12.14   The two common choices for tagged points.

To simplify things further, it helps to choose a strategic tagged point in the decomposition of the dynamics into translation and rotational parts. As mentioned earlier, we consider two possible choices.

The first choice is to locate $\mathbf{R}$ at the center of mass of the rigid body (see Figure 12.14(a)):

$$\mathbf{R} = \mathbf{R}_{\text{cm}}. \tag{12.112}$$

We then have

$$T_1 = \frac{1}{2}M\mathbf{V}_{\text{cm}}^2, \tag{12.113}$$

where $\mathbf{V}_{\text{cm}}$ is the linear velocity of the center of mass. Then

$$T_2 = M\mathbf{V} \cdot (\boldsymbol{\omega} \times \mathbf{R}_{\text{cm}}). \tag{12.114}$$

Often the center of mass dynamics will be trivial and we can change frames to the CM frame while remaining inertial. In this case we have

$$\mathbf{V}_{\text{cm}} = 0 \Rightarrow T_1 = T_2 = 0. \tag{12.115}$$

Alternatively, locate $\mathbf{R}$ at a *fixed pivot point* (see Figure 12.14(b)):

$$\mathbf{R} = \text{constant} \Rightarrow \mathbf{V} = 0. \tag{12.116}$$

We then immediately have

$$T_1 = T_2 = 0. \tag{12.117}$$

In both scenarios we see that $T_1 = T_2 = 0$. We are then left with T_3, where

$$T_3 \equiv T = \frac{1}{2}\int dm\,(\boldsymbol{\omega} \times \boldsymbol{r}) \cdot (\boldsymbol{\omega} \times \boldsymbol{r}), \tag{12.118}$$

where we have also taken the continuum limit by replacing the sum over bits by an integral. Naturally, we call it the **spin kinetic energy**. To simplify this expression further, we make use of the two identities

$$\mathbf{a} \cdot (\mathbf{b} \times \mathbf{c}) = \mathbf{b} \cdot (\mathbf{c} \times \mathbf{a}) = \mathbf{c} \cdot (\mathbf{a} \times \mathbf{b}) \tag{12.119}$$

and

$$\mathbf{a} \times \mathbf{b} = -\mathbf{b} \times \mathbf{a}, \tag{12.120}$$

which allow us to write

$$(\boldsymbol{\omega} \times \boldsymbol{r}_i) \cdot (\boldsymbol{\omega} \times \boldsymbol{r}_i) = -\boldsymbol{r}_i \cdot (\boldsymbol{\omega} \times (\boldsymbol{\omega} \times \boldsymbol{r}_i)). \tag{12.121}$$

We can then use the BAC–CAB rule

$$\mathbf{a} \times (\mathbf{b} \times \mathbf{c}) = (\mathbf{a} \cdot \mathbf{c})\mathbf{b} - (\mathbf{a} \cdot \mathbf{b})\mathbf{c} \tag{12.122}$$

to simplify things further:

$$\begin{aligned}(\boldsymbol{\omega} \times \boldsymbol{r}_i) \cdot (\boldsymbol{\omega} \times \boldsymbol{r}_i) &= -\boldsymbol{r}_i \cdot ((\boldsymbol{\omega} \cdot \boldsymbol{r}_i)\boldsymbol{\omega} - \omega^2\boldsymbol{r}_i) \\ &= -(\boldsymbol{\omega} \cdot \boldsymbol{r}_i)^2 + \omega^2 r_i^2.\end{aligned} \tag{12.123}$$

Inserting this expression back into the original expression for T, we have

$$
\begin{aligned}
T &= \frac{1}{2} \int dm \left(\omega^2 r^2 - (\boldsymbol{\omega} \cdot \boldsymbol{r})^2 \right) \\
&= \frac{1}{2} \omega^a \left[\int dm \left(r^2 \delta_{ab} - r^a r^b \right) \right] \omega^b \\
&= \frac{1}{2} \omega^a I_{ab} \, \omega^b = \frac{1}{2} \boldsymbol{\omega}^T \cdot \hat{\mathbf{I}} \cdot \boldsymbol{\omega},
\end{aligned}
\tag{12.124}
$$

where $\hat{\mathbf{I}}$ is the same moment of inertia matrix encountered earlier in Eq. (12.57).

Note that when this expression is expanded in component form we first need to choose a coordinate system. For example, the expression above references unprimed coordinates implying that the components ω^a and I_{ab} are written in the lab frame. If written in terms of components in a body frame, we would instead write

$$
T = \frac{1}{2} \omega^{a'} I_{a'b'} \, \omega^{b'}.
\tag{12.125}
$$

The two sets of components are of course related to each other by the Euler matrix $\hat{\mathcal{R}}$ from (12.23). Notice also that T is unchanged in value when written in either form; this is in contrast to angular momentum which, being a vector, will have its components change when we transition between the lab and body frames. To emphasize this point, we say that the spin kinetic energy is a *scalar* under rotations.

Another way to appreciate the scalar aspect of spin kinetic energy is to consider what happens to it when we switch between two body frames; say, going from one arbitrary body frame to a special principal axes one that diagonalizes the moment of inertia matrix. As we know, this corresponds to a transformation of the components of $\hat{\mathbf{I}}$ of the form

$$
\hat{\mathbf{I}} \to \hat{\mathcal{A}} \cdot \hat{\mathbf{I}} \cdot \hat{\mathcal{A}}^T,
\tag{12.126}
$$

where $\hat{\mathcal{A}}$ is some rotation matrix (not the Euler matrix $\hat{\mathcal{R}}$ that connects lab and body frames!). As a result of this, the components of the angular velocity vector undergo the change

$$
\boldsymbol{\omega} \to \hat{\mathcal{A}} \cdot \boldsymbol{\omega}.
\tag{12.127}
$$

We can then see that these transformations will have changed the spin kinetic energy, since

$$
T = \frac{1}{2} \boldsymbol{\omega}^T \cdot \hat{\mathbf{I}} \cdot \boldsymbol{\omega} \to \frac{1}{2} \boldsymbol{\omega}^T \cdot \hat{\mathcal{A}}^T \cdot \hat{\mathcal{A}} \cdot \hat{\mathbf{I}} \cdot \hat{\mathcal{A}}^T \cdot \hat{\mathcal{A}} \cdot \boldsymbol{\omega} = \frac{1}{2} \boldsymbol{\omega}^T \cdot \hat{\mathbf{I}} \cdot \boldsymbol{\omega}.
\tag{12.128}
$$

This means that choosing a body frame that diagonalizes the moment of inertia matrix does not change the value of the spin kinetic energy. In particular, if we were to use an $\hat{\mathbf{A}}$ that takes us to the principal axis frame, the spin kinetic energy takes the simpler form

$$T = \frac{1}{2}I_1\omega_1^2 + \frac{1}{2}I_2\omega_2^2 + \frac{1}{2}I_3\omega_3^2 = \frac{L_1^2}{2I_1} + \frac{L_2^2}{2I_2} + \frac{L_3^2}{2I_3},$$ (12.129)

where I_i, ω_i, and L_i with $i = 1, 2, 3$ refer to components in the principal axes frame as introduced earlier in the discussion of angular momentum.

Example 12.8

Kinetic Energy of a Hoop

Let us compute the spin kinetic energy of the hoop we introduced earlier using its center of mass as the tagged point. The tagged point is at the center and the hoop is positioned as in the left picture of Figure 12.11. This shall be our body frame, where we expect that the components of the moment of inertia matrix are constant in time. We know that this is a principal axes system and we have

$$I_1 = I_2 = \frac{1}{2}MR^2, \quad I_3 = MR^2.$$ (12.130)

To write the spin kinetic energy we now need to be careful: we need to express the components of the angular velocity vector $\boldsymbol{\omega}$ in the same *body frame* as well. Looking back at Eq. (12.40), we have the angular velocity components in the body frame in terms of the Euler angles

$$\omega_1 = \dot{\varphi}\sin\theta\sin\psi + \dot{\theta}\cos\psi, \quad \omega_2 = \dot{\varphi}\sin\theta\cos\psi - \dot{\theta}\sin\psi, \quad \omega_3 = \dot{\varphi}\cos\theta + \dot{\psi}.$$ (12.131)

Putting things together in Eq. (12.129), we get

$$T = \frac{1}{4}MR^2(\dot{\theta}^2 + \dot{\varphi}^2\sin^2\theta) + \frac{1}{2}MR^2(\dot{\psi} + \dot{\varphi}\cos\theta)^2,$$ (12.132)

where we used $\cos^2 + \sin^2 = 1$ to simplify the expression. Notice that if the hoop is only spinning about the 3 axis, then $\dot{\varphi} = \dot{\theta} = 0$, and the spin kinetic energy takes the more familiar form

$$T_{\text{rot}} = \frac{1}{2}MR^2\dot{\psi}^2.$$ (12.133)

∎

12.10 Potential Energy

Now we need to find the potential energy of a rigid body. We focus upon a uniform gravitational potential energy, but the procedure is similar for any other potential energy. In general, we divide the rigid body into bits as before and write the total potential energy as the sum of the potential energies of the small bits:

$$U = \sum_i \Delta U_i.$$ (12.134)

For uniform gravity we have

$$U = \sum_i \Delta m_i g\, h_i = \sum_i \Delta m_i g\, \mathbf{r}_i \cdot \hat{\mathbf{z}} = Mg\, \mathbf{R}_{\text{cm}} \cdot \hat{\mathbf{z}} = MgH,$$ (12.135)

where H is the height of the center of mass. So as far as uniform gravity is concerned, the rigid body behaves as if all its mass were concentrated at its center of mass. We say that gravity pulls on the rigid body at its center of mass. This is partly why splitting the dynamics of a rigid body into the translation dynamics of the center of mass plus a rotational one about the center of mass can sometimes be very useful.

A common situation in mechanics involves a force acting at a single point in the rigid body. For example, imagine the center of a wooden plank resting on a rock, with a child sitting on each end of the plank. If the plank is the object of interest, we have three contact forces acting on it. Some contact forces can be viewed as constraining the dynamics and can be included in a Lagrangian approach using Lagrange multipliers. For example, the force from the rock that is the pivot is of this type. However, the effect of the two children sitting on the plank is more usefully accounted for by the appearance of their potential energies in the Lagrangian.

Example 12.9   **A Hoop Hanging on a Spring**

Consider the rather disturbingly complicated situation in which a hoop hangs on a spring, whose other end is a pivot, as in Figure 12.15. The hoop has mass M and radius R, and the spring has spring constant k and zero unstretched length. The system is hanging in uniform gravity, and we want to begin by finding the Lagrangian. There are a total of six degrees of freedom: the location of the center of mass of the hoop X, Y, and Z; and the orientation of the hoop described through the three Euler angles φ, θ, and ψ. The Lagrangian is $\mathcal{L} = T - U$, as usual, where we have used the symbol $\mathcal{L}$ for the Lagrangian to avoid confusion with angular momentum L. We first need the total kinetic energy. Separating the motion of the hoop into the translation of its center of mass plus a rotation about the center of mass, we have

$$T_{\text{tot}} = \frac{1}{2}M(\dot{X}^2 + \dot{Y}^2 + \dot{Z}^2) + \frac{1}{4}MR^2(\dot{\theta}^2 + \dot{\varphi}^2\sin^2\theta) + \frac{1}{2}MR^2(\dot{\psi} + \dot{\varphi}\cos\theta)^2, \quad (12.136)$$

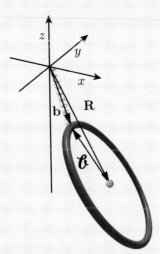

Fig. 12.15   A hoop hanging from a spring.

where we used Eq. (12.132). The interesting part is the potential energy. For the contribution from gravitational potential energy, we have

$$U_{\text{grav}} = MgZ, \tag{12.137}$$

tracking the location of the center of mass as just described. For the spring force, we realize we have a contact force with the spring attached to a fixed point on the rim of the hoop. The potential energy contribution is then

$$U_{\text{spring}} = \frac{1}{2}kb^2, \tag{12.138}$$

where **b** is the vector pointing from the origin to the point on the rim of the hoop where the spring is anchored. The vector **b**, however, depends upon all six degrees of freedom: $X, Y, Z, \varphi, \theta,$ and ψ. We can write

$$\mathbf{b} = \mathbf{R} + \boldsymbol{\mathscr{b}}', \tag{12.139}$$

where **R** is the position vector of the center of mass and $\boldsymbol{\mathscr{b}}'$ points from the center of mass to the rim point where the spring is attached. The prime on the latter vector reminds us that this vector is fixed with respect to the rigid body. We can then write its constant components in the body frame as

$$\boldsymbol{\mathscr{b}}' = (\mathscr{b}^{x'}, \mathscr{b}^{y'}, \mathscr{b}^{z'}) = (0, R, 0), \tag{12.140}$$

where we aligned the y' axis of the body frame with the contact point of the spring. But we need the components of this vector in the lab frame $(\mathscr{b}^x, \mathscr{b}^y, \mathscr{b}^z)$, where we know already that

$$\mathbf{R} = (R^x, R^y, R^z) = (X, Y, Z), \tag{12.141}$$

so that we can write, from Eq. (12.139):

$$\mathbf{b} = (b^x, b^y, b^z) = (X + \mathscr{b}^x, Y + \mathscr{b}^y, Z + \mathscr{b}^z). \tag{12.142}$$

To find $(\mathscr{b}^x, \mathscr{b}^y, \mathscr{b}^z)$, we then use once again Eq. (12.28), which gives

$$\begin{aligned}
\boldsymbol{\mathscr{b}} &= (\mathscr{b}^x, \mathscr{b}^y, \mathscr{b}^z) \\
&= [R(\cos\theta\cos\varphi\sin\psi + \cos\psi\sin\varphi), \\
&\quad R(\cos\theta\cos\varphi\cos\psi - \sin\varphi\sin\psi), \\
&\quad - R(\cos\varphi\sin\theta)].
\end{aligned} \tag{12.143}$$

From all this, we can now construct the spring potential energy (12.138):

$$\begin{aligned}
U_{\text{spring}} &= \frac{1}{2}k\left(X^2 + Y^2 + Z^2 + R^2 + 2RX(\cos\theta\cos\varphi\sin\psi + \cos\psi\sin\varphi)\right. \\
&\quad \left. + 2RY(\cos\theta\cos\varphi\cos\psi - \sin\varphi\sin\psi) - 2RZ\cos\varphi\sin\theta\right).
\end{aligned} \tag{12.144}$$

The full Lagrangian is then

$$\mathcal{L} = T_{\text{tot}} - U_{\text{grav}} - U_{\text{spring}}. \tag{12.145}$$

We leave it as an exercise for the reader to write out the six equations of motion for the system! ∎

Example 12.10

The Angular Momentum of the Hoop Hanging on a Spring

A hoop of mass M and radius R is attached to a spring of force constant k at its rim, as shown in Figure 12.15. We want to find the total angular momentum of the hoop with respect to the fixed point at which the other end of the spring is attached. We know that we can split the angular momentum into two pieces:

$$\mathbf{L}_{\text{tot}} = \mathbf{L}_{\text{cm}} + \mathbf{L}, \tag{12.146}$$

where $\mathbf{L}_{\text{cm}}$ is the orbital angular momentum of the center of mass and $\mathbf{L}$ is the spin angular momentum about the center of mass. Using the same six degrees of freedom from the previous example – X, Y, Z, φ, θ, and ψ – we can immediately write

$$\mathbf{L}_{\text{cm}} = \mathbf{R} \times M\dot{\mathbf{R}}, \tag{12.147}$$

where $\mathbf{R} = (X, Y, Z)$. For the spin part we need to do a little more work. We first write

$$\mathbf{L} = (L_1, L_2, L_3) = (I_1\omega_1, I_2\omega_2, I_3\omega_3), \tag{12.148}$$

where we express the spin angular momentum in the body principal axes frame. Using Eqs. (12.203) and (12.130), we then get

$$
\begin{aligned}
\mathbf{L} = (L_1, L_2, L_3) \\
= \Big(\frac{1}{2}MR^2(\dot{\varphi}\sin\theta\sin\psi + \dot{\theta}\cos\psi), \\
\frac{1}{2}MR^2(-\dot{\varphi}\sin\theta\cos\psi + \dot{\theta}\sin\psi), \\
MR^2(\dot{\varphi}\cos\theta + \dot{\psi}) \Big).
\end{aligned}
\tag{12.149}
$$

In order to add this to $\mathbf{L}_{\text{cm}}$ from Eq. (12.147) to obtain $\mathbf{L}_{\text{tot}}$ as in Eq. (12.146), we first need to compute the components of $\mathbf{L}$ in the *lab frame*. We do this using Eq. (12.28):

$$(L^x, L^y, L^z) = \hat{\mathscr{R}} \cdot (L_1, L_2, L_3). \tag{12.150}$$

We leave the remaining algebra as an exercise. One interesting aspect of the result, however, can be seen without much effort. Because of the form of Eq. (12.148), whenever at least one of I_1, I_2, or I_3 is different from the other two – as in the case for the hoop – then $\boldsymbol{\omega}$ is *not* proportional to $\mathbf{L}$; that is, the angular momentum vector is not parallel to the angular velocity vector. This implies that even if the angular momentum is conserved, and therefore fixed in space, the axis of rotation of the rigid body determined by the direction of $\boldsymbol{\omega}$ is not necessarily fixed. We can have complex wobbling even when no torques act upon the rigid body! ∎

Example 12.11

Rolling, Fixed-Axis Rotation

Suppose that a rigid body is restricted to rotate about a fixed axis which is also a principal axis of the object. This could be a barrel rolling down an incline or a two-dimensional pendulum, as shown in Figure 12.16. In these scenarios, the rotational dynamics is significantly simplified. We need to track the evolution of only a single Euler angle. We may align the principal axes such that the body z' axis is along the fixed axis of rotation, and describe the rotation of the body by the angle ψ. The rotational angular velocity is then always aligned along the fixed axis and has a magnitude $\dot{\psi}$. The angular momentum is also aligned with the same axis:

$$L_1 = L_2 = 0, \quad L_3 = I_3 \omega_3 = I_3 \dot{\psi}. \tag{12.151}$$

The spin kinetic energy is simply

$$T = \frac{1}{2} I_3 \omega_3^2 = \frac{1}{2} I_3 \dot{\psi}^2 = \frac{L_3^2}{2 I_3}. \tag{12.152}$$

The only component of the moment of inertia matrix that matters is I_3, the inertia along the fixed axis. Dynamics then changes the angular momentum in magnitude but not in direction. The torque is

$$\mathbf{N}_{\text{tot}} = \frac{d\mathbf{L}_{\text{tot}}}{dt}, \tag{12.153}$$

which lies along the same fixed axis, and for the rotational piece we have

$$\frac{dL_3}{dt} = I_3 \dot{\omega}_3 = I_3 \alpha, \tag{12.154}$$

where $\alpha = \ddot{\psi}$ is the *angular acceleration*. The torque still needs to be determined by the forces on the rigid body.

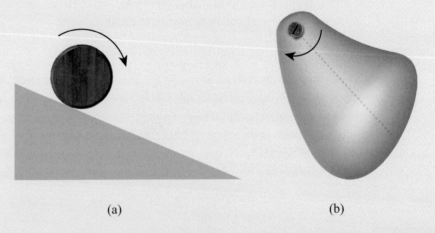

(a)   (b)

Fig. 12.16   Examples of fixed axis rotations: (a) a barrel rolling down an incline, where the fixed axis is the symmetry axis of the barrel; (b) an irregular pendulum swinging in a plane, about a fixed axis directed through the pendulum.   ∎

12.11 Torque-Free Dynamics Using Euler Angles

We now consider a deceptively simple scenario: torque-free motion. You might throw a spinning vase of mass M at a friend as a gesture of discontent. Or a space station might deploy a new satellite in the shape of a flying saucer, spinning about its axis of symmetry. In neither case is there a net torque on the body due to gravity or anything else. Decomposing the motion about the center of mass, the motion of the CM is a simple problem in trajectory physics: in the absence of air resistance the CM of the vase would follow a classic parabolic path, and the CM of the saucer would follow an elliptical path about the earth. The interesting dynamics, however, is the rotational motion *about* the center of mass. Take the vase for example: let us assume that it has cylindrical symmetry so that we can write the components of the moment of inertia matrix about its center of mass and in a principal axes frame in terms of two variables

$$I_1 = I_2 \equiv I, \quad I_3. \tag{12.155}$$

Hence, the x' and y' directions are transverse to the axis of symmetry of the vase, as shown in Figure 12.17.

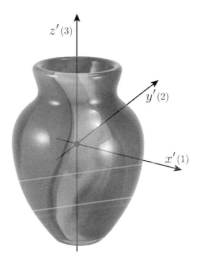

Fig. 12.17 The principal axes frame for the vase.

We want to find the Lagrangian of the vase, decomposing its motion as a translation of the center of mass plus a rotation about the center of mass. Denoting the position of the center of mass by X, Y, and Z, and labeling the vase's orientation using the Euler angles, we write the kinetic energy as

$$T = \frac{1}{2}M\left(\dot{X}^2 + \dot{Y}^2 + \dot{Z}^2\right) + \frac{1}{2}I(\dot{\theta}^2 + \dot{\varphi}^2\sin^2\theta) + \frac{1}{2}I_3(\dot{\psi} + \dot{\varphi}\cos\theta)^2, \tag{12.156}$$

where we obtain the spin kinetic energy employing the same technique used in the case of the hoop in Eq. (12.132). The potential energy is entirely due to gravity, which acts in effect at the vase's center of mass

$$U = MgZ, \tag{12.157}$$

where Z is chosen as the vertical direction. We can now see that the Lagrangian $\mathcal{L} = T - U$ separates into two decoupled systems: the center of mass dynamics describing the evolution of X, Y, and Z and the rotational dynamics described by the Euler angles. The center of mass dynamics is rather trivial, so we will focus on the angular degrees of freedom to determine how the vase tumbles as its center of mass traces a parabolic path.

We have a total of three equations of motion. For ψ, we have

$$I_3 \frac{d}{dt} \left(\dot{\psi} + \dot{\varphi} \cos\theta \right) = 0, \tag{12.158}$$

which is simply the conservation statement associated with the symmetry of rotating about the ψ coordinate

$$p_\psi = \frac{\partial \mathcal{L}}{\partial \dot{\psi}} = I_3 \left(\dot{\psi} + \dot{\varphi} \cos\theta \right) = \text{constant.} \tag{12.159}$$

Similarly, for φ, we get another conservation statement

$$\frac{d}{dt} \left(I\dot{\varphi} \sin^2\theta + I_3 \left(\dot{\psi} + \dot{\varphi} \cos\theta \right) \cos\theta \right) = 0, \tag{12.160}$$

which implies

$$p^\varphi = \frac{\partial L}{\partial \dot{\varphi}} = I\dot{\varphi} \sin^2\theta + p^\psi \cos\theta = \text{constant.} \tag{12.161}$$

Again, φ is a cyclic coordinate of the system. Finally, for θ we get a more complicated statement

$$I\ddot{\theta} = I\dot{\varphi}^2 \sin\theta \cos\theta - p^\psi \dot{\varphi} \sin\theta. \tag{12.162}$$

However, we have an additional conservation law: energy conservation, since the system has time-translational invariance. The Hamiltonian becomes simply the kinetic energy:

$$H = \frac{1}{2}I\dot{\theta}^2 + \frac{1}{2I_3}(p^\psi)^2 + \frac{1}{2}I\sin^2\theta\dot{\varphi}^2 = \text{constant,} \tag{12.163}$$

which means that we have three first-order differential equations for three angular variables. Therefore, the problem can be reduced to quadrature, expressing all variables in terms of integrals. First, we use Eq. (12.161) to solve for $\dot{\varphi}$:

$$\dot{\varphi} = \frac{p^\varphi}{I\sin^2\theta} - \frac{p^\psi \cos\theta}{I \sin^2\theta}. \tag{12.164}$$

We then substitute this into Eq. (12.159) and solve for $\dot{\psi}$:

$$\dot{\psi} = \frac{p^{\psi}}{I_3} - \frac{p_{\varphi} \cos\theta}{I \sin^2\theta} - \frac{I_3}{I} p^{\psi} \frac{\cos^2\theta}{\sin^2\theta}. \qquad (12.165)$$

Finally, we substitute these into Eq. (12.163) for H:

$$H = \frac{1}{2} I \dot{\theta}^2 + \frac{1}{2 I_3} (p^{\psi})^2 + \frac{1}{2} I \sin^2\theta \left(\frac{p^{\varphi}}{I \sin^2\theta} - \frac{p^{\psi}}{I} \frac{\cos\theta}{\sin^2\theta} \right)^2, \qquad (12.166)$$

which we can solve for $\dot{\theta}$, then integrate to find $\theta(t)$. From this, we work backward to find $\varphi(t)$ and $\psi(t)$ using Eqs. (12.164) and (12.165). We will come back to this treatment later. For now, we will proceed to analyze the system using more physical techniques.

First, note that the spin angular momentum of the vase must be conserved: this is because gravity acts at its center of mass, and hence there is no torque about this point. This implies that the spin angular momentum vector $\mathbf{L}$ about the center of mass always points in the same direction and has the same magnitude. For convenience, let us align our lab coordinate system so that $\mathbf{L}$ points along the z axis, and the orientation of the vase is then described using the Euler angles starting from this base configuration, as shown in Figure 12.18.

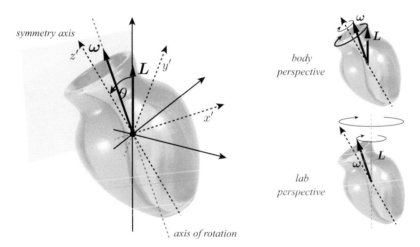

Fig. 12.18 The angular momentum and angular velocity vectors of the tumbling vase. z', $\boldsymbol{\omega}$, and τ lie in a plane.

In body frame coordinates we have

$$L_1 = I \omega_1, \quad L_2 = I \omega_2, \quad L_3 = I_3 \omega_3. \qquad (12.167)$$

Note that the proportionality constants I between $\mathbf{L}$ and $\boldsymbol{\omega}$ in the x' and y' directions are equal, which implies that these two vectors lie in the same plane as the z' axis of the body frame. However, $\mathbf{L}$ and $\boldsymbol{\omega}$ do not point in the same direction: depending on whether $I_3 > I$ or $I_3 < I$, the two vectors change relative position while lying in the same plane with the z' axis. Next, note from Eq. (12.159) that

p_ψ is simply $I_3\omega_3$. Therefore, L_3 is constant and so is the z' component of the angular velocity. But $L_3 = L\cos\theta$, where L is constant as well, so θ must be a constant in time. These observations imply that from the body-frame perspective $\mathbf{L}$ and $\boldsymbol{\omega}$ can only rotate about the z' axis while remaining coplanar with it. In contrast, since $\mathbf{L}$ should appear fixed from the perspective of the lab, this implies that – from the perspective of the lab – the z' axis of the body and $\boldsymbol{\omega}$ can only spin around $\mathbf{L}$ while remaining coplanar. We have thus determined that the torque-free motion may involve a tumble, *with the body's symmetry axis spinning about its fixed angular momentum vector* (see Figure 12.18).

Let us summarize the analysis. With our choice of aligning $\mathbf{L}$ with the lab's z axis, we have chosen our coordinate system in the lab judiciously so that θ is a constant in time, and we have just argued that this is always possible because of the conservation of angular momentum and the axial symmetry of the rigid body. If the initial conditions are such that the axis of symmetry of the vase is not aligned with its initial angular momentum, there ensues a tumbling motion about the fixed angular momentum vector at a constant inclination angle. We now want to determine (i) how fast the symmetry axis of the vase spins about $\mathbf{L}$ as seen from the lab perspective and (ii) how fast $\mathbf{L}$ spins about the vase's symmetry axis as seen from the body frame's perspective.

If θ is a constant fixed by the initial conditions, Eqs. (12.164) and (12.165) immediately imply that both $\dot\psi$ and $\dot\varphi$ must be constants. Hence, the tumbling happens at a constant spin rate. Using Eq. (12.28) and the fact that $\dot\theta = 0$, we can write the components of the angular velocity vector in the body frame as

$$\boldsymbol{\omega} = (\omega_1, \omega_2, \omega_3) = (\dot\varphi\sin\theta\sin\psi, \dot\varphi\sin\theta\cos\psi, \dot\varphi\cos\theta + \dot\psi). \tag{12.168}$$

This implies that $\boldsymbol{\omega}$ is spinning about the z' axis – as seen in the body frame – with a spin rate $\dot\psi$ (since the x' and y' components come with $\sin\psi$ and $-\cos\psi$ factors, respectively). To find $\dot\psi$, we need to look back at Eq. (12.165). Constant θ implies, from Eq. (12.162), that

$$I\dot\varphi\cos\theta = p^\psi. \tag{12.169}$$

Using this in Eq. (12.164), we quickly get

$$p^\varphi = \frac{p^\psi}{\cos\theta}, \tag{12.170}$$

which we can then use in Eq. (12.165) to find that

$$\dot\psi = \left(1 - \frac{I_3}{I}\right)\frac{p^\psi}{I_3} = \left(1 - \frac{I_3}{I}\right)\omega_3. \tag{12.171}$$

We have therefore determined the rate at which both $\mathbf{L}$ and $\boldsymbol{\omega}$ spin about the z' axis in the body frame in terms of the spin rate of the rigid body about its symmetry axis.

This effect can be seen for the earth itself. Treating the earth as a rigid oblate spheroid[3] with axial symmetry $\hat{I}_1 = \hat{I}_2 = I$, the symmetry axis of the planet is slightly tilted away from its spin axis. The torque on the earth due to the moon and sun are small, so from our perspective on the earth – the body frame's perspective – we should see the spin axis tumble or nutate around the symmetry axis at a rate given by Eq. (12.171). Calculating values for I and I_3 using measured radii of the earth, this gives

$$\dot{\psi} \simeq \frac{\omega_3}{300}. \tag{12.172}$$

Now $2\pi/\omega^{z'}$ is one day long, so we find a predicted period of tumbling $P = 2\pi/\dot{\psi} = 300$ days. It turns out, however, that the earth is not really *rigid*: its shape deforms slightly due to its elasticity. The measured tumbling period is roughly $P = 430$ days instead, and is called the "Chandler wobble." The difference between the predicted 300 days and this measured value is accounted for by the malleability of the planet.

It is also interesting to find the spin rate at which the symmetry axis tumbles around the angular momentum vector as seen from the *lab's perspective*. For this, we need to look at the components of $\boldsymbol{\omega}$ in the lab frame, given by Eq. (12.40) with $\dot{\theta} = 0$:

$$\boldsymbol{\omega} = (\omega^x, \omega^y, \omega^z) = \left(\dot{\psi} \sin\theta \sin\varphi, \dot{\psi} \sin\theta \cos\varphi, \dot{\psi} \cos\theta + \dot{\varphi} \right), \tag{12.173}$$

which immediately tells us that this spin rate is given by $\dot{\varphi}$ (given the sine and cosine factors in the x and y components). From Eq. (12.169), we have

$$\dot{\varphi} = \frac{1}{I} \frac{p^{\psi}}{\cos\theta}. \tag{12.174}$$

But it is possible to write this in a simpler form. The magnitude of $\mathbf{L}$ is given by

$$L^2 = I^2 \left(\dot{\varphi}^2 \sin^2\theta + \dot{\theta}^2 \right) + I_3^2 \left(\dot{\psi} + \dot{\varphi} \cos\theta \right)^2, \tag{12.175}$$

where we evaluate it in the body frame: L^2 being the squared length of a vector, it is a scalar quantity that does not change between the body and lab perspectives. Using the conservation laws from above, we can then see that

$$L^2 = 2HI + (p^{\psi})^2 \left(1 - \frac{I}{I_3} \right), \tag{12.176}$$

confirming that the magnitude of $\mathbf{L}$ is constant. Using the fact that $\mathbf{L}$ is aligned along the z axis, one can show that this simplifies to

$$L = \frac{I_3 p^{\psi}}{\cos\theta}. \tag{12.177}$$

---

[3] A shape in which a sphere is somewhat compressed along its axis of rotation.

We leave this step as an exercise to the reader. We then simply get

$$\dot{\varphi} = \frac{L}{I},$$

(12.178)

which is the spin rate at which the symmetry axis of the rigid body tumbles around the angular momentum vector – as seen from the perspective of the lab.

12.12 Euler's Equations of Motion and Stability

Now suppose we have a set of principal axes fixed in the body, with no restrictions on the relative sizes of the principal moments of inertia. The components of angular momentum of the rigid body for arbitrary rotations are then

$$L = I_1\omega_1 + I_2\omega_2 + I_3\omega_3.$$

(12.179)

Now we can take the time derivative of $\mathbf{L}$, which will be different in the rotating "body frame," the frame in which the rigid body is instantaneously at rest, than in the inertial "lab frame" or "space frame," in which the rigid body is instantaneously rotating with angular velocity $\boldsymbol{\omega}$. In Chapter 9 we showed how to translate the time derivative of any vector $\mathbf{A}$ between these two reference frames. The result was

$$\left(\frac{d\mathbf{A}}{dt}\right)_{\text{lab}} = \left(\frac{d\mathbf{A}}{dt}\right)_{\text{body}} + \boldsymbol{\omega} \times \mathbf{A},$$

(12.180)

so translating the angular momentum $\mathbf{L}$ of the rigid body, and using the fact that the time rate of change $d\mathbf{L}/dt$ is equal to the net torque $\mathbf{N}_{\text{tot}}$ acting on the body, then

$$\mathbf{N}_{\text{tot}} = \left(\frac{d\mathbf{L}}{dt}\right)_{\text{lab}} = \left(\frac{d\mathbf{L}}{dt}\right)_{\text{body}} + \boldsymbol{\omega} \times \mathbf{L}.$$

(12.181)

Now along the first principal axis, $(dL_1/dt)_{\text{body}} = I_1\dot{\omega}_1$ and

$$(\boldsymbol{\omega} \times \mathbf{L})_1 = (\omega_2 L_3 - \omega_3 L_2) = (\omega_2 I_3\omega_3 - \omega_3 I_2\omega_2) = (I_3 - I_2)\omega_2\omega_3,$$

(12.182)

with similar results along the other two axes, so altogether

$$N_1 = I_1\dot{\omega}_1 + (I_3 - I_2)\omega_2\omega_3,$$
$$N_2 = I_2\dot{\omega}_2 + (I_1 - I_3)\omega_3\omega_1,$$
$$N_3 = I_3\dot{\omega}_3 + (I_2 - I_1)\omega_1\omega_2,$$

(12.183)

which are *Euler's equations* for rigid-body motion.

An interesting application of the Euler equations is to investigate the stability of rotation about the principal axes in the case where no external torque is being applied. It is easy to see from the equations that in this case, if $\omega_1 \neq 0, \omega_2 = \omega_3 = 0$, then ω_1 remains constant and nonzero, and ω_2 and ω_3 remain zero. Nothing changes. But this could never be done perfectly; inevitably if we try to spin the rigid body only about the first principal axis, it will have some small components of rotation about the other principal axes as well. The question is: Is such a realistic spin stable? That is, do the other components of rotation remain small, in which case the rotation is stable, or do they grow, so that the rotation about the first principal axis is unstable?

Let us suppose no net torque acts, that all three principal moments of inertia I_1, I_2, I_3 are different, and that ω_1 is large and ω_2 and ω_3 are both small. Then the product $\omega_2 \omega_3$ in the first Euler equation is *very* small, so we can safely neglect it. In that case $I_1 \dot{\omega}_1 \simeq 0$, so $\omega_1 \simeq$ constant. This simplifies the remaining Euler equations, converting them to

$$I_2 \dot{\omega}_2 + [(I_1 - I_3)\omega_1]\omega_3 \simeq 0,$$
$$I_3 \dot{\omega}_3 + [(I_2 - I_1)\omega_1]\omega_2 \simeq 0, \qquad (12.184)$$

which are coupled linear equations. An easy way to decouple them is to differentiate the first of these equations and substitute in the second, and to differentiate the second and substitute in the first. After rearranging, the results are

$$\ddot{\omega}_2 + \Omega^2 \omega_2 = 0 \quad \text{and} \quad \ddot{\omega}_3 + \Omega^2 \omega_3 = 0, \qquad (12.185)$$

where

$$\Omega^2 \equiv \frac{[(I_1 - I_2)(I_1 - I_3)]\omega_1^2}{I_2 I_3}. \qquad (12.186)$$

Now notice that Ω^2 can be positive or negative, depending upon the relative sizes of the principal moments of inertia. We have assumed (without loss of generality) that the rigid body is rotating primarily about ω_1, with small amounts of ω_2 and ω_3. So if I_1 is the *largest* moment of inertia or the *smallest* moment of inertia, then either both $(I_1 - I_2)$ and $(I_1 - I_3)$ are positive or both are negative, so Ω^2 is positive and both the ω_2 and ω_3 equations are simple harmonic oscillator equations, with solutions of the form $\sin \Omega t$ or $\cos \Omega t$. This means that the axis of rotation moves around the surface of a cone whose symmetry axis is the first principal axis, and the half-angle of the cone is the small angle

$$\alpha = \arctan\left(\frac{(\omega_2)_{max}}{\omega_1}\right) = \arctan\left(\frac{(\omega_3)_{max}}{\omega_1}\right). \qquad (12.187)$$

This means the motion of the rigid body is stable, in that the angle α remains small, and the spin axis of the rigid body stays always close to what we have called the first principal axis. There is of course a third possibility, that I_1 is the *intermediate* moment of inertia, corresponding to either $I_2 > I_1 > I_3$ or $I_3 > I_1 > I_2$. In that case $\Omega^2 < 0$, so that the solutions of the equations of motion are real exponentials. In this case the angular velocities ω_2 and ω_3 will necessarily grow with time, and so rotation about ω_1 will be *unstable*. This can easily be seen by tossing an object like a book or tennis racquet having three different principal moments of inertia. If the book or racquet is spun about an axis which has the largest or smallest moment of inertia, the motion will be quite stable. It keeps spinning about this axis or close to it. But if the book or racquet is thrown with a spin about its intermediate moment of inertia axis the body axes depart quickly away from this initial axis, so that its motion is unstable, and the object appears to tumble.

Example 12.12

A Graphical Picture of Torque-Free Motion

Let us revisit the torque-free rigid-body problem with $I_1 \neq I_2 \neq I_3$ using a very interesting graphical approach. Once again, the angular momentum and kinetic energy are conserved. The squared magnitude of $\mathbf{L}$ can be written in body-frame components as

$$L^2 = L_1^2 + L_2^2 + L_3^2, \tag{12.188}$$

using the usual principal axes decomposition. The kinetic energy can be written as

$$T = \frac{L_1^2}{2 I_1} + \frac{L_2^2}{2 I_2} + \frac{L_3^2}{2 I_3}. \tag{12.189}$$

Both of these quantities must be constants in time. If we were to visualize the problem in angular momentum space as in Figure 12.19, a constant L^2 describes a sphere of radius L where the axes are $L_1, L_2,$ and L_3. The kinetic energy, in contrast, describes an ellipsoid with radii $2 T I_1, 2 T I_2,$ and $2 T I_3$. Without loss of generality, we can arrange that

$$I_1 \geq I_2 \geq I_3, \tag{12.190}$$

which implies that

$$\sqrt{2 T I_3} \leq L \leq \sqrt{2 T I_1}. \tag{12.191}$$

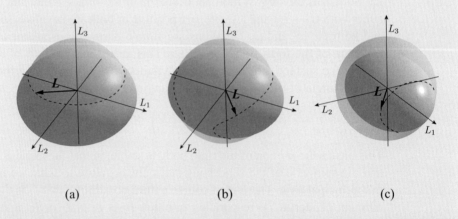

(a)   (b)   (c)

Fig. 12.19 The surfaces of constant angular momentum and kinetic energy for a rigid body. For torque-free dynamics, the angular momentum traces the intersection of the two surfaces as seen from the body frame. We see that starting out a spin along the body y' axis (see the dot in (b)) is unstable; whereas the tumbling about the x' and y' is stable in that the angular momentum circles the x' and y' axes, respectively.

To see this, eliminate L_1 or L_3 between Eq. (12.188) and Eq. (12.189), and then use Eq. (12.190). Therefore the minimum radius of the kinetic energy ellipsoid is always smaller than the radius of the angular momentum sphere, while its maximum radius is always greater. This means that the two surfaces necessarily intersect. The intersection is a curve (or a point in special cases) describing the allowed values for the components of **L**. Therefore in general the angular momentum vector tumbles around as seen from the body perspective – tracing interesting closed curves in this angular momentum space (see Figure 12.19). We can say even more from this qualitative picture. Looking at the intersection curves between the two surfaces near each of the three axes x', y', and z', we see that starting with a configuration where the angular momentum is aligned near the y' axis leads to an unstable tumble! The tip of the angular momentum vector traces a large curve *away* from the y' axis, whereas for the x' and z' axes we get intersection curves *encircling* the corresponding axis – describing a precession of the angular momentum vector. That is, the axis with moment of inertia inbetween the smallest and largest – axis y' – is unstable for tumbling under torque-free dynamics. You can test this easily by throwing a book in the air – preferably not this one – starting a spin about each of three possible symmetry axes of the book. The axis perpendicular to the book's face is x' – the one with largest angular momentum – since the mass of the book is spread away from this axis most; the short side is perpendicular to z' – the one with the smallest angular momentum; and the long side is perpendicular to the unstable axis y'. This qualitative analysis confirms that the tumbling we explored quantitatively for a cylindrically symmetric object is rather general: whenever the initial angular momentum vector is not aligned with the symmetry axes, we should expect a tumbling motion.   ∎

12.13 Gyroscopes

We are now ready to take on the full problem of a rigid body under the influence of *nonzero* torques. This more complicated scenario lends itself to the powerful technology of Lagrangian mechanics. The setup is as shown in Figure 12.20:

A cylindrically symmetric top is pivoted at one endpoint while spinning about its axis of symmetry. Uniform gravity pulls on it as though it were all located at its center of mass, and so causes a nonzero torque about the pivot point. We now have a *gyroscope*.

The problem is surprisingly similar to the torque-free system we discussed before. The difference is simply that we have translated the pivot from the center of mass to the endpoint, so that the full motion is naturally described about the fixed pivot. By the parallel axis theorem, I_3 remains unchanged while I shifts:

$$I = I_0 + M l^2, \tag{12.192}$$

where I_0 is the moment of inertia about the direction transverse to the symmetry axis about the center of mass, and l is the distance from the pivot to the center of mass. The orientation of the body is described by the same Euler angles, and the kinetic energy about the new pivot is simply given by Eq. (12.156):

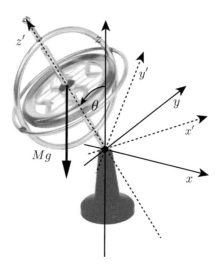

Fig. 12.20   A pivoted gyroscope under the influence of gravity.

$$T = \frac{1}{2}I(\dot{\theta}^2 + \dot{\varphi}^2 \sin^2 \theta) + \frac{1}{2}I_3(\dot{\psi} + \dot{\varphi}\cos\theta)^2, \qquad (12.193)$$

without the translational piece – which we also dropped in the previous discussion, since the center of mass dynamics decoupled from the rotational motion. The contact force at the pivot does no work and so, as usual, is left out of the Lagrangian. It is accounted for implicitly by the fact that the pivot is fixed in the lab frame. The total potential energy is once again entirely gravitational:

$$U = MgZ, \qquad (12.194)$$

where Z is the height of the center of mass of the gyroscope. As a function of the orientation angle:

$$Z = -l \cos \theta, \qquad (12.195)$$

as can be seen from the figure. So as far as rotational dynamics is concerned, the only change between the Lagrangian considered in the torque-free example earlier and the current situation is the addition of the term $+Mgl \cos\theta$ to the Lagrangian. Therefore the only equation of motion that changes is the θ equation, and we still have the three conservation laws: one from each of Eq. (12.159) and Eq. (12.161):

$$p^{\psi} = I_3\left(\dot{\psi} + \dot{\varphi}\cos\theta\right), \quad p^{\varphi} = I\dot{\varphi}\sin^2\theta + p^{\psi}\cos\theta, \qquad (12.196)$$

and a third from energy conservation, now including the contribution from $Mgl \cos\theta$:

$$H = \frac{1}{2}I\dot{\theta}^2 + \frac{1}{2I_3}(p^{\psi})^2 + \frac{1}{2}I\sin^2\theta\dot{\varphi}^2 + Mgl \cos\theta. \qquad (12.197)$$

We can now proceed using the same strategy, writing a decoupled differential equation in θ by eliminating $\dot{\psi}$ and $\dot{\varphi}$ from the energy equation:

$$\frac{1}{2}\dot{\theta}^2 + \frac{Mgl}{I}\cos\theta + \frac{\left(p^\varphi - p^\psi\cos\theta\right)^2}{2\,I^2\sin^2\theta} - \frac{2HI_3 - \left(p^\psi\right)^2}{2\,II_3} = 0. \qquad (12.198)$$

The $g \to 0$ limit takes us back to the torque-free case given by Eq. (12.166). One must then solve this differential equation for $\theta(t)$, from which one can determine $\psi(t)$ and $\varphi(t)$ using the conservation equations (12.196).

To proceed, it is useful to change variables to

$$u = \cos\theta, \qquad (12.199)$$

so the differential equation for $\theta(t)$ then becomes

$$\frac{1}{2}\dot{u}^2 + \frac{1}{2}\frac{\left(p^\varphi - up^\psi\right)^2}{I^2} + \frac{1}{2}\left(\frac{2HI_3 - \left(p^\psi\right)^2}{II_3} - \frac{2Mgl}{I}u\right)\left(u^2 - 1\right) = 0. \quad (12.200)$$

If we turn off gravity, with $g \to 0$, one can show that $\dot{u} = 0$ is a possible solution: i.e., θ is constant, as already determined in the torque-free case, and we can also align **L** with the lab z axis. Therefore, the new interesting physics coming from the nonzero torque has to do with making the angle θ, the angle between the angular momentum and the axis of symmetry, change with time. This phenomenon is known as a **nutation** of the symmetry axis. As the gyroscope precesses around the pivot point, its center of mass can also rise or fall.

The full solution to Eq. (12.200) can be written in terms of elliptic integrals. However, it is more instructive to provide a qualitative analysis using energy considerations. We can think of the fictitious problem of a particle of unit mass moving in one dimension and whose position is denoted by $-1 \le u \le 1$, while Eq. (12.200) provides an effective potential

$$U_{\text{eff}}(u) = \frac{1}{2}\frac{\left(p^\varphi - up^\psi\right)^2}{I^2} + \frac{1}{2}\left(\frac{2HI_3 - \left(p^\psi\right)^2}{II_3} - \frac{2Mgl}{I}u\right)\left(u^2 - 1\right), \quad (12.201)$$

where the total "energy' is zero, as implied by the right-hand side of Eq. (12.200). Therefore, the physics is hidden within the shape of this effective potential, in its maxima and minima. Figure 12.21 shows a generic profile of this effective potential. We have a polynomial cubic in u, and when $u \to \pm\infty$ we find that

$$U_{\text{eff}}(u) \to -\frac{Mgl}{I}u^3, \qquad (12.202)$$

from which we deduce that $U_{\text{eff}}(u)$ slopes downward for large positive u, and slopes upward for large negative u, so gives the shape depicted in the figure. Since the total energy is zero, the zeroes of the potential are turning points. The polynomial can have one, two, or three zeroes. The physical region is given by $-1 \le u \le 1$ along the u axis, since the total "energy" is zero. Different scenarios are explored by adjusting the sizes of the various constant coefficients of the polynomial p^φ, p^ψ, H, I, I_3, M, and l. But only four parameters matter, since we can write the effective potential as $U_{\text{eff}}(u) = d((a - u)^2 + (b - cu)(u^2 - 1))$ for constants a, b, c, and d.

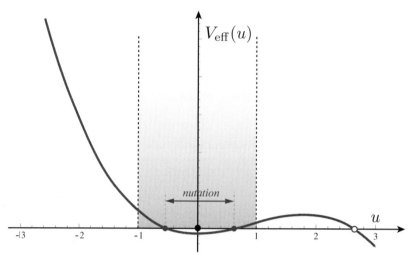

Fig. 12.21   The effective potential of the gyroscope's declination angle θ in terms of the variable $u = \cos\theta$. In the case shown, two zeroes of the potential lie in the physical region $-1 \le u \le 1$ and correspond to turning points: the gyroscope's z' axis nutates between two values of θ determined by these zeroes. Meanwhile, φ and ψ describe the gyroscope's spin around the lab's z axis and the gyroscope's z' axis, respectively.

Figure 12.22 shows different possible nutation patterns hat can result depending on the values of a, b, c, and d.

Interesting special cases can be investigated in more detail. One is the $u = $ constant scenario for all t. Another is to start off with $\dot{u} = 0$ at $t = 0$, i.e., the gyroscope is released from rest. Yet another is for stable solutions near $u = 1$, when $\theta \simeq 0$. We leave these various special cases to the Problems section at the end of this chapter. The main message is always the same: the gyroscope precesses about the vertical, instead of falling down; it hovers and nutates under the influence of the gravitational torque. Another way to view this behavior is to say that the gyroscope "wants" to always point in the same direction $\theta = 0$; and when perturbed by (say) nudging the pivot, it will oscillate about this direction instead of losing its balance. Hence, gyroscopes can be used in navigation, to help encode in memory a direction in space of particular interest, a reference ray.

12.14 Summary

In this chapter we took on the difficult problems of tumbling rigid bodies and gyroscopes. We found that the potentially complex dynamics can readily be

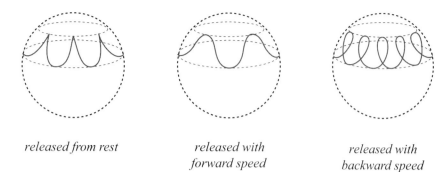

released from rest   *released with*   *released with*
                             forward speed   *backward speed*

Fig. 12.22   The nutation pattern of a gyroscope traced out by the gyroscope's z' axis. The spinning about the vertical axis is the familiar tumble, or *precession*. But now we also have a superimposed nutation as gravity tries to pull the z' axis downward.

understood using the powerful tools of Lagrangian mechanics, conservation laws, and Euler angles. The key is to focus on identifying the degrees of freedom, then divide the motion into translation and rotational parts, using a judicious choice for a tagged point. From that point on, the problem becomes algorithmic: write the Lagrangian using Euler angles, determine the equations of motion and conservation laws, and hope that the system is integrable. If not, one falls back on diagrammatic, approximation, and/or numerical techniques, as usual. We showed that rigid bodies generically have an axis about which they can tumble in a stable pattern that typically involves both a precession and a nutation. And we also described the motions of gyroscopes, which can be used to help orient objects on earth and in space. They are used in submarines, surface vehicles, aircraft, and in space for purposes of orientation and stability. The Hubble Space Telescope, for example, has six onboard gyroscopes, of which three must be operating at any given time to provide the stability required to take time images of distant objects anywhere in the sky. If the telescope turns slightly, one or more of the gyroscopes can detect the turn and signal other devices to bring the telescope back into line.

Problems

★   **Problem 12.1** Three objects, starting from rest at the same altitude, roll without slipping down an inclined plane. One is a ring of mass M and radius R; another is a uniform-density disk of mass $2M$ and radius R, and the last is a uniform-density sphere of mass M and radius $2R$. In what order do they reach the bottom of the inclined plane?

★   **Problem 12.2** A cylindrical pole is inserted into a frozen lake so the pole stands vertically. One end of a rope is attached to a point on the surface of the pole near where it enters the ice, and the rope is then laid out in a straight line on the surface. An ice skater with initial velocity v_0 approaches the opposite end of the rope, moving perpendicular to the rope. As she reaches the rope she grabs it and holds on, so the rope winds up around the pole. (a) When the rope is half wound up, what is her speed, assuming there is no friction between the rope or herself and the ice? (b) Has there been any change in her kinetic energy? If so, identify what positive or negative work has been done on her, and by what source. (c) Has there been any change in her angular momentum? If so, identify the source of the torque done upon her.

★   **Problem 12.3** Humanity collectively uses energy at the average rate of about 18 terawatts. (a) At that rate, after 1 year how long would the length of the day have increased if during that year we were able to power our activities purely by harnessing the rotational kinetic energy of the earth? (The earth has moment of inertia $I = 0.33MR^2$ about the north–south polar axis.) (b) The moment of inertia of a sphere is often given as $I = (2/5)MR^2$. What is the primary reason why this is incorrect for the earth?

★   **Problem 12.4** In a supernova explosion, the core of a heavy star collapses and the outer layers are blown away. Before collapse, suppose the core of a given star has twice the mass of the sun and the same radius as the sun, and rotates with period 20 days. The core collapses in a few seconds to become a neutron star of radius 20 km. (a) Estimate its new period of rotation. (b) Estimate the ratio of the final rotational kinetic energy to the initial rotational kinetic energy of the core. What could account for the change?

★   **Problem 12.5** In some theoretical models of pulsars, which are rotating neutron stars, the braking torque slowing the pulsar's spin rate is proportional to the nth power of the pulsar's angular velocity Ω; that is, $\dot{\Omega} = -K\Omega^n$, where K is a constant. (a) Find a formula for the time rate of change of the pulsar period $\dot{P}$ in terms of P itself and the constants n and K. (b) For the Vela and Crab pulsars, at least, the product $P\dot{P} =$ constant. What is their braking index n?

★★   **Problem 12.6** Tidal effects of the moon on the earth have caused the earth's rotation rate to slow, thus reducing the spin angular momentum of the earth, leading to an increase in earth's day by 0.1 s in the past 3800 years. This reduction has been made up for by an increase in the orbital angular momentum of the moon around the earth. Therefore how much farther is the moon now from the earth than it was 3800 years ago? (Note that the moment of inertia of the earth about its axis is $I = 0.33MR^2$ and that the mean earth–moon distance is 3.8×10^5 km.)

★   **Problem 12.7** Compute the moment of inertia matrix of a solid circular cylinder of height H and base radius R, and of uniform mass density $\rho = \rho_0$. In this expression, the cylinder is arranged so that its symmetry axis is along the z axis and its top cap sits on the x, y plane; i.e., the cylinder extends from $z = -H$ to $z = 0$. Compute all entries of the moment of inertia matrix with respect to the origin in this configuration.

Problem 12.8 A rod of length ℓ and mass m is attached to a pivot on one end. The rim of a disk of radius R and mass M is attached to its other end in such a way that the disk can pivot in the same plane in which the rod is restricted to swing. Find the Lagrangian and equations of motion.

Problem 12.9 (a) Using Euler angles, write the constraint of rolling without slipping for a sphere of radius R moving on a flat surface. (b) Write the Lagrangian and equations of motion using Lagrange multipliers. (c) Show that the rotational and translational kinetic energies are independently conserved.

Problem 12.10 (a) Find the moment of inertia I_{zz} of a thin disk of mass m and radius R about an axis through its center and perpendicular to the plane of the disk. (b) What are I_{xx} and I_{yy} in this case? (c) A solid cylinder of mass M, radius R, and length L can be considered to be a stack of disks. Use the parallel axis theorem to help find the principal moments of inertia of the cylinder whose origin is at the center of the cylinder.

Problem 12.11 A private plane has a single propeller in front, which rotates in the clockwise sense as seen by the pilot. Flying horizontally, the pilot causes the tail rudder to extend out to the left from the plane's flight direction. (a) If the plane is ultralight and the propeller is large, heavy, and rotates fast, what is the primary response of the plane? (b) If instead the plane is heavy and the propeller is small, light, and rotates slowly, what then is the plane's primary response?

Problem 12.12 A uniform-density cone has mass M, base radius R, and height H. Find its inertia matrix if the origin is at the center of the circular base in the x, y plane, the axis of symmetry is along the z axis, and the apex of the cone is at positive z.

Problem 12.13 The Crab Nebula is a bright, reddish nebula consisting of the debris from a supernova explosion observed on earth in 1054 AD. The estimated total power it emits, mostly in X-rays, UV, and visible light, is of order 10^{31} W. The nebula harbors a pulsar in its center, which emits a pulsed light signal. Pulsars are rotating neutron stars, having a mass comparable to that of the sun (2.0×10^{30} kg) but a radius of only about 10 km. The period between successive pulses (the rotational period of the star) is $P = 0.033091$ s, which slowly increases with time, $dP/dt = 4.42 \times 10^{-13}$ s/s. Is it possible that the decreasing rotational energy of the star is the ultimate source of the energy observed in radiation?

Problem 12.14 A cylindrical space station is a hollow cylinder of mass M, radius R, and length D, and endcaps of negligible mass. It spins about its symmetry axis (z axis) with angular velocity ω_0. (a) Find its inertia matrix about its center. A meteor of mass m and velocity v_0, moving in the x direction, strikes the station very near one of the endcaps, and bounces directly back with velocity $-v_0/2$. After the collision, find the station's (b) CM velocity, (c) angular momentum, both magnitude and direction. (d) Show that subsequently the symmetry axis of the station rotates about the angular momentum vector, so the station wobbles as seen by an outside inertial observer. Find the period of this rotation.

★★ **Problem 12.15** (a) Find all elements of the principal moment of inertia matrix for a thin uniform rod of mass Δm and length D if the rod is oriented along the x axis and the origin of coordinates is at the center of the rod. (b) Use the parallel axis theorem, the perpendicular-axis theorem for thin lamina, and the result of part (a) to find the principal moment of inertia matrix for a thin square of side D and total mass ΔM that is perpendicular to the z axis, with the origin of coordinates at the center of the square. (c) Find the principal moment of inertia matrix for a cube of edge length D and mass M. (d) Find the moment of inertia for the cube about an axis parallel to one of the axes in part (c) and which is oriented along the middle of one face of the cube. (e) Find the moment of inertia for the cube about an axis parallel to one of the axes in part (c) and which is oriented along the length of one corner of the cube.

★★ **Problem 12.16** (a) Find the principal moments of inertia for a thin disk of mass Δm and radius R, if its mass density is uniform, the origin of coordinates is at the center of the disk, the x and y axes are in the plane of the disk, and the z axis is perpendicular to the disk. (b) Use this result to help find the principal moments of inertia of a uniform-density sphere of mass M and radius R_0, with origin at the center of the sphere. (c) The moment of inertia for rotation about the symmetry axis of a ring of mass Δm and radius r is $I = \Delta m r^2$. Use this fact to help find the moment of inertia about a symmetry axis for a thin spherical shell of mass ΔM and radius R, with origin at the center of the shell. (d) Use the result of part (c) to find the principal moments of inertia of a solid, uniform-density sphere of mass M and radius R_0. Compare with the result of part (b).

★ **Problem 12.17** Prove that none of the principal moments of inertia of a rigid body can be larger than the sum of the other two.

★ **Problem 12.18** (a) Find all elements of the moment of inertia matrix for a cube of mass M and edge length ℓ using its principal axes. (b) Then find all elements of the moment of inertia matrix for the cube if the axes have been turned by $45°$ about the original z axis.

★ **Problem 12.19** If the entire human race were to leave their current habitats, estimate how much the length of the day would be changed if (a) they gathered at the equator, (b) they gathered at the poles.

★ **Problem 12.20** Consider a square plane lamina with coordinate axes x, y in the plane with origin at the center of the square and which are perpendicular to edges of the square. If the moment of inertia about each of these two axes is I_0, what are the moments of inertia about axes x' and y' in the plane turned about the z axis by a $30°$ angle relative to the original two axes?

★★ **Problem 12.21** In the text we found the total angular velocity vector in the body frame of a rigid body in terms of the Euler angles and their time derivatives:

$$\omega = (\omega^{x'}, \omega^{y'}, \omega^{z'})$$
$$= (\dot\varphi \sin\theta \sin\psi + \dot\theta \cos\psi, \dot\varphi \sin\theta \cos\psi - \dot\theta \sin\psi, \dot\varphi \cos\theta + \dot\psi).$$

Show then that in the laboratory frame:

$$\omega = (\omega^{x}, \omega^{y}, \omega^{z})$$
$$= (\dot\psi \sin\theta \sin\varphi + \dot\theta \cos\varphi, -\dot\psi \sin\theta \cos\varphi + \dot\theta \sin\varphi, \dot\psi \cos\theta + \dot\varphi).$$

★★★   **Problem 12.22** An equilateral triangle of mass M and sidelength L is cut from uniform-density sheet metal. (a) Draw the triangle along with the three perpendicular bisectors, each of which extends from the middle of a side to the opposite vertex. Show that each bisector has length $\sqrt{3}L/2$. (b) Explain why the center of mass of the triangle must be located at the point where the three perpendicular bisectors intersect. Let this point be the origin. (c) Let the z axis be perpendicular to the triangle, the y axis be along one of the perpendicular bisectors, and the x axis be perpendicular to both. Find the moment of inertia matrix for the triangle in these coordinates.

★★   **Problem 12.23** Suppose for a given set of axes the moment of inertia matrix is

$$\left(\frac{m\ell^2}{24}\right) \begin{pmatrix} 1 & -1 & 0 \\ -1 & 1 & 0 \\ 0 & 0 & 2 \end{pmatrix}. \tag{12.203}$$

(a) Find the principal moments of inertia. (b) About what axis, and by what angle, should the original coordinate axes be turned to arrive at the principal axes?

★★   **Problem 12.24** Using the rotation matrices appropriate for each of the three Euler angles, find the overall 3×3 rotation matrix for arbitrary rotations in terms of the angles φ, θ, and ψ, applied in the prescribed order.

★★★   **Problem 12.25** A rigid body has principal moments of inertia $I_{xx} = I_0, I_{yy} = I_{zz} = 2I_0/3$. (a) Find all elements of the moment of inertia matrix in a reference frame that has been rotated by $30°$ about the z axis in the counterclockwise sense relative to the initial axes. (b) In this new (primed) frame the moment of inertia matrix has the form

$$\begin{pmatrix} I'_{xx} & I'_{xy} & I'_{xz} \\ I'_{yx} & I'_{yy} & I'_{yz} \\ I'_{zx} & I'_{zy} & I'_{zz} \end{pmatrix},$$

where the nine entries were found in part (a). Now pretending that you do not already know the answer, diagonalize this matrix to find the principal moments of inertia (that is, subtract I from each of the diagonal elements in the matrix, and then set the determinant of the resulting matrix equal to zero). This will give a cubic equation in I, which when solved will give the three principal moments of inertia.

★   **Problem 12.26** Show that any antisymmetric part of the moment of inertia matrix of a rigid body does not contribute to the body's equations of motion. Therefore we may safely assume that the moment of inertia matrix is symmetric.

★   **Problem 12.27** (a) Write the Lagrangian for the Euler problem of a rigid body undergoing torque-free precession. (b) Write the equations of motion and show that they agree with those in the text.

★★   **Problem 12.28** Write the six equations of motion for the Lagrangian of a hoop attached to a spring from the example in the text.

★★   **Problem 12.29** Show that the magnitude of the angular momentum vector for the torque-free rigid-body dynamics case is given by $L = I_{3'}p_\psi / \cos\theta$.

★   **Problem 12.30** Show that $\dot{u} = 0$ if $g = 0$ from Eq. (12.200).

★   **Problem 12.31** Show that if

$$I_1 \geq I_2 \geq I_3$$

for a torque-free rigid body, then we have

$$\sqrt{2TI_3} \leq L \leq \sqrt{2TI_1}$$

★   **Problem 12.32** Show that $u \to 1$ is a stable point for the gyroscope, and find the corresponding nutation. Show that there is a critical angular momentum $p_\psi = 2\sqrt{MglI}$.

★★   **Problem 12.33** If we start a gyroscope at an angle $u(0) = u_0$ with $\dot{u}(0) = 0$ and nonzero $\dot{\psi}$ but $\varphi = 0$, (a) show that the gyroscope nutates and find the maximum angle u_1 it reaches before bouncing back up. (b) Consider the can of a fast spin, where $p_\psi^2 \gg 2MglI$; find approximate forms for the two nutation angles and nutation frequency.

★★   **Problem 12.34** A rigid body has an axis of symmetry, which we designate as axis 1. The principal moment of inertia about this axis is I_1, while the principal moments of inertia about the remaining two principal axes are $I_2 = I_3 \equiv I_0 \neq I_1$. (a) Write the Euler equations of rotational dynamics in terms of I_1, I_0, and the three angular velocities $\omega_1, \omega_2, \omega_3$. (b) Show that ω_1 is constant. (c) Find a second-order linear differential equation for ω_2 and another for ω_3. (d) Does either the magnitude or direction of this precession depend upon whether the rigid body is prolate (like an American football or rifle bullet) or oblate (like a saucer or frisbee)? (e) Prove that the symmetry axis of the rigid body is coplanar with the angular velocity vector ω and with the angular momentum vector **L**.

★★   **Problem 12.35** Using Euler's equations, show that a rigid body rotating without applied torque has a total angular momentum whose magnitude is constant.

★★   **Problem 12.36** Find the product of two rotation matrices corresponding to successive rotations about (i) the x axis by angle α and (ii) the z axis by angle β, with (a) the *x* axis-rotation first, (b) the *z* axis-rotation first. (c) Then subtract the two results, to illustrate the fact that rotations do not generally commute. (d) By expanding sines and cosines for small angles up through terms of second order, illustrate the fact that infinitesimal rotations *do* commute if second-order effects are counted as negligible.

★★ **Problem 12.37** In 2004 a satellite was launched into a circular polar orbit, 642 km above the earth's surface, containing an experiment called "Gravity Probe B." The satellite remained in orbit for 16 months, flying successively over the north pole, the south pole, back over the north pole, etc. Four gyroscopes were on board, consisting of nearly perfectly spherical, uniform-density, fused-quartz balls 1.9 cm in radius. All were made to spin at about 4200 rpm. Strenuous efforts were made to reduce any torques on the gyroscopes to near zero, so any precession or other drifts caused by them could be minimized. Then if Newton's theory of gravity were correct, the spin direction of the gyros would always point toward the same place in the sky, some particular distant star, for example. According to general relativity, however, there should be two very small drifts in the gyro spin directions. First, there is the "geodetic effect" in which the spin direction should drift slightly *forward* in the orbit (*i.e.*, in the north–south direction), still in the plane of the orbit. Second, there is the "frame dragging" effect, in which the spin direction of the gyro should slowly drift in a direction *perpendicular* to the plane of the orbit (*i.e.*, in the east–west direction). It is only in polar orbit where the predicted geodetic and frame-dragging effects are perpendicular to one another, allowing both to be measured. The predictions from general relativity were that the geodetic effect should lead to a drift of 6.6061 arcseconds/year, while the frame-dragging effect should be 0.0392 arcseconds/year. The data showed a drift in the plane of the orbit of 6.60 ± 0.0183 arcseconds/year and a perpendicular drift of 0.0372 ± 0.0072 arcseconds/year, in good agreement with the predictions. (a) The density of fused quartz is 2.2 g/cm^3. What were the principal moments of inertia of each gyro? (b) When spun at 4200 rpm, what was the angular momentum of each gyro? (c) What was its kinetic energy? (This was sufficient to destroy the entire experiment if the gyro had touched its housing.) (d) Suppose a tiny torque acted on one of the gyros, causing it to precess. Estimate the maximum torque allowable to keep the precession within the quoted errors in drift rates given above.

13

Coupled Oscillators

In the preceding chapter on rigid-body motion we took a step beyond single-particle mechanics to explore the behavior of a more complex system containing many particles bonded rigidly together. Now we will explore additional sets of many-particle systems in which the individual particles are connected by linear, Hooke's-law springs. These have some interest in themselves, but more generally they serve as a model for a large number of coupled systems that oscillate harmonically when disturbed from their natural state of equilibrium, such as elastic solids, electric circuits, and multi-atom molecules. We will begin with the oscillations of a few coupled masses and end with the behavior of a continuum of masses described by a linear mass density. The mathematical techniques required to analyze such coupled oscillators are used throughout physics, including linear algebra and matrices, normal modes, eigenvalues and eigenvectors, and Fourier series and Fourier transforms.

13.1 Linear Systems of Masses and Springs

Suppose we set two blocks, each of mass m, upon a frictionless horizontal surface. Each block is attached to a stationary wall by a spring of force constant k, and the two springs are attached together by a spring of force constant k', all in a straight line, as shown in Figure 13.1.

When the blocks are at rest the springs are unstretched. Let the displacements of block 1 and block 2 from their equilibrium positions be denoted by x_1 and x_2, respectively, each positive to the right. Our goal is to find the differential equations of motion of each block, and then solve them to find $x_1(t)$ and $x_2(t)$. The Lagrangian of the system is

$$L = T(\dot{x}_1, \dot{x}_2) - U(x_1, x_2)$$

$$= \frac{1}{2}m(\dot{x}_1^2 + \dot{x}_2^2) - \frac{1}{2}kx_1^2 - \frac{1}{2}k'(x_2 - x_1)^2 - \frac{1}{2}kx_2^2, \qquad (13.1)$$

taking into account the potential energy stored in each of the three springs (note that the stretch in the middle spring is $x_2 - x_1$). Lagrange's equations give

$$m\ddot{x}_1 = -kx_1 + k'(x_2 - x_1),$$
$$m\ddot{x}_2 = -kx_2 - k'(x_2 - x_1), \qquad (13.2)$$

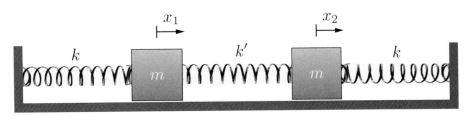

Fig. 13.1 A system of two blocks connected with springs and two walls.

which we could also have written down using $F = ma$ for each block.

We want to solve these coupled equations to find $x_1(t)$ and $x_2(t)$, given the initial conditions. The problem is that each equation involves both x_1 and x_2, so we have to begin by decoupling them. Looking closely at the two equations, we note that if we *sum* them we eliminate the terms with the difference $x_2 - x_1$, giving

$$m(\ddot{x}_1 + \ddot{x}_2) = -k(x_1 + x_2) \tag{13.3}$$

in terms of the single variable $x_1 + x_2$. If instead we *subtract* the first equation from the second, every resulting term contains only the difference $x_2 - x_1$:

$$m(\ddot{x}_2 - \ddot{x}_1) = -k(x_2 - x_1) - 2k'(x_2 - x_1)$$
$$= -(k + 2k')(x_2 - x_1). \tag{13.4}$$

That is, in terms of the new composite coordinates $\xi_1 \equiv x_2 + x_1$ and $\xi_2 \equiv x_2 - x_1$, the equations become

$$m\ddot{\xi}_1 + k\xi_1 = 0 \quad \text{and} \quad m\ddot{\xi}_2 + (k + 2k')\xi_2 = 0, \tag{13.5}$$

which are two *decoupled* simple harmonic oscillator equations. We then have sinusoidal solutions with two angular frequencies ω_1 and ω_2. A mathematically convenient way to write the solution is to extend ξ_1 and ξ_2 to the complex plane, so that

$$\xi_1 = c_1 e^{i\omega_1 t}, \quad \xi_2 = c_2 e^{i\omega_2 t}, \tag{13.6}$$

where c_1 and c_2 are arbitrary complex constants. This is possible since Eq. (13.5) are real and linear. Hence, we can take $\xi_1 = \xi_1^R + i\xi_1^I$ and $\xi_2 = \xi_2^R + i\xi_2^I$ as complex functions – implying that their real and imaginary parts $\{\xi_1^R, \xi_2^R\}$ and $\{\xi_1^I, \xi_2^I\}$ independently satisfy Eq. (13.5). Thus, we can write the solutions of (13.5) through the complex functions given by (13.6) as long as we remember that the physical solutions consist of the real parts only:

$$\xi_1 = \Re\left(c_1 e^{i\omega_1 t}\right), \quad \xi_2 = \Re\left(c_2 e^{i\omega_2 t}\right), \tag{13.7}$$

where the symbol $\Re$ means that we take the real part of what follows. Given that c_1 and c_2 are complex, we achieve the requisite four arbitrary constants for solutions of two second-order differential equations. Writing $c_1 \equiv A_1 - iB_1$ and $c_2 \equiv A_2 - iB_2$ (using minus signs for later convenience), and recalling Euler's identity $e^{i\theta} = \cos\theta + i\sin\theta$, the physical solutions take the form

$$\xi_1 = \Re(\xi_1) = A_1 \cos\omega_1 t + B_1 \sin\omega_1 t \qquad (13.8)$$

and

$$\xi_2 = \Re(\xi_2) = A_2 \cos\omega_2 t + B_2 \sin\omega_2 t, \qquad (13.9)$$

with four arbitrary (real) constants, which can be determined by the initial positions and velocities. As is obvious, these are much more cumbersome to write than the compact forms given by (13.6). We will henceforth work with solutions in the complex plane and remember that, at the end of any algebraic manipulations we can get to the physical solution by simply taking the real parts of ξ_1 and ξ_2.

Substituting (13.6) into the differential equations of motion (13.5), we find that

$$(k - m\omega_1^2)c_1 = 0 \quad \text{and} \quad ((k + 2k') - m\omega_2^2)c_2 = 0, \qquad (13.10)$$

implying

$$\omega_1 = \sqrt{\frac{k}{m}}, \quad \omega_2 = \sqrt{\frac{k + 2k'}{m}} \qquad (13.11)$$

if $c_1 \neq 0$ and $c_2 \neq 0$ in general.

Now consider two special cases.

1. Suppose that $c_2 = 0$, so $\xi_2 = x_2 - x_1 = 0$. In that case $x_1 = x_2 = (c_1/2)e^{i\omega_1 t}$, so each block oscillates with the same frequency ω_1, and with the same amplitude $c_1/2$. They oscillate *in phase* with one another, sliding back and forth on the table together, so that the middle spring is never stretched or compressed, as illustrated in Figure 13.2(a).

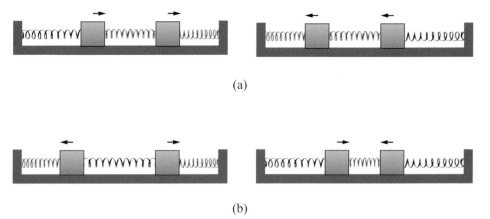

(a)

(b)

Fig. 13.2   The two normal modes of oscillation for the two-block system.

That is why the oscillation frequency $\omega_1 = \sqrt{k/m}$ is independent of k', and why it is the same as the frequency each block would have if it were simply oscillating on the end of a single spring of spring constant k. This motion is called a **normal mode of oscillation**, in which both blocks oscillate with the same frequency. In fact, it is said to be the *first* normal mode, in which the blocks have the same amplitude and the same phase.

2. Suppose instead that $c_1 = 0$, so $\xi_1 = x_1 + x_2 = 0$. In that case $x_2 = -x_1 = -(c_2/2)e^{i\omega_2 t}$, so each block oscillates with the same frequency $\omega_2 = \sqrt{(k + 2k')/m}$ and with equal but *opposite* amplitudes. That is, they move alternately apart and together, with the center of the middle spring remaining fixed, as illustrated in Figure 13.2(b). In effect, each block oscillates on the end of an outer spring plus *half* of the middle spring. The force constant of a half-spring is *twice* that of a full spring (because a half-spring is twice as stiff as the corresponding full spring, since it stretches only half as much for a given applied force). That is why the frequency in this case is $\omega_2 = \sqrt{(k + 2k')/m}$. This motion is the *second* normal mode of oscillation. The blocks move with the same frequency in opposition to one another, with equal but opposite amplitudes (i.e., they are $180°$ out of phase with one another).

With more general initial conditions with general c_1 and c_2, the evolution of x_1 and x_2 is a linear combination of these two normal modes. This method of solving the differential equations of motion, to find the normal-mode frequencies and relative amplitudes, involved the slightly clever guess that adding or subtracting the original $F = ma$ equations decouples them.

In this analysis, in decoupling the original differential equations we effectively *diagonalized* a system of linear equations. This procedure is general and can be developed best with the language of matrices in linear algebra. We will now discuss the same problem but use linear algebra techniques instead that we can later employ when generalizing to more complicated systems.

In a normal mode, *by definition each block oscillates with the same frequency*, so we can simply try the solutions

$$x_1 = b_1 e^{i\omega t}, \qquad x_2 = b_2 e^{i\omega t} \tag{13.12}$$

with the same frequency for each; and once again working in the complex plane provided we remember to take the real parts to connect to the physical solutions. Hence, b_1 and b_2 are complex constants. Substituting these into the original equations of motion (13.2), we get

$$(-m\omega^2 + (k + k'))b_1 - k'b_2 = 0,$$

$$-k'b_1 + (-m\omega^2 + (k + k'))b_2 = 0, \tag{13.13}$$

which in matrix form becomes

$$\begin{pmatrix} (k + k') - m\,\omega^2 & -k' \\ -k' & (k + k') - m\,\omega^2 \end{pmatrix} \begin{pmatrix} b_1 \\ b_2 \end{pmatrix} = 0, \tag{13.14}$$

forming a pair of homogeneous equations in the unknown amplitudes. For arbitrary frequencies, this symmetric matrix can be inverted, giving only the trivial solution $b_1 = b_2 = 0$, where both blocks remain at rest at their equilibrium positions. But from linear algebra we know that with just the *right* choice(s) of ω there are also *non*trivial solution(s) if and only if the determinant of the coefficients vanishes, *i.e.*, if and only if

$$\begin{vmatrix} (k+k') - m\omega^2 & -k' \\ -k' & (k+k') - m\omega^2 \end{vmatrix} = (-m\omega^2 + (k+k'))^2 - k'^2$$
$$= 0, \tag{13.15}$$

since a symmetric matrix with zero determinant cannot be inverted. From this so-called **secular equation** it follows that

$$-m\omega^2 + (k+k') = \pm k', \tag{13.16}$$

with the two solutions

$$\omega_1 = \sqrt{\frac{k}{m}} \quad \text{and} \quad \omega_2 = \sqrt{\frac{k+2k'}{m}}, \tag{13.17}$$

the same results we found for the frequencies using the previous approach. We have once again found the normal-mode frequencies, which are also called the **characteristic frequencies**, **eigenfrequencies**, or **frequency eigenvalues**.

Now we can substitute the normal-mode frequencies one at a time into the original differential equations of motion to find the relative amplitudes of the two blocks. First, let $\omega = \omega_1 = \sqrt{k/m}$. Then Eq. (13.14) become

$$\begin{pmatrix} k' & -k' \\ -k' & k' \end{pmatrix} \begin{pmatrix} b_1 \\ b_2 \end{pmatrix} = 0. \tag{13.18}$$

These two equations are linearly dependent (one is simply a multiple of the other in this case). Therefore the determinant of the matrix vanishes, ensuring that the matrix cannot be inverted. The single independent equation is then solved by $b_1 = b_2$. The two blocks slide back and forth in phase with equal amplitudes, just as we found earlier. If instead we substitute $\omega = \omega_2 = \sqrt{(k+2k')/m}$, then Eqs. (13.14) become

$$\begin{pmatrix} -k' & -k' \\ -k' & -k' \end{pmatrix} \begin{pmatrix} b_1 \\ b_2 \end{pmatrix} = 0, \tag{13.19}$$

so that $b_1 = -b_2$, where the two blocks slide in and out with equal but opposite amplitudes. Note that we can find a single relation between b_1 and b_2 for each normal-mode frequency – leading to $2 \times 1 = 2$ complex degrees of freedom in total, or equivalently 4 *real* degrees of freedom. That makes sense, because we originally had two second-order differential equations with the associated freedom of two real boundary conditions per equation.

We can write normalized **eigenvectors** corresponding to each of the eigenfrequencies in the form

$$\mathbf{e}_{(1)} \equiv \frac{1}{\sqrt{2}} \begin{pmatrix} 1 \\ 1 \end{pmatrix} \tag{13.20}$$

for the first normal mode by choosing $b_2 = 1/\sqrt{2}$ so that $b_1^2 + b_2^2 = 1$, and

$$\mathbf{e}_{(2)} \equiv \frac{1}{\sqrt{2}} \begin{pmatrix} -1 \\ 1 \end{pmatrix} \qquad (13.21)$$

for the second normal mode, choosing $b_2 = 1/\sqrt{2}$ once again. As a result, we have

$$\mathbf{e}_{(1)}^T \cdot \mathbf{e}_{(1)} = \mathbf{e}_{(2)}^T \cdot \mathbf{e}_{(2)} = 1, \quad \mathbf{e}_{(1)}^T \cdot \mathbf{e}_{(2)} = \mathbf{e}_{(2)}^T \cdot \mathbf{e}_{(1)} = 0, \qquad (13.22)$$

where "T" stands for transpose. Note that each eigenvector has the correct relative amplitudes of the two blocks for the corresponding normal mode. Each is also normalized to unity, and the eigenvectors are mutually orthogonal. That we were able to arrange for this is a general feature, as long as the matrix appearing in Eq. (13.14) is real and symmetric.

The differential equations are linear, so any linear combination of the two normal-mode solutions is also a solution. The most general solution of the equations is an arbitrary linear combination of the two normal modes. We can write

$$\begin{pmatrix} x_1(t) \\ x_2(t) \end{pmatrix} = c_1 \mathbf{e}_{(1)} e^{i\omega_1 t} + c_2 \mathbf{e}_{(2)} e^{i\omega_2 t}, \qquad (13.23)$$

where c_1 and c_2 are complex constants determined by the initial conditions. We then say the eigenvectors form an **orthonormal** set and span the space of solutions.

Let us look at a particular case and see how we can build a full solution given initial conditions. Say we have

$$\Re x_1(0) = \Re x_2(0) = 0, \quad \Re \dot{x}_1(0) = v_1, \quad \Re \dot{x}_2(0) = v_2 \qquad (13.24)$$

for some initial speeds v_1 and v_2. Note how to apply the boundary conditions on the real parts of x_1 and x_2 only. Hence both blocks begin at their respective origins, but each is given an initial kick. The subsequent motion of the two blocks will involve *both* normal-mode frequencies and will appear to be rather erratic, in spite of the fact that the motion is completely determined by the initial conditions. To find the solution, we use Eqs. (13.23)) and (13.24) and arrive at

$$0 = \Re(c_1)\mathbf{e}_{(1)} + \Re(c_2)\mathbf{e}_{(2)}, \quad \begin{pmatrix} v_1 \\ v_2 \end{pmatrix} = \omega_1 \Im(c_1)\mathbf{e}_{(1)} + \omega_2 \Im(c_2)\mathbf{e}_{(2)}, \quad (13.25)$$

where the second set of equations is obtained by differentiating (13.23) before setting t to zero, and where the symbol $\Im$ means that we take only the imaginary part of what follows. We then have

$$\Re(c_1) = \Re(c_2) = 0 \qquad (13.26)$$

from the first set of equations, since $\mathbf{e}_{(1)}$ and $\mathbf{e}_{(2)}$ are linearly independent, and

$$\begin{pmatrix} v_1 \\ v_2 \end{pmatrix} = \omega_1 \Im(c_1)\frac{1}{\sqrt{2}} \begin{pmatrix} 1 \\ 1 \end{pmatrix} + \omega_2 \Im(c_2)\frac{1}{\sqrt{2}} \begin{pmatrix} -1 \\ 1 \end{pmatrix}$$
$$= \frac{1}{\sqrt{2}} \begin{pmatrix} \omega_1 \Im(c_1) - \omega_2 \Im(c_2) \\ \omega_1 \Im(c_1) + \omega_2 \Im(c_2) \end{pmatrix} \qquad (13.27)$$

from the second set. This leads us to the solution

$$\Im(c_1) = \frac{1}{\sqrt{2}} \frac{v_2 + v_1}{\omega_1}, \quad \Im(c_2) = \frac{1}{\sqrt{2}} \frac{v_2 - v_1}{\omega_2}, \qquad (13.28)$$

where again the symbol $\Im$ means that we take only the imaginary part of what follows. We have uniquely determined the complex constants c_1 and c_2 from the given initial conditions. Putting all these together into Eq. (13.23), and taking the real part to obtain the physical solution, we get

$$x_1(t) \to \Re(x_1(t)) = \frac{1}{2}\left[\left(\frac{v_2 + v_1}{\omega_1}\right)\sin\omega_1 t - \left(\frac{v_2 - v_1}{\omega_2}\right)\sin\omega_2 t\right], \quad (13.29)$$

$$x_2(t) \to \Re(x_2(t)) = \frac{1}{2}\left[\left(\frac{v_2 + v_1}{\omega_1}\right)\sin\omega_1 t + \left(\frac{v_2 - v_1}{\omega_2}\right)\sin\omega_2 t\right]. \quad (13.30)$$

Hence we obtain the full erratic evolution of the system in terms of the initial conditions. In particular, if we give only block 1 an initial velocity, *i.e.*, if $v_2 = 0$, we have

$$x_1(t) = \left(\frac{v_1}{2}\right)\left[\frac{\sin\omega_1 t}{\omega_1} + \frac{\sin\omega_2 t}{\omega_2}\right] \quad (13.31)$$

and

$$x_2(t) = \left(\frac{v_1}{2}\right)\left[\frac{\sin\omega_1 t}{\omega_1} - \frac{\sin\omega_2 t}{\omega_2}\right]. \quad (13.32)$$

It is easy to show that these equations for x_1 and x_2 can also be written as

$$x_{1,2} = \frac{v_1}{2}\left[\frac{\sin(\omega - \Delta\omega)t}{\omega - \Delta\omega} \pm \frac{\sin(\omega + \Delta\omega)t}{\omega + \Delta\omega}\right], \quad (13.33)$$

where $\omega \equiv (\omega_1 + \omega_2)/2$ is the average of the two frequencies, and $\Delta\omega \equiv (\omega_2 - \omega_1)/2$ is half their difference. The upper (plus) sign corresponds to x_1 and the lower (minus) sign to x_2.

Example 13.1

Weak Coupling and Strong Coupling

Suppose that the middle spring in the system just described is much *weaker* than the other two, with $k' \ll k$, so that the coupling between the two blocks is *weak*. In that case we can approximate ω_2 using the binomial expansion (see Appendix F):

$$\omega_2 = \left(\frac{k + 2k'}{m}\right)^{1/2} = \sqrt{\frac{k}{m}}\left(1 + \frac{2k'}{k}\right)^{1/2} \cong \omega_1\left(1 + \frac{k'}{k}\right), \quad (13.34)$$

so $\Delta\omega \equiv (\omega_2 - \omega_1)/2 \cong \omega_1 k'/2k \cong \omega k'/2k$.

Now because $\Delta\omega$ is small, the phases $(\omega \pm \Delta\omega)t$ in

$$x_{1,2} = \frac{v_1}{2}\left[\frac{\sin(\omega - \Delta\omega)t}{\omega - \Delta\omega} \pm \frac{\sin(\omega + \Delta\omega)t}{\omega + \Delta\omega}\right] \quad (13.35)$$

differ from ωt by only a small amount initially, but as time goes by the $\Delta\omega t$ terms can change the phases in the two terms entirely, so that although for small t the two terms in $x_{1,2}$ are pretty much in phase, after a while they can differ by π or more. The $\Delta\omega$ terms in the denominators, however, which affect the amplitudes of each term, do not change with time, so never build up and are therefore much less important. So for $k'/k \ll 1$, we will neglect the $\Delta\omega$ terms in the denominators, giving

$$x_{1,2} \cong \frac{v_1}{2\omega}[\sin(\omega - \Delta\omega)t \ \pm \ \sin(\omega + \Delta\omega)t], \tag{13.36}$$

which can also be written in the more enlightening form

$$x_1 = \frac{v_1}{\omega}\cos(\Delta\omega\, t)\sin\omega t, \tag{13.37}$$

$$x_2 = -\frac{v_1}{\omega}\sin(\Delta\omega\, t)\cos\omega t, \tag{13.38}$$

where we have used the trig identity

$$\sin A \pm \sin B = 2\cos\left(\frac{A \mp B}{2}\right)\sin\left(\frac{A \pm B}{2}\right). \tag{13.39}$$

Note that both x_1 and x_2 are products of a rapidly oscillating part $\sin\omega t$ or $\cos\omega t$ and a slowly oscillating "envelope" $\cos(\Delta\omega\, t)$ or $\sin(\Delta\omega\, t)$. From the slowly oscillating factors we can see that x_1 starts with large amplitude and x_2 begins with small amplitude, but by the time that $\Delta\omega t = \pi/2$ these have reversed, so that now x_2 has large swings and x_1 only small ones. Then by the time $\Delta\omega\, t = \pi$ the oscillations have returned to where they began. The period of this motion is therefore

$$T = \frac{\pi}{\Delta\omega} = \frac{2\pi k}{\omega k'}. \tag{13.40}$$

That is, for weakly coupled oscillators the energy of oscillation is gradually exchanged from one to the other and back again. This interesting passing of the torch from once oscillator to the other occurs for many weakly coupled mechanical systems.

Now suppose the center spring is much *stronger* than the others, with $k' \gg k$, corresponding to *strong coupling* between the two blocks. Then $\omega_2 \cong \omega_1\sqrt{2\,k'/k}$, so from Eq. (13.33) we have

$$x_{1,2} = \frac{v_1}{2}\left[\frac{\sin\omega_1 t}{\omega_1} \pm \frac{\sin\omega_2 t}{\omega_2}\right] \cong \frac{v_1}{2\omega_1}\left[\sin\omega_1 t \pm \sqrt{\frac{k}{2k'}}\sin\omega_1\sqrt{\frac{2k'}{k}}\,t\right]. \tag{13.41}$$

Compared with the first term, the second term has much smaller amplitude and higher frequency. So in the strong-coupling case, the two blocks basically slide back and forth together in phase with frequency ω_1, but superimposed on this motion is a small-amplitude, high-frequency oscillation in which the blocks move in opposition to one another, caused by the stiff spring between them. ∎

Example 13.2 **Coupled Pendulums**

Two balls, each of mass m, are attached to two strings of equal length ℓ to form side-by-side pendulums of equal period. A weak spring k' is attached to the two balls, as shown in Figure 13.3. When the pendula hang down, the spring is unstretched. All oscillations are assumed to be in the same plane.

We want to find the motion of each ball in the small-amplitude limit, in which the angles θ_1 and θ_2 are both very small. In that case the spring stretch is very nearly $\ell(\theta_2 - \theta_1)$, and the gravitational potential energy of the first ball is $mgh = mg\ell(1 - \cos\theta_1) \cong mg\ell[1 - (1 - \theta_1^2/2)] = mg\ell\theta_1^2/2$, with a similar expression for the second ball. The Lagrangian of the small-amplitude system is therefore

$$L = \frac{1}{2}m\ell^2\dot{\theta}_1^2 + \frac{1}{2}m\ell^2\dot{\theta}_1^2 - \frac{1}{2}mg\ell\theta_1^2 - \frac{1}{2}mg\ell\theta_2^2 - \frac{1}{2}k'\ell^2(\theta_2 - \theta_1)^2. \qquad (13.42)$$

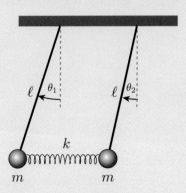

Fig. 13.3   Two pendulums connected with springs.

Note that this Lagrangian is identical to the Lagrangian of the two-mass, three-spring problem we just discussed, if we replace $\ell\theta_1$ and $\ell\theta_2$ by x_1 and x_2, and replace the constant quantity mg/ℓ by k. That is, for small displacements from equilibrium the gravitational force on each mass acts like a spring of force constant mg/ℓ. Therefore the behavior of the small-amplitude coupled pendulum is just like that of the two-mass, three-spring problem. In particular, if the coupling spring k' is very weak, we can start the first pendulum mass swinging back and forth while the second pendulum is initially at rest; then after a while the motion (and energy) is gradually transferred from the first pendulum to the second, so that the second pendulum eventually swings back and forth while the first pendulum comes instantaneously to rest. This alternating behavior would continue indefinitely were it not for friction, which eventually robs the system of its energy, so that both pendulums come to rest.

In contrast, if $k \gg mg/\ell$ the two pendulums are strongly coupled: they swing back and forth together, upon which may be superimposed a small-amplitude, high-frequency oscillation between the two balls. ∎

Example 13.3   **Three Blocks and Four Springs**

Now suppose there are *three* blocks attached to *four* springs. Again, the springs are unstretched in the equilibrium position, and the blocks are free to move in the horizontal direction only. The far end of each outer spring is attached to a stationary wall. The displacements of the three blocks from equilibrium are x_1, x_2, x_3, positive to the right. For simplicity, suppose the blocks have equal mass m and all four springs have the same force constant k, as shown in Figure 13.4. Now the Lagrangian is

$$L = \frac{1}{2}m(\dot{x}_1^2 + \dot{x}_2^2 + \dot{x}_3^2) - \frac{1}{2}kx_1^2 - \frac{1}{2}k(x_2 - x_1)^2 - \frac{1}{2}k(x_3 - x_2)^2 - \frac{1}{2}kx_3^2, \qquad (13.43)$$

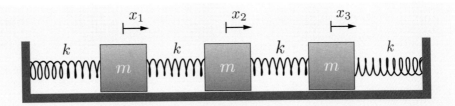

Fig. 13.4 Three blocks attached to two springs and walls.

taking into account the potential energy stored in each of the four springs. Lagrange's equations then give

$$m\ddot{x}_1 = -kx_1 + k(x_2 - x_1),$$
$$m\ddot{x}_2 = -k(x_2 - x_1) + k(x_3 - x_2),$$
$$m\ddot{x}_3 = -kx_3 - k(x_3 - x_2),$$

(13.44)

which again could have been written using $F = ma$ for each block.

As in the two-block case, these can be solved by writing $x_1 = b_1 e^{j\omega t}, x_2 = b_2 e^{j\omega t}, x_3 = b_3 e^{j\omega t}$, resulting in the matrix equation

$$\begin{pmatrix} -m\omega^2 + 2k & -k & 0 \\ -k & -m\omega^2 + 2k & -k \\ 0 & -k & -m\omega^2 + 2k \end{pmatrix} \begin{pmatrix} b_1 \\ b_2 \\ b_3 \end{pmatrix} = 0,$$

(13.45)

which has a nontrivial solution only if the determinant of the coefficient matrix is zero. That is, the secular equation is

$$\begin{vmatrix} -m\omega^2 + 2k & -k & 0 \\ -k & -m\omega^2 + 2k & -k \\ 0 & -k & -m\omega^2 + 2k \end{vmatrix} = 0.$$

(13.46)

Expanding about the top row:

$$(-m\omega^2 + 2k)[(-m\omega^2 + 2k)^2 - k^2] + k(-k(-m\omega^2 + 2k)) = 0.$$

(13.47)

Factoring:

$$(-m\omega^2 + 2k)[(-m\omega^2 + 2k)^2 - 2k^2] = 0,$$

(13.48)

the product of a linear and a quadratic equation in ω^2, with altogether three solutions for ω^2. The first factor is zero if

$$\omega = \omega_1 = \sqrt{\frac{2k}{m}}.$$

(13.49)

The *second* factor is zero if $(-m\omega^2 + 2k)^2 = 2k^2$, i.e., $-m\omega^2 + 2k = \pm\sqrt{2}k$, which gives the other two eigenfrequencies

$$\omega_2 = \sqrt{\frac{(2 - \sqrt{2})k}{m}} \quad \text{and} \quad \omega_3 = \sqrt{\frac{(2 + \sqrt{2})k}{m}}. \tag{13.50}$$

As with the two-block problem, we can find the three normal-mode motions by solving for the eigenvectors: we substitute $\omega = \omega_1, \omega_2, \omega_3$ in (13.45), giving us two linearly independent equations for three unknowns: b_1, b_2, and b_3. We find that

$$\begin{aligned}
b_2 &= 0, & b_3 &= -b_1 & \text{for } \omega = \omega_1, \\
b_3 &= b_1, & b_2 &= \sqrt{2}\,b_1 & \text{for } \omega = \omega_2, \\
b_3 &= b_1, & b_2 &= -\sqrt{2}\,b_1 & \text{for } \omega = \omega_3.
\end{aligned} \tag{13.51}$$

It is clear that ω_1 corresponds to the frequency when the center block remains at rest and the outer blocks oscillate oppositely to one another, both moving outwards and then both moving inwards, etc. Each is connected to two springs whose opposite ends stay at rest, so the frequency should be $\omega = \sqrt{2k/m}$. The eigenfrequency ω_2 corresponds to the two outer blocks moving together in phase, with the middle block moving in the same direction with a different amplitude; and the eigenfrequency ω_3 corresponds to the two outer blocks moving together in phase, with the middle block moving always in the opposite direction, and with a different amplitude.

Finally, to write the orthonormal eigenvectors, we can use the normalization condition $b_1^2 + b_2^2 + b_3^2 = 1$, giving

$$\mathbf{e}_{(1)} = \frac{1}{\sqrt{2}}\begin{pmatrix} 1 \\ 0 \\ -1 \end{pmatrix}, \quad \mathbf{e}_{(2)} = \frac{1}{2}\begin{pmatrix} 1 \\ \sqrt{2} \\ 1 \end{pmatrix}, \quad \mathbf{e}_{(3)} = \frac{1}{2}\begin{pmatrix} 1 \\ -\sqrt{2} \\ 1 \end{pmatrix} \tag{13.52}$$

for the corresponding eigenvalues ω_1, ω_2, and ω_3, respectively. The general solution is then given by

$$\begin{pmatrix} x_1(t) \\ x_2(t) \\ x_3(t) \end{pmatrix} = c_1\mathbf{e}_{(1)}e^{j\omega_1 t} + c_2\mathbf{e}_{(2)}e^{j\omega_2 t} + c_3\mathbf{e}_{(3)}e^{j\omega_3 t}, \tag{13.53}$$

where c_1, c_2, and c_3 are complex constants to be determined from initial conditions, remembering that the physical solution is obtained by taking the real part of the corresponding complex function. ∎

Example 13.4

No Walls

Now we return to the case of three equal-mass blocks, except that we remove the walls and outer springs, so the three blocks are connected together linearly with only *two* springs, as shown in Figure 13.5.

The Lagrangian for this system is

$$L = T - U = \frac{1}{2}m(\dot{x}_1^2 + \dot{x}_2^2 + \dot{x}_3^2) - \frac{1}{2}k(x_2 - x_1)^2 - \frac{1}{2}k(x_3 - x_2)^2, \tag{13.54}$$

from which Lagrange's equations give

$$m\ddot{x}_1 = k(x_2 - x_1),$$
$$m\ddot{x}_2 = -k(x_2 - x_1) + k(x_3 - x_2) = k(x_1 + x_3) - 2kx_2,$$
$$m\ddot{x}_3 = -k(x_3 - x_2). \tag{13.55}$$

Substituting

$$x_1 = b_1 e^{j\omega t}, \quad x_2 = b_2 e^{j\omega t}, \quad x_3 = b_3 e^{j\omega t} \tag{13.56}$$

gives the set of algebraic equations

$$\begin{pmatrix} -m\omega^2 + k & -k & 0 \\ -k & -m\omega^2 + 2k & -k \\ 0 & -k & -m\omega^2 + k \end{pmatrix} \begin{pmatrix} b_1 \\ b_2 \\ b_3 \end{pmatrix} = 0, \tag{13.57}$$

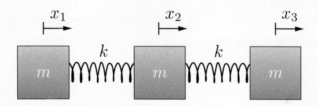

Fig. 13.5 Three blocks attached with two springs.

so the secular determinant is

$$\begin{vmatrix} -m\omega^2 + k & -k & 0 \\ -k & -m\omega^2 + 2k & -k \\ 0 & -k & -m\omega^2 + k \end{vmatrix} = 0. \tag{13.58}$$

Expanding about the top row, we get

$$(-m\omega^2 + k)[(-m\omega^2 + 2k)(-m\omega^2 + k) - k^2] + k[-k(-m\omega^2 + k)] = 0, \tag{13.59}$$

which can be factored to give

$$-m\omega^2(-m\omega^2 + k)(-m\omega^2 + 3k) = 0, \tag{13.60}$$

with the three solutions

$$\omega_1 = 0, \quad \omega_2 = \sqrt{\frac{k}{m}}, \quad \omega_3 = \sqrt{\frac{3k}{m}}. \tag{13.61}$$

Substituting $\omega_1 = 0$ into (13.57), we find that $b_1 = b_2 = b_3$, which means that all three blocks have zero oscillation frequency and the same amplitude at all times: that is, there is no oscillation, and all blocks either are at rest or are moving together at the same velocity. This is called a *translational* mode. Substituting in ω_2 instead gives $b_3 = -b_1$ and $b_2 = 0$, so that the middle block remains at rest while the outer blocks move with equal amplitudes in opposition to one another. The frequency $\omega_2 = \sqrt{k/m}$ makes sense in this case, because in effect each outer block oscillates at the end of a single spring. Substituting in ω_3, we find that $b_2 = -2\,b_1 = -2\,b_3$; that is, the outer blocks move in phase with one another, with the same amplitude, while the center block moves in the opposite direction with twice the amplitude. Note that this motion ensures that the center of mass always remains at rest. The frequency $\omega_3 > \omega_1$, because for a given amplitude of an outer block, the spring is stretched the most if the central block moves in opposition. Finally, the corresponding normalized eigenvectors become

$$\mathbf{e}_{(1)} = \frac{1}{\sqrt{3}} \begin{pmatrix} 1 \\ 1 \\ 1 \end{pmatrix}, \quad \mathbf{e}_{(2)} = \frac{1}{\sqrt{2}} \begin{pmatrix} 1 \\ 0 \\ -1 \end{pmatrix}, \quad \mathbf{e}_{(3)} = \frac{1}{\sqrt{6}} \begin{pmatrix} -1 \\ 2 \\ -1 \end{pmatrix}, \qquad (13.62)$$

in terms of which one can write the most general solution.   ∎

13.2 More Realistic Bound Systems

So far we have dealt only with masses attached to Hooke's-law springs, which of course are highly idealized systems. More generally, for macroscopic one-dimensional mechanical motions there is often some sort of potential energy $U(x)$ between any two masses. If there is a minimum in $U(x)$ at x_0, as illustrated in Figure 13.6, then in equilibrium the two masses are separated by the distance x_0. If they are slightly disturbed, they will oscillate back and forth about the equilibrium point. Most often $U(x)$ rises quadratically from the minimum, which is to say that a parabola can be fit into the bottom of the potential well. In that case the small oscillations will be harmonic, and it is as if a Hooke's-law spring were attached to the two masses.

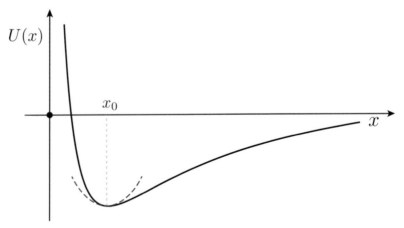

Fig. 13.6 A general shape for a potential with a minimum. In the region near the minimum, the potential can be well approximated by the dashed curve, corresponding to the leading Taylor-series terms.

To see this, the Taylor-series expansion of $U(x)$ about x_0 is (see Appendix F)

$$U(x) = U(x_0) + \left.\frac{dU(x)}{dx}\right|_{x_0} (x - x_0) + \frac{1}{2} \left.\frac{d^2U(x)}{dx^2}\right|_{x_0} (x - x_0)^2 + \ldots, \quad (13.63)$$

where $x - x_0$ is small. Since $U(x)$ has a local minimum at x_0, the second term vanishes, and so $U(x) - U(x_0)$ has the form

$$U(x) - U(x_0) = \frac{1}{2}k(x - x_0)^2 + \ldots \quad (13.64)$$

of a Hooke's-law spring potential for small-amplitude oscillations, where the effective spring constant is the second derivative

$$k = \left.\frac{d^2U(x)}{dx^2}\right|_{x_0}, \quad (13.65)$$

evaluated at the potential energy minimum. If the second derivative happens to be zero at the minimum in $U(x)$, then we cannot model the system by Hooke's-law springs, and the motion is not harmonic.

More generally, suppose we have a coupled system in which the generalized coordinates are q_i and the generalized velocities are $\dot{q}_i$. If the transformation equations from Cartesian to generalized coordinates are not an explicit function of time, then the kinetic energy can be written

$$T = \frac{1}{2}\hat{\mathbf{M}}_{jk}\dot{q}_j\dot{q}_k, \quad (13.66)$$

summed on j and k from 1 to N, where $\hat{\mathbf{M}}_{jk}$ is the symmetric $N \times N$ **mass matrix**.[1] Furthermore, we choose our generalized coordinates so that $q_i = 0$ for all i at a minimum of the potential $U(q_1, q_2, \ldots, q_N)$. Then, the potential energy for the small displacements q_i from equilibrium has the general form

$$U(q_1, q_2, \ldots, q_N) = U_0 + \left.\left|\frac{\partial U}{\partial q_j}\right|\right|_0 q_j + \frac{1}{2}\left.\left|\frac{\partial^2 U}{\partial q_j \partial q_k}\right|\right|_0 q_j q_k + \cdots$$

$$\simeq U_0 + \frac{1}{2}\hat{\mathbf{K}}_{jk}q_j q_k, \quad (13.67)$$

where again we are using the Einstein summation convention on repeated indices. The $N \times N$ **spring matrix** $\hat{\mathbf{K}}_{jk}$, also a symmetric matrix, is then defined as

$$\hat{\mathbf{K}}_{jk} \equiv \frac{\partial^2 U}{\partial q_j \partial q_k}. \quad (13.68)$$

The Lagrangian takes the form

$$L = T - U = \frac{1}{2}\hat{\mathbf{M}}_{jk}\dot{q}_j\dot{q}_k - \frac{1}{2}\hat{\mathbf{K}}_{jk}q_j q_k, \quad (13.69)$$

---

[1] See the proof below.

so

$$\frac{\partial K}{\partial q_i} = -\frac{1}{2}\hat{\mathbf{K}}_{ik}q_k - \frac{1}{2}\hat{\mathbf{K}}_{ki}q_k = -\hat{\mathbf{K}}_{ik}q_k \tag{13.70}$$

since $\hat{\mathbf{K}}_{ik}$ is symmetric. The calculation of $\partial L/\partial\dot{q}_i$ is similar, so the corresponding Lagrange equations are

$$\hat{\mathbf{K}}_{jk}q_j + \hat{\mathbf{M}}_{jk}\ddot{q}_j = 0. \tag{13.71}$$

These are N coupled second-order differential equations, with $k = 1, 2, \ldots, N$. To find the normal modes, we try solutions $q_j = b_j e^{i\omega t}$, which converts the differential equations into the algebraic equations

$$(\hat{\mathbf{K}}_{jk} - \omega^2\hat{\mathbf{M}}_{jk})b_j = 0 \Rightarrow (\hat{\mathbf{K}} - \omega^2\hat{\mathbf{M}})\,\mathbf{b} = 0, \tag{13.72}$$

where $\hat{\mathbf{K}}$ and $\hat{\mathbf{M}}$ are $N \times N$ matrices. The sums and products implied by so-called **index notation** on the left, together with the Einstein sum rule for repeated indices, is exactly the process we go through in multiplying matrices in the matrix form of the equation on the right. As in previous examples, there is a nontrivial solution only if the secular equation $|\hat{\mathbf{K}} - \omega^2\hat{\mathbf{M}}| = 0$ is satisfied, which gives N roots ω_a^2, with $a = 1, 2, \ldots, N$. In some cases two or more of these roots are equal, in which case the system is said to be **degenerate**. Hence, we have solved the problem

$$(\hat{\mathbf{K}} - \omega_a^2\hat{\mathbf{M}})\,\mathbf{e}_{(a)} = 0 \tag{13.73}$$

for the N eigenvalues ω_a and corresponding eigenvectors $\mathbf{e}_{(a)}$, for $a = 1, \ldots, N$.

Note that $\hat{\mathbf{K}}$ and $\hat{\mathbf{M}}$ are necessarily symmetric matrices. To see this, consider for example the term

$$\hat{\mathbf{K}}_{jk}q_jq_k \tag{13.74}$$

that appears in the Lagrangian. We can always write

$$\hat{\mathbf{K}}_{jk} = \frac{1}{2}(\hat{\mathbf{K}}_{jk} + \hat{\mathbf{K}}_{kj}) + \frac{1}{2}(\hat{\mathbf{K}}_{jk} - \hat{\mathbf{K}}_{kj}) \equiv \hat{\mathbf{K}}_{jk}^S + \hat{\mathbf{K}}_{jk}^A, \tag{13.75}$$

where $\hat{\mathbf{K}}_{jk}^S$ and $\hat{\mathbf{K}}_{jk}^A$ are the symmetric and antisymmetric parts of the spring matrix

$$\hat{\mathbf{K}}_{jk}^S = \hat{\mathbf{K}}_{kj}^S, \quad \hat{\mathbf{K}}_{jk}^A = -\hat{\mathbf{K}}_{kj}^A. \tag{13.76}$$

But we may then write

$$\hat{\mathbf{K}}_{jk}q_jq_k = \hat{\mathbf{K}}_{jk}^S q_jq_k + \hat{\mathbf{K}}_{jk}^A q_jq_k = \hat{\mathbf{K}}_{jk}^S q_jq_k, \tag{13.77}$$

showing that the antisymmetric part of $\hat{\mathbf{K}}_{jk}$ drops out of the Lagrangian. This is because

$$\hat{\mathbf{K}}_{jk}^A q_jq_k = \hat{\mathbf{K}}_{kj}^A q_kq_j = -\hat{\mathbf{K}}_{jk}^A q_kq_j = -\hat{\mathbf{K}}_{jk}^A q_jq_k, \tag{13.78}$$

where in the first step we exchanged the labeling $j \leftrightarrow k$, in the second step we used the antisymmetry of $\hat{\mathbf{K}}_{jk}^A$, and in the last step we used the commutativity of multiplication to write $q_k q_j$ as $q_j q_k$: all this amounts to showing that this antisymmetric piece is the negative of itself, implying it is zero. The same trick can be employed with the term in the Lagrangian involving the mass matrix $\hat{\mathbf{M}}_{jk}$. Hence, without loss of generality, we may assume that both $\hat{\mathbf{K}}_{jk}$ and $\hat{\mathbf{M}}_{jk}$, or the corresponding matrices $\hat{\mathbf{K}}$ and $\hat{\mathbf{M}}$, are symmetric.

One can then show that since both $\hat{\mathbf{K}}$ and $\hat{\mathbf{M}}$ are real, symmetric matrices, and because the kinetic energy is $T \geq 0$, it follows that:

1. All eigenvalues ω_a^2 of (13.73) are real and positive. These can be found by finding the roots of the polynomial that arises from the vanishing determinant $|(\hat{\mathbf{K}} - \omega^2 \hat{\mathbf{M}})| = 0$.
2. The eigenvectors $\mathbf{e}_{(a)}$, one for each ω_a^2, are necessarily real and orthogonal with respect to the mass matrix $\hat{\mathbf{M}}$ in the following sense:

$$\mathbf{e}_{(a)}^T \cdot \hat{\mathbf{M}} \cdot \mathbf{e}_{(b)} = 0, \tag{13.79}$$

   for two or more nondegenerate eigenvalues, *i.e.*, in those cases where the eigenvalues are different, $\omega_a^2 \neq \omega_b^2$. Note that "$T$" stands for matrix transposition.
3. One can also show that, even when $\omega_a^2 = \omega_b^2$ for two eigenvalues, one can *choose* the corresponding eigenvectors $\mathbf{e}_{(a)}$ and $\mathbf{e}_{(b)}$ to be orthogonal. In general, we can also choose to normalize the eigenvectors such that one may write

$$\mathbf{e}_{(a)}^T \cdot \hat{\mathbf{M}} \cdot \mathbf{e}_{(b)} = \delta_{ab}, \tag{13.80}$$

   where δ_{ab} is the Kronecker delta (equal to 0 when $a \neq b$, and equal to 1 when $a = b$). The N eigenvectors are found by solving (13.73) for each eigenvalue. With this normalization of the eigenvectors, they can be determined from the spring matrix $\hat{\mathbf{K}}$ and the eigenvalues ω_a^2 using

$$\hat{\mathbf{K}} \cdot \mathbf{e}_{(a)} = \omega_a^2 \, \mathbf{e}_{(a)}. \tag{13.81}$$

4. This set of orthonormal eigenvectors is guaranteed to span the space of all possible solutions of the oscillating system. This means that one may write the most general solution as a linear combination of the normal modes

$$q_k(t) = \sum_a c_k^a \, \mathbf{e}_{(a)}^k \, e^{i \omega_a t}, \tag{13.82}$$

summing over a, but with no sum on the fixed component k. The c_k^a are complex constants to be determined from initial conditions, remembering that the physical solution is obtained by taking the real part $\Re(q_k(t))$.

The previous examples we tackled were essentially applications of this procedure in disguise. In all these previous cases the mass matrix was proportional to the identity

$$\hat{\mathbf{M}} = m \, \hat{\mathbf{1}}, \tag{13.83}$$

where m was a common mass parameter. As a result, the secular equation took the form $|\hat{\mathbf{K}} - m \omega^2 \hat{\mathbf{1}}| = 0$. For finding the eigenvectors, we used

$$\left(\hat{\mathbf{K}} - m\,\omega^2\hat{\mathbf{1}}\right)\mathbf{e} = 0. \qquad (13.84)$$

Acting on this from the left by $\mathbf{e}^T\cdot$, and using the simpler normalization

$$\mathbf{e}^T \cdot \mathbf{e} = 1, \qquad (13.85)$$

we ended up solving for the eigenvector $\mathbf{e}$ using the equation

$$\mathbf{e}^T \cdot \hat{\mathbf{K}} \cdot \mathbf{e} = m\,\omega^2 \Rightarrow \frac{\hat{\mathbf{K}}}{m} \cdot \mathbf{e} = \omega^2\mathbf{e}, \qquad (13.86)$$

which is the problem of diagonalizing the spring matrix $\hat{\mathbf{K}}/m$, or equivalently $\hat{\mathbf{K}}$. All this corresponded to a special class of oscillation problems involving trivial mass matrices. Our treatment in this section allows us now to handle more general scenarios where the mass matrix is nontrivial. Let's now see how to put this technology into use in such cases.

<table>
<tr><td>Example 13.5</td><td>

Longitudinal Vibrations of a Carbon Dioxide Model

Carbon dioxide is a linear molecule, with the carbon atom (of mass m and coordinate x_2) in the middle and the oxygen atoms (of mass M at coordinates x_1 and x_3) at the two ends, as shown in Figure 13.7. In a classical model of the system one can find a potential energy between the carbon atom and an oxygen atom, and neglect any interaction between the two oxygen atoms, which are relatively far apart. The potential energies are quadratic near the minimum, so there is an effective force constant k, and the Lagrangian takes the effective form

$$L = T - U = \frac{1}{2}M(\dot{x}_1^2 + \dot{x}_3^2) + \frac{1}{2}m\dot{x}_2^2 - \frac{1}{2}k(x_2 - x_1)^2 - \frac{1}{2}k(x_3 - x_2)^2. \qquad (13.87)$$

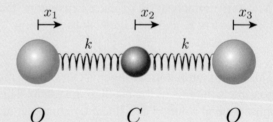

</td></tr>
</table>

Fig. 13.7   The layout of the carbon dioxide molecule.

At the scale of molecules, the motion of the atoms can, of course, only be understood properly using quantum mechanics. However, the use of classical mechanics to describe oscillations of CO_2 is still interesting, as we shall see. Note that we have used Cartesian coordinates as the generalized coordinates such that equilibrium corresponds to $x_1 = x_2 = x_3 = 0$. We can now identify the mass and spring matrices from (13.87) and (13.69):

$$\hat{\mathbf{M}} = \begin{pmatrix} M & 0 & 0 \\ 0 & m & 0 \\ 0 & 0 & M \end{pmatrix}, \quad \hat{\mathbf{K}} = \begin{pmatrix} k & -k & 0 \\ -k & 2k & -k \\ 0 & -k & k \end{pmatrix}. \qquad (13.88)$$

We then need to solve the eigenvalue problem given by

$$(\hat{\mathbf{K}} - \omega^2\hat{\mathbf{M}})\,\mathbf{b} = \begin{pmatrix} k - \omega^2 M & -k & 0 \\ -k & 2k - \omega^2 m & -k \\ 0 & -k & k - \omega^2 M \end{pmatrix}\begin{pmatrix} b_1 \\ b_2 \\ b_3 \end{pmatrix} = 0. \qquad (13.89)$$

The secular determinant is

$$\begin{vmatrix} -M\omega^2 + k & -k & 0 \\ -k & -m\omega^2 + 2k & -k \\ 0 & -k & -M\omega^2 + k \end{vmatrix} = 0 \qquad (13.90)$$

for nontrivial solutions **b**. Expanding about the top row, we get

$$(-M\omega^2 + k)[(-m\omega^2 + 2k)(-M\omega^2 + k) - k^2] + k[-k(-M\omega^2 + k)] = 0, \qquad (13.91)$$

which can be factored to give

$$-m\omega^2(-M\omega^2 + k)\left(-M\omega^2 + k\left(1 + \frac{2M}{m}\right)\right) = 0, \qquad (13.92)$$

with the three solutions

$$\omega_1 = 0, \quad \omega_2 = \sqrt{\frac{k}{M}}, \quad \omega_3 = \sqrt{\frac{k}{M}}\sqrt{1 + \frac{2M}{m}}. \qquad (13.93)$$

Substituting $\omega = \omega_1 = 0$ into Eq. (13.89), we find

$$b_1 = b_2 = b_3. \qquad (13.94)$$

Hence, ω_1 corresponds to *no oscillation at all*: the atoms all remain at rest or move together at constant velocity. This solution corresponds to a *translational* degree of freedom, as we also saw in an earlier example. Substitution of $\omega = \omega_2 = \sqrt{k/M}$ into Eq. (13.89) yields

$$b_3 = -b_1, \quad b_2 = 0. \qquad (13.95)$$

This is a normal mode of oscillation in which the carbon atom remains at rest at the center while the oxygen atoms move in and out, in opposing directions, so the center of mass of the system remains at rest. This explains why the carbon mass m does not appear in the frequency; its mass is irrelevant. Finally, substitution of $\omega = \omega_3 = \sqrt{k/M}\sqrt{1 + (2M/m)}$ into Eq. (13.89) gives

$$b_3 = b_1, \quad \frac{b_2}{b_1} = -\frac{2M}{m}. \qquad (13.96)$$

This corresponds to the two oxygen atoms moving back and forth together, *i.e.*, in phase, while the carbon atom moves in the opposite direction by the distance required to keep the center of mass of the system at rest: we have $Mx_1 + mx_2 + Mx_3 = Mx_1 + m(-2M/m)x_1 + Mx_1 = 0$.

Putting things together, we can write three orthonormal eigenvectors using the normalization condition (13.80). This now involves the mass matrix and amounts to the statement

$$M(b_1^2 + b_3^2) + m\,b_2^2 = 1 \qquad (13.97)$$

instead of $b_1^2 + b_2^2 + b_3^2 = 1$. Therefore we find the normalized eigenvectors

$$\mathbf{e}_{(1)} = \frac{1}{\sqrt{m+2M}} \begin{pmatrix} 1 \\ 1 \\ 1 \end{pmatrix},$$

$$\mathbf{e}_{(2)} = \frac{1}{\sqrt{2M}} \begin{pmatrix} 1 \\ 0 \\ -1 \end{pmatrix}, \quad \mathbf{e}_{(3)} = \frac{1}{\sqrt{m+2M}} \begin{pmatrix} \sqrt{\frac{m}{2M}} \\ -\sqrt{\frac{2M}{m}} \\ \sqrt{\frac{m}{2M}} \end{pmatrix}. \tag{13.98}$$

The most general vibrational solution can then be written in terms of the last two eigenvectors

$$\begin{pmatrix} x_1(t) \\ x_2(t) \\ x_2(t) \end{pmatrix} = \frac{1}{\sqrt{2M}} \begin{pmatrix} c_1^2 \\ 0 \\ -c_3^2 \end{pmatrix} e^{j\omega_2 t} + \frac{1}{\sqrt{m+2M}} \begin{pmatrix} c_1^3 \sqrt{\frac{m}{2M}} \\ -c_2^3 \sqrt{\frac{2M}{m}} \\ c_3^3 \sqrt{\frac{m}{2M}} \end{pmatrix} e^{j\omega_3 t}, \tag{13.99}$$

where the c_k^q are complex constants to be determined from initial conditions. Again, remember that we can connect to the physical solution by taking the real part of this expression.

Carbon dioxide is one of the most important greenhouse gases in our atmosphere. Light from the sun, peaking in the visible range of the spectrum, enters the atmosphere; most of it then strikes earth's surface and is absorbed. The warm surface emits electromagnetic radiation characteristic of its temperature, in the infrared rather than the visible. This radiation makes its way back up through the atmosphere, while some of it is absorbed by atmospheric gases, including carbon dioxide, which have characteristic vibration frequencies in the infrared. A CO_2 molecule can absorb infrared photons with frequencies close to one of the normal-mode frequencies of the molecule, exciting it into a higher-energy state. This absorbed energy is radiated away once again, some reaching the ground, further heating the earth.

Electromagnetic radiation couples most strongly to oscillations in which the electric dipole moment of the molecule changes with time. In a CO_2 molecule the oxygen atoms at each end are slightly negative while the carbon atom in the middle is slightly positive, which implies that in equilibrium the molecule has a *quadrupole* moment but no *dipole* moment. In the first vibrational mode of CO_2 the oxygen atoms move in and out in opposition to one another, while the carbon atom remains at rest, which changes the *quadrupole* moment of the molecule, but not the dipole moment, which remains zero. Therefore the first normal mode of oscillation is only a very weak absorber (or emitter) of radiation. However, in the second normal mode the oxygen atoms move in the *same* direction while the carbon atom moves in the *opposite* direction. Therefore while the negative charge moves to the right (say) the positive charge moves to the left, and vice versa, giving the molecule an ever-changing dipole moment. This second normal mode couples strongly to electromagnetic radiation, and it is the most important absorber of infrared light in our atmosphere, and the most important contributor to global warming.[a]   ∎

---

[a] Methane (CH_4) molecules are many times more effective infrared absorbers than CO_2, but there are fewer methane molecules in the atmosphere, and they degrade over time. Release of large quantities of natural gas (which is about 95% methane) into the atmosphere can nevertheless cause important greenhouse effects.

Example 13.6

Crossed Kinetic and Potential Energy Terms

As a next example, we consider a system that has a nondiagonal mass matrix and a nondiagonal spring-constant matrix. We begin with a single particle of mass m connected to a network of springs so that the equilibrium position of the particle is at the origin of our coordinate system. The Lagrangian of the particle is given in Cartesian coordinates as

$$L = \frac{1}{2}m\dot{x}^2 + \frac{1}{2}m\dot{y}^2 + \frac{1}{2}m\dot{z}^2 - \frac{1}{2}k\left(x^2 + y^2 + \alpha z^2\right),\tag{13.100}$$

where k and α are constants. Note that the spring constants are not the same in all three directions of space: there is an asymmetry between the x–y plane and the z direction. This might, for example, be the result of modeling an atom in a non-isotropic crystal lattice in which the effective springs in the z direction are stiffer than the springs in the x and y directions, which would make $\alpha > 1$. Furthermore, let us suppose that the atom is subject to conditions that confine its motion to the plane $z = x + y$, perhaps due to lattice defects. Therefore our particle has only two degrees of freedom, requiring only two generalized coordinates, which we take to be x and y. We then need to write also $\dot{z} = \dot{x} + \dot{y}$. Substituting into the Lagrangian, we find that

$$L = \frac{m}{2}\left(2\dot{x}^2 + 2\dot{y}^2 + 2\dot{x}\dot{y}\right) - \frac{k}{2}\left(\beta x^2 + \beta y^2 + 2\alpha xy\right),\tag{13.101}$$

where we define $\beta = (1 + \alpha)$. We see that the mass term now involves a nondiagonal term $\dot{x}\dot{y}$. We can then identify the mass matrix as

$$\hat{\mathbf{M}} = \begin{pmatrix} 2m & m \\ m & 2m \end{pmatrix}.\tag{13.102}$$

We have chosen to define the mass matrix without the customary $1/2$ factor in front of the kinetic energy term, and the off-diagonal terms add up to $2m$, as required. The spring matrix takes the form

$$\hat{\mathbf{K}} = \begin{pmatrix} \beta k & \alpha k \\ \alpha k & \beta k \end{pmatrix},\tag{13.103}$$

also off-diagonal, so the eigenvalue problem is given by

$$|\hat{\mathbf{K}} - \omega^2\hat{\mathbf{M}}| = 0,\tag{13.104}$$

which leads to the equation

$$(\beta k - 2m\omega^2)^2 - (\alpha k - m\omega^2)^2 = 0.\tag{13.105}$$

This equation results in the two eigenvalues

$$\omega_1^2 = \frac{k}{m} \quad \text{and} \quad \omega_2^2 = \frac{k}{3m}(1 + 2\alpha).\tag{13.106}$$

For the first eigenvalue we substitute in Eq. (13.73), writing $\mathbf{e}_{(1)}$ as (b_1, b_2), and so find that

$$(\alpha - 1)k\,b_1 + (\alpha - 1)k\,b_2 = 0 \Rightarrow b_1 = -b_2.\tag{13.107}$$

For the second eigenvalue, we write again $\mathbf{e}_{(2)}$ as (b_1, b_2), giving

$$\frac{k}{3}(1-\alpha)b_1 + \frac{k}{3}(\alpha-1)b_2 = 0 \Rightarrow b_1 = b_2. \tag{13.108}$$

We now need to normalize the two eigenvectors using the mass matrix. We first have $\mathbf{e}_{(1)} = (b, -b)$, so

$$\mathbf{e}_{(1)}^T \cdot \hat{\mathbf{M}} \cdot \mathbf{e}_{(1)} = 1 \Rightarrow 2\,m\,b^2 = 1 \Rightarrow b = \frac{1}{\sqrt{2\,m}}. \tag{13.109}$$

For the second eigenvector, we substitute $\mathbf{e}_{(2)} = (b, b)$ in

$$\mathbf{e}_{(2)}^T \cdot \hat{\mathbf{M}} \cdot \mathbf{e}_{(2)} = 1 \Rightarrow 6\,m\,b^2 = 1 \Rightarrow b = \frac{1}{\sqrt{6\,m}}. \tag{13.110}$$

We can now summarize the results. We have two normal modes of vibration. The first has frequency and eigenvector

$$\omega_1^2 = \frac{k}{m} \quad \text{and} \quad \mathbf{e}_{(1)} = \frac{1}{\sqrt{2\,m}}(1, -1); \tag{13.111}$$

the second has frequency and eigenvector

$$\omega_2^2 = \frac{k}{3\,m}(1 + 2\alpha) \quad \text{and} \quad \mathbf{e}_{(2)} = \frac{1}{\sqrt{6\,m}}(1, 1). \tag{13.112}$$

Therefore, if $\alpha > 1$, corresponding to stiffer springs in the z direction, it follows that $\omega_2 > \omega_1$, whereas if $\alpha < 1$, corresponding to weaker springs in the z direction, we have $\omega_2 < \omega_1$. If $\alpha = 1$ the two normal-mode frequencies are equal, and the two normal modes are said to be "degenerate." ∎

13.3 Vibrational Degrees of Freedom

How many normal modes of oscillation are there for a given system? Why does carbon dioxide have *two* vibrational normal modes of oscillation? And why are there *three* normal modes for three masses connected in a straight line by four springs, with the outer two connected to stationary walls?

Suppose there are N particles, free to move only in one dimension. Then we say there are N *degrees of freedom*, $x_1, x_2, \ldots, x_N$, which are the positions measured along the single dimension of each particle. If instead the N particles are allowed to move in two dimensions, there are then $2N$ degrees of freedom, $x_1, y_1; x_2, y_2, \ldots, x_N, y_N$; and in three dimensions we have $3N$ degrees of freedom. So for a carbon dioxide molecule free to move only along one dimension, there are three degrees of freedom. Rather than count by the x coordinate of each atom, we could alternatively say that the one-dimensional motion of the *center of mass* of the molecule counts for one degree of freedom, leaving two more corresponding to possible relative motions of the atoms, in particular the normal-mode oscillations. Thus for one-dimensional motion of three particles:

3 degrees of freedom = (1 *translational* + 2 *vibrational*) degrees of freedom.

Any arbitrary motion of the molecule can be written as some linear combination of these three motions, just as any arbitrary motion can be specified by $x_1(t), x_2(t), x_3(t)$.

Now suppose instead that the CO_2 molecule is free to move in a *plane*. Including now two coordinates for each atom, there are six degrees of freedom. There are two translational degrees of freedom of the center of mass, in the x and y directions, plus now a single *rotational* degree of freedom, corresponding to a rigid rotation of the molecule in the x, y plane (*i.e.*, about a z axis through the center of mass), plus the vibrational modes. So for two-dimensional motion there must be *three* vibrational modes:

6 degrees of freedom =
    (2 *translational* + 1 *rotational* + 3 *vibrational*) degrees of freedom.

We have now learned that there must be a *third* vibrational mode for the CO_2 molecule moving in a plane. This is a vibration in which the molecule bends, with (say) the carbon atom moving in one direction while the two oxygen atoms move in the opposite direction, while keeping the center of mass at rest and keeping the angular momentum equal to zero (see Figure 13.8).

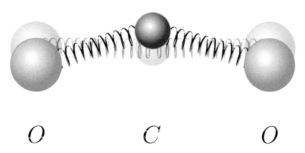

O C O

Fig. 13.8   The third vibrational mode of the CO_2 molecule.

That is, any motion of the center of mass enters exclusively into the translational degrees of freedom, and any net rotation (any rotation with a nonzero angular momentum) enters into the rotational degree of freedom. The vibrational degrees of freedom are what is left over.

If, finally, the same CO_2 model is free to move in all three dimensions, there must be *nine* degrees of freedom. Three of these are translational, in the x, y, and z directions. We might think that there are now also three rotational degrees of freedom, corresponding to rotation about the x, y, or z axis. However, there are really only two: any rotation about an axis passing through all three atoms is unobservable, so does not count. That leaves *four* vibrational degrees of freedom:

the same two we found for one-dimensional motion, plus two bending oscillations. If the molecule is strung out along the x axis, for example, in one bending oscillation the carbon atom moves in the positive y direction while the two oxygen atoms move in the negative y direction (keeping the CM at rest and keeping angular momentum equal to zero), and then vice versa, oscillating back and forth. In the other bending oscillation the carbon atom moves in the positive z direction while the oxygens move in the negative z direction, and then vice versa, oscillating back and forth. Any other bending oscillation is a linear combination of these two. So for the CO_2 molecule in three dimensions:

9 degrees of freedom =
    (3 *translational* + 2 *rotational* + 4 *vibrational*) degrees of freedom.

Now return to the special case of three masses with four springs, with the outer springs attached to rigid walls. How many vibrational modes do we expect? We take the motion to be one-dimensional, not considering any bending motion of the springs. There are then three degrees of freedom, x_1, x_2, x_3. There are obviously no rotational degrees of freedom; *there is also no translational degree of freedom* in this case, because the center of mass of the system is not free to move as it likes, due to the constraint of the walls. So in this case

3 degrees of freedom = 3 *vibrational* degrees of freedom.

That is, any motion x_1, x_2, x_3 permitted by the constraints in this case can be written as a linear combination of the three normal modes of oscillation alone.

13.4 The Continuum Limit

Now we imagine an *infinite number* of masses m connected by an *infinite number* of springs k, as illustrated in Figure 13.9, and allow only longitudinal motions. We can find the equations for such one-dimensional arrays, and then take the limit of an infinite number of infinitesimal masses held together by an infinite number of infinitesimal springs: that is, we can take the continuum limit of our system. This is a good approximation to what happens in longitudinal oscillations of an elastic material, which we would find by banging on one end of a long metal rod with a hammer. We seek to find the sorts of oscillations set up in the rod.

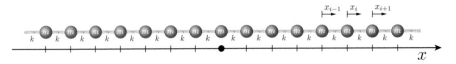

Fig. 13.9   An infinite array of masses and springs.

As before, let the coordinate of mass m be x_i, measured positive to the right from its equilibrium position. Then the system Lagrangian is

$$L = T - U = \sum_i \frac{1}{2} m \dot{x}_i^2 - \frac{1}{2} \sum_i k(x_{i+1} - x_i)^2, \qquad (13.113)$$

since the stretch in the spring connecting mass i with mass $i + 1$ is $x_{i+1} - x_i$. The Lagrange equation for mass m is then

$$m\ddot{x}_i - k(x_{i+1} - x_i) + k(x_i - x_{i-1}) = 0, \qquad (13.114)$$

or

$$k(x_{i+1} - x_i) - k(x_i - x_{i-1}) = m\ddot{x}_i, \qquad (13.115)$$

equivalent to $F = ma$ for m. To take the continuum limit, we need the analogues of force and mass for an essentially continuous elastic material like a thin metal rod. The analogue of mass is the *mass per unit length* μ of the rod. So if a is the equilibrium distance between any two consecutive masses in the spring–mass array, the mass per unit length of the system is

$$\mu = \frac{m}{a}. \qquad (13.116)$$

The analogue of force for the rod is the tension F in the rod, which can be related to the rod's *Young's modulus* Y, defined as

$$Y = \frac{\text{stress}}{\text{strain}}, \qquad (13.117)$$

where the stress in the spring between the $i + 1$th and ith masses is the tension $F = k(x_{i+1} - x_i)$, and the strain is the extension of the spring per unit length, equal to $(x_{i+1} - x_i)/a$. That is:

$$Y = \frac{k(x_{i+1} - x_i)}{(x_{i+1} - x_i)/a} = k\,a. \qquad (13.118)$$

We define the continuum limit as

$$a \to 0 \ \text{ while } \ m \to 0 \text{ and } k \to \infty \ \text{ such that } \mu \text{ and } Y \text{ remain finite.} \qquad (13.119)$$

Therefore dividing Eq. (13.114) by a and using $m = \mu a$ and $k = Y/a$, we find

$$\mu\ddot{x}_i - (Y/a)[(x_{i+1} - x_i) - (x_i - x_{i-1})] = 0. \qquad (13.120)$$

In the limit of a continuous one-dimensional rod, the displacement of mass points in the rod from their equilibrium positions depends upon both the time t and the distance x of the point from some fixed origin within the rod. Let this displacement from equilibrium be given by the continuous function $\eta(t, x)$. That is, if every point in the rod remains in its equilibrium position, then $\eta(t, x) = 0$. In the limit we can identify

$$\ddot{x}_i \to \left. \frac{\partial^2 \eta}{\partial t^2} \right|_{x_i}, \qquad (13.121)$$

the second derivative of $\eta(t, x)$ with respect to time alone, evaluated at the position x_i. Also, in the limit as $a \to 0$:

$$\frac{(x_{i+1} - x_i)}{a} \to \left. \frac{\partial \eta}{\partial x} \right|_{x_i + a} \quad \text{and} \quad \frac{(x_i - x_{i-1})}{a} \to \left. \frac{\partial \eta}{\partial x} \right|_{x_i}, \tag{13.122}$$

the first derivatives of $\eta(t, x)$ with respect to x alone, evaluated at two points separated by the infinitesimal distance a. Then by Taylor's series (see Appendix F):

$$\left. \frac{\partial \eta}{\partial x} \right|_{x_i + a} = \left. \frac{\partial \eta}{\partial x} \right|_{x_i} + a \left. \frac{\partial^2 \eta}{\partial x^2} \right|_{x_i} + \cdots, \tag{13.123}$$

so that in the limit $a \to 0$, Eq. (13.120) becomes the **wave equation**

$$\frac{\partial^2 \eta}{\partial x^2} - \frac{\mu}{Y} \frac{\partial^2 \eta}{\partial t^2} = 0. \tag{13.124}$$

Note that all parameters appearing in this equation, μ and Y, remain finite in the continuum limit (13.119). The standard form for a one-dimensional wave equation is

$$\frac{\partial^2 \eta}{\partial x^2} - \frac{1}{v^2} \frac{\partial^2 \eta}{\partial t^2} = 0, \tag{13.125}$$

where v is the phase velocity of the wave. Therefore longitudinal oscillations in a continuous rod are solutions of a wave equation, in which the phase velocity of waves is

$$v = \sqrt{Y/\mu} \tag{13.126}$$

in terms of the Young's modulus and linear mass density of the rod. Physically, these oscillations represent sound waves in the rod. The waves may be traveling waves, as we could generate by banging on one end of the rod, or they may be standing waves with the usual nodes and crests. This derivation of the wave equation is however rather generic. It arises in many settings where a continuum system is perturbed by a small amount. Much like the Taylor expansion of any potential about its minimum leads generically to the simple harmonic oscillator, the dynamics of small perturbations of many continuum systems leads to the wave equation. For example, instead of restricting ourselves to longitudinal oscillations of the continuous rod, we could also consider a long array of springs and masses and pluck the array in the transverse direction. This leads to exactly the same wave equation for small-amplitude motions, where now $\eta(t, x)$ represents transverse amplitudes such as one would find by plucking a guitar string.

Now we seek the eigenfrequencies and normal-mode solutions for oscillations of a continuous one-dimensional rod. Instead of solving a set of coupled *ordinary* differential equations, one for each of the finite number of masses connected to other masses by springs, we have a single *partial* differential equation with the two independent variables t and x.

This single partial differential equation can be solved by *separation of variables*: that is, we try solutions of Eq. (13.125) with the product form

$$\eta(t,x) = T(t)X(x), \tag{13.127}$$

where $T(t)$ and $X(x)$ are arbitrary functions of their respective independent variable t or x. Substituting this form into the differential equation gives

$$T(t)X(x)'' = \frac{1}{v^2}\ddot{T}X(x), \tag{13.128}$$

where primes are derivatives with respect to x and dots are derivatives with respect to t. Dividing the equation by $\eta = T(t)X(x)$ and multiplying by v^2 gives

$$v^2\frac{X(x)''}{X(x)} = \frac{\ddot{T}}{T}, \tag{13.129}$$

so that the variables t and x have been separated. The left-hand side can be a function of x but not t, while the right-hand side can be a function of t but not x. The two variables are independent of one another, so we could vary t (say) without changing x. In that case the right-hand side might change, but the left-hand side cannot change. However, the two sides are equal to one another, so it is impossible for only one of the two sides to change. The only way to resolve this difficulty is for each side of the equation to be equal to the same *constant*. Let this so-called separation constant be given as $-q$, where $q > 0$. (The reason for the inequality will soon become clear.) This leaves us now with two ordinary differential equations

$$\ddot{T} + qT = 0 \quad \text{and} \quad X'' + (q/v^2)X = 0, \tag{13.130}$$

each of which has the form of a simple harmonic oscillator equation. The $T(t)$ and $X(x)$ equations have solutions

$$T(t) = Ae^{i\omega t} + Be^{-i\omega t} \quad \text{and} \quad X(x) = Ce^{ikx} + De^{-ikx}, \tag{13.131}$$

where A, B, C, D are arbitrary complex constants, $\omega = \sqrt{q}$ and $k = \omega/v$. Note that if we had allowed $q < 0$ we would have had *real* exponential solutions, so that amplitudes η would diverge as $t \to \pm\infty$ and as $x \to \pm\infty$. These we then reject on physical grounds. To construct a solution with a given k, we can write

$$\eta_k(t,x) = T(t)X(x) = A\,Ce^{i(kx+\omega t)} + B\,De^{-i(kx+\omega t)}$$
$$+ A\,De^{-i(kx-\omega t)} + B\,Ce^{i(kx-\omega t)}. \tag{13.132}$$

The most general solution would be a linear superposition of modes like these with various values of k – given that the wave equation is linear in η and hence the sum of any two solutions is still a solution. We may then write in general

$$\eta(t,x) = \sum_k \eta_L(k)e^{ik(x+vt)} + \eta_R(k)e^{ik(x-vt)}, \tag{13.133}$$

where $\eta_L(k)$ and $\eta_R(k)$, called left and right moving modes respectively, are complex functions. As we shall soon see, if the system size in infinite, the sum over k is typically an integral instead. The corresponding representation is then called a **Fourier integral**; the discrete version as given in Eq. (13.133) is instead called a **Fourier series**.[2] Here $\eta_L(k)$ and $\eta_R(k)$ are to be determined from initial conditions[3] – i.e., by specifying $\eta(t = 0, x)$ and $\partial\eta(t = 0, x)/\partial x$. Note however that, to connect to a physical solution, we need to take the real part of Eq. (13.133). We then end up with a superposition of sine and cosine functions with the same arguments, using the Euler identity $e^{i\theta} = \cos\theta + i\sin\theta$.

Finally, notice that Eq. (13.133) has the form

$$\eta(t, x) = f_L(x + vt) + f_R(x - vt) \tag{13.134}$$

with *general* functions f_L and f_R. You can verify this by substituting (13.134) into the wave equation (13.125). We will next show how to find these functions f_L and f_R in terms of initial conditions, using Fourier series or Fourier integrals. That is the wave equation essentially tells us that, given an initial profile of $\eta(t, x)$, the disturbances propagate left and right at speed v.

Periodic Boundary Conditions

We now describe in some detail the technique of Fourier series. Fourier's idea was that many functions can be constructed by adding together different sine (or cosine) waves with the appropriate wavelengths.

Suppose in particular that a real function $f(\theta)$ is *periodic*, with $f(\theta + 2\pi) = f(\theta)$ for all θ. Suppose also that $f(\theta)$ is Riemann integrable on every bounded interval, meaning that f is bounded and is also piecewise continuous (*i.e.*, continuous except possibly at a finite number of points in any bounded interval). Then $f(\theta)$ can be expanded in the Fourier series

$$f(\theta) = \sum_{n=-\infty}^{\infty} a_n e^{in\theta}, \tag{13.135}$$

where n can take all integer values (positive, negative, zero) and the a_n are numbers, generally complex with $a_n = a_{-n}^*$ to assure that f is real. Alternatively:

$$f(\theta) = b_0/2 + \sum_{n=1}^{\infty}(b_n \cos n\theta + c_n \sin n\theta), \tag{13.136}$$

where the b_n and c_n are *real* constants, with $n = 1, 2, 3, \ldots$ We have translated between the sine, cosine, and complex exponential form using $e^{in\theta} = \cos n\theta + i\sin n\theta$. The coefficients a_n are related to b_n and c_n by $a_0 = b_0/2$, $a_n = (b_n - ic_n)/2$, and $a_{-n} = (b_n + ic_n)/2$, where now $n = 1, 2, 3, \ldots$

---

[2] Jean Baptiste Joseph Fourier (1768–1830) was a French mathematician and physicist who was also a government and academic administrator.

[3] The wave equation satisfies the *existence and uniqueness theorem* of partial differential equations: given enough initial conditions, a unique solution is guaranteed.

We can calculate the a_n (and from them also b_n and c_n) by multiplying the $e^{in\theta}$ series by $e^{-im\theta}$ and integrating from $-\pi$ to π, where m is a positive or negative integer, or zero. That is:

$$\int_{-\pi}^{\pi} d\theta\, f(\theta)e^{-im\theta} = \sum_{n=-\infty}^{\infty} a_n \int_{-\pi}^{\pi} d\theta\, e^{i(n-m)\theta} = 2\pi a_m, \qquad (13.137)$$

since

$$\int_{-\pi}^{\pi} d\theta\, e^{i(n-m)\theta} = 2\pi\, \delta_{nm} = \begin{cases} 0 & \text{if } m \neq n \\ 2\pi & \text{if } m = n \end{cases}. \qquad (13.138)$$

Here δ_{nm} is the Kronecker delta, equal to unity if $n = m$ and zero otherwise. Therefore we have found that

$$a_n = \frac{1}{2\pi} \int_{-\pi}^{\pi} d\theta\, f(\theta)e^{-in\theta}, \qquad (13.139)$$

so also

$$b_n = \frac{1}{\pi} \int_{-\pi}^{\pi} d\theta\, f(\theta) \cos n\theta \quad (n \geq 0) \qquad (13.140)$$

and

$$c_n = \frac{1}{\pi} \int_{-\pi}^{\pi} d\theta\, f(\theta) \sin n\theta \quad (n \geq 1). \qquad (13.141)$$

Note that we have assumed that $f(\theta)$ is periodic in θ. However, this formalism goes through as long as we are dealing with a finite system. If our function depends upon a distance x rather than an angle θ, we can identify $\theta \equiv 2\pi x/L$, where L is some fixed size of the system. This is suited for example for a rod of length L in the case of longitudinal waves, or a string of length L in the case of transverse waves, where fixed boundary conditions are imposed at the two ends. That is, we can let L be the periodicity of the solution, so even though the Fourier series solves the wave equation for (hypothetical) previous or subsequent rods or strings of length L, they really are not there, and we can ignore these repetitions. The boundary conditions also have to be periodic, so only a countable infinity of frequencies or wave numbers can satisfy them.

To reprise the exponential form of the series, if $f(\theta)$ is periodic and piecewise continuous, with $f(\theta + 2\pi) = f(\theta)$, then we can write

$$f(\theta) = \sum_{n=-\infty}^{\infty} a_n e^{in\theta} \quad \text{where} \quad a_n = \frac{1}{2\pi} \int_{-\pi}^{\pi} d\theta\, f(\theta)e^{-in\theta}. \qquad (13.142)$$

Example 13.7   **Waves on a Clamped Rod**

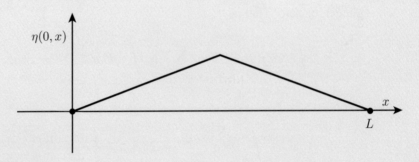

$\eta(0, x)$

L

x

Fig. 13.10   An initial displacement function in a rod.

We can use Fourier series to find longitudinal wave solutions on a thin metal rod of finite length L, with one end at $x = 0$ and the other at $x = L$. The displacement function $\eta(t, x)$ along the rod must satisfy the wave equation (13.125) subject to certain boundary conditions and initial conditions. We specify these as follows. We clamp the ends of the rod, so that $\eta(t, x)$ remains zero at each end, so the boundary conditions are $\eta(t, 0) = 0$ and $\eta(t, L) = 0$. Furthermore, we initially distort the rod longitudinally so that $\eta(0, x) = \alpha x$ for $0 \le x \le L/2$, and $\alpha(L - x)$ for $L/2 \le x \le L$, where α is a constant. In other words, $\eta = 0$ initially at both ends, increasing linearly to a maximum in the middle, as shown in Figure 13.10. We then release the rod from rest, so that $\partial \eta(t, x)/\partial t|_{t=0} = 0$. We could certainly express the general solution in terms of waves traveling to the right and to the left, where $f_L(x + vt)$ and $f_R(x - vt)$ are expanded in Fourier modes as in (13.142). However, it is simpler and more natural here to use sums over *standing* waves rather than traveling waves, since from the boundary conditions there are fixed nodes at $x = 0$ and $x = L$. Standing wave solutions are made from linear combinations of right and left-traveling waves. For example, using the identity $\sin(A + B) = \sin A \cos B + \cos A \sin B$, the sum

$$f_L + f_R = \sin(kx + \omega t) + \sin(kx - \omega t) = (\sin kx \cos \omega t + \cos kx \sin \omega t)$$
$$+ (\sin kx \cos \omega t - \cos kx \sin \omega t) = 2 \sin kx \cos \omega t, \tag{13.143}$$

which is a standing wave, oscillating in place. Here k and ω are related by $\omega/k = v$, where v is fixed by the wave equation. Other combinations of sines and cosines of position and time can be found by adding traveling waves with different phases. For the problem at hand, however, the example we just found is exactly what is needed, because the spatial dependence must involve sines, not cosines, since $\eta(t, x) = 0$ at $x = 0$. Also the time dependence must involve cosines, not sines, because the amplitude is a maximum at $t = 0$, which follows from the initial condition that the distortion is released from rest, so the distortion can only decrease, not increase. Putting this all together in the framework of Fourier series, the solution must have the form

$$\eta(t, x) = \sum_{n=1}^{\infty} c_n \sin nk_0 x \cos nk_0 vt. \tag{13.144}$$

where k_0 is the smallest possible wave number allowed by the boundary conditions, corresponding to a half-wavelength between $x = 0$ and $x = L$. In fact, there are two boundary conditions, $\eta(t, x = 0) = 0$

and also $\eta(t, x = L) = 0$, so $\sin nk_0 L = 0$, fixing $k_0 \equiv 2\pi/\lambda_0 = \pi/L$. The solution can therefore be written

$$\eta(t, x) = \sum_{n=1}^{\infty} c_n \sin \frac{n\pi x}{L} \cos \frac{n\pi vt}{L}. \tag{13.145}$$

where all we have left to do is find the Fourier coefficients c_n needed to match the shape of the initial distortion of the rod, $\eta(0, x) = \alpha x$ for $0 \le x \le L/2$, and $\alpha(L - x)$ for $L/2 \le x \le L$. That is, at $t = 0$:

$$\eta(0, x) = \sum_{n=1}^{\infty} c_n \sin \frac{n\pi x}{L} = \begin{cases} \alpha x & 0 \le x \le L/2 \\ \alpha(L - x) & L/2 \le x \le L \end{cases}. \tag{13.146}$$

To find how much of each wavelength (or wave number $k = 2\pi/\lambda$) is required, multiply both sides of this equation by $\sin(m\pi x/L)$, where m is an arbitrary positive integer, and then integrate from 0 to L. That is:

$$\sum_{n=1}^{\infty} c_n \int_0^L dx \, \sin \frac{m\pi x}{L} \sin \frac{n\pi x}{L}$$

$$= \alpha \int_0^{L/2} dx \, x \sin \frac{m\pi x}{L} + \alpha \int_{L/2}^{L} dx \, (L - x) \sin \frac{m\pi x}{L}. \tag{13.147}$$

All of these integrals are readily performed. The integral on the left is 0 if $m \ne n$ and equal to $L/2$ if $m = n$. That is, the given sine functions are orthogonal, as we would expect for normal mode solutions. The integrals on the right containing the product $x \sin m\pi x/L$ can be evaluated by parts.[a] The final result is

$$c_m = \left(\frac{4\alpha L}{\pi^2 m^2}\right) \times \begin{cases} 0 & m = 2, 4, 6, \ldots \\ 1 & m = 1, 5, 9, \ldots \\ -1 & m = 3, 7, 11 \ldots \end{cases} \tag{13.148}$$

Note that for the nonzero (odd m) modes, their amplitudes decrease like $1/m^2$. So if we add up the first few modes we come quite close to reproducing the shape of the initial conditions. The more modes we add in, the better fit we get.   ∎

[a] See the Problems section at the end of this chapter on how to do integrals like the one on the left.

Beyond Periodic Boundary Conditions

Now abandon the assumption that the longitudinal waves on a metal rod (or transverse waves on a stretched string) are necessarily periodic. A hypothetical stretched string might be (effectively) infinite,[4] with a displacement function $\eta(t, x)$ for transverse waves. Or we can admit that a rod or string has finite length, but we do not establish fixed boundary conditions at the ends, so there are now no restrictions on frequencies or wave numbers like those we have seen so far.

[4] "Effectively infinite" means here that any transverse displacement function $\eta(t, x)$ on the string never reaches either end of it in any time of interest to us.

For example, we might distort a stretched string into the initial shape $\eta(0, x) \equiv f(x)$, and then release the string from rest; we could then try to find $\eta(t, x)$ for all $t > 0$. It is very useful in problems like these to use **Fourier transforms** to solve problems with a continuum of frequencies or wave numbers. No longer do the frequencies of oscillation form a *discrete* infinity, but a *continuous* infinity instead; all frequencies or wave numbers must be considered. In the discrete case we had 13.142 or (in terms of a distance x)

$$f(x) = \sum_{n=-\infty}^{\infty} a_n e^{2\pi i n x / L} \quad \text{and} \quad a_n = \frac{1}{2\pi} \int_{-L/2}^{L/2} dx\, f(x) e^{-2\pi i n x / L}, \quad (13.149)$$

which include a countable $(n = 0, \pm 1, \pm 2, \ldots)$ infinity of modes. Now in the continuous case we have[5]

$$f(x) = \int_{-\infty}^{\infty} dk\, a(k) e^{ikx} \quad \text{and} \quad a(k) = \frac{1}{2\pi} \int_{-\infty}^{\infty} dx\, f(x) e^{-ikx}. \quad (13.150)$$

The wave number $k = 2\pi/\lambda$, a continuous variable, has replaced the discrete counting index n, and an integral over the continuous variable k has replaced the sum over the discrete index n.

This pair of equations is called a *Fourier transform* pair. That is, a function $f(x)$ in position space is the Fourier transform of a function $a(k)$ in wave-number space, and vice versa. If we know either function we can find the other by performing an integration. One can think of $a(k)$ as a measure of the amplitude of sinusoidal waves of wave number k required to sum up to the given $f(x)$. In this case of Fourier integrals we are adding up a continuous infinity of different wave numbers, which is what allows us to sum them up to functions that are not periodic, as we shall see.

Example 13.8 **A Single Square Bump**

Consider a square bump of width A:

$$f(x) = \begin{cases} 1 & -A/2 < x < A/2 \\ 0 & |x| > A/2 \end{cases} \quad (13.151)$$

shown in Figure 13.11(a). Obviously $f(x)$ is *not* periodic. The Fourier transform of $f(x)$ is

$$a(k) = \frac{1}{2\pi} \int_{-A/2}^{A/2} dx\, e^{-ikx} = \frac{1}{2\pi} \left(\frac{e^{-ikx}}{-ik} \right) \Big|_{-A/2}^{A/2}$$

$$= \frac{1}{2\pi} \left(\frac{e^{ikA/2} - e^{-ikA/2}}{ik} \right) = \frac{A}{2\pi} \left(\frac{\sin kA/2}{kA/2} \right), \quad (13.152)$$

shown in Figure 13.11(b).

---

[5] Fourier transforms are often written in an alternative form, with the factor $1/\sqrt{2\pi}$ in front of both the k and x integrals. As long as the product of these factors is $1/2\pi$, any choice is possible.

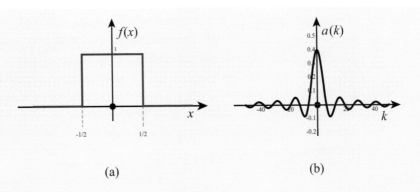

(a)   (b)

Fig. 13.11 (a) A square bump $f(x)$. (b) Its Fourier transform $a(k)$.

The pair of functions $f(x)$ and $a(k)$ have very nice physical interpretations in this case. Suppose that a plane wave of sound moving vertically upward, strikes a solid horizontal plate at altitude $h = 0$, with a slit of width A removed, as shown in Figure 13.12(a). Let x be the horizontal coordinate, and $f(x)$ be the amplitude of the wave at the position of the plate. This amplitude will have the form of a square bump, because it will have a constant nonzero value where it passes through the slit, and zero where it is blocked by the plate. After striking the slit, the emitted wave has $a(\boldsymbol{k}) \neq 0$ given by (13.152) for $\boldsymbol{k}$ transverse to the direction of propagation. These nonzero horizontal modes mean that the waves diffract to the left and right, and this diffraction pattern can be observed on the screen. The parts of the wave with large k will deflect more to the sides than parts with small k.

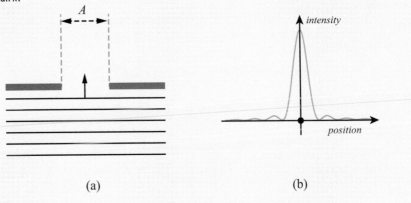

(a)   (b)

Fig. 13.12 (a) A plane sound wave moving upward toward a plate with a slit removed. (b) Wave intensity on a distant screen.

The intensity of the sound wave on the screen is proportional to the square of the amplitude; that is, the intensity is proportional to

$$I \sim a(k)^2 \sim \frac{\sin^2 kA/2}{(kA/2)^2},\qquad(13.153)$$

as illustrated in Figure 13.12(b). Let us measure how far the diffraction pattern has spread out by finding the width of the central diffraction peak. The *central* peak extends from $kA/2 = -\pi$ to $kA/2 = +\pi$, so $\Delta k = 2 \times 2\pi/A$, which is a measure of the width of the wave in k-space. The width of the wave in position space is the width of the slit, which is equal to A. So if we were to decrease A, the width of the slit, the position space width decreases while the k space width *increases*. Note that the product is $\Delta k \Delta x = 2 \times 2\pi/A \times A = 4\pi$, independent of A. The more one tries to confine the beam in position space (by narrowing the slit), the less confined the beam is in wave-number space, and vice versa. Narrow slits cause more diffraction than wide ones.   ∎

Example 13.9 | **Dirac Delta Function**

Let us find the Fourier transform of the Dirac "delta function"

$$f(x) = \delta(x - x_0) = \infty \quad (x = x_0)$$
$$= 0 \quad (x \neq x_0), \tag{13.154}$$

with

$$\int_{-\infty}^{\infty} dx \, \delta(x - x_0)g(x) = g(x_0). \tag{13.155}$$

Note that for $g(x) = 1$, we have

$$\int_{-\infty}^{\infty} dx \, \delta(x - x_0) = 1. \tag{13.156}$$

The Fourier transform of $f(x) = \delta(x - x_0)$ is then

$$a(k) = \frac{1}{2\pi} \int_{-\infty}^{\infty} dx \, e^{-ikx} \, \delta(x - x_0) = \frac{1}{2\pi} e^{-ikx_0}, \tag{13.157}$$

and so the inverse Fourier transform returns us back to

$$\delta(x - x_0) = \frac{1}{2\pi} \int_{-\infty}^{\infty} dx \, e^{ikx} \, e^{-ikx_0} = \frac{1}{2\pi} \int_{-\infty}^{\infty} dx \, e^{ik(x - x_0)}, \tag{13.158}$$

a very useful *representation* of the delta function.

In particular, this expression for the delta function demonstrates the consistency of the Fourier transform pair of integrals: as seen earlier, an arbitrary function $f(x)$ can be expressed as the integral

$$f(x) = \int_{-\infty}^{\infty} dk \, a(k)e^{ikx} \tag{13.159}$$

and

$$a(k) = \frac{1}{2\pi} \int_{-\infty}^{\infty} dx' \, f(x')e^{-ikx'}, \tag{13.160}$$

using a variable x' to avoid confusion with the variable x we will use in the next equation. Now substitute this latter expression back into (13.159). This gives

$$f(x) = \frac{1}{2\pi} \int_{-\infty}^{\infty} dk\, e^{ikx} \int_{-\infty}^{\infty} dx'\, f(x')e^{-ikx'}$$

$$= \frac{1}{2\pi} \int_{-\infty}^{\infty} dx'\, f(x') \int_{-\infty}^{\infty} dk\, e^{ik(x-x')}, \tag{13.161}$$

exchanging the order of integration, which we assume is valid here.[a] However, using (13.158), we get

$$f(x) = \int_{-\infty}^{\infty} dx'\, f(x')\delta(x - x') = f(x), \tag{13.162}$$

demonstrating self-consistency.   ∎

[a] A theorem named for the Italian mathematician Guido Fubini shows when this exchange is valid.

The Wave Equation, Revisited

Now we can apply Fourier transforms to find longitudinal waves on a thin rod or transverse waves on a stretched string. Transverse waves $y(t, x)$ are easier to visualize, so we will discuss those here; the mathematics of longitudinal waves is exactly the same. First, we take the Fourier transform of the entire wave equation

$$\frac{\partial^2 y}{\partial x^2} = \frac{1}{v^2} \frac{\partial^2 y}{\partial t^2}. \tag{13.163}$$

by multiplying through by e^{ikx} and integrating over x. That is:

$$\int_{-\infty}^{\infty} dx \frac{\partial^2 y(t, x)}{\partial x^2}\, e^{ikx} = \frac{1}{v^2} \int_{-\infty}^{\infty} dx \frac{\partial^2 y(t, x)}{\partial t^2}\, e^{ikx}. \tag{13.164}$$

The integral on the left can be integrated twice by parts (each time integrating one of the partial derivatives with respect to x); the integrated parts vanish because we assume the string is so long that no disturbances $y(t, x)$ reach the ends in times of interest. The wave equation in k space then becomes

$$(-ik)^2 Y(t, k) = \frac{1}{v^2} \frac{\partial^2 Y(t, k)}{\partial t^2}, \tag{13.165}$$

where $Y(t, k)$ is the Fourier transform of $y(t, x)$:

$$Y(t, k) = \frac{1}{2\pi} \int_{-\infty}^{\infty} dx\, y(t, x)e^{ikx}. \tag{13.166}$$

The differential equation for the Fourier transform $Y(t, k)$ is an *ordinary* differential equation in the dependent variable Y and independent variable t, which is a great simplification we shall now exploit.[6] Equation (13.165) is just a simple harmonic oscillator equation, with solution

$$Y(t, k) = A(k)e^{ivkt} + B(k)e^{-ivkt}, \tag{13.167}$$

where $A(k)$ and $B(k)$ are arbitrary functions of the wave number k. These can be determined from initial conditions as usual.

At time $t = 0$, and using both of the two preceding equations, the initial values of Y and $\partial Y/\partial t$ are

$$Y(0, k) = A(k) + B(k) = \frac{1}{2\pi} \int_{-\infty}^{\infty} dx \, y(0, x)e^{ikx}, \tag{13.168}$$

$$\frac{\partial Y}{\partial t}(0, k) = (ivk)[A(k) - B(k)] = \frac{1}{2\pi} \int_{-\infty}^{\infty} dx \, \frac{\partial y(0, x)}{\partial t}e^{ikx}. \tag{13.169}$$

Let the initial conditions be the initial position and the initial velocity of every point on the string:

$$y(0, x) \equiv f(x) \quad \text{and} \quad \frac{\partial y(0, x)}{\partial t} \equiv g(x). \tag{13.170}$$

Therefore

$$A(k) + B(k) = \frac{1}{2\pi} \int_{-\infty}^{\infty} dx \, f(x)e^{ikx}. \tag{13.171}$$

and

$$(ivk)[A(k) - B(k)] = \frac{1}{2\pi} \int_{-\infty}^{\infty} dx \, g(x)e^{ikx}, \tag{13.172}$$

which are two equations for the two unknown functions $A(k)$ and $B(k)$ in terms of the initial positions $f(x)$ and velocities $g(x)$. Solving for $A(k)$ and $B(k)$, we find

$$A(k) = \frac{1}{4\pi} \int_{-\infty}^{\infty} dx \left(f(x) + \frac{g(x)}{ikv} \right) e^{ikx} \tag{13.173}$$

and

$$B(k) = \frac{1}{4\pi} \int_{-\infty}^{\infty} dx \left(f(x) - \frac{g(x)}{ikv} \right) e^{ikx}. \tag{13.174}$$

We now have $Y(t, k)$ in terms of the initial conditions $f(x)$ and $g(x)$:

$$Y(t, k) = A(k)e^{ivkt} + B(k)e^{-ivkt}$$

$$= \frac{1}{4\pi} \int_{-\infty}^{\infty} dx \left[f(x) \left(e^{ik(x+vt)} + e^{ik(x-vt)} \right) \right.$$

$$\left. + \frac{g(x)}{ikv} \left(e^{ik(x+vt)} - e^{ik(x-vt)} \right) \right]. \tag{13.175}$$

---

[6] Note that the Fourier transform $Y(t, k)$ of the displacement $y(t, x)$ is unrelated to the Young's modulus Y of an elastic material, introduced earlier in this chapter.

The final step is to find $y(t,x)$ itself, by taking the inverse Fourier transform of $Y(t,k)$. That is:

$$y(t,x) = \int_{-\infty}^{\infty} dk\ Y(t,k)e^{-ikx}. \tag{13.176}$$

The quantity $Y(t,k)$ contains an integral over x, which we will now relabel as x', to avoid confusion with the x in $y(t,x)$. Therefore

$$y(t,x) = \int_{-\infty}^{\infty} dk\ Y(t,k)e^{-ikx}$$

$$= \frac{1}{4\pi} \int_{-\infty}^{\infty} dk \int_{-\infty}^{\infty} dx' \left[f(x') \left(e^{ik(x'+vt)} + e^{ik(x'-vt)} \right) e^{-ikx} \right.$$

$$\left. + \frac{g(x')}{ikv} \left(e^{ik(x'+vt)} - e^{ik(x'-vt)} \right) e^{-ikx} \right]. \tag{13.177}$$

Now invert the order of integration:

$$y(t,x) = \frac{1}{4\pi} \int_{-\infty}^{\infty} dx'\ f(x') \int_{-\infty}^{\infty} dk\ \left(e^{ik(x'-(x-vt))} + e^{ik(x'-(x+vt))} \right)$$

$$+ \frac{1}{4\pi iv} \int_{-\infty}^{\infty} dx'\ g(x') \int_{-\infty}^{\infty} \frac{dk}{k} \left(e^{ik(x'-(x-vt))} - e^{ik(x'-(x+vt))} \right),$$

which is the sum of two expressions we will call y_f and y_g. The first term y_f is greatly simplified using Eq. (13.158). It becomes

$$y_f = \frac{1}{4\pi} \int_{-\infty}^{\infty} dx'\ f(x') \int_{-\infty}^{\infty} dk\ \left(e^{ik(x'-(x-vt))} + e^{ik(x'-(x+vt))} \right)$$

$$= \frac{1}{2} \int_{-\infty}^{\infty} dx'\ f(x') \left[\delta(x' - (x - vt)) + \delta(x' - (x + vt)) \right]$$

$$= \frac{1}{2}[f(x - vt) + f(x + vt)], \tag{13.178}$$

the sum of two terms, each of which has exactly the same shape (but half the height) of the original shape $f(x)$ of the string. The term $f(x - vt)$ represents this shape moving to the *right* with speed v, while the term $f(x + vt)$ represents this same shape moving to the *left* with speed v. In fact, if the string is released from rest, then $g(x) = 0$, so these are the *only* terms present in the solution. Therefore if we begin by pulling the very long string into a Gaussian shape, for example, and then releasing it from rest, we will see two Gaussian shapes emerge, one moving to the right and one to the left, each half the height of the original.

But now suppose we do *not* release the string from rest, so that $g(x) \neq 0$. Then we have to evaluate the second set of terms:

$$y_g = \frac{1}{4\pi iv} \int_{-\infty}^{\infty} dx' g(x') \int_{-\infty}^{\infty} \frac{dk}{k} \left(e^{ik(x'-(x-vt))} - e^{ik(x'-(x+vt))} \right). \tag{13.179}$$

We can evaluate the k integrals as follows. We know that $e^{i\theta} = \cos\theta + i\sin\theta$, where the cosine and sine are even and odd functions of their arguments, respectively. These arguments are proportional to $1/k$ in each case, so the integrands containing sine functions divided by k are *even* functions of k, while those containing cosine functions divided by k are *odd* functions of k. The k integrals extend from $-\infty$ to $+\infty$, which is an even interval, so those integrals containing cosine functions will vanish (because the integral from $-\infty \to 0$ will cancel the integral from $0 \to +\infty$.) This leaves only the sine integrals. Therefore

$$y_g = \frac{1}{4\pi v} \int_{-\infty}^{\infty} dx' g(x') \int_{-\infty}^{\infty} \frac{dk}{k} \left(\sin kX_1' - \sin kX_2' \right),$$

where

$$X_1' \equiv x' - (x - vt) \quad \text{and} \quad X_2' \equiv x' - (x + vt). \tag{13.180}$$

The remaining integrals are somewhat nontrivial. With some effort, or looking it up in tables, we find

$$\int_{-\infty}^{\infty} dk \, \frac{\sin ak}{k} = \begin{cases} \pi & a > 0 \\ 0 & a = 0 \\ -\pi & a < 0 \end{cases}. \tag{13.181}$$

Therefore, we get

$$4\pi v y_g = \pi \int_{x-vt}^{\infty} dx' \, g(x') - \pi \int_{-\infty}^{x-vt} dx' \, g(x')$$

$$- \pi \int_{x+vt}^{\infty} dx' \, g(x') + \pi \int_{-\infty}^{x+vt} dx' \, g(x')$$

$$= 2\pi \int_{x-vt}^{\infty} dx' \, g(x') - 2\pi \int_{x+vt}^{\infty} dx' \, g(x')$$

$$= 2\pi \int_{x-vt}^{x+vt} dx' \, g(x'). \tag{13.182}$$

In summary, we have found that at any time $t > 0$, the shape of the string is

$$y(t, x) = y_f + y_g$$

$$= \frac{1}{2}[f(x - vt) + f(x + vt)] + \frac{1}{2v} \int_{x-vt}^{x+vt} dx' \, g(x'), \tag{13.183}$$

given that $f(x)$ is the initial shape of the string and that $g(x)$ is the initial time derivative of the shape. With the help of Fourier transform methods, we have found a general expression for transverse waves on a very long string, in terms of the initial conditions $y(0, x) \equiv f(x)$ and $\partial y(t, x)/\partial t|_{t=0} \equiv g(x)$. The solution is equally valid for longitudinal waves on a long, thin rod, which we have modeled by the continuum limit of an infinite number of springs connecting an infinite number of infinitesimal masses.

Finally, let us note that Eq. (13.183) can be written as

$$
y(t,x) = \left(\frac{1}{2}f(x - vt) + \frac{1}{2v} \int_{x-vt}^{0} dx' \, g(x') \right)
$$

$$
+ \left(\frac{1}{2}f(x + vt)] + \frac{1}{2v} \int_{0}^{x+vt} dx' \, g(x') \right). \tag{13.184}
$$

The first term is some function of $x - vt$ while the second is a function of $x + vt$. Hence, the general solution of the wave equation is a sum of left and right-moving profiles, as verified earlier.

We have illustrated the very powerful methods of Fourier series and Fourier transforms, the first used for finite rods or strings or for periodic displacements, and the second used for infinite rods or strings, or for displacements that are not periodic.

13.5 Summary

In this chapter, we developed general techniques that can be used to analyze a great many systems, with arbitrary numbers of degrees of freedom, when perturbed near an equilibrium point. We saw that the general solution is a linear superposition of normal modes of vibrations, which can in turn be determined by solving a well-defined eigenvalue problem. It is often possible to intuitively guess at these normal modes and hence quickly unravel what otherwise might be complex dynamics. It is difficult to overemphasize the importance of such harmonic motion in physics. We also developed techniques to extend the oscillations of a discrete number of coupled oscillators to the realm of continuous systems, which leads to waves and wave equations. Wave equations of course are very useful and even essential in understanding quantum-mechanical systems as well as classical systems. Small oscillations, whether of discrete or continuous systems, play central roles in virtually every branch of modern physics.

Problems

★★   **Problem 13.1** Two blocks, of masses m and M, are connected by a single spring of force constant k. The blocks are free to slide on a frictionless table. Beginning with the Lagrangian, find the oscillation frequency of the system in terms of k and the reduced mass $\mu \equiv mM/(m + M)$. Show that for the special case $M = m$, the frequency is what you would expect when the center of the spring remains at rest.

★★   **Problem 13.2** Two blocks, of masses m and $2m$, are connected together linearly by three springs of equal force constants k. The outer springs are also attached to stationary walls, while the middle spring connects the two masses. Find the normal-mode frequencies.

★★   **Problem 13.3** Reconsider the problem of two equal-mass blocks and three springs, in a straight line with the outer springs attached to stationary walls. Now suppose the outer springs have the same force constant k, while the central spring has force constant $2k$. Find the eigenfrequencies and eigenvectors.

★   **Problem 13.4** A hypothetical linear molecule of four atoms is free to move in three dimensions. How many degrees of freedom are there? How many translational modes? How many rotational modes? How many vibrational modes? Then suppose instead that the four atoms are all in the same plane but not lined up, still free to move in three dimensions. How many degrees of freedom are there in this case, and how many are there of each kind of translational, rotational, and vibrational modes?

★★   **Problem 13.5** Find the normal modes of oscillation for small-amplitude motions of a double pendulum (a lower mass m hanging from an upper mass M), where the pendulum lengths are equal. Find the normal-mode frequencies and the amplitude ratios of M and m in each case. Let the generalized coordinates be θ_1, the angle of the upper mass relative to the vertical, and θ_2, the angle of the lower mass relative to the vertical.

★★   **Problem 13.6** A uniform horizontal rod of mass m and length ℓ is supported against gravity by two identical springs, one at each end of the rod. Assuming the motion is confined to the vertical plane, find the normal modes and frequencies of the system. Then find the motion in case just one end of the rod is displaced from equilibrium and released from rest.

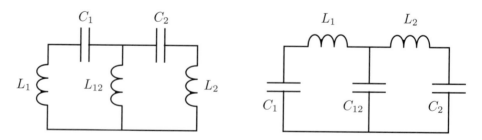

★★   **Problem 13.7** The voltage across a capacitor is $V_C = q/C$, where C is the capacitance and q is the charge on the capacitor. The voltage across an inductor is $V_L = dI/dt$, where L is the inductance and I is the current through the inductor. A wire attached to a capacitor whose charge is changing carries a current $I = dq/dt$. The net voltage drop around any closed circuit is zero, so a simple electrical L, C circuit obeys $L\ddot{q} + q/C = 0$, and so oscillates with frequency $\omega = 1/\sqrt{LC}$. Find the normal-mode oscillation frequencies and eigenvectors for each of the two-loop circuits shown above, in the case $C_1 = C_2 \equiv C, L_1 = L_2 \equiv L, C_{12} = 2C$, and $L_{12} = 2L$.

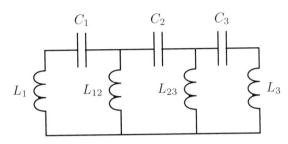

Problem 13.8 The voltage across a capacitor is $V_C = q/C$, where C is the capacitance and q is the charge on the capacitor. The voltage across an inductor is $V_L = dI/dt$, where L is the inductance and I is the current through the inductor. A wire attached to a capacitor whose charge is changing carries a current $I = dq/dt$. The net voltage drop around any closed circuit is zero, so a simple electrical L, C circuit obeys $L\ddot{q} + q/C = 0$, and so oscillates with frequency $\omega = 1/\sqrt{LC}$. Find the normal-mode oscillation frequencies of the three-loop circuit shown above, for the case $C_1 = C_2 = C_3 = C$ and $L_1 = L_2 = L_{12} = L_{23} = L$.

Problem 13.9 A block of mass M can move without friction on a horizontal rail. A simple pendulum of mass m and length ℓ hangs from the block. Find the normal-mode frequencies for small-amplitude oscillations.

Problem 13.10 A block of mass M can move without friction on a horizontal rail. A horizontal spring of force constant k connects one end of the block to a stationary wall. A simple pendulum of mass m and length ℓ hangs from the block. Find the normal-mode frequencies for small-amplitude oscillations.

Problem 13.11 The techniques used in this chapter can be extended to two- and three-dimensional systems. For example, we can find the normal-mode oscillations of a system of three equal masses m and three equal springs k in the configuration of an equilateral triangle, as shown in Figure 13.13(a).

We will suppose the masses are free to move only in the plane of the triangle, so there are $3 \times 2 = 6$ degrees of freedom for this system. (a) How many of these modes are translational? rotational? vibrational? (b) Show that the mass matrix is given by

$$
\hat{\mathbf{M}} = \begin{pmatrix} m & 0 & 0 & 0 & 0 & 0 \\ 0 & m & 0 & 0 & 0 & 0 \\ 0 & 0 & m & 0 & 0 & 0 \\ 0 & 0 & 0 & m & 0 & 0 \\ 0 & 0 & 0 & 0 & m & 0 \\ 0 & 0 & 0 & 0 & 0 & m \end{pmatrix},
$$

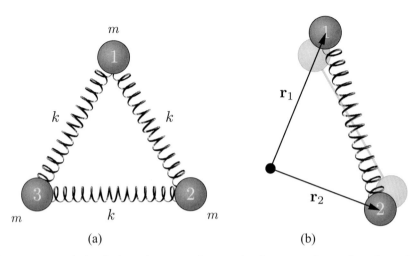

Fig. 13.13   (a) Three masses attached with identical springs and arranged at the corners of an equilateral triangle. (b) Depiction of the stretching of one of the springs, along with two position vectors.

while the spring matrix takes the form

$$\hat{\mathbf{K}} = \begin{pmatrix} 2k & -k & -k & 0 & 0 & 0 \\ -k & 2k & -k & 0 & 0 & 0 \\ -k & -k & 2k & 0 & 0 & 0 \\ 0 & 0 & 0 & 2k & -k & -k \\ 0 & 0 & 0 & -k & 2k & -k \\ 0 & 0 & 0 & -k & -k & 2k \end{pmatrix}.$$

(c) Find the normal modes of vibrations; note that the matrices are block diagonal in that 3×3 sub-blocks do not mix. Note that, due to degeneracies, you will need to make choices for picking orthonormal eigenvectors. Show that, for one choice, the normal modes take the form shown in Figure 13.14.

★   **Problem 13.12** In the previous problem, three degenerate normal modes were derived for the case of three equal masses at the vertices of an equilateral triangle, where the springs form the sides of the triangle. Any other oscillation in which the CM remains at rest and the system has no angular momentum must be a linear combination of these three modes. In particular, consider an oscillation identical to that of the second normal mode of the previous problem, except that it has been rotated to the right by $120°$, so for example mass no. 2 at the lower right now oscillates directly toward and away from the CM, rather than mass no. 1. By symmetry with the second normal mode, this mode should be possible, and should have the same frequency. Find the linear combination of the normal modes of the previous problem which is equal to the oscillation described above.

★★★   **Problem 13.13** Starting from the matrix equation

$$(\hat{\mathbf{K}} - \omega_i^2 \hat{\mathbf{M}})\mathbf{b_i} = 0, \tag{13.185}$$

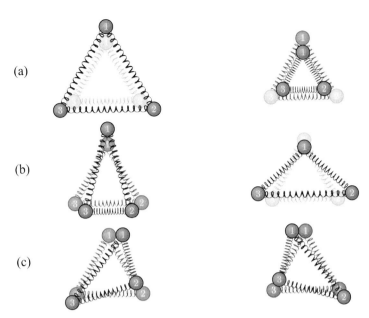

Fig. 13.14 The three vibrational modes of the triangular molecule.

for the ith eigenvector, and knowing that $\hat{\mathbf{K}}$ and $\hat{\mathbf{M}}$ are symmetric and real, prove the following statements: (a) $\mathbf{b_j}^\dagger \hat{\mathbf{M}} \mathbf{b_j}$ is real and positive definite (no sum over j implied); (b) the eigenvalues ω_i^2 are real; (c) $\mathbf{b_j}^\dagger \hat{\mathbf{K}} \mathbf{b_j}$ is also real and positive definite if the system is stable (no sum over j implied); (d) $\mathbf{b_i}$ and $\mathbf{b_j}$ are orthogonal if $\omega_i^2 \neq \omega_j^2$; (e) the eigenvectors $\mathbf{b_i}$ are real up to a multiplicative constant.

★ **Problem 13.14** (a) The "6–12" potential energy $U(r) = -2a/r^6 + b/r^{12}$, where a and b are positive constants, is sometimes used to approximate the potential energy between two atoms in a diatomic molecule, where the atoms are separated by a distance r. Find the effective force constant and the angular frequency of small oscillations of a classical atom of mass m about the equilibrium point. (b) Repeat part (a) for the "Morse" potential energy $U(r) = D_e(1 - e^{a(r-r_0)})^2$, where D_e, a, and r_0 are constants.

★★ **Problem 13.15** Consider an infinite number of masses m connected in a linear array to an infinite number of springs k. In equilibrium the masses are separated by distance a. Now allow small-amplitude transverse displacements of the masses, and take the limit as $a \to 0$, with an infinite number of infinitesimal masses and an infinite number of infinitesimal spring constants, so that the shape of the array as a function of time and space is given by $\eta(t, x)$, where η is transverse to the direction of the array in equilibrium. Show that if the amplitude is very small, then $\eta(t, x)$ obeys a linear wave equation, whose solutions can be traveling or standing transverse waves.

★★ **Problem 13.16** (a) With the help of the trig identities

$$\sin(A \pm B) = \sin A \cos B \pm \cos A \sin B,$$
$$\cos(A \pm B) = \cos A \cos B \mp \sin A \sin B,$$

prove the following results, often useful in Fourier analysis, where m and n are positive integers with $m \neq n$:

$$\text{(i)} \int_{-\pi}^{\pi} d\theta \sin m\theta \sin n\theta = 0,$$

$$\text{(ii)} \int_{-\pi}^{\pi} d\theta \sin m\theta \cos n\theta = 0,$$

$$\text{(iii)} \int_{-\pi}^{\pi} d\theta \cos m\theta \cos n\theta = 0.$$

(b) Evaluate the same three integrals for the case $m = n$. (c) Evaluate the same three integrals, for both $m \neq n$ and $m = n$, if the range of integration is $(0, \pi)$ instead of $(-\pi, \pi)$.

Problem 13.17 A rod of length L is clamped at both ends $x = 0, L$ so that the displacement function obeys $\eta(t, 0) = \eta(t, L) = 0$. Initially the displacement function is $\eta(0, x) = b \sin^2(\pi x / L)$ and $\partial \eta(t, x) / \{\text{partial deriv}\} t|_0 = 0$, where b is a positive constant. Find a Fourier-series representation of the solution of the wave equation at all future times.

Problem 13.18 A rod of length L, with ends at $(x = 0, L)$, has an initial displacement function $\eta(0, x) = b$ for $0 \leq x \leq L/2$ and $\eta(0, x) = -b$ for $L/2 \leq x \leq L$, where b is a positive constant. At time $t = 0$ the derivative of $\eta(t, x)$ is $\partial \eta(t, x) / \partial t|_0 = -v_0$ for $0 \leq x \leq L/2$ and equal to $+v_0$ for $L/2 \leq x \leq L$, where v_0 is a positive constant. Find a Fourier-series representation of the solution of the wave equation at all future times using the doubling trick.

Problem 13.19 A rod of length L, with ends at $(x = 0, L)$, has an initial displacement function $\eta(0, 0) = \eta(0, L) = 0$ and $\eta(0, x) = b$ for $0 < x < L$, where b is a positive constant. (That is, η is discontinuous at the ends.) At time $t = 0$ the derivative of η with respect to time is $\partial \eta(t, x) / \partial t = 0$ for all x. Find a Fourier-series representation of the solution of the wave equation at all future times, for points $0 < x < L$.

Problem 13.20 One end of a rod of length L is held at $x = 0$ while the other end is stretched from $x = L$ to $x = (1 + a)L$, where a is a constant. In this way an arbitrary point x in the rod is moved to $(1 + a)x$. Then at time $t = 0$ the rod is released. (a) What is the initial value of the displacement function $\eta(0, x)$? (b) Find $\eta(t, x)$. (c) Show that the velocity at the left end of the rod is either $2av$ or $-2av$, alternating between these values with a time interval L/v, where v is the wave velocity in the rod. You might want to use the doubling trick from the text.

★ **Problem 13.21** An infinite rod has an initial square-pulse displacement function $\eta(0,x) = C$, a constant, for $|x| \leq b$ and $\eta(0,x) = 0$ for $|x| > b$. (a) Find the displacement function $\eta(t,x)$ at later times, assuming all mass points in the rod are initially at rest. (b) Carry out a Fourier transform of $\eta(0,x)$ to determine $g(0,k)$, which shows the degree to which various wavelengths make up the square pulse.

★★ **Problem 13.22** An infinite rod has an initial triangular-pulse displacement function $\eta(0,x) = C - |x|$ for $x < C$, where C is a constant, and zero otherwise. (a) Find the displacement function $\eta(t,x)$ at later times, assuming all mass points in the rod are initially at rest. (b) Carry out a Fourier transform of $\eta(0,x)$ to determine $g(0,k)$, which shows the degree to which various wavelengths make up the triangular pulse.

★★ **Problem 13.23** An infinite rod has an initial Gaussian displacement function $\eta(0,x) = Ae^{-x^2/b^2}$, where A and b are constants. (a) Carry out a Fourier transform of $\eta(0,x)$, and show that the result is a Gaussian function in k space. (b) Then show that if the Gaussian in position space is narrow (with b small), then the Gaussian in k space is wide, and vice versa. (c) Define Δx as the distance between the two points on the position-space Gaussian for which $\eta(0,x)$ is half its maximum value. Similarly, define Δk as the distance in k space between the two points on $g(0,k)$ for which $g(0,k)$ is half its maximum value. Then find the product $\Delta x \cdot \Delta k$, and show that it is independent of b. *Hint*: The Fourier integrals can be evaluated by completing the square in the exponents.

★ **Problem 13.24** At the end of the chapter we derived a general expression for waves $y(t,x)$ on a long string, in terms of the initial displacement $y(0,x) \equiv f(x)$ and velocity $\partial y(0,x)/\partial t \equiv g(x)$. Suppose that the initial displacement is $y(0,x) = f(x)$, where $f(x)$ is some given function. (a) What $g(x)$ would be required, in terms of $f(x)$, so that for any time $t > 0$, there is *only* a wave traveling to the right: $y(t,x) = f(x - vt)$? (b) Find this $g(x)$ in the special case that $f(x)$ is the Gaussian function $f(x) = Ae^{-x^2/b^2}$, where A and b are constants.

★★★ **Problem 13.25** In the text we saw an example involving a nondiagonal mass matrix arising in the case of a single particle. In this problem, we will look at a similar scenario for two particles. Consider two interacting particles of mass m_1 and m_2 constrained to move in one dimension described by the Lagrangian

$$L = \frac{1}{2}m_1\,\dot{q}_1^2 + \frac{1}{2}m_2\,\dot{q}_2^2 - U(q_1 + q_2).$$

The coordinates of the two particles are represented by q_1 and q_2 and the potential energy function is given by $U(Q) = \alpha Q^2/2$ for some constant α. The novelty here is that the potential between the two particles is not translationally invariant; it does not depend on the distance between the particles, $q \equiv q_1 - q_2$. Instead, the potential depends on the *sum* of the two coordinates $Q \equiv q_1 + q_2$. As a result, the usual coordinate transformation from q_1 and q_2 to the center of mass coordinate $Q_{cm} = (m_1q_1 + m_2q_2)/(m_1 + m_2)$ and the relative distance $q = q_1 - q_2$ is not very useful. Instead, we want to transform to $Q = q_1 + q_2$ and $q = q_1 - q_2$. (a) Show that the Lagrangian in the Q and q coordinates takes the form

$$L = \frac{1}{8}M\dot{Q}^2 + \frac{1}{8}M\dot{q}^2 + \frac{1}{4}m\dot{Q}\dot{q} - U(Q),$$

where $M \equiv m_1 + m_2$ and $m \equiv m_1 - m_2$. (b) Find the eigenvalues and eigenvectors of small oscillations. Elaborate briefly on the meaning of each normal mode.

★★   **Problem 13.26** Consider a particle of mass m moving in three dimensions but constrained to the surface of the paraboloid $z = \alpha\left((x-1)^2 + (y-1)^2\right)$. The particle is also subject to the spring potential $U(x, y, z) = (1/2)k\left(x^2 + y^2\right)$. (a) Show that the Lagrangian of the system is given by

$$L = \frac{1}{2}m\dot{x}^2\left(1 + 4\alpha^2\right) + \frac{1}{2}m\dot{y}^2\left(1 + 4\alpha^2\right) + \frac{1}{2}8m\alpha^2\dot{x}\dot{y} - \frac{1}{2}k\left(x^2 + y^2\right)$$

to quadratic order in x, y, $\dot{x}$, and $\dot{y}$ – assuming the displacement from the origin is small. (b) Find the normal modes, eigenfrequencies, and eigenvectors.

Deterministic evolution is a hallmark of classical mechanics. Given a set of exact initial conditions, differential equations evolve the trajectories of particles into the future and can exactly predict the location of every particle at any instant in time. As famously written by the French mathematician and astronomer Pierre Simon de Laplace (1749–1827):

> We may regard the present state of the universe as the effect of its past and the cause of its future. An intellect which at a certain moment would know all forces that set nature in motion, and all positions of all items of which nature is composed, if this intellect were also vast enough to submit these data to analysis, it would embrace in a single formula the movements of the greatest bodies of the universe and those of the tiniest atom; for such an intellect nothing would be uncertain and the future just like the past would be present before its eyes.[1]

Laplace's universe was *clock-like*. Of course we now know that quantum mechanics makes a precise and complete specification of initial conditions impossible, and that the equations needed to solve for the future positions are fundamentally probabilistic in nature. But even if the universe operated strictly according to Newton's laws, just as Laplace believed, *in the real world it would never be possible, with infinite precision, to know either the initial conditions or the exact future motion.* Neither measurements nor computer calculations can be perfectly precise!

So what happens if our uncertainties in the initial position or velocity of a particle are tiny? Does that mean that our uncertainties about the subsequent motion of the particle are necessarily tiny as well? Or are there situations in which a very slight change in initial conditions leads to huge changes in the later motion? For example, can you really balance a pencil on its point? What has been learned in relatively recent years is that, in contrast to Laplace's vision of a clock-like universe, *deterministic systems are not necessarily predictable.* Motion which is so sensitive to initial conditions that it is hopeless to try to predict future motion is said to be *chaotic*. For example, in the long run it turns out that even the future of the solar system is unpredictable: the Newtonian gravitational forces between the various planets and moons and asteroids and the sun add up to complex nonlinear dynamics with all the necessary ingredients for **chaos**. But how then can we have

---

[1] Laplace should have written: "... and all positions *and velocities* of all items of which nature is composed, ..." so as to specify the complete set of initial conditions the intellect would need for the analysis.

stable planetary orbits in such a system? What are the attributes of chaos and how can we quantify it? We begin our discussion with the notion of **integrability**, which ensures the *absence* of chaos.

14.1 Integrability

Consider a simple two-dimensional system consisting of a bob of mass m attached to a spring of spring constant k and zero equilibrium length – as depicted in Figure 14.1. The other end of the spring is kept fixed. For now, no gravity acts on the bob.

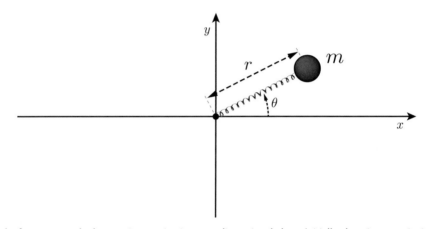

Fig. 14.1   A bob of mass m attached to a spring moving in a two-dimensional plane. Initially, there is no gravity in the problem.

The two degrees of freedom of the system can conveniently be chosen as the polar coordinates r and θ of the bob. The Lagrangian is

$$L = \frac{1}{2}m\left(\dot{r}^2 + r^2\dot{\theta}^2\right) - \frac{1}{2}kr^2. \tag{14.1}$$

The system is then symmetric under rotations as well as displacements in time. From time displacement invariance, the energy is conserved:

$$E = \frac{1}{2}m\left(\dot{r}^2 + r^2\dot{\theta}^2\right) + \frac{1}{2}kr^2. \tag{14.2}$$

Also, rotational invariance leads to angular momentum conservation:

$$\ell = mr^2\dot{\theta}. \tag{14.3}$$

This is a familiar system that we analyzed already in Chapter 7. Switching to the Hamiltonian picture, the system is described in phase space by the coordinates r, p_r, θ, p_θ, where

$$p_\theta = m\,r^2\dot{\theta} = \ell \tag{14.4}$$

and

$$p_r = m\,\dot{r}. \tag{14.5}$$

The Hamiltonian is the energy written in phase-space coordinates:

$$H = \frac{p_r^2}{2\,m} + \frac{p_\theta^2}{2\,m\,r^2} + \frac{1}{2}k\,r^2 = E. \tag{14.6}$$

This system is said to be **integrable**, where integrability is defined as follows. If a system with N degrees of freedom – that is, with a $2\,N$-dimensional phase space – has N conserved quantities that are in **involution**, the system is integrable.

So what is meant by "involution"? Let the conserved quantities of a system be denoted by

$$C_i = C_i(q,p) \tag{14.7}$$

for $i = 1, \ldots, N$, each being a function of all $2\,N$ phase-space coordinates in general. These conserved quantities are said to be in involution if the Poisson brackets vanish:

$$\{C_i, C_j\} = 0, \tag{14.8}$$

for all i and j.

Let us apply this idea to the current example. We have two degrees of freedom and a four-dimensional phase space. Our required two conserved quantities can be $C_1 = H(r, p_r, \theta, p_\theta)$ and $C_2 = p_\theta$. We can then check that $\{C_i, C_j\} = 0$. That is:

$$\{H, p_\theta\} = \frac{\partial H}{\partial r}\frac{\partial p_\theta}{\partial p_r} - \frac{\partial H}{\partial p_r}\frac{\partial p_\theta}{\partial r} + \frac{\partial H}{\partial \theta}\frac{\partial p_\theta}{\partial p_\theta} - \frac{\partial H}{\partial p_\theta}\frac{\partial p_\theta}{\partial \theta} = 0, \tag{14.9}$$

since, as is easy to show, each term vanishes. Note also that we have $\{H, H\} = \{p_\theta, p_\theta\} = 0$: the Poisson bracket of a function with itself is zero. Therefore our system is integrable. This implies that we are guaranteed to succeed in writing the solution to the equations of motion in the form of integrals. For example, we can find $r(t), p_r(t), \theta(t), p_\theta(t)$ for our system as follows. From Hamilton's equations

$$\dot{\theta} = \frac{\partial H}{\partial p_\theta} = \frac{p_\theta}{m\,r^2}, \quad \dot{p}_\theta = -\frac{\partial H}{\partial \theta} = 0, \tag{14.10}$$

so that

$$\int d\theta = \theta = \int \frac{p_\theta}{m\,r^2}\,dt, \tag{14.11}$$

$$p_\theta = \text{constant.} \tag{14.12}$$

Also

$$\dot{r} = \frac{\partial H}{\partial p_r}, \quad \dot{p}_r = -\frac{\partial H}{\partial r},$$ (14.13)

implying

$$\int \frac{m\,r}{\sqrt{2\,E\,m\,r^2 - p_\theta^2 - k\,m\,r^4}}\,dr = \int dt = t$$ (14.14)

and

$$\int dp_r = p_r = \int \left(\frac{p_\theta^2}{m\,r^3} - k\,r \right)\,dt.$$ (14.15)

We can start by integrating Eq. (14.14), which then allows us to find $p_r(t)$ and $\theta(t)$ using Eqs. (14.15) and (14.11), respectively. And of course p_θ is a constant throughout, as implied by Eq. (14.12).

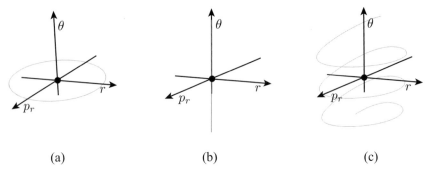

(a)   (b)   (c)

Fig. 14.2 (a) The r–p_r–θ subspace of the phase space of the spring–mass system. A trajectory with $p_\theta = 0$ is shown. Note that the $\theta = 0$ plane is to be identified with the $\theta = 2\pi$ plane. (b) A trajectory with $\dot{\theta} = \sqrt{k/m}$, which is preserved by the evolution, is shown. (c) A trajectory for a generic initial condition is shown.

We now want to demonstrate how integrability implies *nonchaotic* evolution. In phase space, the evolution of a mechanical system traces out a curve described by $q_i(t)$ and $p_i(t)$. At every instant in time, the state of the system is a point in $2N$-dimensional phase space. In our case, the phase space is four-dimensional, so it is difficult to visualize from our three-dimensional point of view. However, we can always take two- or three-dimensional cross-sections of the full four-dimensional phase space. For example, Figure 14.2 shows a projection of the trajectory of the bob into the r–p_r–θ subspace. In this scenario, the picture is quite clear: the fourth dimension of phase space is p_θ, which remains constant through the evolution of the bob's motion. Hence, the trajectory is already confined to the $p_\theta = \text{constant}$ subspace and the projection in question is trivial. The figure shows trajectories for three different initial conditions. For (a), we have $p_\theta = 0$, implying that the bob oscillates radially. For (b), we have chosen $\dot{\theta} = \sqrt{k/m}$ initially, which leads to circular motion, in which the spring force causes a purely centripetal acceleration. Finally, for (c) we have the case of generic and arbitrary initial conditions. Hence, we see that the trajectory is always confined to the surface of the cylinder shown in the figure. However, notice that the z axis is an angular direction: $\theta = 2\pi$ is to be identified with $\theta = 0$. This implies that we need to glue the two ends of the cylinder so that the shape of the surface where the trajectory lives is in fact a *torus* (see Figure 14.3).

Fig. 14.3   A torus can be constructed by gluing the endpoints of a cylinder together.

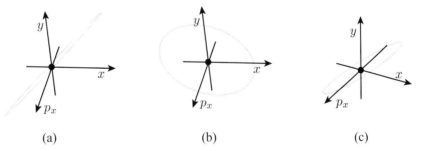

       (a)   (b)   (c)

Fig. 14.4   The trajectory of the bob in the x–p_x–y subspace. The initial conditions depicted correspond to the $p_\theta = 0$ case in (a), where the motion lies along a line in the x–y plane; $\dot{\theta} = \sqrt{k/m}$ case in (b), where the motion is a circle in the x–y plane; and the generic case in (c).

We can see this a little better if we switch to Cartesian coordinates and write the Hamiltonian instead in terms of x, p_x, y, and p_y:

$$H = \frac{p_x^2}{2m} + \frac{1}{2}kx^2 + \frac{p_y^2}{2m} + \frac{1}{2}ky^2, \tag{14.16}$$

which makes it clear that we simply have two decoupled harmonic oscillators with the same angular frequency $\sqrt{k/m}$. Figure 14.4 shows the corresponding phase-space picture in the x–p_x–y plane. We see the three different trajectories discussed above as curves, once again, on a torus. Note that a torus has two noncontractible cycles, as indicated in the figure.

The perspective in Cartesian coordinates also makes it clear how to think of more general situations. In the current example, the two oscillators have the same angular frequencies. Hence, when the trajectory winds once around the major cycle of the torus, it also winds once around the minor cycle. The resulting trajectory is then closed. However, imagine a more general Hamiltonian of the form

$$H = \frac{p_x^2}{2\,m} + \frac{1}{2} m\,\omega_1^2 x^2 + \frac{p_y^2}{2\,m} + \frac{1}{2} m\,\omega_2^2\, y^2, \tag{14.17}$$

where $\omega_1 \neq \omega_2$. We still have two decoupled harmonic oscillators, but now they have *different* frequencies. The system remains integrable, however; the energy of each one-dimensional oscillator is separately conserved and thus we have two conserved quantities for two total degrees of freedom. We can easily check that the two energies are indeed in involution, as required for an integrable system. Now, however, the trajectory in phase space can be much more interesting.

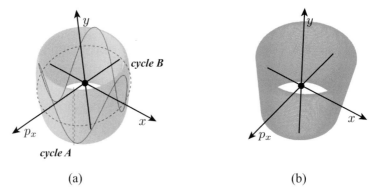

cycle B

cycle A

(a)   (b)

Fig. 14.5 (a) The trajectory of the double oscillator in the x–p_x–y subspace when $\nu = \omega_1/\omega_2 = 1/3$; the trajectory lies on the surface of a torus as shown. Note that, in this case, cycle A of the torus is essentially flat. (b) The scenario where ν is irrational.

Figure 14.5(a) shows the case where $\omega_1/\omega_2 = 1/3$. Each time the trajectory winds three times around cycle A, it winds only once around cycle B. The full trajectory is again closed, and it still lies on a torus. We can see that this pattern will persist for any case where the ratio

$$\nu = \frac{\omega_1}{\omega_2} \tag{14.18}$$

is a rational number, *i.e.*, a ratio of two integers. This ratio ν is called the **winding number** of the trajectory. Conventionally one chooses to write ν such that $\omega_1 < \omega_2$, so that $0 \leq \nu \leq 1$. Figure 14.5(b) shows a qualitatively different behavior if the winding number happens to be *irrational*: the trajectory never closes and densely covers the surface of the torus as it evolves in time. Given enough time, the trajectory can be shown to come as close as needed to any given point on the torus.

For an integrable system in $2\,N$-dimensional phase space, the pattern we just depicted is a general one: the trajectory of the system in phase space is confined to the surface of an N dimensional torus. Hence the intersection of the N-subspaces defined by the conserved quantities $C_i = C_i(q, p)$, with $i = 1, \ldots, N$, is guaranteed to be a torus in a $2N - N = N$-dimensional subspace. Such a torus, called an N-torus, has N noncontractible cycles and can be imagined as an N-dimensional hypercube with its opposing sides identified. The system is then characterized by N angular frequencies, one for each noncontractible cycle. And depending on the relation between these frequencies, the trajectory may close or may densely cover the torus. The special coordinate system where the N-torus can be seen as an N-dimensional hypercube is called the **action angle** perspective. This coordinate system can always be constructed, provided the system is integrable. We will discuss this topic in more detail later.

There is another technique for analyzing phase-space trajectories that is particularly useful when the system under consideration is *not* integrable. Since a $2N$-dimensional phase-space picture is difficult to visualize for $N > 1$, one takes snapshots of the trajectory in two-dimensional cross-sections of the full space.

Figure 14.6 illustrates the idea for constructing x–p_x maps for various cases of the winding number ν. For our two-oscillator example, a point is drawn on the two-dimensional map every time the trajectory crosses the $y = 0$ plane. For rational ν, we end up with a discrete set of points circling the origin.

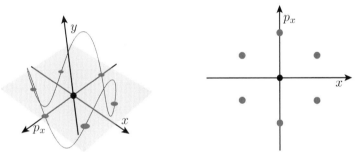

Fig. 14.6 The construction of the x–p_x map (Poincaré section) for the double oscillator. We see the process of constructing the map from the perspective of the x–p_x–y subspace of phase space.

Figure 14.7(a) shows how the pattern of such points can be used to read off the winding number ν. However, when ν is irrational, the points coalesce into a closed curve as shown in Figure 14.7(b). Such curves are called **KAM tori**[2] (even though they are cross-sections of higher-dimensional tori), while the two-dimensional maps are called **Poincaré sections**. For integrable systems, Poincaré sections consist of collections of closed loops and discrete points when we scan over different initial conditions, as shown in Figure 14.8.

Something much more interesting happens when the mechanical system is not integrable and hence is prone to chaos.

---

[2] The name derives from the mathematicians Kolmogorov, Arnol'd, and Moser. We shall shortly describe an important theorem they developed.

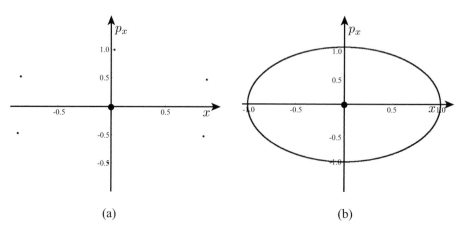

Fig. 14.7 (a) A Poincaré section for rational winding number $\nu = a/b$. Here, a and b can be read off from the number of times we circle the central point and the number of points, respectively. (b) A Poincaré section for irrational winding number.

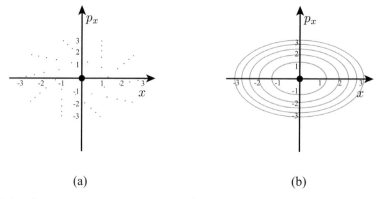

Fig. 14.8 A typical Poincaré section as we map out various initial conditions. (a) The case with rational winding numbers for various initial conditions corresponding to different total energies. (b) The case with irrational winding for similar initial conditions.

Consider a modified setup where we couple two equal-frequency harmonic oscillators to each other as follows:

$$H = \frac{p_1^2}{2\,m} + \frac{1}{2} m\,\Omega^2 q_1^2 + \frac{p_2^2}{2\,m} + \frac{1}{2} m\,\Omega^2\,q_2^2 + \frac{1}{2\,m^{3/2}} q_1 q_2^2 - \frac{1}{6\,m^{3/2}} q_1^3. \quad (14.19)$$

This system can be used to model the dynamics of a star in the axisymmetric potential of a galaxy – with q_1 being related to the radial location of the star from the center of the galaxy while q_2 is related to its angle of declination. This is named the **Hénon–Heiles** system, and is known to be chaotic.

Figure 14.9(a) show a Poincaré section of the **Hénon–Heiles** dynamics. We see that for certain initial conditions, the KAM tori have degenerated – implying that the trajectory in the full phase space is no longer on the surface of a torus. To

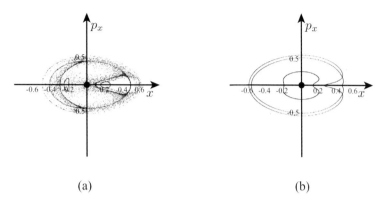

(a)   (b)

Fig. 14.9 (a) The $x-p_x$ Poincaré section for the Hénon–Heiles system. (b) The same map when the system is rendered integrable by a flip of a sign in the Hamiltonian.

contrast this with integrability, Figure 14.9(b) shows what happens if we flip a single sign in the Hamiltonian of the **Hénon–Heiles** system. If we consider instead

$$H = \frac{p_1^2}{2\,m} + \frac{1}{2}m\,\Omega^2 q_1^2 + \frac{p_2^2}{2\,m} + \frac{1}{2}m\,\Omega^2\,q_2^2 + \frac{1}{2\,m^{3/2}}q_1 q_2^2 + \frac{1}{6\,m^{3/2}}q_1^3 \quad (14.20)$$

(note the sign flip in the last term!), the dynamics becomes in fact integrable, although that is far from obvious. Poincaré sections provide us with a rather elegant technique to determine whether a system is chaotic (and hence non-integrable). It turns out to be much more difficult to show that a system is integrable.

Our goal is to analyze what happens as an integrable system is turned into a chaotic one. There are several ways to achieve chaos starting from an integrable system. One technique is to add some significant nonlinear time-independent couplings in the Hamiltonian all at once, as in the **Hénon–Heiles** case. A more controlled approach is to add such a coupling, but to do so gradually. That is, add a *small* nonlinear perturbation to an integrable system in the hope of observing the onset of chaos as the perturbation is gradually increased. These all lead to examples of **conservative chaos**: the Hamiltonian remains independent of time, and energy is conserved. Alternatively, one can achieve chaotic dynamics by adding dissipation and external time-dependent forces to a system. The latter corresponds to **dissipative chaos** and is described in detail in Section 14.3. For now we explore the case of conservative chaos.

14.2 Conservative Chaos

Consider our example from the previous section, the bob at the end of a spring, with two modifications. We set the rest length of the spring equal to r_0 instead of zero, and we add uniform gravity so that the system is more like a pendulum:

$$L = \frac{1}{2}m\left(\dot{r}^2 + r^2\dot{\theta}^2\right) - \frac{1}{2}k(r - r_0)^2 + mgr\cos\theta, \tag{14.21}$$

where we have returned to polar coordinates. We plan to turn on the gravitational effect gradually, tuning g from zero to higher values as we study the response of the system through Poincaré sections. The energy

$$E = \frac{1}{2}m\left(\dot{r}^2 + r^2\dot{\theta}^2\right) + \frac{1}{2}k(r - r_0)^2 - mgr\cos\theta \tag{14.22}$$

is still conserved, since the Hamiltonian is time independent. However, we have lost rotational invariance, so the angular momentum is no longer conserved. We therefore have only one of the two required conserved quantities needed for an integrable system with two degrees of freedom. Indeed, a second conserved quantity does *not* exist and the system is known to be chaotic.

To demonstrate the non-integrability of this system, we need to solve the equations of motion *numerically*. Section 14.6 presents a brief discussion of handy numerical techniques – including a fourth-order Runge–Kutta algorithm – to study such a mechanical system on a computer. The first task is to write the dynamics in terms of dimensionless variables and to identify the relevant external tunable parameters. In this case, we first note that the natural scale for the radial coordinate r is given by the spring's rest length r_0, so that we will let

$$R = \frac{r}{r_0}, \tag{14.23}$$

where the new radial coordinate R is dimensionless. We can use two of the other parameters, k and m, to define a dimensionless time

$$T = \sqrt{\frac{k}{m}}t, \tag{14.24}$$

where T is dimensionless and measures the number of oscillation periods. Time derivatives then convert as

$$\frac{d}{dt} = \sqrt{\frac{k}{m}}\frac{d}{dT}. \tag{14.25}$$

We can then rewrite the Lagrangian in terms of the dimensionless radius and dimensionless time:

$$L = \frac{1}{2}\left(\dot{R}^2 + R^2\dot{\theta}^2\right) - \frac{1}{2}(R - 1)^2 + \varepsilon R\cos\theta, \tag{14.26}$$

where we have dropped an overall multiplicative constant (which does not affect the equations of motion), where the overdots now mean d/dT, and where we have defined the dimensionless perturbation parameter

$$\varepsilon \equiv \frac{mg}{kr_0}. \tag{14.27}$$

Now the idea is to begin with $\varepsilon = 0$, and then gradually increase this "control parameter" ε to see how the dynamics is affected. From now on we will assume that $\varepsilon \ll 1$.

We start by considering a special set of initial conditions where θ is small and R is near its static equilibrium length $R_{eq} = 1 + \varepsilon$. So we let

$$R = R_{eq} + x = 1 + \varepsilon + x, \quad \cos\theta \simeq 1 - \frac{\theta^2}{2}, \tag{14.28}$$

where $x \ll R_{eq}$ and $\theta \ll 1$. Substituting these into the Lagrangian, and expanding to quadratic order in the small parameters x and θ, we find

$$L = \frac{1}{2}\dot{x}^2 - \frac{1}{2}x^2 + \frac{1}{2}(1 + \varepsilon)^2\dot{\theta}^2 - \frac{1}{2}\varepsilon(1 + \varepsilon)\theta^2. \tag{14.29}$$

Therefore, as we may have expected, we have in effect *two* harmonic oscillators with angular frequencies

$$\omega_x = 1, \quad \omega_\theta = \sqrt{\frac{\varepsilon}{1 + \varepsilon}}. \tag{14.30}$$

Within this restricted set of initial conditions – small angles and small radial displacements – the curves on the Poincaré sections would be simply KAM tori. The winding number

$$\nu = \frac{\omega_\theta}{\omega_x} = \sqrt{\frac{\varepsilon}{1 + \varepsilon}} \tag{14.31}$$

is in general irrational, except for special values of ε. Note that the treatment is self-consistent since starting with small x and θ assures that x and θ stay small.

If we look beyond this very restricted set of initial conditions, the Poincaré sections become much more interesting. For small but fixed perturbation ε, the winding numbers of KAM tori for different initial conditions become functions of the initial conditions, as depicted in Figure 14.10.

Shown are two KAM tori with different irrational winding numbers. Their shapes are found to be slightly deformed from what they would have been had we not turned on the perturbation. At the center is a single point – a **KAM torus** of winding number $\nu = 1$. For the θ–p_θ section, this corresponds to stable vertical oscillations at fixed $\theta = 0$. Such central points on Poincaré sections are stable tori, often called fixed points or **elliptic points**, around which other KAM tori are arranged. As we scan over initial conditions moving between the two irrational KAM tori shown in the figure, we will come upon the occasional rational winding number case. In these regions of the Poincaré section, we discover the onset of chaos.

The celebrated *KAM theorem* of Kolmogorov, Arnol'd, and Moser proves that if a small perturbation is added to an integrable system, initial conditions corresponding to irrational KAM tori in Poincaré sections are affected by minimal deformations – with the dynamics remaining nonchaotic. However, tori with rational winding numbers are expected to disintegrate into chaotic motion. The "more

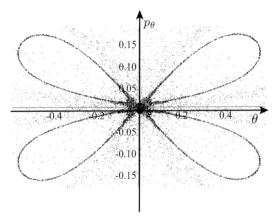

Fig. 14.10 The θ–p_θ Poincaré section for the bead–spring system with gravity. The winding numbers become functions of the initial conditions.

rational" an irrational number, the more unstable it becomes as the perturbation is increased. To understand this subtlety, we need to know that any irrational number ν can be approximated by a continued fraction expansion, as in

$$\nu = c_1 + \cfrac{1}{c_2 + \cfrac{1}{c_3 + \cfrac{1}{c_4 + \cdots}}}, \tag{14.32}$$

where the c_is are unique integers. The more c_is we need to achieve a prescribed precision for approximating ν, the "less rational" is ν. The larger the c_is, the faster the sequence converges to the desired precision. In this sense, the most irrational number corresponds to the minimum value for the c_is: $c_i = 1$. This corresponds to $\nu = (\sqrt{5} - 1)/2 \simeq 0.618033988\ldots$, which is the inverse of the *golden mean* or *golden ratio* in mathematics.[3] It would correspond to the most stable KAM torus under perturbation.

There is an intuitive way to understand why rational winding numbers are seeds for chaotic dynamics: when the winding number is a rational number for given initial conditions, the pattern in the Poincaré sections consists of a sequence of disconnected dots, as shown earlier in Figure 14.8. This means that, as time evolves, a small perturbation can constructively accumulate its effect on the trajectory – as the latter traces over and over the same trajectory. Hence, the effect of the perturbation grows with time and eventually the nonlinear effects from the perturbation take over and destroy the potential KAM torus. However in a sense, irrational winding numbers diffuse the effect of the perturbation as the trajectory never traces back on itself, hence leading to stable tori.

---

[3] The golden ratio $\phi \equiv 1.618033988\ldots$ is found as follows. Given two numbers a and b with $a > b > 0$, their ratio is "golden" if $a/b = (a + b)/a$. This ratio was studied by both Pythagoras and Euclid, and has been applied in architecture and elsewhere.

The KAM theorem leads to very interesting conclusions. Most initial conditions correspond to stable KAM tori – in the sense that most numbers between 0 and 1 are irrational. In contrast, there are an infinite number of rational numbers between 0 and 1, and hence an infinite set of initial conditions that can lead to chaos. In phase space, and for small perturbations, there are large basins or regions of stable deterministic evolution. But even for the tiniest perturbation, we are guaranteed that there also exists regions of phase space corresponding to initial conditions leading to chaotic dynamics. In our example, for the smallest gravitational perturbation of the pendulum, there are initial conditions at large angles and large radial extents where the motion of the bead becomes chaotic. As the nonlinear perturbation is tuned to even larger values, chaos takes over more and more of the phase space, until we reach *macroscopic chaos*, where most of the phase space describes chaotic evolution.

We can say a little more about the onset of chaos in regions of phase space where rational winding numbers accumulate the effects of a weak nonlinear perturbation.

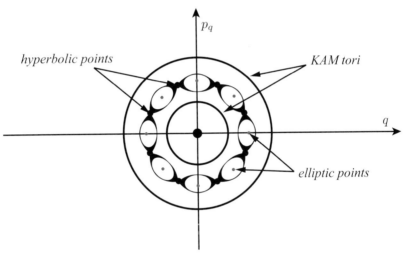

Fig. 14.11 The onset of chaos according to the Poincaré–Birkhoff theorem. A typical curve with rational winding number $\nu = a/b$ is shown as it disintegrates into $2b$ points, alternating between elliptic and hyperbolic ones. The **elliptic points** seed to KAM tori around them, while the **hyperbolic points** seed chaos in their vicinity.

Figure 14.11 shows a close-up of a chaotic region, and describes the pattern of chaotic dynamics. This pattern is shown to be a general one through the Poincaré–Birkhoff theorem: for a rational winding number of $\nu = a/b$, one gets an integer multiple of $2b$ points along the would-be KAM torus – alternating between elliptic points and *hyperbolic points*. The elliptic points are encircled by new tori as the would-be KAM torus degenerates. For, say, $2b$ total points, we get b elliptic points, called also *period b resonances*. In contrast, the hyperbolic points seed fully chaotic evolution. As one zooms onto hyperbolic regions, more elliptic regions may be found in a self-similar pattern. These patterns can be accorded **fractal dimension**, defined and described in Appendix D.

So far, we have defined chaos as the absence of integrability. In this sense, the small perturbation of our pendulum system renders the entire setup chaotic. However, by looking at the details of the onset of chaos, we also saw that, for different initial conditions, we may encounter basins of stability – specially when the non-integrable nonlinearities are small. Thus, we have been visually identifying regions of chaos in Poincaré sections as well as regions of KAM stability. It is hence useful to introduce a more restrictive yet quantitative definition of chaos that applies only in regions of phase space where KAM tori disintegrate and the trajectory of the system does not live on the surface of any torus. A quantifiable new definition of chaos goes as follows. Say we shift a given set of initial conditions by a tiny amount and compute the coordinate distance d between the two trajectories

$$d^2(t) = \sum_i \left(q_i'(t) - q_i(t) \right)^2 , \tag{14.33}$$

where $q_i(t)$ is the trajectory resulting from the unshifted initial conditions and the $q_i'(t)$ arises from the shifted one. If we find that, at long times, we have

$$d(t) \propto e^{\lambda t} \tag{14.34}$$

with $\lambda > 0$, we say that the corresponding initial conditions lead to chaotic evolution. The quantity λ can then be computed and is known as the **Lyapunov exponent**. Hence, regions of chaos in a Poincaré section can be labeled by Lyapunov exponents. Extending this notion, if one finds that $\lambda \leq 0$, we say that we have **attractor** behavior – where the dynamics is attracted towards a stable torus in phase space. Hence, you can imagine a contour map of λ onto a Poincaré section, identifying chaotic initial conditions as well as nonchaotic ones.

Finally, let us connect our discussion of chaos to equilibrium statistical mechanics – the microscopic framework that underlies thermodynamics. In equilibrium statistical mechanics, there exists the central notion of **ergodic evolution**. One typically has a system with a very large number of degrees of freedom (say of the order of Avogadro's number) plus typically some nonlinear interactions between these degrees of freedom (whether weak or strong). The evolution of the system in phase space is said to be ergodic if, given enough time, the trajectory of the system comes arbitrarily close to any point in phase space. Indeed, chaotic dynamics shares this feature with ergodic evolution: in the domains of phase space where chaos ensues, the trajectory comes as close as you like to any point in the region, given enough time. Hence, chaos implies ergodicity. The reverse is not true however. One can have an integrable system with irrational winding numbers that lead to trajectories on the N-torus that exhibit ergodicity on the surface of the torus. While chaos is sufficient and often present in statistical mechanics systems, it is not necessary.

14.3 Dissipative Chaos

So far we have looked only at energy-conserving systems. However, chaos is also observed in systems in which energy is *not* conserved. As a first step in constructing an example, we begin with a simple pendulum, a nonlinear, *energy-conserving* system, in which a bob of mass m swings back and forth in a plane, on the end of a string (or massless rod) of length ℓ. The equation of motion is

$$\frac{d^2\theta}{dt^2} + \frac{g}{\ell} \sin \theta = 0, \qquad (14.35)$$

which is linear in the limit of small amplitudes θ, but is generally nonlinear yet integrable (given that there exists one conserved quantity in a two-dimensional phase space). We have taken $\theta = 0$ when the bob is at its lowest point. The mass appeared in both terms, so has dropped out. Such nonlinear dynamical equations can generate complicated solutions. However, this simple pendulum does not exhibit chaos as expected; it is not extraordinarily sensitive to initial conditions – small changes in initial conditions generally lead to small changes in the resulting solution.

Now write the simple pendulum equation using a dimensionless time T, by letting $t = \sqrt{\ell/g}\, T$. The equation then becomes

$$\frac{d^2\theta}{dT^2} + \sin \theta = 0. \qquad (14.36)$$

The corresponding motion in two-dimensional θ, p_θ phase space is shown in Figure 14.12. The paths are closed. We can solve the pendulum equation analytically, although not in terms of simple functions. The first integral of motion is the dimensionless energy

$$E(\theta, \dot{\theta}) = \frac{1}{2}\dot{\theta}^2 + (1 - \cos \theta). \qquad (14.37)$$

For small E the motion is represented by circles centered about the points with $\theta = \pm 2n\pi$, with $n = 0, 1, 2, \ldots$, and the periods are independent of the amplitude. The circles morph into ovals for somewhat larger energies, and the periods increase with amplitude. When $E > 2$ the oscillations are replaced by continuous rotation about the origin (the pendulum is then swinging round and round either clockwise or counterclockwise). All these curves lie on a torus as we now know. Each of the two curves forming the boundary between the oscillating and rotating regimes is called a *separatrix.*

The first integral of motion can itself be integrated in terms of elliptic functions. The result is a solution $\theta(T)$ determined by the initial conditions, which is what we expect; but also the motion is not particularly sensitive to initial conditions except near the separatrix. Even here we know the motion will turn out to be either very large-amplitude oscillations or rotations that barely make it all the way around

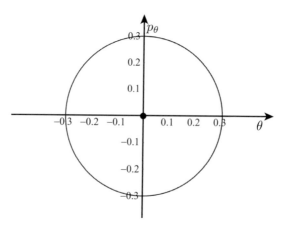

Fig. 14.12 Phase-space diagram for a simple pendulum.

in the same direction. The system is not yet complicated enough to exhibit truly chaotic behavior.

Now add linear damping to the system, so that the equation of motion becomes

$$\frac{d^2\theta}{dT^2} + 2\gamma\frac{d\theta}{dT} + \sin\theta = 0. \tag{14.38}$$

The equation can be solved exactly, and a typical phase-space plot of the solution is shown in Figure 14.13(a). Note that there is a point "attractor" at the origin; all solutions with $\gamma > 0$ eventually become zero-amplitude, zero-velocity solutions, approaching the origin in phase space, which is the attractor. Energy of course is no longer conserved, the system is not integrable, but there is still no sign of chaotic behavior.

Now add a sinusoidal forcing function, which might be supplied by causing the support point of the pendulum to move up and down periodically. The equation of motion then becomes

$$\frac{d^2\theta}{dT^2} + 2\gamma\frac{d\theta}{dT} + \sin\theta = \xi\cos\omega t = \xi\cos\Omega T, \tag{14.39}$$

where ω is the driving frequency and $\Omega = \omega\sqrt{\ell/g}$ is a dimensionless driving frequency. ξ measures the strength of the oscillating forcing function. A typical solution for a weak driving force (with ξ very small) is shown in Figure 14.13(b). It approaches what is called a *limit cycle*, which is the dotted phase-plane path in the figure. In this case the attractor is the dotted path, which is a one-dimensional curve.

There are now three parameters, γ, ξ, and Ω, with solutions depending upon all three. The equation is nonlinear, so except in restricted ranges of parameter space the solutions can be found only by numerical methods. First, suppose that ξ is small, and that θ and $\dot{\theta}$ are both small initially. Then we expect that θ will remain small, and so the differential equation becomes approximately

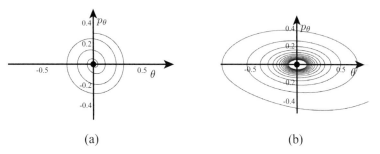

(a)   (b)

Fig. 14.13 Phase-space diagram for (a) a linearly damped pendulum, with a point attractor at the origin, (b) a damped oscillator with a small driving force, with a limit cycle attractor corresponding to the steady-state solution, after transients have died out.

$$\frac{d^2\theta}{dT^2} + 2\gamma\frac{d\theta}{dT} + \theta = \xi\cos\Omega T, \tag{14.40}$$

which is a *linear* damped, driven oscillator equation, which is exactly solvable. The solution is the sum of a damped oscillator (the "transient" solution) and a steady-state solution, which oscillates at frequency Ω but is generally out of phase with the driving force. In the long run the transient dies out, leaving a pendulum that swings back and forth with low amplitude and the same frequency as the driver, but out of phase with it, so that

$$\theta(t) = A\cos(\Omega T - \delta), \tag{14.41}$$

where δ is the phase angle between the driver and the bob's motion. There is no chaotic behavior. The amplitude and phase angle are

$$A = \frac{\xi}{\sqrt{\Omega^4 - 2(1 - 2\gamma^2)\Omega^2 + 1}} \quad \text{and} \quad \delta = \tan^{-1}\frac{2\gamma\Omega}{1 - \Omega^2}. \tag{14.42}$$

Now suppose we gradually increase the driver strength ξ. Eventually there will be departures from linearity; that is, with $\sin\theta \neq \theta$. At first one finds changes in the amplitude *shape* and also small admixtures of higher-frequency modes of oscillation, still with no chaotic behavior. But as the driving strength increases still further, eventually very interesting behaviors take place. For example, the solutions may begin to exhibit **bifurcations** in frequency, in which the pendulum may oscillate alternately at one frequency and another, as illustrated in Figure 14.14. As the driving force is cranked up further, there may be further bifurcations. The solutions are still not chaotic, however; although not as regular as for very small driving strengths, there is not the *apparent* complete randomness observed in chaotic solutions. Eventually, however, as the driving force parameter ξ is raised still further, the solutions become chaotic.

As we learned earlier, one can use Poincaré sections or maps to characterize the motion of nonlinear systems and the onset of chaos. In the Poincaré section for the pendulum, one draws the θ axis horizontally, and the p_θ axis vertically.

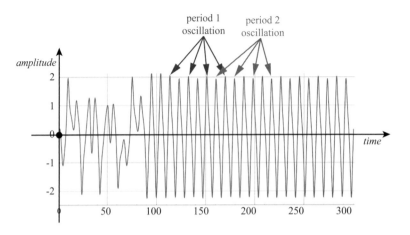

Steady-state motion of a driven, damped pendulum for somewhat larger driving parameters ξ, showing period doubling, in which the pendulum alternates between two frequencies, called a bifurcation.

Then *once each cycle of the driver* one plots the state of the pendulum on this diagram. That is, at that particular cycle the pendulum will be at some particular angle and will have some particular value of p_θ, so the state of the pendulum is mapped onto a single point in phase space. This process is repeated once each cycle. Figure 14.15 shows the Poincaré map exhibiting the hallmark features of chaos, now in a dissipative setting. As we can see, if the motion (after transients have died out) becomes perfectly regular (as it will in the linear regime), the entire plot becomes a single point in phase space, repeated *ad infinitum* at this same single point. But in the nonlinear regime there will be some scatter in the points, which may repeat in several or many points. In the chaotic regime the points may seem to cover vast regions in phase space, while other regions are still free of points. As shown in the diagram, the points seem to cover a wandering, two-dimensional portion of the plane. Closer inspection reveals that they do not cover the two-dimensional region quite densely, however. It turns out that the dimension of the points is neither one-dimensional (as in a line) or two-dimensional (as in a region of the plane) but somewhere in between. The dimensions are then said to be **fractal**, as discussed in Appendix D. Note the regions in which the points tend to cluster; these are **attractors**. However, these points have fractal dimensions, so are said to be *strange attractors*.

Alternatively, we can study the system through a new analysis tool known as a **bifurcation diagram**. Such a diagram is equally interesting, but serves a quite different purpose. In a bifurcation diagram one plots the strength of the *control parameter* (we have been using the driving strength parameter ξ as the control parameter in our discussion of the pendulum) on the horizontal axis, and the long-term pendulum amplitude on the vertical axis. In this way the bifurcation diagram shows the linear regime, and where the solutions begin to bifurcate (alternating between one frequency and another), and where chaos sets in. Figure 14.16

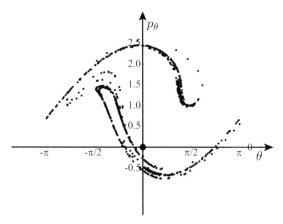

Fig. 14.15 Typical Poincaré map for a driven, damped pendulum that has become chaotic. There are still attractors, regions where the Poincaré points cluster, but they do not have integer dimensions, being somewhere between one-dimensional (as in a limit cycle) and two-dimensional (densely covering a region of the two-dimensional diagram).

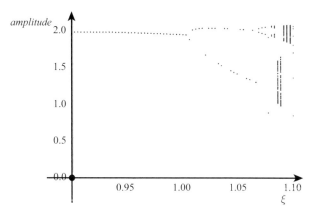

Fig. 14.16 A bifurcation diagram for a driven, damped pendulum, for certain initial conditions. The horizontal axis represents the strength ξ of the driver, and the vertical axis measures the long-term response (amplitude) of the pendulum. For very weak driving strengths the pendulum responds with a single steady-state amplitude. As ξ is increased, there are eventually "pitchfork bifurcations" in which the pendulum alternates between two frequencies. Additional bifurcations take place as ξ is increased still further. Eventually chaos develops, where the amplitudes become unpredictable. There are, however, occasionally small regions of relative stability (nonchaotic behavior) for various special values of ξ.

illustrates a bifurcation diagram for the driven pendulum. Note that each point in the space is the result of a numerical solution of the equations with different control parameter ξ, with both of the other parameters kept constant, and every point corresponding to the same initial conditions for the pendulum. In the figure we have used the initial conditions $(\theta_0) = 0$, $(p_\theta)_0 = 0$.

14.4 The Logistic Map

So far we have explored properties of *nonlinear* differential equations, such as $\mathbf{F} = m\mathbf{a}$ for nonlinear forces $\mathbf{F}$. Chaos is also observed in solutions of nonlinear **difference equations**, which often exhibit many of the same properties and are easier to solve. Indeed, we will see in Section 14.6 that we can map a large class of differential equations arising in dynamics onto difference equations – in an attempt to solve a system numerically using a computer. A particularly interesting example of a difference equation is the **logistic equation**, or **logistic map**

$$x_{n+1} = \alpha x_n (1 - x_n), \tag{14.43}$$

where $0 \le x_n \le 1$ and where the constant "control parameter" α is in the range $0 < \alpha < 4$. The map is nonlinear due to the x_n^2 term on the right. It is also iterative: by choosing an initial value for x_1 for the right-hand side we can calculate x_2 from the equation, and then from x_2 we can calculate x_3, and so on. So the logistic equation becomes dynamical if we suppose that x changes once in every standard time interval. The map has been used to model the population growth and decline of a colony of animals, for example. If members of the colony reproduce once each year, then after each year we let n advance by unity. Note that for very small x_n we have $x_{n+1} = \alpha x_n$, so early on, x grows approximately exponentially, as it might if there is plenty of food. But then the colony eventually outgrows its food supply, so begins to die off in accord with the second term on the right, $-\alpha x_n^2$. In the population model the quantity x can be taken to be the ratio of the number of animals in the colony to the "carrying capacity," the largest number of animals possible in the colony, consistent with the limitation $0 \le x_n \le 1$. The question is, how does x_n evolve, and is there eventually a stable population as $n \to \infty$?

We can solve the logistic equation graphically with the help of Figure 14.17(a), which plots x_{n+1} on the vertical axis and x_n on the horizontal axis, for the special

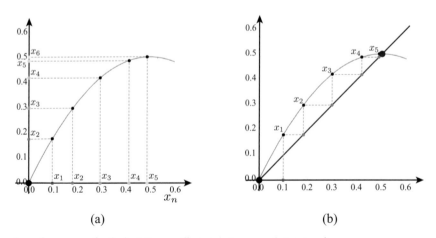

(a)   (b)

Fig. 14.17   The values of x_{n+1} vs. x_n for the logistic map. Shown are two ways to iterate values.

case $\alpha = 2$. Shown is the continuous parabola $y = \alpha x(1 - x)$; every iteration of the logistical map must have a value x_n that is somewhere on this parabola. So we begin with an arbitrary (small) value of x_1, say $x_1 = 0.1$, as shown. We note the value of x_2 on the vertical axis where the line from x_1 intersects the parabola, which is $x_2 = 0.18$. Then we start over with this value of x_2 on the horizontal axis, project it up to the parabola to find $x_3 = 0.2952$, $x_4 = 0.4161$, $x_5 = 0.4859$, $x_6 = 0.4996$, etc., asymptotically approaching $x_\infty = 0.5$. This technique works, but there is a faster and more visual way to find the successive values of x_n, as illustrated in Figure 14.17(b). In addition to the parabola, we draw the straight line $x_{n+1} = x_n$, tilted at $45°$. Now as before, we draw the vertical line denoting x_1, which intersects the parabola at vertical height x_2. Then we "reflect" this original, vertical line off the parabola into a horizontal line which remains at height x_2. So then the horizontal line intersects the tilted $45°$ line at x_2. There follow alternate vertical and horizontal "reflections" off the tilted line; every time a horizontal line intersects the tilted line, we can read off a new value of x_n. In this case, the value of x eventually settles down to a steady-state "fixed point" at $x_{n+1} = x_n = 0.5$, which is the intersection point between parabola and line. Note that the population represented by x_n grows steadily at first, and then levels out, asymptotically approaching $x_\infty = 0.5$.

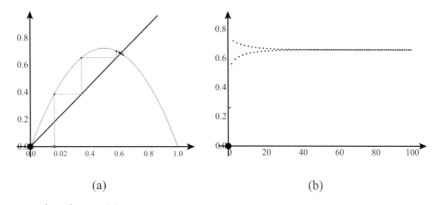

(a)   (b)

Fig. 14.18   Faster growth, with $\alpha = 2.9$.

Now let the initial population growth be faster, with $\alpha = 2.9$. We begin again with the same initial population, $x_1 = 0.1$, as shown in Figure 14.18(a). After the faster initial growth, the plot now develops a "cobweb" appearance, with x_n varying from larger to smaller values and back again, decreasing in amplitude to an "attractor" at $x = 0.655\ldots$, which is where the tilted $45°$ line intersects the parabola. (An "attractor" is a point, line, region, $\ldots$ toward which solutions of an equation evolve.) Therefore in this case, as shown in Figure 14.18(b), the population begins by growing quickly, and then overshoots where it will eventually settle, and subsequently oscillates about this equilibrium value with ever-decreasing amplitude.

It is very interesting to make similar plots for various values of α. If there is a sufficiently small control parameter, $\alpha < 1$, corresponding to a low reproduction rate in a population, then $x_n \to 0$ as $n \to \infty$: the animal colony ultimately goes extinct. Graphically, when the slope of the parabola near the origin is less than one:

$$\frac{d}{dx}\left(\alpha x \left(1 - x\right)\right)_0 = \alpha < 1, \tag{14.44}$$

the only intersection point between the $45°$ line and the parabola is at the origin. If $1 < \alpha < 3$, the population grows and ultimately reaches a nonzero steady-state fixed point, as we showed above for the cases $\alpha = 2$ and $\alpha = 2.9$. The solution oscillates for a time before settling down. Graphically, we now have an nontrivial intersection point between the parabola and the $45°$ line. In general, the intersection point is given by

$$\alpha x \left(1 - x\right) = x \Rightarrow x = 1 - \frac{1}{\alpha}, \tag{14.45}$$

and hence the sequence converges eventually to this value.

Now consider a case for which $\alpha > 3$, say $\alpha = 3.1$. Then the graph of x_{n+1} vs. x_n is shown in Figure 14.19(a). The solution never does settle down to a fixed value of x_n, but oscillates forever between $x = 0.558$ and $x = 0.765$, as shown in Figure 14.19(b). In this case x_n repeats itself every *second* oscillation, so is said to be in a *period-2 cycle*. There are alternating boom and bust years for the population being described. Graphically, one needs to consider $f(f(x))$, where $f(x) = \alpha x \left(a - x\right)$ is the right-hand side of Eq. (14.43).

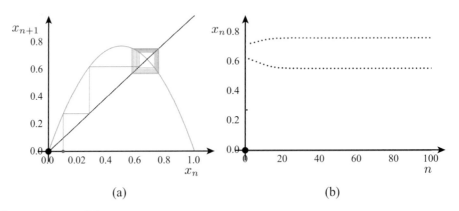

(a)   (b)

Fig. 14.19 The case with $\alpha = 3.1$.

More complicated behaviors await us for still larger values of α. If $\alpha = 3.5$, for example, the comparable diagrams are shown in Figure 14.20(a) and (b). Now the cycle repeats itself every *four* generations, so is a *period-4 cycle*. This kind of doubling behavior keeps occurring as α is gradually increased, to give a *period-8 cycle*, a *period-16 cycle*, and so on.

Computer simulations reveal that the value of α for which a new doubling, called a *bifurcation*, occurs is as given in the table below:

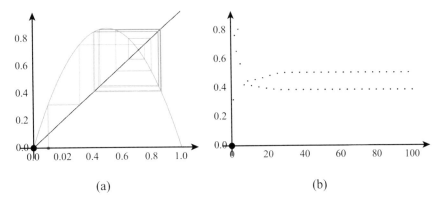

(a)   (b)

Fig. 14.20   The case with $\alpha = 3.5$.

$$\alpha_1 = 3.0000 \qquad \text{period 2 begins}$$
$$\alpha_2 = 3.449\ldots \qquad \text{period 4 begins}$$
$$\alpha_3 = 3.54409\ldots \quad \text{period 8 begins}$$
$$\alpha_4 = 3.5644\ldots \quad \text{period 16 begins}$$
$$\ldots$$
$$\alpha_\infty = 3.569946 \qquad \text{period } \infty \text{ begins}$$

The spacing between successive bifurcations becomes smaller and smaller, so that the α_k converge to a limiting value α_∞. Now if we take the ratios

$$r_k \equiv \frac{\alpha_k - \alpha_{k-1}}{\alpha_{k+1} - \alpha_k}, \tag{14.46}$$

we find from the above numbers that the ratio is 4.722 for $k = 2$, 4.682 for $k = 3$, and so on; as $k \to \infty$, the ratio is

$$r_\infty = \lim_{k \to \infty} \frac{\alpha_k - \alpha_{k-1}}{\alpha_{k+1} - \alpha_k} \equiv \delta = 4.669\ 201\ 609\ldots, \tag{14.47}$$

called **Feigenbaum's number**. The remarkable thing about Feigenbaum's number $\delta = 4.669\ldots$ is its **universal property**. The same number is obtained for any system that undergoes the same kind of period-doubling behavior for maps with a quadratic maximum. In other ways the systems may be quite different, but this number is independent of the specific map involved.

Another remarkable thing happens if $3.569946 < \alpha < 4$. We then enter a regime in which there is a mixture of order and *chaos*. In the chaotic regions the value of x no longer repeats itself at all. Figure 14.21 shows the plot of x_{n+1} vs. x_n for $\alpha = 3.9$, for example.

Finally, we can construct a bifurcation diagram for the logistic map by carrying out a large number of computer calculations for various control parameters α, as shown in Figure 14.22(a). The horizontal axis is the value of α, and the vertical axis shows the long-term value of x for that value of α. So for $\alpha < 3.0$ there is a unique long-term value of x; above $\alpha = 3$ the bifurcations begin. Above $\alpha = 3.569946$

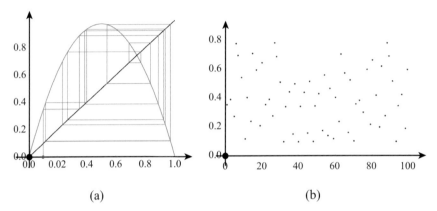

(a)   (b)

Fig. 14.21 Nonrepeating "populations" x for a control parameter $\alpha = 3.9$.

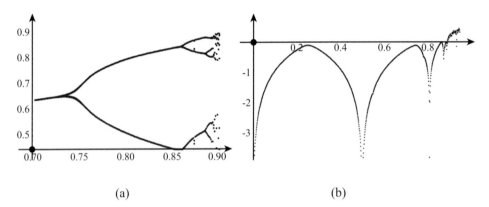

(a)   (b)

Fig. 14.22 (a) Bifurcation diagram for the logistic map, plotting the "population" x as a function of the control parameter α. (b) The Lyapunov exponents as a function of α.

the diagram displays a chaotic regime. In this regime there are no fixed long-term values of x. Also the slightest change in the initial value of x will likely make an enormous difference in the final outcome. Interpreting the diagram as an application of population growth and decline, we can say that for a moderate growth rate of $1 \leq \alpha \leq 3.0$, there is only a single ultimate number of animals in the colony. For $3.0 < \alpha < 3.449\ldots$, the ultimate number of animals keeps changing each year between two possible results, called *period doubling*. In one year the number of animals might be at the high end, too high for all to survive, so the following year the number falls to the lower value. But then the food is adequate for the animals to not only survive but to have offspring, so the number bounces back to the higher value. As α is increased in the range $3.449 < \alpha < 3.569946$, there are further bifurcations, so there are always an even number 4, 8, 16, ... of possible numbers of animals, cycling from one to another from year to year. Then for most of the range $\alpha > 3.569946$ there is no repetition whatever from year to year. This is the chaotic regime. There are still "windows" of periodicity within this region, which can be seen from the figure. Notice also that the special points at which bifurcation emerges are also where the Lyapunov exponents vanish, as can be seen from Figure 14.22(b).

As we see, the complex visual patterns that one often generates in studying chaos can be quantitatively analyzed and classified using various techniques. Another useful notion that commonly arises in the analysis of chaotic systems is that of *fractal dimension*. For this, the reader is referred to Appendix D for further reading.

14.5 Perturbation Techniques

In many situations, the dynamics of a physical system divides into two sectors with qualitatively different roles: one sector involves well-understood and integrable dynamics; the other is a *small* nonlinear effect that perturbs the otherwise tractable situation. For example, consider a one-dimensional simple harmonic oscillator perturbed by a small quadratic term as in

$$\ddot{q} + \omega^2 q + \epsilon q^3 = 0, \tag{14.48}$$

written in terms of the degree of freedom $q(t)$. The two constants ω and ϵ parameterize the equation. In the limit where $\epsilon \to 0$, the system describes a harmonic oscillator with angular frequency ω. The term proportional to ϵ adds a nonlinearity, and we want to focus on the regime where the effect of ϵ is small. More concretely, we want

$$\omega^2 \gg \epsilon q(t)^2, \tag{14.49}$$

so that the last term in Eq. (14.48) is negligible. This is achieved by taking ϵ small, but also by ensuring that $q(t)$ does not become too big. If this scenario is realized, we can study the dynamics of the system using a myriad of approximation techniques that are often collectively and broadly referred to as classical **perturbation theory**. Here, we describe a couple of the most common approaches.

In case we have $\epsilon = 0$ in Eq. (14.48), we know that $q(t)$ would be a combination of sines and cosines. Taking the boundary condition

$$q(0) = A, \quad \dot{q}(0) = 0, \tag{14.50}$$

we get in particular

$$q(t) = q_0(t) = A \cos \omega t, \tag{14.51}$$

where we have labeled the solution with a zero subscript to emphasize that it corresponds to the special case without the nonlinear force term in Eq. (14.48), *i.e.*, when $\epsilon = 0$. The next step is to devise a controlled scheme where one computes small corrections to (14.51), induced by the ϵ-dependent nonlinear term in the equation of motion. For this purpose, it is very helpful if we were to identify a *dimensionless* small parameter that controls the importance of these corrections. As it currently stands, a statement like $\epsilon \ll 1$ is meaningless as ϵ is not dimensionless: indeed, ϵq^2 has units of frequency squared, as can be seen from (14.48). First, we rescale time to absorb the angular frequency w: $s \equiv wt$. s is then dimensionless and the equation of motion becomes

$$\ddot{q} + q + \frac{\epsilon}{w^2} q^3 = 0, \tag{14.52}$$

where $\ddot{q}$ now stands for d^2q/ds^2 instead of d^2q/dt^2. Next, we write $Q \equiv q/A$, where A is the amplitude of the harmonic oscillation from Eq. (14.51). Q is then dimensionless and the equation of motion becomes

$$\ddot{Q} + Q + \frac{\epsilon A^2}{w^2} Q^3 = 0. \tag{14.53}$$

Here both Q and time have been turned into dimensionless variables. So we now see that we should define

$$\varepsilon \equiv \frac{\epsilon A^2}{w^2} \tag{14.54}$$

as the small dimensionless perturbation parameter. We then have the rescaled equation of motion

$$\ddot{Q} + Q + \varepsilon Q^3 = 0, \tag{14.55}$$

where we are interested in the regime with $\varepsilon \ll 1$, so that the last term of the equation is considered a small perturbation to the harmonic oscillator. We still need to make sure that the resulting $Q(s)$ does not grow much more than order unity, since this could make the last term large even for $\varepsilon \ll 1$. In terms of dimensionless variables, the solution given by Eq. (14.51) takes the form

$$Q_0(s) = \cos s, \tag{14.56}$$

with $Q_0(0) = 1$ and $\dot{Q}_0(0) = 0$. Assuming $\varepsilon \ll 1$, we might expect that the solution to (14.55) is close to Eq. (14.56), differing from it by a small amount. In this spirit, we write

$$Q(s) = Q_0(s) + \varepsilon Q_1(s) + \varepsilon^2 Q_2(s) + \cdots = \sum_{n=0}^{\infty} \varepsilon^n Q_n(s), \tag{14.57}$$

where $Q_n(s)$ for $n > 0$ are functions of time yet to be determined. These are meant to be successively more refined corrections to the zeroth-order solution $Q_0(s)$. We then substitute (14.57) into the equation of motion (14.55) and expand. The key to this technique is that we expect that the resulting expression holds for *any* value of ε. This in turn implies that all terms in the resulting equation with no powers of ε should cancel among themselves, and that all terms with a single power of ε should also cancel among themselves, and the same for terms involving ε^2, and so forth... We saw a similar approach in Chapter 10 where we studied the precession of planetary orbits due to small general relativistic corrections to Newtonian gravity. In the current situation, the terms without ε give

$$\ddot{Q}_0 + Q_0 = 0. \tag{14.58}$$

Then canceling the terms multiplied by a single power of ε leads to

$$\ddot{Q}_1 + Q_1 = -Q_0^2. \tag{14.59}$$

For order ε^2 we find

$$\ddot{Q}_2 + Q_2 = -2 Q_0 Q_1, \tag{14.60}$$

and we can continue in this manner for higher powers of ε. Thus we see a pattern emerging: at each order in ε we must solve an equation of the form

$$\ddot{Q}_n + Q_n = F_n(s) \tag{14.61}$$

for $Q_n(s)$, with $n = 1, 2, 3, \ldots$; $F_n(s)$ is always constructed from the previous solutions in the sequence, $Q_m(s)$ with $m = 0, \ldots, n-1$. For example, we have $F_1(s) = -Q_0(s)^2$ from (14.59); and $F_2(s) = -2 Q_0(s) Q_1(s)$ from (14.60). Equation (14.61) is known as the forced harmonic oscillator, where $F_n(s)$ plays the role of an external time-dependent force. This differential equation can readily be solved using Green function methods,[4] and leads to the general solution

$$Q_n(s) = \int_0^s F_n(s') \sin(s - s') \, ds', \tag{14.62}$$

which corresponds to the boundary conditions

$$Q_n(0) = 0, \quad \dot{Q}_n(0) = 0, \tag{14.63}$$

so that we still preserve the original boundary conditions on the full solution $Q(0) = 1$ and $\dot{Q}(0) = 0$. Hence, we now have an algorithm for successively computing more refined corrections to $Q_0(s)$, building up towards the solution of the full equation of motion given by (14.55).

Using (14.62), we then get from (14.59)

$$Q_1(s) = -\int_0^s Q_0(s')^2 \sin(s - s') \, ds' = \frac{1}{6}(-3 + 2\cos s + \cos 2s). \tag{14.64}$$

From this, using (14.57), we can reconstruct a solution to our original equation (14.48) to first order in ε as

$$q(t) = A \cos \omega t - \epsilon \frac{A^3}{2\omega^2}\left(1 - \frac{2}{3}\cos \omega t - \frac{1}{3}\cos 2\omega t\right), \tag{14.65}$$

where we have restored the result in terms of the original variables.

[4] Green functions.

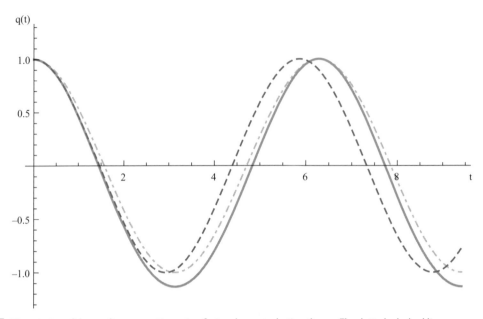

Fig. 14.23 The solution of the nonlinear equation using first-order perturbation theory. The dotted–dashed line describes the unperturbed scenario. The dashed line is the accurate solution obtained through numerical integration. The solid line depicts the result from first-order perturbation theory discussed in the text. We see how the errors accumulate as time progresses and the perturbation technique eventually falters.

Figure 14.23 shows a graph of $q(t)$, comparing the first-order solution given by (14.65) to a high-accuracy numerical solution. We see that our technique yields decent results for small enough times, but errors accumulate over time, and perhaps higher-order corrections in ε are needed to do better. However, this technique does not guarantee that any truncated approximation of the series (14.57) will provide sufficient accuracy. In many cases, as exemplified in the Problems section at the end of this chapter, the scheme falters and leads to poor approximations for sufficiently long times. There are various improvements that one can apply to the method to make things more robust and useful. One common improvement involves expanding other parameters in the original equation of motion in powers of ε as well. In this example, we might consider solutions with a shifted angular frequency

$$\Omega = \omega + \omega_1 \epsilon + \omega_2 \epsilon + \cdots , \tag{14.66}$$

instead of ω. The suggestion is that a good approximate solution is a function of Ωt. The constants $\omega_1, \omega_2, \ldots$ are then determined by substituting the series (14.57) in (14.55) as before – but replacing s with $\Omega s/\omega$; then expanding and matching powers of ε as usual. This approach can work quite well and will be explored in the Problems section at the end of this chapter.

14.6 Numerical Techniques

There are relatively few mechanics problems that can be solved exactly, problems that can be associated with closed-form analytical solutions. The vast majority of interesting physical situations involve complex nonlinear dynamics, like the ones encountered in this chapter. It is often jokingly said that physicists only know how to solve the harmonic oscillator – everything else they either treat as a perturbation of the oscillator or handle through numerical computational techniques. This is not far from reality, so it is important for a modern physicist to be fluent in the use of basic numerical methods.

When using a numerical approach to solve a physics problem, it can be easy to lose sight of the physics. As modern numerical techniques get more involved and sophisticated, we can have a "tail wagging the dog" scenario, where more effort and time is spent on technique than on the underlying physics. At the other extreme, a "black box" phenomenon can emerge where the physicist uses the technology without knowing how it works under the hood – and this can lead to incorrect physical interpretations of numerical results. All this can also be present in the context of cherished analytical approaches, but it is more likely to be an issue with numerical approaches because of the complexities involved and the additional elaborate electronic layer between results and input. Hence, while it is important to have a command of powerful numerical techniques, one needs to use them with care – as yet another tool out of many, always keeping a focus on the physics. For example, given a particular system, one should start by taking certain simplifying limits in which one can solve these cases analytically. Then, once analytical methods have been exhausted, implement carefully crafted and reliable numerical techniques – keeping track of numerical errors in the results. The numerical results must then be subjected to (1) sanity checks against the analytical special limits, (2) repeated computations with different hardware systems, and (3) internal numerical testing that for example checks against conservation laws and other physical constraints on the solution. Computational physics lies, in a sense, between theory and experiment: on the one hand, it builds upon a theoretical hypothesis to understand a physical system; on the other hand, it generates data just as experiments do and hence should be controlled for reproducibility, robustness, and systematic and statistical errors.

In this section we collect several key numerical techniques useful in mechanics and physics in general. We focus on techniques for computing the roots and extrema of functions, integration methods, and strategies for solving differential equations. Each topic can be handled by a wide array of methods and approaches, but we will focus in each case on discussing a single technique that is both powerful and suitable for use in many situations. The techniques will be described in some detail, but to keep things concise the mathematical proofs will not be presented. The reader is referred to the Further Reading section.

Roots and Extremization

A common problem that one encounters in physics involves finding the roots of a set of functions

$$F_i(\mathbf{r}) = 0, \quad \text{for } i = 1, \ldots, N, \tag{14.67}$$

where we consider a system of N equations that depend on N variables $\mathbf{r} = (x^1, x^2, \ldots, x^N)$. Solutions to this set of equations are points $\mathbf{r}_0$, if any, where all the N functions vanish. For the simplest $N = 1$ case, one wants to find the roots of a single function depending on a single variable. One of the most versatile algorithms for solving this problem numerically is known as the *Newton–Raphson method*. This requires that we compute, analytically or numerically, the derivatives of the functions

$$J_{ij} = \frac{\partial F_i}{\partial x^j}. \tag{14.68}$$

The approach relies on starting with a guess for $\mathbf{r}_0$ and iteratively zeroing onto the solution by perturbing the guess. The method generally works best if the initial guess is close to the final solution, and there are methods to improve the quality of the initial guess. These might involve bracketing techniques, for example, where one coarsely scans over $\mathbf{r}$ to find windows where the functions F_i flip signs. Let us assume that a good guess has been identified, and let us call it $\mathbf{r}_0$. The Newton–Raphson algorithm then proceeds as follows:

1. Solve the linear algebra problem

$$\hat{\mathbf{J}}_0 \cdot \delta\mathbf{x} = -\mathbf{F}_0 \tag{14.69}$$

   for $\delta\mathbf{x}$ – where we have written the derivatives from (14.68) evaluated at $\mathbf{r}_0$ as an $N \times N$ matrix $\hat{\mathbf{J}}_0$, and $\mathbf{F}_0$ is the vector whose components are the functions F_i evaluated at the guess $\mathbf{r}_0$. To find $\delta\mathbf{x}$, one needs to invert $\hat{\mathbf{J}}$, hence this matrix cannot be singular. A common approach to finding $\delta\mathbf{x}$ is to use a numerical **LU decomposition** of $\hat{\mathbf{J}}$.[5]

2. Modify your guess as follows:

$$\mathbf{r}_0 \rightarrow \mathbf{r}_0 + \delta\mathbf{x}. \tag{14.70}$$

3. Iterate over steps 1 and 2 until $\mathbf{F}_0$ approaches zero to the desired level of accuracy.

Figure 14.24 shows the algorithm at work for the case where $N = 1$. As you can see, the idea is to use the local slope of the function to shoot toward a point where the function crosses zero.

   In certain special circumstances this algorithm can fail miserably. For example, if a guess lands near an extremum during an iteration, the $\hat{\mathbf{J}}$ would become very small or vanish!

---

[5] From linear algebra, a matrix $\hat{\mathbf{M}}$ can be LU decomposed as $\hat{\mathbf{M}} = \hat{\mathbf{L}} \cdot \hat{\mathbf{U}}$, where $\hat{\mathbf{L}}$ is lower triangular and $\hat{\mathbf{U}}$ is upper triangular.

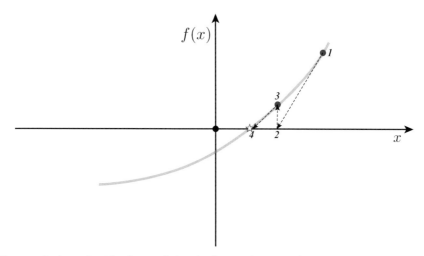

Fig. 14.24 The Newton–Raphson algorithm for root finding for the one-dimensional case. The root, indicated by a star, is zeroed onto by successive steps 1, 2, 3, etc.

There is a particular root-finding problem that arises commonly enough in physics for special attention to be warranted: the problem of finding the roots of a polynomial. Consider the polynomial of degree N of the form

$$p(x) = \sum_{n=0}^{N} a_n x^n. \tag{14.71}$$

To find the roots $p(x) = 0$ of this polynomials, it can be shown that one needs to solve the linear algebra problem

$$\det\left[\hat{\mathbf{A}} - x\,\hat{\mathbf{1}}\right] = 0, \tag{14.72}$$

where the $N \times N$ matrix $\hat{\mathbf{A}}$ is given by

$$\hat{\mathbf{A}} = \begin{pmatrix} -\frac{a_{N-1}}{a_N} & -\frac{a_{N-2}}{a_N} & -\frac{a_{N-32}}{a_N} & \cdots & -\frac{a_{11}}{a_N} & -\frac{a_0}{a_N} \\ 1 & 0 & 0 & \cdots & 0 & 0 \\ 0 & 1 & 0 & \cdots & 0 & 0 \\ & & & \ddots & 0 & 0 \\ 0 & 0 & 0 & \cdots & 1 & 0 \end{pmatrix}. \tag{14.73}$$

The eigenvalues of $\hat{\mathbf{A}}$ are the roots of the polynomial. There are many well-tested efficient algorithms to find the eigenvalues of a matrix numerically, so the problem is straightforward to tackle.

Another class of problems that we often encounter in physics involves finding the extrema of a function instead of its roots. Suppose we have a function $f(\mathbf{r})$ of N dependent variables, and we want to find where the function has minima. The straightforward approach is to start with a guess $\mathbf{r}_0$ and move in the direction of the local slope of f so as to decrease the value of the function. And we want to repeat this process as we slide down valleys of f and land in a minimum. The challenge is that the convergence to a minimum might be slow. For example, if the function has a long narrow valley, a naive algorithm might slalom back and forth in the valley – instead of heading down the narrow ditch towards the minimum more directly. The algorithm known as the conjugate gradient method is an attempt at moving down the slopes of the function in a smarter way so as to avoid long zigzags toward the minima. The method starts with a good guess, one that lies in a narrow window where we know we will find a minimum. As in the discussion of root finding, bracketing techniques might be used to start with a decent guess. In this case, bracketing would involve scanning over $\mathbf{r}$ to find islands where the function's value somewhere within the island is less than all of its values on the boundaries. We then call the guess $\mathbf{r}_0$ and define

$$\mathbf{g}_0 = \mathbf{h}_0 = -\nabla f(\mathbf{r}_0). \tag{14.74}$$

We then proceed as follows:

1. We compute step $i + 1$ of the algorithm from step i, starting from $i = 0$:

$$\mathbf{r}_{i+1} = \mathbf{r}_i + \lambda \mathbf{h}_i, \tag{14.75}$$

where we find the value of λ that minimizes f along $\mathbf{h}_i$. We can do this by scanning upward over λ starting from 0 until the function starts increasing instead of decreasing.
2. We then compute the next values of $\mathbf{g}$ and $\mathbf{h}$ as follows:

$$\mathbf{g}_{i+1} = -\nabla f(\mathbf{r}_{i+1}) \quad \text{and} \quad \mathbf{h}_{i+1} = \mathbf{g}_{i+1} + \gamma_i \mathbf{h}_i, \tag{14.76}$$

where we define

$$\gamma_i = \frac{\mathbf{g}_{i+1} \cdot (\mathbf{g}_{i+1} - \mathbf{g}_i)}{\mathbf{g}_i \cdot \mathbf{g}_i}. \tag{14.77}$$

3. Steps 1 and 2 are then repeated until a value of i is reached that identifies the minimum to the desired accuracy. This can be assessed by comparing the change in subsequent steps $f(\mathbf{r}_{i+1}) - f(\mathbf{r}_i)$.

Notice that at each step one moves toward lower values of f *almost* along the decreasing slope of the function – hence the "gradient" in the name of the method. The movement is done along a vector $\mathbf{h}_i$ that is the gradient of the function but shifted "sideways" to avoid being trapped zigzagging in narrow valleys. The $\mathbf{g}_i$ and $\mathbf{h}_i$ vectors are chosen so that they satisfy

$$\mathbf{g}_i \cdot \mathbf{g}_j = \mathbf{g}_i \cdot \mathbf{h}_j = 0 \tag{14.78}$$

for $j < i$; that is, they are mutually orthogonal. So from (14.76), we see that one always moves in a direction orthogonal to the gradient in the *previous* step. Figure 14.76 shows how this technique accelerates convergence to the minimum in the dangerous scenario in which one lands in a narrow valley.

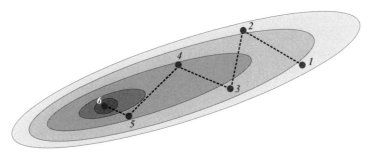

Fig. 14.25 The conjugate gradient algorithm in a narrow valley in successive steps 1, 2, 3, etc.

Integration

Integration is ubiquitous in physics, and one often has to resort to numerical methods to handle complicated integrals. We start with the case of one-dimensional integration. While there are many numerical techniques that we might discuss, we focus on one that is powerful and well-established, that of *Gaussian quadratures*. Consider an integral of the form

$$I = \int_a^b W(x)f(x)dx, \tag{14.79}$$

where we have written the integrand as a product of two functions, $W(x)$ and $f(x)$. The function $W(x)$ is meant to be a factor that we will specify to identify different algorithmic classes, and $f(x)$ is whatever is left to make the integrand what it is supposed to be. The idea is that $W(x)$ can be the singular part of the integrand, and for given $W(x)$ we can employ a method of discretizing the integral into a sum that is best adapted to the singular behavior of the integrand. The Gaussian quadrature method writes the numerical result of the integral as

$$I \simeq \sum_{i=0}^{N-1} w_i f(x_i), \tag{14.80}$$

where the w_is are called weights and the x_is are called the abscissa. Here N is the order of the approximation. As we shall see, all these parameters are determined from special polynomial functions. It can be shown mathematically that the Gaussian quadrature method, given by expression (14.80) at fixed order N, can approximate the integral of interest well if the function $f(x)$ can be approximated to a desired level of accuracy by a polynomial of degree $2N - 1$ in the integration interval (a, b).

To determine the weights, abscissa, and order, we proceed as follows. Let $p_k(x)$ be a set of orthonormal polynomials such that their orthogonality is given by

$$\int_a^b W(x)p_k^*(x)p_l(x)dx = 0 \tag{14.81}$$

for $k \neq l$. Note that the limits of the integration a and b, and the function $W(x)$, are all the same as in the integral of interest given by Eq. (14.79). Each polynomial is of degree k:

$$p_k(x) = \sum_{n=0}^{k} a_n^k x^n, \qquad (14.82)$$

where a_n^k are constants. We start by fixing an order N for the integration, determined by trial and error so as to achieve a desired level of numerical accuracy for the integral (14.79). That is, we use the algorithm for N and $N + 1$, and we look at the difference in the results; if the difference is less than the desired accuracy, we can pick N; otherwise, we compare $N + 1$ and $N + 2$, and so forth... Given N, the abscissa x_i are the N roots of the polynomial $p_N(x) = 0$.

The weights w_i are then given by

$$w_i = \frac{a_N^N}{a_{N-1}^{N-1}} \frac{\int_a^b W(x) p_{N-1}^*(x) p_{N-1}(x) dx}{p_{N-1}(x_i) p_N'(x_i)}. \qquad (14.83)$$

Therefore, we now have a full accounting of all the parameters appearing in Eq. (14.80).

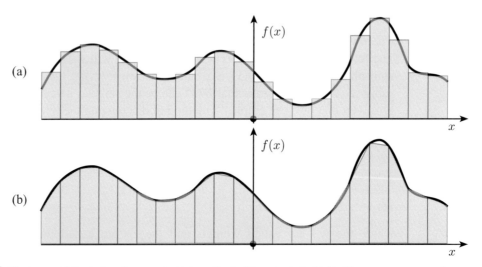

Fig. 14.26 The trapezoidal technique for numerical integration in (b) contrasted with the Riemann-sum approach in (a).

For example, if the integral of interest has no singularities to remove, and if the integration limits are $a = -1$ and $b = 1$, we might choose the polynomials to be the Legendre polynomials, with $W(x) = 1$, since the Legendre polynomials do satisfy the orthonormality relation (14.81) for given $W(x)$, a, and b (see Appendix E for a brief discussion of Legendre polynomials and other special polynomials we mention below). This method is significantly better than, for example, the trapezoidal technique where almost all the w_is are taken equal and the x_is are evenly spaced in the interval (a, b) (see Figure 14.26). One often finds that $N = 5$ or $N = 6$ is enough to achieve good results. The technique is most powerful when the integrand has singularities yet the integral is finite. For example, if the singularity in the integrand can be shifted to $x = \pm 1$ with a coordinate transformation, and we have $(a, b) = (-1, 1)$ once again, we might use

$$W(x) = \frac{1}{\sqrt{1 - x^2}} \; ; \tag{14.84}$$

and we would then correspondingly choose the Chebyshev polynomials for the roots and weights. For more general singularities so that

$$W(x) = (1 - x)^\alpha (1 + b)^\beta \tag{14.85}$$

with $\alpha, \beta > -1$ is better suited, one can use the Jacobi polynomials. Note that all these assume an integration interval $(a, b) = (-1, 1)$ which we are typically able to arrange by a proper coordinate transformation. In contrast, for $(a, b) = (0, \infty)$, one can use $W(x) = \exp(-x)$ along with Laguerre polynomials; and for $(a, b) = (-\infty, \infty)$, take $W(x) = \exp(-x^2)$ with Hermite polynomials. In general, given a desired $W(x)$, one can numerically construct a set of orthonormal polynomials that work well; but this can sometimes be computationally complex and costly.

For multi-dimensional integrals, the most interesting algorithm is based on a Monte-Carlo technique. Consider the N-dimensional integral

$$\int_{a_1}^{b_1} \int_{a_2}^{b_2} \cdots \int_{a_N}^{b_N} dx_1 \, dx_2 \cdots dx_N \, f(x_1, x_2, \ldots, x_N). \tag{14.86}$$

Using a combination of one-dimensional integration techniques is usually prohibitively costly and error prone. Instead, one can use the following efficient technique. The integration region describes an N-dimensional hypercube delineated by $(a_1, b_1), (a_2, b_2), \ldots, (a_N, b_N)$. The integral is computing an $N + 1$-dimensional volume that lies under the surface f. Find the minimum and maximum of f, f_{min}, and f_{max}, within the integration intervals; imagine an $N + 1$-dimensional hypercube constructed from the N-dimensional one plus an $N + 1$th direction constrained between f_{min} and f_{max}. The function f fits nicely in this $N + 1$-dimensional hypercube. Proceed then as follows:

- Generate M random points in the $N + 1$-dimensional hypercube.
- Let us label each of these M points as x_i^a, where $i = 1, 2, \ldots, M$ and $a = 1, 2, \ldots, N + 1$. For each point, evaluate the function f, the integrand of the integral, at x_i^a for $a = 1, 2, \ldots, N$ and call it f_i. Find the fraction of points for which $f_i > x_i^{N+1}$. Call this fraction κ.
- The integral is then approximated by $\kappa \times V$, where V is the volume of the $N + 1$-dimensional hypercube.

The algorithm is really a simple one. We find the fraction of random points that would fall under the function f, then estimate the volume under the function f by multiplying this fraction by the volume of the hypercube which encloses the function. This is an efficient algorithm that works pretty well in many cases.

Differential Equations

Time evolution in mechanics is generically described by a set of differential equations. In the Lagrangian formulation, these equations are solved by the path that extremizes the action given boundary conditions. Consequently, one approach to numerically solve for the time evolution of a physical system is to find the extremum of the action functional numerically. While this approach can work, in practice integrating the differential equations of motion is often more fruitful. So, we will focus on a class of computational techniques used to integrate differential equations efficiently.

Lagrange equations of motion are second-order differential equations. These are generally more difficult to handle than first-order ones. Fortunately, the Hamiltonian formulation writes the time evolution of a system in terms of first-order differential equations – at the cost of double the degrees of freedom. Hence, a typical problem encountered in physics is that of solving a set of coupled first-order differential equations. We write these as follows:

$$\frac{d\mathbf{z}}{dt} = \mathbf{f}(t, \mathbf{z}). \tag{14.87}$$

where $\mathbf{z}(t)$ is a $2N$-dimensional vector whose components are the canonical coordinates and momenta; $\mathbf{f}$ is typically constructed from derivatives of the Hamiltonian. We will describe the Rosenbrock method of integrating this set of coupled equations. The idea is to iteratively advance in time in steps Δt from an initial time t_0; at each step, we compute $\mathbf{z}(t)$ from data from the previous step and the form of $\mathbf{f}$. The most naive thing we can do is write $\Delta \mathbf{z} = \mathbf{f} \, \Delta t$ and advance $\mathbf{z} \rightarrow \mathbf{z} + \Delta \mathbf{z}$. This typically results in large errors as we advance in time, and requires extremely small time steps Δt that render the approach computationally inefficient. The idea of the Rosenbrock method is to sample the $\mathbf{f}$ at various moments in time between time t and $t + \Delta t$ in a manner to minimize accumulated error. The result is a class of algorithm that can be summarized by the following:

$$\mathbf{z}(t + \Delta t) = \mathbf{z}(t) + \sum_{i=1}^{N} a_i \, \mathbf{k}_i, \tag{14.88}$$

where N is the order of the algorithm, the a_is are constants to be determined, and the $\mathbf{k}_i$s are given by

$$\mathbf{k}_i = \Delta t \, \mathbf{f} \left(t + \alpha_i \Delta t, \mathbf{z} + \sum_{j=1}^{i-1} \alpha_{ij} \mathbf{k}_j \right), \tag{14.89}$$

where the α_{ij} are constants to be determined and

$$\alpha_i = \sum_{j=1}^{i-1} \alpha_{ij}. \tag{14.90}$$

The constants a_i and α_{ij} parameterize the algorithm. The essence of the technique is to sample $\mathbf{f}$ at a sequence of N points near time t and position $\mathbf{z}$, and construct a weighted sum of these samples to advance $\mathbf{z}$ to the next time step. Depending on the values of N, a_i, and α_{ij}, we obtain different algorithms. These values are typically computed using various mathematical approximation methods and they are tabulated in the literature. One typically sees tables arranged as follows:

$$
\begin{array}{c|ccccc}
0 & 0 & 0 & 0 & \cdots \\
\alpha_2 & \alpha_{21} & 0 & 0 & \cdots \\
\alpha_3 & \alpha_{31} & \alpha_{32} & 0 & \cdots \\
\vdots & \vdots & \vdots & \vdots & \ddots \\
\alpha_N & \alpha_{N1} & \alpha_{N2} & \cdots & \\
\hline
 & a_1 & a_2 & \cdots & a_N
\end{array}
$$

The simplest case is known as the Euler method, given by the boring table

$$
\begin{array}{c|c}
0 & 0 \\
\hline
 & 1
\end{array}
$$

which is in practice very error prone. A workhorse of mechanics is the fourth-order Runge–Kutta, given by the table

$$
\begin{array}{c|cccc}
0 & 0 & 0 & 0 & 0 \\
1/2 & 1/2 & 0 & 0 & 0 \\
1/2 & 0 & 1/2 & 0 & 0 \\
1 & 0 & 0 & 1 & 0 \\
\hline
 & 1/6 & 1/3 & 1/3 & 1/6
\end{array}
$$

There are numerous other alternatives that the reader may look up in the literature; but for many practical purposes, fourth-order Runge–Kutta works very well. One important aspect of using any of these algorithms is to estimate the error that accumulates in each iteration. You can do that by computing a time step with two different algorithms, and comparing the results. If the difference is less than the error you can tolerate, you can continue or even try increasing the time step to speed things up; otherwise, you want to decrease the time step and repeat the test until the accuracy is acceptable. This technique is often termed "adaptive" and there are pairs of tables of coefficients designed to be used in this approach.

Finally, one issue that one sometimes encounters is poor convergence or stiffness of the evolution, where the time steps get stuck near domains of $\mathbf{f}$ where $\mathbf{f}$ might not be very well behaved. In these situations, one can use the *implicit* algorithms that generalize (14.89) to

$$
\mathbf{k}_i = \Delta t\, \mathbf{f} \left(t + \alpha_i \Delta t, \mathbf{z} + \sum_{j=1}^{N} \alpha_{ij} \mathbf{k}_j \right). \tag{14.91}
$$

Note the change in the upper limit of the sum. This means we now have nonzero α_{ij} for $j > i$, which makes solving for $\mathbf{k}_i$ more tricky. For example, the implicit trapezoidal algorithm is given by

$$
\begin{array}{c|cc}
0 & 0 & 0 \\
1 & 1/2 & 1/2 \\
\hline
 & 1/2 & 1/2
\end{array}
$$

and works decently in many situations. In all these algorithms, a critical role is played by the size of Δt, the time step one uses in each iteration. The larger this step, the faster the algorithm and the more error prone it is. The smaller the time step, the slower the algorithm and the more accurate it becomes. Hence, there is an art in choosing the right Δt that one learns by trial and error; as mentioned above, employing adaptive time steps can be a powerful and efficient approach.

14.7 Summary

In this chapter, we explored the concepts of integrability and chaos. We developed techniques to analyze dynamics in phase space that allow one to unravel and quantify complexity in dynamics. For integrable systems, trajectories in phase space live on tori. Once, however, nonlinear effects kick in, integrability is lost and dynamics becomes chaotic and highly sensitive to initial conditions. We learned to analyze such situations using Poincaré maps and Lyapunov exponents. We demonstrated interesting phenomena such as period doubling, bifurcations, and fractal dimension. In the process, we connected to an important modern tool in analyzing mechanics, that of computer simulations of complex systems.

Problems

★   **Problem 14.1** Consider the equation of motion

$$
\frac{d^2u}{d\varphi^2} + u - \frac{1}{p} = 3\,\lambda\,u^2, \tag{14.92}
$$

where p and λ are constants. Find the solution using perturbation theory to first order in the small parameter λ. Assess whether your approximate solution is a good one by solving the problem using numerical techniques and comparing with your result from the first-order perturbation method.

★   **Problem 14.2** Using numerical methods, solve the differential equation

$$
\frac{dq}{dt} = -\alpha\,q + \beta. \tag{14.93}
$$

Compare your results with the exact solution as a function of the discrete time step you use, and the order/method of the algorithm you adopt. Try in particular contrasting fourth-order Runge–Kutta with another algorithm of your choosing.

★ **Problem 14.3** Consider the logistic map discussed in the text. To gauge the density of bifurcations, one uses a measure of distance between fixed points as follows. Define $d = x^* - (1/2)$ as the distance between the fixed point $1/2$ and the nearest fixed point to it – labeled x_n^*. That is, you first find the r value corresponding to a convergence at value $x = 1/2$, then you identify the closest fixed point x^* to $1/2$ at this value of r, and compute the distance d. For example, at first period doubling, we have $d_1 = 0.3090\ldots$; then, after the second, we have $d_2 = -0.1164\ldots$ We then define the parameter γ as

$$\gamma = \lim_{n \to \infty} -\frac{d_n}{d_{n+1}}. \tag{14.94}$$

Using numerical methods, compute γ and verify that it is given by $\gamma = 2.502907\ldots$

★★ **Problem 14.4** Consider a particle in a Newtonian potential $V(r) = -k/r + \epsilon/r^n$ for some integer n. Using the alternate variable $u = 1/r$, (a) show that the radial equation of motion can be put into the form

$$\frac{d^2u}{d\varphi^2} + u = \frac{1}{p} + n\,\kappa\,u^{n+1}m, \tag{14.95}$$

where

$$p \equiv \frac{\ell^2}{\mu k} \quad \text{and} \quad \kappa \equiv \frac{\mu k}{\ell^2}. \tag{14.96}$$

Here, μ is the reduced mass and ℓ is the angular momentum. (b) We want to find $u(\varphi)$, the shape of the trajectory. Write numerical code that solves this equation, taking as input μ, k, ϵ, n, and ℓ. Study various scenarios, including (1) $n = 2$ and ϵ large, and (2) $n = 3$ with ϵ small.

★★ **Problem 14.5** Consider the one-dimensional harmonic oscillator with angular frequency ω perturbed by the small nonlinear potential ϵq^4. (a) Find the solution using the perturbation technique introduced in the text to first order in the small perturbation. (b) Improve your solution from part (a) by implementing the technique outlined at the end of Section 14.5, writing a solution with angular frequency $\Omega = \omega + \epsilon \omega_1$. That is, your solution now depends on $s = \Omega t$ instead of $s = \omega t$. Fix ω_1 so that you cancel a term in the solution that is *not* periodic.

★ **Problem 14.6** The celebrated Lorenz attractor is described by the differential equations

$$\frac{dx}{dt} = -\sigma x + \sigma y, \quad \frac{dy}{dt} = -xz + \alpha x - y, \quad \frac{dz}{dt} = xy - \beta z \tag{14.97}$$

and is used to described chaotic fluid dynamics involving heat flow. It is parameterized by α, β, and σ. (a) Solve this system of equations numerically and plot, for example, x vs. y and z vs. y. Determine the onset of chaos by testing supersensitivity to initial conditions.

★★   Problem 14.7 Consider the recursion relation

$$x_{n+1} = x_n + y_n, \quad y_{n+1} = a y_n - b \cos(x_n + y_n), \tag{14.98}$$

where a and b are constants; this system is known as the *standard map*. Analyze the system as we did for the logistic map in the text. In particular, explore regions of the parameter space where (1) $a = 1$ and (2) $a = 1/2$ while varying b: (3) $b = 6$ near point $(3, 3)$.

★   Problem 14.8 Show that the standard map of the previous problem described by

$$x_{n+1} = x_n + y_n, \quad y_{n+1} = a y_n - b \cos(x_n + y_n) \tag{14.99}$$

can be obtained by discretizing time in the Hamiltonian equations of motion of the following physical system: a planar pendulum in the absence of gravity that is periodically kicked in a fixed direction with a fixed force. The phase space would be described by $\theta(t)$, the pendulum's angle from the vertical, and its canonical momentum $p_\theta(t)$ – which you will need to map to the discrete sequence given by x_n and y_n.

★★   Problem 14.9 Consider the variant of the standard map described by the recursion relation

$$y_{n+1} = y_n + k \sin x_n \quad x_{n+1} = x_n + y_{n+1}, \tag{14.100}$$

where k is a constant. (a) Study the distortion of the KAM tori as k is taken from $k = 0$ to $k = 0.6$. (b) Analyze the system when $k = 0.9716$. Compute the "winding number" $\Omega \equiv \lim_{n \to \infty} (x_n - x_1)/n$ as a function of k.

★★   Problem 14.10 Consider the map given by

$$x_{n+1} = x_n e^{\alpha (1 - x_n)} \tag{14.101}$$

used to study population growth limited by disease. Analyze the system as done for the logistic map in the text, identifying the onset of chaos and bifurcations, if any. Consider in particular values of $\alpha = 1.5, 2, 2.7$.

★★   Problem 14.11 Consider the map given by

$$x_{n+1} = \alpha \sin(\pi x_n), \tag{14.102}$$

where $0 < \alpha < 1$. Analyze the system as was done for the logistic map in the text, identifying the onset of chaos and bifurcations, if any. Compute the parameter δ introduced in the text.

★★★   Problem 14.12 Consider the two-dimensional recursion

$$x_{n+1} = y_n + 1 - \alpha x_n^2, \quad y_{n+1} = \beta x_n \tag{14.103}$$

introduced by Hénon to describe chaotic behavior in the trajectories of asteroids. Study the sequence for $\alpha = 1.4$ and $\beta = 0.3$ with the initial condition $x_0 = 0.63135448$ and $y_0 = 0.18940634$. Explore the parameter space for interesting features.

★   **Problem 14.13** Consider the two-dimensional map described by the recursions

$$\theta_{n+1} = \theta_n + 2\pi \frac{\beta^{3/2}}{r_n^{3/2}}, \quad r_{n+1} = 2 r_n - r_{n-1} - \alpha \frac{\cos \theta_n}{(r_n - \beta)^2} \tag{14.104}$$

parameterized by the constants α and β. This system arises in analyzing Saturn's rings produced by the influence of one of its moons, Mimas. θ and r refer to the angular position and radius of a particle in the ring – averaged over a period of Mimas. (a) Verify numerically that a volume element in r–θ is preserved by the recursion. (b) Plot $(r_n \cos \theta_n, r_n \sin \theta_n)$ and identify bands of r_n where one has stable trajectories.

★★ **Problem 14.14** Consider the logistic map analyzed in the text. Divide the range $(0, 1)$ into N equal intervals and numerically compute the probabilities p_k that the recursion lands in the kth interval. Then compute the *entropy*

$$S \equiv - \sum_{k=1}^{N} p_k \ln p_k.$$

Do this so as to build up the function $S(\alpha)$ for the range $2.8 < \alpha < 4$. Plot the function and correlate with the conclusions in the text.

★★ **Problem 14.15** Consider a magnetic compass needle with moment of inertia I and magnetic dipole moment μ, free to rotate in the x–y plane. Denote the polar angle by θ. A time-dependent external magnetic field $\mathbf{B} = B_0 \cos \omega t \hat{x}$ applies a torque given by $\mu \times \mathbf{B}$ on the needle. (a) Write the equation of motion for θ. (b) Solve for $\theta(t)$ numerically and generate a Poincaré map by plotting discrete points $\theta(t = 2\pi n/\omega)$ for integer n. Verify the onset of chaos for $2 B_0 \mu / I > \omega^2$.

★★ **Problem 14.16** Consider the two-dimensional map

$$x_{n+1} = \alpha \left(x_n - \frac{1}{4}(x_n + y_n)^2 \right), \quad y_{n+1} = \frac{1}{\alpha} \left(y_n + \frac{1}{4}(x_n + y_n)^2 \right) \tag{14.105}$$

that approximates the chaotic scattering behavior of a projectile off a region near the origin where it collides with a bunch of targets. Fix $\alpha = 5$ and y_0 to some small value near the origin. Then start with a bunch of values for x_0 near the origin but positive, and compute the number of steps $S(x_0)$ it takes for the projectile to leave the collision basin, say when $x_n < -5$. Plot $S(x_0)$.

★★★ **Problem 14.17** Consider the so-called *circle map* for the angular variable

$$\theta_{n+1} = \theta_n + r - \kappa \sin \theta_n. \tag{14.106}$$

Note that we have $\theta \sim \theta + 2\pi$. This system can approximately describe a damped driven pendulum with angle θ. (a) First consider the case where $\kappa = 0$. Using $\theta_0 = 2\pi \times 0.2$ and $r = 2\pi k$ where k is the ratio of two integers, check that the motion is periodic. Then try $r = 2\pi/\sqrt{2}$ and check periodicity. You can study periodicity by computing the "winding number"

$$w = \lim_{k \to \infty} \frac{1}{2\pi k} \sum_{n=0}^{k-1} (\theta_{n+1} - \theta_n).$$

For periodic or almost periodic motion, we would have $w \to r/(2\pi)$. (b) Now consider $\kappa = 1/2$ with $0 < r < 2\pi$, and explore what happens to the periodicity of the motion by computing w. (c) Study the case where $\kappa = 1$. Plot $w(r)$.

Seeds of Quantization

While the Lagrangian formulation describes classical dynamics through second-order differential equations, and the Hamiltonian formulation describes it through first-order differential equations, a third formalism known as the Hamilton–Jacobi technique encapsulates the dynamics in a single *partial* differential equation. This method is conceptually very useful, albeit not necessarily advantageous for solving problems. Even though these three formalisms are essentially equivalent, each has its own advantages and disadvantages in problem solving, for insight into classical mechanics, and for how classical mechanics and quantum mechanics are related.

In this final chapter we introduce Hamilton–Jacobi theory along with its special insights into classical mechanics, and then go on to show how Erwin Schrödinger used the Hamilton–Jacobi equation to learn how to write his famous quantum-mechanical wave equation. In doing so we will have introduced the reader to two of the ways classical mechanics served as a stepping stone to the world of quantum mechanics. Back in Chapter 5 we showed how Feynman's sum-over-paths method is related to the principle of least action and the Lagrangian, and here we will show how Schrödinger used the Hamilton–Jacobi equations to invent wave mechanics. These two approaches, along with a third approach developed by Werner Heisenberg called "matrix mechanics," turn out to be quantum-mechanical analogues of the classical mechanical theories of Newton, Lagrange, Hamilton, and Hamilton and Jacobi, in that they are describing the same thing in different ways, each with its own advantages and disadvantages.

15.1 Hamilton–Jacobi Theory

Recall that in Chapter 11 we introduced *canonical transformations*, in which one set of canonical coordinates q_i and momenta p_i is transformed to a new set Q_i and P_i. Such a transformation is the key to developing the Hamilton–Jacobi technique. In particular, the strategy for finding the single partial differential equation describing a system is to find a canonical transformation such that the transformed Hamiltonian $\widetilde{H}$ *vanishes*:

$$\widetilde{H}(Q,P,t) = H(q,p,t) + \frac{\partial F_2}{\partial t} = 0 \tag{15.1}$$

for all values of the variables. Here, $F_2(q, P, t)$ is the generator of the transformation, as introduced in Chapter 11.

In Hamilton–Jacobi theory it is traditional to use the symbol S as the name of the generator, rather than F_2, and to call it **Hamilton's principle function**. Then the old and new qs and ps are related by

$$p_i = \frac{\partial S}{\partial q_i}, \quad Q_i = \frac{\partial S}{\partial P_i}, \tag{15.2}$$

so $S(q, P, t)$ satisfies the partial differential equation

$$H\left(q, \frac{\partial S}{\partial q_i}, t\right) + \frac{\partial S}{\partial t} = 0. \tag{15.3}$$

This is the **Hamilton–Jacobi equation**. To form the Hamilton–Jacobi equation for a particular problem, we begin by finding the Hamiltonian, and then replace any momentum p_i that occurs within H by $\partial S/\partial q_i$, where $S = S(q, P, t)$.

Now, given our condition that $\widetilde{H} = 0$ *identically*, it follows that the Q_is and P_is are *constants*, since from Hamilton's equations

$$\dot{Q}_i = \frac{\partial \widetilde{H}}{\partial P_i} = 0, \quad \dot{P}_i = -\frac{\partial \widetilde{H}}{\partial Q_i} = 0. \tag{15.4}$$

Example 15.1

The One-Dimensional Simple Harmonic Oscillator

We illustrate the Hamilton–Jacobi method by finding the motion of a one-dimensional simple harmonic oscillator, for which the original Hamiltonian is

$$H(q, p, t) = \frac{p^2}{2m} + \frac{1}{2}k q^2. \tag{15.5}$$

Hamilton's principle function S must then satisfy the Hamilton–Jacobi equation

$$\frac{1}{2m}\left(\frac{\partial S}{\partial q}\right)^2 + \frac{1}{2}k q^2 + \frac{\partial S}{\partial t} = 0. \tag{15.6}$$

A partial differential equation of the form (15.3) for a function S of N q_is and time t leads to a solution with $N + 1$ constants of integration. In the example of the one-dimensional harmonic oscillator we have $N = 1$, so we expect two constants of integration in the solution to Eq. (15.6). Since shifting S by a constant does not change the Hamilton–Jacobi equation (15.3), one of the $N + 1$ integration constants is simply this freedom to shift S by a constant. Hence, we are left with N nontrivial integration constants. In the example at hand, this is just a single constant of integration.

We can solve Eq. (15.6) by separation of variables:

$$S = f(q) + g(t), \tag{15.7}$$

giving

$$\frac{1}{2m}\left(\frac{df}{dq}\right)^2 + \frac{1}{2}kq^2 + \frac{dg}{dt} = 0. \tag{15.8}$$

Separating the terms that can depend only upon space from those that can depend only upon time, this implies that

$$\frac{1}{2m}\left(\frac{df}{dq}\right)^2 + \frac{1}{2}kq^2 = \text{constant} = -\frac{dg}{dt}. \tag{15.9}$$

Labeling this single integration constant as C_1, we find that

$$g(t) = -C_1 t,$$

$$f(q) = \frac{\sqrt{km}}{2}\left(q\sqrt{\frac{2C_1}{k} - q^2} + \frac{2C_1}{k}\arcsin\frac{q}{\sqrt{2C_1/k}}\right), \tag{15.10}$$

where we have dropped the integration constant corresponding to the shift freedom in S. This gives

$$S(q,P,t) = -C_1 t$$
$$+ \frac{\sqrt{km}}{2}\left(q\sqrt{\frac{2C_1}{k} - q^2} + \frac{2C_1}{k}\arcsin\frac{q}{\sqrt{2C_1/k}}\right). \tag{15.11}$$

Then, using Eqs. (15.2) we have

$$p = \frac{\partial S}{\partial q} = \sqrt{km}\sqrt{\frac{2C_1}{k} - q^2}, \quad \text{so} \quad \frac{p^2}{2m} + \frac{k}{2}q^2 = C_1. \tag{15.12}$$

Now we know that the new momentum P is a constant, and that C_1 is a constant as well. Therefore it is possible to simply *identify* them. That is, we choose $P = C_1$. Then it is clear from the second equation above that the orbits in phase space are closed ellipses, as expected for a simple harmonic oscillator, and also that $C_1 = P$ is the *energy* E of the system. That is, we have identified the constant of integration C_1 with the new (constant) canonical momentum, and found that these quantities also turn out to be the conserved energy.

The other relation in Eqs. (15.2) gives

$$Q = \frac{\partial S}{\partial P} = \left(\sqrt{\frac{m}{k}}\arcsin\frac{q}{\sqrt{2E/k}} - t\right) = \text{constant}. \tag{15.13}$$

If we let

$$t_0 \equiv -Q, \tag{15.14}$$

then

$$q(t) = \sqrt{\frac{2E}{k}}\sin\sqrt{\frac{k}{m}}(t - t_0), \tag{15.15}$$

so that $q = 0$ at time $t = t_0$, and we have the sinusoidal motion we expected. Using Eq. (15.12) we can also find,

$$p(t) = \sqrt{2mE}\cos\sqrt{\frac{k}{m}}(t - t_0). \qquad (15.16)$$

This is a rather perverse way of solving the simple harmonic oscillator problem! Yet as we shall now see, the techniques give us a new and useful perspective on mechanics.   ∎

One illustration of the conceptual usefulness of the Hamilton–Jacobi approach can be seen by taking the total derivative of $S(q, P, t)$ with respect to time:

$$\frac{dS}{dt} = \sum_i \frac{\partial S}{\partial q_i}\dot{q}_i + \frac{\partial S}{\partial t} = \sum_i p_i\dot{q}_i - H, \qquad (15.17)$$

where we used Eqs. (15.2) and (15.3). Note that the right-hand side is nothing but the Lagrangian

$$\frac{dS}{dt} = L \Rightarrow S = \int dt\, L, \qquad (15.18)$$

implying that it is the *action* which is the generator of canonical transformations that take us from any given coordinates (q_i, p_i) to new ones where the Hamiltonian simply vanishes. Note also that this S is the action evaluated *at the equation of motion*. For the example of the one-dimensional harmonic oscillator, we can write

$$S = \int dt\, L = \int dt\left(\frac{1}{2}m\dot{q}^2 - \frac{1}{2}kq^2\right) = \int \frac{dq}{\dot{q}}\left(\frac{1}{2}m\dot{q}^2 - \frac{1}{2}kq^2\right)$$
$$= \int dq\,\frac{m}{p}\left(\frac{1}{2m}p^2 - \frac{1}{2}kq^2\right) \qquad (15.19)$$

and use Eq. (15.12) to eliminate p and obtain (15.11).

Clearly any well-behaved function of the C_is is also a constant, so we can write the solution to Eq. (15.3) in terms of an alternative set of constants $C_i' = C_i'(C)$. Since the new constant momenta P_i are necessarily functions of the C_is (or equivalently the C_i's), we see that we can always *choose* our new momenta to be any functions we like of the constants of integration. For example, we can choose them identically equal to a particular set of constants of integration $P_i = C_i'$.

In general, the partial differential equation (15.3) is difficult to solve analytically. Yet a solution may exist in principle. Whenever it is possible to write the solution analytically, there exists a coordinate system (q_i, p_i) in which Hamilton's principle function S takes the form

$$S = S^{(1)}(q_1, C_1, C_2, \ldots, t) + S^{(2)}(q_2, C_1, C_2, \ldots, t)$$
$$+ S^{(3)}(q_3, C_1, C_2, \ldots, t) + \cdots + S^{(N)}(q_N, C_1, C_2, \ldots, t), \qquad (15.20)$$

leading to a **separable system**. Each of the $S^{(i)}(q_i, C_1, C_2, \ldots, t)$ is then found by direct integration, where the C_is are the constants of integration, just as we saw in the example above. We then have N relations

$$p_i = \frac{\partial S^{(i)}}{\partial q_i} = p_i(q_i, C_1, C_2, \ldots, t) \tag{15.21}$$

describing curves in each q_i–p_i plane. If we choose the P_is such that $P_i = C_i$, this implies that the constant P_is are algebraic functions of the q_is and p_is. And since these are canonical momenta, they are constants of motion that satisfy $\{P_i, P_j\} = 0$. Hence, we have N conserved quantities that are said to be "in involution." So separable Hamilton–Jacobi equations imply that the system at hand is integrable, as discussed in Chapter 14.

Integrable systems are delicate and rare, so in practice the Hamilton–Jacobi method is seldom useful in finding the trajectory of a system. In fact, even when a system is known to be integrable, finding the particular coordinate system in which the Hamilton–Jacobi partial differential equation is separable may be prohibitively difficult.

15.2 Hamilton's Characteristic Function

Now specialize to problems in which the Hamiltonian is not an explicit function of time. This of course is not always the case, but it was true for the simple harmonic oscillator problem solved in the preceding section. Then the Hamilton–Jacobi equation can be written

$$H\left(q, \frac{\partial S}{\partial q_i}\right) = -\frac{\partial S}{\partial t}, \tag{15.22}$$

so we can separate the spatial and time parts of S; that is, we can write

$$S(q_i, t) = W(q_i) - Et, \tag{15.23}$$

where E is the separation constant and $W(q_i)$, *which depends on coordinates only*, is called *Hamilton's characteristic function.*

This then leads to the *time-independent* equation

$$H\left(\frac{\partial W}{\partial q_i}, q_i\right) = E, \tag{15.24}$$

which is a first-order partial differential equation in which the momenta p_i in H are replaced by $\partial W/\partial q_i$. The separation constant E is in fact the energy of the system. This differential equation is often called the *time-independent* Hamilton–Jacobi equation.

Example 15.2   **Plumb Bob on a Spring**

As an illustration of using Hamilton's characteristic function W, consider the problem from Chapter 14, a plumb bob on the end of a spring confined to oscillate and swing in two dimensions. The bob has mass m and is attached by a spring of force constant k to a fixed point. The rest length of the spring is zero, and we suppose that there is no gravity. The Hamiltonian is given by

$$H = \frac{p_r^2}{2m} + \frac{p_\theta^2}{2mr^2} + \frac{1}{2}kr^2, \tag{15.25}$$

where we adopt polar coordinates and describe phase space with the coordinates r, p_r, θ, and p_θ. Since the Hamiltonian does not depend on time explicitly, the energy of this system is conserved. To obtain the partial differential equation satisfied by Hamilton's characteristic function $W(r, P_R, \theta, P_\Theta)$, we write

$$p_r \to \frac{\partial W}{\partial r}, \quad p_\theta \to \frac{\partial W}{\partial \theta} \tag{15.26}$$

in the Hamiltonian and so find that

$$\frac{1}{2m}\left(\frac{\partial W}{\partial r}\right)^2 + \frac{1}{2mr^2}\left(\frac{\partial W}{\partial \theta}\right)^2 + \frac{1}{2}kr^2 = C_1 \equiv E, \tag{15.27}$$

where it is clear that the first constant C_1 is in fact the conserved energy E. This equation is separable in the given coordinate system. To see this, write

$$W(r, P_R, \theta, P_\Theta) = W^{(1)}(r, P_R, P_\Theta) + W^{(2)}(\theta, P_R, P_\Theta), \tag{15.28}$$

so that

$$r^2\left(\frac{dW^{(1)}}{dr}\right)^2 + \left(\frac{dW^{(2)}}{d\theta}\right)^2 + mkr^4 = 2mr^2E. \tag{15.29}$$

Then separation of the variables r and θ gives two ordinary differential equations

$$\left(\frac{dW^{(1)}}{dr}\right)^2 + \frac{l^2}{r^2} + mkr^2 = 2mE \quad \text{and} \quad \left(\frac{dW^{(2)}}{d\theta}\right)^2 = l^2 = \text{constant}. \tag{15.30}$$

Note that we have labeled the new constant of integration l – suggestively identifying it as the angular momentum. The second equation of (15.30) can be integrated immediately, giving

$$W^{(2)}(\theta, P_R, P_\Theta) = l\theta + \text{shift constant}. \tag{15.31}$$

As usual, the new constant momenta P_R and P_Θ are generally functions of the constants of integration, which we have labeled E and l. And we are free to choose $P_R(E, l)$ and $P_\Theta(E, l)$. From the first differential equation of (15.30), we find that

$$W^{(1)}(r, P_R, P_\Theta) = \int \sqrt{2mE - mkr^2 - \frac{l^2}{r^2}}\, dr, \tag{15.32}$$

which is a rather unpleasant integral that can be evaluated in terms of inverse sines and inverse tangents. But we need not worry about its explicit form if we are interested in the trajectory of the bob in phase space. From (15.26), we find

$$p_r = \sqrt{2mE - mkr^2 - \frac{l^2}{r^2}}, \quad p_\theta = l. \tag{15.33}$$

Therefore, the orbit of the plumb bob in phase space is identified in the r–p_r and θ–p_θ planes. To find the time dependences $r(t), p_r(t), \theta(t)$, and $p_\theta(t)$, let us choose $P_R = E$ and $P_\Theta = l$ so that we have:

$$R = \frac{\partial W}{\partial P_R} = \frac{\partial W}{\partial E} = \omega_1(t + t_1), \quad \Theta = \frac{\partial W}{\partial P_\Theta} = \frac{\partial W}{\partial l} = \omega_2(t + t_2), \tag{15.34}$$

where ω_1, ω_2, t_1, and t_2 are constants.   ∎

Now let us further specialize the time-independent Hamilton–Jacobi equation to a single particle of mass m whose Cartesian coordinates are x, y, z, and whose potential energy is $U(x, y, z)$. Then we have

$$\frac{1}{2m}\left[\left(\frac{\partial W}{\partial x}\right)^2 + \left(\frac{\partial W}{\partial y}\right)^2 + \left(\frac{\partial W}{\partial z}\right)^2\right] + U(x, y, z) = E, \tag{15.35}$$

so

$$(\nabla W)^2 = 2m(E - U). \tag{15.36}$$

As we shall see later in the chapter, *this form of the time-independent Hamilton–Jacobi equation proves to be extremely useful in making the transition to quantum mechanics.*

There is yet another but related approach to tackling dynamics. It is sometimes useful when dealing with dynamical systems which have constant energy. In this approach, we look for a generator $G_2(q, P, t)$ of a canonical transformation such that the new Hamiltonian $\widetilde{H}$ does not contain any of the new coordinates Q_i – instead of finding $F_2(q, P, T)$ that leads to a Hamiltonian that identically vanishes. This implies that we will have

$$\dot{P}_i = -\frac{\partial \widetilde{H}}{\partial Q_i} = 0 \tag{15.37}$$

and thus the new momenta P_i are constants. We then get

$$\dot{Q}_i = \frac{\partial \widetilde{H}}{\partial P_i} = \text{constant} \equiv \omega_i(P) \Rightarrow Q_i(t) = \omega_i(P)t + t_i, \tag{15.38}$$

with N integration constants t_i. We therefore have the relation

$$\widetilde{H}(P) = H(q, p) + \frac{\partial G_2}{\partial t} = \text{constant}, \tag{15.39}$$

and since $H(q,p)$ is a constant, G_2 must be of the form

$$G_2(q,P,t) = -\alpha t + W(q,P) \tag{15.40}$$

for a constant α. We also have the transformation relations

$$p_i = \frac{\partial G_2}{\partial q_i} = \frac{\partial W}{\partial q_i}, \quad Q_i = \frac{\partial G_2}{\partial P_i} = \frac{\partial W}{\partial P_i}. \tag{15.41}$$

Therefore $W(q,P)$ must satisfy the partial differential equation

$$H\left(q, \frac{\partial W}{\partial q_i}, t\right) - \alpha = \tilde{H}(P) = \text{constant.} \tag{15.42}$$

Alternatively, we can write

$$H\left(q, \frac{\partial W}{\partial q_i}\right) = C_1, \tag{15.43}$$

where C_1 is a constant. Solving this partial differential equation leads to $N-1$ additional constants of integration $C_2, \ldots, C_N$ (one of the N total constants is simply a constant shift of W). Along with C_1, these consist of N integration constants. As before, we also have the freedom to choose the new momenta P_i as any functions of the C_is. Note also that C_1 is simply the conserved value of the Hamiltonian, often being the energy. Taking the direct time derivative of Hamilton's characteristic function W, we can write

$$\frac{dW}{dt} = \sum_i \frac{\partial W}{\partial q_i} \dot{q}_i = \sum_i p_i \dot{q}_i. \tag{15.44}$$

15.3 Action Angle Variables

As described in Chapter 14, a theorem by Liouville states that if a system with N degrees of freedom is integrable, its phase-space trajectory is necessarily confined to the surface of an N-dimensional torus. The dynamics involves N-angular frequencies of the trajectory along each of the N noncontractible cycles of the torus. We saw in Chapter 14 that irrational winding numbers, *i.e.*, irrational ratios of angular frequencies, lead to dense tracks of evolution on the torus. The method of Hamilton's characteristic function discussed above can be a handy tool to determine these frequencies *without the need to solve for the details of the dynamics*.

To apply the technique, imagine we are to employ a canonical transformation generator $W(q,P)$ that leads to the Hamiltonian $\tilde{H}(P_i)$, with all the Q_is being cyclic as described above. Furthermore, we assume separability in the coordinates (q_i, p_i). As argued above, separability implies that the trajectory in each q_i–p_i plane decouples and is described by functions

$$p_i = p_i(q_i, C_1, C_2, \ldots, C_N), \tag{15.45}$$

where the C_is are constants of integration. These curves in each q_i–p_i plane are *closed*, since they are projections of a trajectory confined to the surface of an N-torus. We now choose the P_is as follows:

$$P_i = \frac{1}{2\pi} \oint p_i(q_i, C_1, C_2, \ldots, C_N)\, dq_i, \tag{15.46}$$

where the integrals are taken along the closed trajectories in each q_i–p_i plane. This gives a specific choice of functions $P_i(C_i)$ as our new momenta. Conventionally, these momenta are labeled J_i instead of P_i and they are called **action angle variables**. Their canonical pairs, the Q_is, are labeled Θ_i to emphasize they are angular in character, since their evolution is given by

$$\Theta_i(t) = \omega_i(J_1, J_2, \ldots)t + t_i. \tag{15.47}$$

Hence, in a time $2\pi/(\omega_i)$, each Θ_i increases by 2π and hence must wrap like an angle. We then have

$$\omega_i(J_1, J_2, \ldots) = \dot{\Theta}_i(t) = \frac{\partial \widetilde{H}}{\partial J_i}. \tag{15.48}$$

If we can write the constant Hamiltonian in terms of the action angle variables J_i, we can then read off the angular frequencies of the integrable trajectory by simple differentiation.

Example 15.3 **Plumb Bob on a Spring Revisited**

To demonstrate these ideas we return to the example of the bob on a spring. The action angle variables are

$$P_R = J_R = \frac{1}{2\pi} \oint p_r dr, \quad P_\Theta = J_\Theta = \frac{1}{2\pi} \oint p_\theta d\theta. \tag{15.49}$$

The integrals are along closed paths in the r–p_r and θ–p_θ planes. From the second expression, we immediately get

$$J_\Theta = p_\theta, \tag{15.50}$$

since p_θ is constant. The first expression requires more effort:

$$J_R = \frac{1}{2\pi} \oint \sqrt{2mE - mkr^2 - \frac{l^2}{r^2}}\, dr, \tag{15.51}$$

where the limits are between a maximum and a minimum value of r if we multiply the integral by two. These can be obtained from the turning points, the points where $p_r = 0$ in Eq. (15.33):

$$r_{min,max} = \frac{E}{k} \sqrt{1 \mp \sqrt{1 - \frac{k p_\theta^2}{m E^2}}}. \tag{15.52}$$

We can then write

$$J_R = 2 \times \frac{1}{2\pi} \int_{r_{min}}^{r_{max}} \sqrt{2mE - mkr^2 - \frac{l^2}{r^2}} \, dr. \qquad (15.53)$$

This definite integral is easy to evaluate if we use the alternate variable

$$u = \frac{k^2 r^2/E^2) - 1}{1 - (k p_\theta^2/m E^2)}. \qquad (15.54)$$

After some algebra, we find that

$$J_R = \sqrt{km} \left(1 - \sqrt{\frac{k}{m} \frac{l}{E}} \right). \qquad (15.55)$$

Equations (15.50) and (15.55) allow us to write the Hamiltonian in terms of the action angle variables J_R and J_Θ:

$$\widetilde{H} = \sqrt{\frac{k}{m}} (J_R + J_\Theta). \qquad (15.56)$$

Using Eq. (15.48) we then find the two angular frequencies of motion on the phase-space torus

$$\omega_R = \omega_\Theta = \sqrt{\frac{k}{m}}, \qquad (15.57)$$

as expected. The corresponding angle variables R and Θ are the natural angle coordinates along the two noncontractible cycles of the two-dimensional torus. Since the winding number, the ratio of the frequencies, is an integer (in this case simply unity), the orbits of the bead are closed. ∎

15.4 Adiabatic Invariants

Action angle variables are particularly useful in studying a class of mechanics problems involving **adiabatic dynamics**. We already introduced the idea of adiabatic dynamics and adiabatic invariants back in Chapter 8, where we discussed the orbits of charged particles in a slowly varying magnetic field. To illustrate adiabatic dynamics here, let us focus on cases with a single degree of freedom, and hence a two-dimensional phase space. We consider a system that exhibits periodic motion, then make one of the constant parameters, call it A, vary very slowly over a duration of a period of the system, $A \to A(t)$. For example, our system might be a pendulum consisting of a plumb bob at the end of a cord threaded through a hole, and, as the pendulum oscillates, the cord is slowly pulled up (see Figure 15.1). The parameter $A(t)$ is then the time-varying length of the pendulum. If T denotes the period of the unperturbed system, *slowly varying* can be defined through the relation

$$\frac{\dot{A}}{A} \ll \frac{1}{T}. \qquad (15.58)$$

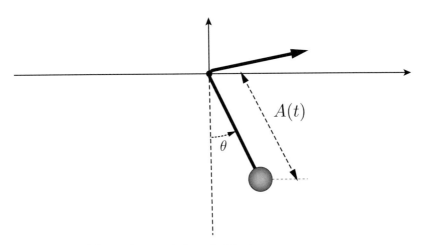

Fig. 15.1 A two-dimensional pendulum with its rope being pulled slowly.

Hence, the length of the pendulum cord does not change much over the period of a single oscillation of the bob. Note that this system is not conservative: energy is *not* conserved, as the external agent does work in varying the parameter $A(t)$ (*i.e.*, the work done by the person pulling the cord in our example). We can still formally consider a canonical transformation that takes us from the original (θ, p_θ) phase-space coordinates to a new angle and action angle coordinates Θ, J_Θ. Of course, we do not expect that Θ will be linear in time and that J_Θ is constant – given the time dependence in the Hamiltonian. We *define*

$$J_\Theta \equiv \frac{1}{2\pi} \oint p_\theta \, d\theta \tag{15.59}$$

as usual, and employ Hamilton's characteristic function $W(q, J_\Theta, A(t))$, where we have now explicitly indicated that W depends on time through its dependence on $A(t)$. We still have the general relations

$$p_\theta = \frac{\partial W}{\partial \theta}, \quad \Theta = \frac{\partial W}{\partial J_\Theta}. \tag{15.60}$$

It is useful to switch pictures and go between the same set of coordinates $(\theta, p_\theta) \to (\Theta, J_\Theta)$, employing instead a generator of the first kind $F_1(\theta, \Theta, A(t))$ instead of W, which is of the second kind $W = F_2(\theta, J_\Theta, A(t))$. W and F_1 are related by a Legendre transformation

$$F_1 = W - \Theta J_\Theta. \tag{15.61}$$

Note also that Hamilton's characteristic function is given by

$$W = \int p_\theta \, d\theta. \tag{15.62}$$

We then observe the following: if the system undergoes a full oscillation, $\Theta \to \Theta + 2\pi$, W must change as $W \to W + 2\pi J_\Theta$ given (15.60). We then see from Eq. (15.61) that $F_1 \to F_1$, *i.e.*, F_1 must be a periodic function in Θ. This fact will help us prove that, in the adiabatic regime, J_Θ remains approximately constant. To see this, note that the new Hamiltonian is given by the standard canonical transformation relation

$$\widetilde{H} = H(p_\theta(\Theta, J_\Theta), \theta(\Theta, J_\Theta), A(t)) + \frac{\partial F_1}{\partial t} = H(J_\Theta, A(t)) + \dot{A}\frac{\partial F_1}{\partial A}. \qquad (15.63)$$

In writing this expression, we noted that the Hamiltonian, when expressed in the Θ and J_Θ coordinates, depends only on J_Θ – this functional property of H being unchanged by the time dependence of the external parameter $A(t)$. We can then write

$$\dot{J}_\Theta = -\frac{\partial \widetilde{H}}{\partial \Theta} = -\dot{A}\frac{\partial^2 F_1}{\partial\Theta\partial A}. \qquad (15.64)$$

Therefore, due to the time dependence of $A(t)$, the action angle variable is not strictly constant in time. However, if we look at the change in J_Θ over a full period of the system, we can write

$$\langle \dot{J}_\Theta \rangle \equiv \frac{1}{2\pi}\int_0^{2\pi} \dot{J}_\Theta\, d\Theta = -\frac{1}{2\pi}\int_0^{2\pi} \dot{A}\frac{\partial^2 F_1}{\partial\Theta\partial A}\, d\Theta, \qquad (15.65)$$

where we have defined the average of $\dot{J}_\Theta$ over a period of oscillation as $\langle \dot{J}_\Theta \rangle$.

Since, by assumption, $\dot{A}$ does not vary much over a period of the system, we can take it out of the integral and we have instead

$$\langle \dot{J}_\Theta \rangle \simeq -\frac{\dot{A}}{2\pi}\int_0^{2\pi} \frac{\partial}{\partial\Theta}\left(\frac{\partial F_1}{\partial A}\right) d\Theta = 0, \qquad (15.66)$$

since F_1 is periodic in Θ. Hence, we conclude that the action angle variable is on average approximately constant in an adiabatic scenario. We say that J_Θ is an **adiabatic invariant**.

Let us apply this conclusion to tackling the problem of a pendulum swinging in a plane, whose cord is being slowly pulled up through a hole. The Lagrangian for small angles θ is given by

$$L = \frac{1}{2}m\,l^2\dot{\theta}^2 - \frac{1}{2}m\,g\,l\,\theta^2, \qquad (15.67)$$

yielding the Hamiltonian

$$H = \frac{p_\theta^2}{2\,m\,l^2} + \frac{1}{2}m\,g\,l\,\theta^2. \qquad (15.68)$$

We then make the length of the cord vary *slowly* with time, $l \to l(t)$:

$$H = \frac{p_\theta^2}{2\,m\,l(t)^2} + \frac{1}{2}m\,g\,l(t)\,\theta^2. \qquad (15.69)$$

The action angle variable is given by

$$J_\Theta = \oint p_\theta \, d\theta = \sqrt{2H(t)m} \, l(t) \oint \sqrt{1 - \frac{m \, g \, l(t)}{2 \, H(t)} \theta^2} \, d\theta, \tag{15.70}$$

where we have explicitly indicated that $H(t)$ is not a constant in time. We can evaluate this integral using a change of variable $\theta \to \alpha$, so that $m \, g \, l(t)\theta^2/2 \, H(t) = \sin^2 \alpha$, with

$$J_\Theta = 2 H(t) \sqrt{l/g} \oint \cos^2 \alpha \, d\alpha = 2 \pi H(t) \sqrt{l(t)/g}, \tag{15.71}$$

where we used the fact that $\oint \cos^2 \alpha \, d\alpha = \int_0^{2\pi} \cos^2 \alpha \, d\alpha = \pi$, corresponding to one full cycle. This action angle variable is then approximately constant when $l(t)$ changes adiabatically:

$$H(t) \sqrt{\frac{l(t)}{g}} = \text{constant}. \tag{15.72}$$

Noting that the angular frequency of the pendulum is $\omega = \sqrt{g/l}$, we conclude that

$$\frac{H(t)}{\omega(t)} = \text{constant}. \tag{15.73}$$

As one pulls the cord very slowly, that is, adiabatically, $l(t)$ becomes smaller and thus $\omega(t)$ becomes larger. Equation (15.73) implies that the energy of the system increases correspondingly. This means that the amplitude of oscillations increases, since the energy of a harmonic pendulum is proportional to the square of the oscillation amplitude.[1]

Particle in Slowly Varying Magnetic Field

As another illustration of the technique of adiabatic invariants, consider the example of a particle of mass m and electric charge q moving in a uniform magnetic field B pointing in the z direction. In Chapter 11 we tackled this problem in the Hamiltonian formalism. Using cylindrical coordinates ρ, φ, and z, we have the Hamiltonian

$$H = \frac{p_\rho^2}{2m} + \frac{p_\varphi^2}{2m\rho^2} + \frac{p_z^2}{2m} - \frac{q^2 B^2}{8m}\rho^2. \tag{15.74}$$

---

[1] In the quantum version of this problem, the cord is pulled over timescales much longer than the timescale associated with transition between the levels of the simple harmonic oscillator. The adiabatic process then amounts to having the pendulum locked in a given level of the quantum oscillator. The system does not transition in or out of the initial level as long as the process is slow enough, shifting the energy of the system gradually with the shifting of the space between energy levels that results from an adiabatic change in frequency.

For simplicity, let us suppress the trivial dynamics in the z direction. The action angle variable associated with φ is

$$J = \oint p_\varphi \, d\varphi. \tag{15.75}$$

Without loss of generality, we know that the particle will orbit at constant radius ρ around the z axis. Its angular frequency is

$$\omega = -\frac{q\,B}{m} = \dot{\varphi}. \tag{15.76}$$

This implies that

$$\dot{\varphi} = -\frac{\partial H}{\partial p_\varphi} = \frac{p_\varphi}{m\,\rho^2} = \omega = -\frac{q\,B}{m}. \tag{15.77}$$

Hence, we have

$$p_\varphi = -q\,B\rho^2 = \text{constant}. \tag{15.78}$$

The action angle variable then becomes

$$J = -2\,\pi\,q\,B\,\rho^2. \tag{15.79}$$

Next, consider the scenario where we adiabatically tune the external magnetic field $B \rightarrow B(t)$. We then expect J to be approximately constant:

$$q\,B(t)\rho(t)^2 = \text{constant} = \frac{q\,B(t)}{m}m\,\rho(t)^2 = \rho(t) \times m\,\rho(t)\omega(t) = L, \tag{15.80}$$

where here L is the angular momentum of the particle about the origin. So as we tune $B(t)$, the radius and angular frequency of the particle change in such a way as to keep the angular momentum constant. For example, if $B(t)$ increases, so will $\omega(t)$, which implies the radius of the circular trajectory decreases. Another way to look at this is to say

$$B(t)\,\pi\,\rho(t)^2 = \text{constant} = \Phi, \tag{15.81}$$

where Φ is the magnetic flux through the circular trajectory of the particle. Therefore *the flux Φ is an adiabatic invariant.*   ∎

15.5 Early Quantum Theory

As mentioned in Chapter 5, in 1905 Einstein proposed that the photoelectric effect suggests that light consists of "quanta," later called photons, which have both a *particle* nature and a *wave* nature. The wave nature of photons is described by properties like wavelength λ and frequency ν, while their particle nature is described by properties like momentum p and energy E. The two worlds of particles and waves for photons are related by $E = h\nu$ and $p = h/\lambda$, where h is Planck's constant, introduced by Max Planck in 1901.

Then in 1924, in his doctoral dissertation, the French student Louis de Broglie proposed that if light (long considered a wave) could have particle properties, why shouldn't massive particles (electrons, protons, chairs, etc.) have wave properties? In fact, he proposed that the same relationships $E = h\nu$ and $p = h/\lambda$ should apply to massive particles as to massless photons. Einstein for one was intrigued by de Broglie's idea, and over the next few years, the wave nature of particles was confirmed for electrons by the experiments of Davisson, Germer, and others.

When it came to atoms, it had already been shown by Niels Bohr in 1915 that to explain atomic spectra, which consist of discrete wavelengths, one needs to postulate discrete energies for the electrons in atoms. Bohr had suggested that electrons can only orbit the nuclei at particular radii, and that the discrete energies needed are associated with these radii. Using his particle waves, de Broglie showed that the particular radii in hydrogen atoms correspond to circular orbits for which there is a whole number of electron wavelengths. In the ground state, there is one complete wavelength in traversing the circular orbit, in the first excited state there are two complete wavelengths, etc. And the energy difference between these two orbits, calculated from classical physics, corresponds to the correct frequency of a photon radiated as the electron falls from the first excited state to the ground state, using $\Delta E = h\nu$.

Another important step was the statement of the "indeterminacy principle" or "uncertainty principle" by the German physicist Werner Heisenberg, that if Δx is the uncertainty in a particle's position at some time, and Δp is the uncertainty in its momentum at the same time, then $\Delta x \, \Delta p \geq \hbar/2$, where $\hbar = h/2\pi$. In classical physics one can in principle learn both x and p to any degree of accuracy at the same time, but not so in the new quantum physics.

Physicists soon developed what is now called the "old quantum theory," using a mixture of classical ideas and de Broglie wavelengths. In particular, they used phase-space pictures and Liouville's theorem. Recall that in Section 11.7, we learned from Liouville's theorem that evolution in phase space preserves phase-space volume. For example, if a small phase-space area in two-dimensional phase space were to be squeezed along the position axis through time evolution, it would then have to expand along the momentum axis: the area must remain constant in time. This suggests that there is a natural "incompressibility" of information in phase space, respected by Hamiltonian evolution, and obeying Heisenberg's principle.

Quantum evolution necessitates Liouville's theorem, and Liouville's theorem may be viewed as hinting at phase-space quantization. This does not mean that Liouville's theorem necessarily *implies* a notion of discretization of phase space; but it certainly would be consistent with such an idea. These observations did not evade the founders of quantum mechanics. Indeed, a semiclassical quantization scheme was developed based on such observations by Bohr first, and then later by Wentzel, Kramers, Brillouin, Jeffreys, and Sommerfeld. To understand the idea, let us go back to the one-dimensional simple harmonic oscillator example of Chapter 11. We have a Hamiltonian of the form

$$H = E = \frac{p^2}{2m} + \frac{1}{2}kx^2 \tag{15.82}$$

and trajectories in phase space are ellipses, as shown in Figure 11.4. Consider the area of such an ellipse:

$$\text{Area} = \pi \sqrt{\frac{2E}{k}} \sqrt{2mE} = 2\pi E \sqrt{\frac{m}{k}} = \frac{2\pi E}{\omega}. \tag{15.83}$$

Let us now propose that phase space must be discretized, so that

$$\text{Area} \simeq n \times h, \tag{15.84}$$

where n is an integer and h is Planck's constant. We would then have

$$E = n\hbar\omega. \tag{15.85}$$

Comparing this to the quantum-mechanical expression for the energy states:

$$E = \hbar\omega\left(n + \frac{1}{2}\right) \simeq n\hbar\omega, \tag{15.86}$$

where we have taken the regime $n \gg 1$. We see that, at large quantum level number n, the semiclassical proposition given by Eq. (15.84) leads to the correct quantization of energy. We do however miss the ground-state energy $\hbar\omega/2$, and need to consider only large values of n.

Emboldened by this success, we may then propose the general semiclassical quantization scheme

$$\text{Area} = \oint p\,dq = nh, \tag{15.87}$$

where the integral is over any of the cycles of the phase-space torus, assuming periodic trajectories. This is a rather good approximation for quantum systems and quantization, as long as we focus on large quantum numbers. For example, let us look at the example of the particle on a parabola from Chapter 11. The energy is given by

$$E = \frac{p_x^2}{2m(1 + x^2)} + \frac{mg}{2}x^2 \tag{15.88}$$

and the phase-space trajectories are shown Figure 11.6. We can then write

$$p_x = m(1 + x^2)\dot{x} \tag{15.89}$$

and construct the area of a closed curve as

$$\text{Area} = 4 \int_0^{x_0} p_x \, dx = 4 \, m \int_0^{x_0} (1 + x^2) \dot{x} \, dx$$

$$= 4 \int_0^{x_0} \sqrt{E - \frac{m g x^2}{2}} \sqrt{2 m \, (1 + x^2)} \, dx. \tag{15.90}$$

But note that the energy and the largest distance x_0 are related:

$$E = \frac{m g}{2} x_0^2 \Rightarrow x_0 = \sqrt{\frac{2 E}{m g}}. \tag{15.91}$$

Applying now the quantization scheme (15.87), we get

$$n h = 4 \int_0^{\sqrt{2 E/m g}} \sqrt{E - \frac{m g x^2}{2}} \sqrt{2 m \, (1 + x^2)} \, dx. \tag{15.92}$$

We can express this result in terms of **elliptic functions** $\mathscr{E}$ and $\mathscr{K}$:

$$n h = \frac{1}{3 \sqrt{2 m g}} \times$$
$$\left[(2 E - m g) \mathscr{E}(-2 E/m g) + (2 E + m g) \mathscr{K}(-2 E/m g) \right]. \tag{15.93}$$

This is useful because these elliptic functions satisfy many properties and have well-known asymptotic expansions. Expecting that large quantum level number corresponds to large energy, we can then take the large E limit of this expression, which gives

$$\frac{2}{3 \, m g} E^{3/2} \simeq n h \Rightarrow E \sim n^{2/3} \tag{15.94}$$

for $n \gg 1$. A plot of this function is shown in Figure 15.2.

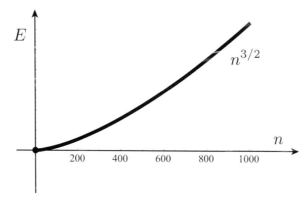

Fig. 15.2   Energy levels of a particle on a parabola.

While we expect this result to be only an approximation, the method did allow us to tackle a rather complicated quantum mechanics problem with relatively little effort. More importantly, the connection between Liouville's theorem and phase-space quantization is rather beautiful and good for developing our intuition of quantum mechanics.

15.6 Optics: From Waves to Rays

Although it might at first seem irrelevant to Hamilton–Jacobi methods in mechanics, and to quantum ideas as well, we will now look into the relationship between two well-known approaches to *optics*, namely *ray* optics (or "geometrical" optics) and *wave* optics. *Ray* optics is very useful in tracing the behavior of light as it passes through lenses or bounces off mirrors, for example; the rays obey Fermat's principle of stationary time, as discussed in Chapter 3. In contrast, *wave* optics is required to understand such phenomena as interference and diffraction. What is the relationship between ray optics and wave optics?

In Chapter 5 we showed that small-amplitude transverse waves on a stretched rope obey the wave equation

$$\frac{\partial^2 y}{\partial x^2} - \frac{1}{v^2}\frac{\partial^2 y}{\partial t^2} = 0, \tag{15.95}$$

where $y(x, t)$ is the transverse displacement of the rope and v is the speed of the wave, which is related to the rope's tension T and mass per unit length μ by $v = \sqrt{T/\mu}$. A solution of the equation for fixed wave number $k = 2\pi/\lambda$, frequency $\omega = 2\pi\nu$, and amplitude A is

$$y = Ae^{i(kx-\omega t)}, \tag{15.96}$$

which describes a wave traveling toward positive x with speed $v = \omega/k = \lambda\nu$. As always, we can retrieve a physical wave by taking only the real part of the solution.

Maxwell's equations of electromagnetism can be used to derive the wave equation for what is called the scalar potential φ, which for one dimension and in vacuum is

$$\frac{\partial^2 \varphi}{\partial x^2} - \frac{1}{c^2}\frac{\partial^2 \varphi}{\partial t^2} = 0, \tag{15.97}$$

where c is the speed of light. That is, according to Maxwell, light is an electromagnetic wave traveling at speed c. He thus unified electricity, magnetism, and optics, the great synthesis of the nineteenth century.

In three dimensions, the linear wave equation becomes

$$\nabla^2 \varphi - \frac{1}{v^2}\frac{\partial^2 \varphi}{\partial t^2} = 0, \tag{15.98}$$

where in Cartesian coordinates

$$\nabla^2 \equiv \frac{\partial^2}{\partial x^2} + \frac{\partial^2}{\partial y^2} + \frac{\partial^2}{\partial z^2}. \tag{15.99}$$

A solution of the three-dimensional wave equation is the plane wave

$$\varphi = \varphi_0 e^{i(\mathbf{k_0 \cdot r} - \omega t)} = A e^{i(k_{0x}x + k_{0y}y + k_{0z}z - \omega t)}, \tag{15.100}$$

where now the wave number $\mathbf{k_0}$ has magnitude $|\mathbf{k_0}| = 2\pi/\lambda_0$, and also a direction, the direction in which the wave is propagating.

If φ represents a light wave in a medium with fixed index of refraction n, then the velocity of the wave is $v = c/n$, and the wave equation becomes

$$\nabla^2 \varphi - \frac{n^2}{c^2} \frac{\partial^2 \varphi}{\partial t^2} = 0, \tag{15.101}$$

with solutions

$$\varphi = \varphi_0 e^{i(\mathbf{k \cdot r} - \omega t)}, \tag{15.102}$$

where $\mathbf{k}$ points in the direction of the wave propagation, and has magnitude $|\mathbf{k}| = n k_0$. For the same frequency ω as in vacuum, the wavelength is shorter in the medium, with $\lambda = 2\pi/|\mathbf{k}| = \lambda_0/n$.

More generally, the index of refraction $n(\mathbf{r})$ depends upon position, as for example in earth's atmosphere, where n is larger near the ground than at higher altitudes. If $n(\mathbf{r})$ changes only slowly with position, we have the case of *geometrical optics*, the language of light rays we discussed already in Chapter 3. So in that case we try a modified solution of the form

$$\varphi = e^{A(\mathbf{r}) + i k_0 (L(\mathbf{r}) - ct)}. \tag{15.103}$$

The real, slowly varying functions $A(\mathbf{r})$ and $L(\mathbf{r})$ are related to the amplitude and phase of the wave, respectively. The function $L(\mathbf{r})$ is called the *optical path length* or *eikonal* of the wave. It is the actual physical distance covered by the wave weighted by the local index of refraction. So now we substitute this trial solution into the wave equation to find equations governing the amplitude and phase functions $A(\mathbf{r})$ and $L(\mathbf{r})$. Applying the gradient operator to the given expression for φ:

$$\nabla \varphi = \varphi \nabla (A + i k_0 L), \tag{15.104}$$

so then taking the divergence of this gradient, we have the Laplacian

$$\nabla^2 \varphi \equiv \nabla \cdot \nabla \varphi = \varphi [\nabla^2 (A + i k_0 L) + (\nabla (A + i k_0 L)^2]. \tag{15.105}$$

Substituting this into the wave equation, we find terms that are real and terms that are imaginary. However, we chose both $A(\mathbf{r})$ and $L(\mathbf{r})$ to be real, so to satisfy the wave equation the real terms and the imaginary terms must vanish separately. These terms require, respectively, that

$$\nabla^2 A + (\nabla A)^2 + k_0^2 (n^2 - (\nabla L))^2) = 0 \tag{15.106}$$

and

$$\nabla^2 L + 2\nabla A \cdot \nabla L = 0. \tag{15.107}$$

These equations are *exact*, but they are also coupled, nonlinear equations that we have no realistic hope of solving analytically. However, we can now make the *geometrical optics approximation*. We suppose that the wavelength of light is very small compared with the distances the light travels or distances over which n changes appreciably, That is certainly the case for visible light passing through earth's atmosphere, for example, but not so much for radio waves with wavelengths many kilometers long. This means that the two terms proportional to $k_0^2 = 4\pi^2/\lambda_0^2$ threaten to dominate all others, making that equation invalid, *unless* the quantity $n^2 - (\nabla L)^2 \simeq 0$. That is, in the geometrical optics limit we require that the eikonal L obey

$$(\nabla L)^2 = n^2, \tag{15.108}$$

called the *eikonal equation*. Given an index of refraction $n(\mathbf{r})$, we can try to solve this nonlinear first-order differential equation to find the eikonal function $L(\mathbf{r})$. Surfaces of constant L are the wave fronts of the wave; the wave itself propagates in a direction normal to the wave fronts.

Example 15.5   **The Eikonal for a Flat Atmosphere**

A plane wave of light from a distant star approaches the earth. Let us find the eikonal for this wave, entering the atmosphere at high altitude, and descending to the ground. Without loss of generality we can orient coordinates so that the wave lies in the x, y plane, where x is the horizontal axis and y the vertical axis, which is zero at high altitude where $n = 1$; then n increases as the wave moves downward. We will assume the atmosphere is essentially flat. Then the eikonal equation becomes

$$(\nabla L)^2 = \left(\frac{\partial L}{\partial x}\right)^2 + \left(\frac{\partial L}{\partial y}\right)^2 = n^2(y). \tag{15.109}$$

We try separation of variables, with $L(x, y) = f(x) + g(y)$. Substituting, we find

$$\left(\frac{df}{dx}\right)^2 = n^2(y) - \left(\frac{dg}{dy}\right)^2, \tag{15.110}$$

where the left side can depend only on x, and the right side can depend only on y. Therefore, since they are equal, each side must equal the same *constant*, which we will call C^2. Therefore

$$\left(\frac{df}{dx}\right)^2 = C^2, \quad \text{with solution } f = Cx \tag{15.111}$$

and

$$\left(\frac{dg}{dy}\right)^2 = n^2 - C^2, \quad \text{with solution } g = \int \sqrt{n^2 - C^2}\, dy \tag{15.112}$$

where we have neglected additive constants, since they will play no role in the physics. Our solution for the eikonal is therefore

$$L(x, y) = Cx + \int dy \sqrt{n^2(y) - C^2},$$
(15.113)

where we can in principle evaluate the integral if we know $n(y)$. However, as we shall now see, we can understand everything about the direction of the associated light ray without performing this integration at all.

In the language of geometrical optics, light paths are represented by light "rays," which trace paths that are perpendicular to the wave front of the light waves in the small-wavelength limit. Then since the gradient of the wave fronts is perpendicular to the wave fronts themselves, the direction of a light ray is in the direction of the gradient of these wave-front eikonals L. In the example we have been discussing, this is in the direction of ∇L, where

$$\nabla L = \hat{x} \left(\frac{\partial L}{\partial x} \right) + \hat{y} \left(\frac{\partial L}{\partial y} \right) = C\hat{x} + \sqrt{n^2(y) - C^2}\,\hat{y}.$$
(15.114)

Now it is easy to find the angle θ of a light ray relative to the vertical. Simply note that this angle is determined by

$$\tan \theta = \frac{|\nabla L_x|}{|\nabla L_y|} = \frac{C}{\sqrt{n^2(y) - C^2}} = \frac{C/n(y)}{\sqrt{1 - C^2/n^2(y)}},$$
(15.115)

but also

$$\tan \theta = \frac{\sin \theta}{\cos \theta} = \frac{\sin \theta}{\sqrt{1 - \sin^2 \theta}},$$
(15.116)

so clearly

$$n(y) \sin \theta = C = \text{constant},$$
(15.117)

which is just Snell's law of geometrical optics applied to successive horizontal slices of the atmosphere. We already noted this result in Chapter 3, in discussing minimum-time light rays. In particular, if a ray enters the top of the atmosphere at angle θ_0 relative to the vertical, then it will strike the ground at the steeper angle $\sin \theta_{\text{ground}} = \sin \theta_0 / n_{\text{ground}}$. ∎

15.7 Schrödinger's Wave Mechanics

In 1924 the particle–wave ideas of de Broglie became a hot topic of discussion all over Europe, including physicists in both of the physics departments in Zürich. One of these departments was at the University of Zürich, whose chair was Peter Debye, and the other was at the E.T.H., the same department in which Einstein had earned his doctorate many years before.[2] The two departments held a weekly joint colloquium. Felix Bloch was a graduate student at the University of Zürich, and Erwin Schrödinger was a faculty member at the E.T.H. As Bloch wrote many years later:[3]

> Once at the end of a colloquium I heard Debye saying something like: "Schrödinger, you are not working right now on very important problems anyway. Why don't you tell us some time about that thesis of de Broglie, which seems to have attracted some attention?" So in one of the next colloquia, Schrödinger gave a beautifully clear account of how de Broglie associated a wave with a particle and how he could obtain the quantization rules of Niels Bohr and Sommerfeld by demanding that an integer number of waves should be fitted along a stationary orbit. When he had finished, Debye casually remarked that this way of talking was rather childish. As a student of Sommerfeld he had learned that, to deal properly with waves, one had to have a wave equation. It sounded quite trivial and did not seem to make a great impression, but Schrödinger evidently thought a bit more about the idea afterwards. Just a few weeks later he gave another talk in the colloquium which he started by saying: "My colleague Debye suggested that one should have a wave equation; well I have found one!" And then he told us essentially what he was about to publish under the title "Quantization as Eigenvalue Problem" as the first paper of a series in the *Annalen der Physik*. I was still too green to really appreciate the significance of this talk, but from the general reaction of the audience I realized that something rather important had happened, and I need not tell you what the name of Schrödinger has meant from then on. Many years later, I reminded Debye of his remark about the wave equation; interestingly enough he claimed that he had forgotten about it and I am not quite sure whether this was not the subconscious suppression of his regret that he had not done it himself. In any event, he turned to me with a broad smile and said: "Well, wasn't I right?"

So how can one go about finding a wave equation for de Broglie waves?

The first step is to show that de Broglie wavelengths for typical classical particles are exceedingly small. Take a 1-g particle moving at 1 cm/s, for example. Its momentum is $p = mv = 10^{-5}$ kg m/s, so its de Broglie wavelength is $\lambda = h/p = 6.6 \times 10^{-34}$ J · s/ 10^{-5} kg m/s $= 6.6 \times 10^{-29}$ m, much, much smaller than the size of an atom. This immediately reminds us of the eikonal approximation in optics. For small wavelengths of light, small compared with any physical barriers in a situation or of distances over which the index of refraction changes appreciably, we can replace the wave nature of light with the trajectory of light *rays*, which travel on definite paths. Such rays remind us of the paths of classical particles, yet they arise from a wave theory. Is it possible that *classical particles* are to de Broglie waves as *light rays* are to light waves? In other words, is it possible that classical mechanics is the eikonal approximation to a wave theory of matter?

There is an encouraging sign that this idea might work. As shown in the preceding section, the eikonal equation of geometrical optics is $(\nabla L)^2 = n^2$, relating the phase L to the index of refraction. There is the strikingly similar version

---

[2] E.T.H. stands for *Eidgenössische Technische Hochschule*, rendered in English as the Swiss Federal Institute of Technology or simply ETH Zürich.
[3] Reproduced from *Physics Today* 29(12), 23 (1976), with the permission of the American Institute of Physics.

of the Hamilton–Jacobi equation, $(\nabla W)^2 = 2m(E - U)$, for a particle of mass m, total energy E, and potential energy U, related to Hamilton's characteristic function W for the case where the Hamiltonian is not an explicit function of time. Here W is analogous to L and $2m(E - U)$ is analogous to n.

To derive ray optics from wave optics, we began with the wave equation

$$\nabla^2 \varphi - \frac{n^2}{c^2} \frac{\partial^2 \varphi}{\partial t^2}, \tag{15.118}$$

then wrote a general solution of the form

$$\varphi = e^{A(\mathbf{r}) + ik_0(L(\mathbf{r}) - ct)}. \tag{15.119}$$

Finally, after making the eikonal approximation, we found a differential equation for $L(\mathbf{r})$:

$$(\nabla L)^2 = n^2, \tag{15.120}$$

called the eikonal equation.

So now for *wave mechanics*, let us try working *backwards*. We already have an analogue for the eikonal equation, the time-independent Hamilton–Jacobi equation

$$(\nabla W)^2 = 2m(E - U) \tag{15.121}$$

in terms of Hamilton's characteristic function W.

Now backing up a step, the solution ψ of some hoped-for wave equation for particles would have the form

$$\psi = e^{A(\mathbf{r}) + i\alpha(W(\mathbf{r}) - \beta t)}, \tag{15.122}$$

where $A(\mathbf{r})$ is related to the amplitude of the wave and α and β are constants. Requiring dimensional consistency shows that β must have the dimensions of energy, since W has the dimensions of energy $\times$ time, and α must have dimensions of $1/(\text{energy} \times \text{time})$, since exponents must be dimensionless overall. As a trial, we will therefore try setting $\beta = E$, the energy of the particle, and $\alpha = 1/\hbar$, where at this point $\hbar$ is some unknown constant with dimensions of action, i.e., energy $\times$ time. So we rewrite the wave function as

$$\psi = e^{A(\mathbf{r}) + \frac{i}{\hbar}(W(\mathbf{r}) - Et)}. \tag{15.123}$$

Now we take the final step backward, by trying to find a wave equation which has the solution we have just written. First, take derivatives of the trial solution with respect to x:

$$\frac{\partial \psi}{\partial x} = \left(\frac{\partial A}{\partial x} + \frac{i}{\hbar} \frac{\partial W}{\partial x} \right) \psi, \tag{15.124}$$

$$\frac{\partial^2 \psi}{\partial x^2} = \left(\frac{\partial^2 A}{\partial x^2} + \frac{i}{\hbar} \frac{\partial^2 W}{\partial x^2} \right) \psi + \left(\frac{\partial A}{\partial x} + \frac{i}{\hbar} \frac{\partial W}{\partial x} \right) \frac{\partial \psi}{\partial x} \tag{15.125}$$

$$= \left[\frac{\partial^2 A}{\partial x^2} + \frac{i}{\hbar} \frac{\partial^2 W}{\partial x^2} + \left(\frac{\partial A}{\partial x} + \frac{i}{\hbar} \frac{\partial W}{\partial x} \right)^2 \right] \psi.$$

Now if we use the eikonal approximation, where (as with light waves) the most important term is $(\partial W/\partial x)^2 \psi$ in the case of very short wavelengths, we have

$$\frac{\partial^2 \psi}{\partial x^2} = -\frac{1}{\hbar^2} \left(\frac{\partial W}{\partial x} \right)^2 \psi. \tag{15.126}$$

Adding in similar terms for derivatives in the y and z directions:

$$\frac{\partial^2 \psi}{\partial x^2} + \frac{\partial^2 \psi}{\partial y^2} + \frac{\partial^2 \psi}{\partial z^2} \equiv \nabla^2 \psi \approx -\frac{1}{\hbar}^2 (\nabla W)^2 \psi = -\frac{2m}{\hbar^2} (E - U(\mathbf{r})) \psi. \tag{15.127}$$

The derivative of our trial solution with respect to *time* is

$$\frac{\partial \psi}{\partial t} = -\frac{i}{\hbar} E \psi. \tag{15.128}$$

So if we now combine the spatial derivatives and time derivative of ψ to obtain a nonrelativistic wave equation for particles, note there is only one way to do it. Multiplying the result for $\nabla^2 \psi$ by $-\hbar^2/2m$, we have

$$-\frac{\hbar^2}{2m} \nabla^2 \psi = (E - U(\mathbf{r})) \psi \tag{15.129}$$

and multiplying the time derivative by $i\hbar$, we have

$$i\hbar \frac{\partial \psi}{\partial t} = E \psi. \tag{15.130}$$

Combining these results, we get

$$-\frac{\hbar^2}{2m} \nabla^2 \psi + U(\mathbf{r}) \psi = i\hbar \frac{\partial \psi}{\partial t}, \tag{15.131}$$

which is Schrödinger's equation, first presented at that colloquium in Zürich.

Note some interesting features of the equation. First, unlike the wave equations we have met before, it contains only a single derivative with respect to time, but second derivatives with respect to space. We can trace this fact back to the relation $E = p^2/2m$ for a nonrelativistic particle; E came from the time derivative and p came from spatial derivatives.

Second, the equation contains the imaginary number "i" *explicitly*. We have often used imaginary numbers for mathematical convenience in *solutions* of the wave equation or other fundamental equations of physics, but this is the first time an imaginary number has appeared explicitly in a fundamental equation itself.

Third, what is the meaning of the wave function ψ? What is it that is waving? Clearly ψ is an *amplitude*, but an amplitude of what exactly? This question was to worry Schrödinger for the rest of his life, and Einstein and others as well, and even today is a matter of some controversy. We have developed a standard version that works superbly well in practice, but there remain physicists who question it.

Let us at least test the equation in a simple case, that of a free particle (with $U(\mathbf{r}) = 0$) moving in one dimension. Then the equation becomes

$$-\frac{\hbar^2}{2m}\frac{\partial^2 \psi}{\partial x^2} = i\hbar\frac{\partial \psi}{\partial t}. \tag{15.132}$$

We try the plane wave $\psi(x,t) = \psi_0 e^{i(kx-\omega t)}$, which solves the equation if $\hbar^2 k^2/2m = \hbar\omega$. For a wave, $\omega = 2\pi\nu$, so the right-hand side is $2\pi\hbar\nu$, which is the energy $E = h\nu$ of the particle according to de Broglie, *if we set the constant* $\hbar \equiv h/2\pi$. Then the left-hand side is

$$\frac{\hbar^2 k^2}{2m} = \frac{\hbar^2(2\pi/\lambda)^2}{2m} = \frac{(h/\lambda)^2}{2m} = \frac{p^2}{2m} \tag{15.133}$$

using the other de Broglie relation $p = h/\lambda$. This is of course the kinetic energy of a classical free particle, the only energy it has. So the solution is consistent with both the de Broglie relations and classical particle mechanics.

Now let us begin with quantum mechanics according to Schrödinger, and see how classical mechanics emerges from it as a special case. We will substitute the very general wave form

$$\psi = e^{A(\mathbf{r})+\frac{i}{\hbar}(W(\mathbf{r})-Et)} \tag{15.134}$$

into the Schrödinger equation, and see what happens. The derivative of ψ with respect to time is

$$\frac{\partial \psi}{\partial t} = -\frac{i}{\hbar}E\psi. \tag{15.135}$$

The gradient of ψ is

$$\nabla\psi = \left(\nabla A + \frac{i}{\hbar}\nabla W\right)\psi \tag{15.136}$$

so then the Laplacian is

$$\nabla^2\psi = \nabla \cdot \nabla\psi \tag{15.137}$$

$$= \left[\nabla^2 A + \frac{i}{\hbar}\nabla^2 W + \left(\nabla A + \frac{i}{\hbar}\nabla W\right)^2\right]\psi$$

$$= \left(\nabla^2 A + \frac{i}{\hbar}\nabla^2 W + (\nabla A)^2 - \frac{1}{\hbar^2}(\nabla W)^2 + \frac{2i}{\hbar}\nabla A \cdot \nabla W\right)\psi.$$

Substituting into Schrödinger's equation, we find

$$(\nabla W)^2 + 2m(U - E) = i\hbar(\nabla^2 W + 2\nabla A \cdot \nabla W) + \hbar^2(\nabla^2 A + (\nabla A)^2). \tag{15.138}$$

Now note that if we let $\hbar \to 0$, the entire right-hand side goes away, leaving us with the Hamilton–Jacobi equation! That is, if $\hbar$ were equal to zero, Schrödinger's equation would reduce to classical mechanics. However, even though Planck's constant is very small by everyday standards, it is *not* zero, so Schrödinger's equation is describing a more general theory than classical mechanics.

There is an alternative way to set the right-hand side equal to zero. With the same sort of arguments used earlier in this section, we can show that the right-hand side becomes negligible in the short-wavelength limit. That is, beginning with quantum mechanics as expressed by the Schrödinger equation, *classical mechanics is the short-wavelength limit of quantum mechanics*. Said another way, *classical mechanics is the eikonal approximation to quantum mechanics*.

15.8 Quantum Operators and the Bracket

There is yet another way to approach quantum mechanics, beginning with a formalism of classical mechanics introduced back in Chapter 11. The Poisson bracket was a convenient tool we introduced then to make Hamilton's equations of motion and canonical transformations more concise and transparent. We defined the bracket as

$$\{A, B\}_{q,p} \equiv \frac{\partial A}{\partial q_k}\frac{\partial B}{\partial p_k} - \frac{\partial A}{\partial p_k}\frac{\partial B}{\partial q_k}, \qquad (15.139)$$

i.e., a certain combination of derivatives with respect to the canonical coordinates and momenta. Hamilton's equations then took the form

$$\dot{q}_k = \{q_k, H\}_{q,p}, \quad \dot{p}_k = \{p_k, H\}_{q,p}. \qquad (15.140)$$

Conservation laws and canonical transformations could also be written in terms of this bracket, as shown in Eqs. (11.136) and (11.140). We also learned there that the Poisson bracket satisfies the following four identities.

1. Anticommutativity: $\{A, B\}_{q,p} = -\{B, A\}_{q,p}$
2. Distributivity: $\{A, b\,B + c\,C\}_{q,p} = b\,\{A, B\}_{q,p} + c\,\{A, C\}_{q,p}$
3. A modified notion of *associativity*: $\{AB, C\}_{q,p} = \{A, C\}_{q,p}\,B + A\,\{B, C\}_{q,p}$
4. The Jacobi identity: $\{A, \{B, C\}_{q,p}\}_{q,p} + \{B, \{C, A\}_{q,p}\}_{q,p} + \{C, \{A, B\}_{q,p}\}_{q,p} = 0$

Let us now think of the Poisson bracket as some abstract bilinear mathematical operation involving two inputs – with the associated four properties of the operator listed above as *defining* properties. That is, we are not to think of the bracket as necessarily taking functions of canonical coordinates and momenta as inputs and then computing some derivatives of these functions as in Eq. (15.139). Instead, we allow for any other realization of the operator – as long as its four properties above hold. We will then represent this more abstract notion of the Poisson bracket by *square* brackets $[\cdot, \cdot]$ instead of the curly brackets with subscripts $\{\cdot, \cdot\}_{p,q}$. As an example, we can propose another representation of the bracket as a bi-linear operator that takes as inputs *matrices*. That is:

$$[\hat{\mathbf{A}}, \hat{\mathbf{B}}] = \hat{\mathbf{A}} \cdot \hat{\mathbf{B}} - \hat{\mathbf{B}} \cdot \hat{\mathbf{A}}, \tag{15.141}$$

where $\hat{\mathbf{A}}$ and $\hat{\mathbf{B}}$ are matrices and the dot implies matrix multiplication. The bracket operation is now the so-called *commutator* of the matrices. One can show that this realization of the bracket operation satisfies all four defining properties (see the Problems section at the end of this chapter), so we say it is a valid representation of the bracket. And we can indeed find other valid representations as well – all satisfying the four defining properties.

The reason this is interesting is the following: the abstract structure of the bracket, a bi-linear operator with the four properties listed above and nothing more, captures the essence of dynamics when used to describe time evolution as in Eq. (15.140). As we saw, a great deal of mechanics, from equations of motion to canonical transformations of phase space, can be neatly written using these brackets. For example, inspired by Eq. (15.140), suppose we posit a crazy new theory of the time evolution of some matrices based on the representation of the bracket given by Eq. (15.141). That is:

$$\frac{d\hat{\mathbf{A}}}{dt} \sim \left[\hat{\mathbf{A}}, \hat{\mathbf{H}}\right]. \tag{15.142}$$

Then this new theory will inherit a lot of the hallmark features of mechanics: evolution would be dictated by a Hamiltonian (some matrix $\hat{\mathbf{H}}$ in this case) that acts as the generator of time translations, constants of motion would correlate with symmetries of this Hamiltonian, a familiar phase-space picture might be developed, and many attributes of dynamics that we often take for granted would be encoded in our new theory of matrices for free. This might not sound terribly useful or too hypothetical, but it is indeed how the process of **canonical quantization** led some physicists from classical mechanics to quantum mechanics.

The canonical quantization prescription dictates that classical coordinates and momenta should be elevated to the stature of abstract *operators*: mathematical entities that can "act" on a vector space that represents the different states of the physical system. We can think of these operators as matrices, but they don't have to be. For a single particle, the classical variables $q(t)$ and $p(t)$ are then thought of as the $\hat{q}(t)$ and $\hat{p}(t)$ operators[4]

$$q(t) \to \hat{q}(t), \quad p(t) \to \hat{p}(t), \tag{15.143}$$

where we use hats on the variables to emphasize that they are now general operators. The vector space these operators act upon is proposed to be the space of all states of the single particle.[5] When an operator acts on such a vector state it

---

[4] This is known as the Heisenberg picture, where operators are time dependent instead of the vectors they act on.

[5] If it helps, one can think of these operators as $n \times n$ matrices, and the vectors they act upon as n-component column matrices representing the various states of the particle. Note however that, for a matrix representation of position and momentum in this case, n must be strictly infinite.

gives another vector state. Then we assume the following central prescription that helps convert classical equations to quantum equations:

$$\{\cdot,\cdot\}_{p,q} \to \frac{1}{i\hbar}[\cdot,\cdot], \tag{15.144}$$

where $i = \sqrt{-1}$ is there by convention, and $\hbar$ is Planck's constant we encountered in the first capstone chapter (Chapter 5). In particular, this implies that

$$\{q,p\}_{p,q} = 1 \Rightarrow [\hat{q},\hat{p}] = i\hbar. \tag{15.145}$$

We then take the bracket of two operators to be their commutator, as given by Eq. (15.141) for matrices. For example, we have $[\hat{q},\hat{p}] = \hat{q}\hat{p} - \hat{p}\hat{q}$. As mentioned earlier, this representation of the bracket can be shown to be a valid one. The time evolution of the particle, classically given by Eq. (15.140), will then be described by the evolution of the operators

$$i\hbar\frac{d\hat{q}}{dt} = [\hat{q},\hat{H}], \quad i\hbar\frac{d\hat{p}}{dt} = [\hat{p},\hat{H}], \tag{15.146}$$

where the Hamiltonian operator $\hat{H}$ is to be constructed from the classical Hamiltonian of the particle by replacing qs and ps with their corresponding operator incarnations.[6] The next step is to connect how the time evolution of these operators translates to evolution of the particle state – in particular, the evolution of its position and momentum. This in turn requires us to define the physical meaning of "acting" with an operator upon a state vector of the particle. Once these potholes are filled in, the road to quantizing a classical system is paved through the procedure just outlined.

In this story, the canonical quantization prescription motivates and provides a roadmap for transitioning from the classical to the quantum. However, it is also important to appreciate that classical mechanics does *not* imply quantum mechanics: there are a lot of additional new ideas not inherent to classical mechanics – from the introduction of abstract operators for position and momentum, to the use of a vector space describing states of a physical system. These new ideas need to be added to make quantum mechanics emerge from classical mechanics. Yet, it is very instructive to see how the seeds of quantization are already sown in the mathematical landscape of classical mechanics; and they are made more transparent with the language of Poisson brackets and Hamiltonian evolution.

It is worthwhile noting that the quantum commutator that finds its origin and inspiration in the classical Poisson bracket is related to one of the most celebrated taglines of quantum mechanics: the **Heisenberg uncertainty principle**. In its most common incarnation, this principle asserts that the uncertainties in knowing the position and momentum of a particle are correlated. In the quantum world of competing realities, the results of measurements of position and momentum randomly fluctuate and one can at most know the likelihood of one measurement

---

[6] Note that there can be ambiguities in this prescription given that $\hat{q}$ and $\hat{p}$ do not commute – but this is a matter beyond this textbook.

outcome over another. Hence, a measurement of the position q of a particle is associated with an uncertainty denoted by Δq, which is essentially the standard deviation of multiple attempts at measuring the position of the particle with the same setup. Similarly, one has a quantum uncertainty for the particle's momentum p, denoted as Δp. The Heisenberg uncertainty principle notes that we must then have

$$\Delta q \, \Delta p \geq \frac{\hbar}{2}, \qquad (15.147)$$

correlating the two uncertainties. This statement can be *derived* from the commutator $[\hat{q}, \hat{p}] = i\hbar$ in quantum mechanics, which we learned in turn is inspired from the classical Poisson bracket $\{q, p\} = 1$. If we were to imagine *importing* (15.147) into classical phase space, we might think of it as saying that area in phase space is "quantized" in units of $\hbar/2$. That is, one cannot locate a point in phase space that is smaller than a cell of area $\hbar/2$. This fits nicely with the classical Liouville theorem in which phase-space area is preserved under time evolution. The Liouville theorem hints at a notion of "incompressibility" of area in phase space, which agrees well with Heisenberg's principle of minimal phase-space area. Once again, we see that there are well-hidden imprints of the quantum in Hamiltonian classical mechanics.

Example 15.6   **The Quantum Harmonic Oscillator**

As an example of the prescription of canonical quantization, let us develop the case of the simple harmonic oscillator. We start with a classical Hamiltonian with a single degree of freedom

$$H = \frac{p^2}{2m} + \frac{1}{2}k\,q^2 \qquad (15.148)$$

for a particle of mass m attached to a spring of spring constant k. To "quantize" this system, we elevate q and p to $\hat{q}$ and $\hat{p}$ satisfying

$$[\hat{q}, \hat{p}] = i\,\hbar, \quad [\hat{q}, \hat{q}] = 0, \quad [\hat{p}, \hat{p}] = 0, \qquad (15.149)$$

using the $\{\}_{p,q} \to [\,]/i\hbar$ prescription. And we write the new Hamiltonian operator as

$$\hat{H} = \frac{\hat{p}^2}{2m} + \frac{1}{2}k\,\hat{q}^2. \qquad (15.150)$$

We can then describe the evolution of these operators by

$$i\,\hbar\frac{d\hat{q}}{dt} = [\hat{q}, \hat{H}] = i\,\hbar\,\frac{\hat{p}}{m} \qquad (15.151)$$

for $\hat{q}(t)$, where in the last equality we used the first three properties of the Poisson bracket: anticommutation, distributivity, and modified associativity. For example, we have

$$\left[\hat{q}, \frac{\hat{p}^2}{2m}\right] = \frac{1}{2m}[\hat{q}, \hat{p}\hat{p}] = -\frac{1}{2m}[\hat{p}\hat{p}, \hat{q}] = -\frac{1}{2m}[\hat{p}, \hat{q}]\hat{p} - \frac{1}{2m}\hat{p}[\hat{p}, \hat{q}] = i\,\hbar\,\frac{\hat{p}}{m}. \qquad (15.152)$$

Notice that we did not need any particular representation of the bracket operator; its abstract properties and (15.149) was all that we used. Similarly, we get

$$i\hbar\frac{d\hat{p}}{dt} = [\hat{p},\hat{H}] = -i\hbar k\hat{q}.$$ (15.153)

At the operator level, this looks like the familiar restoring effect of a spring force. To describe the actual evolution of the quantum particle under the influence of the spring force we would next need to define the state-vector space of the particle on which these operators act. This vector space is to describe the various possible states of the particle as vectors – and hence the particle's position and momentum are to be extracted from such a state vector. We would also need to define the meaning of an operator acting on a state – which is related to the postulate of measurement in quantum mechanics. The Poisson bracket machinery gets us a long way into the quantum world of this particle, but we still have a long way to go to complete the quantum picture. And that will require a different textbook.   ∎

15.9 A Hint at Quantum Time Evolution

There is a way to understand classical time evolution of physical systems in a manner that ties nicely into their quantum counterpart. As we learned earlier, Hamilton's equations of motion take the form

$$\dot{q} = \{q,H\}, \quad \dot{p} = \{p,H\},$$ (15.154)

written in terms of the Poisson bracket. For simplicity, we focus on the case of a two-dimensional phase space but the discussion can easily be extended to general $2N$-dimensional phase space. These equations can be solved formally and in general with a series expansion in time. We start by writing

$$q(t) = q(t_0 + \Delta t) = q(t_0) + \dot{q}(t_0)\Delta t + \frac{1}{2!}\ddot{q}(t_0)\Delta t^2 + \cdots,$$ (15.155)

Taylor expanding $q(t)$ about an initial time t_0 (see Appendix F). Assuming convergence of this series for given Δt, we then need to compute the time derivatives of $q(t)$ evaluated at $t = t_0$. From (15.154), we can write

$$\dot{q}(t_0) = \{q,H\}_0,$$ (15.156)

where the result of the bracket is to be evaluated at $t = t_0$. But we then also have

$$\ddot{q} = \frac{d}{dt}\{q,H\} = \{\dot{q},H\} + \{q,\dot{H}\}.$$ (15.157)

We also know that

$$\frac{dH}{dt} = \frac{\partial H}{\partial t}.$$ (15.158)

For simplicity, let us assume that the Hamiltonian does not depend on time explicitly, so that $\partial H/\partial t = 0$. We then have

$$\ddot{q} = \{\dot{q}, H\} = \{\{q, H\}, H\}, \tag{15.159}$$

where we used (15.154) to eliminate $\dot{q}$. This implies that we have

$$\ddot{q}(t_0) = \{\{q, H\}, H\}_0, \tag{15.160}$$

with the subscript once again indicating that the expression is to be evaluated at $t = t_0$ after the Poisson brackets are computed. Similarly, one can derive

$$\frac{d^3 q(t_0)}{dt} = \{\{\{q, H\}, H\}, H\}_0, \tag{15.161}$$

and more generally

$$\frac{d^n q(t_0)}{dt} = \{\{\{q, H\}, H\}, \ldots, H\}_0, \tag{15.162}$$

where we would have n nested Poisson brackets and n occurrences of H. We now see that one can find $q(t)$ using Eq. (15.155) by computing the coefficients of the Taylor expansion in terms of nested Poisson brackets with the Hamiltonian – all evaluated at $t = t_0$ (see Appendix F). We then write

$$q(t_0 + \Delta t) = q(t_0) + \{q, H\}_0 \Delta t + \frac{1}{2!}\{\{q, H\}, H\}_0 \Delta t^2 + \cdots \tag{15.163}$$

A similar treatment can be undertaken for computing the canonical momentum $p(t_0 + \Delta t)$. In principle, Eq. (15.163) is a complete solution in the form of an infinite series, but in practice it is not a very useful expression for solving Hamiltonian time evolution. However, the expression allows us to define how a system time-evolves in quantum mechanics. As we transition from the classical to the quantum, we learn that we need to replace Poisson brackets with commutators, as in $\{.,.\} \to [.,.]/i\hbar$. This translates to

$$\hat{q}(t_0 + \Delta t) = \hat{q}(t_0) + (-i)\frac{\Delta t}{\hbar}[\hat{q}, \hat{H}] + (-i)^2 \frac{1}{2!}\frac{\Delta t^2}{\hbar^2}[[\hat{q}, \hat{H}], \hat{H}] + \cdots \tag{15.164}$$

Formally, this is written as

$$\hat{q}(t_0 + \Delta t) = e^{\frac{i}{\hbar}\hat{H}\Delta t}\hat{q}(t_0)e^{-\frac{i}{\hbar}\hat{H}\Delta t}, \tag{15.165}$$

where one defines the exponential operator as

$$e^{-\frac{i}{\hbar}\hat{H}\Delta t} = \sum_{n=0}^{\infty} \frac{1}{n!}\left(\frac{-i\hat{H}}{\hbar}\right)^n \tag{15.166}$$

through a Taylor-like expansion. One can then show (see the Problems section at the end of this chapter) that Eqs. (15.165) and (15.166) indeed lead to (15.164). Hence, we see a direct connection between evolution in classical phase space, and time evolution of operators in quantum mechanics. Once again, the Lagrangian and Hamiltonian formulations of classical mechanics provide the foundations and proper guidance to lead us to quantum mechanics.

15.10 Summary

In this concluding chapter we began by developing Hamilton–Jacobi theory, in which motion of systems is found not by solving second-order or first-order ordinary differential equations, as in Lagrangian or Hamiltonian mechanics, but by solving a partial differential equation. We then explored the method of adiabatic invariants, which is very useful in practice, as there is often a hierarchical separation of time scales in physical situations, with a fast motion superimposed on a slow one. This was followed by a brief exploration of some of the early semiclassical ideas and schemes to address quantum phenomena, as in the so-called "old quantum theory," ideas developed and promoted by such physicists as Bohr, de Broglie, Wentzel, Kramers, Brilliouin, Jeffreys, and Sommerfeld, which helped prepare the way for the full quantum mechanics. We then showed at some length how de Broglie's proposal that particles have a wave nature led Schrödinger, with his thorough understanding of optics and Hamilton–Jacobi theory, to develop his famous wave equation. Finally, we described the canonical quantization prescription for developing the quantum version of a classical system – where we saw that the Hamiltonian formulation and Poisson brackets play a central role. This quantization procedure was due initially to Heisenberg and then developed further by him and many others as well. Historically, Heisenberg's so-called matrix mechanics came first, soon followed by Schrödinger's wave mechanics – it was Schrödinger who subsequently showed that the two formulations were different ways of understanding the same theory. Feynman's path-integration approach, as already discussed in Chapter 5, was a third and unique way of viewing quantum theory; it came a quarter-century later. Each of the three fundamental approaches to quantum mechanics has its own advantages and disadvantages for problem solving and for insight into the theory, just as the various formulations of classical mechanics, due to Newton, Lagrange, Hamilton, Jacobi, and others, have each provided their own insights and techniques. In this book we have tried to demonstrate that all three fundamental approaches to quantum mechanics have emerged in large part from different formulations of classical mechanics, and that in turn our understanding of the limits and meaning of classical mechanics have benefited from the quantum theory to which they all helped give rise.

Problems

★ **Problem 15.1** A ball of mass m is dropped from rest above the surface of an airless moon in essentially uniform gravity g. (a) If y is the vertical axis, show that the Hamilton–Jacobi equation for the ball is

$$\frac{1}{2m}\left(\frac{\partial S}{\partial y}\right)^2 + mgy + \frac{\partial S}{\partial t} = 0,$$

where S is Hamilton's principal function. (b) Then show that

$$S = \pm\frac{2\sqrt{2}}{3g\sqrt{m}}(C - mgy)^{3/2} - Ct,$$

where C is a constant. (c) Then show, using a judicious choice of constants, that the equation of motion of the ball can be written as

$$y = y_0 - \frac{1}{2}g(t - t_0)^2.$$

★★ **Problem 15.2** Starting from rest at time $t = 0$ and at altitude h_0, a block of mass M slides down a frictionless plane inclined at angle α to the horizontal. There is a uniform gravitational field g directed vertically downward. (a) Write the Hamilton–Jacobi equation for the block. (b) Solve the equation to find Hamilton's principal function S. (c) Find the equation of motion $s(t)$ for the block, where s is the distance along the incline, measured from the top of the incline.

★ **Problem 15.3** A spaceship drifts in gravity-free space. If its velocity is v_0 in the positive x direction at time $t = 0$, find its subsequent motion using the Hamilton–Jacobi method.

★★ **Problem 15.4** A projectile is fired in a uniform gravitational field g with initial speed v_0 and angle θ_0 relative to the horizontal. Note that the Hamiltonian is

$$H = \frac{p_x^2}{2m} + \frac{p_y^2}{2m} + mgy.$$

Find the projectile's motion $x(t)$ and $y(t)$ using the Hamilton–Jacobi method.

★★ **Problem 15.5** A block m can slide along a frictionless tabletop in the x, y plane, subject to the forces exerted by one spring that lies along the x axis and has force constant k_1, and another spring that lies along the y axis and has force constant k_2. Assume the motions of m are so small that the springs remain essentially perpendicular to one another. Write the Hamiltonian, the Hamilton–Jacobi equation, and solve the equation to find the motions $x(t)$ and $y(t)$ of the block.

★★ **Problem 15.6** A thin, stiff metal ring of radius R is placed in a vertical plane, and made to spin with constant angular velocity ω about a vertical axis that passes through the center of the ring. A bead of mass m is free to slide around the ring, with its position defined by its angle θ up from the bottom of the ring. (a) Find the Hamiltonian of the bead and write out the corresponding Hamilton–Jacobi equation. (b) Show that Hamilton's principal function can be separated into $S = S_\theta(\theta) + S_t(t)$, and find $S_t(t)$ explicitly and $S_\theta(\theta)$ as an integral over a function of θ.

★★★   **Problem 15.7** The Hamiltonian for a particle of mass m, with arbitrary initial position and velocity, and subject to an inverse-square attractive force, can be written

$$H = \frac{p_r^2}{2m} + \frac{p_\theta^2}{2mr^2} - \frac{k}{r},$$

where k is a positive constant. (a) Write the Hamilton–Jacobi equation for the particle. (b) By separating variables, show that Hamilton's principal function can be written in the form

$$S = S_r + S_\theta + S_t = S_r + C_1 t + C_2 \theta,$$

where C_1 and C_2 are constants, and S_r depends only upon r. (c) Write an expression for S_r in the form of an integral over r. (d) The new coordinate $Q_r = \partial S/\partial C_2 = \partial(S_r + S_\theta)/\partial C_2 \alpha$, a constant, since the new coordinates in Hamilton–Jacobi theory are necessarily constants. Take this partial derivative (right through the integral sign!) to show that (with an appropriate choice of signs)

$$\theta - \alpha = \int \frac{C_2 dr}{r^2 \sqrt{-2mC_1 + 2mk/r - C_2^2/r^2}}.$$

(e) Evaluate the integral with the help of the substitution $u = 1/r$; then find an expression for $r(\theta)$. This gives the possible orbital shapes: circles, ellipses, parabolas, and hyperbolas.

★   **Problem 15.8** Using the action angle variables approach, find the oscillation frequency of a one-dimensional simple harmonic oscillator of mass m and force constant k.

★★★   **Problem 15.9** Using the action angle variables approach, find the frequencies of oscillation of a planet orbiting the sun, for both the radial and angular motions. What is the consequence of the fact that these frequencies turn out to be the same? You might need to use contour integration to evaluate an integral.

★   **Problem 15.10** A particle of mass m moves in one dimension subject to a force F of constant magnitude, but directed toward the left for positive x and to the right for negative x. Thus the potential energy has the form $U = k|x|$ for some constant k. Using action angle variables, find the frequency of oscillation as a function of the particle's energy.

★★   **Problem 15.11** As an example of adiabatic invariance, consider the preceding problem in the case where the magnitude of the force F is slowly changed, i.e., slow relative to the oscillation period of the particle. Which (if any) of the following quantities remain constant in the adiabatic limit: (a) the oscillation amplitude? (b) the frequency of oscillation? (c) the energy?

★ **Problem 15.12** A particle of mass m can move along the positive x axis only, subject to a constant force to the left. That is, there is an impenetrable wall at $x = 0$ preventing it from reaching negative x, and for positive x there is a constant force attracting the particle back towards the origin. Find the possible energy levels of the particle according to the "old quantum theory."

★ **Problem 15.13** A particle of mass m can move along the positive x axis only, subject to a Hooke's-law spring force $F = -kx$. That is, there is an impenetrable wall at $x = 0$ preventing it from reaching negative x, and for positive x there is a spring force attracting the particle back towards the origin. Find the allowed energies of the particle according to the "old quantum theory." Compare these energy levels with those for a particle moving *anywhere* on the x axis subject to a spring force $F = -kx$ attracting it to $x = 0$.

★ **Problem 15.14** A particle of mass m is confined to move inside a cubic box of side length L, with potential energy zero. Find the allowed energies of the particle according to the "old quantum theory."

★★ **Problem 15.15** One end of a spring of rest length zero and force constant k is attached to a fixed point while the other end is attached to a ball of mass m, which is otherwise free to move as it likes in three-dimensional space. Find the allowed energies of the system according to the "old quantum theory."

★★ **Problem 15.16** From the classical point of view, the electron in a hydrogen atom moves under the influence of a central attractive force $F = -e^2/r^2$ caused by the proton nucleus. According to "old quantum theory," the phase integrals over r and θ are given by

$$\oint p_r \, dr = n_1 h \quad \text{and} \quad \oint p_\theta \, d\theta = n_2 h,$$

where p_r and p_θ are the classical canonical momenta, h is Planck's constant, and n_1 and n_2 are positive integers. Show that according to old quantum theory there are only a discrete set of possible energy levels, given by

$$E_n = -\frac{me^4}{2n^2\hbar^2},$$

where $\hbar \equiv h/2\pi$ and $n = n_1 + n_2$.

★★ **Problem 15.17** If a quantum-mechanical particle has definite energy E we can write its wave function in the form $\Psi(\mathbf{r}, t) = \psi(\mathbf{r})e^{-iEt/\hbar}$. (a) Substitute this into the full Schrödinger equation to show that the *time-independent Schrödinger equation* for $\psi(\mathbf{r})$ may be written

$$\nabla^2\psi + \frac{2m}{\hbar^2}(E - U)\psi = 0,$$

where both ψ and U are functions of position. (b) A particle of mass m is trapped inside a one-dimensional box of width L. The potential energy of the particle is zero for $0 < x < L$ and infinite otherwise. The wave function of the particle is zero outside the box and at $x = 0, L$. According to the one-dimensional Schrödinger equation, the lowest-energy "eigenfunction" ψ_1 for the particle is $\psi_1 = A \sin \pi x/L$, where A is a constant. Find the corresponding energy eigenvalue E_1. (c) Find all other energy eigenfunctions ψ_n of the particle, and the corresponding energy eigenvalues E_n.

★★ **Problem 15.18** A particle of mass m is trapped inside a three-dimensional box of side length L. The potential energy of the particle is zero for $0 < x < L, 0 < y < L, 0 < z < L$ and infinite otherwise. Possible energy "eigenfunctions" $\psi_n(x, y, z)$ of the particle are zero outside the box and at every face. The lowest-energy eigenfunction ψ_1 for the particle is $\psi_1 = A \sin(\pi x/L) \sin(\pi y/L) \sin(\pi z/L)$, where A is a constant. (a) Find the corresponding energy eigenvalue E_1. (b) Find all other energy eigenfunctions ψ_n and corresponding energy eigenvalues E_n in terms of the "quantum number" n. (c) Compare these allowed energy values with those predicted by the "old quantum theory," showing that they agree in the limit of large n.

★★★ **Problem 15.19** The lowest-energy (*i.e.*, "ground state") eigenfunction of the electron in a hydrogen atom is spherically symmetric. (a) Using the time-independent Schrödinger equation, find this eigenfunction in terms of the radius r of the electron from the nucleus as origin. (b) Find also the electron's corresponding energy eigenvalue.

★★ **Problem 15.20** The ground-state wave function of a one-dimensional simple harmonic oscillator of mass m and force constant k is the Gaussian function $\psi(x) = Ae^{-\alpha x^2}$, where A is a normalization constant (adjusted to make $\int \psi^* \psi dx = 1$) and α is also a constant. (a) Using Schrödinger's equation, find α in terms of m, k, and $\hbar \equiv h/2\pi$. (b) Find the energy eigenvalue for this wave function, in terms of the same constants. (c) Compare with the energy predicted for this ground state by the "old quantum theory."

★★ **Problem 15.21** The angular momentum vector of a particle of mass m is written as $\mathbf{L} = \mathbf{r} \times \mathbf{p}$. Find the Poisson brackets of any two components of the angular momentum vector in Cartesian coordinates. Do this using the explicit representation of the Poisson bracket as derivatives with respect to canonical coordinates and momenta. Show that the result is given by $\{L_x, L_y\} = L_z$, $\{L_y, L_z\} = L_x$, and $\{L_z, L_x\} = L_y$ (*i.e.*, cyclic permutations of (xyz)). This is known as the angular momentum algebra.

★ **Problem 15.22** (a) Repeat the previous problem but instead use only the four properties of the Poisson bracket and the facts that $\{x, p_x\} = \{y, p_y\} = \{z, p_z\} = 1$ while the other brackets of positions and momenta vanish. (b) Write the quantum version of this algebra using the quantization scheme described in the chapter.

★   **Problem 15.23** Using Eq. (11.140), find the generator of the transformation that can translate a function of the canonical momentum $f(p)$ by $p \rightarrow p + \epsilon$.

★★   **Problem 15.24** Using Eq. (11.140), show that if we use $G = \epsilon L_x$ as a generator of a transformation (where L_x is the x component of the angular momentum), we end up rotating the components of the position vector $\mathbf{r}$ by an infinitesimal angle ϵ about the x axis. Show this by applying the generator onto an arbitrary function of position $A(\mathbf{r})$. Similarly, find the generators that rotate the position vector about the y and z axes.

★★   **Problem 15.25** Inspired by the previous problem, find the generators that rotate the momentum vector $\mathbf{p}$ about the x, y, and z axes by infinitesimal angles.

★★   **Problem 15.26** Compute the Poisson bracket of any components of position or momentum with any component of angular momentum. Use the Poisson bracket representation as derivatives with respect to canonical coordinates and momenta.

★   **Problem 15.27** Repeat the previous problem but use only the four properties of the Poisson bracket and the particular Poisson brackets between the components of the position and momentum vectors. From this, deduce the corresponding commutation relations in quantum mechanics.

★★   **Problem 15.28** Find the generator that performs a Galilean boost in the x direction by an infinitesimal speed ϵ. Do this using Eq. (11.140), working backwards and considering expected effects of the transformation on the position and momentum of a particle.

★★★   **Problem 15.29** (a) Find the generator that performs an infinitesimal scale transformation, where $\mathbf{r}' = (1 + \epsilon)\mathbf{r}$, and similarly for momentum. (b) Find the brackets of this generator with the components of angular momentum of a particle.

★★★   **Problem 15.30** Show that

$$\hat{q}(t_0 + \Delta t) = e^{\frac{i}{\hbar}\hat{H}\Delta t}\hat{q}(t_0)e^{-\frac{i}{\hbar}\hat{H}\Delta t}, \tag{15.167}$$

where we define

$$e^{-\frac{i}{\hbar}\hat{H}\Delta t} = \sum_{n=0}^{\infty} \frac{1}{n!}\left(\frac{-i\hat{H}}{\hbar}\right)^n, \tag{15.168}$$

implies

$$\hat{q}(t_0 + \Delta t) = \hat{q}(t_0) + (-i)\frac{\Delta t}{\hbar}[\hat{q}, \hat{H}] + (-i)^2\frac{1}{2!}\frac{\Delta t^2}{\hbar^2}[[\hat{q}, \hat{H}], \hat{H}] + \cdots \tag{15.169}$$

Do this by showing the pattern for the first few terms only.

APPENDICES

A Coordinate Systems

In solving physics problems, it is often necessary to make a judicious choice of coordinate system from the outset. Using the "wrong" coordinate system that does not take advantage of the symmetries offered by the problem can easily lead to unnecessarily complicated algebra and intractable differential equations. Three coordinate systems arise most frequently in physics problems: Cartesian coordinates – most useful in problems with planar symmetry, cylindrical coordinates – most useful in problems with cylindrical symmetry, and spherical coordinates – for problems with spherical symmetry. In this appendix, we describe each of these three coordinate systems.

A.1 Cartesian Coordinates

We label a point in space in Cartesian coordinates by x, y, and z, as shown in Figure A.1. The position vector then takes the form

$$\mathbf{r} = x\,\hat{\mathbf{x}} + y\,\hat{\mathbf{y}} + z\,\hat{\mathbf{z}}, \tag{A.1}$$

where $\hat{\mathbf{x}}$, $\hat{\mathbf{y}}$, and $\hat{\mathbf{z}}$ are mutually orthogonal unit-normalized basis vectors that span the vector space in three dimensions. Any vector $\mathbf{W}$ can be decomposed into components as follows:

$$\mathbf{W} = \mathrm{W}^x\hat{\mathbf{x}} + \mathrm{W}^y\hat{\mathbf{y}} + \mathrm{W}^z\hat{\mathbf{z}}. \tag{A.2}$$

The Cartesian basis vectors are constant as we move around in x, y, and z. The line element takes the form

$$d\mathbf{r} = dx\,\hat{\mathbf{x}} + dy\,\hat{\mathbf{y}} + dz\,\hat{\mathbf{z}}. \tag{A.3}$$

The area element is written as

$$d\mathbf{A} = \pm dydz\,\hat{\mathbf{x}} \pm dxdz\,\hat{\mathbf{y}} \pm dxdy\,\hat{\mathbf{z}}, \tag{A.4}$$

while the volume element is

$$d\mathrm{Vol} = dxdydz. \tag{A.5}$$

The gradient of a function $f(x,y,z)$ takes the form

$$\nabla f = \frac{\partial f}{\partial x}\hat{\mathbf{x}} + \frac{\partial f}{\partial y}\hat{\mathbf{y}} + \frac{\partial f}{\partial z}\hat{\mathbf{z}}. \tag{A.6}$$

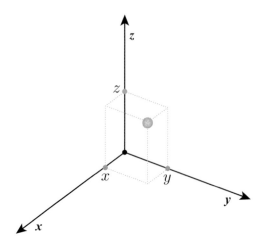

Cartesian coordinate system.

Given that the basis vectors are constant, the divergence, curl, and Laplacian take very simple forms. We get

$$\mathrm{div}\mathbf{W} = \nabla \cdot \mathbf{W} = \frac{\partial W^x}{\partial x} + \frac{\partial W^y}{\partial y} + \frac{\partial W^z}{\partial z}, \tag{A.7}$$

$$\mathrm{curl}\mathbf{W} = \nabla \times \mathbf{W} = \left(\frac{\partial W^z}{\partial y} - \frac{\partial W^y}{\partial z}\right)\hat{\mathbf{x}} + \left(\frac{\partial W^x}{\partial z} - \frac{\partial W^z}{\partial x}\right)\hat{\mathbf{y}} + \left(\frac{\partial W^y}{\partial x} - \frac{\partial W^x}{\partial y}\right)\hat{\mathbf{z}}, \tag{A.8}$$

where $\mathbf{A}(x, y, z)$ is an arbitrary vector field. And

$$\nabla^2 f = \frac{\partial^2 f}{\partial x^2} + \frac{\partial^2 f}{\partial y^2} + \frac{\partial^2 f}{\partial z^2}. \tag{A.9}$$

A.2 Cylindrical Coordinates

We label a point in space in cylindrical coordinates by ρ, φ, and z, as shown in Figure A.2.

These are related to the Cartesian coordinates by

$$x = \rho \cos \varphi, \quad y = \rho \sin \varphi \tag{A.10}$$

or

$$\rho = \sqrt{x^2 + y^2}, \quad \tan \varphi = \frac{y}{x} \tag{A.11}$$

with z unchanged. The position vector then takes the form

$$\mathbf{r} = \rho \,\hat{\boldsymbol{\rho}} + z \,\hat{\mathbf{z}}. \tag{A.12}$$

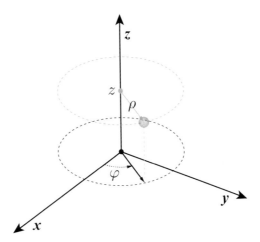

Fig. A.2 Cylindrical coordinate system.

We write $\hat{\rho}$, $\hat{\varphi}$, and $\hat{z}$ for the mutually orthogonal unit-normalized basis vectors that span the vector space. In particular, we have

$$\hat{\rho} \times \hat{\varphi} = \hat{z}, \quad \hat{\varphi} \times \hat{z} = \hat{\rho}, \quad \hat{z} \times \hat{\rho} = \hat{\varphi}. \tag{A.13}$$

We can also relate these basis vectors to those used in Cartesian coordinates:

$$\hat{\rho} = \cos\varphi\,\hat{x} + \sin\varphi\,\hat{y}, \quad \hat{\varphi} = -\sin\varphi\,\hat{x} + \cos\varphi\,\hat{y} \tag{A.14}$$

or

$$\hat{x} = \cos\varphi\,\hat{\rho} - \sin\varphi\,\hat{\varphi}, \quad \hat{y} = \sin\varphi\,\hat{\rho} + \cos\varphi\,\hat{\varphi}. \tag{A.15}$$

This means that any vector $\mathbf{W}$ can be decomposed into components as follows:

$$\mathbf{W} = W^\rho\hat{\rho} + W^\varphi\hat{\varphi} + W^z\hat{z}. \tag{A.16}$$

The basis vectors in cylindrical coordinates are *not* constant as we change φ; they remain constant only when we move in ρ and z. We then have the nonzero derivatives

$$\frac{\partial\hat{\rho}}{\partial\varphi} = \hat{\varphi} \tag{A.17}$$

and

$$\frac{\partial\hat{\varphi}}{\partial\varphi} = -\hat{\rho}. \tag{A.18}$$

The line element takes the form

$$d\mathbf{r} = d\rho\,\hat{\rho} + \rho d\varphi\,\hat{\varphi} + dz\,\hat{z}. \tag{A.19}$$

The area element is

$$d\mathbf{A} = \pm\rho\,d\varphi dz\,\hat{\rho} \pm d\rho dz\,\hat{\varphi} \pm \rho\,d\rho d\varphi\,\hat{z}, \tag{A.20}$$

while the volume element is

$$dVol = \rho\, d\rho d\varphi dz. \tag{A.21}$$

The gradient of a function $f(\rho, \varphi, z)$ is

$$\nabla f = \frac{\partial f}{\partial \rho}\hat{\boldsymbol{\rho}} + \frac{1}{\rho}\frac{\partial f}{\partial \varphi}\hat{\boldsymbol{\varphi}} + \frac{\partial f}{\partial z}\hat{\mathbf{z}}. \tag{A.22}$$

The divergence, curl, and Laplacian take more complicated forms arising from (A.17) and (A.18). We get

$$\mathrm{div}\mathbf{W} \equiv \nabla \cdot \mathbf{W} = \frac{1}{\rho}\frac{\partial}{\partial \rho}\left(\rho W^\rho\right) + \frac{1}{\rho}\frac{\partial W^\varphi}{\partial \varphi} + \frac{\partial W^z}{\partial z}, \tag{A.23}$$

$$\mathrm{curl}\mathbf{W} \equiv \nabla \times \mathbf{W} = \left(\frac{1}{\rho}\frac{\partial W^z}{\partial \varphi} - \frac{\partial W^\varphi}{\partial z}\right)\hat{\boldsymbol{\rho}} + \left(\frac{\partial W^\rho}{\partial z} - \frac{\partial W^z}{\partial \rho}\right)\hat{\boldsymbol{\varphi}}$$
$$+ \left(\frac{1}{\rho}\frac{\partial}{\partial \rho}\left(\rho W^\varphi\right) - \frac{1}{\rho}\frac{\partial W^\rho}{\partial \varphi}\right)\hat{\mathbf{z}}, \tag{A.24}$$

where $\mathbf{W}(\rho, \varphi, z)$ is an arbitrary vector field. Finally:

$$\nabla^2 f = \frac{1}{\rho}\frac{\partial}{\partial \rho}\left(\rho \frac{\partial f}{\partial \rho}\right) + \frac{1}{\rho^2}\frac{\partial^2 f}{\partial \varphi^2} + \frac{\partial^2 f}{\partial z^2}. \tag{A.25}$$

In two dimensions, the cylindrical coordinate system can be collapsed onto the ρ–φ subspace to represent the **Polar coordinate system**. All formulae in this section go through by simply dropping the z coordinate dependence throughout. To accord with common conventions, two-dimensional polar coordinates are denoted in this text by r and θ, instead of ρ and φ.

A.3 Spherical Coordinates

We label a point in space in spherical coordinates by r, ϕ, and θ, as shown in Figure A.3.

These are related to the Cartesian coordinates by

$$x = r \sin \theta \cos \phi, \quad y = r \sin \theta \sin \phi, \quad z = r \cos \theta \tag{A.26}$$

or

$$r = \sqrt{x^2 + y^2 + z^2}, \quad \tan \theta = \frac{\sqrt{x^2 + y^2}}{z}, \quad \tan \phi = \frac{y}{x}. \tag{A.27}$$

The position vector then takes the simple form

$$\mathbf{r} = r\hat{\mathbf{r}}. \tag{A.28}$$

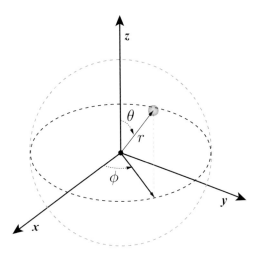

Spherical coordinate system.

We write $\hat{\mathbf{r}}$, $\hat{\boldsymbol{\theta}}$, and $\hat{\boldsymbol{\phi}}$ for the mutually orthogonal unit-normalized basis vectors that span the vector space. In particular, we have

$$\hat{\mathbf{r}} \times \hat{\boldsymbol{\theta}} = \hat{\boldsymbol{\phi}}, \quad \hat{\boldsymbol{\theta}} \times \hat{\boldsymbol{\phi}} = \hat{\mathbf{r}}, \quad \hat{\boldsymbol{\phi}} \times \hat{\mathbf{r}} = \hat{\boldsymbol{\theta}}. \tag{A.29}$$

We can also relate these basis vectors to those used in Cartesian coordinates:

$$\hat{\mathbf{r}} = \sin\theta\cos\phi\,\hat{\mathbf{x}} + \sin\theta\sin\phi\,\hat{\mathbf{y}} + \cos\theta\,\hat{\mathbf{z}}, \tag{A.30}$$

$$\hat{\boldsymbol{\theta}} = \cos\theta\cos\phi\,\hat{\mathbf{x}} + \cos\theta\sin\phi\,\hat{\mathbf{y}} - \sin\theta\,\hat{\mathbf{z}}, \tag{A.31}$$

$$\hat{\boldsymbol{\phi}} = -\sin\phi\,\hat{\mathbf{x}} + \cos\phi\,\hat{\mathbf{y}} \tag{A.32}$$

or

$$\hat{\mathbf{x}} = \sin\theta\cos\phi\,\hat{\mathbf{r}} + \cos\theta\cos\phi\,\hat{\boldsymbol{\theta}} - \sin\phi\,\hat{\boldsymbol{\phi}}, \tag{A.33}$$

$$\hat{\mathbf{y}} = \sin\theta\sin\phi\,\hat{\mathbf{r}} + \cos\theta\sin\phi\,\hat{\boldsymbol{\theta}} + \cos\phi\,\hat{\boldsymbol{\phi}}, \tag{A.34}$$

$$\hat{\mathbf{z}} = \cos\theta\,\hat{\mathbf{r}} - \sin\theta\,\hat{\boldsymbol{\theta}}. \tag{A.35}$$

This means that any vector $\mathbf{W}$ can be decomposed into components according to

$$\mathbf{W} = \mathrm{W}^r\hat{\mathbf{r}} + \mathrm{W}^\phi\hat{\boldsymbol{\phi}} + \mathrm{W}^\theta\hat{\boldsymbol{\theta}}. \tag{A.36}$$

The basis vectors in spherical coordinates are *not* constant as we move in space. We have the nonzero derivatives

$$\frac{\partial\hat{\mathbf{r}}}{\partial\theta} = \hat{\boldsymbol{\theta}}, \quad \frac{\partial\hat{\mathbf{r}}}{\partial\phi} = \sin\theta\,\hat{\boldsymbol{\phi}} \tag{A.37}$$

$$\frac{\partial \hat{\boldsymbol{\theta}}}{\partial \theta} = -\hat{\mathbf{r}}, \quad \frac{\partial \hat{\boldsymbol{\theta}}}{\partial \phi} = \cos \theta \, \hat{\boldsymbol{\phi}} \tag{A.38}$$

and

$$\frac{\partial \hat{\boldsymbol{\phi}}}{\partial \phi} = -\sin \theta \, \hat{\mathbf{r}} - \cos \theta \, \hat{\boldsymbol{\theta}}. \tag{A.39}$$

All other derivatives are zero. The line element then takes the form

$$d\mathbf{r} = dr \, \hat{\mathbf{r}} + r \, d\theta \, \hat{\boldsymbol{\theta}} + r \sin \theta \, d\phi \, \hat{\boldsymbol{\phi}}. \tag{A.40}$$

The area element is written as

$$d\mathbf{A} = \pm r^2 \sin \theta \, d\theta d\phi \, \hat{\mathbf{r}} \pm r \sin \theta \, dr d\phi \, \hat{\boldsymbol{\theta}} \pm r \, dr d\theta \, \hat{\boldsymbol{\phi}}, \tag{A.41}$$

while the volume element is

$$d\mathrm{Vol} = r^2 \sin \theta \, dr d\theta d\phi. \tag{A.42}$$

The gradient of a function $f(r, \theta, \phi)$ is

$$\nabla f = \frac{\partial f}{\partial r}\hat{\mathbf{r}} + \frac{1}{r}\frac{\partial f}{\partial \theta}\hat{\boldsymbol{\theta}} + \frac{1}{r \sin \theta}\frac{\partial f}{\partial \phi}\hat{\boldsymbol{\phi}}. \tag{A.43}$$

The divergence, curl, and Laplacian take more complicated forms arising from (A.37), (A.38), and (A.39). We get

$$\mathrm{div}\mathbf{W} \equiv \nabla \cdot \mathbf{W} = \frac{1}{r^2}\frac{\partial}{\partial r}\left(r^2 \mathrm{W}^r\right) + \frac{1}{r \sin \theta}\frac{\partial}{\partial \theta}\left(\sin \theta \, \mathrm{W}^\theta\right) + \frac{1}{r \sin \theta}\frac{\partial \mathrm{W}^\phi}{\partial \phi}, \tag{A.44}$$

$$\begin{aligned}
\mathrm{curl}\mathbf{W} \equiv \nabla \times \mathbf{W} = {} & \left(\frac{1}{r \sin \theta}\frac{\partial}{\partial \theta}\left(\sin \theta \, \mathrm{W}^\phi\right) - \frac{\partial \mathrm{W}^\theta}{\partial \phi}\right)\hat{\mathbf{r}} \\
& + \frac{1}{r}\left(\frac{1}{\sin \theta}\frac{\partial \mathrm{W}^r}{\partial \phi} - \frac{\partial}{\partial r}\left(r \, \mathrm{W}^\phi\right)\right)\hat{\boldsymbol{\theta}} \\
& + \frac{1}{r}\left(\frac{\partial}{\partial r}\left(r \mathrm{W}^\theta\right) - \frac{\partial \mathrm{W}^r}{\partial \theta}\right)\hat{\boldsymbol{\phi}},
\end{aligned} \tag{A.45}$$

where $\mathbf{W}(r, \theta, \phi)$ is an arbitrary vector field. Finally, the Laplacian takes the form

$$\nabla^2 f = \frac{1}{r^2}\frac{\partial}{\partial r}\left(r^2 \frac{\partial f}{\partial r}\right) + \frac{1}{r^2 \sin \theta}\frac{\partial}{\partial \theta}\left(\sin \theta \frac{\partial f}{\partial \theta}\right) + \frac{1}{r^2 \sin^2 \theta}\frac{\partial^2 f}{\partial \phi^2}. \tag{A.46}$$

Integral Theorems

When analyzing dynamics in three space dimensions, we often encounter integrals over volumes, areas, and curves and relations between them. These geometrical integral relations come in three flavors that we will elaborate on here. All three are closely related, and can be unified within the proper mathematical language of differential geometry. Here we take each on its own and outline the proofs. These proofs help us understand the physical and geometrical meaning of the integral relations between volumes, areas, and curves.

B.1 Green's Theorem

Consider the line integral over a closed two-dimensional curve $\mathscr{C}$:

$$\oint_{\mathscr{C}} U\,dx + V\,dy, \tag{B.1}$$

for arbitrary functions $U(x,y)$ and $V(x,y)$. We start by assuming the curve has the shape shown in Figure B.1, where the left and right edges are vertical. We will now show that

$$\oint_{\mathscr{C}} U\,dx + V\,dy = \int\int_{\mathscr{A}} \left(-\frac{\partial U}{\partial y} + \frac{\partial V}{\partial x} \right) dx\,dy. \tag{B.2}$$

Notice that this equation relates the integral around the curve to an integral over the area enclosed by the curve. For now, we make the statement regarding curves of the form depicted in Figure B.1; and then we will show that this generalizes to a curve of any shape.

Let us start with the first term in (B.2) expressed as an integral over dx. On the left-hand side we get

$$\oint_{\mathscr{C}} U\,dx = \int -U(x,y_2(x))\,dx + U(x,y_1(x))\,dx, \tag{B.3}$$

where the minus sign in the first term comes from the fact that the top part of the curve is oriented in the negative x direction. We also note that there are no contributions from the vertical segments of the curve, since $dx = 0$ on these parts. Now let us look at the first term of the right-hand side of (B.2). We have

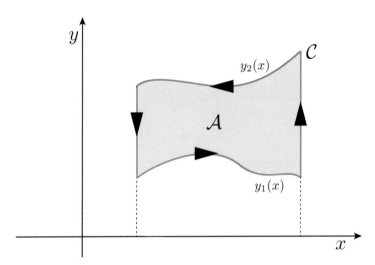

Proof of Green's theorem for a curve with vertical sides.

$$\int\int -\frac{\partial U}{\partial y} dy\, dx = \int \left(-U(x,y_2(x)) + U(x,y_1(x)) \right) dx, \qquad \text{(B.4)}$$

where we performed the integral over dy. We see that this equals the expression in (B.3). Similarly, we can show that

$$\oint_{\mathcal{C}} V\, dy = \int\int \frac{\partial V}{\partial x} dx dy, \qquad \text{(B.5)}$$

this time by performing the integral over dx on the right-hand side. We thus have shown the relation (B.2) for the type of curves that have vertical left and right edges. For more general curves, we can slice the curve into parts that do have vertical edges, as shown in Figure B.2.

Applying our integral relation to each closed curve, and adding the results, we get (B.2) for *any* curve: this is because the vertical segments that splice the larger area come in pairs oriented in opposite directions. This means that they would cancel on the left-hand side of (B.2), so that we are left with the line integral only on the outer rim. And obviously the sum of the areas on the right add up for the full enclosed area by the general curve. Hence, we have a general statement in (B.2) that applies to any closed two-dimensional curve. This relation is known as **Green's theorem**.

B.2 Stokes's Theorem

Consider a general vector field $\mathbf{B}(x,y,z)$ in three dimensions. Stokes's theorem states that

$$\oint_{\mathcal{C}} \mathbf{B} \cdot d\mathbf{l} = \int\int_{\mathcal{A}} (\nabla \times \mathbf{B}) \cdot d\mathbf{A}, \qquad \text{(B.6)}$$

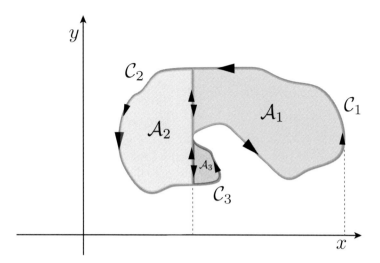

Fig. B.2 When the curve is more complicated, we can divide the relevant enclosed area into parts where each obeys Green's theorem, as proven in the text. The sum of the integrals reproduces Green's theorem for the full area as contributions from vertical edges cancel.

relating the line integral of the vector field over an arbitrary curve to the integral of its curl over any surface area bounded by the curve – as shown in Figure B.3.

The orientation of the curve determines the direction of the area element $d\mathbf{A}$ by the right-hand rule: curl the finger of your right hand along the chosen orientation of the curve, then your thumb tells you how to orient $d\mathbf{A}$ for the surface.

To prove this theorem, let us start by assuming that $\mathbf{B} = B^z\hat{\mathbf{z}}$. We also assume that the surface bounded by the curve can be described by one-to-one functions $x(y,z)$, $y(x,z)$, and $z(x,y)$. We will come back to this restriction at the end and show that we can extend the theorem to more general surface shapes.

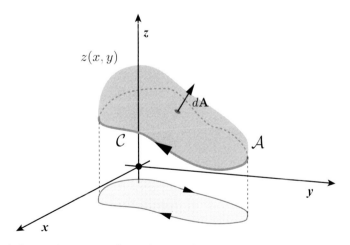

Fig. B.3   Proof of Stokes's theorem for a curve and area of a particular well-behaved form.

We start with the left-hand side of (B.6). We have

$$\oint_{\mathscr{C}} B^z(x,y,z)dz = \oint B^z(x,y,z(x,y)) \left(\frac{\partial z}{\partial x} + \frac{\partial z}{\partial y} \right), \qquad \text{(B.7)}$$

where we used the chain rule

$$dz = \frac{\partial z}{\partial x} + \frac{\partial z}{\partial y}. \qquad \text{(B.8)}$$

Note that the curve over which we integrate the right-hand side is the projection of the three-dimensional curve $\mathscr{C}$ onto the x–y plane. Now, let us look at the right-hand side of (B.6). We have

$$\int\!\!\int_{\mathscr{A}} (\nabla \times \mathbf{B}) \cdot d\mathbf{A} = \int\!\!\int \left(-\frac{\partial B^z}{\partial y} \frac{\partial z}{\partial x} + \frac{\partial B^z}{\partial x} \frac{\partial z}{\partial y} \right) dxdy, \qquad \text{(B.9)}$$

where we used the outward normal area element $d\mathbf{A}$:

$$d\mathbf{A} = -\frac{\partial z}{\partial x}\hat{\mathbf{x}} - \frac{\partial z}{\partial y}\hat{\mathbf{y}} + dxdy\hat{\mathbf{z}}. \qquad \text{(B.10)}$$

We can now make use of Green's theorem (B.2) with

$$U = B^z(x,y,z(x,y))\frac{\partial z}{\partial x}, \quad V = B^z(x,y,z(x,y))\frac{\partial z}{\partial y}. \qquad \text{(B.11)}$$

We then have

$$\frac{\partial U}{\partial y} = \left(\frac{\partial B^z}{\partial y} + \frac{\partial B^z}{\partial z}\frac{\partial z}{\partial y} \right)\frac{\partial z}{\partial x} + B^z \frac{\partial^2 z}{\partial x \partial y} \qquad \text{(B.12)}$$

and

$$\frac{\partial V}{\partial x} = \left(\frac{\partial B^z}{\partial x} + \frac{\partial B^z}{\partial z}\frac{\partial z}{\partial x} \right)\frac{\partial z}{\partial y} + B^z \frac{\partial^2 z}{\partial y \partial x}. \qquad \text{(B.13)}$$

When we subtract this second relation from the first, all terms cancel except the two shown in (B.9). By Green's theorem, we then have (B.9) equal to

$$\int\!\!\int_{\mathscr{A}} (\nabla \times \mathbf{B}) \cdot d\mathbf{A} = \oint B^z(x,y,z(x,y))\frac{\partial z}{\partial x}dx + B^z(x,y,z(x,y))\frac{\partial z}{\partial y}dy, \quad \text{(B.14)}$$

where the curve over which we integrate the right-hand side is once again the projection of the three-dimensional curve $\mathscr{C}$ onto the x–y plane. Therefore, we have proven Stokes's theorem for $\mathbf{B} = B^z\hat{\mathbf{z}}$. We can repeat this procedure for $\mathbf{B} = B^x\hat{\mathbf{x}}$, and then again for $\mathbf{B} = B^y\hat{\mathbf{y}}$. Since Stokes's theorem is linear in $\mathbf{B}$, we can then add all three realizations of the theorem in the x, y, and z directions to conclude that (B.6) is valid for any arbitrary vector field $\mathbf{B}$.

We now come to the assumption that the shape of the surface in Stokes's theorem is one-to-one in all three directions. Physically, this means that the shadows of the surface onto the x–y, y–z, and z–x planes are "clean" or "simple": that is, the shadows of different parts of the surface do not overlap. If we have a more complicated surface like the one shown in Figure B.4, where the surface folds on itself, all we need to do is divide up the surface in "simple" parts that satisfy the one-to-one criterion we assumed, and we can apply the theorem to each segment. We can then add up the segments, since, as in the case of Green's theorem, boundaries between segments come in oppositely oriented pairs that cancel – so that we recover a line integral over the full curve; and the area integrals just sum up to the full area integral as needed. We thus conclude that Stokes's theorem works for any arbitrary vector field, any arbitrary three-dimensional curve, and any arbitrarily shaped surfaces as long as the boundary of the surface is the corresponding curve on the left-hand side of (B.6).

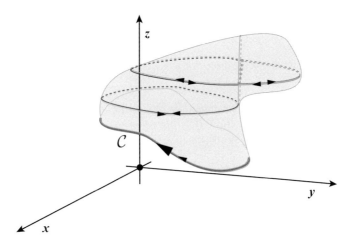

Fig. B.4　When the area bounded by the curve is more complicated, we can divide it into parts for which our proof of Stokes's theorem applies. The sum of the integrals then implies Stokes's theorem for the general area, since the contributions from the additional bounding curves cancel.

B.3 Divergence Theorem

The divergence theorem relates the integral of any vector field $\mathbf{B}(x,y,z)$ over a closed surface to the volume integral of its divergence

$$\oint_{\mathcal{A}} \mathbf{B} \cdot d\mathbf{A} = \int\int\int_{\mathcal{V}} \nabla \cdot \mathbf{B}\, d\text{Vol}, \tag{B.15}$$

where the volume $\mathcal{V}$ is the one enclosed by the surface $\mathcal{A}$ as shown in Figure B.5.

The convention is that $d\mathbf{A}$ is oriented such that it points outward from the enclosed volume.

To prove this theorem, let us start by assuming that $\mathbf{B} = B^z\hat{\mathbf{z}}$. We also assume that the closed surface is such that any lines parallel to the x, y, or z axes intersect the surface only at two points at most. This basically means that the surface does

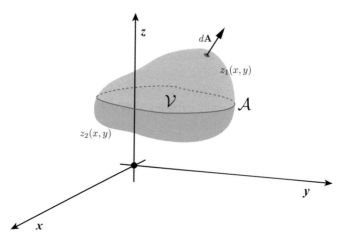

Proof of the divergence theorem of a class of simple volumes.

not have strange protrusions. We will come back at the end and remove these two assumptions and show that the theorem applies in general.

We start from the right-hand side of (B.15):

$$\iiint_{\mathcal{V}} \nabla \cdot \mathbf{B} \, d\text{Vol} = \iiint \frac{\partial B^z}{\partial z} \, dxdydz$$

$$= \iint \left(B^z(x, y, z_1(x, y)) - B^z(x, y, z_2(x, y)) \right) \, dxdy, \quad \text{(B.16)}$$

where we performed the integration over dz as shown in Figure B.5. The left-hand side of (B.15) is

$$\oint_{\mathcal{A}} \mathbf{B} \cdot d\mathbf{A} = \oint B^z dA^z, \quad \text{(B.17)}$$

with

$$d\mathbf{A} = \mp \frac{\partial z}{\partial x} \hat{\mathbf{x}} \mp \frac{\partial z}{\partial y} \hat{\mathbf{y}} \pm dxdy\hat{\mathbf{z}}, \quad \text{(B.18)}$$

where the upper sign is for cap 1 with its normal pointing upward, and the lower sign is for cap 2 with its normal pointing downward. We then have

$$\oint_{\mathcal{A}} \mathbf{B} \cdot d\mathbf{A} = \int \left(B^z(x, y, z_1(x, y)) - B^z(x, y, z_2(x, y)) \right) \, dxdy, \quad \text{(B.19)}$$

accounting for both caps. This matches with what we had in (B.16), and hence we have proved the theorem. We can now repeat the procedure for $\mathbf{B} = B^x\hat{\mathbf{x}}$ and $\mathbf{B} = B^y\hat{\mathbf{y}}$; and in the case of Stokes's theorem, we can add the three statements for three different directions to conclude that the theorem works for any vector field $\mathbf{B}$.

Finally, we can address the assumption that the surface is "simple," with no protrusions. For a general surface like the one shown in Figure B.6, we divide up the surface into parts that fit our criteria, as shown. We then apply the divergence theorem to each part and sum. On the left side of (B.15), areas that separate different regions come in pairs with opposing normal vectors and hence will cancel; this leaves the area integral over the outside area only. On the right-hand side, the different volume integrals add up to the full volume integral. We thus conclude that the divergence theorem holds for any surface and its enclosed volume, and for any vector field.

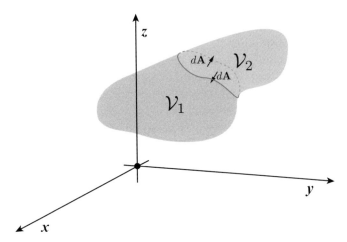

Fig. B.6   When the volume has a more complicated shape, we can divide it up into regions for which our proof of the divergence theorem applies. Adding up the integrals reproduces the divergence theorem for the general volume, since contributions for areas that slice the volume cancel.

Dimensional reasoning is a powerful tool that can help us learn how one quantity depends upon others. The secret is that in classical mechanics, both sides of an equation must have the same dimensions of mass M, length L, and time T. All other quantities can be expressed in terms of these three. For example, the dimensions of momentum (which we will write as $[p]$, with square brackets) are ML/T, and the dimensions of energy are $[E] = ML^2/T^2$.

Suppose we hold up a ball, drop it from rest, and then seek to find its momentum when it strikes the ground. The first step is to ask "what would the momentum likely depend upon?" Using physical intuition, it seems reasonable that the momentum might depend upon the ball's mass m, the height h from which it is dropped, and the acceleration of gravity g. We are not sure *how* it depends upon these quantities, however. The next step is to compare dimensions. The dimensions are $[p] = ML/T$, $[m] = [M]$, $[g] = L/T^2$, and $[h] = L$. The only way to get the "M" in momentum is to suppose that p is directly proportional to m, because neither g nor h contains a dimension of mass. Then the only way to get the $1/T$ in momentum is to suppose that p is proportional to $\sqrt{g}$. The product $m\sqrt{g}$ has dimensions $(M/T)\sqrt{L}$, which only needs to be multiplied by $\sqrt{h}$ to achieve the correct dimensions for momentum. That is, the momentum when the ball strikes the ground must have the dependence

$$p = km\sqrt{gh}, \tag{C.1}$$

where k is some *dimensionless* constant. Dimensional reasoning alone cannot give us this constant, so in fact we still do not know what the momentum of the ball is when it reaches the ground. What we do know, however, is that if the momentum at the ground of a particular dropped ball is p_0, the momentum at the ground of a similar ball dropped from twice the height will be $\sqrt{2}\, p_0$, or the momentum of a ball dropped on the moon from the original height will be $p_0/\sqrt{6}$, since gravity on the moon is only one-sixth that on earth.

Note that this particular problem is easily solved exactly using $\mathbf{F} = m\mathbf{a}$, giving the same equation while providing the value $k = \sqrt{2}$. Dimensional analysis, however, also works in much more complicated problems where the proportionality constant may be more difficult to find.

In general, we can solve such problems by writing a general relation such as

$$p = km^{\alpha}g^{\beta}h^{\gamma}, \tag{C.2}$$

where α, β, and γ are constants to be determined. We then expand everything in terms of mass, length, and time:

$$\frac{ML}{T} \sim M^{\alpha} \frac{L^{\beta}}{T^{2\beta}} L^{\gamma}, \tag{C.3}$$

yielding simple equations for α, β, and γ:

$$1 = \alpha, \quad 1 = \beta + \gamma, \quad -1 = -2\beta, \tag{C.4}$$

confirming that $\alpha = 1$, $\beta = 1/2$, and $\gamma = 1/2$.

Example C.1

Flow Rate of Molasses Through a Narrow Pipe

By *flow rate*, we mean the volume/second (with dimensions [flow rate] $= L^3/T$) that passes though a pipe. We expect that this depends upon the radius of the pipe, with $[r] = L$, since a wider pipe should allow more fluid to flow than a narrower one. It should also depend upon friction within the fluid itself, and between the fluid and sides of the pipe. Friction in a fluid is characterized by its **viscosity** η, with dimensions $[\eta] = M/LT$, and with values that can be found in tables.[a] The greater the viscosity, the greater the friction, and the lower the flow rate should be: molasses or honey (with high viscosity) should flow more slowly than a light oil (with low viscosity). Finally, the flow rate should also depend upon how hard one pushes on the fluid; *i.e.*, the pressure difference ΔP between one end of the pipe and the other. More precisely, it should depend upon the pressure difference/unit length of pipe, since it makes sense that the viscous friction must be overcome by the pressure gradient within the pipe.[b] The dimensions of pressure are [force/area] $= (ML/T^2)/L^2 = M/(LT^2)$, so the dimensions of pressure per unit length are $[\Delta P/\ell] = M/(L^2T^2)$.

Now we can formally calculate, using dimensional analysis, how the volume per second of the flow depends upon r, η, and $\Delta P/\ell$, by taking arbitrary powers of each and finding the powers by matching dimensions on both sides. That is:

$$\text{flow volume}/s = k\, r^{\alpha} \eta^{\beta} (\Delta P/\ell)^{\gamma}, \tag{C.5}$$

where k is a dimensionless constant. Therefore dimensionally:

$$\frac{L^3}{T} = L^{\alpha} \left(\frac{M}{LT}\right)^{\beta} \left(\frac{M}{L^2T^2}\right)^{\gamma}. \tag{C.6}$$

We match exponents in turn for M, L, and T. That is:

$$\text{mass: } 0 = \beta + \gamma, \quad \text{length: } 3 = \alpha - \beta - 2\gamma, \quad \text{time: } -1 = -\beta - 2\gamma. \tag{C.7}$$

From the first of these we learn that $\gamma = -\beta$, so then from the third equation we find that $\gamma = -\beta = 1$. Finally, the second equation tells us that $\alpha = 3 + \beta + 2\gamma = 4$. Thus the equation for the flow rate through a pipe is

$$\text{flow volume}/s = k \left(\frac{\Delta P/\ell}{\eta}\right) r^4. \tag{C.8}$$

Again, dimensional analysis alone cannot tell us the numerical value of the dimensionless number k. However, we have learned a lot. Most spectacularly, we have learned that the flow rate of a highly viscous fluid is not proportional to the cross-sectional area of the pipe, but to the *fourth power* of the radius: a pipe of twice the radius will transport 16 times the volume of fluid. This formula corresponds to what is called **Poiseuille flow**, and an exact analytic calculation shows that the constant $k = 6\pi$.

We have carried out the dimensional analysis here in a rather formal way; one can often speed up the process without using arbitrary powers like α, β, and γ. Note from Eq. (C.8) that the flow rate must depend upon the ratio $(\Delta P/\ell)/\eta$ to cancel out the dimension of mass, so we can rewrite Eq. (C.8) as

$$\frac{L^3}{T} = L^\alpha \left(\frac{M}{L^2 T^2} \times \frac{LT}{M} \right)^\gamma = L^\alpha \left(\frac{1}{LT} \right)^\gamma, \tag{C.9}$$

from which it is clear that $\gamma = 1$ to obtain the needed $1/T$ dimension, and so then $\alpha = 4$ to obtain L^3. ■

[a] The viscosity η of a fluid can be measured in principle by placing the fluid between two parallel metal plates of area A that are separated by a distance d. When one plate is kept fixed while the other is moved parallel to the fixed plate with constant velocity v, the drag force on the moving plate is observed to have magnitude $F = \eta Av/d$. From this formula one can see that the dimensions of η are M/LT.

[b] In this problem we are assuming smooth, so-called **laminar flow**, which is nonturbulent. High-viscosity fluids (like molasses) that move slowly in narrow pipes are less likely to become turbulent. Turbulent flow is more complicated and depends on additional parameters.

C.1 A Few Exercises

1. The velocity of some waves on the surface of a lake depends upon gravity g and the wavelength λ as long as the depth h of the lake satisfies $h \gg \lambda$, corresponding to what is called "deep water waves." If we were to increase the wavelength by a factor of two, by what factor would the wave velocity be changed?

2. Capillary waves on the surface of a liquid come about because of the liquid's surface tension σ, which has dimensions M/T^2. The velocity of capillary waves depends upon σ and also upon the wavelength λ and mass density ρ of the liquid. Two capillary waves on the same liquid have wavelengths λ_1 and $\lambda_2 = 2\lambda_1$. What is the ratio of their velocities?

3. The Planck length ℓ_p depends upon Planck's constant $\hbar$ (with dimensions of energy $\times$ time), Newton's constant of gravity G, and the speed of light c. If Planck's constant were twice as large as it actually is, how would that affect ℓ_p? Taking the proportionality constant to be unity, how large is ℓ_p numerically in SI units?

4. Two very flat parallel metal plates, with a vacuum between and surrounding them, are attracted to one another by what is called the *Casimir force*. The force is proportional to the area A of each plate, and also depends upon the distance d between the plates, the speed of light c, and Planck's constant $\hbar$, which has dimensions of energy $\times$ time. If the distance d were halved, would the Casimir force increase or decrease? By what factor?

5. When a nuclear explosion takes place, the resulting fireball expands quickly through the surrounding air, heating more and more air with time t. The radius of the fireball depends on the energy E_0 of the explosion, the density ρ of the air, and time t. Given data for one fireball (say, the first such fireball at the Trinity site near Alamogordo, New Mexico on July 16, 1945), if in a subsequent explosion near ground level the fireball radius had become twice as large in the same time t, by what factor would the energy E of the second explosion exceed E_0?

Fractal Dimension

A single point in a $2N$-dimensional phase space has zero dimensions, a curve in phase space has dimension $d = 1$, a surface has dimension $d = 2$, and so on. For example, the attractor of a damped pendulum is a point at $\theta = 0, p_\theta = 0$, so has $d = 0$ dimensions. If the attractor for a system is a limit cycle, that attractor has $d = 1$, etc. Now it is interesting to find the dimensionality of the set of points in a Poincaré map corresponding to a chaotic system. Surprisingly, it turns out that not all attractors have integer dimensionality. They may instead have *fractal dimensions*, which means simply that they have non-integer dimensions.

To show this, we first need a general definition of dimensionality. The definition uses what is called "box counting." Given a set of points in $2N$ dimensions, we begin by covering the points with M small $2N$-dimensional "cubes" of edge length ϵ, where $2N$ is the phase-space dimension. Then using natural logarithms, we can define the dimension d of the set of points by

$$d = \lim_{\epsilon \to 0} \frac{\ln M(\epsilon)}{\ln(1/\epsilon)}. \tag{D.1}$$

We will test this definition for two well-known special cases.

1. First, suppose we have a point attractor. We only need a single small box to cover a point, so $M(\epsilon) = 1$ for any ϵ, *i.e.*, for a box of any size. Then $\ln(M) = \ln 1 = 0$, which gives $d = 0$, which is correct for a zero-dimensional set.
2. Now suppose we have an attractor which is a limit cycle. Covering the limit cycle in squares of side ϵ, as illustrated in Figure D.1, we require $N(\epsilon) \sim \ell/\epsilon$, where ℓ is the length of the closed curve. In that case

$$d = \lim_{\epsilon \to 0} \frac{\ln(\ell/\epsilon)}{\ln(1/\epsilon)} = \frac{\ln(\ell) + \ln(1/\epsilon)}{\ln(1/\epsilon)} = 1, \tag{D.2}$$

since as $\epsilon \to 0$ the first term in the numerator becomes negligible compared with the second term. So far, so good.

A more exotic case is that of the **Cantor set**, shown in Figure D.2. Begin with the points in a line segment extending from 0 to 1, as shown in the topmost picture. Then in the next step remove the middle third of the line, leaving two line segments; then the middle third of each of these two, and so on. We label the sets of points by $n = 0$ (the original line segment), $n = 1$ (just below it), $n = 2$, and so on. Note

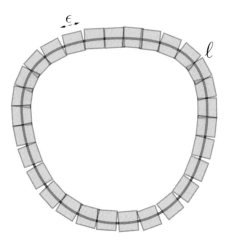

Covering a limit cycle with squares of side ϵ.

The Cantor set, in which successive middle thirds of lines are removed. In the limit of an infinite number of steps, the dimension of the remaining points is not an integer.

then that at the nth step $M(\epsilon) = 2^n$ and $\epsilon = 1/3^n$. Therefore the dimension of the set of points remaining in the limit $n \to \infty$ is

$$d = \lim_{\epsilon \to 0} \frac{\ln 2^n}{\ln 3^n} = \lim_{\epsilon \to 0} \frac{n \ln 2}{n \ln 3} = \frac{\ln 2}{\ln 3} = \frac{0.693147...}{1.09861...} = 0.63092..., \quad (D.3)$$

so this set of points has a fractal dimension, greater than that of a point, but less than that of a line.

In many areas of physics, one often encounters various sets of special polynomials that satisfy interesting and useful properties. These typically arise in trying to solve certain differential equations, and they often have deep connections to various disciplines of mathematics such as group theory. One celebrated class of special polynomials is named after its discoverer, the French mathematician Adrien-Marie Legendre, and arises as solutions to the differential equation

$$\frac{d}{dx}\left((1-x^2)\frac{dP_n(x)}{dx}\right) + n(n+1)P_n(x) = 0, \tag{E.1}$$

which in turn is encountered when solving Poisson's equation in spherical coordinates. Here n is an integer, $n = 0, 1, 2, \ldots$, and the solutions $P_n(x)$ are known as **Legendre polynomials** – a set of polynomials that truncate to order n. The Legendre polynomials are orthogonal to one another, and with conventional normalization satisfy

$$\int_{-1}^{1} P_m(x)P_n(x)\,dx = \frac{2}{2n+1}\delta_{mn}. \tag{E.2}$$

Hence, we say that P_m is orthogonal to P_n for $m \neq n$. Here are the first few polynomials in the sequence:

$$P_0(x) = 1, \quad P_1(x) = x, \quad P_2(x) = \frac{1}{2}\left(3x^2 - 1\right)$$

$$P_3(x) = \frac{1}{2}\left(5x^3 - 3x\right), \quad P_4(x) = \frac{1}{8}\left(35x^4 - 30x^2 + 3\right). \tag{E.3}$$

More generally, we can generate the sequence using the Rodrigues formula

$$P_n(x) = \frac{1}{2^n n!}\frac{d^n}{dx^n}(x^2 - 1)^n. \tag{E.4}$$

Legendre polynomials satisfy numerous other relations and form a *complete set*; this means that, with the help of the orthogonality relation (E.2), one can express any real function as a (possibly infinite) sum of Legendre polynomials with real coefficients, much like one can generate a Fourier decomposition of a function in terms of sines and cosines. This works for any function $f(x)$ with finitely many discontinuities in the interval $-1 < x < 1$, with coefficients a_n given by

$$f(x) = \sum_{n=0}^{\infty} a_n P_n(x), \quad a_n = \frac{2n+1}{2} \int_{-1}^{1} f(x)\, P_n(x)\, dx, \tag{E.5}$$

as long as the integral is finite.

There are other sets of special polynomials like the Legendre polynomials with similar useful properties: orthogonality, completeness, satisfying numerous mathematical identities, and arising as solutions to interesting and useful differential equations. Amongst this set and in addition to the Legendre polynomials, one encounters in physics the **Chebyshev polynomials** (e.g. Poisson equation in elliptic coordinates), the **Jacobi polynomials** (e.g. combining quantum spins), the **Laguerre polynomials** (e.g. the hydrogen atom in quantum mechanics), the **Hermite polynomials** (e.g. the quantum harmonic oscillator), the **Bessel polynomials** (e.g. Poisson equation in cylindrical coordinates, electromagnetism), and variations of these. As we saw in the text, all these also arise in quadrature algorithms for numerical integration. The reader is referred to the Further Reading section for more information on this rich subject.

Taylor Series

A Taylor series is a technique for writing a function as a series expansion in its derivatives about a point. It was introduced by English mathematician Brook Taylor, and is used extensively in physics as an approximation technique. Generally, a Taylor series of a function takes the form

$$f(x) = \sum_{n=0}^{\infty} \frac{1}{n!} \frac{d^n f}{dx}\bigg|_{x=x_0} (x - x_0)^n, \tag{F.1}$$

where the expansion is carried out around a point of our choosing, say $x = x_0$. This infinite series is not guaranteed to converge to $f(x)$, however. If there is a finite value r such that the infinite Taylor series converges to $f(x)$ for $|x - x_0| < r$, we say that $f(x)$ is *analytic* at x_0 with radius of convergence r. If the radius of convergence is infinite, the function $f(x)$ is said to be *entire*. A celebrated entire function is the exponential

$$e^x = 1 + x + \frac{1}{2!}x^2 + \frac{1}{3!}x^3 + \cdots \tag{F.2}$$

The logarithm, in contrast, is analytic about $x = 1$ but not entire – it has a radius of convergence equal to one:

$$\ln x = \ln(1 - y) = -y - \frac{1}{2}y^2 - \frac{1}{3}y^3 + \cdots, \tag{F.3}$$

where we have written the expansion in terms of $y = 1 - x$ for convenience. Therefore one must be careful when using Taylor series. Even when a Taylor series is formally convergent, it might not equal the function being expanded. Consider $f(x) = e^{-1/x^2}$, for example. Its Taylor series about the origin is well defined, but is equal to zero! This $f(x)$ is therefore not analytic, even though it is smooth.

In physics we often use the Taylor series as an approximation scheme by truncating the infinite series to the first few terms. For example, we write

$$e^x \simeq 1 + x + \mathcal{O}(x^2) \tag{F.4}$$

by dropping terms with powers x^2 or higher – in a regime where presumably $|x| \ll 1$ so that higher-order terms are smaller. A particularly useful truncated expansion of this type that we commonly encounter in physics problems involves

$$f(x) = (1 + x)^n, \tag{F.5}$$

where $|x| \ll 1$ and n is a constant. For integer n, we have an exact finite polynomial known as the **binomial expansion**:

$$(1+x)^n = \sum_{k=0}^{n} \binom{n}{k} x^k, \tag{F.6}$$

where

$$\binom{n}{k} = \frac{n!}{(n-k)!\,k!}. \tag{F.7}$$

We could think of this as a special Taylor-series expansion about $x = 0$ that truncates after a finite number of terms. For any n – even when non-integer, the Taylor-series expansion would still be well defined with radius of convergence equal to one. And for $|x| \ll 1$, we can often keep the first few terms to a very good approximation, and write

$$(1+x)^n \simeq 1 + nx + \frac{n(n-1)}{2!}x^2 + \mathcal{O}(x^3). \tag{F.8}$$

In most physical situations, these first three terms are often enough to achieve a desired level of accuracy. In many situations, even keeping only the first two terms can be adequate. In general, the idea is to truncate the expansion at the point where one captures the leading small effects from x, and drop any subleading contributions.

F.1 A Few Exercises

1. (a) Find the first three terms in the series expansion of $\sin\theta$ about $\theta = 0$, using the Taylor series. (b) Find this series instead given the series for e^x and also Euler's formula for e^x in terms of $\sin x$ and $\cos x$. (c) Find $\sin 0.025$, including three terms in the expansion.
2. Repeat all three parts of Exercise 1 for $\cos\theta$.
3. Find approximate values for $(1.004)^{3.5}$ and $(0.994)^{-3}$.
4. Find a power series for $4(1 - x^3)/\sqrt{2 + x^2}$ for small x, valid through terms of order x^3.
5. Using three terms in its expansion, find an approximate value of $\ln(x)$ for $x = 1.025$.

Further Reading

- V. I. Arnold, *Mathematical Methods of Classical Mechanics*, Springer, 1997. This is an advanced, graduate-level textbook that emphasizes the connections of mechanics to mathematics and mathematical physics. It is an excellent resource for the more mathematically oriented.

- G. B. Arfken, H. J. Weber, and F. E. Harris, *Mathematical Methods for Physicists*, Academic Press, 2012. A good reference on special functions and many other topics in mathematical physics.

- R. P. Feynman and A. R. Hibbs, *Quantum Mechanics and Path Integrals*, McGraw Hill, 1965. An introduction to quantum mechanics via the path-integral approach.

- H. Goldstein, C. P. Poole, Jr., and J. L. Safko, *Classical Mechanics*, Pearson, 2002. An excellent, albeit a bit advanced, exposition on the subject of mechanics, with a traditional approach.

- H. Gould, J. Tobochnik, and W. Christian, *An Introduction to Computer Simulation Methods*, Addison-Wesley, 2006. A good place to start for methods in computational physics.

- L. N. Hand and J. D. Finch, *Analytical Mechanics*, Cambridge University Press, 1998. An interesting exposition on the subject of mechanics at the undergraduate level following a more or less standard and traditional approach, but with interesting and unique examples.

- J. Hartle, *Gravity – An Introduction to Einstein's General Relativity*, Addison-Wesley, 2003. A clear and insightful exposition at the undergraduate level.

- J. D. Jackson, *Classical Electrodynamics*, Wiley, 1999. A classic and advanced exposition of electromagnetism, the standard graduate text on the subject.

- D. Kleppner and R. Kolenkow, *An Introduction to Mechanics*, Cambridge University Press, 2014. An excellent textbook that can serve as a solid first course on mechanics preceding the current one.

- L. D. Landau and E. M. Lifshitz, *Mechanics*, Elsevier Butterworth-Heinemann, 1976. A classic exposition with conciseness and depth, yet not terribly pedagogical. The book serves as an excellent reference on a physicist's bookshelf.

- D. Lemons, *An Introduction to Stochastic Processes in Physics*, The Johns Hopkins University Press, 2002. A very good introductory textbook on stochastic dynamics in physics.

- E. Merzbacher, *Quantum Mechanics*, Wiley, 1998. A rather complete and advanced discussion of modern quantum mechanics with many unique elements; excellent reference book on the subject.
- T. A. Moore, *A General Relativity Workbook*, University Science Books, 2013. A concise overview of general relativity at the undergraduate level, guiding the reader through hundreds of problems.
- W. H. Press, S. A. Teukolsky, W. T. Vetterling, and B. P. Flannery, *Numerical Recipes*, Cambridge University Press, 2007. All that is numerical and computational physics in a beautiful and exhaustive exposition.
- E. Purcell and D. Morin, *Electricity and Magnetism*, Cambridge University Press, 2013. An excellent undergraduate textbook on electromagnetism.
- T. Tél and M. Gruiz, *Chaotic Dynamics*, Cambridge University Press, 2006. Excellent exposition on the topic of chaos.
- J. S. Townsend, *A Modern Approach to Quantum Mechanics*, University Science Books, 2012. Excellent exposition on undergraduate quantum mechanics with a unique motivational development of Feynman path integration.
- R. M. Wald, *General Relativity*, University of Chicago Press, 1984. An excellent and modern exposition on the subject of general relativity at the graduate level.

Index